NASA'S CONTRIBUTIONS TO AERONAUTICS

VOLUME 2

FLIGHT ENVIRONMENT

OPERATIONS

FLIGHT TESTING AND RESEARCH

DR. RICHARD P. HALLION, EDITOR

National Aeronautics and Space Administration

Headquarters

300 E St SW

Washington, DC 20546

2010

NASA/SP-2010-570-Vol 2

www.nasa.gov

Contents

Foreword

AS THIS BOOK GOES TO PRESS, the National Aeronautics and Space Administration (NASA) has passed beyond the half century mark, its longevity a tribute to how essential successive Presidential administrations—and the American people whom they serve—have come to regard its scientific and technological expertise. In that half century, flight has advanced from supersonic to orbital velocities, the jetliner has become the dominant means of intercontinental mobility, astronauts have landed on the Moon, and robotic spacecraft developed by the Agency have explored the remote corners of the solar system and even passed into interstellar space.

Born of a crisis—the chaotic aftermath of the Soviet Union's space triumph with Sputnik—NASA rose magnificently to the challenge of the emergent space age. Within a decade of NASA's establishment, teams of astronauts would be planning for the first lunar landings, accomplished with Neil Armstrong's "one small step" on July 20, 1969. Few events have been so emotionally charged, and none so publicly visible or fraught with import, as his cautious descent from the spindly little Lunar Module Eagle to leave his historic boot-print upon the dusty plain of Tranquillity Base.

In the wake of Apollo, NASA embarked on a series of space initiatives that, if they might have lacked the emotional and attention-getting impact of Apollo, were nevertheless remarkable for their accomplishment and daring. The Space Shuttle, the International Space Station, the Hubble Space Telescope, and various planetary probes, landers, rovers, and flybys speak to the creativity of the Agency, the excellence of its technical personnel, and its dedication to space science and exploration.

But there is another aspect to NASA, one that is too often hidden in an age when the Agency is popularly known as America's space agency and when its most visible employees are the astronauts who courageously rocket into space, continuing humanity's quest into the unknown. That hidden aspect is aeronautics: lift-borne flight within the atmosphere, as distinct from the ballistic flight of astronautics, out into space. It is the first "A" in the Agency's name, and the oldest-rooted of the Agency's technical competencies, dating to the formation, in 1915, of NASA's lineal predecessor, the National Advisory Committee for Aeronautics (NACA). It was the NACA that largely restored America's aeronautical primacy in the interwar years after 1918, deriving the airfoil profiles and configuration concepts that defined successive generations of ever-more-capable aircraft as America progressed from the subsonic piston era into the transonic and supersonic jet age. NASA, succeeding the NACA after the shock of Sputnik, took American aeronautics across the hypersonic frontier and onward into the era of composite structures, electronic flight controls and energy-efficient flight.

As with the first in this series, this second volume traces contributions by NASA and the post–Second World War NACA to aeronautics. The surveys, cases, and biographical examinations presented in this work offer just a sampling of the rich legacy of aeronautics research having been produced by the NACA and NASA. These include

- Atmospheric turbulence, wind shear, and gust research, subjects of crucial importance to air safety across the spectrum of flight, from the operations of light general-aviation aircraft through large commercial and supersonic vehicles.
- Research to understand and mitigate the danger of lightning strikes upon aerospace vehicles and facilities.
- The quest to make safer and more productive skyways via advances in technology, cross-disciplinary integration of developments, design innovation, and creation of new operational architectures to enhance air transportation.
- Contributions to the melding of human and machine, via the emergent science of human factors, to increase the safety, utility, efficiency, and comfort of flight.
- The refinement of free-flight model testing for aerodynamic research, the anticipation of aircraft behavior, and design validation and verification, complementing traditional wind tunnel and full-scale aircraft testing.

- The evolution of the wind tunnel and expansion of its capabilities, from the era of the slide rule and subsonic flight to hypersonic excursions into the transatmosphere in the computer and computational fluid dynamics era.
- The advent of composite structures, which, when coupled with computerized flight control systems, gave aircraft designers a previously unknown freedom enabling them to design aerospace vehicles with optimized aerodynamic and structural behavior.
- Contributions to improving the safety and efficiency of general-aviation aircraft via better understanding of their unique requirements and operational circumstances, and the application of new analytical and technological approaches.
- Undertaking comprehensive flight research on sustained supersonic cruise aircraft—with particular attention to their aerodynamic characteristics, airframe heating, use of integrated flying and propulsion controls, and evaluation of operational challenges such as inlet "unstart," aircrew workload—and blending them into the predominant national subsonic and transonic air traffic network.
- Development and demonstration of Synthetic Vision Systems, enabling increased airport utilization, more efficient flight deck performance, and safer air and ground aircraft operations.
- Confronting the persistent challenge of atmospheric icing and its impact on aircraft operations and safety.
- Analyzing the performance of aircraft at high angles of attack and conducting often high-risk flight-testing to study their behavior characteristics and assess the value of developments in aircraft design and flight control technologies to reduce their tendency to depart from controlled flight.
- Undertaking pathbreaking flight research on VTOL and V/STOL aircraft systems to advance their ability to enter the mainstream of aeronautical development.
- Conducting a cooperative international flight-test program to mutually benefit understanding of the potential, behavior, and performance of large supersonic cruise aircraft.

As this sampling—far from a complete range—of NASA work in aeronautics indicates, the Agency and its aeronautics staff spread across the Nation maintain a lively interest in the future of flight, benefitting NASA's reputation earned in the years since 1958 as a national repository of aerospace excellence and its legacy of accomplishment in the 43-year history of the National Advisory Committee for Aeronautics, from 1915 to 1958.

As America enters the second decade of the second century of winged flight, it is again fitting that this work, like the volume that precedes it, be dedicated, with affection and respect, to the men and women of NASA, and the NACA from whence it sprang.

Dr. Richard P. Hallion
August 25, 2010

THIS PAGE INTENTIONALLY BLANK

NASA 515, Langley Research Center's Boeing 737 testbed, is about to enter a microburst wind shear. The image is actual test footage, reflecting the murk and menace of wind shear. NASA.

CASE 1

Eluding Aeolus: Turbulence, Gusts, and Wind Shear

Kristen Starr

Since the earliest days of American aeronautical research, NASA has studied the atmosphere and its influence upon flight. Turbulence, gusts, and wind shears have posed serious dangers to air travelers, forcing imaginative research and creative solutions. The work of NASA's researchers to understand atmospheric behavior and NASA's derivation of advanced detection and sensor systems that can be installed in aircraft have materially advanced the safety and utility of air transport.

BEFORE WORLD WAR II, the National Advisory Committee for Aeronautics (NACA), founded in 1915, performed most of America's institutionalized and systematic aviation research. The NACA's mission was "to supervise and direct the scientific study of the problems of flight with a view to their practical solution." Among the most serious problem it studied was that of atmospheric turbulence, a field related to the Agency's great interest in fluid mechanics and aerodynamics in general. From the 1930s to the present, the NACA and its successor—the National Aeronautics and Space Administration (NASA), formed in 1958—concentrated rigorously on the problems of turbulence, gusts, and wind shear. Midcentury programs focused primarily on gust load and boundary-layer turbulence research. By the 1980s and 1990s, NASA's atmospheric turbulence and wind shear programs reached a level of sophistication that allowed them to make significant contributions to flight performance and aircraft reliability. The aviation industry integrated this NASA technology into planes bought by airlines and the United States military. This research has resulted in an aviation transportation system exponentially safer than that envisioned by the pioneers of the early air age.

An Unsettled Sky

When laypeople think of the words "turbulence" and "aviation" together, they probably envision the "bumpy air" that passengers are often subjected to on long-duration plane flights. But the term "turbulence" has a particular technical meaning. Turbulence describes the motion of a fluid (for, our purposes, air) that is characterized by chaotic, seemingly random property changes. Turbulence encompasses fluctuations in diffusion, convection, pressure, and velocity. When an aircraft travels through air that experiences these changes, its passengers feel the turbulence buffeting the aircraft. Engineers and scientists characterize the degree of turbulence with the Reynolds number, a scaling parameter identified in the 1880s by Osborne Reynolds at the University of Manchester. Lower numbers denote laminar (smooth) flows, intermediate values indicate transitional flows, and higher numbers are characteristic of turbulent flow.[1]

A kind of turbulent airflow causes drag on all objects, including cars, golf balls, and planes, which move through the air. A boundary layer is "the thin reaction zone between an airplane [or missile] and its external environment." The boundary layer is separated from the contour of a plane's airfoil, or wing section, by only a few thousandths of an inch. Air particles change from a smooth laminar flow near the leading edge to a turbulent flow toward the airfoil's rear.[2] Turbulent flow increases friction on an aircraft's skin and therefore increased surface heat while slowing the speed of the aircraft because of the drag it produces.

Most atmospheric circulation on Earth causes some kind of turbulence. One of the more common forms of atmospheric turbulence experienced by aircraft passengers is clear air turbulence (CAT), which is caused by the mixing of warm and cold air in the atmosphere by wind, often via the process of wind shear. Wind shear is a difference in wind speed and direction over a relatively short distance in Earth's atmosphere. One engineer describes it as "any situation where wind velocity varies sharply from point to point."[3] Wind shears can have both horizontal and vertical components. Horizontal wind shear is usually encountered near coastlines and along fronts, while vertical wind shear appears closer to Earth's surface and sometimes at higher levels in the atmosphere, near frontal zones and upper-level air jets.

1. James R. Hansen, *Engineer in Charge: a History of the Langley Aeronautical Laboratory, 1917–1958*, NASA SP-4305 (Washington, DC: GPO, 1987), p. 76.

2. Theodore von Kármán, *Aerodynamics* (New York: Dover Publications, 2004 ed.), pp. 86–91.

3. Terry Zweifel, "Optimal Guidance during a Windshear Encounter," *Scientific Honeyweller* (Jan. 1989), p. 110.

Large-scale weather events, such as weather fronts, often cause wind shear. Weather fronts are boundaries between two masses of air that have different properties, such as density, temperature, or moisture. These fronts cause most significant weather changes. Substantial wind shear is observed when the temperature difference across the front is 9 degrees Fahrenheit (°F) or more and the front is moving at 30 knots or faster. Frontal shear is seen both vertically and horizontally and can occur at any altitude between surface and tropopause, which is the lowest portion of Earth's atmosphere and contains 75 percent of the atmosphere's mass. Those who study the effects of weather on aviation are concerned more with vertical wind shear above warm fronts than behind cold fronts because of the longer duration of warm fronts.[4]

The occurrence of wind shear is a microscale meteorological phenomenon. This means that it usually develops over a distance of less than 1 kilometer, even though it can emerge in the presence of large weather patterns (such as cold fronts and squall lines). Wind shear affects the movement of soundwaves through the atmosphere by bending the wave front, causing sounds to be heard where they normally would not. A much more violent variety of wind shear can appear near and within downbursts and microbursts, which may be caused by thunderstorms or weather fronts, particularly when such phenomena occur near mountains. Vertical shear can form on the lee side of mountains when winds blow over them. If the wind flow is strong enough, turbulent eddies known as "rotors" may form. Such rotors pose dangers to both ascending and descending aircraft.[5]

The microburst phenomenon, discovered and identified in the late 1970s by T. Theodore Fujita of the University of Chicago, involves highly localized, short-lived vertical downdrafts of dense cool air that impact the ground and radiate outward toward all points of the compass at high speed, like a water stream from a kitchen faucet impacting a basin.[6]

4. Integrated Publishing, "Meteorology: Low-Level Wind Shear," *http://www.tpub.com/weather3/6-15.htm*, accessed July 25, 2009.
5. National Center for Atmospheric Research, "T-REX: Catching the Sierra's Waves and Rotors," *http://www.ucar.edu/communications/quarterly/spring06/trex.jsp*, accessed July 21, 2009.
6. T. Theodore Fujita, "The Downburst, Microburst, and Macroburst," Satellite and Mesometeorology Research Project [SMRP] Research Paper 210, Dept. of Geophysical Sciences, University of Chicago, NTIS Report PB-148880 (1985).

Speed and directional wind shear result at the three-dimensional boundary's leading edge. The strength of the vertical wind shear is directly proportional to the strength of the outflow boundary. Typically, microbursts are smaller than 3 miles across and last fewer than 15 minutes, with rapidly fluctuating wind velocity.[7]

Wind shear is also observed near radiation inversions (also called nocturnal inversions), which form during rapid cooling of Earth's surface at night. Such inversions do not usually extend above the lower few hundred feet in the atmosphere. Favorable conditions for this type of inversion include long nights, clear skies, dry air, little or no wind, and cold or snow-covered surfaces. The difference between the inversion layer and the air above the inversion layer can be up to 90 degrees in direction and 40 knots. It can occur overnight or the following morning. These differences tend to be strongest toward sunrise.[8]

The troposphere is the lowest layer of the atmosphere in which weather changes occur. Within it, intense vertical wind shear can slow or prevent tropical cyclone development. However, it can also coax thunderstorms into longer life cycles, worsening severe weather.[9]

Wind shear particularly endangers aircraft during takeoff and landing, when the aircraft are at low speed and low altitude, and particularly susceptible to loss of control. Microburst wind shear typically occurs during thunderstorms but occasionally arises in the absence of rain

7. For microbursts and NASA research on them, see the recommended readings at the end of this paper by Roland L. Bowles, Kelvin K. Droegemeier, Fred H. Proctor, Paul A. Robinson, Russell Targ, and Dan D. Vicroy.

8. NASA has undertaken extensive research on wind shear, as evidenced by numerous reports listed in the recommended readings section following this study. For introduction to the subject, see NASA Langley Research Center, "Windshear," *http://oea.larc.nasa.gov/PAIS/Windshear.html*, accessed July 30, 2009; Integrated Publishing, "Meterology: Low-Level Wind Shear," *http://www.tpub.com/weather3/6-15.htm*, accessed July 25, 2009; Amos A. Spady, Jr., Roland L. Bowles, and Herbert Schlickenmaier, eds., *Airborne Wind Shear Detection and Warning Systems, Second Combined Manufacturers and Technological Conference*, two parts, NASA CP-10050 (1990); U.S. National Academy of Sciences, Committee on Low-Altitude Wind Shear and Its Hazard to Aviation, *Low Altitude Wind Shear and Its Hazard to Aviation* (Washington, DC: National Academy Press, 1983); and Dan D. Vicroy, "Influence of Wind Shear on the Aerodynamic Characteristics of Airplanes," NASA TP-2827 (1988).

9. Department of Atmospheric Sciences, University of Illinois-Champaign, "Jet Stream," *http://ww2010.atmos.uiuc.edu/%28Gh%29/guides/mtr/cyc/upa/jet.rxml*, accessed July 25, 2009. Lightning aspects of the thunderstorm risk are addressed in an essay by Barrett Tillman and John Tillman in this volume.

near the ground. There are both "wet" and "dry" microbursts. Before the developing of forward-looking detection and evasion strategies, it was a major cause of aircraft accidents, claiming 26 aircraft and 626 lives, with over 200 injured, between 1964 and 1985.[10]

Another macro-level weather event associated with wind shear is an upper-level jetstream, which contains vertical and horizontal wind shear at its edges. Jetstreams are fast-flowing, narrow air currents found at certain areas of the tropopause. The tropopause is the transition between the troposphere (the area in the atmosphere where most weather changes occur and temperature decreases with height) and the stratosphere (the area where temperature increases with height).[11] A combination of atmospheric heating (by solar radiation or internal planetary heat) and the planet's rotation on its axis causes jetstreams to form. The strongest jetstreams on Earth are the polar jets (23,000–39,000 feet above sea level) and the higher and somewhat weaker subtropical jets (33,000–52,000 feet). Both the northern and southern hemispheres have a polar jet and a subtropical jet. Wind shear in the upper-level jetstream causes clear air turbulence. The cold-air side of the jet, next to the jet's axis, is where CAT is usually strongest.[12]

Although most aircraft passengers experience clear air turbulence as a minor annoyance, this kind of turbulence can be quite hazardous to aircraft when it becomes severe. It has caused fatalities, as in the case of United Airlines Flight 826.[13] Flight 826 took off from Narita International Airport in Japan for Honolulu, HI, on December 28, 1997.

10. Statistic from Emedio M. Bracalente, C.L. Britt, and W.R. Jones, "Airborne Doppler Radar Detection of Low Altitude Windshear," AIAA Paper 88-4657 (1988); see also Joseph R. Chambers, *Concept to Reality: Contributions of the NASA Langley Research Center to U.S. Civil Aircraft of the 1990s*, NASA SP-2003-4529 (Washington, DC: GPO, 2003), p. 185; NASA Langley Research Center, "Windshear," *http://oea.larc.nasa.gov/PAIS/Windshear.html*, accessed July 30, 2009.
11. U.S. Department of Energy, "Ask a Scientist," *http://www.newton.dep.anl.gov/aas.htm*, accessed Aug. 5, 2009.
12. BBC News, "Jet Streams in the UK," *http://www.bbc.co.uk/weather/features/understanding/jetstreams_uk.shtml*, accessed July 30, 2009; M.P. de Villiers and J. van Heerden, "Clear Air Turbulence Over South Africa," *Meteorological Applications*, vol. 8 (2001), pp. 119–126; T.L. Clark, W.D. Hall, et al., "Origins of Aircraft-Damaging Clear-Air Turbulence During the 9 December 1992 Colorado Downslope Windstorm: Numerical Simulations and Comparison with Observations," *Journal of Atmospheric Sciences*, vol. 57 (Apr. 2000), p. 20.
13. National Transportation Safety Board, "Aircraft Accident Investigation Press Release: United Airlines Flight 826," *http://www.ntsb.gov/Pressrel/1997/971230.htm*, accessed July 30, 2009.

At 31,000 feet, 2 hours into the flight, the crew of the plane, a Boeing 747, received warning of severe clear air turbulence in the area. A few minutes later, the plane abruptly dropped 100 feet, injuring many passengers and forcing an emergency return to Tokyo, where one passenger subsequently died of her injuries.[14] A low-level jetstream is yet another phenomenon causing wind shear. This kind of jetstream usually forms at night, directly above Earth's surface, ahead of a cold front. Low-level vertical wind shear develops in the lower part of the low-level jet. This kind of wind shear is also known as nonconvective wind shear, because it is not caused by thunderstorms.

The term "jetstream" is often used without further modification to describe Earth's Northern Hemisphere polar jet. This is the jet most important for meteorology and aviation, because it covers much of North America, Europe, and Asia, particularly in winter. The Southern Hemisphere polar jet, on the other hand, circles Antarctica year-round.[15] Commercial use of the Northern Hemisphere polar jet began November 18, 1952, when a Boeing 377 Stratocruiser of Pan American Airlines first flew from Tokyo to Honolulu at an altitude of 25,000 feet. It cut the trip time by over one-third, from 18 to 11.5 hours.[16] The jetstream saves fuel by shortening flight duration, since an airplane flying at high altitude can attain higher speeds because it is passing through less-dense air. Over North America, the time needed to fly east across the continent can be decreased by about 30 minutes if an airplane can fly with the jetstream but can increase by more than 30 minutes it must fly against the jetstream.[17]

Strong gusts of wind are another natural phenomenon affecting aviation. The National Weather Service reports gusts when top wind speed reaches 16 knots and the variation between peaks and lulls reaches 9 knots.[18] A gust load is the wind load on a surface caused by gusts.

14. Aviation Safety Network, "ASN Aircraft accident Boeing 747 Tokyo," *http://aviation-safety.net/database/record.php?id=19971228-0*, accessed July 4, 2009.

15. U.S. Department of Energy, "Ask a Scientist," *http://www.newton.dep.anl.gov/aas.htm*, accessed Aug. 20, 2009.

16. M.D. Klaas, "Stratocruiser: Part Three," *Air Classics* (June 2000), at *http://findarticles.com/p/articles/mi_qa3901/is_200006/ai_n8911736/pg_2/*, accessed July 8, 2009.

17. Ned Rozell, Alaska Science Forum, "Amazing flying machines allow time travel," *http://www.gi.alaska.edu/ScienceForum/ASF17/1727.html*, accessed July 8, 2009.

18. U.S. Weather Service, "Wind Gust," *http://www.weather.gov/forecasts/wfo/definitions/defineWindGust.html*, accessed Aug. 1, 2009.

Otto Lilienthal, the greatest of pre-Wright flight researchers, in flight. National Air and Space Museum.

The more physically fragile a surface, the more danger a gust load will pose. As well, gusts can have an upsetting effect upon the aircraft's flightpath and attitude.

Initial NACA–NASA Research

Sudden gusts and their effects upon aircraft have posed a danger to the aviator since the dawn of flight. Otto Lilienthal, the inventor of the hang glider and arguably the most significant aeronautical researcher before the Wright brothers, sustained fatal injuries in an 1896 accident, when a gust lifted his glider skyward, died away, and left him hanging in a stalled flight condition. He plunged to Earth, dying the next day, his last words reputedly being "Opfer müssen gebracht werden"—or "Sacrifices must be made."[19]

NASA's interest in gust and turbulence research can be traced to the earliest days of its predecessor, the NACA. Indeed, the first NACA

19. Richard P. Hallion, *Taking Flight: Inventing the Aerial Age from Antiquity Through the First World War* (New York: Oxford University Press, 2003), p. 161.

technical report, issued in 1917, examined the behavior of aircraft in gusts.[20] Over the first decades of flight, the NACA expanded its interest in gust research, looking at the problems of both aircraft and lighter-than-air airships. The latter had profound problems with atmospheric turbulence and instability: the airship Shenandoah was torn apart over Ohio by violent stormwinds; the Akron was plunged into the Atlantic, possibly from what would now be considered a microburst; and the Macon was doomed when clear air turbulence ripped off a vertical fin and opened its gas cells to the atmosphere. Dozens of airmen lost their lives in these disasters.[21]

During the early part of the interwar years, much research on turbulence and wind behavior was undertaken in Germany, in conjunction with the development of soaring, and the long-distance and long-endurance sailplane. Conceived as a means of preserving German aeronautical skills and interest in the wake of the Treaty of Versailles, soaring evolved as both a means of flight and a means to study atmospheric behavior. No airman was closer to the weather, or more dependent upon an understanding of its intricacies, than the pilot of a sailplane, borne aloft only by thermals and the lift of its broad wings. German soaring was always closely tied to the nation's excellent technical institutes and the prestigious aerodynamics research of Ludwig Prandtl and the Prandtl school at Göttingen. Prandtl himself studied thermals, publishing a research paper on vertical air currents in 1921, in the earliest years of soaring development.[22] One of the key figures in German sailplane development was Dr. Walter Georgii, a wartime meteorologist who headed the postwar German Research Establishment for Soaring Flight (Deutsche Forschungsanstalt für Segelflug ([DFS]). Speaking before

20. J.C. Hunsaker and Edwin Bidwell Wilson, "Report on Behavior of Aeroplanes in Gusts," NACA TR-1 (1917); see also Edwin Bidwell Wilson, "Theory of an Airplane Encountering Gusts," pts. II and III, NACA TR-21 and TR-27 (1918).

21. For an example of NACA research, see C.P. Burgess, "Forces on Airships in Gusts," NACA TR-204 (1925). These—and other—airship disasters are detailed in Douglas A. Robinson, *Giants in the Sky: A History of the Rigid Airship* (Seattle: University of Washington Press, 1973).

22. Ludwig Prandtl, "Some Remarks Concerning Soaring Flight," NACA Technical Memorandum No. 47 (Oct. 1921), a translation of a German study; Howard Siepen, "On the Wings of the Wind," *The National Geographic Magazine*, vol. 55, no. 6 (June 1929), p. 755. For an example of later research, see Max Kramer, "Increase in the Maximum Lift of an Airplane Wing due to a Sudden Increase in its Effective Angle of Attack Resulting from a Gust," NACA TM-678 (1932), a translation of a German study.

Britain's Royal Aeronautical Society, he proclaimed, "Just as the master of a great liner must serve an apprenticeship in sail craft to learn the secret of sea and wind, so should the air transport pilot practice soaring flights to gain wider knowledge of air currents, to avoid their dangers and adapt them to his service."[23] His DFS championed weather research, and out of German soaring, came such concepts as thermal flying and wave flying. Soaring pilot Max Kegel discovered firsthand the power of storm-generated wind currents in 1926. They caused his sailplane to rise like "a piece of paper that was being sucked up a chimney," carrying him almost 35 miles before he could land safely.[24] Used discerningly, thermals transformed powered flight from gliding to soaring. Pioneers such as Gunter Grönhoff, Wolf Hirth, and Robert Kronfeld set notable records using combinations of ridge lift and thermals. On July 30, 1929, the courageous Grönhoff deliberately flew a sailplane with a barograph into a storm, to measure its turbulence; this flight anticipated much more extensive research that has continued in various nations.[25]

The NACA first began to look at thunderstorms in the 1930s. During that decade, the Agency's flagship laboratory—the Langley Memorial Aeronautical Laboratory in Hampton, VA—performed a series of tests to determine the nature and magnitude of gust loadings that occur in storm systems. The results of these tests, which engineers performed in Langley's signature wind tunnels, helped to improve both civilian and military aircraft.[26] But wind tunnels had various limitations, leading to use of specially instrumented research airplanes to effectively use the sky as a laboratory and acquire information unobtainable by traditional tunnel research. This process, most notably associated with the post–World War II X-series of research airplanes, led in time to such future NASA research aircraft as the Boeing 737 "flying laboratory" to study wind shear. Over subsequent decades, the NACA's successor, NASA,

23. Walter Georgii, "Ten Years' Gliding and Soaring in Germany," *Journal of the Royal Aeronautical Society,* vol. 34, no. 237 (Sept. 1930), p. 746.

24. Siepen, "On the Wings of the Wind," p. 771.

25. Ibid., pp. 735–741; see also B.S. Shenstone and S. Scott Hall's "Glider Development in Germany: A Technical Survey of Progress in Design in Germany Since 1922," NACA TM No. 780 (Nov. 1935), pp. 6–8.

26. See also James R. Hansen, *Engineer in Charge: A History of the Langley Aeronautical Laboratory, 1917–1958,* NASA SP-4305 (Washington, DC: GPO, 1987), p. 181; and Hansen, *The Bird is on the Wing: Aerodynamics and the Progress of the American Airplane* (College Station, TX: Texas A&M University Press, 2003), p. 73.

would perform much work to help planes withstand turbulence, wind shear, and gust loadings.

From the 1930s to the 1950s, one of the NACA's major areas of research was the nature of the boundary layer and the transition from laminar to turbulent flow around an aircraft. But Langley Laboratory also looked at turbulence more broadly, to include gust research and meteorological turbulence influences upon an aircraft in flight. During the previous decade, experimenters had collected measurements of pressure distribution in wind tunnels and flight, but not until the early 1930s did the NACA begin a systematic program to generate data that could be applied by industry to aircraft design, forming a committee to oversee loads research. Eventually, in the late 1930s, Langley created a separate structures research division with a structures research laboratory. By this time, individuals such as Philip Donely, Walter Walker, and Richard V. Rhode had already undertaken wideranging and influential research on flight loads that transformed understanding about the forces acting on aircraft in flight. Rhode, of Langley, won the Wright Brothers Medal in 1935 for his research of gust loads. He pioneered the undertaking of detailed assessments of the maneuvering loads encountered by an airplane in flight. As noted by aerospace historian James Hansen, his concept of the "sharp edge gust" revised previous thinking of gust behavior and the dangers it posed, and it became "the backbone for all gust research."[27] NACA gust loads research influenced the development of both military and civilian aircraft, as did its research on aerodynamic-induced flight-surface flutter, a problem of particular concern as aircraft design transformed from the era of the biplane to that of the monoplane. The NACA also investigated the loads and stresses experienced by combat aircraft when undertaking abrupt rolling and pullout maneuvers, such as routinely occurred in aerial dogfighting and in dive-bombing.[28] A dive bomber encountered particularly punishing aerodynamic and structural loads as the pilot executed a pullout: abruptly recovering the airplane from a dive and resulting in it

27. Ibid., p. 73; for Rhode's work on maneuver loads, see R.V. Rhode, "The Pressure Distribution over the Horizontal and Vertical Tail Surfaces of the F6C-4 Pursuit Airplane in Violent Maneuvers," NACA TR-307 (1929).

28. For example, C.H. Dearborn and H.W. Kirschbaum, "Maneuverability Investigation of the F6C-3 Airplane with Special Flight Instruments," NACA TR-369 (1932); and Philip Donely and Henry A. Pearson, "Flight and Wind-Tunnel Tests of an XBM-1 Dive Bomber," NACA TN-644 (1938).

swooping back into the sky. Researchers developed charts showing the relationships between dive angle, speed, and the angle required for recovery. In 1935, the Navy used these charts to establish design requirements for its dive bombers. The loads program gave the American aeronautics community a much better understanding of load distributions between the wing, fuselage, and tail surfaces of aircraft, including high-performance aircraft, and showed how different extreme maneuvers "loaded" these individual surfaces.

In his 1939 Wilbur Wright lecture, George W. Lewis, the NACA's legendary Director of Aeronautical Research, enumerated three major questions he believed researchers needed to address:

- What is the nature or structure of atmospheric gusts?
- How do airplanes react to gusts of known structure?
- What is the relation of gusts to weather conditions?[29]

Answering these questions, posed at the close of the biplane era, would consume researchers for much of the next six decades, well into the era of jet airliners and supersonic flight.

The advent of the internally braced monoplane accelerated interest in gust research. The long, increasingly thin, and otherwise unsupported cantilever wing was susceptible to load-induced failure if not well-designed. Thus, the stresses caused by wind gusts became an essential factor in aircraft design, particularly for civilian aircraft. Building on this concern, in 1943, Philip Donely and a group of NACA researchers began design of a gust tunnel at Langley to examine aircraft loads produced by atmospheric turbulence and other unpredictable flow phenomena and to develop devices that would alleviate gusts. The tunnel opened in August 1945. It utilized a jet of air for gust simulation, a catapult for launching scaled models into steady flight, curtains for catching the model after its flight through the gust, and instruments for recording the model's responses. For several years, the gust tunnel was useful, "often [revealing] values that were not found by the best known methods of calculation . . . in one instance, for example, the gust tunnel tests showed that it would be safe to design the airplane for load increments 17 to 22 percent less than the previously accepted

29. George W. Gray, *Frontiers of Flight: the Story of NACA Research* (New York: Alfred A. Knopf, 1948), p. 173.

The experimental Boeing XB-15 bomber was instrumented by the NACA to acquire gust-induced structural loads data. NASA.

values."[30] As well, gust researchers took to the air. Civilian aircraft—such as the Aeronca C-2 light, general-aviation airplane, Martin M-130 flying boat, and the Douglas DC-2 airliner—and military aircraft, such as the Boeing XB-15 experimental bomber, were outfitted with special loads recorders (so-called "v-g recorders," developed by the NACA). Extensive records were made on the weather-induced loads they experienced over various domestic and international air routes.[31]

This work was refined in the postwar era, when new generations of long-range aircraft entered air transport service and were also instrumented to record the loads they experienced during routine airline

30. Ibid., p. 174; Hansen, *Engineer in Charge*, p. 468. NACA researchers created the gust tunnel to provide information to verify basic concepts and theories. It ultimately became obsolete because of its low Reynolds and Mach number capabilities. After being used as a low-velocity instrument laboratory and noise research facility, the gust tunnel was dismantled in 1965.
31. Philip Donely, "Effective Gust Structure at Low Altitudes as Determined from the Reactions of an Airplane," NACA TR-692 (1940); Walter G. Walker, "Summary of V-G Records Taken on Transport Airplanes from 1932 to 1942," NACA WRL-453 (1942); Donely, "Frequency of Occurrence of Atmospheric Gusts and of Related Loads on Airplane Structures," NACA WRL-121 (1944); Walker, "An Analysis of the Airspeeds and Normal Accelerations of Martin M-130 Airplanes in Commercial Transport Operation," NACA TN-1693 (1948); and Walker, "An Analysis of the Airspeed and Normal Accelerations of Douglas DC-2 Airplanes in Commercial Transport Operations," NACA TN-1754 (1948).

operation.[32] Gust load effects likewise constituted a major aspect of early transonic and supersonic aircraft testing, for the high loads involved in transiting from subsonic to supersonic speeds already posed a serious challenge to aircraft designers. Any additional loading, whether from a wind gust or shear, or from the blast of a weapon (such as the over-pressure blast wave of an atomic weapon), could easily prove fatal to an already highly loaded aircraft.[33] The advent of the long-range jet bomber and transport—a configuration typically having a long and relatively thin swept wing, and large, thin vertical and horizontal tail surfaces—added further complications to gust research, particularly because the penalty for an abrupt gust loading could be a fatal structural failure. Indeed, on one occasion, while flying through gusty air at low altitude, a Boeing B-52 lost much of its vertical fin, though fortunately, its crew was able to recover and land the large bomber.[34]

The emergence of long-endurance, high-altitude reconnaissance aircraft such as the Lockheed U-2 and Martin RB-57D in the 1950s and the long-range ballistic missile further stimulated research on high-altitude gusts and turbulence. Though seemingly unconnected, both the high-altitude jet airplane and the rocket-boosted ballistic missile required understanding of the nature of upper atmosphere turbulence and gusts. Both transited the upper atmospheric region: the airplane cruising in the high stratosphere for hours, and the ballistic missile

32. Donely, "Summary of Information Relating to Gust Loads on Airplanes," NACA TR-997 (1950); Walker, "Gust Loads and Operating Airspeeds of One Type of Four-Engine Transport Airplane on Three Routes from 1949 to 1953," NACA TN-3051 (1953); and Kermit G. Pratt and Walker, "A Revised Gust-Load Formula and a Re-Evaluation of V-G Data Taken on Civil Transport Airplanes from 1933 to 1950," NACA TR-1206 (1954).

33. For example, E.T. Binckley and Jack Funk, "A Flight Investigation of the Effects of Compressibility on Applied Gust Loads," NACA TN-1937 (1949); and Harvard Lomax, "Lift Developed on Unrestrained Rectangular Wings Entering Gusts at Subsonic and Supersonic Speeds," NACA TN-2925 (1953).

34. Jack Funk and Richard H. Rhyne, "An Investigation of the Loads on the Vertical Tail of a Jet-Bomber Airplane Resulting from Flight Through Rough Air," NACA TN-3741 (1956); Philip Donely, "Safe Flight in Rough Air," NASA TMX-51662 (1964); W.H. Andrews, S.P. Butchart, T.R. Sisk, and D.L. Hughes, "Flight Tests Related to Jet-Transport Upset and Turbulent-Air Penetration," and R.S. Bray and W.E. Larsen, "Simulator Investigations of the Problems of Flying a Swept-Wing Transport Aircraft in Heavy Turbulence," both in NASA LRC, Conference on Aircraft Operating Problems, NASA SP-83 (1965); M. Sadoff, R.S. Bray, and W.H. Andrews, "Summary of NASA Research on Jet Transport Control Problems in Severe Turbulence," AIAA Paper 65-330 (1965); and Richard J. Wasicko, "NASA Research Experience on Jet Aircraft Control Problems in Severe Turbulence," NASA TM-X-60179 (1966).

or space launch vehicle transiting through it within seconds on its way into space. Accordingly, from early 1956 through December 1959, the NACA, in cooperation with the Air Weather Service of the U.S. Air Force, installed gust load recorders on Lockheed U-2 strategic reconnaissance aircraft operating from various domestic and overseas locations, acquiring turbulence data from 20,000 to 75,000 feet over much of the Northern Hemisphere. Researchers concluded that the turbulence problem would not be as severe as previous estimates and high-altitude balloon studies had indicated.[35]

High-altitude loitering aircraft such as the U-2 and RB-57 were followed by high-altitude, high-Mach supersonic cruise aircraft in the early to mid-1960s, typified by Lockheed's YF-12A Blackbird and North American's XB-70A Valkyrie, both used by NASA as Mach 3+ Supersonic Transport (SST) surrogates and supersonic cruise research testbeds. Test crews found their encounters with high-altitude gusts at supersonic speeds more objectionable than their exposure to low-altitude gusts at subsonic speeds, even though the given g-loading accelerations caused by gusts were less than those experienced on conventional jet airliners.[36] At the other extreme of aircraft performance, in 1961, the Federal Aviation Agency (FAA) requested NASA assistance to document the gust and maneuver loads and performance of general-aviation aircraft. Until the program was terminated in 1982, over 35,000 flight-hours of data were assembled from 95 airplanes, representing every category of general-aviation airplane, from single-engine personal craft to twin-engine business airplanes and including such specialized types as crop-dusters and aerobatic aircraft.[37]

---

35. Thomas L. Coleman and Emilie C. Coe, "Airplane Measurements of Atmospheric Turbulence for Altitudes Between 20,000 and 55,000 Feet Over the Western part of the United States," NACA RM-L57G02 (1957); and Thomas L. Coleman and Roy Steiner, "Atmospheric Turbulence Measurements Obtained from Airplane Operations at Altitudes Between 20,000 and 75,000 Feet for Several Areas in the Northern Hemisphere," NASA TN-D-548 (1960).

36. Eldon E. Kordes and Betty J. Love, "Preliminary Evaluation of XB-70 Airplane Encounters with High-Altitude Turbulence," NASA TN-D-4209 (1967); L.J. Ehernberger and Betty J. Love, "High Altitude Gust Acceleration Environment as Experienced by a Supersonic Airplane," NASA TN-D-7868 (1975). NASA's supersonic cruise flight test research is the subject of an accompanying essay in this volume by William Flanagan, a former Air Force Blackbird navigator.

37. Joseph W. Jewel, Jr., "Tabulations of Recorded Gust and Maneuver Accelerations and Derived Gust Velocities for Airplanes in the NSA VGH General Aviation Program," NASA TM-84660 (1983).

Along with studies of the upper atmosphere by direct measurement came studies on how to improve turbulence detection and avoidance, and how to measure and simulate the fury of turbulent storms. In 1946–1947, the U.S. Weather Bureau sponsored a study of turbulence as part of a thunderstorm study project. Out of this effort, in 1948, researchers from the NACA and elsewhere concluded that ground radar, if properly used, could detect storms, enabling aircraft to avoid them. Weather radar became a common feature of airliners, their once-metal nose caps replaced by distinctive black radomes.[38] By the late 1970s, most wind shear research was being done by specialists in atmospheric science, geophysical scientists, and those in the emerging field of mesometeorology—the study of small atmospheric phenomena, such as thunderstorms and tornadoes, and the detailed structure of larger weather events.[39] Although turbulent flow in the boundary layer is important to study in the laboratory, the violent phenomenon of microburst wind shear cannot be sufficiently understood without direct contact, investigation, and experimentation.[40]

Microburst loadings constitute a threat to aircraft, particularly during approach and landing. No one knows how many aircraft accidents have been caused by wind shear, though the number is certainly considerable. The NACA had done thunderstorm research during World War II, but its instrumentation was not nearly sophisticated enough to detect microburst (or thunderstorm downdraft) wind shear. NASA would join with the FAA in 1986 to systematically fight wind shear and would only have a small pool of existing wind shear research data from which to draw.[41]

---

38. Robert W. Miller, "The Use of Airborne Navigational and Bombing Radars for Weather-Radar Operations and Verifications," *Bulletin of the American Meteorological Society*, vol. 28, no. 1 (Jan. 1947), pp. 19–28; H. Press and E.T. Binckley, "A Preliminary Evaluation of the Use of Ground Radar for the Avoidance of Turbulent Clouds," NACA TN-1864 (1948).

39. W. Frost and B. Crosby, "Investigations of Simulated Aircraft Flight Through Thunderstorm Outflows," NASA CR-3052 (1978); Norbert Didden and Chi-Minh Ho, Department of Aerospace Engineering, University of Southern California, "Unsteady Separation in a Boundary Layer Produced by an Impinging Jet," *Journal of Fluid Mechanics*, vol. 160 (1985), pp. 235–236.

40. See, for example, Paul A. Robinson, Roland L. Bowles, and Russell Targ, "The Detection and Measurement of Microburst Wind Shear by an Airborne Lidar System," NASA LRC, NTRS Report 95A87798 (1993); Dan D. Vicroy, "A Simple, Analytical, Axisymmetric Microburst Model for Downdraft Estimation," NASA TM-104053 (1991); and Vicroy, "Assessment of Microburst Models for Downdraft Estimation," AIAA Paper 91-2947 (1991).

41. Hansen, *The Bird is on the Wing*, p. 207.

The Lockheed L-1011 TriStar uses smoke generators to show its strong wing vortex flow patterns in 1977. NASA.

A revealing view taken down the throat of a wingtip vortex, formed by a low-flying crop-duster. NASA.

Wind Shear Emerges as an Urgent Aviation Safety Issue

In 1972, the FAA had instituted a small wind shear research program, with emphasis upon developing sensors that could plot wind speed and direction from ground level up to 2,000 feet above ground level (AGL). Even so, the agency's major focus was on wake vortex impingement. The powerful vortexes streaming behind newer-generation wide-body aircraft could—and sometimes did—flip smaller, lighter aircraft out of control. Serious enough at high altitude, these inadvertent excursions could be disastrous if low over the ground, such as during landing and takeoff, where a pilot had little room to recover. By 1975, the FAA had developed an experimental Wake Vortex Advisory System, which it installed later that year at Chicago's busy O'Hare International Airport. NASA undertook a detailed examination of wake vortex studies, both in tunnel tests and with a variety of aircraft, including the Boeing 727 and 747, Lockheed L-1011, and smaller aircraft, such as the Gates Learjet, helicopters, and general-aviation aircraft.

But it was wind shear, not wake vortex impingement, which grew into a major civil aviation concern, and the onset came with stunning and deadly swiftness.[42] Three accidents from 1973 to 1975 highlighted the extreme danger it posed. On the afternoon of December 17, 1973, while making a landing approach in rain and fog, an Iberia Airlines McDonnell-Douglas DC-10 wide-body abruptly sank below the glide-slope just seconds before touchdown, impacting amid the approach lights of Runway 33L at Boston's Logan Airport. No one died, but the crash seriously injured 16 of the 151 passengers and crew. The subsequent National Transportation Safety Board (NTSB) report determined "that the captain did not recognize, and may have been unable to recognize an increased rate of descent" triggered "by an encounter with a low-altitude wind shear at a critical point in the landing approach."[43] Then, on June 24, 1975, Eastern Air Lines' Flight 66, a Boeing 727, crashed on approach to John F. Kennedy International Airport's Runway 22L. This time, 113 of the 124 passengers and crew perished. All afternoon, flights had encountered and reported wind shear conditions, and at least one pilot had recommended closing the runway. Another Eastern captain, flying a Lockheed L-1011 TriStar, prudently abandoned his approach and landed instead at Newark. Shortly after the L-1011 diverted, the EAL Boeing 727 impacted almost a half mile short of the runway threshold, again amid the approach lights, breaking apart and bursting into flames. Again, wind shear was to blame, but the NTSB also faulted Kennedy's air traffic controllers for not diverting the 727 to another runway, after the EAL TriStar's earlier aborted approach.[44]

Just weeks later, on August 7, Continental Flight 426, another Boeing 727, crashed during a stormy takeoff from Denver's Stapleton

42. William J. Cox, "The Multi-Dimensional Nature of Wind Shear Investigations," in *Society of Experimental Test Pilots, 1976 Report to the Aerospace Profession: Proceedings of the Twentieth Symposium of The Society of Experimental Test Pilots, Beverly Hills, CA, Sept. 22–25, 1976,* vol. 13, no. 2 (Lancaster, CA: Society of Experimental Test Pilots, 1976).

43. National Transportation Safety Board, "Aircraft Accident Report: Iberia Lineas Aereas de España (Iberian Airlines), McDonnell-Douglas DC-10-30, EC CBN, Logan International Airport, Boston, Massachusetts, December 17, 1973," Report NTSB-AAR-74-14 (Nov. 8, 1974).

44. "Aviation: A Fatal Case of Wind Shear," *Time* (July 7, 1975); National Transportation Safety Board, "Aircraft Accident Report: Eastern Air Lines Inc. Boeing 727-225, N8845E, John F. Kennedy International Airport, Jamaica, New York, June 14, 1975," Report NTSB-AAR-76-8 (Mar. 12, 1976); Edmund Preston, *Troubled Passage: The Federal Aviation Administration during the Nixon-Ford Term, 1973–1977* (Washington, DC: FAA, 1987), p. 197.

International Airport. Just as the airliner began its climb after lifting off the runway, the crewmembers encountered a wind shear so severe that they could not maintain level flight despite application of full power and maintenance of a flight attitude that ensured the wings were producing maximum lift.[45] The plane pancaked in level attitude on flat, open ground, sustaining serious damage. No lives were lost, though 15 of the 134 passengers and crew were injured.

Less than a year later, on June 23, 1976, Allegheny Airlines Flight 121, a Douglas DC-9 twin-engine medium-range jetliner, crashed during an attempted go-around at Philadelphia International Airport. The pilot, confronting "severe horizontal and vertical wind shears near the ground," abandoned his landing approach to Runway 27R. As controllers in the airport tower watched, the straining DC-9 descended in a nose-high attitude, pancaking onto a taxiway and sliding to a stop. The fact that it hit nose-high, wings level, and on flat terrain undoubtedly saved lives. Even so, 86 of the plane's 106 passengers and crew were seriously injured, including the entire crew.[46]

In these cases, wind shear brought about by thunderstorm downdrafts (microbursts), rather than the milder wind shear produced by gust fronts, caused these accidents. This led to a major reinterpretation of the wind shear–causing phenomena that most endangered low-flying planes. Before these accidents, meteorologists believed that gust fronts, or the leading edge of a large dome of rain-cooled air, provided the most dangerous sources of wind shear. Now, using data gathered from the planes that had crashed and from weather radar, scientists, engineers, and designers came to realize that the small, focused, jet-like downdraft columns characteristic of microbursts produced the most threatening kind of wind shear.[47]

Microburst wind shear poses an insidious danger for an aircraft. An aircraft landing will typically encounter the horizontal outflow of a microburst as a headwind, which increases its lift and airspeed, tempting

45. U.S. National Transportation Safety Board, "Aircraft Accident Report: Continental Airlines Inc, Boeing 727-224, N88777, Stapleton International Airport, Denver, Colorado, August 7, 1975," Report NTSB-AAR-76-14 (May 5, 1976).
46. National Transportation Safety Board, "Aircraft Accident Report: Allegheny Airlines, Inc., Douglas DC-9, N994VJ, Philadelphia, Pennsylvania, June 23, 1976," Report NTSB-AAR-78-2 (Jan. 19, 1978).
47. For various perspectives on the multiagency research spawned by these accidents, see Amos A. Spady, Jr., Roland L. Bowles, and Herbert Schlickenmaier, eds., *Airborne Wind Shear Detection and Warning Systems, Second Combined Manufacturers and Technological Conference,* two parts, NASA CP-10050 (1990).

1

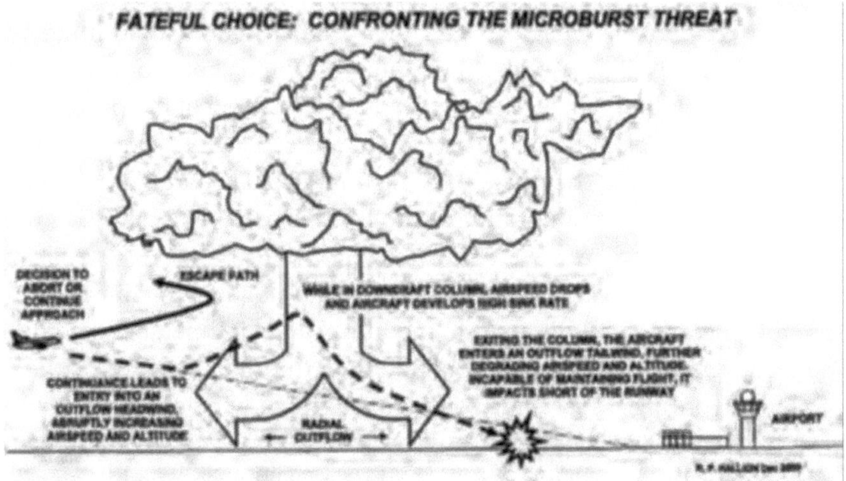

Fateful choice: confronting the microburst threat. Richard P. Hallion.

the pilot to reduce power. But then the airplane encounters the descending vertical column as an abrupt downdraft, and its speed and altitude both fall. As it continues onward, it will exit the central downflow and experience the horizontal outflow, now as a tailwind. At this point, the airplane is already descending at low speed. The tailwind seals its fate, robbing it of even more airspeed and, hence, lift. It then stalls (that is, loses all lift) and plunges to Earth. As NASA testing would reveal, professional pilots generally need between 10 to 40 seconds of warning to avoid the problems of wind shear.[48]

Goaded by these accidents and NTSB recommendations that the FAA improve its weather advisory and runway selection procedures, "step up research on methods of detecting the [wind shear] phenomenon," and develop aircrew wind shear training process, the FAA mandated installation at U.S. airports of a new Low-Level Windshear Alert System (LLWAS), which employed acoustic Doppler radar, technically similar to the FAA's Wake Vortex Advisory System installed at O'Hare.[49] The LLWAS incorporated a variety of equipment that measured wind velocity (wind speed and direction). This equipment included a master station, which had a main computer and system console to monitor LLWAS performance, and a transceiver, which transmitted signals

48. NASA Langley Research Center, "Windshear," *http://oea.larc.nasa.gov/PAIS/Windshear.html*, accessed July 30, 2009.
49. Preston, *Troubled Passage*, p. 197.

to the system's remote stations. The master station had several visual computer displays and auditory alarms for aircraft controllers. The remote stations had wind sensors made of sonic anemometers mounted on metal pipes. Each remote station was enclosed in a steel box with a radio transceiver, power supplies, and battery backup. Every airport outfitted with this system used multiple anemometer stations to effectively map the nature of wind events in and around the airport's runways.[50]

At the end of March 1981, over 70 representatives from NASA, the FAA, the military, the airline community, the aerospace industry, and academia met at the University of Tennessee Space Institute in Tullahoma to explore weather-related aviation issues. Out of that came a list of recommendations for further joint research, many of which directly addressed the wind shear issue and the need for better detection and warning systems. As the report summarized:

> 1. There is a critical need to increase the data base for wind and temperature aloft forecasts both from a more frequent updating of the data as well as improved accuracy in the data, and thus, also in the forecasts which are used in flight planning. This will entail the development of rational definitions of short term variations in intensity and scale length (of turbulence) which will result in more accurate forecasts which should also meet the need to improve numerical forecast modeling requirements relative to winds and temperatures aloft.
> 2. The development of an on-board system to detect wind induced turbulence should be beneficial to meeting the requirement for an investigation of the subjective evaluation of turbulence "feel" as a function of motion drive algorithms.
> 3. More frequency reporting of wind shift in the terminal area is needed along with greater accuracy in forecasting.
> 4. There is a need to investigate the effects of unequal wind components acting across the span of an airfoil.

50. Ibid., pp. 197–198; Cox, "Multi-Dimensional Nature," pp. 141–142. Anemometers are tools that originated in the late Middle Ages and measure wind speed. The first anemometer, a deflection anemometer, was developed by Leonardo da Vinci. Several new varieties, including cup, pressure, and sonic anemometers, have emerged in the intervening centuries.

5. The FAA Simulator Certification Division should monitor the work to be done in conjunction with the JAWS project relative to the effects of wind shear on aircraft performance.
6. Robert Steinberg's ASDAR effort should be utilized as soon as possible, in fact it should be encouraged or demanded as an operational system beneficial for flight planning, specifically where winds are involved.
7. There is an urgent need to review the way pilots are trained to handle wind shear. The present method, as indicated in the current advisory circular, of immediately pulling to stick shaker on encountering wind shear could be a dangerous procedure. It is suggested the circular be changed to recommend the procedure to hold at whatever airspeed the aircraft is at when the pilot realizes he is encountering a wind shear and apply maximum power, and that he not pull to stick shaker except to flare when encountering ground effect to minimize impact or to land successfully or to effect a go-around.
8. Need to develop a clear non-technical presentation of wind shear which will help to provide improved training for pilots relative to wind shear phenomena. Such training is of particular importance to pilots of high performance, corporate, and commercially used aircraft.
9. Need to develop an ICAO type standard terminology for describing the effects of windshear on flight performance.
10. The ATC system should be enhanced to provide operational assistance to pilots regarding hazardous weather areas and in view of the envisioned controller workloads generated, perfecting automated transmissions containing this type of information to the cockpit as rapidly and as economically as practicab1e.
11. In order to improve the detection in real time of hazardous weather, it is recommended that FAA, NOAA, NWS, and DOD jointly address the problem of fragmental meteorological collection, processing, and dissemination pursuant to developing a system dedicated to making effective use of perishable weather information. Coupled with this would be the need to conduct a cost

benefit study relative to the benefits that could be realized through the use of such items as a common winds and temperature aloft reporting by use of automated sensors on aircraft.

12. Develop a capabi1ity for very accurate four to six minute forecasts of wind changes which would require terminal reconfigurations or changing runways.
13. Due to the inadequate detection of clear air turbulence an investigation is needed to determine what has happened to the promising detection systems that have been reported and recommended in previous workshops.
14. Improve the detection and warning of windshear by developing on-board sensors as well as continuing the development of emerging technology for ground-based sensors.
15. Need to collect true three and four dimensional wind shear data for use in flight simulation programs.
16. Recommend that any systems whether airborne or ground based that can provide advance or immediate alert to pilots and controllers should be pursued.
17. Need to continue the development of Doppler radar technology to detect the wind shear hazard, and that this be continued at an accelerated pace.
18. Need for airplane manufacturers to take into consideration the effect of phenomena such as microbursts which produce strong periodic longitudinal wind perturbations at the aircraft phugoid frequency.
19. Consideration should be given, by manufacturers, to consider gust alleviation devices on new aircraft to provide a softer ride through turbulence.
20. Need to develop systems to automatically detect hazardous weather phenomena through signature recognition algorithms and automatically data linking alert messages to pilots and air traffic controllers.[51]

51. Dennis W. Camp, Walter Frost, and Pamela D. Parsley, *Proceedings: Fifth Annual Workshop on Meteorological and Environmental Inputs to Aviation Systems, Mar. 31–Apr. 2, 1981*, NASA CP-2192 (1981).

Given the subsequent history of NASA's research on the wind shear problem (and others), many of these recommendations presciently forecast the direction of Agency and industry research and development efforts.

Unfortunately, that did not come in time to prevent yet another series of microburst-related accidents. That series of catastrophes effectively elevated microburst wind shear research to the status of a national air safety emergency. By the early 1980s, 58 U.S. airports had installed LLWAS. Although LLWAS constituted a great improvement over verbal observations and warnings by pilots communicated to air traffic controllers, LLWAS sensing technology was not mature or sophisticated enough to remedy the wind shear threat. Early LLWAS sensors were installed without fullest knowledge of microburst characteristics. They were usually installed in too-few numbers, placed too close to the airport (instead of farther out on the approach and departure paths of the runways), and, worst, were optimized to detect gust fronts (the traditional pre-Fujita way of regarding wind shear)—not the columnar downdrafts and horizontal outflows characteristic of the most dangerous shear flows. Thus, wind shear could still strike, and viciously so.

On July 9, 1982, Clipper 759, a Pan American World Airways Boeing 727, took off from the New Orleans airport amid showers and "gusty, variable, and swirling" winds.[52] Almost immediately, it began to descend, having attained an altitude of no more than 150 feet. It hit trees, continued onward for almost another half mile, and then crashed into residential housing, exploding in flames. All 146 passengers and crew died, as did 8 people on the ground; 11 houses were destroyed or "substantially" damaged, and another 16 people on the ground were injured. The NTSB concluded that the probable cause of the accident was "the airplane's encounter during the liftoff and initial climb phase of flight with a microburst-induced wind shear which imposed a downdraft and a decreasing headwind, the effects of which the pilot would have had difficulty recognizing and reacting to in time for the airplane's descent to be arrested before its impact with trees." Significantly, it also noted, "Contributing to the accident was the limited capability of current ground based low level wind shear detection technology [the LLWAS] to provide

52. National Transportation Safety Board, "Aircraft Accident Report: Pan American World Airways, Clipper 759, N4737, Boeing 727-235, New Orleans International Airport, Kenner, Louisiana, July 9, 1982," Report NTSB-AAR-83-02 (Mar. 21, 1983).

definitive guidance for controllers and pilots for use in avoiding low level wind shear encounters."[53] This tragic accident impelled Congress to direct the FAA to join with the National Academy of Sciences (NAS) to "study the state of knowledge, alternative approaches and the consequences of wind shear alert and severe weather condition standards relating to take off and landing clearances for commercial and general aviation aircraft."[54]

As the FAA responded to these misfortunes and accelerated its research on wind shear, NASA researchers accelerated their own wind shear research. In the late 1970s, NASA Ames Research Center contracted with Bolt, Baranek, and Newman, Inc., of Cambridge, MA, to perform studies of "the effects of wind-shears on the approach performance of a STOL aircraft . . . using the optimal-control model of the human operator." In laymen's terms, this meant that the company used existing data to mathematically simulate the combined pilot/aircraft reaction to various wind shear situations and to deduce and explain how the pilot should manipulate the aircraft for maximum safety in such situations. Although useful, these studies did not eliminate the wind shear problem.[55] Throughout the 1980s, NASA research into thunderstorm phenomena involving wind shear continued. Double-vortex thunderstorms and their potential effects on aviation were of particular interest. Double-vortex storms involve a pair of vortexes present in the storm's dynamic updraft that rotate in opposite directions. This pair forms when the cylindrical thermal updraft of a thunderstorm penetrates the upper-level air and there is a large amount of vertical wind shear between the lower- and upper-level air layers. Researchers produced a numerical tornado prediction scheme based on the movement of the double-vortex thunderstorm. A component of this scheme was the Energy-Shear Index (ESI), which researchers calculated from radiosonde measurements. The index integrated parameters that were representative of thermal instability and the blocking effect. It indicated

53. Ibid., p. ii.

54. "Wind Shear Study: Low-Altitude Wind Shear," *Aviation Week & Space Technology* (Mar. 28, 1983), p. 32. One outcome was a seminal report completed before the end of the year by the National Academy's Committee on Low-Altitude Wind Shear and Its Hazard to Aviation, *Low Altitude Wind Shear and Its Hazard to Aviation* (Washington, DC: National Academy Press, 1983).

55. Sheldon Baron, Bolt Baranek, et al., *Analysis of Response to Wind-Shears using the Optimal Control Model of the Human Operator*, NASA Ames Research Center Technical Paper NAS2-0652 (Washington, DC: NASA, 1979).

1

NASA 809, a Martin B-57B flown by Dryden research crews in 1982 for gust and microburst research. NASA.

environments appropriate for the development of double-vortex thunderstorms and tornadoes, which would help pilots and flight controllers determine safe flying conditions.[56]

In 1982, in partnership with the National Center for Atmospheric Research (NCAR), the University of Chicago, the National Oceanic Atmospheric Administration (NOAA), the National Science Foundation (NSF), and the FAA, NASA vigorously supported the Joint Airport Weather Studies (JAWS) effort. NASA research pilots and flight research engineers from the Ames-Dryden Flight Research Facility (now the NASA Dryden Flight Research Center) participated in the JAWS program from mid-May through mid-August 1982, using a specially instrumented Martin B-57B jet bomber. NASA researchers selected the B-57B for its strength, flying it on low-level wind shear research flights around the Sierra Mountains near Edwards Air Force Base (AFB), CA, about the Rockies near Denver, CO, around Marshall Space Flight Center, AL, and near Oklahoma City, OK. Raw data were digitally collected on microbursts, gust fronts, mesocyclones, torna-

56. J.R. Connell, et al., "Numeric and Fluid Dynamic Representation of Tornadic Double Vortex Thunderstorms," NASA CR-171023 (1980).

does, funnel clouds, and hail storms; converted into engineering format at the Langley Research Center; and then analyzed at Marshall Space Flight Center and the University of Tennessee Space Institute at Tullahoma. Researchers found that some microbursts recorded during the JAWS program created wind shear too extreme for landing or departing airliners to survive if they encountered it at an altitude less than 500 feet.[57] In the most severe case recorded, the B-57B experienced an abrupt 30-knot speed increase within less than 500 feet of distance traveled and then a gradual decrease of 50 knots over 3.2 miles, clear evidence of encountering the headwind outflow of a microburst and then the tailwind outflow as the plane transited through the microburst.[58]

At the same time, the Center for Turbulence Research (CTR), run jointly by NASA and Stanford University, pioneered using an early parallel computer, the Illiac IV, to perform large turbulence simulations, something previously unachievable. CTR performed the first of these simulations and made the data available to researchers around the globe. Scientists and engineers tested theories, evaluated modeling ideas, and, in some cases, calibrated measuring instruments on the basis of these data. A 5-minute motion picture of simulated turbulent flow provided an attention-catching visual for the scientific community.[59]

In 1984, NASA and FAA representatives met at Langley Research Center to review the status of wind shear research and progress toward developing sensor systems and preventing disastrous accidents. Out of this, researcher Roland L. Bowles conceptualized a joint NASA–FAA

57. National Academy of Sciences, Committee on Low-Altitude Wind Shear and Its Hazard to Aviation, *Low Altitude Wind Shear and Its Hazard to Aviation* (Washington, DC: National Academy Press, 1983), pp. 14–15; Roland L. Bowles, "Windshear Detection and Avoidance: Airborne Systems Survey," *Proceedings of the 29th IEEE Conference on Decision and Control, Honolulu, HI* (New York: IEEE Publications, 1990), p. 708; H. Patrick Adamson, "Development of the Advance Warning Airborne System (AWAS)," paper presented at the *Fourth Combined Manufacturers' and Technologists' Airborne Windshear Review Meeting, Turbulence Prediction Systems, Boulder, CO, Apr. 14, 1992*. JAWS program research continued into the 1990s.
58. John McCarthy, "The Joint Airport Weather Studies (JAWS) Project," in Camp, Frost, and Parsley, *Proceedings: Fifth Annual Workshop on Meteorological and Environmental Inputs to Aviation*, pp. 91–95; and Weneth D. Painter and Dennis W. Camp, "NASA B-57B Severe Storms Flight Program," NASA TM-84921 (1983).
59. Center for Turbulence Research, Stanford University, "About the Center for Turbulence Research (CTR)," *http://www.stanford.edu/group/ctr/about.html*, accessed Oct. 3, 2009. For Illiac IV and its place in computing history, see Paul E. Ceruzzi, *A History of Modern Computing* (Cambridge: The MIT Press, 1999), pp. 196–197.

program to develop an airborne detector system, perhaps one that would be forward-looking and thus able to furnish real-time warning to an airline crew of wind shear hazards in its path. Unfortunately, before this program could yield beneficial results, yet another wind shear accident followed the dismal succession of its predecessors: the crash of Delta Flight 191 at Dallas-Fort Worth International Airport (DFW) on August 2, 1985.[60]

Delta Flight 191 was a Lockheed L-1011 TriStar wide-body jumbo jet. As it descended toward Runway 17L amid a violent turbulence-producing thunderstorm, a storm cell produced a microburst directly in the airliner's path. The L-1011 entered the fury of the outflow when only 800 feet above ground and at a low speed and energy state. As the L-1011 transitioned through the microburst, a lift-enhancing headwind of 26 knots abruptly dropped to zero and, as the plane sank in the downdraft column, then became a 46-knot tailwind, robbing it of lift. At low altitude, the pilots had insufficient room for recovery, and so, just 38 seconds after beginning its approach, Delta Flight 191 plunged to Earth, a mile short of the runway threshold. It broke up in a fiery heap of wreckage, slewing across a highway and crashing into some water tanks before coming to a rest, burning furiously. The accident claimed the lives of 136 passengers and crewmembers and the driver of a passing automobile. Just 24 passengers and 3 of its crew survived: only 2 were without injury.[61] Among the victims were several senior staff members from IBM, including computer pioneer Don Estridge, father of the IBM PC. Once again, the NTSB blamed an "encounter at low altitude with a microburst-induced, severe wind shear" from a rapidly developing thunderstorm on the final approach course. But the accident illustrated as well the immature capabilities of the LLWAS at that time; only after Flight 191 had crashed did the DFW LLWAS detect the fatal microburst.[62]

60. Chambers, *Concept to Reality*, p. 188.

61. National Transportation Safety Board, "Aircraft Accident Report: Delta Air Lines, Inc., Lockheed L-1011-385-1, N726DA, Dallas/Fort Worth International Airport, Texas, August 2, 1985," Report NTSB-AAR-86-05 (Aug. 15, 1986). See also James Ott, "Inquiry Focuses on Wind Shear As Cause of Delta L-1011 Crash," *Aviation Week & Space Technology* (Aug. 12, 1985), pp. 16–19; F. Caracena, R. Ortiz, and J. Augustine, "The Crash of Delta Flight 191 at Dallas-Fort Worth International Airport on 2 August 1985: Multiscale Analysis of Weather Conditions," National Oceanic and Atmospheric Report TR ERL 430-ESG-2 (1987); T. Theodore Fujita, "DFW Microburst on August 2, 1985," Satellite and Mesometeorology Research Project Research Paper 217, Dept. of Geophysical Sciences, University of Chicago, NTIS Report PB-86-131638 (1986).

62. Chambers, *Concept to Reality*, p. 188.

The Dallas accident resulted in widespread shock because of its large number of fatalities. It particularly affected airline crews, as American Airlines Capt. Wallace M. Gillman recalled vividly at a NASA-sponsored 1990 meeting of international experts in wind shear:

> About one week after Delta 191's accident in Dallas, I was taxiing out to take off on Runway 17R at DFW Airport. Everybody was very conscience of wind shear after that accident. I remember there were some storms coming in from the northwest and we were watching it as we were in a line of airplanes waiting to take off. We looked at the wind socks. We were listening to the tower reports from the LLWAS system, the winds at various portions around the airport. I was number 2 for takeoff and I said to my co-pilot, "I'm not going to go on this runway." But just at that time, the number 1 crew in line, Pan Am, said, "I'm not going to go." Then the whole line said, "We're not going to go" then the tower taxies us all down the runway, took us about 15 minutes, down to the other end. By that time the storm had kind of passed by and we all launched to the north.[63]

Taming Microburst: NASA's Wind Shear Research Effort Takes Wing

The Dallas crash profoundly accelerated NASA and FAA wind shear research efforts. Two weeks after the accident, responding to calls from concerned constituents, Representative George Brown of California requested a NASA presentation on wind shear and subsequently made a fact-finding visit to the Langley Research Center. Dr. Jeremiah F. Creedon, head of the Langley Flight Systems Directorate, briefed the Congressman on the wind shear problem and potential technologies that might alleviate it. Creedon informed Brown that Langley researchers were running a series of modest microburst and wind shear modeling projects, and that an FAA manager, George "Cliff" Hay, and NASA Langley research engineer Roland L. Bowles had a plan underway for a comprehensive airborne wind shear detection research program. During the briefing, Brown asked how much money it would take; Creedon estimated several million dollars. Brown remarked the amount was "nothing"; Creedon

63. Wallace M. Gillman, "Industry Terms of Reference," in Spady, et al., eds., *Airborne Wind Shear Detection and Warning Systems*, pt. 1, p. 16.

1

replied tellingly, "It's a lot of money if you don't have it." As the Brown party left the briefing, one of his aides confided to a Langley manager "NASA [has] just gotten itself a wind shear program." The combination of media attention, public concern, and congressional interest triggered the development of "a substantial, coordinated interagency research effort to address the wind shear problem."[64]

On July 24, 1986, NASA and the FAA mandated the National Integrated Windshear Plan, an umbrella project overseeing several initiatives at different agencies.[65] The joint effort responded both to congressional directives and National Transportation Safety Board recommendations after documentation of the numerous recent wind shear accidents. NASA Langley Research Center's Roland L. Bowles subsequently oversaw a rigorous plan of wind shear research called the Airborne Wind Shear Detection and Avoidance Program (AWDAP), which included the development of onboard sensors and pilot training. Building upon earlier supercomputer modeling studies by Michael L. Kaplan, Fred H. Proctor, and others, NASA researchers developed the Terminal Area Simulation System (TASS), which took into consideration a variety of storm parameters and characteristics, enabling numerical simulation of microburst formation. Out of this came data that the FAA was able to use to build standards for the certification of airborne wind shear sensors. As well, the FAA created a flight

64. Lane E. Wallace, *Airborne Trailblazer: Two Decades with NASA Langley's 737 Flying Laboratory*, NASA SP 4216 (Washington, DC: GPO, 1994), p. 41.

65. NASA Langley Research Center, "NASA Facts On-line: Making the Skies Safe from Windshear," *http://oea.larc.nasa.gov/PAIS/Windshear.html*, accessed July 15, 2009. For subsequent research, see for example Roland L. Bowles, "Windshear Detection and Avoidance: Airborne Systems Survey," *Proceedings of the 29th IEEE Conference on Decision and Control, Honolulu, HI* (New York: IEEE Publications, 1990); E.M. Bracalente, C.L. Britt, and W.R. Jones, "Airborne Doppler Radar Detection of Low Altitude Windshear," AIAA Paper 88-4657 (1988); Dan D. Vicroy, "Investigation of the Influence of Wind Shear on the Aerodynamic Characteristics of Aircraft Using a Vortex-Lattice Method," NASA LRC, NTRS Report 88N17619 (1988); Vicroy, "Influence of Wind Shear on the Aerodynamic Characteristics of Airplanes," NASA TP-2827 (1988); "Wind Shear Study: Low-Altitude Wind Shear," *Aviation Week & Space Technology* (Mar. 28, 1983); Terry Zweifel, "Optimal Guidance during a Windshear Encounter," *Scientific Honeywell* (Jan. 1989); Zweifel, "Temperature Lapse Rate as an Adjunct to Windshear Detection," paper presented at the *Airborne Wind Shear Detection and Warning Systems Third Combined Manufacturer's and Technologist's Conference, Hampton, VA, Oct. 16–18, 1990*; Zweifel, "The Effect of Windshear During Takeoff Roll on Aircraft Stopping Distance" NTRS Report 91N11699 (1990); Zweifel, "Flight Experience with Windshear Detection," NTRS Report 91N11684 (1990).

safety program that supported NASA development of wind shear detection technologies.[66]

At NASA Langley, the comprehensive wind shear studies started with laboratory analysis and continued into simulation and flight evaluation. Some of the sensor systems that Langley tested work better in rain, while others performed more successfully in dry conditions.[67] Most were tested using Langley's modified Boeing 737 systems testbed.[68] This research airplane studied not only microburst and wind shear with the Airborne Windshear Research Program, but also tested electronic and computerized control displays ("glass cockpits" and Synthetic Vision Systems) in development, microwave landing systems in development, and Global Positioning System (GPS) navigation.[69]

NASA's Airborne Windshear Research Program did not completely resolve the problem of wind shear, but "its investigation of microburst detection systems helped lead to the development of onboard monitoring systems that offered airliners another way to avoid potentially lethal situations."[70] The program achieved much and gave confidence to those pursuing practical applications. The program had three major goals. The first was to find a way to characterize the wind shear threat in a way that would indicate the hazard level that threatened aircraft. The second was to develop airborne remote-sensor technology to provide accurate, forward-looking wind shear detection. The third was to design flight management systems and concepts to transfer this information to pilots in such a way that they could effectively respond to a wind shear threat. The program had to pursue these goals under tight time constraints.[71] Time was of the essence, partly because the public had demanded a solution to the scourge of microburst wind shear and because a proposed FAA regulation stipulated that any "forward-looking" (predictive) wind shear detection technology produced by NASA be swiftly transferred to the airlines.

An airborne technology giving pilots advanced warning of wind shear would allow them the time to increase engine power, "clean up"

66. Chambers, *Concept to Reality*, p. 189.

67. NASA Langley Research Center, "NASA Facts On-line: Making the Skies Safe from Windshear," *http://oea.larc.nasa.gov/PAIS/Windshear.html*, accessed July 15, 2009.

68. Chambers, *Concept to Reality*, p. 192; Wallace, *Airborne Trailblazer*, ch. 5.

69. For SVS research, see the accompanying essay in this volume by Robert Rivers.

70. Hansen, *The Bird is on the Wing*, p. 211.

71. Wallace, *Airborne Trailblazer*, ch. 5.

the aircraft aerodynamically, increase penetration speed, and level the airplane before entering a microburst, so that the pilot would have more energy, altitude, and speed to work with or to maneuver around the microburst completely. But many doubted that a system incorporating all of these concepts could be perfected. The technologies offering most potential were microwave Doppler radar, Doppler Light Detecting and Ranging (LIDAR, a laser-based system), and passive infrared radiometry systems. However, all these forward-looking technologies were challenging. Consequently, developing and exploiting them took a minimum of several years. At Langley, versions of the different detection systems were "flown" as simulations against computer models, which re-created past wind shear accidents. However, computer simulations could only go so far; the new sensors had to be tested in actual wind shear conditions. Accordingly, the FAA and NASA expanded their 1986 memorandum of understanding in May 1990 to support flight research evaluating the efficacy of the advanced wind shear detection systems integrating airborne and ground-based wind shear measurement methodologies. Researchers swiftly discovered that pilots needed as much as 20 seconds of advance warning if they were to avert or survive an encounter with microburst wind shear.[72]

Key to developing a practical warning system was deriving a suitable means of assessing the level of threat that pilots would face, because this would influence the necessary course of action to avoid potential disaster. Fortunately, NASA Project Manager Roland Bowles devised a hazard index called the "F-Factor." The F-Factor, as ultimately refined by Bowles and his colleagues Michael Lewis and David Hinton, indicated how much specific excess thrust an airplane would require to fly through wind shear without losing altitude or airspeed.[73] For instance, a typical twin-engine jet transport plane might have engines capable

72. P. Douglas Arbuckle, Michael S. Lewis, and David A. Hinton, "Airborne Systems Technology Application to the Windshear Threat," Paper 96-5.7.1, *20th Congress of the International Council of the Aeronautical Sciences, Sorrento, Italy, 1996*; see also Wallace, *Airborne Trailblazer*, ch. 5.
73. Fred H. Proctor, David A. Hinton, and Roland L. Bowles, "A Windshear Hazard Index," NASA LRC NTRS Report 200.001.16199 (2000). Specific excess thrust is thrust minus the drag of the airplane, divided by airplane's weight. It determines the climb gradient (altitude gain vs. horizontal distance), which is expressed as $\gamma = (T - D) / W$, where γ is the climb gradient, T is thrust, D is drag, and W is weight. See Roger D. Schaufele, *The Elements of Aircraft Preliminary Design* (Santa Ana: Aries Publications, 2000), p. 18, and Arbuckle, Lewis, and Hinton, "Airborne Systems Technology Application," p. 2.

of producing 0.17 excess thrust on the F-Factor scale. If a microburst wind shear registered higher than 0.17, the airplane would not be able to fly through it without losing airspeed or altitude. The F-Factor provided a way for information from any kind of sensor to reach the pilot in an easily recognizable form. The technology also had to locate the position and track the movement of dangerous air masses and provide information on the wind shear's proximity and volume.[74] Doppler-based wind shear sensors could only measure the first term in the F-Factor equation (the rate of change of horizontal wind). This limitation could result in underestimation of the hazard. Luckily, there were several ways to measure changes in vertical wind from radial wind measurements, using equations and algorithms that were computerized. Although error ranges in the device's measurement of the F-Factor could not be eliminated, these were taken into account when producing the airborne system.[75] The Bowles team derivation and refinement of the F-Factor constituted a major element of NASA's wind shear research, to some, "the key contribution of NASA in the taming of the wind-shear threat." The FAA recognized its significance by incorporating F-Factor in its regulations, directing that at F-Factors of 0.13 or greater, wind shear warnings must be issued.[76]

In 1988, NASA and researchers from Clemson University worked on new ways to eliminate clutter (or data not related to wind shear) from information received via Doppler and other kinds of radar used on an airborne platform. Such methods, including antenna steering and adaptive filtering, were somewhat different from those used to eliminate clutter from information received on a ground-based platform. This was

74. Roland L. Bowles's research is enumerated in the recommended readings; for a sample, see his "Reducing Wind Shear Risk Through Airborne Systems Technology," *Proceedings of the 17th Congress of the Int'l Congress of Aeronautical Sciences, Stockholm, Sweden, Sept. 1990*; and Roland L. Bowles and Russell Targ, "Windshear Detection and Avoidance—Airborne Systems Perspective," NASA LRC, NTRS Report 89A13506 (1988).

75. Bowles, "Reducing Wind Shear Risk Through Airborne Systems Technology," *Proceedings of the 17th Congress of the International Congress of Aeronautical Sciences, Stockholm, Sweden, Sept. 1990*; Wallace, *Airborne Trailblazer, http://oea.larc.nasa.gov/trailblazer/SP-4216/chapter5/ch5.html*, accessed Aug. 1, 2009, "Vertical Wind Estimation from Horizontal Wind Measurements: General Questions and Answers," *NASA–FAA Wind Shear Review Meeting, Sept. 28, 1993.*

76. Chambers, *Concept to Reality*, pp. 190, 197. Bowles's subsequently received numerous accolades for his wind shear research, including, fittingly, the Langley Research Center H.J.E. Reid Award for 1993 (shared with Fred Proctor) and AIAA Engineer of the Year Award for 1994.

because the airborne environment had unique problems, such as large clutter-to-signal ratios, ever-changing range requirements, and lack of repeatability.[77]

The accidents of the 1970s and 1980s stimulated research on a variety of wind shear predictive technologies and methodologies. Langley's success in pursuing both enabled the FAA to decree in 1988 that all commercial airline carriers were required to install wind shear detection devices by the end of 1993. Most airlines decided to go with reactive systems, which detect the presence of wind shear once the plane has already flown into it. For American, Northwest, and Continental—three airlines already testing predictive systems capable of detecting wind shear before an aircraft flew into it—the FAA extended its deadline to 1995, to permit refinement and certification of these more demanding and potentially more valuable sensors.[78]

From 1990 onwards, NASA wind shear researchers were particularly energetic, publishing and presenting widely, and distributing technical papers throughout the aerospace community. Working with the FAA, they organized and sponsored well-attended wind shear conferences that drew together other researchers, aviation administrators, and—very importantly—airline pilots and air traffic controllers. Finally, cognizant of the pressing need to transfer the science and technology of wind shear research out of the laboratory and onto the flight line, NASA and the FAA invited potential manufacturers to work with the agencies in pursuing wind shear detector development.[79]

The invitations were welcomed by industry. Three important avionics manufacturers—Allied Signal, Westinghouse, and Rockwell Collins—sent engineering teams to Langley. These teams followed NASA's wind shear effort closely, using the Agency's wind shear simulations to enhance the capabilities of their various systems. In 1990, Lockheed introduced its Coherent LIDAR Airborne Shear Sensor (CLASS), developed under contract to NASA Langley. CLASS was a predictive system allowing pilots to avoid hazards of low-altitude wind shear under all weather conditions. CLASS would detect thunderstorm downburst early in its development

77. Ernest G. Baxa, "Clutter Filter Design Considerations for Airborne Doppler Radar Detection of Wind Shear," 527468, N91-11690, Oct. 19, 1988.

78. "Technology for Safer Skies," *http://er.jsc.nasa.gov/SEH/pg56s95.html*, accessed Dec. 11, 2009, p. 3.

79. Ibid.

and emphasize avoidance rather than recovery. After consultation with airline and military pilots, Lockheed engineers decided that the system should have a 2- to 4-kilometer range and should provide a warning time of 20 to 40 seconds. A secondary purpose of the system would be to provide predictive warnings of clear air turbulence. In conjunction with NASA, Lockheed conducted a 1-year flight evaluation program on Langley's 737 during the following year to measure line-of-sight wind velocities from many wind fields, evaluating this against data obtained via air- and ground-based radars and accelerometer-based systems and thus acquiring a comparative database.[80]

Also in 1990, using technologies developed by NASA, Turbulence Prediction Systems of Boulder, CO, successfully tested its Advance Warning Airborne System (AWAS) on a modified Cessna Citation small, twin-jet research aircraft operated by the University of North Dakota. Technicians loaded AWAS into the luggage compartment in front of the pilot. Pilots intentionally flew the plane into numerous wind shear events over the course of 66 flights, including several wet microbursts in Orlando, FL, and a few dry microbursts in Denver. On the Cessna, AWAS measured the thermal characteristics of microbursts to predict their presence during takeoff and landing. In 1991, AWAS units were flown aboard three American Airlines MD-80s and three Northwest Airlines DC-9s to study and improve the system's nuisance alert response. Technicians also installed a Honeywell Windshear Computer in the planes, which Honeywell had developed in light of NASA research. The computer processed the data gathered by AWAS via external aircraft measuring instruments. AWAS also flew aboard the NASA Boeing 737 during summer 1991. Unfortunately, results from these research flights were not conclusive, in part because NASA conducted research flights outside AWAS's normal operating envelope, and in an attempt to compensate for differences in airspeed, NASA personnel sometimes overrode automatic features. These complications did not stop the development of more sophisticated versions of the system and ultimate FAA certification.[81]

80. Russell Targ, "CLASS: Coherent Lidar Airborne Shear Sensor and Windshear Avoidance," Electro-Optical Sciences Directorate, Lockheed Missiles and Space Company, Oct. 1990.
81. H. Patrick Adamson, "Development of the Advance Warning Airborne System (AWAS)," *Fourth Combined Manufacturers and Technologists' Airborne Windshear Review Meeting, Turbulence Prediction Systems, Boulder, CO, Apr. 14, 1992.*

After analyzing data from the Dallas and Denver accidents, Honeywell researchers had concluded that temperature lapse rate, or the drop in temperature with the increase in altitude, could indicate wind shear caused by both wet and dry microbursts. Lapse rate could not, of course, communicate whether air acceleration was horizontal or vertical. Nonetheless, this lapse rate could be used to make reactive systems more "intelligent," "hence providing added assurance that a dangerous shear has occurred." Because convective activity was often associated with turbulence, the lapse rate measurements could also be useful in warning of impending "rough air." Out of this work evolved the first-generation Honeywell Windshear Detection and Guidance System, which gained wide acceptance.[82]

Supporting its own research activities and the larger goal of air safety awareness, NASA developed a thorough wind shear training and familiarization program for pilots and other interested parties. Flightcrews "flew" hundreds of simulated wind shears. Crews and test personnel flew rehearsal flights for 2 weeks in the Langley and Wallops areas before deploying to Orlando or Colorado for actual in-flight microburst encounters in 1991 and 1992.

The NASA Langley team tested three airborne systems to predict wind shear. In the creation of these systems, it was often assisted by technology application experts from the Research Triangle Institute of Triangle Park, NC.[83] The first system tested was a Langley-sponsored Doppler microwave radar, whose development was overseen by Langley's Emedio "Brac" Bracalente and the Langley Airborne Radar Development Group. It sent a microwave radar signal ahead of the plane to detect raindrops and other moisture in the air. The returning signal provided information on the motion of raindrops and moisture particles, and it translated this information into wind speed. Microwave radar was best in damp or wet conditions, though not in dry conditions. Rockwell International's Collins Air Transport Division in Cedar Rapids, IA, made the radar transmitter, extrapolated from the standard Collins 708 weather radar. NASA's Langley Research Center in Hampton, VA, developed

82. Terry Zweifel, "Temperature Lapse Rate as an Adjunct to Windshear Detection," paper presented at the *Airborne Wind Shear Detection and Warning Systems Third Combined Manufacturers' and Technologists' Conference, Hampton, VA, Oct. 16–18, 1990.*

83. "Technology for Safer Skies," *http://er.jsc.nasa.gov/SEH/pg56s95.html*, accessed Dec. 11, 2009.

the receiver/detector subsystem and the signal-processing algorithms and hardware for the wind shear application. So enthusiastic and confident were the members of the Doppler microwave test team that they designed their own flight suit patch, styling themselves the "Burst Busters," with an international slash-and-circle "stop" sign overlaying a schematic of a microburst.[84]

The second system was a Doppler LIDAR. Unlike radio beam-transmitting radar, LIDAR used a laser, reflecting energy from aerosol particles rather than from water droplets. This system had fewer problems with ground clutter (interference) than Doppler radar did, but it did not work as well as the microwave system does in heavy rain. The system was made by the Lockheed Corporation's Missiles and Space Company in Sunnyvale, CA; United Technologies Optical Systems, Inc., in West Palm Beach, FL; and Lassen Research of Chico, CA.[85] Researchers noted that an "inherent limitation" of the radar and LIDAR systems was their inability to measure any velocities running perpendicular to the system's line of sight. A microburst's presence could be detected by measuring changes in the horizontal velocity profile, but the inability to measure a perpendicular downdraft could result in an underestimation of the magnitude of the hazard, including its spatial size.[86]

The third plane-based system used an infrared detector to find temperature changes in the airspace in front of the plane. It monitored carbon dioxide's thermal signatures to find cool columns of air, which often indicate microbursts. The system was less expensive and less complex than the others but also less precise, because it could not directly measure wind speed.[87]

84. Emedio M. Bracalente, C.L. Britt, and W.R. Jones, "Airborne Doppler Radar Detection of Low Altitude Windshear," AIAA Paper 88-4657 (1988); and David D. Aalfs, Ernest G. Baxa, Jr., and Emedio M. Bracalente, "Signal Processing Aspects of Windshear Detection," *Microwave Journal*, vol. 96, no. 9 (Sept. 1993), pp. 76, 79, 82–84, available as NTRS Report 94A12361 (1993); and Chambers, *Concept to Reality*, pp. 193, 195. Radar details are in S.D. Harrah, E.M. Bracalente, P.R. Schaffner, and E.G. Baxa, "Description and Availability of Airborne Doppler Radar Data," in NASA Jet Propulsion Laboratory, *JPL Progress in Electromagnetics Research Symposium (PIERS)* (Pasadena: JPL, 1993), p. 262, NTIS ID N94-20403 05-32.

85. "Technology for Safer Skies."

86. D. Vicroy, "Vertical Wind Estimation from Horizontal Wind Measurements," *NASA–FAA Wind Shear Review Meeting, NASA Langley Research Center, Sept. 28, 1993.*

87. "Making the Skies Safe from Windshear."

NASA 515, the Langley Boeing 737, on the airport ramp at Orlando, FL, during wind shear sensor testing. NASA.

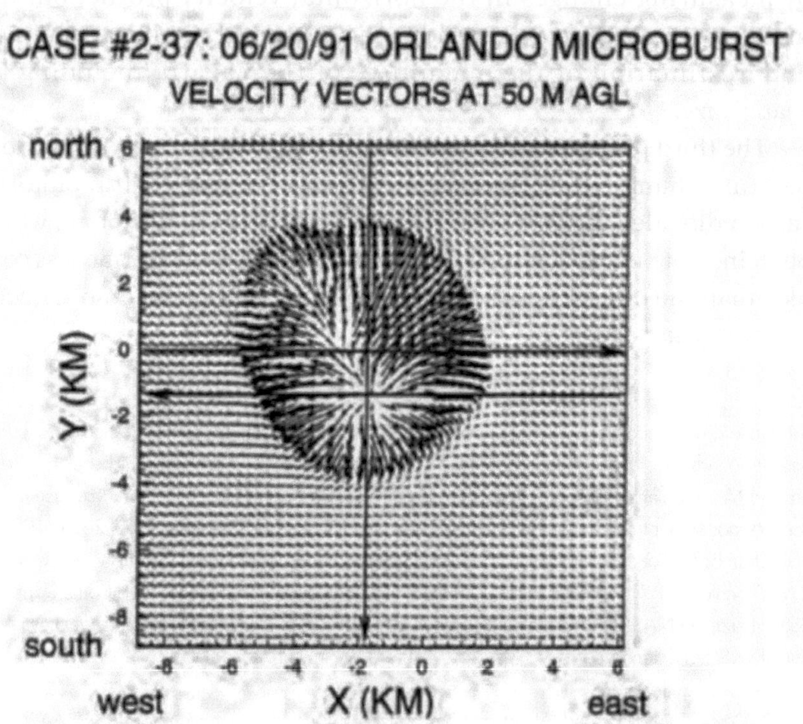

A June 1991 radar plot of a wind shear at Orlando, showing the classic radial outflow. This one is approximately 5 miles in diameter. NASA.

In 1990–1992, Langley's wind shear research team accumulated and evaluated data from 130 sensor-evaluation research flights made using the Center's 737 testbed.[88] Flight-test crews flew research missions in the Langley local area, Philadelphia, Orlando, and Denver. Risk mitigation was an important program requirement. Thus, wind shear investigation flights were flown at higher speeds than airliners typically flew, so that the 737 crew would have better opportunity to evade any hazard it encountered. As well, preflight ground rules stipulated that no penetrations be made into conditions with an F-Factor greater than 0.15. Of all the systems tested, the airborne radar functioned best. Data were accumulated during 156 weather runs: 109 in the turbulence-prone Orlando area. The 737 made 15 penetrations of microbursts at altitudes ranging from 800 to 1,100 feet. During the tests, the team evaluated the radar at various tilt angles to assess any impact from ground clutter (a common problem in airborne radar clarity) upon the fidelity of the airborne system. Aircraft entry speed into the microburst threat region had little effect on clutter suppression. All together, the airborne Doppler radar tests collected data from approximately 30 microbursts, as well as 20 gust fronts, with every microburst detected by the airborne radar. F-Factors measured with the airborne radar showed "excellent agreement" with the F-Factors measured by Terminal Doppler Weather Radar (TDWR), and comparison of airborne and TDWR data likewise indicated "comparable results."[89] As Joseph Chambers noted subsequently, "The results of the test program demonstrated that Doppler radar systems offered the greatest promise for early introduction to airline service. The Langley forward-looking Doppler radar detected wind shear consistently and at longer ranges than other systems, and it was able to provide 20 to 40 seconds warning of upcoming microburst."[90] The Burst Busters clearly had succeeded. Afterward, forward-looking Doppler radar was adopted by most airlines.

88. The team consisted of wind shear Program Manager Roland Bowles, Deputy Program Manager Michael S. Lewis, research engineers Emedio "Brac" Bracalente and David Hinton, research pilots Lee H. Person, Jr., and Kenneth R. Yenni, crew chief Michael Basnett, and lead electronics technician Artie D. Jessup, supported by others.

89. Emedio Bracalente, "Doppler Radar Results," and Charles L. Britt and Emedio Bracalente, "NASA Airborne Radar Wind Shear Detection Algorithm and the Detection of Wet Microbursts in the Vicinity of Orlando, Florida," both presented at the *4th Combined Manufacturers' and Technologists' Airborne Wind Shear Review Meeting, Williamsburg, VA, Apr. 4–16, 1992,* NTIS Reports Nos. N93-19595 and N93-19611 (1992).

90. Chambers, *Concept to Reality*, p. 195.

NASA Langley's wind shear team at Orlando in the cockpit of NASA 515. Left to right: Program Manager Roland Bowles, research pilot Lee Person, Deputy Program Manager Michael Lewis, research engineer David Hinton, and research engineer Emedio Bracalente. Note Bracalente's "Burst Buster" shoulder patch. NASA.

Assessing NASA's Wind Shear Research Effort

NASA's wind shear research effort involved complex, cooperative relationships between the FAA, industry manufacturers, and several NASA Langley directorates, with significant political oversight, scrutiny, and public interest. It faced many significant technical challenges, not the least of which were potentially dangerous flight tests and evaluations.[91] Yet, during a 7-year effort, NASA, along with industry technicians and researchers, had risen to the challenge. Like many classic NACA research projects, it was tightly focused and mission-oriented, taking "a proven,

91. Ibid., p. 198. See also "Report of the Committee on Low-Altitude Wind Shear and Its Hazard to Aviation," *Low Altitude Wind Shear and Its Hazard to Aviation* (Washington, DC: National Academy Press, 1983), pp. 14–15; Roland L. Bowles, "Windshear Detection and Avoidance: Airborne Systems Survey," *Proceedings of the 29th IEEE Conference on Decision and Control, Honolulu, HI* (New York: IEEE Publications, 1990), p. 708; Michael S. Lewis, et al., "Design and Conduct of a Windshear Detection Flight Experiment," AIAA Paper 92-4092 (1992).

significant threat to aviation and air transportation and [developing] new technology that could defeat it."[92] It drew on technical capabilities and expertise from across the Agency—in meteorology, flight systems, aeronautics, engineering, and electronics—and from researchers in industry, academia, and agencies such as the National Center for Atmospheric Research. This collaborative effort spawned several important breakthroughs and discoveries, particularly the derivation of the F-Factor and the invention of Langley's forward-looking Doppler microwave radar wind shear detector. As a result of this Government-industry-academic partnership, the risk of microburst wind shear could at last be mitigated.[93]

In 1992, the NASA–FAA Airborne Windshear Research Program was nominated for the Robert J. Collier Trophy, aviation's most prestigious honor. Industry evaluations described the project as "the perfect role for NASA in support of national needs" and "NASA at its best." Langley's Jeremiah Creedon said, "we might get that good again, but we can't get any better."[94] In any other year, the program might easily have won, but it was the NASA–FAA team's ill luck to be competing that year with the revolutionary Global Positioning System, which had proven its value in spectacular fashion during the Gulf War of 1991. Not surprisingly, then, it was GPS, not the wind shear program, which was awarded the Collier Trophy. But if the wind shear team members lost their shot at this prestigious award, they could nevertheless take satisfaction in knowing that together, their agencies had developed and demonstrated a "technology base" enabling the manufacture of many subsequent wind shear detection and prediction systems, to the safety and undoubted benefit of the traveling public, and airmen everywhere.[95]

NASA engineers had coordinated their research with commercial manufacturers from the start of wind shear research and detector development, so its subsequent transfer to the private sector occurred quickly and effectively. Annual conferences hosted jointly by NASA Langley and the FAA during the project's evolution provided a ready forum for manufacturers to review new technology and for NASA researchers to obtain a better understanding of the issues that manufacturers were

92. Wallace, *Airborne Trailblazer.*

93. "Airborne Wind Shear Detection and Warning Systems," *Second Combined Manufacturers and Technological Conference*, NASA CP-10050.

94. Wallace, *Airborne Trailblazer*, pp. 5–48.

95. "Technology for Safer Skies."

encountering as they developed airborne equipment to meet FAA certification requirements. The fifth and final combined manufacturers' and technologists' airborne wind shear conference was held at NASA Langley on September 28–30, 1993, marking an end to what NASA and the FAA jointly recognized as "the highly successful wind shear experiments conducted by government, academic institutions, and industry." From this point onward, emphasis would shift to certification, regulation, and implementation as the technology transitioned into commercial service.[96] There were some minor issues among NASA, the airlines, and plane manufacturers about how to calibrate and where to place the various components of the system for maximum effectiveness. Sometimes, the airlines would begin testing installed systems before NASA finished its testing. Airline representatives said that they were pleased with the system, but they noted that their pilots were highly trained professionals who, historically, had often avoided wind shear on their own. Pilots, who of course had direct control over plane performance, wished to have detailed information about the system's technical components. Airline representatives debated the necessity of considering the performance specifications of particular aircraft when installing the airborne system but ultimately went with a single Doppler radar system that could work with all passenger airliners.[97] Through all this, Langley researchers worked with the FAA and industry to develop certification standards for the wind shear sensors. These standards involved the wind shear hazard, the cockpit interface, alerts given to flight crews, and sensor performance levels. NASA research, as it had in other aspects of aeronautics over the history of American civil aviation, formed the basis for these specifications.[98]

Although its airborne sensor development effort garnered the greatest attention during the 1980s and 1990s, NASA Langley also developed several ground-based wind shear detection systems. One was the

96. V.E. Delnore, ed., *Airborne Windshear Detection and Warning Systems: Fifth and Final Combined Manufacturers' and Technologists' Conference*, NASA CP-10139, pts. 1–2 (1994).
97. Vicroy, *NASA–FAA Wind Shear Review Meeting*, "Vertical Wind Estimation from Horizontal Wind Measurements: Results of American in-service Evaluations," Sept. 28, 1993.
98. G.F. Switzer, J.V. Aanstoos, F.H. Proctor, and D.A. Hinton, "Windshear Database for Forward-Looking Systems Certification," NASA TM-109012 (1993); and Charles L. Britt, George F. Switzer, and Emedio M. Bracalente, "Certification Methodology Applied to the NASA Experimental Radar System," paper presented at the *Airborne Windshear Detection and Warning Systems' 5th and Final Combined Manufacturers' and Technologists' Conference*, pt. 2, pp. 463–488, NTIS Report 95N13205 (1994).

low-level wind shear alert system installed at over 100 United States airports. By 1994, ground-based radar systems (Terminal Doppler Weather Radar) were in place at hundreds of airports that could predict when such shears would come, but plane-based systems continue to be necessary because not all of the thousands of airports around the world had such systems. Of plane-based systems, NASA's forward-looking predictive radar worked best.[99]

The end of the tyranny of microburst did not come without one last serious accident that had its own consequences for wind shear alleviation. On July 2, 1994, US Air Flight 1016, a twin-engine Douglas DC-9, crashed and burned after flying through a microburst during a missed approach at Charlotte-Douglas International Airport. The crew had realized too late that conditions were not favorable for landing on Runway 18R, had tried to go around, and had been caught by a violent microburst that sent the airplane into trees and a home. Of the 57 passengers and crew, 37 perished, and the rest were injured, 16 seriously. The NTSB faulted the crew for continuing its approach "into severe convective activity that was conducive to a microburst," for "failure to recognize a windshear situation in a timely manner," and for "failure to establish and maintain the proper airplane attitude and thrust setting necessary to escape the windshear." As well, it blamed a "lack of real-time adverse weather and windshear hazard information dissemination from air traffic control."[100] Several factors came together to make the accident more tragic. In 1991, US Air had installed a Honeywell wind shear detector in the plane that could furnish the crew with both a visual warning light and an audible "wind shear, wind shear, wind shear" warning once an airplane entered a wind shear. But it failed to function during this encounter. Its operating algorithms were designed to minimize "nuisance alerts," such as routine changes in aircraft motions induced by flap movement. When Flight 1016 encountered its fatal shear, the plane's landing flaps were in transition as the crew executed its missed approach, and this likely played a role in its failure to function. As well, Charlotte had been scheduled to be the fifth airport to receive Terminal Doppler Weather Radar, a highly sensitive and precise wind shear

99. "Making the Skies Safer from Windshear."

100. Quotes from National Transportation Safety Board, "Aircraft Accident Report: Flight Into Terrain During Missed Approach: US Air Flight 1016, DC-9-31, N954VJ, Charlotte-Douglas International Airport, Charlotte, North Carolina, July 2, 1994," Report NTSB-AAR-95-03 (Apr. 4, 1995), p. vi.

detection system. But a land dispute involving the cost of property that the airport was trying to purchase for the radar site bumped it from 5th to 38th on the list to get the new TDWR. Thus, when the accident occurred, Charlotte only had the far less capable LLWAS in service.[101] Clearly, to survive the dangers of wind shear, airline crews needed aircraft equipped with forward-looking predictive wind shear warning systems, airports equipped with up-to-date precise wind shear Doppler radar detection systems, and air traffic controllers cognizant of the problem and willing to unhesitatingly shift flights away from potential wind shear threats. Finally, pilots needed to exercise extreme prudence when operating in conditions conducive to wind shear formation.

Not quite 5 months later, on November 30, 1994, Continental Airlines Flight 1637, a Boeing 737 jetliner, lifted off from Washington-Reagan Airport, Washington, DC, bound for Cleveland. It is doubtful whether any passengers realized that they were helping usher in a new chapter in the history of aviation safety. This flight marked the introduction of a commercial airliner equipped with a forward-looking sensor for detecting and predicting wind shear. The sensor was a Bendix RDR-4B developed by Allied Signal Commercial Avionic Systems of Fort Lauderdale, FL. The RDR-4B was the first of the predictive Doppler microwave radar wind shear detection systems based upon NASA Langley's research to gain FAA certification, achieving this milestone on September 1, 1994. It consisted of an antenna, a receiver-transmitter, and a Planned Position Indicator (PPI), which displayed the direction and distance of a wind shear microburst and the regular weather display. Since then, the number of wind shear accidents has dropped precipitously, reflecting the proliferation and synergistic benefits accruing from both air- and land-based advanced wind shear sensors.[102]

In the mid-1990s, as part of NASA's Terminal Area Productivity Program, Langley researchers used numerical modeling to predict weather in the area of airport terminals. Their large-eddy simulation (LES) model had a meteorological framework that allowed the prediction and depiction of the interaction of the airplane's wake vortexes (the rotating turbulence that streams from an aircraft's wingtips when it passes through the air) with environments containing crosswind shear,

101. Ibid., pp. 15 and 85. As the NTSB report makes clear, cockpit transcripts and background signals confirmed the failure of the Honeywell system to alert the crew.
102. "Technology for Safer Skies"; "Making the Skies Safer From Windshear."

stratification, atmospheric turbulence, and humidity. Meteorological effects can, to a large degree, determine the behavior of wake vortexes. Turbulence can gradually decay the rotation of the vortex, robbing it of strength, and other dynamic instabilities can cause the vortex to collapse. Results from the numerical simulations helped engineers to develop useful algorithms to determine the way aircraft should be spaced when aloft in the narrow approach corridors surrounding the airport terminal, in the presence of wake turbulence. The models utilized both two and three dimensions to obtain the broadest possible picture of phenomena interaction and provided a solid basis for the development of the Aircraft Vortex Spacing System (AVOSS), which safely increased airport capacity.[103]

In 1999, researchers at NASA's Goddard Space Flight Center in Greenbelt, MD, concluded a 20-year experiment on wind-stress simulations and equatorial dynamics. The use of existing datasets and the creation of models that paired atmosphere and ocean forecasts of changes in sea surface temperatures helped the researchers to obtain predictions of climatic conditions of large areas of Earth, even months and years in advance. Researchers found that these conditions affect the speed and timing of the transition from laminar to turbulent airflow in a plane's boundary layer, and their work contributed to a more sophisticated understanding of aerodynamics.[104]

In 2008, researchers at NASA Goddard compared various NASA satellite datasets and global analyses from the National Centers for Environmental Protection to characterize properties of the Saharan Air Layer (SAL), a layer of dry, dusty, warm air that moves westward off the Saharan Desert of Africa and over the tropical Atlantic. The researchers also examined the effects of the SAL on hurricane development. Although the SAL causes a degree of low-level vertical wind shear that pilots have to be cognizant of, the researchers concluded that the SAL's effects on hurricane and microburst formation were negligible.[105]

103. Fred H. Proctor, "The NASA-Langley Wake Vortex Modeling Effort in Support of an Operational Aircraft Spacing System," AIAA Paper 98-0589 (1998).

104. Julio T. Bacmeister and Max J. Suarez, "Wind-Stress Simulations and Equatorial Dynamics in an AGCM [Atmospheric-land General Circulation Model]," NASA Goddard Earth Sciences and Technology Center, NASA Seasonal-to-Interannual Prediction Project, pts. I–II (June 6, 1999), NTIS CASI ID 200.101.00385.

105. Scott Braun and Chung-Lin Shie, "Improving Our Understanding of Atlantic Tropical Cyclones Through Knowledge of the Saharan Air Layer: Hope or Hype?" *Bulletin of the American Meteorological Society* (Aug. 14, 2008).

Advanced research into turbulence will be a vital part of the aerospace sciences as long as vehicles move through the atmosphere. Since 1997, Stanford has been one of five universities sponsored by the U.S. Department of Energy as a national Advanced Simulation and Computing Center. Today, researchers at Stanford's Center for Turbulence use computer clusters, which are many times more powerful than the pioneering Illiac IV. For large-scale turbulence research projects, they also have access to cutting-edge computational facilities at the National Laboratories, including the Columbia computer at NASA Ames Research Center, which has 10,000 processors. Such advanced research into turbulent flow continues to help steer aerodynamics developments as the aerospace community confronts the challenges of the 21st century.[106]

In 2003, President George W. Bush signed the Vision 100 Century of Aviation Reauthorization Act.[107] This initiative established within the FAA a joint planning and development office to oversee and manage the Next Generation Air Transportation System (NextGen). NextGen incorporated seven goals:

1. Improve the level of safety, security, efficiency, quality, and affordability of the National Airspace System and aviation services.
2. Take advantage of data from emerging ground-based and space-based communications, navigation, and surveillance technologies.
3. Integrate data streams from multiple agencies and sources to enable situational awareness and seamless global operations for all appropriate users of the system, including users responsible for civil aviation, homeland security, and national security.
4. Leverage investments in civil aviation, homeland security, and national security and build upon current air traffic management and infrastructure initiatives to meet system performance requirements for all system uses.

106. Stanford University, "About the Center for Turbulence Research," *http://www.stanford.edu/group/ctr/about.html*, accessed Oct. 4, 2009.
107. Public Law 108-176 (2003).

5. Be scalable to accommodate and encourage substantial growth in domestic and international transportation and anticipate and accommodate continuing technology upgrades and advances.
6. Accommodate a range of aircraft operations, including airlines, air taxis, helicopters, general-aviation, and unmanned aerial vehicles.
7. Take into consideration, to the greatest extent practicable, design of airport approach and departure flight paths to reduce exposure of noise and emissions pollution on affected residents.[108]

NASA is now working with the FAA, industry, the academic community, the Departments of Commerce, Defense, Homeland Security, and Transportation, and the Office of Science and Technology Policy to turn the ambitious goals of NextGen into air transport reality. Continual improvement of Terminal Doppler Weather Radar and the Low-Level Windshear Alert System are essential elements of the reduced weather impact goals within the NextGen initiatives. Service life extension programs are underway to maintain and improve airport TDWR and the older LLWAS capabilities.[109] There are LLWAS at 116 airports worldwide, and an improvement plan for the program was completed in 2008, consisting of updating system algorithms and creating new information/alert displays to increase wind shear detection capabilities, reduce the number of false alarms, and lower maintenance costs.[110]

FAA and NASA researchers and engineers have not been content to rest on their accomplishment and have continued to perfect the wind shear prediction systems they pioneered in the 1980s and 1990s. Building upon this fruitful NASA–FAA turbulence and wind shear partnership effort, the FAA has developed Graphical Turbulence Guidance (GTG), which provides clear air turbulence forecasts out to 12 hours in advance for planes flying at altitudes of 20,000 feet and higher. An improved system, GTG-2, will enable forecasts out to 12 hours for planes flying at lower altitudes down to 10,000 feet.[111] As of 2010, forward-looking

108. Ibid.
109. Section 3, DOT 163.
110. Section 3, DOT 171.
111. Section 3, DOT 171.

predictive Doppler microwave radar systems of the type pioneered by Langley are installed on most passenger aircraft.

This introduction to NASA research on the hazards of turbulence, gusts, and wind shear offers but a glimpse of the detailed work undertaken by Agency staff. However brief, it furnishes yet another example of how NASA, and the NACA before it, has contributed to aviation safety. This is due, in no small measure, to the unique qualities of its professional staff. The enthusiasm and dedication of those who worked NASA's wind shear research programs, and the gust and turbulence studies of the NACA earlier, have been evident throughout the history of both agencies. Their work has helped the air traveler evade the hazards of wild winds, turbulence, and storm, to the benefit of all who journey through the world's skies.

Recommended Additional Readings

Reports, Papers, Articles, and Presentations:

David D. Aalfs, "Real-Time Processing of Radar Return on a Parallel Computer," NASA CR-4456 (1992).

David D. Aalfs, Ernest G. Baxa, Jr., and Emedio M. Bracalente, "Signal Processing Aspects of Windshear Detection," *Microwave Journal*, vol. 96, no. 9 (Sept. 1993), pp. 76, 79, 82–84, available as NTRS Report 94A12361 (1993).

H. Patrick Adamson, "Development of the Advance Warning Airborne System (AWAS)," paper presented at the *Fourth Combined Manufacturers and Technologists' Airborne Windshear Review Meeting, Turbulence Prediction Systems, Boulder, CO, Apr. 14, 1992.*

W.H. Andrews, S.P. Butchart, T.R. Sisk, and D.L. Hughes, "Flight Tests Related to Jet-Transport Upset and Turbulent-Air Penetration," NASA LRC, *Conference on Aircraft Operating Problems*, NASA SP-83 (1965).

P. Douglas Arbuckle, Michael S. Lewis, and David A. Hinton, "Airborne Systems Technology Application to the Windshear Threat," Paper 96-5.7.1, presented at *20th Congress of the International Council of the Aeronautical Sciences, Sorrento, Italy, 1996.*

Aviation Safety Network, "ASN Aircraft accident Boeing 747 Tokyo," *http://aviation-safety.net/database/record.php?id=19971228-0,* accessed July 4, 2009.

Julio T. Bacmeister and Max J. Suarez, "Wind-Stress Simulations and Equatorial Dynamics in an AGCM [Atmospheric-land General Circulation Model]," NASA Goddard Earth Sciences and Technology Center, NASA Seasonal-to-Interannual Prediction Project, pts. I–II (June 6, 1999), NTIS CASI ID 200.101.00385.

Ernest G. Baxa, Jr., "Clutter Filter Design Considerations for Airborne Doppler Radar Detection of Wind Shear," 527468, NTRS Report N91-11690 (Oct. 19, 1988).

Ernest G. Baxa, Jr., "Signal processing for Airborne Doppler Radar Detection of Hazardous Wind Shear as Applied to NASA 1991 Radar Flight Experiment Data," NTRS Report 93N19612 (1992).

Ernest G. Baxa, Jr., "Windshear Detection Radar Signal Processing Studies," NASA CR-194615 (1993).

Ernest G. Baxa, Jr., and Manohar Deshpande, "Signal Processing Techniques for Clutter Filtering and Wind Shear Detection," NTRS 91N24144 (1991).

Ernest G. Baxa, Jr., and Jonggil Lee, "The Pulse-Pair Algorithm as a Robust Estimator of Turbulent Weather Spectral Parameters Using Airborne Pulse Doppler Radar," NASA CR-4382 (1991).

Sheldon Baron, Bolt Baranek, et al., "Analysis of Response to Wind-Shears using the Optimal Control Model of the Human Operator," NASA Ames Research Center Technical Paper NAS2-0652 (1979).

G.M. Bezos, R.E. Dunham, Jr., G.L. Gentry, Jr., and W. Edward Melson, Jr., "Wind Tunnel Test Results of Heavy Rain Effects on Airfoil Performance," AIAA Paper 87-0260 (1987).

E.T. Binckley and Jack Funk, "A Flight Investigation of the Effects of Compressibility on Applied Gust Loads," NACA TN-1937 (1949).

Roland L. Bowles, "Reducing Wind Shear Risk Through Airborne Systems Technology," *Proceedings of the 17th Congress of the International Congress of Aeronautical Sciences, Stockholm, Sweden, Sept. 1990.*

Roland L. Bowles, "Response of Wind Shear Warning Systems to Turbulence with Implication of Nuisance Alerts," NASA LRC, NTRS Report 88N17618 (1988).

Roland L. Bowles, "Windshear Detection and Avoidance: Airborne Systems Survey," *Proceedings of the 29th IEEE Conference on Decision and Control, Honolulu, HI* (New Yok: IEEE Publications, 1990).

Roland L. Bowles, "Wind Shear and Turbulence Simulation." NASA LRC, NTRS Report 87N25274 (1987).

Roland L. Bowles, "Program Overview: 1991 Flight Test Objectives," NASA LRC, NTRS Report 93N19591 (1992).

Roland L. Bowles and Bill K. Buck, "A Methodology for Determining Statistical Performance Compliance for Airborne Doppler Radar with Forward-Looking Turbulence Detection Capability," NASA CR-2009-215769 (2009).

Roland L. Bowles and David A. Hinton, "Windshear Detection—Airborne System Perspective," NASA LRC, NTRS Report 91A19807 (1990).

Roland L. Bowles, Tony R. Laituri, and George Trevino, "A Monte Carlo Simulation Technique for Low-Altitude, Wind-Shear Turbulence," AIAA Paper 90-0564 (1990).

Roland L. Bowles and Russell Targ, "Windshear Detection and Avoidance—Airborne Systems Perspective," NASA LRC, NTRS Report 89A13506 (1988).

Emedio M. Bracalente, "Doppler Radar Results," paper presented at the *4th Combined Manufacturers' and Technologists' Airborne Wind Shear Review Meeting, Williamsburg, VA, Apr. 4–16, 1992*, NTIS Report N93-19595 (1992).

Emedio M. Bracalente, C.L. Britt, and W.R. Jones, "Airborne Doppler Radar Detection of Low Altitude Windshear," AIAA Paper 88-4657 (1988).

Scott Braun and Chung-Lin Shie, "Improving our understanding of Atlantic tropical cyclones through knowledge of the Saharan Air Layer: Hope or Hype?" *Bulletin of the American Meteorological Society* (Aug. 14, 2008).

R.S. Bray and W.E. Larsen, "Simulator Investigations of the Problems of Flying a Swept-Wing Transport Aircraft in Heavy Turbulence," NASA LRC, *Conference on Aircraft Operating Problems*, NASA SP-83 (1965).

Charles L. Britt and Emedio Bracalente, "NASA Airborne Radar Wind Shear Detection Algorithm and the Detection of Wet Microbursts in the Vicinity of Orlando, Florida," paper presented at the *4th Combined Manufacturers' and Technologists' Airborne Wind Shear Review Meeting, Williamsburg, VA, Apr. 4–16, 1992*, NTIS Report N93-19611 (1992).

Charles L. Britt, George F. Switzer, and Emedio M. Bracalente, "Certification Methodology Applied to the NASA Experimental Radar System," paper presented at the *Airborne Windshear Detection and Warning Systems' 5th and Final Combined Manufacturers' and Technologists' Conference*, pt. 2, pp. 463–488, NTIS Report 95N13205 (1994).

C.P. Burgess, "Forces on Airships in Gusts," NACA TR-204 (1925).

Dennis W. Camp, Walter Frost, and Pamela D. Parsley, *Proceedings: Fifth Annual Workshop on Meteorological and Environmental Inputs to Aviation Systems, Mar. 31–Apr. 2, 1981*, NASA CP-2192 (1981).

F. Caracena, R. Ortiz, and J. Augustine, "The Crash of Delta Flight 191 at Dallas-Fort Worth International Airport on 2 August 1985: Multiscale Analysis of Weather Conditions," National Oceanic and Atmospheric Report TR ERL 430-ESG-2 (1987).

T.L. Clark, W.D. Hall, et al., "Origins of Aircraft-Damaging Clear-Air Turbulence During the 9 December 1992 Colorado Downslope Windstorm: Numerical Simulations and Comparison with Observations," *Journal of Atmospheric Sciences*, vol. 57 (Apr. 2000).

Thomas L. Coleman and Emilie C. Coe, "Airplane Measurements of Atmospheric Turbulence for Altitudes Between 20,000 and 55,000 Feet Over the Western part of the United States," NACA RM-L57G02 (1957).

Thomas L. Coleman and Roy Steiner, "Atmospheric Turbulence Measurements Obtained from Airplane Operations at Altitudes Between 20,000 and 75,000 Feet for Several Areas in the Northern Hemisphere," NASA TN-D-548 (1960).

J.R. Connell, et al., "Numeric and Fluid Dynamic Representation of Tornadic Double Vortex Thunderstorms," NASA Technical Paper CR-171023 (1980).

William J. Cox, "The Multi-Dimensional Nature of Wind Shear Investigations," in Society of Experimental Test Pilots, 1976 Report to the Aerospace Profession: *Proceedings of the Twentieth Symposium of The Society of Experimental Test Pilots, Beverly Hills, CA, Sept. 22–25, 1976*, vol. 13, no. 2 (Lancaster, CA: Society of Experimental Test Pilots, 1976).

M.P. de Villiers and J. van Heerden, "Clear air turbulence over South Africa," *Meteorological Applications,* vol. 8 (2001), pp. 119–126.

V.E. Delnore, ed., *Airborne Windshear Detection and Warning Systems: Fifth and Final Combined Manufacturers' and Technologists' Conference*, NASA CP-10139, pts. 1–2 (1994).

"Delta Accident Report Focuses On Wind Shear Research," *Aviation Week & Space Technology* (Nov. 24, 1986).

Norbert Didden and Chi-Minh Ho, "Unsteady Separation in a Boundary Layer Produced by an Impinging Jet," *Journal of Fluid Mechanics,* vol. 160 (1985), pp. 235–236.

Philip Donely, "Effective Gust Structure at Low Altitudes as Determined from the Reactions of an Airplane," NACA TR-692 (1940).

Philip Donely, "Frequency of Occurrence of Atmospheric Gusts and of Related Loads on Airplane Structures," NACA WRL-121 (1944).

Philip Donely, "Safe Flight in Rough Air," NASA TM-X-51662 (1964).

Philip Donely, "Summary of Information Relating to Gust Loads on Airplanes," NACA TR-997 (1950).

Kelvin K. Droegemeier and Terry Zweifel, "A Numerical Field Experiment Approach for Determining Probabilities of Microburst Intensity," NTRS Report 93N19599 (1992).

R.E. Dunham, Jr., "Low-Altitude Wind Measurements from Wide-Body Jet Transports," NASA TM-84538 (1982).

R.E. Dunham, Jr., "Potential Influences of Heavy Rain on General Aviation Airplane Performance," AIAA Paper 86-2606 (1986).

R.E. Dunham, Jr., G.M. Bezos, and B.A. Campbell, "Heavy Rain Effects on Airplane Performance," NTRS Report 91N11686 (1990).

R.E. Dunham, Jr., and J.W. Usry, "Low Altitude Wind Shear Statistics Derived from Measured and FAA Proposed Standard Wind Profiles," AIAA Paper 84-0114 (1984).

L.J. Ehernberger and Betty J. Love, "High Altitude Gust Acceleration Environment as Experienced by a Supersonic Airplane," NASA TN-D-7868 (1975).

W. Frost and B. Crosby, "Investigations of Simulated Aircraft Flight Through Thunderstorm Outflows," NASA CR-3052 (1978).

T. Theodore Fujita, "DFW Microburst on August 2, 1985," Satellite and Mesometeorology Research Project Research Paper 217, Dept. of Geophysical Sciences, University of Chicago, NTIS Report PB-86-131638 (1986).

T. Theodore Fujita, "The Downburst, Microburst, and Macroburst," Satellite and Mesometeorology Research Project Research Paper 210, Dept. of Geophysical Sciences, University of Chicago, NTIS Report PB-85-148880 (1985).

T. Theodore Fujita and F. Caracena, "An Analysis of Three Weather Related Aircraft Accidents," *Bulletin of the American Meteorological Society,* vol. 58 (1977), pp. 1164–1181.

Jack Funk and Richard H. Rhyne, "An Investigation of the Loads on the Vertical Tail of a Jet-Bomber Airplane Resulting from Flight Through Rough Air," NACA TN-3741 (1956).

Walter Georgii, "Ten Years' Gliding and Soaring in Germany," *Journal of the Royal Aeronautical Society,* vol. 34, no. 237 (Sept. 1930).

S.D. Harrah, E.M. Bracalente, P.R. Schaffner, and E.G. Baxa, "Description and Availability of Airborne Doppler Radar Data," in NASA Jet Propulsion Laboratory, *JPL Progress in Electromagnetics Research Symposium (PIERS)* (Pasadena: JPL, 1993), NTIS ID N94-20403 05-32.

J.C. Hunsaker and Edwin Bidwell Wilson, "Report on Behavior of Aeroplanes in Gusts," NACA TR-1 (1917).

Joseph W. Jewel, Jr., "Tabulations of Recorded Gust and Maneuver Accelerations and Derived Gust Velocities for Airplanes in the NSA VGH General Aviation Program," NASA TM-84660 (1983).

W.A. Kilgore, S. Seth, N.L. Crabill, S.T. Shipley, and J. O'Neill, "Pilot Weather Advisor," NASA CR-189723 (1992).

Eldon E. Kordes and Betty J. Love, "Preliminary Evaluation of XB 70 Airplane Encounters with High-Altitude Turbulence," NASA TN-D-4209 (1967).

Jonggil Lee and Ernest G. Baxa, Jr., "Phase Noise Effects on Turbulent Weather Radar Spectrum Parameter Estimation," NTRS Report 91A25458 (1990).

Michael S. Lewis, et al., "Design and Conduct of a Windshear Detection Flight Experiment," AIAA Paper 92-4092 (1992).

Harvard Lomax, "Lift Developed on Unrestrained Rectangular Wings Entering Gusts at Subsonic and Supersonic Speeds," NACA TN-2925 (1953).

G.A. Lucchi, "Commercial Airborne Weather Radar Technology," paper presented at the *IEEE International Radar Conference, 1980.*

Robert W. Miller, "The Use of Airborne Navigational and Bombing Radars for Weather-Radar Operations and Verifications," *Bulletin of the American Meteorological Society,* vol. 28, no. 1 (Jan. 1947), pp. 19–28.

NASA Langley Research Center, "Windshear," *http://oea.larc.nasa.gov/PAIS/Windshear.html,* accessed July 30, 2009.

NASA Langley Research Center, "NASA Facts On-line: Making the Skies Safe from Windshear," *http://oea.larc.nasa.gov/PAIS/Windshear.html,* accessed July 15, 2009.

National Center for Atmospheric Research, "T-REX: Catching the Sierra's Waves and Rotors," *http://www.ucar.edu/communications/quarterly/spring06/trex.jsp,* accessed July 21, 2009.

Rosa M. Oseguera and Roland L. Bowles, "A Simple, Analytical 3-Dimensional Downburst Model Based on Boundary Layer Stagnation Flow," NASA TM-100632 (1988).

Rosa M. Oseguera, Roland L. Bowles, and Paul A. Robinson, "Airborne In Situ Computation of the Wind Shear Hazard Index," AIAA Paper 92-0291 (1992).

Weneth D. Painter and Dennis W. Camp, "NASA B-57B Severe Storms Flight Program," NASA TM-84921 (1983).

Ludwig Prandtl, "Some Remarks Concerning Soaring Flight," NACA Technical Memorandum No. 47 (Oct. 1921).

Kermit G. Pratt and Walter G. Walker, "A Revised Gust-Load Formula and a Re-Evaluation of V-G Data Taken on Civil Transport Airplanes from 1933 to 1950," NACA TR-1206 (1954).

Fred H. Proctor, "NASA Wind Shear Model—Summary of Model Analyses," NASA CP-100006 (1988).

H. Press and E.T. Binckley, "A Preliminary Evaluation of the Use of Ground Radar for the Avoidance of Turbulent Clouds," NACA TN-1864 (1948).

Fred H. Proctor, "The NASA-Langley Wake Vortex Modeling Effort in Support of an Operational Aircraft Spacing System," AIAA Paper 98-0589 (1998).

Fred H. Proctor, "Numerical Simulations of an Isolated Microburst, Part I: Dynamics and Structure," *Journal of the Atmospheric Sciences,* vol. 45, no. 21 (Nov. 1988), pp. 3137–3160.

Fred H. Proctor, "Numerical Simulations on an Isolated Microburst. Part II: Sensitivity Experiments," *Journal of the Atmospheric Sciences,* vol. 46, no. 14 (July 1989), pp. 2143–2165.

Fred H. Proctor, David W. Hamilton, and Roland L. Bowles, "Numerical Simulation of a Convective Turbulence Encounter," AIAA Paper 2002-0944 (2002).

Fred H. Proctor, David A. Hinton, and Roland L. Bowles, "A Windshear Hazard Index," NASA LRC NTRS Report 200.001.16199 (2000).

Paul A. Robinson, Roland L. Bowles, and Russell Targ, "The Detection and Measurement of Microburst Wind Shear by an Airborne Lidar System," NASA LRC, NTRS Report 95A87798 (1993).

M. Sadoff, R.S. Bray, and W.H. Andrews, "Summary of NASA Research on Jet Transport Control Problems in Severe Turbulence," AIAA Paper 65-330 (1965).

"Safety Board Analyzes Responses to Weather Data," *Aviation Week & Space Technology* (Dec. 15, 1986).

B.S. Shenstone and S. Scott Hall, "Glider Development in Germany: A Technical Survey of Progress in Design in Germany Since 1922," NACA TM-780 (Nov. 1935).

Howard Siepen, "On the Wings of the Wind," *The National Geographic Magazine,* vol. 55, no. 6 (June 1929), p. 755.

Amos A. Spady, Jr., Roland L. Bowles, and Herbert Schlickenmaier, eds., *Airborne Wind Shear Detection and Warning Systems, Second Combined Manufacturers and Technological Conference,* two parts, NASA CP-10050 (1990).

Stanford University, Center for Turbulence Research, "About the Center for Turbulence Research (CTR)," *http://www.stanford.edu/group/ctr/about.html,* accessed Oct. 3, 2009.

G.F. Switzer, J.V. Aanstoos, F.H. Proctor, and D.A. Hinton, "Windshear Database for Forward-Looking Systems Certification," NASA TM-109012 (1993).

Russell Targ, "CLASS: Coherent Lidar Airborne Shear Sensor and Windshear Avoidance," Electro-Optical Sciences Directorate, Lockheed Missiles and Space Company, Oct. 1990.

Russell Targ and Roland L. Bowles, "Investigation of Airborne Lidar for Avoidance of Windshear Hazards," AIAA Paper 88-4658 (1988).

Russell Targ and Roland L. Bowles, "Lidar Windshear Detection for Commercial Aircraft," NASA LRC, NTRS Report 93A17864 (1991).

Russell Targ and Roland L. Bowles, "Windshear Avoidance: Requirements and Proposed System for Airborne Lidar Detection," NASA LRC, NTRS Report 89A15876 (1988).

Russell Targ, Michael J. Kavaya, R. Milton Huffaker, and Roland L. Bowles, "Coherent Lidar Airborne Windshear Sensor—Performance Evaluation," NASA LRC, NTRS Report 91A39873 (1991).

U.S. Department of Energy, "Ask a Scientist," *http://www.newton.dep.anl.gov/aas.htm,* accessed Aug. 5, 2009.

U.S. Department of Transportation, Federal Aviation Administration, Advisory Circular (AC) 25-12, "Airworthiness Criteria for the Approval of Airborne Windshear Warning Systems in Transport Category Airplanes" (Washington, DC: DOT–FAA, Nov. 2, 1987).

U.S. Department of Transportation, Federal Aviation Administration, "Aircraft Certification Service, Airborne Windshear Warning and Escape Guidance Systems for Transport Airplanes," TSO-C117 (Washington, DC: DOT–FAA, July 24, 1990).

U.S. National Academy of Sciences, Committee on Low-Altitude Wind Shear and Its Hazard to Aviation, *Low Altitude Wind Shear and Its Hazard to Aviation* (Washington, DC: National Academy Press, 1983).

U.S. National Transportation Safety Board, "Aircraft Accident Report: Iberia Lineas Areas de España (Iberian Airlines), McDonnell-Douglas DC-10-30, EC CBN, Logan International Airport, Boston, Massachusetts, December 17, 1973," Report NTSB-AAR-74-14 (Nov. 8, 1974).

U.S. National Transportation Safety Board, "Aircraft Accident Report: Eastern Air Lines Inc., Boeing 727-225, N8845E, John F. Kennedy International Airport, Jamaica, New York, June 14, 1975," Report NTSB-AAR-76-8 (Mar. 12, 1976).

U.S. National Transportation Safety Board, "Aircraft Accident Report: Continental Airlines Inc, Boeing 727-224, N88777, Stapleton International Airport, Denver, Colorado, August 7, 1975," Report NTSB-AAR-76-14 (May 5, 1976).

U.S. National Transportation Safety Board, "Aircraft Accident Report: Allegheny Airlines, Inc., Douglas DC-9, N994VJ, Philadelphia, Pennsylvania, June 23, 1976," Report NTSB-AAR-78-2 (Jan. 19, 1978).

U.S. National Transportation Safety Board, "Aircraft Accident Report: Pan American World Airways, Clipper 759, N4737, Boeing 727-235, New Orleans International Airport, Kenner, Louisiana, July 9, 1982," Report NTSB-AAR-83-02 (Mar. 21, 1983).

U.S. National Transportation Safety Board, "Aircraft Accident Report: United Airlines Flight 663, Boeing 727-222, N7647U, Denver, Colorado, May 31, 1984," Report NTSB AAR-85-05 (Mar. 21, 1985).

U.S. National Transportation Safety Board, "Aircraft Accident Report: Delta Air Lines, Inc., Lockheed L-1011-385-1, N726DA, Dallas/Fort Worth International Airport, Texas, August 2, 1985," Report NTSB-AAR-86-05 (Aug. 15, 1986).

U.S. National Transportation Safety Board, "Aircraft Accident Report: Flight Into Terrain During Missed Approach: US Air Flight 1019, DC-9-31, N954VJ, Charlotte-Douglas International Airport, Charlotte, North Carolina, July 2, 1994," Report NTSB-AAR-95-03 (Apr. 4, 1995).

U.S. National Transportation Safety Board, "Aircraft Accident Investigation Press Release: United Airlines Flight 826," *http://www.ntsb.gov/Pressrel/1997/971230.htm,* accessed July 30, 2009.

United States Weather Service, "Wind Gust," *http://www.weather.gov/forecasts/wfo/definitions/defineWindGust.html,* accessed Aug. 1, 2009.

University of Illinois-Champaign, Department of Atmospheric Sciences, "Jet Stream," *http://ww2010.atmos.uiuc.edu/%28Gh%29/guides/mtr/cyc/upa/jet.rxml,* accessed July 25, 2009.

"Vertical Wind Estimation from Horizontal Wind Measurements: General Questions and Answers," NASA/FAA Wind Shear Review Meeting, Sept. 28, 1993.

Dan D. Vicroy, "The Aerodynamic Effect of Heavy Rain on Airplane Performance," AIAA Paper 90-3131 (1990).

Dan D. Vicroy, "Assessment of Microburst Models for Downdraft Estimation," AIAA Paper 91-2947 (1991).

Dan D. Vicroy, "Influence of Wind Shear on the Aerodynamic Characteristics of Airplanes," NASA TP-2827 (1988).

Dan D. Vicroy, "Investigation of the Influence of Wind Shear on the Aerodynamic Characteristics of Aircraft Using a Vortex-Lattice Method," NASA LRC, NTRS Report 88N17619 (1988).

Dan D. Vicroy, "Microburst Vertical Wind Estimation from Horizontal Wind Measurements," NASA TP-3460 (1994).

Dan D. Vicroy, "A Simple, Analytical, Axisymmetric Microburst Model for Downdraft Estimation," NASA TM-104053 (1991).

Dan D. Vicroy, "Vertical Wind Estimation from Horizontal Wind Measurements: General Questions and Answers," Report of the *NASA/FAA Wind Shear Review Meeting* (Washington, DC: NASA, Sept. 28, 1993).

Dan D. Vicroy and Roland L. Bowles, "Effect of Spatial Wind Gradients on Airplane Aerodynamics," AIAA Paper 88-0579 (1988).

Walter G. Walker, "An Analysis of the Airspeed and Normal Accelerations of Douglas DC-2 Airplanes in Commercial Transport Operations," NACA TN-1754 (1948).

Walter G. Walker, "An Analysis of the Airspeeds and Normal Accelerations of Martin M-130 Airplanes in Commercial Transport Operation," NACA TN-1693 (1948).

Walter G. Walker, "Gust Loads and Operating Airspeeds of One Type of Four-Engine Transport Airplane on Three Routes from 1949 to 1953," NACA TN-3051 (1953).

Richard J. Wasicko, "NASA Research Experience on Jet Aircraft Control Problems in Severe Turbulence," NASA TM-X-60179 (1966).

Edwin Bidwell Wilson, "Theory of an Airplane Encountering Gusts," pts. II and III, NACA TR-21 and TR-27 (1918).

"Wind Shear Study: Low-Altitude Wind Shear," *Aviation Week & Space Technology* (Mar. 28, 1983).

Terry Zweifel, "The Effect of Windshear During Takeoff Roll on Aircraft Stopping Distance," NTRS Report 91N11699 (1990).

Terry Zweifel, "Flight Experience with Windshear Detection," NTRS Report 91N11684 (1990).

Terry Zweifel, "Optimal Guidance during a Windshear Encounter," *Scientific Honeyweller* (Jan. 1989).

Terry Zweifel, "Temperature Lapse Rate as an Adjunct to Windshear Detection", paper presented at the *Airborne Wind Shear Detection and Warning Systems Third Combined Manufacturers' and Technologists' Conference, Hampton, VA, Oct. 16–18, 1990.*

Books and Monographs:

Joseph R. Chambers, *Concept to Reality: Contributions of the NASA Langley Research Center to U.S. Civil Aircraft of the 1990s,* NASA SP-2003-4529 (Washington, DC: GPO, 2003).

George W. Gray, *Frontiers of Flight: The Story of NACA Research* (New York: Alfred A. Knopf, 1948).

James R. Hansen, *The Bird is on the Wing: Aerodynamics and the Progress of the American Airplane* (College Station, TX: Texas A&M University Press, 2003).

James R. Hansen, *Engineer in Charge: A History of the Langley Aeronautical Laboratory, 1917–1958,* NASA SP-4305 (Washington, DC: GPO, 1987).

Edwin P. Hartman, *Adventures in Research: A History of the Ames Research Center, 1940–1965,* SP-4302 (Washington, DC: GPO, 1970).

Edmund Preston, *Troubled Passage: The Federal Aviation Administration during the Nixon-Ford Term, 1973–1977* (Washington, DC: FAA, 1987).

Douglas A. Robinson, *Giants in the Sky: A History of the Rigid Airship* (Seattle: University of Washington Press, 1973).

Theodore von Kármán, *Aerodynamics* (New York: Dover Publications, 2004 ed.).

Lane E. Wallace, *Airborne Trailblazer: Two Decades with NASA Langley's 737 Flying Laboratory,* NASA SP-4216 (Washington, DC: GPO, 1994).

THIS PAGE INTENTIONALLY BLANK

A lightning strike reveals the breadth, power, and majesty of this still mysterious electromagnetic phenomenon. NOAA.

CASE 2

Coping With Lightning: A Lethal Threat to Flight

Barrett Tillman and John L. Tillman

The beautiful spectacle and terrible power of lightning have always inspired fear and wonder. In flight, it has posed a significant challenge. While the number of airships, aircraft, and occupants lost to lightning have been few, they offer sobering evidence that lightning is a hazard warranting intensive study and preventative measures. This is an area of NASA research that crosses between the classic fields of aeronautics and astronautics, and that has profound implications for both.

"I LEARNED MORE ABOUT LIGHTNING from flying at night over Bosnia while wearing night vision goggles than I ever learned from a meteorologist. You'd occasionally see a green flash as a bolt discharged to the ground, but that was nothing compared to what was happening inside the clouds themselves. Even a moderate-sized cloud looked like a bubbling witches' cauldron, with almost constant green discharges left and right, up and down. You'd think, "Bloody hell! I wouldn't want to fly through that!" But of course you do, all the time. You just don't notice if you don't have the goggles."[1]

So stated one veteran airman of his impressions with lightning. Lightning is an electrical discharge in the atmosphere usually generated by thunderstorms but also by dust storms and volcanic eruptions. Because only about a fourth of discharges reach the ground, lightning represents a disproportionate hazard to aviation and rocketry. In any case, lightning is essentially an immense spark that can be many miles long.[2]

1. Statement of Air Commodore Andrew P.N. Lambert, RAF, to Richard P. Hallion, Nov. 15, 2009, referring to his experiences on Operation Deny Flight, in 1993, when he was Officer Commanding 23 Squadron and former OC of the RAF Phantom Top Gun school.

2. M.A. Uman, *The Lightning Discharge* (New York: Academic Press, Inc., 1987); Franklin A. Fisher and J. Anderson Plumer, "Lightning Protection of Aircraft," NASA RP-1008 (1977); Michael J. Rycroft, R. Giles Harrison, Keri A. Nicoll, and Evgeny A. Mareev, "An Overview of Earth's Global Electric Circuit and Atmospheric Conductivity," *Space Science Reviews*, vol. 137, no. 104 (June 2008).

Lightning generates radio waves. Scientists at the National Aeronautics and Space Administration (NASA) discovered that very low frequency (VLF) waves cause a gap between the inner and outer Van Allen radiation belts surrounding Earth. The gap offers satellites a potential safe zone from solar outburst particle streams. But, as will be noted, protection of spacecraft from lightning and electromagnetic pulses (EMPs) represents a lasting concern.

There are numerous types of lightning. By far the most common is the streak variety, which actually is the return stroke in open air. Most lightning occurs inside clouds and is seldom witnessed inside thunderstorms. Other types include: ball (spherical, semipersistent), bead (cloud to ground), cloud-to-cloud (aka, sheet or fork lightning), dry (witnessed in absence of moisture), ground-to-cloud, heat (too distant for thunder to be heard), positive (also known as high-voltage lightning), ribbon (in high crosswinds), rocket (horizontal lightning at cloud base), sprites (above thunderstorms, including blue jets), staccato (short cloud to ground), and triggered (caused by aircraft, volcanoes, or lasers).

Every year, some 16 million thunderstorms form in the atmosphere. Thus, over any particular hour, Earth experiences over 1,800. Estimates of the average global lightning flash frequency vary from 30 to 100 per second. Satellite observations produce lower figures than did prior scientific studies yet still record more than 3 million worldwide each day.[3] Between 1959 and 1994, lightning strikes in the United States killed 3,239 people and injured a further 9,818, a measure of the lethality of this common phenomenon.[4]

Two American regions are notably prone to ground strikes: Florida and the High Plains, including foothills of the Rocky Mountains. Globally, lightning is most common in the tropics. Therefore, Florida records the most summer lightning strikes per day in the U.S. Heat differentials between land and water on the three sides of peninsular Florida, over its lakes and swamps and along its panhandle coast, drive air circulations that spin off thunderstorms year-round, although most intensely in summer.

3. Data from weather archive at *http://www.newton.dep.anl.gov/askasci/wea00/wea00239.htm*, accessed Nov. 30, 2009.

4. Joseph R. Chambers, *Concept to Reality: Contributions of the NASA Langley Research Center to U.S. Civil Aircraft of the 1990s*, NASA SP-2003 (Washington, DC: GPO, 2003), p. 173.

Lightning: What It Is, What It Does

Despite recent increases in understanding, scientists are still somewhat mystified by lightning. Modern researchers might concur with stone age shaman and bronze age priests that it partakes of the celestial.

Lightning is a form of plasma, the fourth state of matter, after solids, liquids, and gases. Plasma is an ionized gas in which negatively charged electrons have been stripped by high energy from atoms and molecules, creating a cloud of electrons, neutrons, and positively charged ions.

As star stuff, plasma is by far the most common state of matter in the universe. Interstellar plasmas, such as solar wind particles, occur at low density. Plasmas found on Earth include flames, the polar auroras, and lightning.

Lightning is like outer space conditions coming fleetingly to Earth. The leader of a bolt might zip at 134,000 miles per hour (mph). The energy released heats air instantaneously around the discharge from 36,000 to 54,000 degrees Fahrenheit (°F), or more than three to five times the Sun's surface temperature. The sudden, astronomical increase in local pressure and temperature causes the atmosphere within and around a lightning bolt to expand rapidly, compressing the surrounding clear air into a supersonic shock wave, which decays to the acoustic wave perceived as thunder. Ranging from a sharp, loud crack to a long, low rumble, the sound of a thunderclap is determined by the hearer's distance from the flash and by the type of lightning.

Lightning originates most often in cumulonimbus thunderclouds. The bases of such large, anvil-shaped masses may stretch for miles. Their tops can bump up against, spread out along, and sometimes blast through the tropopause: the boundary between the troposphere (the lower portion of the atmosphere, in which most weather occurs) and the higher stratosphere. The altitude of the lower stratosphere varies with season and latitude, from about 5 miles above sea level at the poles in winter to 10 miles near the equator. The tropopause is not a "hard" ceiling. Energetic thunderstorms, particularly from the tropics, may punch into the lower stratosphere and oscillate up and down for hours in a multicycle pattern.

A Lightning Primer

The conditions if not the mechanics that generate lightning are now well known. In essence, this atmospheric fire is started by rubbing particles together. But there is still no agreement on which processes ignite lightning. Current hypotheses focus on the separation of electric charge and generation of an electric field within a thunderstorm. Recent studies further suggest that lightning initiation requires ice, hail, and semifrozen water droplets, called "graupel." Storms that do not produce large quantities of ice usually do not develop lightning.[5] Graupel forms when super-cooled water droplets condense around a snowflake nucleus into a sphere of rime, from 2 to 5 millimeters across. Scientific debate continues as experts grapple with the mysteries of graupel, but the stages of lightning creation in thunderstorms are clear, as outlined by the National Weather Service of the National Oceanic and Atmospheric Administration (NOAA).

First comes charge separation. Thunderstorms are turbulent, with strong updrafts and downdrafts regularly occurring close to one another. The updrafts lift water droplets from warmer lower layers to heights between 35,000 and 70,000 feet, miles above the freezing level. Simultaneously, downdrafts drag hail and ice from colder upper layers. When the opposing air currents meet, water droplets freeze, releasing heat, which keeps hail and ice surfaces slightly warmer than the surrounding environment, so that graupel, a "soft hail," forms.

Electrons carry a negative charge. As newly formed graupel collides with more water droplets and ice particles, electrons are sheared off the ascending particles, charging them positively. The stripped electrons collect on descending bits, charging them negatively. The process results in a storm cloud with a negatively charged base and positively charged top.

Once that charge separation has been established, the second step is generation of an electrical field within the cloud and, somewhat like a mirror image, an electrical field below the storm cloud. Electrical opposites attract, and insulators inhibit current flow. The separation of positive and negative charges within a thundercloud generates an electric field between its top and base. This field strengthens with further separation of these charges into positive and negative pools. But the atmosphere acts as an insulator, inhibiting electric flow, so an enormous charge must build up before lightning can occur. When that high charge threshold is finally crossed, the strength of the electric field overpowers atmospheric insulation, unleashing lightning. Another electrical field develops with Earth's surface below negatively charged storm base,

5. NOAA Online School for Weather, "How Lightning is Created," at *http://www.srh.noaa.gov/jetstream/lightning/lightning.htm*, accessed Nov. 30, 2009.

where positively charged particles begin to pool on land or sea. Whither the storm goes, the positively charged field—responsible for cloud-to-ground lightning—will follow it. Because the electric field within the storm is much stronger than the shadowing positive charge pool, most lightning (about 75 to 80 percent) remains within the clouds and is thus not attracted groundward.

The third phase is the building of the initial stroke that shoots between the cloud and the ground. As a thunderstorm moves, the pool of positively charged particles traveling with it along the ground gathers strength. The difference in charge between the base of the clouds and ground grows, leading positively charged particles to climb up taller objects like houses, trees, and telephone poles. Eventually a "stepped leader," a channel of negative charge, descends from the bottom of the storm toward the ground. Invisible to humans, it shoots to the ground in a series of rapid steps, each happening quicker than the blink of an eye. While this negative leader works its way toward Earth, a positive charge collects in the ground and in objects resting upon it. This accumulation of positive charge "reaches out" to the approaching negative charge with its own channel, called a "streamer." When these channels connect, the resulting electrical transfer appears to the observer as lightning.

Finally, a return stroke of lightning flows along a charge channel about 0.39 inches wide between the ground and the cloud. After the initial lightning stroke, if enough charge is left over, additional strokes will flow along the same channel, giving the bolt its flickering appearance.

Land struck by a bolt may reach more than 3,300 °F, hot enough to almost instantly melt the silica in conductive soil or sand, fusing the grains together. Within about a second, the fused grains cool into fulgurites, or normally hollow glass tubes that can extend some distance into the ground, showing the path of the lightning and its dispersion over the surface.

The tops of trees, skyscrapers, and mountains lie closer to the base of storm clouds than does low-lying ground, so such objects are commonly struck by lightning. The less atmospheric insulation that lightning must burn through, the easier falls its strike. The tallest object beneath a storm will not necessarily suffer a hit, however, because the opposite charges may not accumulate around the highest local point or in the clouds above it. Lightning can strike an open field rather than a nearby line of trees.

Lightning leader development depends not only upon the electrical breakdown of air, which requires about 3 million volts per meter, but on prior channel carving. Ambient electric fields required for lightning leader propagation can be one or two orders of magnitude less than the electrical breakdown strength. The potential gradient inside a developed return stroke channel is on the order of hundreds of volts per meter because of intense channel ionization, resulting in a power output on the order of a megawatt per meter for a vigorous return stroke current of 100,000 amperes (100 kiloamperes, kA).

Negative, Positive, Helpful, and Harmful

Most lightning forms in the negatively charged region under the base of a thunderstorm, whence negative charge is transferred from the cloud to the ground. This so-called "negative lightning" accounts for over 95 percent of strikes. An average bolt of negative lightning carries an electric current of 30 kA, transferring a charge of 5 coulombs, with energy of 500 megajoules (MJ). Large lightning bolts can carry up to 120 kA and 350 coulombs. The voltage is proportional to the length of the bolt.[6]

Some lightning originates near the top of the thunderstorm in its cirrus anvil, a region of high positive charge. Lightning formed in the upper area behaves similarly to discharges in the negatively charged storm base, except that the descending stepped leader carries a positive charge, while its subsequent ground streamers are negative. Bolts thus created are called "positive lightning," because they deliver a net positive charge from the cloud to the ground. Positive lightning usually consists of a single stroke, while negative lightning typically comprises two or more strokes. Though less than 5 percent of all strikes consist of positive lightning, it is particularly dangerous. Because it originates in the upper levels of a storm, the amount of air it must burn through to reach the ground is usually much greater. Therefore, its electric field typically is much stronger than a negative strike would be and generates enormous amounts of extremely low frequency (ELF) and VLF waves. Its flash duration is longer, and its peak charge and potential are 6 to 10 times greater than a negative strike, as much as 300 kA and 1 billion volts!

Some positive lightning happens within the parent thunderstorm and hits the ground beneath the cloud. However, many positive strikes occur near the edge of the cloud or may even land more than 10 miles away, where perhaps no one would recognize risk or hear thunder.

6. Richard Hasbrouck, "Mitigating Lightning Hazards," Science & Technology Review (May 1996), p. 7.

Such positive lightning strikes are called "bolts from the blue." Positive lightning may be the main type of cloud-to-ground during winter months or develop in the late stages of a thunderstorm. It is believed to be responsible for a large percentage of forest fires and power-line damage, and poses a threat to high-flying aircraft. Scientists believe that recently discovered high-altitude discharges called "sprites" and "elves" result from positive lightning. These phenomena occur well above parent thunderstorms, at heights from 18 to 60 miles, in some cases reaching heights traversed only by transatmospheric systems such as the Space Shuttle.

Lightning is by no means a uniformly damaging force. For example, fires started by lightning are necessary in the life cycles of some plants, including economically valuable tree species. It is probable that, thanks to the evolution and spread of land plants, oxygen concentrations achieved the 13-percent level required for wildfires before 420 million years ago, in the Paleozoic Era, as evinced by fossil charcoal, itself proof of lightning-caused range fires.

In 2003, NASA-funded scientists learned that lightning produces ozone, a molecule composed of three oxygen atoms. High up in the stratosphere (about 6 miles above sea level at midlatitudes), ozone shields the surface of Earth from harmful ultraviolet radiation and makes the land hospitable to life, but low in the troposphere, where most weather occurs, it's an unwelcome byproduct of manmade pollutants. NASA's researchers were surprised to find that more low-altitude ozone develops naturally over the tropical Atlantic because of lightning than from the burning of fossil fuels or vegetation to clear land for agriculture.

Outdoors, humans can be injured or killed by lightning directly or indirectly. No place outside is truly safe, although some locations are more exposed and dangerous than others. Lightning has harmed victims in improvised shelters or sheds. An enclosure of conductive material does, however, offer refuge. An automobile is an example of such an elementary Faraday cage.

Property damage is more common than injuries or death. Around a third of all electric power-line failures and many wildfires result from lightning. (Fires started by lightning are, however, significant in the natural life cycle of forests.) Electrical and electronic devices, such as telephones, computers, and modems, also may be harmed by lightning, when overcurrent surges fritz them out via plug-in outlets, phone jacks, or Ethernet cables.

The Lightning Hazard in Aeronautics and Astronautics: A Brief Synopsis

Since only about one-fourth of discharges reach Earth's surface, lightning presents a disproportionate hazard to aviation and rocketry. Commercial aircraft are frequently struck by lightning, but airliners are built to reduce the hazard, thanks in large part to decades of NASA research. Nevertheless, almost every type of aircraft has been destroyed or severely damaged by lightning, ranging from gliders to jet airliners. The following is a partial listing of aircraft losses related to lightning:

- August 1940: a Pennsylvania Central Airlines Douglas DC-3A dove into the ground near Lovettsville, VA, killing all 25 aboard (including Senator Ernest Lundeen of Minnesota), after "disabling of the pilots by a severe lightning discharge in the immediate neighborhood of the airplane, with resulting loss of control."[7]
- June 1959: a Trans World Airlines (TWA) four-engine Lockheed Starliner with 68 passengers and crew was destroyed near Milan, Italy.
- August 1963: a turboprop Air Inter Vickers Viscount crashed on approach to Lyon, France, killing all 20 on board plus 1 person on the ground.
- December 1963: a Pan American Airlines Boeing 707 crashed at night when struck by lightning over Maryland. All 82 aboard perished.
- April 1966: Abdul Salam Arif, President of Iraq, died in a helicopter accident, reportedly in a thunderstorm that could have involved lightning.
- April 1967: an Iranian Air Force C-130B was destroyed by lightning near Mamuniyeh. The 23 passengers and crew all died.
- Christmas Eve 1971: a Lockheed Electra of Líneas Aéreas Nacionales Sociedad Anónima (LANSA) was destroyed over Peru with 1 survivor among 92 souls on board.
- May 1976: an Iranian Air Force Boeing 747 was hit during descent to Madrid, Spain, killing all 17 aboard.

7. Civil Aeronautics Board, Accident Investigation Report on Loss of DC-3A NC21789, Aug. 31, 1940, p. 84; Donald R. Whitnah, *Safer Skyways: Federal Control of Aviation, 1926–1966* (Ames, IA: The Iowa State University Press, 1966), p. 157; "Disaster: Death in the Blue Ridge," *Time,* Sept. 9, 1940.

- November 1978: a U.S. Air Force (USAF) C-130E was struck by lightning near Charleston, SC, and fatally crashed, with six aboard.
- September 1980: a Kuwaiti C-130 crashed after a lightning strike near Montelimar, France. The eight-man crew was killed.
- February 1988: a Swearingen Metro operated by Nürnberger Flugdienst was hit near Mulheim, Germany, with all 21 aboard killed.
- January 1995: a Super Puma helicopter en route to a North Sea oil platform was struck in the tail rotor, but the pilot autorotated to a water landing. All 16 people aboard were safely recovered.
- April 1999: a British glider was struck, forcing both pilots to bail; they landed safely.

Additionally, lightning posed a persistent threat to rocket-launch operations, forcing extensive use of protective systems such as lightning rods and "tripwire" devices. These devices included small rockets trailing conductive wires that can trigger premature cloud-to-ground strokes, reducing the risk of more powerful lightning strokes. The classic example was the launch of Apollo 12, on November 14, 1969. "The flight of Apollo 12," NASA historian Roger E. Bilstein has written, "was electrifying, to say the least."[8]

During its ascent, it built up a massive static electricity charge that abruptly discharged, causing a brief loss of power. It had been an exceptionally close call. Earlier, the launch had been delayed while technicians dealt with a liquid hydrogen leak. Had a discharge struck the fuel-air mix of the leak, the conflagration would have been disastrous. Of course, three decades earlier, a form of lightning (a brush discharge, commonly called "St. Elmo's fire") that ignited a hydrogen gas-air mix was blamed by investigators for the loss of the German airship Hindenburg in 1937 at Lakehurst, NJ.[9]

8. Roger E. Bilstein, *Stages to Saturn: A Technological History of the Apollo/Saturn Launch Vehicles*, NASA SP-4206 (Washington, DC: NASA, 1980), p. 374.

9. U.S. Department of Commerce, Bureau of Air Commerce, Robert W. Knight, *The Hindenburg Accident: A Comparative Digest of the Investigations and Findings, with the American and Translated German Reports Included*, Report No. 11 (Washington, DC: GPO, 1938).

Flight Research on Lightning

Benjamin Franklin's famous kite experiments in the 1750s constituted the first application of lightning's effect upon "air vehicles." Though it is uncertain that Franklin personally conducted such tests, they certainly were done by others who were influenced by him. But nearly 200 years passed before empirical data were assembled for airplanes.[10]

Probably the first systematic study of lightning effects on aircraft was conducted in Germany in 1933 and was immediately translated by NASA's predecessor, the National Advisory Committee on Aeronautics (NACA). German researcher Heinrich Koppe noted diverse opinions on the subject. He cited the belief that any aircraft struck by lightning "would be immediately destroyed or at least set on fire," and, contrarily, that because there was no direct connection between the aircraft and the ground, "there could be no force of attraction and, consequently, no danger."[11]

Koppe began his survey detailing three incidents in which "the consequences for the airplanes were happily trivial." However, he expanded the database to 32 occasions in 6 European nations over 8 years. (He searched for reports from America but found none at the time.) By discounting incidents of St. Elmo's fire and a glider episode, Koppe had 29 lightning strikes to evaluate. All but 3 of the aircraft struck had extended trailing antennas at the moment of impact. His conclusion was that wood and fabric aircraft were more susceptible to damage than were metal airframes, "though all-metal types are not immune." Propellers frequently attracted lightning, with metal-tipped wooden blades being more susceptible than all-metal props. While no fatalities occurred with the cases in Koppe's studies, he did note disturbing effects upon aircrew, including temporary blindness, short-term stunning, and brief paralysis; in each case, fortunately, no lingering effects occurred.[12]

Koppe called for measures to mitigate the effects of lightning strikes, including housing of electrical wires in metal tubes in wood airframes and "lightning protection plates" on the external surfaces. He said radio

10. E. Philip Krider, "Benjamin Franklin and the First Lightning Conductors," *Proceedings of the International Commission on History of Meteorology* (2004).

11. Heinrich Koppe, "Practical Experiences with Lightning Discharges to Airplanes," *Zeitschrift für Flugtechnik und Motorluftschiffahrt*, vol. 24, no. 21, translated and printed as NACA Technical Memorandum No. 730 (Nov. 4, 1933), p. 1.

12. Ibid., p. 7.

masts and the sets themselves should be protected. One occasionally overlooked result was "electrostriction," which the author defined as "very heavy air pressure effect." It involved mutual attraction of parallel tracks into the area of the current's main path. Koppe suggested a shield on the bottom of the aircraft to attract ionized air. He concluded: "airplanes are not 'hit' by lightning, neither do they 'accidentally' get into the path of a stroke. The hits to airplanes are rather the result of a release of more or less heavy electrostatic discharges whereby the airplane itself forms a part of the current path."[13]

American studies during World War II expanded upon prewar examinations in the United States and elsewhere. A 1943 National Bureau of Standards (NBS, now the National Institute for Standards and Technology, NIST) analysis concluded that the power of a lightning bolt was so enormous—from 100 million to 1 billion volts—that there was "no possibility of interposing any insulating barrier that can effectively resist it." Therefore, aircraft designers needed to provide alternate paths for the discharge via "lightning conductors."[14] Postwar evaluation reinforced Koppe's 1933 observations, especially regarding lightning effects upon airmen: temporary blindness (from seconds to 10 minutes), momentary loss of hearing, observation of electrical effects ranging from sparks to "a blinding blue flash," and psychological effects. The latter were often caused more by the violent sensations attending the entrance of a turbulent storm front rather than a direct result of lightning.[15]

Drawing upon British data, the NACA's 1946 study further detailed atmospheric discharges by altitude bands from roughly 6,500 to 20,500 feet, with the maximum horizontal gradient at around 8,500 feet. Size and configuration of aircraft became recognized factors in lightning, owing to the greater surface area exposed to the atmosphere. Moisture and dust particles clinging to the airframe had greater potential for drawing a lightning bolt than on a smaller aircraft. Aircraft speed also was considered, because the ram-air effect naturally forced particles closer together.[16]

---

13. Ibid., p. 14.
14. National Bureau of Standards, "Protection of Nonmetallic Aircraft from Lightning," High Voltage Laboratory, Advance Report 3110 (Sept. 1943).
15. National Bureau of Standards, "Electrical Effects in Glider Towlines," High Voltage Laboratory, Advance Restricted Report 4C20 (Mar. 1944), p. 47.
16. L.P. Harrison, "Lightning Discharges to Aircraft and Associated Meteorological Conditions," U.S. Weather Bureau, Washington, DC (May 1946), pp. 58–60.

A Weather Bureau survey of more than 150 strikes from 1935 to 1944 defined a clear "danger zone": aircraft flying at or near freezing temperatures and roughly at 1,000 to 2,000 feet above ground level (AGL). The most common factors were 28–34 °F and between 5,000 and 8,000 feet AGL. Only 15 percent of strikes occurred above 10,000 feet.[17]

On February 19, 1971, a Beechcraft B90 King Air twin-turboprop business aircraft owned by Marathon Oil was struck by a bolt of lightning while descending through 9,000 feet preparatory to landing at Jackson, MI. The strike caused "widespread, rather severe, and unusual" damage. The plane suffered "the usual melted metal and cracked nonmetallic materials at the attachments points" but in addition suffered a local structural implosion on the inboard portions of the lower right wing between the fuselage and right engine nacelle, damage to both flaps, impact-and-crush-type damage to one wingtip at an attachment point, electrical arc pitting of flap support and control rod bearings, a hole burned in a ventral fin, missing rivets, and a brief loss of power. "Metal skins were distorted," NASA inspectors noted, "due to the 'magnetic pinch effect' as the lightning current flowed through them." Pilots J.R. Day and J.W. Maxie recovered and landed the aircraft safely. Marathon received a NASA commendation for taking numerous photographs of record and contacting NASA so that a much more detailed examination could be performed.[18]

The jet age brought greater exposure to lightning, prompting further investigation by NOAA (created in 1970 to succeed the Environmental Science Services Administration, which had replaced the Weather Bureau in 1965). The National Severe Storms Laboratory conducted Project Rough Rider, measuring the physical characteristics and effects of thunderstorms, including lightning. The project employed two-seat F-100F and T-33A jets to record the intensity of lightning strikes over Florida and Oklahoma in the mid-1960s and later. The results of the research flights were studied and disseminated to airlines, providing safety guidelines for flight in the areas of thunderstorms.[19]

17. Ibid., pp. 91–95.

18. Quotes from Paul T. Hacker, "Lightning Damage to a General Aviation Aircraft: Description and Analysis," NASA TN-D-7775 (1974).

19. Edward Miller, "1964 Rough Rider Summary of Parameters Recorded, Test Instrumentation, Flight Operations, and Aircraft Damage," USAF Aeronautical Systems Division (1965), DTIC AD 0615749, at *http://oai.dtic.mil/oai/oai?verb=getRecord&metadataPrefix=html&identifier=AD0615749*, accessed Nov. 30, 2009.

In December 1978, two Convair F-106A Delta Dart interceptors were struck within a few minutes near Castle Air Force Base (AFB), CA. Both had lightning protection kits, which the Air Force had installed beginning in early 1976. One Dart was struck twice, with both jets sustaining "severe" damage to the Pitot booms and area around the radomes. The protection kits prevented damage to the electrical systems, though subsequent tests determined that the lightning currents well exceeded norms, in the area of 225 kA. One pilot reported that the strike involved a large flash, and that the impact felt "like someone hit the side of the aircraft with a sledgehammer." The second strike a few minutes later exceeded the first. The report concluded that absent the protection kits, damage to electrical and avionic systems might have been extensive.[20]

Though rare, other examples of dual aircraft strikes have been recorded. In January 1982, a Grumman F-14A Tomcat was en route to the Grumman factory at Calverton, NY, flown by CDR Lonny K. McClung from Naval Air Station (NAS) Miramar, CA, when it was struck by lightning. The incident offered a dramatic example of how a modern, highly sophisticated aircraft could be damaged, and its safety compromised, by a lightning strike. As CDR McClung graphically recalled:

> We were holding over Calverton at 18,000 waiting for a rainstorm to pass. A lightning bolt went down about half a mile in front of us. An arm reached out and zapped the Pitot probe on the nose. I saw the lightning bolt go down and almost as if a time warp, freeze frame, an arm of that lightning came horizontal to the nose of our plane. It shocked me, but not badly, though it fried every computer in the airplane—Grumman had to replace everything. Calverton did not open in time for us to recover immediately so we had to go to McGuire AFB (112 miles southwest) and back on the "peanut gyro" since all our displays were fried. With the computers zapped, we had a bit of an adventure getting the plane going again so we could go to Grumman and get it fixed. When we got back to Calverton, one of the linemen told us that the

20. J. Anderson Plumer, "Investigation of Severe Lightning Strike Incidents to Two USAF F-106A Aircraft," NASA CR-165794 (1981).

2

same lightning strike hit a news helo below us. Based on the time, we were convinced it was the same strike that got us. An eerie feeling.[21]

The 1978 F-106 Castle AFB F-106 strikes stimulated further research on the potential danger of lightning strikes on military aircraft, particularly as the Castle incidents involved currents beyond the strength usually encountered.

Coincidentally, the previous year, the National Transportation Safety Board had urged cooperative studies among academics, the aviation community, and Government researchers to address the dangers posed to aircraft operations by thunderstorms. Joseph Stickle and Norman Crabill of the NASA Langley Research Center, strongly supported by Allen Tobiason and John Enders at NASA Headquarters, structured a comprehensive program in thunderstorm research that the Center could pursue. The next year, Langley researchers evaluated a lightning location detector installed on an Agency light research aircraft, a de Havilland of Canada DHC-6 Twin Otter. But the most extensive and prolonged study NASA undertook involved, coincidentally, the very sort of aircraft that had figured so prominently in the Castle AFB strikes: a two-seat NF-106B Delta Dart, lent from the Air Force to NASA for research purposes.[22]

The NASA Langley NF-106B lightning research program began in 1980 and continued into 1986. Extensive aerial investigations were undertaken after ground testing, modeling, and simulation.[23] Employing the NF-106B, Langley researchers studied two subjects in particular: the mechanisms influencing lightning-strike attachments on aircraft and the electrical and physical effects of those strikes. Therefore, the Dart was fitted with sensors in 14 locations: 9 in the fuselage plus 3 in the wings and 2 in the vertical stabilizer. In all, the NF-106B sustained 714 strikes during 1,496 storm penetrations at altitudes from 5,000 to 50,000 feet, typically

21. Capt. Lonny K. McClung, USN (ret.), e-mail to authors, May 2009.

22. Chambers, *Concept to Reality*, p. 175.

23. Literature on NASA's NF-106B program is understandably extensive. The following are particularly recommended: J.H. Helsdon, "Atmospheric Electrical Modeling in Support of the NASA F-106 Storm Hazards Project," NASA CR-179801 (1986); V. Mazur, B.D. Fisher, and J.C. Gerlach, "Lightning Strikes to a NASA Airplane Penetrating Thunderstorms at Low Altitudes," AIAA Paper 86-0021 (1986); R.M. Winebarger, "Loads and Motions of an F-106B Flying Through Thunderstorms," NASA TM-87671 (1986).

flying within a 150-mile radius of its operating base at Langley.[24] One NASA pilot—Bruce Fisher—experienced 216 lightning strikes in the two-seat Dart. Many test missions involved multiple strikes; during one 1984 research flight at an altitude of 38,000 feet through a thunderstorm, the NF-106B was struck 72 times within 45 minutes, and the peak recorded on that particular test mission was an astounding 9 strikes per minute.[25]

NASA's NF-106B lightning research program constituted the single most influential flight research investigation undertaken in atmospheric electromagnetic phenomena by any nation. The aircraft, now preserved in an aviation museum, proved one of the longest-lived and most productive of all NASA research airplanes, retiring in 1991. As a team composed of Langley Research Center, Old Dominion University, and Electromagnetic Applications, Inc., researchers reported in 1987:

> This research effort has resulted in the first statistical quantification of the electromagnetic threat to aircraft based on in situ measurements. Previous estimates of the in-flight lightning hazard to aircraft were inferred from ground-based measurements. The electromagnetic measurements made on the F-106 aircraft during these strikes have established a statistical basis for determination of quantiles and "worst-case" amplitudes of electromagnetic parameters of rate of change of current and the rate of change of electric flux density. The 99.3 percentile of the peak rate of change of current on the F-106 aircraft struck by lightning is about two and a half times that of previously accepted airworthiness criteria. The findings are at present being included in new criteria concerning protection of aircraft electrical and

24. Rosemarie L. McDowell, "Users Manual for the Federal Aviation Administration Research and Development Electromagnetic Database (FRED) for Windows: Version 2.0," Department of Transportation, Federal Aviation Administration, Report DOT/FAA/AR-95/18 (1998), p. 41; and R.L. McDowell, D.J. Grush, D.M. Cook, and M.S. Glynn, "Implementation of the FAA Research and Development Electromagnetic Database," in NASA KSC, *The 1991 International Aerospace and Ground Conference on Lightning and Static Electricity*, vol. 2 (1991). Fittingly, the NASA Langley NF-106B is now a permanent exhibit at the Virginia Air and Space Museum, Hampton.

25. Chambers, *Concept to Reality*, p. 181; NASA News Release, "NASA Lightning Research on ABC 20/20," Dec. 11, 2007, at *http://www.nasa.gov/topics/aeronautics/features/fisher-2020.html*, accessed Nov. 30, 2009.

> electronic systems against lightning. Since there are at present no criteria on the rate of change of electric flux density, the new data can be used as the basis for new criteria on the electric characteristics of lightning-aircraft electrodynamics. In addition to there being no criteria on the rate of change of electric flux density, there are also no criteria on the temporal durations of this rate of change or rate of change of electric current exceeding a prescribed value. Results on pulse characteristics presented herein can provide the basis for this development. The newly proposed lightning criteria and standards are the first which reflect actual aircraft responses to lightning measured at flight altitudes.[26]

The data helped shape international certification and design standards governing how aircraft should be shielded or hardened to minimize damage from lightning. Recognizing its contributions to understanding the lightning phenomena, its influence upon design standards, and its ability to focus the attention of lightning researchers across the Federal Government, the Flight Safety Foundation accorded the NF-106B program recognition as an Outstanding Contribution to Flight Safety for 1989. This did not mark the end of the NF-106B's electromagnetic research, however, for it was extensively tested at the Air Force Weapons Laboratory at Kirtland AFB, NM, in a cooperative Air Force–NASA study comparing lightning effects with electromagnetic pulses produced by nuclear explosions.[27]

As well, the information developed in F-106B flights led to extension of "triggered" (aircraft-induced) lightning models applied to other aircraft. Based on scaling laws for triggering field levels of differing airframe sizes and configurations, data were compiled for types as diverse as Lockheed C-130 airlifters and light, business aircraft, such as the Gates (now Bombardier) Learjet. The Air Force operated a Lockheed WC-130 during 1981, collecting data to characterize airborne lightning. Operating in Florida, the Hercules flew at altitudes between 1,500

26. Felix L. Pitts, Larry D. Lee, Rodney A. Perala, and Terence H. Rudolph, "New Methods and Results for Quantification of Lightning-Aircraft Electrodynamics," NASA TP-2737 (1987), p. 18.
27. Chambers, *Concept to Reality*, p. 182. This NF-106B, NASA 816, is exhibited in the Virginia Air and Space Center, Hampton, VA.

The workhorse General Dynamics NF-106B Delta Dart used by NASA for a range of electromagnetic studies and research. NASA.

and 18,000 feet, using 11 sensors to monitor nearby thunderstorms. The flights were especially helpful in gathering data on intercloud and cloud-to-ground strokes. More than 1,000 flashes were recorded by analog and 500 digitally.[28]

High-altitude research flights were conducted in 1982 with instrumented Lockheed U-2s carrying the research of the NF-106B and the WC-130 at lower altitudes well into the stratosphere. After a smaller 1979 project, the Thunderstorm Overflight Program was cooperatively sponsored by NASA, NOAA, and various universities to develop criteria for a lightning mapping satellite system and to study the physics of lightning. Sensors included a wide-angle optical pulse detector, electric field change meter, optical array sensor, broadband and high-resolution Ebert spectrometers, cameras, and tape recorders. Flights recorded data from Topeka, KS, in May and from Moffett Field, CA, in August. The project collected some 6,400 data samples of visible pulses, which were analyzed by NASA and university researchers.[29] NASA expanded the studies to include

28. B.P. Kuhlman, M.J. Reazer, and P.L. Rustan, "WC-130 Airborne Lightning Characterization Program Data Review," USAF Wright Aeronautical Laboratories (1984), DTIC ADA150230, at *http://oai.dtic.mil/oai/oai?verb=getRecord&metadataPrefix=html&identifier=ADA150230*, accessed Nov. 30, 2009.

29. Otha H. Vaughan, Jr., "NASA Thunderstorm Overflight Program—Research in Atmospheric Electricity from an Instrumented U-2 Aircraft," NASA TM-82545 (1983); Vaughn, "NASA Thunderstorm Overflight Program—Atmospheric Electricity Research: An Overview Report on the Optical Lightning Detection Experiment for Spring and Summer 1983," NASA TM-86468 (1984); Vaughn, et al., "Thunderstorm Overflight Program," AIAA Paper 80-1934 (1980).

flights by an Agency Lockheed ER-2, an Earth-resources research aircraft derived from the TR-2, itself a scaled-up outgrowth of the original U-2.[30]

Complementing NASA's lightning research program was a cooperative program of continuing studies at lower altitudes undertaken by a joint American-French study team. The American team consisted of technical experts and aircrew from NASA, the Federal Aviation Administration (FAA), the USAF, the United States Navy (USN), and NOAA, using a specially instrumented American Convair CV-580 twin-engine medium transport. The French team was overseen by the Offices Nationales des Études et Recherchés Aerospatiales (National Office for Aerospace Studies and Research, ONERA) and consisted of experts and aircrew from the Centre d'Essais Aéronautique de Toulouse (Toulouse Aeronautical Test Center, CEAT) and the l'Armée de l'Air (French Air Force) flying a twin-engine medium airlifter, the C-160 Transall. The Convair was fitted with a variety of external sensors and flown into thunderstorms over Florida in 1984 to 1985 and 1987. Approximately 60 strikes were received, while flying between 2,000 and 18,000 feet. The hits were categorized as lightning, lightning attachment, direct strike, triggered strike, intercepted strike, and electromagnetic pulse. Flight tests revealed a high proportion of strikes initiated by the aircraft itself. Thirty-five of thirty-nine hits on the CV-580 were determined to be aircraft-induced. Further data were obtained by the C-160 with high-speed video recordings of channel formation, which reinforced the opinion that aircraft initiate the lightning. The Transall operated over southern France (mainly near the Pyrenees Mountains) in 1986–1988, and CEAT furnished reports from its strike data to the FAA, and thence to other agencies and industry.[31]

---

30. Richard Blakeslee, "ER-2 Investigations of Lightning and Thunderstorms," in NASA MSFC, *FY92 Earth Science and Applications Program Research Review* (Huntsville: NASA MSFC, 1993), NRTS 93-N20088; Doug M. Mach, et al., "Electric Field Profiles Over Hurricanes, Tropical Cyclones, and Thunderstorms with an Instrumented ER-2 Aircraft," paper presented at the *International Conferences on Atmospheric Electricity (ICAE), International Commission on Atmospheric Electricity, Beijing, China, Aug. 13–17, 2007*, NTRS 2007.003.7460.

31. Centre d'Essais Aéronautique de Toulouse, "Measurement of Characteristics of Lightning at High Altitude," a translation of CEAT, "Mesure des caracteristiques de la foudre en altitude," Test No. 76/650000 P.4 (May 1979), NASA TM-76669 (1981); Harold D. Burket, et al., "In-Flight Lightning Characterization Program on a CV-580 Aircraft." Wright-Patterson AFB Flight Dynamics Lab (June 1988); Martin A. Uman, *The Art and Science of Lightning Protection* (Cambridge: Cambridge University Press, 2008), p. 155; McDowell, "User's Manual for FRED," pp. 5, 49.

NASA's Earth-resource research aircraft, a derivative of the Lockheed TR-2 (U-2R) reconnaissance aircraft. NASA.

Electrodynamic Research Using UAVs

Reflecting their growing acceptance for a variety of military missions, unmanned ("uninhabited") aerial vehicles (UAVs) are being increasingly used for atmospheric research. In 1997, a Goddard Space Flight Center space sciences team consisting of Richard Goldberg, Michael Desch, and William Farrell proposed using UAVs for electrodynamic studies. Much research in electrodynamics centered upon the direct-current (DC) Global Electric Circuit (GEC) concept, but Goldberg and his colleagues wished to study the potential upward electrodynamic flow from thunderstorms. "We were convinced there was an upward flow," he recalled over a decade later, "and [that] it was AC."[32] To study upward flows, Goldberg and his colleagues decided that a slow-flying, high-altitude UAV had advantages of proximity and duration that an orbiting spacecraft did not. They contacted Richard Blakeslee at Marshall Space Flight Center, who had a great interest in Earth sciences research. The Goddard-Marshall part-

32. Notes of telephone conversation, Richard P. Hallion with Richard A. Goldberg, NASA Goddard Space Flight Center, Sept. 10, 2009, in author's possession. Goldberg had begun his scientific career studying crystallography but found space science (particularly using sounding rockets) much more exciting. His perception of the upward flow of electrodynamic energy was, as he recalled, "in the pre-sprite days. Sprites are largely insignificant anyway because their duration is so short."

NASA Altus 2 electrodynamic research aircraft, a derivative of the General Atomics Predator UAV, in flight on July12, 2002. NASA.

nership quickly secured Agency support for an electrodynamic UAV research program to be undertaken by the National Space Science and Technology Center (NSSTC) at Huntsville, AL. The outcome was Altus, a modification of the basic General Atomics Predator UAV, leased from the manufacturer and modified to carry a NASA electrodynamic research package. Altus could fly as slow as 70 knots and as high as 55,000 feet, cruising around and above (but never into) Florida's formidable and highly energetic thunderstorms. First flown in 2002, Altus constituted the first time that UAV technology had been applied to study electrodynamic phenomena.[33] Initially, NASA wished to operate the UAV from Patrick AFB near Cape Canaveral, but concerns about the potential dangers of flying a UAV over a heavily populated area resulted in switching its operational location to the more remote Key West Naval Air Station. Altus flights confirmed the suppositions of Goldberg and his colleagues, and it complemented other research methodologies that took electric, magnetic, and optical measurements of thunderstorms, gauging lightning

33. Although this was not the first time drones had been used for measurements in hazardous environments. Earlier, in the heyday of open-atmospheric tests of nuclear weapons, drone aircraft such as Lockheed QF-80 Shooting Stars were routinely used to "sniff" radioactive clouds formed after a nuclear blast and to map their dispersion in the upper atmosphere. Like the electromagnetic research over a quarter century later, these trials complemented sorties by conventional aircraft such as the U-2, another atomic monitor.

The launch of Apollo 12 from the John F. Kennedy Space Center in 1969. NASA.

activity and associated electrical phenomena, including using ground-based radars to furnish broader coverage for comparative purposes.[34]

While not exposing humans to thunderstorms, the Altus Cumulus Electrification Study (ACES) used UAVs to collect data on cloud properties throughout a 3- or 4-hour thunderstorm cycle—not always possible with piloted aircraft. ACES further gathered material for three-dimensional storm models to develop more-accurate weather predictions.

Lightning bolt photographed at the John F. Kennedy Space Center immediately after the launch of Apollo 12 in November 1969. NASA.

Spacecraft and Electrodynamic Effects

With advent of piloted orbital flight, NASA anticipated the potential effects of lightning upon launch vehicles in the Mercury, Gemini, and Apollo programs. Sitting atop immense boosters, the spacecraft were especially vulnerable on their launch pads and in the liftoff phase. One NASA lecturer warned his audience in 1965 that explosive squibs, detonators, vapors, and dust were particularly vulnerable to static electrical

34. For Altus background, see Richard Blakeslee, "The ALTUS Cumulus Electrification Study (ACES): A UAV-Based Investigation of Thunderstorms," paper presented at the *Technical Analysis and Applications Center Unmanned Aerial Vehicle Annual Symposium, Las Cruces, NM, Oct. 30–31, 2001*; and Tony Kim and Richard Blakeslee, "ALTUS Cumulus Electrification Study (ACES)," paper presented at the *Technical Analysis and Applications Center Conference, Santa Fe, NM, Oct. 28–30, 2002.*

detonation; the amount of energy required to initiate detonation was "very small," and, as a consequence, their triggering was "considerably more frequent than is generally recognized."[35]

As mentioned briefly, on November 14, 1969, at 11:22 a.m. EST, Apollo 12, crewed by astronauts Charles "Pete" Conrad, Richard F. Gordon, and Alan L. Bean, thundered aloft from Launch Complex 39A at the Kennedy Space Center. Launched amid a torrential downpour, it disappeared from sight almost immediately, swallowed up amid dark, foreboding clouds that cloaked even its immense flaring exhaust. The rain clouds produced an electrical field, prompting a dual trigger response initiated by the craft. As historian Roger Bilstein wrote subsequently:

> Within seconds, spectators on the ground were startled to see parallel streaks of lightning flash out of the cloud back to the launch pad. Inside the spacecraft, Conrad exclaimed "I don't know what happened here. We had everything in the world drop out." Astronautics Pete Conrad, Richard Gordon, and Alan Bean, inside the spacecraft, had seen a brilliant flash of light inside the spacecraft, and instantaneously, red and yellow warning lights all over the command module panels lit up like an electronic Christmas tree. Fuel cells stopped working, circuits went dead, and the electrically operated gyroscopic platform went tumbling out of control. The spacecraft and rocket had experienced a massive power failure. Fortunately, the emergency lasted only seconds, as backup power systems took over and the instrument unit of the Saturn V launch vehicle kept the rocket operating.[36]

The electrical disturbance triggered the loss of nine solid-state instrumentation sensors, none of which, fortunately, was essential to the safety or completion of the flight. It resulted in the temporary loss of communications, varying between 30 seconds and 3 minutes, depending upon the particular system. Rapid engagement of backup

35. G.J. Bryan, "Static Electricity and Lightning Hazards, Part II," NASA Explosive Safety Executive Lecture Series, June 1965, NTRS N67-15981, pp. 6-10, 6-11.
36. Bilstein, *Stages to Saturn*, pp. 374–375.

systems permitted the mission to continue, though three fuel cells were automatically (and, as subsequently proved, unnecessarily) shut down. Afterward, NASA incident investigators concluded that though lightning could be triggered by the long combined length of the Saturn V rocket and its associated exhaust plume, "The possibility that the Apollo vehicle might trigger lightning had not been considered previously."[37]

Apollo 12 constituted a dramatic wake-up call on the hazards of mixing large rockets and lightning. Afterward, the Agency devoted extensive efforts to assessing the nature of the lightning risk and seeking ways to mitigate it. The first fruit of this detailed study effort was the issuance, in August 1970, of revised electrodynamic design criteria for spacecraft. It stipulated various means of spacecraft and launch facility protection, including

1. Ensuring that all metallic sections are connected electrically (bonded) so that the current flow from a lightning stroke is conducted over the skin without any caps where sparking would occur or current would be carried inside.
2. Protecting objects on the ground, such as buildings, by a system of lightning rods and wires over the outside to carry the lightning stroke to the ground.
3. Providing a cone of protection for the lightning protection plan for Saturn Launch Complex 39.
4. Providing protection devices in critical circuits.
5. Using systems that have no single failure mode; i.e., the Saturn V launch vehicle uses triple-redundant circuitry on the auto-abort system, which requires two out of three of the signals to be correct before abort is initiated.
6. Appropriate shielding of units sensitive to electromagnetic radiation.[38]

37. R. Godfrey, et al., "Analysis of Apollo 12 Lightning Incident," NASA Marshall Space Flight Center, MSC-01540 (Feb. 1970), NTIS N72-73978; L.A. Ferrara, "Analysis of Air-Ground Voice Contacts During the Apollo 12 Launch Phase," NASA CR-110575 (1970).

38. Glenn E. Daniels, "Atmospheric Electricity Criteria Guidelines for Use in Space Vehicle Development," NASA TM-X-64549 (1970), pp. 1–2.

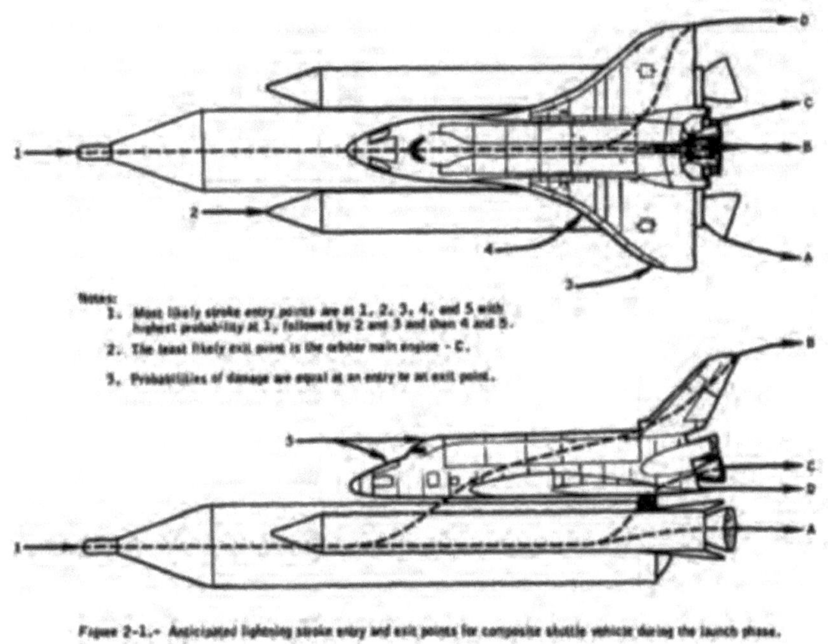

A 1973 NASA projection of likely paths taken by lightning striking a composite structure Space Shuttle, showing attachment and exit points. NASA.

The stakes involved in lightning protection increased greatly with the advent of the Space Shuttle program. Officially named the Space Transportation System (STS), NASA's Space Shuttle was cnvisioned as a routine space logistical support vehicle and was touted by some as a "space age DC-3," a reference to the legendary Douglas airliner that had galvanized air transport on a global scale. Large, complex, and expensive, it required careful planning to avoid lightning damage, particularly surface burnthroughs that could constitute a flight hazard (as, alas, the loss of Columbia would tragically demonstrate three decades subsequently). NASA predicated its studies on Shuttle lightning vulnerabilities on two major strokes, one having a peak current of 200 kA at a current rate of change of 100 kA per microsecond (100 kA / 10^{-6} sec), and a second of 100 kA at a current rate of change of 50 kA / 10^{-6} sec. Agency researchers also modeled various intermediate currents of lower energies. Analysis indicated that the Shuttle and its launch stack (consisting of the orbiter, mounted on a liquid fuel tank flanked by two solid-fuel boosters) would most likely have lightning entry points at the tip of its tankage and boosters, the leading edges of its wings at mid-span and at the wingtip, on its upper nose surface, and (least likely) above the cockpit. Likely exit points were the nozzles of the two solid-fuel boosters, the trailing-edge tip of the vertical fin, the trailing edge of the body flap, the trailing edges of the wing tip, and (least likely) the nozzles of its three liquid-fuel Space Shuttle main engines (SSMEs).[39] Because the Shuttle orbiter was, effectively, a large delta aircraft, data and criteria assembled previously for conventional aircraft furnished a good reference base for Shuttle lightning prediction studies, even studies dating to the early 1940s. As well, Agency researchers undertook extensive tests to guard against inadvertent triggering of the Shuttle's solid rocket boosters (SRBs), because their premature ignition would be catastrophic.[40]

Prudently, NASA ensured that the servicing structure on the Shuttle launch complex received an 80-foot lightning mast plus safety wires to guide strikes to the ground rather than through the launch vehicle. Dramatic proof of the system's effectiveness occurred in August 1983, when lightning struck the launch pad of the Shuttle Challenger before launching mission STS-8, commanded by Richard H. Truly. It was the first Shuttle night launch, and it subsequently proceeded as planned.

The hazards of what lightning could do to a flight control system (FCS) was dramatically illustrated March 26, 1987, when a bolt led to the loss of AC-67, an Atlas-Centaur mission carrying FLTSATCOM 6, a TRW, Inc., communications satellite developed for the Navy's Fleet Satellite Communications system. Approximately 48 seconds after launch, a cloud-to-ground lightning strike generated a spurious signal into the Centaur launch vehicle's digital flight control computer, which then sent a hard-over engine command. The resultant abrupt yaw overstressed the vehicle, causing its virtual immediate breakup. Coming after the weather-related loss of the Space Shuttle Challenger the previous year,

39. NASA JSC Shuttle Lightning Protection Committee, "Space Shuttle Lightning Protection Criteria Document," NASA JSC-07636 (1973); for studies cited by NASA as having particular value, see K.B. McEachron and J.H. Hayenguth, "Effect of Lightning on Thin Metal Surfaces," *Transactions of the American Institute of Electrical Engineers*, vol. 61 (1942), pp. 559–564; and R.O. Brick, L.L. Oh, and S.D. Schneider, "The Effects of Lightning Attachment Phenomena on Aircraft Design," paper presented at the *1970 Lightning and Static Electricity Conference, San Diego, CA, Dec. 1970.*
40. William M. Druen, "Lightning Tests and Analyses of Tunnel Bond Straps and Shielded Cables on the Space Shuttle Solid Rocket Booster," NASA CR-193921 (1993).

the loss of AC-67 was particularly disturbing. In both cases, accident investigators found that the two Kennedy teams had not taken adequate account of meteorological conditions at the time of launch.[41]

The accident led to NASA establishing a Lightning Advisory Panel to provide parameters for determining whether a launch should proceed in the presence of electrical activity. As well, it understandably stimulated continuing research on the electrodynamic environment at the Kennedy Space Center and on vulnerabilities of launch vehicles and facilities at the launch site. Vulnerability surveys extended to in-flight hardware, launch and ground support equipment, and ultimately almost any facility in areas of thunderstorm activity. Specific items identified as most vulnerable to lightning strikes were electronic systems, wiring and cables, and critical structures. The engineering challenge was to design methods of protecting those areas and systems without adversely affecting structural integrity or equipment performance.

To improve the fidelity of existing launch models and develop a better understanding of electrodynamic conditions around the Kennedy Center, between September 14 and November 4, 1988, NASA flew a modified single-seat single-engine Schweizer powered sailplane, the Special Purpose Test Vehicle (SPTVAR), on 20 missions over the spaceport and its reservation, measuring electrical fields. These trials took place in consultation with the Air Force (Detachment 11 of its 4th Weather Wing had responsibility for Cape lightning forecasting) and the New Mexico Institute of Mining and Technology, which selected candidate cloud forms for study and then monitored the real-time acquisition of field data. Flights ranged from 5,000 to 17,000 feet, averaged over an hour in duration, and took off from late morning to as late as 8 p.m. The SPTVAR aircraft dodged around electrified clouds as high as 35,000 feet, while taking measurements of electrical fields, the net airplane charge, atmospheric liquid water content, ice particle concentrations, sky brightness, accelerations, air temperature and

41. H.J. Christian, et al., "The Atlas-Centaur Lightning Strike Incident," *Journal of Geophysical Research*, vol. 94 (Sept. 30, 1989), pp. 13169–13177; John Busse, et al., "AC 67 Investigation Board Final Report," NASA Video VT-200.007.8606 (May 11, 1987); NASA release, "Lightning and Launches," Apr. 22, 2004, *http://www.nasa.gov/audience/foreducators/9-12/features/F_Lightning_and_Launches_9_12.html*, accessed Nov. 30, 2009; Virginia P. Dawson and Mark D. Bowles, *Taming Liquid Hydrogen: The Centaur Upper Stage Rocket, 1958–2002*, NASA SP-2004-4230 (Washington, DC: NASA, 2004), p. 234.

pressure, and basic aircraft parameters, such as heading, roll and pitch angles, and spatial position.[42]

After the Challenger and AC-67 launch accidents, the ongoing Shuttle program remained a particular subject of Agency concern, particularly the danger of lightning currents striking the Shuttle during rollout, on the pad, or upon liftoff. As verified by the SPTVAR survey, large currents (greater than 100 kA) were extremely rare in the operating area. Researchers concluded that worst-case figures for an on-pad strike ran from 0.0026 to 0.11953 percent. Trends evident in the data showed that specific operating procedures could further reduce the likelihood of a lightning strike. For instance, a study of all lightning probabilities at Kennedy Space Center observed, "If the Shuttle rollout did not occur during the evening hours, but during the peak July afternoon hours, the resultant nominal probabilities for a >220 kA and >50 kA lightning strike are 0.04% and 0.21%, respectively. Thus, it does matter 'when' the Shuttle is rolled out."[43] Although estimates for a triggered strike of a Shuttle in ascent were not precisely determined, researchers concluded that the likelihood of triggered strike (one caused by the moving vehicle itself) of any magnitude on an ascending launch vehicle is 140,000 times likelier than a direct hit on the pad. Because Cape Canaveral constitutes America's premier space launch center, continued interest in lightning at the Cape and its potential impact upon launch vehicles and facilities will remain major NASA concerns.

NASA and Electromagnetic Pulse Research

The phrase "electromagnetic pulse" usually raises visions of a nuclear detonation, because that is the most frequent context in which it is used. While EMP effects upon aircraft certainly would feature in a thermonuclear event, the phenomenon is commonly experienced in and around lightning storms. Lightning can cause a variety of EMP radiations, including radio-frequency pulses. An EMP "fries" electrical

42. J.J. Jones, et al., "Aircraft Measurements of Electrified Clouds at Kennedy Space Center," Final Report, parts I and II (Apr. 27, 1990), NTIS N91 14681-2; and J. Weems, et al., "Assessment and Forecasting of Lightning Potential and its Effect on Launch Operations at Cape Canaveral Air Force Station and John F. Kennedy Space Center," in NASA, KSC, *The 1991 International Aerospace and Ground Conference on Lightning and Static Electricity*, vol. 1, NASA CP-3106 (Washington, DC: NASA, 1991).

43. D.L. Johnson and W.W. Vaughan, "Analysis and Assessment of Peak Lightning Current Probabilities at the NASA Kennedy Space Center," NASA TM-2000-210131 (1999), p. 10.

circuits by passing a magnetic field past the equipment in one direction, then reversing in an extremely short period—typically a few nanoseconds. Therefore, the magnetic field is generated and collapses within that ephemeral time, creating a focused EMP. It can destroy or render useless any electrical circuit within several feet of impact.

Any survey of lightning-related EMPs brings attention to the phenomena of "elves," an acronym for Emissions of Light and Very low-frequency perturbations from Electromagnetic pulses. Elves are caused by lightning-generated EMPs, usually occurring above thunderstorms and in the ionosphere, some 300,000 feet above Earth. First recorded on Space Shuttle Mission STS-41 in 1990, elves mostly appear as reddish, expanding flashes that can reach 250 miles in diameter, lasting about 1 millisecond.

EMP research is multifaceted, conducted in laboratories, on airborne aircraft and rockets, and ultimately outside Earth's atmosphere. Research into transient electric fields and high-altitude lightning above thunderstorms has been conducted by sounding rockets launched by Cornell University. In 2000, a Black Brant sounding rocket from White Sands was launched over a storm, attaining a height of nearly 980,000 feet. Onboard equipment, including electronic and magnetic instruments, provided the first direct observation of the parallel electric field within 62 miles horizontal from the lightning.[44]

By definition, NASA's NF-106B flights in the 1980s involved EMP research. Among the overlapping goals of the project was quantification of lightning's electromagnetic effects, and Langley's Felix L. Pitts led the program intended to provide airborne data of lightning-strike traits. Bruce Fisher and two other NASA pilots (plus four Air Force pilots) conducted the flights. Fisher conducted analysis of the information he collected in addition to backseat researchers' data. Those flying as flight-test engineers in the two-seat jet included Harold K. Carney, Jr., NASA's lead technician for EMP measurements.

NASA Langley engineers built ultra-wide-bandwidth digital transient recorders carried in a sealed enclosure in the Dart's missile bay. To acquire the fast lightning transients, they adapted or devised electromagnetic sensors based on those used for measurement of nuclear pulse radiation. To aid understanding of the lightning transients recorded on

44. D.E. Rowland, et al., "Propagation of the Lightning Electromagnetic Pulse Through the E- and F-region Ionosphere and the Generation of Parallel Electric Fields," *American Geophysical Union* (May 2004).

2

the jet, a team from Electromagnetic Applications, Inc., provided mathematical modeling of the lightning strikes to the aircraft. Owing to the extra hazard of lightning strikes, the F-106 was fueled with JP-5, which is less volatile than the then-standard JP-4. Data compiled from dedicated EMP flights permitted statistical parameters to be established for lightning encounters. The F-106's onboard sensors showed that lightning strikes to aircraft include bursts of pulses lasting shorter than previously thought, but they were more frequent. Additionally, the bursts are more numerous than better-known strikes involving cloud-to-Earth flashes.[45]

Rocket-borne sensors provided the first ionospheric observations of lightning-induced electromagnetic waves from ELF through the medium frequency (MF) bands. The payload consisted of a NASA double-probe electric field sensor borne into the upper atmosphere by a Black Brant sounding rocket that NASA launched over "an extremely active thunderstorm cell." This mission, named Thunderstorm III, measured lightning EMPs up to 2 megahertz (MHz). Below 738,000 feet, a rising whistler wave was found with a nose-whistler wave shape with a propagating frequency near 80 kHz. The results confirmed speculation that the leading intense edge of the lightning EMP was borne on 50–125-kHz waves.[46]

Electromagnetic compatibility is essential to spacecraft performance. The requirement has long been recognized, as the insulating surfaces on early geosynchronous satellites were charged by geomagnetic substorms to a point where discharges occurred. The EMPs from such discharges coupled into electronic systems, potentially disrupting satellites. Laboratory tests on insulator charging indicated that discharges could be initiated at insulator edges, where voltage gradients could exist.[47]

45. The global aerospace industry has also pursued such research. For example, British Aerospace modeled lightning strikes and direct and indirect phenomena (including EMPs), current flow through composite material representing a wing or tail, field ingression within the airframe, and coupling to wiring and avionics systems. See BAE Systems, "Lightning, Electromagnetic Pulse (EMP) and Electrostatic Discharge (ESD)," 2009, at *http://www.baesystems.com/ProductsServices/ss_tes_atc_emp_esd.html*, accessed Nov. 30, 2009.

46. M.C. Kelley, et al., "LF and MF Observations of the Lightning Electromagnetic Pulse at Ionospheric Altitudes," *Geophysical Research Letters*, vol. 24, no. 9 (May 1997), p. 1111.

47. N.J. Stevens, et al., "Insulator Edge Voltage Gradient Effects in Spacecraft Charging Phenomena," NASA TM-78988 (1978); Stevens, "Interactions Between Spacecraft and the Charged-Particle Environment," NASA Lewis [Glenn] Research Center, NTRS Report 79N24021 (1979); Stevens, "Interactions Between Large Space Power Systems and Low-Earth-Orbit Plasmas," NASA, NTRS Report 85N22490 (1985).

Apart from observation and study, detecting electromagnetic pulses is a step toward avoidance. Most lightning detections systems include an antenna that senses atmospheric discharges and a processor to determine whether the strobes are lightning or static charges, based upon their electromagnetic traits. Generally, ground-based weather surveillance is more accurate than an airborne system, owing to the greater number of sensors. For instance, ground-based systems employ numerous antennas hundreds of miles apart to detect a lightning stroke's radio frequency (RF) pulses. When an RF flash occurs, electromagnetic pulses speed outward from the bolt to the ground at hyper speed. Because the antennas cover a large area of Earth's surface, they are able to triangulate the bolt's site of origin. Based upon known values, the RF data can determine with considerable accuracy the strength or severity of a lightning bolt.

Space-based lightning detection systems require satellites that, while more expensive than ground-based systems, provide instantaneous visual monitoring. Onboard cameras and sensors not only spot lightning bolts but also record them for analysis. NASA launched its first lightning-detection satellite in 1995, and the Lightning Imaging Sensor, which analyzes lightning through rainfall, was launched 2 years later. From approximately 1993, low-Earth orbit (LEO) space vehicles carried increasingly sophisticated equipment requiring increased power levels. Previously, satellites used 28-volt DC power systems as a legacy of the commercial and military aircraft industry. At those voltage levels, plasma interactions in LEO were seldom a concern. But use of high-voltage solar arrays increased concerns with electromagnetic compatibility and the potential effects of EMPs. Consequently, spacecraft design, testing, and performance assumed greater importance.

NASA researchers noted a pattern wherein insulating surfaces on geosynchronous satellites were charged by geomagnetic substorms, building up to electrical discharges. The resultant electromagnetic pulses can couple into satellite electronic systems, creating potentially disruptive results. Reducing power loss received a high priority, and laboratory tests on insulator charging showed that discharges could be initiated at insulator edges, where voltage gradients could exist. The benefits of such tests, coupled with greater empirical knowledge, afforded greater operating efficiency, partly because of greater EMP protection.[48]

48. G.B. Hillard and D.C. Ferguson, "Low Earth Orbit Spacecraft Charging Design Guidelines," *42nd AIAA Aerospace Sciences Meeting* (Jan. 2004).

Research into lightning EMPs remains a major focus. In 2008, Stanford's Dr. Robert A. Marshall and his colleagues reported on time-modeling techniques to study lightning-induced effects upon VLF transmitter signals called "early VLF events." Marshall explained:

> This mechanism involves electron density changes due to electromagnetic pulses from successive in-cloud lightning discharges associated with cloud-to-ground discharges (CGs), which are likely the source of continuing current and much of the charge moment change in CGs. Through time-domain modeling of the EMP we show that a sequence of pulses can produce appreciable density changes in the lower ionosphere, and that these changes are primarily electron losses through dissociative attachment to molecular oxygen. Modeling of the propagating VLF transmitter signal through the disturbed region shows that perturbed regions created by successive horizontal EMPs create measurable amplitude changes.[49]

However, the researchers found that modeling optical signatures was difficult when observation was limited by line of sight, especially by ground-based observers. Observation was further complicated by clouds and distance, because elves and "sprites" (large-scale discharges over thunderclouds) were mostly seen at ranges of 185 to 500 statute miles. Consequently, the originating lightning usually was not visible. But empirical evidence shows that an EMP from lightning is extremely short-lived when compared to the propagation time across an elve's radius. Observers therefore learned to recognize that the illuminated area at a given moment appears as a thin ring rather than as an actual disk.[50]

In addition to the effects of EMPs upon personnel directly engaged with aircraft or space vehicles, concern was voiced about researchers being exposed to simulated pulses. Facilities conducting EMP tests upon avionics and communications equipment were a logical area of investi-

49. R.A. Marshall, et al. "Early VLF perturbations caused by lightning EMP-driven dissociative attachment," *Geophysical Research Letters*, vol. 35, Issue 21, (Nov. 13, 2008).

50. Michael J. Rycroft, R. Giles Harrison, Keri A. Nicoll, and Evgeny A. Mareev, "An Overview of Earth's Global Electric Circuit and Atmospheric Conductivity," *Space Science Reviews*, vol. 137, no. 104 (June 2008).

gation, but some EMP simulators had the potential to expose operators and the public to electromagnetic fields of varying intensities, including naturally generated lightning bolts. In 1988, the NASA Astrophysics Data System released a study of bioelectromagnetic effects upon humans. The study stated, "Evidence from the available database does not establish that EMPs represent either an occupational or a public health hazard." Both laboratory research and years of observations on staffs of EMP manufacturing and simulation facilities indicated "no acute or short-term health effects." The study further noted that the occupational exposure guideline for EMPs is 100 kilovolts per meter, "which is far in excess of usual exposures with EMP simulators."[51]

NASA's studies of EMP effects benefited nonaerospace communities. The Lightning Detection and Ranging (LDAR) system that enhanced a safe work environment at Kennedy Space Center was extended to private industry. Cooperation with private enterprises enhances commercial applications not only in aviation but in corporate research, construction, and the electric utility industry. For example, while two-dimensional commercial systems are limited to cloud-to-ground lightning, NASA's three-dimensional LDAR provides precise location and elevation of in-cloud and cloud-to-cloud pulses by measuring arrival times of EMPs.

Nuclear- and lightning-caused EMPs share common traits. Nuclear EMPs involve three components, including the "E2" segment, which is similar to lightning. Nuclear EMPs are faster than conventional circuit breakers can handle. Most are intended to stop millisecond spikes caused by lightning flashes rather than microsecond spikes from a high-altitude nuclear explosion. The connection between ionizing radiation and lightning was readily demonstrated during the "Mike" nuclear test at Eniwetok Atoll in November 1952. The yield was 10.4 million tons, with gamma rays causing at least five lightning flashes in the ionized air around the fireball. The bolts descended almost vertically from the cloud above the fireball to the water. The observation demonstrated that, by causing atmospheric ionization, nuclear radiation can trigger a shorting of the natural vertical electric gradient, resulting in a lightning bolt.[52]

51. T.E. Aldrich, et al., "Bioelectromagnetic effects of EMP: Preliminary findings," *The Smithsonian/NASA Astrophysics Data System* (1988); Aldrich, et al., "Bioelectromagnetic Effects of EMP: Preliminary Findings," NASA Scientific and Technical Information, Report 1988STIN 8912791A (June 1988).
52. J.D. Colvin, et al., "An Empirical Study of the Nuclear Explosion-Induced Lightning Seen on Ivy Mike," *Journal of Geophysical Research*, vol. 92, Issue D5 (1987), pp. 5696–5712.

Thus, research overlap between thermonuclear and lightning-generated EMPs is unavoidable. NASA's workhorse F-106B, apart from NASA's broader charter to conduct lightning-strike research, was employed in a joint NASA–USAF program to compare the electromagnetic effects of lightning and nuclear detonations. In 1984, Felix L. Pitts of NASA Langley proposed a cooperative venture, leading to the Air Force lending Langley an advanced, 10-channel recorder for measuring electromagnetic pulses.

Langley used the recorder on F-106 test flights, vastly expanding its capability to measure magnetic and electrical change rates, as well as currents and voltages on wires inside the Dart. In July 1993, an Air Force researcher flew in the rear seat to operate the advanced equipment, when 72 lightning strikes were obtained. In EMP tests at Kirtland Air Force Base, the F-106 was exposed to a nuclear electromagnetic pulse simulator while mounted on a special test stand and during flybys. NASA's Norman Crabill and Lightning Technologies' J.A. Plumer participated in the Air Force Weapons Laboratory review of the acquired data.[53]

With helicopters becoming ever-more complex and with increasing dependence upon electronics, it was natural for researchers to extend the Agency's interest in lightning to rotary wing craft. Drawing upon the Agency's growing confidence in numerical computational analysis, Langley produced a numerical modeling technique to investigate the response of helicopters to both lightning and nuclear EMPs. Using a UH-60A Black Hawk as the focus, the study derived three-dimensional time domain finite-difference solutions to Maxwell's equations, computing external currents, internal fields, and cable responses. Analysis indicated that the short-circuit current on internal cables was generally greater for lightning, while the open-circuit voltages were slightly higher for nuclear-generated EMPs. As anticipated, the lightning response was found to be highly dependent upon the rise time of the injected current. Data showed that coupling levels to cables in a helicopter are 20 to 30 decibels (dB) greater than in a fixed wing aircraft.[54]

53. Chambers, *Concept to Reality*, at *http://oea.larc.nasa.gov/PAIS/Concept2Reality/lightning.html*, accessed Nov. 30, 2009.

54. C.C. Easterbrook and R.A. Perala, "A Comparison of Lightning and Nuclear Electromagnetic Pulse Response of a Helicopter," presented at the *Aerospace and Ground Conference on Lightning and Static Electricity*, NTIS N85-16343 07-47 (Dec. 1984).

Lightning and the Composite, Electronic Airplane

FAA Federal Air Regulation (FAR) 23.867 governs protection of aircraft against lightning and static electricity, reflecting the influence of decades of NASA lightning research, particularly the NF-106B program. FAR 23.867 directs that an airplane "must be protected against catastrophic effects from lightning," by bonding metal components to the airframe or, in the case of both metal and nonmetal components, designing them so that if they are struck, the effects on the aircraft will not be catastrophic. Additionally, for nonmetallic components, FAR 23.867 directs that aircraft must have "acceptable means of diverting the resulting electrical current so as not to endanger the airplane."[55]

Among the more effective means of limiting lightning damage to aircraft is using a material that resists or minimizes the powerful pulse of an electromagnetic strike. Late in the 20th century, the aerospace industry realized the excellent potential of composite materials for that purpose. Aside from older bonded-wood-and-resin aircraft of the interwar era, the modern all-composite aircraft may be said to date from the 1960s, with the private-venture Windecker Eagle, anticipating later aircraft as diverse as the Cirrus SR-20 lightplane, the Glasair III LP (the first composite homebuilt aircraft to meet the requirements of FAR 23), and the Boeing 787. The 787 is composed of 50-percent carbon laminate, including the fuselage and wings; a carbon sandwich material in the engine nacelles, control surfaces, and wingtips; and other composites in the wings and vertical fin. Much smaller portions are made of aluminum and titanium. In contrast, indicative of the rising prevalence of composites, the 777 involved just 12-percent composites.

An even newer composite testbed design is the Advanced Composite Cargo Aircraft (ACCA). The modified twin-engine Dornier 328Jet's rear fuselage and vertical stabilizer are composed of advanced composite materials produced by out-of-autoclave curing. First flown in June 2009, the ACCA is the product of a 10-year project by the Air Force Research Laboratory.[56]

NASA research on lightning protection for conventional aircraft structures translated into use for composite airframes as well. Because experience proved that lightning could strike almost any spot on an

55. U.S. Department of Transportation, Federal Aviation Administration, *Federal Air Regulations* (Washington, DC: FAA, 2009), FAR 23.867.

56. U.S. Patent Olson composite aircraft structure having lightning protection. 4,352,142 (Sept. 28, 1982).

airplane's surface—not merely (as previously believed) extremities such as wings and propeller tips—researchers found a lesson for designers using new materials. They concluded, "That finding is of great importance to designers employing composite materials, which are less conductive, hence more vulnerable to lightning damage than the aluminum allows they replace."[57] The advantages of fiberglass and other composites have been readily recognized: besides resistance to lightning strikes, composites offer exceptional strength for light weight and are resistant to corrosion. Therefore, it was inevitable that aircraft designers would increasingly rely upon the new materials.[58]

But the composite revolution was not just the province of established manufacturers. As composites grew in popularity, they increasingly were employed by manufacturers of kit planes. The homebuilt aircraft market, a feature of American aeronautics since the time of the Wrights, expanded greatly over the 1980s and afterward. NASA's heavy investment in lightning research carried over to the kit-plane market, and Langley released a Small Business Innovation Research (SBIR) contract to Stoddard-Hamilton Aircraft, Inc., and Lightning Technologies, Inc., for development of a low-cost lightning protection system for kit-built composite aircraft. As a result, Stoddard-Hamilton's composite-structure Glasair III LP became the first homebuilt aircraft to meet the standards of FAR 23.[59]

One of the benefits of composite/fiberglass airframe materials is inherent resistance to structural damage. Typically, composites are produced by laying spaced bands of high-strength fibers in an angular pattern of perhaps 45 degrees from one another. Selectively winding the material in alternating directions produces a "basket weave" effect that enhances strength. The fibers often are set in a thermoplastic resin four or more layers thick, which, when cured, produces extremely high strength and low weight. Furthermore, the weave pattern affords excellent resistance to peeling and delamination, even when struck by lightning. Among the earliest aviation uses of composites were engine cowlings, but eventually, structural components and then entire composite airframes were envisioned. Composites can provide additional electromagnetic resistance by winding conductive filaments in a

57. D.C. Ferguson and G.B. Hillard, "Low Earth Orbit Spacecraft Charging Design Guidelines," NASA TP-2003-212287 (2003).

58. The development of the composite aircraft is the subject of a companion essay in this volume.

59. Chambers, *Concept to Reality*, p. 184.

spiral pattern over the structure before curing the resin. The filaments help dissipate high-voltage energy across a large area and rapidly divert the impulses before they can inflict significant harm.[60]

It is helpful to compare the effects of lightning on aluminum aircraft to better understand the advantage of fiberglass structures. Aluminum readily conducts electromagnetic energy through the airframe, requiring designers to channel the energy away from vulnerable areas, especially fuel systems and avionics. The aircraft's outer skin usually offers the path of least resistance, so the energy can be "vented" overboard. Fiberglass is a proven insulator against electromagnetic charges. Though composites conduct electricity, they do so less readily than do aluminum and other metals. Consequently, though it may seem counterintuitive, composites' resistance to EMP strokes can be enhanced by adding small metallic mesh to the external surfaces, focusing unwanted currents away from the interior. The most common mesh materials are aluminum and copper impressed into the carbon fiber. Repairs of lightning-damaged composites must take into account the mesh in the affected area and the basic material and attendant structure. Composites mitigate the effect of a lightning strike not only by resisting the immediate area of impact, but also by spreading the effects over a wider area. Thus, by reducing the energy for a given surface area (expressed in amps per square inch), a potentially damaging strike can be rendered harmless.

Because technology is still emerging for detection and diagnosis of lightning damage, NASA is exploring methods of in-flight and postflight analysis. Obviously, the most critical is in-flight, with aircraft sensors measuring the intensity and location of a lightning strike's current, employing laboratory simulations to establish baseline data for a specific material. Thus, the voltage/current test measurements can be compared with statistical data to estimate the extent of damage likely upon the composite. Aircrews thereby can evaluate the safety of flight risks after a specific strike and determine whether to continue or to land.

NASA's research interests in addressing composite aircraft are threefold:

- Deploying onboard sensors to measure lightning-strike strength, location, and current flow.

60. United States Patent 5132168, "Lightning strike protection for composite aircraft structures."

- Obtaining conductive paint or other coatings to facilitate current flow, mitigating airframe structural damage, and eliminating requirements for additional internal shielding of electronics and avionics.
- Compiling physics-based models of complex composites that can be adapted to simulate lightning strikes to quantify electrical, mechanical, and thermal parameters to provide real-time damage information.

As testing continues, NASA will provide modeling data to manufacturers of composite aircraft as a design tool. Similar benefits can accrue to developers of wind turbines, which increasingly are likely to use composite blades. Other nonaerospace applications can include the electric power industry, which experiences high-voltage situations.[61]

Avionics

Lightning effects on avionics can be disastrous, as illustrated by the account of the loss of AC-67. Composite aircraft with internal radio antennas require fiberglass composite "windows" in the lightning-strike mesh near the antenna. (Fiberglass composites are employed because of their transparency to radio frequencies, unlike carbon fiber.) Lightning protection and avoidance are important for planning and conducting flight tests. Consequently, NASA's development of lightning warning and detection systems has been a priority in furthering fly-by-wire (FBW) systems. Early digital computers in flight control systems encountered conditions in which their processors could be adversely affected by lightning-generated electrical pulses. Subsequently, design processes were developed to protect electronic equipment from lightning strikes. As a study by the North Atlantic Treaty Organization (NATO) noted, such protection is "particularly important on aircraft with composite structures. Although equipment bench tests can be used to demonstrate equipment resistance to lightning strikes and EMP, it is now often considered necessary to perform whole aircraft lightning-strike tests to validate the design and clearance process."[62]

Celeste M. Belcastro of Langley contrasted laboratory, ground-based, and in-flight testing of electromagnetic environmental effects, noting:

61. "Lightning Strike Protection for Composite Aircraft," *NASA Tech Briefs* (June 1, 2009).
62. F. Webster and T.D. Smith, "Flying Qualities Flight Testing of Digital Flight Control Systems," in NATO, *AGARDograph*, No. 300, vol. 21, in the AGARD *Flight Test Techniques Series* (2001), p. 3.

> Laboratory tests are primarily open-loop and static at a few operating points over the performance envelope of the equipment and do not consider system level effects. Full-aircraft tests are also static with the aircraft situated on the ground and equipment powered on during exposure to electromagnetic energy. These tests do not provide a means of validating system performance over the operating envelope or under various flight conditions. . . . The assessment process is a combination of analysis, simulation, and tests and is currently under development for demonstration at the NASA Langley Research Center. The assessment process is comprehensive in that it addresses (i) closed-loop operation of the controller under test, (ii) real-time dynamic detection of controller malfunctions that occur due to the effects of electromagnetic disturbances caused by lightning, HIRF, and electromagnetic interference and incompatibilities, and (iii) the resulting effects on the aircraft relative to the stage of flight, flight conditions, and required operational performance.[63]

A prime example of full-system assessment is the F-16 Fighting Falcon, nicknamed "the electric jet," because of its fly-by-wire flight control system. Like any operational aircraft, F-16s have received lightning strikes, the effects of which demonstrate FCS durability. Anecdotal evidence within the F-16 community contains references to multiple lightning strikes on multiple aircraft—as many as four at a time in close formation. In another instance, the leader of a two-plane section was struck, and the bolt leapt from his wing to the wingman's canopy.

Aircraft are inherently sensor and weapons platforms, and so the lightning threat to external ordnance is serious and requires examination. In 1977, the Air Force conducted tests on the susceptibility of AIM-9 missiles to lightning strikes. The main concern was whether the Sidewinders, mounted on wingtip rails, could attract strobes that could enter the airframe via the missiles. The evaluators concluded that the optical dome of the missile was vulnerable to simulated lightning strikes

63. C.M. Belcastro, "Assessing Electromagnetic Environment Effects on Flight Critical Aircraft Control Computers," NASA Langley Research Center Technical Seminar Paper (Nov. 17, 1997), at *http://www.ece.odu.edu/~gray/research/abstracts.html#Assessing*, accessed Nov. 30, 2009.

even at moderate currents. The AIM-9's dome was shattered, and burn marks were left on the zinc-coated fiberglass housing. However, there was no evidence of internal arcing, and the test concluded that "it is unlikely that lightning will directly enter the F-16 via AIM-9 missiles."[64] Quite clearly, lightning had the potential of damaging the sensitive optics and sensors of missiles, thus rendering an aircraft impotent. With the increasing digitization and integration of electronic engine controls, in addition to airframes and avionics, engine management systems are now a significant area for lightning resistance research.

Transfer of NASA Research into Design Practices

Much of NASA's aerospace research overlaps various fields. For example, improving EMP tolerance of space-based systems involves studying plasma interactions in a high-voltage system operated in the ionosphere. But a related subject is establishing design practices that may have previously increased adverse plasma interactions and recommending means of eliminating or mitigating such reactions in future platforms.

Standards for lightning protection tests were developed in the 1950s, under FAA and Department of Defense (DOD) auspices. Those studies mainly addressed electrical bonding of aircraft components and protection of fuel systems. However, in the next decade, dramatic events such as the in-flight destruction of a Boeing 707 and the triggered responses of Apollo 12 clearly demonstrated the need for greater research. With advent of the Space Shuttle, NASA required further means of lightning protection, a process that began in the 1970s and continued well beyond the Shuttle's inaugural flight, in 1981.

Greater interagency cooperation led to new research programs in the 1980s involving NASA, the Air Force, the FAA, and the government of France. The goal was to develop a lightning-protection design philosophy, which in turn required standards and guidelines for various aerospace vehicles.

NASA's approach to lightning research has emphasized detection and avoidance, predicated on minimizing the risk of strikes, but then, if strikes occur nevertheless, ameliorating their damaging effects. Because early detection enhances avoidance, the two approaches work hand in glove. Translating those related philosophies into research and thence

64. Air Force Flight Dynamics Laboratory, Electromagnetic Hazards Group, "Lightning Strike Susceptibility Tests on the AIM-9 Missile," AFFDL-TR-78-95 (Aug. 1978), p. 23.

to design practices contains obvious benefits. The relationship between lightning research and protective design was noted by researchers for Lightning Technologies, Inc., in evaluating lightning protection for digital engine control systems. They emphasized, "The coordination between the airframe manufacturer and system supplies in this process is fundamental to adequate protection."[65] Because it is usually impractical to perform full-threat tests on fully configured aircraft, lightning protection depends upon accurate simulation using complete aircraft with full systems aboard. NASA, and other Federal agencies and military services, has undertaken such studies, dating to its work on the F-8 DFBW testbed of the early 1970s, as discussed subsequently.

In their Storm Hazards Research Program (SHRP) from 1980 to 1986, Langley researchers found that multiple lightning strikes inject random electric currents into an airframe, causing rapidly changing magnetic fields that can lead to erroneous responses, faulty commands, or other "upsets" in electronic systems. In 1987, the FAA (and other nations' aviation authorities) required that aircraft electronic systems performing flight-critical functions be protected from multiple-burst lightning.

At least from the 1970s, NASA recognized that vacuum tube electronics were inherently more resistant to lightning-induced voltage surges than were solid-state avionics. (The same was true for EMP effects. When researchers in the late 1970s were able to examine the avionics of the Soviet MiG-25 Foxbat, after defection of a Foxbat pilot to Japan, they were surprised to discover that much of its avionics were tube-based, clearly with EMP considerations in mind.) While new microcircuitry obviously was more vulnerable to upset or damage, many new-generation aircraft would have critical electronic systems such as fly-by-wire control systems.

Therefore, lightning represented a serious potential hazard to safety of flight for aircraft employing first-generation electronic flight control architectures and systems. A partial solution was redundancy of flight controls and other airborne systems, but in 1978, there were few if any standards addressing indirect effects of lightning. That time, however, was one of intensive interest in electronic flight controls. New fly-by-wire aircraft such as the F-16 were on the verge of entering squadron service. Even more radical designs—notably highly unstable early stealth aircraft such as the Lockheed XST Have Blue testbed, the Northrop Tacit Blue,

65. M. Dargi, et al., "Design of Lightning Protection for a Full-Authority Digital Engine Control," Lightning Technologies, Inc., NTIS N91-32717 (1991).

the Lockheed F-117, and the NASA–Rockwell Space Shuttle orbiter—were either already flying or well underway down the development path.

NASA's digital fly-by-wire (DFBW) F-8C Crusader afforded a ready means of evaluating lightning-induced voltages, via ground simulation and evaluation of electrodynamic effects upon its flight control computer. Dryden's subsequent research represented the first experimental investigation of lightning-induced effects on any FBW system, digital or analog. A summary concluded:

> Results are significant, both for this particular aircraft and for future generations of aircraft and other aerospace vehicles such as the Space Shuttle, which will employ digital FBW FCSs. Particular conclusions are: Equipment bays in a typical metallic airframe are poorly shielded and permit substantial voltages to be induced in unshielded electrical cabling. Lightning-induced voltages in a typical a/c cabling system pose a serious hazard to modern electronics, and positive steps must be taken to minimize the impact of these voltages on system operation. Induced voltages of similar magnitudes will appear simultaneously in all channels of a redundant system. A single-point ground does not eliminate lightning-induced voltages. It reduces the amount of diffusion-flux induced and structural IR voltage but permits significant aperture-flux induced voltages. Cable shielding, surge suppression, grounding and interface modifications offer means of protection, but successful design will require a coordinated sharing of responsibility among those who design the interconnecting cabling and those who design the electronics. A set of transient control levels for system cabling and transient design levels for electronics, separated by a margin of safety, should be established as design criteria.[66]

66. J.A. Plumer, W.A. Malloy, and J.B. Craft, "The Effects of Lightning on Digital Flight Control Systems," NASA, *Advanced Control Technology and its Potential for Future Transport Aircraft* (Edwards: DFRC, 1976), pp. 989–1008; C.R. Jarvis and K.J. Szalai, "Ground and Flight Test Experience with a Triple Redundant Digital Fly By Wire Control System," in NASA LRC, *Advanced Aerodynamics and Active Controls*, NIST N81-19001 10-01 (1981).

The F-8 DFBW program is the subject of a companion study on electronic flight controls and so is not treated in greater detail here. In brief, a Navy Ling-Temco-Vought F-8 Crusader jet fighter was modified with a digital electronic flight control system and test-flown at the NASA Flight Research Center (later the NASA Dryden Flight Research Center). When the F-8 DFBW program ended in 1985, it had made 210 flights, with direct benefits to aircraft as varied as the F-16, the F/A-18, the Boeing 777, and the Space Shuttle. It constituted an excellent example of how NASA research can prove and refine design concepts, which are then translated into design practice.[67]

The versatile F-106B program also yielded useful information on protection of digital computers and other airborne systems that translated into later design concepts. As NASA engineer-historian Joseph Chambers subsequently wrote: "These findings are now reflected in lightning environment and test standards used to verify adequacy of protection for electrical and avionics systems against lightning hazards. They are also used to demonstrate compliance with regulations issued by airworthiness certifying authorities worldwide that require lightning strikes not adversely affect the aircraft systems performing critical and essential functions."[68]

Similarly, NASA experience at lightning-prone Florida launch sites provided an obvious basis for identifying and implementing design practices for future use. A 1999 lessons-learned study identified design considerations for lightning-strike survivability. Seeking to avoid natural or triggered lightning in future launches, NASA sought improvements in electromagnetic compatibility (EMC) for launch sites used by the Shuttle and other launch systems. They included proper grounding of vehicle and ground-support equipment, bonding requirements, and circuit protection. Those aims were achieved mainly via wire shielding and transient limiters.

In conclusion, it is difficult to improve upon D.L. Johnson and W.W. Vaughn's blunt assessment that "Lightning protection assessment and design consideration are critical functions in the design and development of an aerospace vehicle. The project's engineer responsible for

67. James E. Tomayko, *Computers Take Flight: A History of NASA's Pioneering Digital Fly-By-Wire Project*, NASA SP-2000-4224 (Washington, DC: GPO, 2000).

68. Chambers, *Concept to Reality*, "Lightning Protection and Standards," at *http://oea.larc.nasa.gov/PAIS/Concept2Reality/lightning.html*, accessed Nov. 30, 2009.

2

lightning must be involved in preliminary design and remain an integral member of the design and development team throughout vehicle construction and verification tests."[69] This lesson is applicable to many aerospace technical disciplines and reflects the decades of experience embedded within NASA and its predecessor, the NACA, involving high-technology (and often high-risk) research, testing, and evaluation. Lightning will continue to draw the interest of the Agency's researchers, for there is still much that remains to be learned about this beautiful and inherently dangerous electrodynamic phenomenon and its interactions with those who fly.

69. D.L. Johnson and W.W. Vaughan, "Lightning Strike Peak Current Probabilities as Related to Space Shuttle Operations" (Huntsville: NASA MSFC, 1999), p. 3, at *http://ntrs.nasa.gov/archive/nasa/casi.ntrs.nasa.gov/199.900.09077_199.843.2277.pdf*, accessed Nov. 30, 2009; C.C. Goodloe, "Lightning Protection Guidelines for Aerospace Vehicles," NASA TM-209734 (1999).

Recommended Additional Reading

Reports, Papers, Articles, and Presentations:

T.E. Aldrich, et al., "Bioelectromagnetic effects of EMP: Preliminary findings," in *The Smithsonian/NASA Astrophysics Data System* (Washington, DC: Smithsonian Institution and NASA, 1988).

T.E. Aldrich, et al., "Bioelectromagnetic Effects of EMP: Preliminary Findings," NASA Scientific and Technical Information (June 1988), NRTS N8912791A.

Constantine A. Balanis, et al., "Advanced Electromagnetic Methods for Aerospace Vehicles: Final Report," Telecommunications Research Center, College of Engineering and Applied Science, Arizona State University, Report No. TRC-EM-CAB-0001 (1999).

C.M. Belcastro, "Assessing Electromagnetic Environment Effects on Flight Critical Aircraft Control Computers," NASA Langley Research Center Technical Seminar Paper (Nov. 17, 1997).

Richard Blakeslee, "The Altus Cumulus Electrification Study (ACES): A UAV-Based Investigation of Thunderstorms," paper presented at the *Technical Analysis and Applications Center Unmanned Aerial Vehicle Annual Symposium, Las Cruces, NM, Oct. 30–31, 2001.*

Richard Blakeslee, "ER-2 Investigations of Lightning and Thunderstorms," in NASA MSFC, *FY92 Earth Science and Applications Program Research Review* (Huntsville: NASA MSFC, 1993), NRTS 93N20088.

R.O. Brick, L.L. Oh, and S.D. Schneider, "The Effects of Lightning Attachment Phenomena on Aircraft Design," paper presented at the *1970 Lightning and Static Electricity Conference, San Diego, CA, Dec. 1970.*

Harold D. Burket, et al., "In-Flight Lightning Characterization Program on a CV-580 Aircraft," Wright-Patterson AFB Flight Dynamics Lab (June 1988).

J.R. Busse, et al., *Report of Atlas Centaur 67 / FLTSATCOM F-6 Investigation Board*, vol. 1: final report (July 15, 1987).

Centre d'Essais Aéronautique de Toulouse, "Measurement of Characteristics of Lightning at High Altitude," a translation of CEAT, "Mesure des caracteristiques de la foudre en altitude," Test No. 76/650000 P.4 (May 1979), NASA TM-76669 (1981).

H.J. Christian, et al., "The Atlas-Centaur Lightning Strike Incident," *Journal of Geophysical Research*, vol. 94 (Sept. 30, 1989).

J.D. Colvin, et al., "An Empirical Study of the Nuclear Explosion-Induced Lightning Seen on Ivy Mike," *Journal of Geophysical Research*, vol. 92, no. 5 (1987), pp. 5696–5712.

David Crofts, "Lightning Protection of Full Authority Digital Electronic Systems," in NASA, KSC, *The 1991 International Aerospace and Ground Conference on Lightning and Static Electricity*, vol. 1, NASA CP-3106 (Washington, DC: NASA, 1991).

M. Dargi, et al., "Design of Lightning Protection for a Full-Authority Digital Engine Control," Lightning Technologies, Inc., NRTS N91-32717 (1991).

William M. Druen, "Lightning Tests and Analyses of Tunnel Bond Straps and Shielded Cables on the Space Shuttle Solid Rocket Booster," NASA CR-193921 (1993).

C.C. Easterbrook and R.A. Perala, "A comparison of lightning and nuclear electromagnetic pulse response of a helicopter," presented at the *Aerospace and Ground Conference on Lightning and Static Electricity*, NRTS N85-16343 07-47 (1984).

James Fenner, "Rocket-Triggered Lightning Strikes and Forest Fire Ignition," paper prepared for the *1990 NASA/ASEE Summer Faculty Fellowship Program, NASA Kennedy Space Center and the University of Central Florida*, NRTS N91-20027 (1990).

D.C. Ferguson and G.B. Hillard, "Low Earth Orbit Spacecraft Charging Design Guidelines," NASA TP-2003-212287 (2003).

L.A. Ferrara, "Analysis of Air-Ground Voice Contacts During the Apollo 12 Launch Phase," NASA CR-110575 (1970).

Bruce D. Fisher, G.L. Keyser, Jr., and P.L. Deal, "Lightning Attachment Patterns and Flight Conditions for Storm Hazards," NASA TP-2087 (1982).

Bruce D. Fisher, et al., "Storm Hazards '79: F-106B Operations Summary," NASA TM-81779 (1980).

Franklin A. Fisher and J. Anderson Plumer, "Lightning Protection of Aircraft," NASA RP-1008 (1977).

D.V. Giri, R.S. Noss, D.B. Phuoc, and F.M. Tesche, "Analysis of Direct and Nearby Lightning Strike Data for Aircraft," NASA CR-172127 (1983).

R. Godfrey, et al., "Analysis of Apollo 12 Lightning Incident," NASA Marshall Space Flight Center, MSC-01540 (Feb. 1970), NTIS N72-73978.

C.C. Goodloe, "Lightning Protection Guidelines for Aerospace Vehicles," NASA TM-209734 (1999).

Charles F. Griffin and Arthur M. James, "Fuel Containment, Lightning Protection, and Damage Tolerance in Large Composite Primary Aircraft Structures," NASA CR-3875 (1985).

Paul T. Hacker, "Lightning Damage to a General Aviation Aircraft: Description and Analysis," NASA TN-D-7775 (1974).

L.P. Harrison, "Lightning Discharges to Aircraft and Associated Meteorological Conditions," U.S. Weather Bureau, Washington, DC (May 1946), pp. 58–60.

T.L. Harwood, "An Assessment of Tailoring of Lightning Protection Design Requirements for a Composite Wing Structure on a Metallic Aircraft," NASA, KSC, *The 1991 International Aerospace and Ground Conference on Lightning and Static Electricity*, vol. 1, NASA CP-3106 (Washington, DC: NASA, 1991).

Richard Hasbrouck, "Mitigating Lightning Hazards," *Science & Technology Review* (May 1996).

B.D. Heady and K.S. Zeisel, "NASA F-106B Lightning Ground Tests," McDonnell Co. (1983), DTIC ADP002194.

J.H. Helsdon, Jr., "Atmospheric Electrical Modeling in Support of the NASA F-106 Storm Hazards Project," NASA CR-179801 (1986).

J.H. Helsdon, Jr., "Investigations Into the F-106 Lightning Strike Environment as Functions of Altitude and Storm Phase," NASA CR-180292 (1987).

G.B. Hillard and D.C. Ferguson, "Low Earth Orbit Spacecraft Charging Design Guidelines," paper presented at the *42nd AIAA Aerospace Sciences Meeting, Jan. 2004.*

C.R. Jarvis and K.J. Szalai, "Ground and flight Test Experience with a Triple Redundant Digital Fly By Wire Control System," in NASA LRC, *Advanced Aerodynamics and Active Controls*, NASA CP-2172 (1981).

D.L. Johnson and W.W. Vaughan, "Analysis and Assessment of Peak Lightning Current Probabilities at the NASA Kennedy Space Center," NASA TM-2000-210131 (1999).

D.L. Johnson and W.W. Vaughn, "Lightning Characteristics and Lightning Strike Peak Current probabilities as Related to Aerospace Vehicle Operations," paper presented at the *8th Conference on Aviation, Range, and Aerospace Meteorology, 79th American Meteorological Society Annual Meeting, Dallas, TX, Jan. 10–15, 1999.*

D.L. Johnson and W.W. Vaughan, "Lightning Strike Peak Current Probabilities as Related to Space Shuttle Operations" (Huntsville: NASA MSFC, 1999).

J.J. Jones, et al., "Aircraft Measurements of Electrified Clouds at Kennedy Space Center," two parts, NRTS 91N14681 and 91N14682 (1990).

M.C. Kelley, et al., "LF and MF Observations of the Lightning Electromagnetic Pulse at Ionospheric Altitudes," *Geophysical Research Letters*, vol. 24, no. 9 (May 1997).

Alexander Kern, "Simulation and Measurement of Melting Effects on Metal Sheets Caused by Direct Lightning Strikes," in NASA, KSC, *The 1991 International Aerospace and Ground Conference on Lightning and Static Electricity*, vol. 1, NASA CP-3106 (Washington, DC: NASA, 1991).

Tony Kim and Richard Blakeslee, "ALTUS Cumulus Electrification Study (ACES)," paper presented at the *Technical Analysis and Applications Center Conference, Santa Fe, NM, Oct. 28–30, 2002.*

Heinrich Koppe, "Practical Experiences with Lightning Discharges to Airplanes," *Zeitschrift für Flugtechnik und Motorluftschiffahrt*, vol. 24, no. 21, translated and printed as NACA Technical Memorandum No. 730 (Nov. 4, 1933), p. 1.

E. Philip Krider, "Benjamin Franklin and the First Lightning Conductors," *Proceedings of the International Commission on History of Meteorology*, vol. 1, no. 1 (2004).

B.P. Kuhlman, M.J. Reazer, and P.L. Rustan, "WC-130 Airborne Lightning Characterization Program Data Review," USAF Wright Aeronautical Laboratories (1984), DTIC ADA150230.

D.M. Levine, "Lightning Electric Field Measurements Which Correlate with Strikes to the NASA F-106B Aircraft, 22 July 1980," NASA TM-82142 (1981).

D.M. Levine, "RF Radiation from Lightning Correlated with Aircraft Measurements During Storm Hazards-82," NASA TM-85007 (1983).

"Lightning Strike Protection for Composite Aircraft." *NASA Tech Briefs* (June 1, 2009).

R.A. Marshall, et al., "Early VLF Perturbations Caused by Lightning EMP-driven Dissociative Attachment," *Geophysical Research Letters*, vol. 35, Issue 21 (Nov. 13, 2008).

Vladislav Mazur, "Lightning Threat to Aircraft: Do We Know All We Need to Know?" in NASA, KSC, *The 1991 International Aerospace and Ground Conference on Lightning and Static Electricity*, vol.1, NASA CP-3106 (Washington, DC: NASA, 1991).

V. Mazur, B.D. Fisher, and J.C. Gerlach, "Lightning Strikes to a NASA Airplane Penetrating Thunderstorms at Low Altitudes," AIAA Paper 86-0021 (1986).

Rosemarie L. McDowell, "Users Manual for the Federal Aviation Administration Research and Development Electromagnetic Database (FRED) for Windows: Version 2.0," Department of Transportation, Federal Aviation Administration, Report DOT/FAA/AR-95/18 (1998).

K.B. McEachron and J.H. Hayenguth, "Effect of Lightning on Thin Metal Surfaces," *Transactions of the American Institute of Electrical Engineers*, vol. 61 (1942), pp. 559–564.

Edward Miller, "1964 Rough Rider Summary of Parameters Recorded, Test Instrumentation, Flight Operations, and Aircraft Damage," USAF Aeronautical Systems Division (1965), DTIC AD 0615749.

NASA JSC Shuttle Lightning Protection Committee, "Space Shuttle Lightning Protection Criteria Document," NASA JSC-07636 (1973).

NASA KSC, *The 1991 International Aerospace and Ground Conference on Lightning and Static Electricity*, two vols., NASA CP-3106 (Washington, DC: NASA, 1991).

Poh H. Ng, et al., "Application of Triggered Lightning Numerical Models to the F-106B and Extension to Other Aircraft," NASA CR-4207 (1988).

I.I. Pinkel, "Airplane Design Trends: Some Safety Implications," NASA TM-X-52692 (1969).

Felix L. Pitts, "In-Flight Direct Strike Lightning Research," NASA LRC (Feb. 1, 1981), NTRS Acc. No. 81N19005.

Felix L. Pitts and M.E. Thomas, "The 1980 Direct Strike Lightning Data," NASA TM-81946 (1981).

Felix L. Pitts and M.E. Thomas, "The 1981 Direct Strike Lightning Data," NASA TM-83273 (1982).

Felix L. Pitts, et al., "Aircraft Jolts from Lightning Bolts," *IEEE Spectrum*, vol. 25 (July 1988), pp. 34–38.

Felix L. Pitts and M.E. Thomas, "Initial Direct Strike Lightning Data," NASA TM-81867 (1980).

Felix L. Pitts, et al., "New Methods and Results for Quantification of Lightning-Aircraft Electrodynamics," NASA TP-2737 (1987).

J.A. Plumer, "Investigation of Severe Lightning Strike Incidents to Two USAF F-106A Aircraft," NASA CR-165794 (1981).

J.A. Plumer, W.A. Malloy, and J.B. Craft, "The Effects of Lightning on Digital Flight Control Systems," in NASA DFRC, *Advanced Control Technology and its Potential for Future Transport Aircraft* (Washington, DC: NASA, 1976), pp. 989–1008.

J.A. Plumer, N.O. Rasch, and M.S. Glynn, "Recent Data from the Airlines Lightning Strike Reporting Project," *Journal of Aircraft*, vol. 22 (1985), pp. 429–433.

Vernon J. Rossow, et al., "A Study of Wire-Deploying Devices Designed to Trigger Lightning," NASA TM-X-62085 (1971).

D.E. Rowland, et al., "Propagation of the Lightning Electromagnetic Pulse Through the E- and F-region Ionosphere and the Generation of Parallel Electric Fields," paper presented at the *Spring Meeting of the American Geophysical Union, May 2004*, abstract No. SA33A-03.

T.H. Rudolph, et al., "Interpretation of F-106B and CV-580 In-Flight Lightning Data and Form Factor Determination," NASA CR-4250 (1989).

T.H. Rudolph and R.A. Perala, "Linear and Nonlinear Interpretation of the Direct Strike Lightning Response of the NASA F-106B Thunderstorm Research Aircraft," NASA CR-3746 (1983).

Michael J. Rycroft, R. Giles Harrison, Keri A. Nicoll, and Evgeny A. Mareev, "An Overview of Earth's Global Electric Circuit and Atmospheric Conductivity," *Space Science Reviews*, vol. 137, no. 104 (June 2008).

Olaf Spiller, "Protection of Electrical and Electronic Equipment Against Lightning Indirect Effects on the Airbus A340 Wing," in NASA, KSC, *The 1991 International Aerospace and Ground Conference on Lightning and Static Electricity*, vol. 1, NASA CP-3106 (Washington, DC: NASA, 1991).

N.J. Stevens, et al., "Insulator Edge Voltage Gradient Effects in Spacecraft Charging Phenomena," NASA TM-78988 (1978).

N.J. Stevens, "Interactions Between Large Space Power Systems and Low-Earth-Orbit Plasmas," NASA Lewis [Glenn] Research Center, NTRS Report 85N22490 (1985).

N.J. Stevens, "Interactions Between Spacecraft and the Charged-Particle Environment," NASA Lewis [Glenn] Research Center, NTRS Report 79N24021 (1979).

Gale R. Sundberg, "Civil Air Transport: A Fresh Look at Power-by-Wire and Fly-by-Light," NASA TM-102574 (1990).

C.D. Turner, "External Interaction of the Nuclear EMP with Aircraft and Missiles," *IEEE Transactions on Electromagnetic Compatibility*, vol. EMC-20, no. 1 (Feb. 1978).

U.S. Air Force, Air Force Flight Dynamics Laboratory, Electromagnetic Hazards Group, "Lightning Strike Susceptibility Tests on the AIM-9 Missile," AFFDL-TR-78-95 (Aug. 1978).

U.S. Department of Commerce, "Report of Airship 'Hindenburg' Accident Investigation," *Air Commerce Bulletin*, vol. 9, no. 2 (Aug. 15, 1937), pp. 21–36.

U.S. National Bureau of Standards, "Electrical Effects in Glider Towlines," High Voltage Laboratory, Advance Restricted Report 4C20 (Mar. 1944), p. 47.

U.S. National Bureau of Standards, "Protection of Nonmetallic Aircraft from Lightning," High Voltage Laboratory, Advance Report 3I10 (Sept. 1943).

Otha H. Vaughn, Jr., "NASA Thunderstorm Overflight Program—Atmospheric Electricity Research: An Overview Report on the Optical Lightning Detection Experiment for Spring and Summer 1983," NASA TM-86468 (1984).

Otha H. Vaughan, Jr., "NASA Thunderstorm Overflight Program—Research in Atmospheric Electricity from an Instrumented U-2 Aircraft," NASA TM-82545 (1983).

Otha H. Vaughn, Jr., et al., "Thunderstorm Overflight Program," AIAA Paper 80-1934 (1980).

F. Webster and T.D. Smith, "Flying Qualities Flight Testing of Digital Flight Control Systems," in NATO, *AGARDograph,* vol. 21, no. 300, in the AGARD *Flight Test Techniques Series* (2001).

J. Weems, et al., "Assessment and Forecasting of Lightning Potential and its Effect on Launch Operations at Cape Canaveral Air Force Station and John F. Kennedy Space Center," in NASA, KSC, *The 1991*

International Aerospace and Ground Conference on Lightning and Static Electricity, vol. 1, NASA CP-3106 (Washington, DC: NASA, 1991).

R.M. Winebarger, "Loads and Motions of an F-106B Flying Through Thunderstorms," NASA TM-87671 (1986).

News Releases and Summaries:

John Busse, et al, "Atlas Centaur 67 Investigation Board Final Report," NASA Video VT-200.007.8606 (May 11, 1987).

NASA news release, "Lightning and Launches, 22 April 2004, at *http://www.nasa.gov/audience/foreducators/9-12/features/F_Lightning_and_Launches_9_12.html*, accessed Nov. 30, 2009.

NASA news release, "NASA Lightning Research on ABC 20/20," Dec. 11, 2007, at *http://www.nasa.gov/topics/aeronautics/features/fisher-2020.html*, accessed Nov. 30, 2009.

NASA Marshall Space Flight Center, "Lightning and Atmospheric Electricity Research at the GHCC," 2009, at *http://thunder.msfc.nasa.gov/primer*, accessed Nov. 30, 2009.

Books and Monographs:

Roger E. Bilstein, *Stages to Saturn: A Technological History of the Apollo/Saturn Launch Vehicles*, NASA SP-4206 (Washington, DC: NASA, 1980).

H.R. Byers and R.R. Braham, *The Thunderstorm* (Washington, DC: GPO, 1949).

Joseph R. Chambers, *Concept to Reality: Contributions of the NASA Langley Research Center to U.S. Civil Aircraft of the 1990s*, NASA SP-2003 (Washington, DC: GPO, 2003).

Virginia P. Dawson and Mark D. Bowles, *Taming Liquid Hydrogen: The Centaur Upper Stage Rocket, 1958–2002*, NASA SP-2004-4230 (Washington, DC: NASA, 2004).

James E. Tomayko, *Computers Take Flight: A History of NASA's Pioneering Digital Fly-By-Wire Project*, NASA SP-2000-4224 (Washington, DC: GPO, 2000).

Martin A. Uman, *The Art and Science of Lightning Protection* (Cambridge: Cambridge University Press, 2008).

Martin A. Uman, *The Lightning Discharge* (New York: Academic Press, Inc., 1987).

Donald R. Whitnah, *Safer Skyways: Federal Control of Aviation, 1926–1966* (Ames, IA: The Iowa State University Press, 1966).

More than 87,000 flight take place each day over the United States. The work of NASA and others has helped develop ways to ensure safety in these crowded skies. Richard P. Hallion.

CASE 3

The Quest for Safety Amid Crowded Skies

James Banke

Since 1926 and the passage of the Air Commerce Act, the Federal Government has had a vital commitment to aviation safety. Even before this, however, the NACA championed regulation of aeronautics, the establishment of licensing procedures for pilots and aircraft, and the definition of technical criteria to enhance the safety of air operations. NASA has worked closely with the FAA and other aviation organizations to ensure the safety of America's air transport network.

WHEN THE FIRST AIRPLANE LIFTED OFF from the sands of Kitty Hawk during 1903, there was no concern of a midair collision with another airplane. The Wright brothers had the North Carolina skies all to themselves. But as more and more aircraft found their way off the ground and then began to share the increasing number of new airfields, the need to coordinate movements among pilots quickly grew. As flight technology matured to allow cross-country trips, methods to improve safe navigation between airports evolved as well. Initially, bonfires lit the airways. Then came light towers, two-way radio, omnidirectional beacons, radar, and—ultimately—Global Positioning System (GPS) navigation signals from space.[1]

Today, the skies are crowded, and the potential for catastrophic loss of life is ever present, as more than 87,000 flights take place each day over the United States. Despite repeated reports of computer crashes or bad weather slowing an overburdened national airspace system, air-related fatalities remain historically low, thanks in large part to the technical advances developed by the National Aeronautics and Space Administration (NASA), but especially to the daily efforts of some 15,000 air traffic controllers keeping a close eye on all of those airplanes.[2]

1. Edmund Preston, *FAA Historical Chronology, Civil Aviation and the Federal Government 1926–1996* (Washington, DC: Federal Aviation Administration).
2. NATCA: *A History of Air Traffic Control* (Washington, DC: National Air Traffic Controllers Association, 2009), p. 16.

From an Australian government slide show in 1956, the basic concepts of an emerging air traffic control system are explained to the public. Airways Museum & Civil Aviation Historical Society, Melbourne, Australia (*www.airwaysmuseum.com*).

All of those controllers work for, or are under contract to, the Federal Aviation Administration (FAA), which is the Federal agency responsible for keeping U.S. skyways safe by setting and enforcing regulations. Before the FAA (formed in 1958), it was the Civil Aeronautics Administration (formed in 1941), and even earlier than that, it was the Department of Commerce's Aeronautics Bureau (formed in 1926). That that administrative job today is not part of NASA's duties is the result of decisions made by the White House, Congress, and NASA's predecessor organization, the National Advisory Committee for Aeronautics (NACA), during 1920.[3]

At the time (specifically 1919), the International Commission for Air Navigation had been created to develop the world's first set of rules for governing air traffic. But the United States did not sign on to the convention. Instead, U.S. officials turned to the NACA and other organizations to determine how best to organize the Government for handling

3. Alex Roland, *Model Research: The National Advisory Committee for Aeronautics 1915–1958*, NASA SP-4103 (Washington, DC: NASA, 1985).

all aspects of this new transportation system. The NACA in 1920 already was the focal point of aviation research in the Nation, and many thought it only natural, and best, that the Committee be the Government's all-inclusive home for aviation matters. A similar organizational model existed in Europe but didn't appear to some with the NACA to be an ideal solution. This sentiment was most clearly expressed by John F. Hayford, a charter member of the NACA and a Northwestern University engineer, who said during a meeting, "The NACA is adapted to function well as an advisory committee but not to function satisfactorily as an administrative body."[4]

So, in a way, NASA's earliest contribution to making safer skyways was to shed itself of the responsibility for overseeing improvements to and regulating the operation of the national airspace. With the FAA secure in that management role, NASA has been free to continue to play to its strengths as a research organization. It has provided technical innovation to enhance safety in the cockpits; increase efficiencies along the air routes; introduce reliable automation, navigation, and communication systems for the many air traffic control (ATC) facilities that dot the Nation; and manage complex safety reporting systems that have required creation of new data-crunching capabilities.

This case study will present a survey in a more-or-less chronological order of NASA's efforts to assist the FAA in making safer skyways. An overview of key NASA programs, as seen through the eyes of the FAA until 1996, will be presented first. NASA's contributions to air traffic safety after the 1997 establishment of national goals for reducing fatal air accidents will be highlighted next. The case study will continue with a survey of NASA's current programs and facilities related to airspace safety and conclude with an introduction of the NextGen Air Transportation System, which is to be in place by 2025.

NASA, as Seen by the FAA

Nearly every NASA program related to aviation safety has required the involvement of the FAA. Anything new from NASA that affects—for example, the design of an airliner or the layout of a cockpit panel[5] or the introduction of a modified traffic control procedure that relies on

4. Roland, *Model Research*, p. 57.

5. *Part 21 Aircraft Certification Procedures for Products and Parts*, Federal Aviation Regulations (Washington, DC: FAA, 2009).

new technology[6]—must eventually be certified for use by the FAA, either directly or indirectly. This process continues today, extending the legacy of dozens of programs that came before—not all of which can be detailed here. But in terms of a historical overview through the eyes of the FAA, a handful of key collaborations with NASA were considered important enough by the FAA to mention in its official chronology, and they are summarized in this section.

Partners in the Sky: 1965

The partnership between NASA and the FAA that facilitates that exchange of ideas and technology was forged soon after both agencies were formally created in 1958. With the growing acceptance of commercial jet airliners and the ever-increasing number of passengers who wanted to get to their destinations as quickly as possible, the United States began exploring the possibility of fielding a Supersonic Transport (SST). By 1964, it was suggested that duplication of effort was underway by researchers at the FAA and NASA, especially in upgrading existing jet powerplants required to propel the speedy airliner. The resulting series of meetings during the next year led to the creation in May 1965 of the NASA–FAA Coordinating Board, which was designed to "strengthen the coordination, planning, and exchange of information between the two agencies."[7]

Project Taper: 1965

During that same month, the findings were released of what the FAA's official historical record details as its first joint research project with NASA.[8]

A year earlier, during May and June 1964, two series of flight tests were conducted using FAA aircraft with NASA pilots to study the hazards of light to moderate air turbulence to jet aircraft from several perspectives. The effort was called Project Taper, short for Turbulent Air Pilot Environment Research.[9] In conjunction with ground-based wind tunnel runs and early use of simulator programs, FAA Convair 880 and

6. *Aeronautical Information Manual: Official Guide to Basic Flight Information and ATC Procedures* (Washington, DC: FAA, 2008).
7. Preston, *FAA Chronology,* p. 108.
8. Ibid., p. 109.
9. William H. Andrews, Stanley P. Butchart, Donald L. Hughes, and Thomas R. Sisk, "Flight Tests Related to Jet Transport Upset and Turbulent-Air Penetration," *Conference on Aircraft Operating Problems,* NASA SP-83 (Washington, DC: NASA, 1965).

Boeing 720 airliners were flown to define the handling qualities of aircraft as they encountered turbulence and determine the best methods for the pilot to recover from the upset. Another part of the study was to determine how turbulence upset the pilots themselves and if any changes to cockpit displays or controls would be helpful. Results of the project presented at a 1965 NASA Conference on Aircraft Operating Problems indicated that in terms of aircraft control, retrimming the stabilizer and deploying the spoilers were "valuable tools," but if those devices were to be safely used, an accurate g-meter should be added to the cockpit to assist the pilot in applying the correct amount of control force. The pilots also observed that initially encountering turbulence often created such a jolt that it disrupted their ability to scan the instrument dials (which remained reliable despite the added vibrations) and recommended improvements in their seat cushions and restraint system.[10]

But the true value of Project Taper to making safer skyways may have been the realization that although aircraft and pilots under controlled conditions and specialized training could safely penetrate areas of turbulence—even if severe—the better course of action was to find ways to avoid the threat altogether. This required further research and improvements in turbulence detection and forecasting, along with the ability to integrate that data in a timely manner to the ATC system and cockpit instrumentation.[11]

Avoiding Bird Hazards: 1966

After millions of years of birds having the sky to themselves, it only took 9 years from the time the Wright brothers first flew in 1903 for the first human fatality brought about by a bird striking an aircraft and causing the plane to crash in 1912. Fast-forward to 1960, when an Eastern Air Lines plane went down near Boston, killing 62 people as a result of a bird strike—the largest loss of life from a single bird incident.[12]

With the growing number of commercial jet airplanes, faster aircraft increased the potential damage a small bird could inflict and the larger airplanes put more humans at risk during a single flight. The need to address methods for dealing with birds around airports and in the skies also rose in priority. So, on September 9, 1966, the Interagency Bird

10. Ibid.

11. Philip Donely, "Safe Flight in Rough Air," NASA TM-X-51662 (Hampton, VA: NASA, 1964).

12. Micheline Maynard, "Bird Hazard is Persistent for Planes," *New York Times* (Jan. 19, 2009).

A DeTect, Inc., MERLIN bird strike avoidance radar is seen here in use in South Africa. NASA uses the same system at Kennedy Space Center for Space Shuttle missions, and the FAA is considering its use at airports around the Nation. NASA.

Hazard Committee was formed to gather data, share information, and develop methods for mitigating the risk of collisions between birds and airplanes. With the FAA taking the lead, the Committee included representatives from NASA; the Civil Aeronautics Board; the Department of Interior; the Department of Health, Education, and Welfare; and the U.S. Air Force, Navy, and Army.[13]

Through the years since the Committee was formed, the aviation community has approached the bird strike hazard primarily on three fronts: (1) removing or relocating the birds, (2) designing aircraft components to be less susceptible to damage from bird strikes, and (3) increasing the understanding of bird habitats and migratory patterns so as to alter air traffic routes and minimize the potential for bird strikes. Despite these efforts, the problem persists today, as evidenced by the January 2009 incident involving a US Airways jet that was forced to ditch in the Hudson River. Both of its jet engines failed because of

13. John L. Seubert, "Activities of the FAA Inter-Agency Bird Hazard Committee" (Washington, DC: FAA, 1968).

bird strikes shortly after takeoff. Fortunately, all souls on board survived the water landing thanks to the training and skills of the entire flightcrew.[14]

NASA's contributions in this area include research to characterize the extent of damage that birds might inflict on jet engines and other aircraft components in a bid to make those parts more robust or forgiving of a strike,[15] and the development of techniques to identify potentially harmful flocks of birds[16] and their local and seasonal flight patterns using radar so that local air traffic routes can be altered.[17]

Radar is in use to warn pilots and air traffic controllers of bird hazards at the Seattle-Tacoma International Airport. As of this writing, the FAA plans to deploy test systems at Chicago, Dallas, and New York airports, as the technology still needs to be perfected before its deployment across the country, according to an FAA spokeswoman quoted in a *Wall Street Journal* story published January 26, 2009.[18]

Meanwhile, a bird detecting radar system first developed for the Air Force by DeTect, Inc., of Panama City, FL, has been in use since 2006 at NASA's Kennedy Space Center to check for potential bird strike hazards before every Space Shuttle launch. Two customized marine radars scan the sky: one oriented in the vertical, the other in the horizontal. Together with specialized software, the MERLIN system can detect flocks of birds up to 12 miles from the launch pad or runway, according to a company fact sheet.

In the meantime, airports with bird problems will continue to rely on broadcasting sudden loud noises, shooting off fireworks, flashing strobe lights, releasing predator animals where the birds are nesting, or, in the worst case, simply eliminating the birds.

---

14. Maynard, "Bird Hazard is Persistent for Planes."
15. M.S. Hirschbein, "Bird Impact Analysis Package for Turbine Engine Fan Blades," *23rd Structures, Structural Dynamics and Materials Conference, New Orleans, LA, May 10–12, 1982.*
16. E.B. Dobson, J.J. Hicks, and T.G. Konrad, "Radar Characteristics of Known, Single Birds in Flight," *Science*, vol. 159, no. 3812 (Jan. 19, 1968), pp. 274–280.
17. Bruno Bruderer and Peter Steidinger, "Methods of Quantitative and Qualitative Analysis of Bird Migration with a Tracking Radar," *Animal Orientation and Navigation* (Washington, DC: NASA, 1972), pp. 151–167.
18. Andy Pasztor and Susan Carey, "New Focus Put on Avoiding Bird Strikes," *Wall Street Journal* (Jan. 26, 2009), p. A3.

Applications Technology Satellite 1 (ATS 1): 1966–1967

Aviation's use of actual space-based technology was first demonstrated by the FAA using NASA's Applications Technology Satellite 1 (ATS 1) to relay voice communications between the ground and an airborne FAA aircraft using very high frequency (VHF) radio during 1966 and 1967, with the aim of enabling safer air traffic control over the oceans.[19]

Launched from Cape Canaveral atop an Atlas Agena D rocket on December 7, 1966, the spin-stabilized ATS 1 was injected into geosynchronous orbit to take up a perch 22,300 miles high, directly over Ecuador. During this early period in space history, the ATS 1 spacecraft was packed with experiments to demonstrate how satellites could be used to provide the communication, navigation, and weather monitoring that we now take for granted. In fact, the ATS 1's black and white television camera captured the first full-Earth image of the planet's cloud-covered surface.[20]

Eight flight tests were conducted using NASA's ATS 1 to relay voice signals between the ground and an FAA aircraft using VHF band radio, with the intent of allowing air traffic controllers to speak with pilots flying over an ocean. Measurements were recorded of signal level, signal plus noise-to-noise ratio, multipath propagation, voice intelligibility, and adjacent channel interference. In a 1970 FAA report, the author concluded that the "overall communications reliability using the ATS 1 link was considered marginal."[21]

All together, the ATS project attempted six satellite launches between 1966 and 1974, with ATS 2 and ATS 4 unable to achieve a useful orbit. ATS 1 and ATS 3 continued the FAA radio relay testing, this time including a specially equipped Pan American Airways 747 as it flew a commercial flight over the ocean. Results were better than when the ATS 1 was tested alone, with a NASA summary of the experiments concluding that

> The experiments have shown that geostationary satellites can provide high quality, reliable, un-delayed communications

19. J.N. Sivo, W.H. Robbins, and D.M. Stretchberry, "Trends in NASA Communications Satellites," NASA TM-X-68141 (1972).

20. A.N. Engler, J.F. Nash, and J.D. Strange, "Applications Technology Satellite and Communications Technology/Satellite User Experiments for 1967-1980 Reference Book," NASA CR-165169-VOL-1 (1980).

21. F.W. Jefferson, "ATS-1 VHF Communications Experimentation," FAA 0444707 (1970).

> between distant points on the earth and that they can also be used for surveillance. A combination of un-delayed communications and independent surveillance from shore provides the elements necessary for the implementation of effective traffic control for ships and aircraft over oceanic regions. Eventually the same techniques may be applied to continental air traffic control.[22]

Aviation Safety Reporting System: 1975

On December 1, 1974, a Trans World Airlines (TWA) Boeing 727, on final approach to Dulles airport in gusty winds and snow, crashed into a Virginia mountain, killing all aboard. Confusion about the approach to the airport, the navigation charts the pilots were using, and the instructions from air traffic controllers all contributed to the accident. Six weeks earlier, a United Airlines flight nearly succumbed to the same fate. Officials concluded, among other things, that a safety awareness program might have enabled the TWA flight to benefit from the United flight's experience. In May 1975, the FAA announced the start of an Aviation Safety Reporting Program to facilitate that kind of communication. Almost immediately, it was realized the program would fail because of fear the FAA would retaliate against someone calling into question its rules or personnel. A neutral third party was needed, so the FAA turned to NASA for the job. In August 1975, the agreement was signed, and NASA officially began operating a new Aviation Safety Reporting System (ASRS).[23]

NASA's job with the ASRS was more than just emptying a "big suggestion box" from time to time. The memorandum of agreement between the FAA and NASA proposed that the updated ASRS would have four functions:

> 1. Take receipt of the voluntary input, remove all evidence of identification from the input, and begin initial processing of the data.
> 2. Perform analysis and interpretation of the data to identify any trends or immediate problems requiring action.

22. "VHF Ranging and Position Fixing Experiment using ATS Satellites," NASA CR-125537 (1971).
23. C.E. Billings, E.S. Cheaney, R. Hardy, and W.D. Reynard, "The Development of the NASA Aviation Safety Reporting System," NASA RP-1114 (1986), p. 3.

3. Prepare and disseminate appropriate reports and other data.
4. Continually evaluate the ASRS, review its performance, and make improvements as necessary.

Two other significant aspects of the ASRS included a provision that no disciplinary action would be taken against someone making a safety report and that NASA would form a committee to advise on the ASRS. The committee would be made up of key aviation organizations, including the Aircraft Owners and Pilots Association, the Air Line Pilots Association, the Aviation Consumer Action Project, the National Business Aircraft Association, the Professional Air Traffic Controllers Organization, the Air Transport Association, the Allied Pilots Association, the American Association of Airport Executives, the Aerospace Industries Association, the General Aviation Manufacturers' Association, the Department of Defense, and the FAA.[24]

Now in existence for more than 30 years, the ASRS has racked up an impressive success record of influencing safety that has touched every aspect of flight operations, from the largest airliners to the smallest general-aviation aircraft. According to numbers provided by NASA's Ames Research Center at Moffett Field, CA, between 1976 and 2006, the ASRS received more than 723,400 incident reports, resulting in 4,171 safety alerts being issued and the instigation of 60 major research studies. Typical of the sort of input NASA receives is a report from a Mooney 20 pilot who was taking a young aviation enthusiast on a sightseeing flight and explaining to the passenger during his landing approach what he was doing and what the instruments were telling him. This distracted his piloting just enough to complicate his approach and cause the plane to flare over the runway. He heard his stall alarm sound, then silence, then another alarm with the same tone. Suddenly, his aircraft hit the runway, and he skidded to a stop just off the pavement. It turned out that the stall warning alarm and landing gear alarm sounded alike. His suggestion was to remind the general-aviation community there were verbal alarms available to remind pilots to check their gear before landing.[25]

24. C.E. Billings, "Aviation Safety Reporting System," p. 6.

25. "Horns and Hollers," *CALLBACK From NASA's Aviation Safety Reporting System*, No. 359 (Nov. 2009), p. 2.

Although the ASRS continues today, one negative about the program is that it is passive and only works if information is voluntarily offered. But from April 2001 through December 2004, NASA fielded the National Aviation Operations Monitoring Service (NAOMS) and conducted almost 30,000 interviews to solicit specific safety-related data from pilots, air traffic controllers, mechanics, and other operational personnel. The aim was to identify systemwide trends and establish performance measures, with an emphasis on tracking the effects of new safety-related procedures, technologies, and training. NAOMS was part of NASA's Aviation Safety Program, detailed later in this case study.[26]

With all these data in hand, more coming in every day, and none of them in a standard, computer-friendly format, NASA researchers were prompted to develop search algorithms that recognized relevant text. The first such suite of software used to support ASRS was called QUOROM, which at its core was a computer program capable of analyzing, modeling, and ranking text-based reports. NASA programmers then enhanced QUOROM to provide:

- Keyword searches, which retrieve from the ASRS database narratives that contain one or more user-specified keywords in typical or selected contexts and rank the narratives on their relevance to the keywords in context.
- Phrase searches, which retrieve narratives that contain user-specified phrases, exactly or approximately, and rank the narratives on their relevance to the phrases.
- Phrase generation, which produces a list of phrases from the database that contain a user-specified word or phrase.
- Phrase discovery, which finds phrases from the database that are related to topics of interest.[27]

QUORUM's usefulness in accessing the ASRS database would evolve as computers became faster and more powerful, paving the way for a new suite of software to perform what is now called "data mining." This in turn would enable continual improvement in aviation safety and

26. "NAOMS Reference Report: Concepts, Methods, and Development Roadmap" Battelle Memorial Institute (2007).
27. Michael W. McGreevy, "Searching the ASRS Database Using QUORUM Keyword Search, Phrase Search, Phrase Generation, and Phrase Discovery," NASA TM-2001-210913 (2001), p. 4.

Microwave Landing System hardware at NASA's Wallops Flight Research Facility in Virginia as a NASA 737 prepares to take off to test the high-tech navigation and landing aid. NASA.

find applications in everything from real-time monitoring of aircraft systems[28] to Earth sciences.[29]

Microwave Landing System: 1976

As soon as it was possible to join the new inventions of the airplane and the radio in a practical way, it was done. Pilots found themselves "flying the beam" to navigate from one city to another and lining up with the runway, even in poor visibility, using the Instrument Landing System (ILS). ILS could tell the pilots if they were left or right of the runway centerline and if they were higher or lower than the established glide slope during the final approach. ILS required straight-in approaches and separation between aircraft, which limited the number of landings allowed each hour at the busiest airports. To improve upon this, the FAA, NASA, and the Department of Defense (DOD) in 1971 began developing the Microwave Landing System (MLS), which promised,

28. Glenn Sakamoto, "Intelligent Data Mining Capabilities as Applied to Integrated Vehicle Health Management," *2007 Research and Engineering Annual Report* (Edwards, CA: NASA, 2008), p. 65.
29. Sara Graves, Mahabaleshwa Hegde, Ken Keiser, Christopher Lynnes, Manil Maskey, Long Pham, and Rahul Ramachandran, "Earth Science Mining Web Services," *American Geophysical Union Meeting, San Francisco, Dec. 15–19, 2008.*

among other things, to increase the frequency of landings by allowing multiple approach paths to be used at the same time. Five years later, the FAA took delivery of a prototype system and had it installed at the FAA's National Aviation Facilities Experimental Center in Atlantic City, NJ, and at NASA's Wallops Flight Research Facility in Virginia.[30]

Between 1976 and 1994, NASA was actively involved in understanding how MLS could be integrated into the national airspace system. Configuration and operation of aircraft instrumentation,[31] pilot procedures and workload,[32] air traffic controller procedures,[33] use of MLS with helicopters,[34] effects of local terrain on the MLS signal,[35] and the determination to what extent MLS could be used to automate air traffic control[36] were among the topics NASA researchers tackled as the FAA made plans to employ MLS at airports around the Nation.

But having proven with NASA's Applications Technology Satellite program that space-based communication and navigation were more than feasible (but skipping endorsement of the use of satellites in the FAA's 1982 National Airspace System Plan), the FAA dropped the MLS program in 1994 to pursue the use of GPS technology, which was just beginning to work itself into the public consciousness. GPS signals, when enhanced by a ground-based system known as the Wide Area Augmentation System (WAAS), would provide more accurate position information and do it in a more efficient and potentially less costly manner than by deploying MLS around the Nation.[37]

Although never widely deployed in the United States for civilian use, MLS remains a tool of the Air Force at its airbases. NASA has

30. Preston, *FAA Chronology,* p. 188.

31. D.G. Moss, P.F. Rieder, B.P. Stapleton, A.D. Thompson, and D.B. Walen, "MLS: Airplane System Modeling," NASA CR-165700 (1981).

32. Jon E. Jonsson and Leland G. Summers, "Crew Procedures and Workload of Retrofit Concepts for Microwave Landing System," NASA CR-181700 (1989).

33. S. Hart, J.G. Kreifeldt, and L. Parkin, "Air Traffic Control by Distributed Management in a MLS Environment," *13th Conference on Manual Control, Cambridge, MA, 1977.*

34. H.Q. Lee, P.J. Obrien, L.L. Peach, L. Tobias, and F.M. Willett, Jr., "Helicopter IFR Approaches into Major Terminals Using RNAV, MLS and CDTI," *Journal of Aircraft,* vol. 20 (Aug. 1983).

35. M.M. Poulose, "Terrain Modeling for Microwave Landing System," *IEEE Transactions on Aerospace and Electronic Systems,* vol. 27 (May 1991).

36. M.M. Poulose, "Microwave Landing System Modeling with Application to Air Traffic Control Automation," *Journal of Aircraft,* vol. 29, no. 3 (May–June 1992).

37. "Navigating the Airways," *Spinoff* (Washington, DC: NASA, 1999), p. 50.

employed a version of the system called the Microwave Scan Beam Landing System for use at its Space Shuttle landing sites in Florida and California. Moreover, Europe has embraced MLS in recent years, and an increasing number of airports there are being equipped with the system, with London's Heathrow Airport among the first to roll it out.[38]

NUSAT: 1985

NUSAT, a tiny satellite designed by Weber State College in northern Utah, was deployed into Earth orbit from the cargo bay of the Space Shuttle Challenger on April 29, 1985. Its purpose was to serve as a radar target for the FAA.

The satellite employed three L-band receivers, an ultra high frequency (UHF) command receiver, a VHF telemetry transmitter, associated antennas, a microprocessor, fixed solar arrays, and a power supply to acquire, store, and forward signal strength data from radar. All of that was packed inside a basketball-sized, 26-sided polyhedron that weighed about 115 pounds.[39]

NUSAT was used to optimize ground-based ATC radar systems for the United States and member nations of the International Civil Aviation Organization by measuring antenna patterns.[40]

National Plan for Civil Aviation Human Factors: 1995

In June 1995, the FAA announced its plans for a joint FAA–DOD–NASA initiative called the National Plan for Civil Aviation Human Factors. The plan detailed a national effort to reduce and eliminate human error as the cause of aviation accidents. The plan called for projects that would identify needs and problems related to human performance, guide research programs that addressed the human element, involve the Nation's top scientists and aviation professionals, and report the results of these efforts to the aviation community.[41]

NASA's extensive involvement in human factors issues is detailed in another case study of this volume.

38. Brian Evans, "MLS: Back to the Future?" *Aviation Today* (Apr. 1, 2003).

39. R.G. Moore, "A Proof-of-Principle Getaway Special Free-Flying Satellite Demonstration," *2nd Symposium on Space Industrialization* (Huntsville, AL: NASA, 1984), p. 349.

40. Charles A. Bonsall, "NUSAT Update," *The 1986 Get Away Special Experimenter's Symposium* (Greenbelt, MD: NASA, 1987), p. 63.

41. FAA, "National Plan for Civil Aviation Human Factors: An Initiative for Research and Application" (Washington, DC: FAA, 1990).

Aviation Performance Measuring System: 1996

With the Aviation Safety Reporting System fully operational for two decades, NASA in 1996 once again found itself working with the FAA to gather raw data, process it, and make reports—all in the name of identifying potential problems and finding solutions. In this case, as part of a Flight Operations Quality Assurance program that the FAA was working with industry on, the agency partnered with NASA to test a new Aviation Performance Measuring System (APMS). The new system was designed to convert digital data taken from the flight data recorders of participating airlines into a format that could easily be analyzed.[42]

More specifically, the objectives of the NASA–FAA APMS research project was to establish an objective, scientifically and technically sound basis for performing flight data analysis; identify a flight data analysis system that featured an open and flexible architecture, so that it could easily be modified as necessary; and define and articulate guidelines that would be used in creating a standardized database structure that would form the basis for future flight data analysis programs. This standardized database structure would help ensure that no matter which data-crunching software an airline might choose, it would be compatible with the APMS dataset. Although APMS was not intended to be a nationwide flight data collection system, it was intended to make available the technical tools necessary to more easily enable a large-scale implementation of flight data analysis.[43]

At that time, commercially available software development was not far enough advanced to meet the needs of the APMS, which sought identification and analysis of trends and patterns in large-scale databases involving an entire airline. Software then was primarily written with the needs of flight crews in mind and was more capable of spotting single events rather than trends. For example, if a pilot threw a series of switches out of order, the onboard computer could sound an alarm. But that computer, or any other, would not know how frequently pilots made the same mistake on other flights.[44]

---

42. Preston, *FAA Chronology*, p. 301.

43. Irving Statler, "APMS: An Integrated Set of Tools for Measuring Safety," *ISASI Flight Recorder Working Group Workshop, Santa Monica, CA, Apr. 16–18, 1996.*

44. Statler, "The Aviation Performance Measuring System (APMS): An Integrated Suite of Tools for Measuring Performance and Safety," *World Aviation Congress, Anaheim, CA, Sept. 28–30, 1998.*

The FAA's air traffic control tower facility at the Dallas/Fort Worth International Airport is a popular site that the FAA uses for testing new ATC systems and procedures, including new Center TRACON Automation System tools. FAA.

A particularly interesting result of this work was featured in the 1998 edition of NASA's annual *Spinoff* publication, which highlights successful NASA technology that has found a new home in the commercial sector:

> A flight data visualization system called FlightViz™ has been created for NASA's Aviation Performance Measuring System (APMS), resulting in a comprehensive flight visualization and

> analysis system. The visualization software is now capable of very high-fidelity reproduction of the complete dynamic flight environment, including airport/airspace, aircraft, and cockpit instrumentation. The APMS program calls for analytic methods, algorithms, statistical techniques, and software for extracting useful information from digitally-recorded flight data. APMS is oriented toward the evaluation of performance in aviation systems, particularly human performance. . . . In fulfilling certain goals of the APMS effort and related Space Act Agreements, SimAuthor delivered to United Airlines in 1997, a state-of-the-art, high-fidelity, reconfigurable flight data replay system. The software is specifically designed to improve airline safety as part of Flight Operations Quality Assurance (FOQA) initiatives underway at United Airlines. . . . Pilots, instructors, human factors researchers, incident investigators, maintenance personnel, flight operations quality assurance staff, and others can utilize the software product to replay flight data from a flight data recorder or other data sources, such as a training simulator. The software can be customized to precisely represent an aircraft of interest. Even weather, time of day and special effects can be simulated.[45]

While by no means a complete list of every project NASA and the FAA have collaborated on, the examples detailed so far represent the diverse range of research conducted by the agencies. Much of the same kind of work continued as improved technology, updated systems, and fresh approaches were applied to address a constantly evolving set of challenges.

Aviation Safety Program

After the in-flight explosion and crash of TWA 800 in July 1996, President Bill Clinton established a Commission on Aviation Safety and Security, chaired by Vice President Al Gore. The Commission's emphasis was to find ways to reduce the number of fatal air-related accidents. Ultimately, the Commission challenged the aviation community to lower the fatal aircraft accident rate by 80 percent in 10 years and 90 percent in 25 years.

45. "Improving Airline Safety," *Spinoff* (Washington, DC: NASA, 1998), p. 62.

NASA's response to this challenge was to create in 1997 the Aviation Safety Program (AvSP) and, as seen before, partner with the FAA and the DOD to conduct research on a number of fronts.[46]

NASA's AvSP was set up with three primary objectives: (1) eliminate accidents during targeted phases of flight, (2) increase the chances that passengers would survive an accident, and (3) beef up the foundation upon which aviation safety technologies are based. From those objectives, NASA established six research areas, some having to do directly with making safer skyways and others pointed at increasing aircraft safety and reliability. All produced results, as noted in the referenced technical papers. Those research areas included accident mitigation,[47] systemwide accident prevention,[48] single aircraft accident prevention,[49] weather accident prevention,[50] synthetic vision,[51] and aviation system modeling and monitoring.[52]

Of particular note is a trio of contributions that have lasting influence today. They include the introduction and incorporation of the glass cockpit into the pilot's work environment and a pair of programs to gather key data that can be processed into useful, safety enhancing information.

Glass Cockpit

As aircraft systems became more complex and the amount of navigation, weather, and air traffic information available to pilots grew in abundance, the nostalgic days of "stick and rudder" men (and women) gave way to "cockpit managers." Mechanical, analog dials showing a

46. Jaiwon Shin, "The NASA Aviation Safety Program: Overview," NASA TM-2000-209810 (2000).

47. Lisa E. Jones, "Overview of the NASA Systems Approach to Crashworthiness Program," *American Helicopter Society 58th Annual Forum, Montreal, Canada, June 11–13, 2002.*

48. Doreen A. Comerford, "Recommendations for a Cockpit Display that Integrates Weather Information with Traffic Information," NASA TM-2004-212830 (2004).

49. Roger M. Bailey, Mark W. Frye, and Artie D. Jessup, "NASA-Langley Research Center's Aircraft Condition Analysis and Management System Implementation," NASA TM-2004-213276 (2004).

50. "Proceedings of the Second NASA Aviation Safety Program Weather Accident Review," NASA CP-2003-210964 (2003).

51. Jarvis J. Arthur, III, Randall E. Bailey, Lynda J. Kramer, R.M. Norman, Lawrence J. Prinzel, III, Kevin J. Shelton, and Steven P. Williams, "Synthetic Vision Enhanced Surface Operations With Head-Worn Display for Commercial Aircraft," *International Journal of Aviation Psychology,* vol. 19, no. 2 (Apr. 2009), pp. 158–181.

52. "The Aviation System Monitoring and Modeling (ASMM) Project: A Documentation of its History and Accomplishments: 1999–2005," NASA TP-2007-214556 (2007).

A prototype "glass cockpit" that replaces analog dials and mechanical tapes with digitally driven flat panel displays is installed inside the cabin of NASA's 737 airborne laboratory, which tested the new hardware and won support for the concept in the aviation community. NASA.

single piece of information (e.g., airspeed or altitude) weren't sufficient to give pilots the full status of their increasingly complicated aircraft flying in an increasingly crowded sky. The solution came from engineers at NASA's Langley Research Center in Hampton, VA, who worked with key industry partners to come up with an electronic flight display—what is generally known now as the glass cockpit—that took advantage of powerful, small computers and liquid crystal display (LCD) flat panel technology. Early concepts of the glass cockpit were flight-proven using NASA's Boeing 737 flying laboratory and eventually certified for use by the FAA.[53]

According to a NASA fact sheet,

> The success of the NASA-led glass cockpit work is reflected in the total acceptance of electronic flight displays beginning with the introduction of the Boeing 767 in 1982. Airlines and their passengers, alike, have benefitted. Safety and efficiency of flight have been increased with improved pilot understanding of the airplane's situation relative to its environment.

53. Lane E. Wallace, "Airborne Trailblazer: Two Decades with NASA Langley's 737 Flying Laboratory," NASA SP-4216 (1994).

> The cost of air travel is less than it would be with the old technology and more flights arrive on time.[54]

After developing the first glass cockpits capable of displaying basic flight information, NASA has continued working to make more information available to the pilots,[55] while at the same time being conscious of information overload,[56] the ability of the flight crew to operate the cockpit displays without distraction during critical phases of flight (takeoff and landing),[57] and the effectiveness of training pilots to use the glass cockpit.[58]

Performance Data Analysis and Reporting System

In yet another example of NASA developing a database system with and for the FAA, the Performance Data Analysis and Reporting System (PDARS) began operation in 1999 with the goal of collecting, analyzing, and reporting of performance-related data about the National Airspace System. The difference between PDARS and the Aviation Safety Reporting System is that input for the ASRS comes voluntarily from people who see something they feel is unsafe and report it, while input for PDARS comes automatically—in real time—from electronic sources such as ATC radar tracks and filed flight plans. PDARS was created as an element of NASA's Aviation Safety Monitoring and Modeling project.[59]

From these data, PDARS calculates a variety of performance measures related to air traffic patterns, including traffic counts, travel times between airports and other navigation points, distances flown, general traffic flow parameters, and the separation distance from trailing

---

54. "The Glass Cockpit: Technology First Used in Military, Commercial Aircraft," FS-2000-06-43-LaRC (2000).
55. Marianne Rudisill, "Crew/Automation Interaction in Space Transportation Systems: Lessons Learned from the Glass Cockpit," NASA Langley Research Center (2000).
56. Susan T. Heers and Gregory M. Pisanich, "A Laboratory Glass-Cockpit Flight Simulator for Automation and Communications Research," *Human Factors Society Conference, San Diego, Oct. 9–13, 1995.*
57. Earl L. Wiener, "Flight Training and Management for High-Technology Transport Aircraft," NASA CR-200816 (1996).
58. Wiener, "Flight Training and Management for High-Technology Transport Aircraft," NASA CR-199670 (1995).
59. Thomas R. Chidester, "Aviation Performance Measuring System," Ames Research Center Research and Technology 2000 (Moffett Field: NASA, 2000).

aircraft. Nearly 1,000 reports to appropriate FAA facilities are automatically generated and distributed each morning, while the system also allows for sharing data and reports among facilities, as well as facilitating larger research projects. With the information provided by PDARS, FAA managers can quickly determine the health, quality, and safety of day-to-day ATC operations and make immediate corrections.[60]

The system also has provided input for several NASA and FAA studies, including measurement of the benefits of the Dallas/Fort Worth Metroplex airspace, an analysis of the Los Angeles Arrival Enhancement Procedure, an analysis of the Phoenix Dryheat departure procedure, measurement of navigation accuracy of aircraft using area navigation en route, a study on the detection and analysis of in-close approach changes, an evaluation of the benefits of domestic reduced vertical separation minimum implementation, and a baseline study for the airspace flow program. As of 2008, PDARS was in use at 20 Air Route Traffic Control Centers, 19 Terminal Radar Approach Control facilities, three FAA service area offices, the FAA's Air Traffic Control System Command Center in Herndon, VA, and at FAA Headquarters in Washington, DC.[61]

National Aviation Operations Monitoring Service

A further contribution to the Aviation Safety Monitoring and Modeling project provided yet another method for gathering data and crunching numbers in the name of making the Nation's airspace safer amid increasingly crowded skies. Whereas the Aviation Safety Reporting System involved volunteered safety reports and the Performance Data Analysis and Reporting System took its input in real time from digital data sources, the National Aviation Operations Monitoring Service was a scientifically designed survey of the aviation community to generate statistically valid reports about the number and frequency of incidents that might compromise safety.[62]

---

60. Wim den Braven and John Schade, "Concept and Operation of the Performance Data Analysis and Reporting System (PDARS)," *SAE Conference, Montreal, 2003.*

61. R. Nehl and J. Schade, "Update: Concept and Operation of the Performance Data Analysis and Reporting System (PDARS)," *2007 IEEE Aerospace Conference, Big Sky, MT, Mar. 3–10, 2007.*

62. Battelle Memorial Institute, "NAOMS Reference Report: Concepts, Methods and Development Roadmap" (Moffett Field: NASA, 2007).

After a survey was developed that would gather credible data from anonymous volunteers, an initial field trial of the NAOMS was held in 2000, followed by the launch of the program in 2001. Initially, the surveyors only sought out air carrier pilots who were randomly chosen from the FAA Airman's Medical Database. Researchers characterized the response to the NAOMS survey as enthusiastic. Between April 2001 and December 2004, nearly 30,000 pilot interviews were completed, with a remarkable 83-percent return rate, before the project ran short of funds and had to stop. The level of response was enough to achieve statistical validity and prove that NAOMS could be used as a permanent tool for managers to assess the operational health of the ATC system and suggest changes before they were actually needed. Although NASA and the FAA desired for the project to continue, it was shut down on January 31, 2008.[63]

It's worth mentioning that the NAOMS briefly became the subject of public controversy in 2007, when NASA received a Freedom of Information Act request by a reporter for the data obtained in the NAOMS survey. NASA denied the request, using language that then NASA Administrator Mike Griffin said left an "unfortunate impression" that the Agency was not acting in the best interest of the public. NASA eventually released the data after ensuring the anonymity originally guaranteed to those who were surveyed. In a January 14, 2008, letter from Griffin to all NASA employees, the Administrator summed up the experience by writing: "As usual in such circumstances, there are lessons to be learned, remembered, and applied. The NAOMS case demonstrates again, if such demonstrations were needed, the importance of peer review, scientific integrity, admitting mistakes when they are made, correcting them as best we can, and keeping our word, despite the criticism that can ensue."[64]

An Updated Safety Program

In 2006, NASA's Aeronautics Research Mission Directorate (ARMD) was reorganized. As a result, the projects that fell under ARMD's Aviation Safety Program were restructured as well, with more of a focus on

63. Statler, "The Aviation System Monitoring and Modeling (ASMM) Project: A Documentation of its History and Accomplishments: 1999–2005," NASA TP-2007-214556 (2007).
64. Michael Griffin, "Letter from NASA Administrator Mike Griffin" (Washington, DC: NASA, 2008).

aircraft safety than on the skies they fly through. Air traffic improvements in the new plan now fall almost exclusively within the Airspace Systems Program. The Aviation Safety Program is now dedicated to developing the principles, guidelines, concepts, tools, methods, and technologies to address four project areas: the Integrated Vehicle Health Management Project,[65] the Integrated Intelligent Flight Deck Technologies Project,[66] the Integrated Resilient Aircraft Control Project,[67] and the Aircraft Aging and Durability Project.[68]

Commercial Aviation Safety Team (CAST)

When NASA's Aviation Safety Program was begun in 1997, the agency joined with a large group of aviation-related organizations from Government, industry, and academia in forming a Commercial Aviation Safety Team (CAST) to help reduce the U.S. commercial aviation fatal accident rate by 80 percent in 10 years. During those 10 years, the group analyzed data from some 500 accidents and thousands of safety incidents and helped develop 47 safety enhancements.[69] In 2008, the group could boast that the rate had been reduced by 83 percent, and for that, CAST was awarded aviation's most prestigious honor, the Robert J. Collier Trophy.

NASA's work with improving the National Airspace System has won the Agency two Collier Trophies: one in 2007 for its work with developing the new next-generation ADS-B instrumentation, and one in 2008 as part of the Commercial Aviation Safety Team, which helped improve air safety during the past decade. NASA.

65. Luis Trevino, Deidre E. Paris, and Michael D. Watson, "Intelligent Vehicle Health Management," *41st AIAA–ASME–SAE–ASEE Joint Propulsion Conference and Exhibit, Tucson, July 10–13, 2005.*
66. David B. Kaber and Lawrence J. Prinzel, III, "Adaptive and Adaptable Automation Design: A Critical Review of the Literature and Recommendations for Future Research," NASA TM-2006-214504 (2006).
67. Sanjay Garg, "NASA Glenn Research in Controls and Diagnostics for Intelligent Aerospace Propulsion Systems," *Integrated Condition Management 2006, Anaheim, Nov. 14–16, 2006.*
68. Doug Rohn and Rick Young, "Aircraft Aging and Durability Project: Technical Plan Summary" (Washington, DC: NASA, 2007).
69. Samuel A. Morello and Wendell R. Ricks, "Aviation Safety Issues Database," NASA TM-2009-215706 (2009).

Air Traffic Management Research

The work of NASA's Aeronautics Research Mission Directorate primarily takes place at NASA Field Centers in Virginia, Ohio, and California. It's at the Ames Research Center at Moffett Field, CA, that a large share of the work to make safer skyways has been managed. Many of the more effective programs to improve the safety and efficiency of the Nation's air traffic control system began at Ames and continue to be studied.[70]

Seven programs managed within the divisions of Ames's Air Traffic Management Research office, described in the next section, reveal how NASA research is making a difference in the skies every day.

Airspace Concept Evaluation System

The Airspace Concept Evaluation System (ACES) is a computer tool that allows researchers to try out novel Air Traffic Management (ATM) theories, weed out those that are not viable, and identify the most promising concepts. ACES looks at how a proposed air transportation concept can work within the National Airspace System (NAS), with the aim of reducing delays, increasing capacity, and handling projected growth in air traffic. ACES does this by simulating the major components of the NAS, modeling a flight from gate to gate, and taking into account in its models the individual behaviors of those that affect the NAS, from departure clearance to the traffic control tower, the weather office, navigation systems, pilot experience, type of aircraft, and other major components. ACES also is able to predict how one individual behavior can set up a ripple effect that touches, or has the potential to touch, the entire NAS. This modeling approach isolates the individual models so that they can continue to be enhanced, improved, and modified to represent new concepts without impacting development of the overall simulation system.[71]

Among the variables ACES has been tasked to run through its simulations are environmental impacts when a change is introduced,[72] use

70. Gano Chatterji, Kapil Sheth, and Banavar Sridhar, "Airspace Complexity and its Application in Air Traffic Management," *Second USA/Europe Air Traffic Management R&D Seminar, Dec. 1–4, 1998.*

71. Brian Capozzi, Patrick Carlos, Vikram Manikonda, Larry Meyn, and Robert Windhorst, "The Airspace Concepts Evaluation System Architecture and System Plant," *AIAA Guidance, Navigation, and Control Conference, Keystone, CO, Aug. 21–24, 2006.*

72. Stephen Augustine, Brian Capozzi, John DiFelici, Michael Graham, Raymond M.C. Miraflor, and Terry Thompson, "Environmental Impact Analysis with the Airspace Concept Evaluation System," *6th ATM Research and Development Seminar, Baltimore, June 27–30, 2005.*

of various communication and navigation models,[73] validation of certain concepts under different weather scenarios,[74] adjustments to spacing and merging of traffic around dense airports,[75] and reduction of air traffic controller workload by automating certain tasks.[76]

Future ATM Concepts Evaluation Tool

Another NASA air traffic simulation tool, the Future ATM Concepts Evaluation Tool (FACET), was created to allow researchers to explore, develop, and evaluate advanced traffic control concepts. The system can operate in several modes: playback, simulation, live, or in a sort of hybrid mode that connects it with the FAA's Enhanced Traffic Management System (ETMS). ETMS is an operational FAA program that monitors and reacts to air traffic congestion, and it can also predict when and where congestion might happen. (The ETMS is responsible, for example, for keeping a plane grounded in Orlando because of traffic congestion in Atlanta.) Streaming the ETMS live data into a run of FACET makes the simulation of a new advanced traffic control concept more accurate. Moreover, FACET is able to model airspace operations on a national level, processing the movements of more than 5,000 aircraft on a single desktop computer, taking into account aircraft performance, weather, and other variables.[77]

Some of the advanced concepts tested in FACET include allowing aircraft to have greater freedom in maintaining separation on their own,[78] integrating space launch vehicle and aircraft operations into the

73. Greg Kubat and Don Vandrei, "Airspace Concept Evaluation System, Concept Simulations using Communication, Navigation and Surveillance System Models," *Proceedings of the Sixth Integrated Communications, Navigation and Surveillance Conference & Workshop, Baltimore, May 1–3, 2006.*

74. Larry Meyn and Shannon Zelinski, "Validating the Airspace Concept Evaluation System for Different Weather Days," *AIAA Modeling and Simulation Technologies Conference, Keystone, CO, Aug. 21–24, 2006.*

75. Art Feinberg, Gary Lohr, Vikram Manikonda, and Michel Santos, "A Simulation Testbed for Airborne Merging and Spacing," *AIAA Atmospheric Flight Mechanics Conference, Honolulu, Aug. 18–21, 2008.*

76. Heinz Erzberger and Robert Windhorst, "Fast-time Simulation of an Automated Conflict Detection and Resolution Concept," *6th AIAA Aviation Technology, Integration and Operations Conference, Wichita, Sept. 25–27, 2006.*

77. Banavar Sridhar, "Future Air Traffic Management Concepts Evaluation Tool," Ames Research Center Research and Technology 2000 (Moffett Field: NASA, 2000), p. 5.

78. Karl D. Bilimoria and Hilda Q. Lee, "Properties of Air Traffic Conflicts for Free and Structured Routing," *AIAA GN&C Conference, Montreal, Aug. 2001.*

airspace, and monitoring how efficiently aircraft comply with ATC instructions when their flights are rerouted.[79] In fact, the last of these concepts was so successful that it was deployed into the FAA's operational ETMS. NASA reports that the success of FACET has lead to its use as a simulation tool not only with the FAA, but also with several airlines, universities, and private companies. For example, Flight Dimensions International—the world's leading vendor of aircraft situational displays—recently integrated FACET with its already popular Flight Explorer product. FACET won NASA's 2006 Software of the Year Award.[80]

Surface Management System

Making the skyways safer for aircraft to fly by reducing delays and lowering the stress on the system begins and ends with the short journey on the ground between the active runway and the terminal gate. To better coordinate events between the air and ground sides, NASA developed, in cooperation with the FAA, a software tool called the Surface Management System (SMS), whose purpose is to manage the movements of aircraft on the surface of busy airports to improve capacity, efficiency, and flexibility.[81]

The SMS has three parts: a traffic management tool, a controller tool, and a National Airspace System information tool.[82]

The traffic management tool monitors aircraft positions in the sky and on the ground, along with the latest times when a departing airliner is about to be pushed back from its gate, to predict demand for taxiway and runway usage, with an aim toward understanding where backups might take place. Sharing this information among the traffic control tools and systems allows for more efficient planning. Similarly, the controller tool helps personnel in the ATC and ramp towers to better coordinate the movement of arriving and departing flights and to

---

79. Sarah Stock Patterson, "Dynamic Flow Management Problems in Air Transportation," NASA CR-97-206395 (1997).

80. "Comprehensive Software Eases Air Traffic Management," Spinoff 2007 (Washington, DC: NASA, 2007).

81. Dave Jara and Yoon C. Jung, "Development of the Surface Management System Integrated with CTAS Arrival Tools," *AIAA 5th Aviation Technology, Integration and Operations Forum, Arlington, TX, Sept. 2005.*

82. Katherine Lee, "CTAS and NASA Air Traffic Management Fact Sheets for En Route Descent Advisor and Surface Management System," *NATCA Safety Conference, Fort Worth, Apr. 2004.*

advise pilots on which taxiways to use as they navigate between the runway and the gate.[83] Finally, the NAS information tool allows data from the SMS to be passed into the FAA's national Enhanced Traffic Management System, which in turn allows traffic controllers to have a more accurate picture of the airspace.[84]

Center TRACON Automation System

The computer-based tools used to improve the flow of traffic across the National Airspace System—such as SMS, FACET, and ACES already discussed—were built upon the historical foundation of another set of tools that are still in use today. Rolled out during the 1990s, the underlying concepts of these tools go back to 1968, when an Ames Research Center scientist, Heinz Erzberger, first explored the idea of introducing air traffic control concepts—such as 4-D trajectory synthesis—and then proposed what was, in fact, developed: the Center TRACON Automation System (CTAS), the Traffic Manager Adviser (TMA), the En Route Descent Adviser (EDA), and the Final Approach Spacing Tool (FAST). Each of the tools provides controllers with advice, information, and some amount of automation—but each tool does this for a different segment of the NAS.[85]

CTAS provides automation tools to help air traffic controllers plan for and manage aircraft arriving to a Terminal Radar Approach Control (TRACON), which is the area within about 40 miles of a major airport. It does this by generating air traffic advisories that are designed to increase fuel efficiency and reduce delays, as well as assist controllers in ensuring that there is an acceptable separation between aircraft and that planes are approaching a given airport in the correct order. CTAS's goals also include improving airport capacity without threatening safety or increasing the workload of controllers.[86]

83. Gautam Gupta and Matthew Stephen Kistler, "Effect of Surface Traffic Count on Taxi Time at Dallas-Fort Worth International Airport," NASA ARC-E-DAA-TN286 (2008).

84. John O'Neill and Roxana Wales, "Information Management for Airline Operations," Ames Research Center Research and Technology Report (Moffett Field: NASA, 1998).

85. Heinz Erzberger and William Nedell, "Design of Automation Tools for Management of Descent Traffic," NASA TM-101078 (1988).

86. Dallas G. Denery and Heinz Erzberger, "The Future of Air Traffic Management," *NASA–ASEE Stanford University Seminars, Stanford, CA, 1998.*

Flight controllers test the Traffic Manager Adviser tool at the Denver TRACON. The tool helps manage the flow of air traffic in the area around an airport. National Air and Space Museum.

Traffic Manager Adviser

Airspace over the United States is divided into 22 areas. The skies within each of these areas are managed by an Air Route Traffic Control Center. At each center, there are controllers designated Traffic Management Coordinators (TMCs), who are responsible for producing a plan to deliver aircraft to a TRACON within the center at just the right time, with proper separation, and at a rate that does not exceed the capacity of the TRACON and destination airports.[87]

The NASA-developed Traffic Manager Adviser tool assists the TMCs in producing and updating that plan. The TMA does this by using graphical displays and alerts to increase the TMCs' situational awareness. The program also computes and provides statistics on the undelayed estimated time of arrival to various navigation milestones of an arriving aircraft and even gives the aircraft a runway assignment and scheduled time of arrival (which might later be changed by FAST). This informa-

87. Harry N. Swenson and Danny Vincent, "Design and Operational Evaluation of the Traffic Management Advisor at the Ft. Worth Air Route Traffic Control Center," *United States/Europe Air Traffic Management Research and Development Seminar, Paris, June 16–19, 1997.*

tion is constantly updated based on live radar updates and controller inputs and remains interconnected with other CTAS tools.[88]

En Route Descent Adviser

The National Airspace System relies on a complex set of actions with thousands of variables. If one aircraft is so much as 5 minutes out of position as it approaches a major airport, the error could trigger a domino effect that results in traffic congestion in the air, too many airplanes on the ground needing to use the same taxiway at the same time, late arrivals to the gate, and missed connections. One specific tool created by NASA to avoid this is the En Route Descent Adviser. Using data from CTAS, TMA, and live radar updates, the EDA software generates specific traffic control instructions for each aircraft approaching a TRACON so that it crosses an exact navigation fix in the sky at the precise time set by the TMA tool. The EDA tool does this with all ATC constraints in mind and with maneuvers that are as fuel efficient as possible for the type of aircraft.[89]

Improving the efficient flow of air traffic through the TRACON to the airport by using EDA as early in the approach as practical makes it possible for the airport to receive traffic in a constant feed, avoiding the need for aircraft to waste time and fuel by circling in a parking orbit before taking turn to approach the field. Another benefit: EDA allows controllers during certain high-workload periods to concentrate less on timing and more on dealing with variables such as changing weather and airspace conditions or handling special requests from pilots.[90]

Final Approach Spacing Tool

The last of the CTAS tools, which can work independently but is more efficient when integrated into the full CTAS suite, is the Final Approach Spacing Tool. It assists the TRACON controllers to determine the most efficient sequence, schedule, and runway assignments for aircraft intending to land. FAST takes advantage of information provided by the TMA and EDA tools in making its assessments and displaying advisories to

88. Greg Carr and Frank Neuman, "A Fast-Time Study of Aircraft Reordering in Arrival Sequencing and Scheduling," *AIAA Guidance, Navigation and Control Conference, Boston, Aug. 10–12, 1998.*
89. Lee, "CTAS Fact Sheets," 2004.
90. Steven Green and Robert Vivona, "En Route Descent Advisor Multi-Sector Planning Using Active and Provisional Controller Plans," AIAA Paper 2003-5572 (2003).

the controller, who then directs the aircraft as usual by radio communication. FAST also makes its determinations by using live radar, weather and wind data, and a series of other static databases, such as aircraft performance models, each airline's preferred operational procedures, and standard air traffic rules.[91]

Early tests of a prototype FAST system during the mid-1990s at the Dallas/Fort Worth International Airport TRACON showed immediate benefits of the technology. Using FAST's runway assignment and sequence advisories during more than 25 peak traffic periods, controllers measured a 10- to 20-percent increase in airport capacity, depending on weather and airport conditions.[92]

Simulating Safer Skyways

From new navigation instruments to updated air traffic control procedures, none of the developments intended to make safer skyways that was produced by NASA could be deployed into the real world until it had been thoroughly tested in simulated environments and certified as ready for use by the FAA. Among the many facilities and aircraft available to NASA to conduct such exercises, the Langley-based Boeing 737 and Ames-based complement of air traffic control simulators stand out as major contributors to the effort of improving the National Airspace System.

Langley's Airborne Trailblazer

The first Boeing 737 ever built was acquired by NASA in 1974 and modified to become the Agency's Boeing 737-100 Transport Systems Research Vehicle. During the next 20 years, it flew 702 missions to help NASA advance aeronautical technology in every discipline possible, first as a NASA tool for specific programs and then more generally as a national airborne research facility. Its contributions to the growth in capability and safety of the National Airspace System included the testing of hardware and procedures using new technology, most notably in the cockpit. Earning its title as an airborne trailblazer, it was the Langley 737 that tried out and won acceptance for new ideas such as the glass

91. Christopher Bergh, Thomas J. Davis, and Ken J. Krzeczowski, "The Final Approach Spacing Tool," *IFAC Conference, Palo Alto, CA, Sept. 1994.*

92. Thomas J. Davis, Douglas R. Isaacson, Katharine K. Lee, and John E. Robinson, III, "Operational Test Results of the Final Approach Spacing Tool," *Transportation Systems 1997, Chania, Greece, June 16–18, 1997.*

NASA's Airborne Trailblazer is seen cruising above the Langley Research Center in Virginia. The Boeing 737 served as a flying laboratory for NASA's aeronautics research for two decades. NASA.

cockpit. Those flat panel displays enabled other capabilities tested by the 737, such as data links for air traffic control communications, the microwave landing system, and satellite-based navigation using the revolutionary Global Positioning System.[93]

With plans to retire the 737, NASA Langley in 1994 acquired a Boeing 757-200 to be the new flying laboratory, earning the designation Airborne Research Integrated Experiments System (ARIES). In 2006, NASA decided to retire the 757.[94]

Ames's SimLabs

NASA's Ames Research Center in California is home to some of the more sophisticated and powerful simulation laboratories, which Ames calls SimLabs. The simulators support a range of research, with an emphasis on aerospace vehicles, aerospace systems and operations, human factors, accident investigations, and studies aimed at improving aviation

93. Wallace, "Airborne Trailblazer," 1994.

94. Michael S. Wusk, "ARIES: NASA Langley's Airborne Research Facility," AIAA 2002-5822 (2002).

3

safety. They all have played a role in making work new air traffic control concepts and associated technology. The SimLabs include:

- Future Flight Central, which is a national air traffic control and Air Traffic Management simulation facility dedicated to exploring solutions to the growing problem of traffic congestion and capacity, both in the air and on the ground. The simulator is a two-story facility with a 360-degree, full-scale, real-time simulation of an airport, in which new ideas and technology can be tested or personnel can be trained.[95]
- Vertical Motion Simulator, which is a highly adaptable flight simulator that can be configured to represent any aerospace vehicle, whether real or imagined, and still provide a high-fidelity experience for the pilot. According to a facility fact sheet, existing vehicles that have been simulated include a blimp, helicopters, fighter jets, and the Space Shuttle orbiter. The simulator can be integrated with Future Flight Central or any of the air traffic control simulators to provide real-time interaction.[96]
- Crew-Vehicle Systems Flight Facility,[97] which itself has three major simulators, including a state-of-the-art Boeing 747 motion-based cockpit,[98] an Advanced Concept Flight Simulator,[99] and an Air Traffic Control Simulator consisting of 10 PC-based computer workstations that can be used in a variety of modes.[100]

95. Jim McClenahen, "Virtual Planning at Work: A Tour of NASA Future Flight Central," NASA Tech Server Document ID: 7 (2000).
96. R.A. Hess and Y. Zeyada, "A Methodology for Evaluating the Fidelity of Ground-Based Flight Simulators," AIAA Paper 99-4034 (1999).
97. Durand R. Begault and Marc T, Pittman, "Three Dimensional Audio Versus Head Down TCAS Displays," NASA CR-177636 (1994).
98. Barry Crane, Everett Palmer, and Nancy Smith, "Simulator Evaluation of a New Cockpit Descent Procedure," *9th International Symposium on Aviation Psychology, Columbus, OH, Apr. 27–May 1, 1997.*
99. Thomas J. Davis and Steven M. Green, "Piloted Simulation of a Ground-Based Time-Control Concept for Air Traffic Control," NASA TM-10185 (1989).
100. Sharon Doubek, Richard F. Haines, Stanton Harke, and Boris Rabin, "Information Presentation and Control in a Modern Air Traffic Control Tower Simulator," *7th International Conference on Human Computer Interface, San Francisco, Aug. 24–29, 1997.*

A full-sized Air Traffic Control Simulator with a 360-degree panorama display, called Future Flight Central, is available to test new systems or train controllers in extremely realistic scenarios. NASA.

The Future of ATC

Fifty years of working to improve the Nation's airways and the equipment and procedures needed to manage the system have laid the foundation for NASA to help lead the most significant transformation of the National Airspace System in the history of flight. No corner of the air traffic control operation will be left untouched. From airport to airport, every phase of a typical flight will be addressed, and new technology and solutions will be sought to raise capacity in the system, lower operating costs, increase safety, and enhance the security of an air transportation system that is so vital to our economy.

This program originated from the 2002 Commission on the Future of Aerospace in the United States, which recommended an overhaul of the air transportation system as a national priority—mostly from the concern that air traffic is predicted to double, at least, during the next 20 years. Congress followed up with some money, and President George W. Bush signed into law a plan to create a Next Generation Air Transportation System (NextGen). To manage the effort, a Joint Planning and Development Office (JPDO) was created, with NASA, the FAA, the DOD, and other key aviation organizations as members.[101]

101. Jeremy C. Smith and Kurt W. Neitzke, "Metrics for the NASA Airspace Systems Program," NASA SP-2009-6115 (2009).

NASA then organized itself to manage its NextGen efforts through the Airspace Systems Program. Within the program, NASA's efforts are further divided into projects that are in support of either NextGen Airspace or NextGen Airportal. The airspace project is responsible for dealing with air traffic control issues such as increasing capacity, determining how much more automation can be introduced, scheduling, spacing of aircraft, and rolling out a GPS-based navigation system that will change the way we perceive flying. Naturally, the airportal project is examining ways to improve terminal operations in and around the airplanes, including the possibility of building new airports.[102]

Already, several technologies are being deployed as part of NextGen. One is called the Wide Area Augmentation System, another the Automatic Dependent Surveillance-Broadcast-B (ADS-B). Both have to do with deploying a satellite-based GPS tracking system that would end reliance on radars as the primary means of tracking an aircraft's approach.[103]

WAAS is designed to enhance the GPS signal from Earth orbit and make it more accurate for use in civilian aviation by correcting for the errors that are introduced in the GPS signal by the planet's ionosphere.[104] Meanwhile, ADS-B, which is deployed at several locations around the U.S., combines information with a GPS signal and drives a cockpit display that tells the pilots precisely where they are and where other aircraft are in their area, but only if those other aircraft are similarly equipped with the ADS-B hardware. By combining ADS-B, GPS, and WAAS signals, a pilot can navigate to an airport even in low visibility.[105] NASA was a member of the Government and industry team led by the FAA that conducted an ADS-B field test several years ago with United Parcel Service at its hub in Louisville, KY. This work earned the team the 2007 Collier Trophy.

In these various ways, NASA has worked to increase the safety of the air traveler and to enhance the efficiency of the global air transportation

102. Stephen T. Darr, Katherine A. Lemos, and Wendell R. Ricks, "A NextGen Aviation Safety Goal," *2008 Digital Avionics Systems Conference, St. Paul, MN, Oct. 26–30, 2008.*

103. A. Buige, "FAA Global Positioning System Program," *Global Positioning System for Gen. Aviation: Joint FAA–NASA Seminar, Washington, DC, 1978.*

104. Muna Demitri, Ian Harris, Byron Iijima, Ulf Lindqwister, Anthony Manucci, Xiaoqing Pi, and Brian Wilson, "Ionosphere Delay Calibration and Calibration Errors for Satellite Navigation of Aircraft," *Jet Propulsion Laboratory, Pasadena, CA, 2000.*

105. T. Breen, R. Cassell, C. Evers, R. Hulstrom, and A. Smith, "System-Wide ADS-B Back Up and Validation," *Sixth Integrated Communications, Navigation and Surveillance Conference, Baltimore, May 1–3, 2006.*

network. As winged flight enters its second century, it is a safe bet that the Agency's work in coming years will be as comprehensive and influential as it has been in the past, thanks to the competency, dedication, and creativity of NASA people.

Recommended Additional Readings

Reports, Papers, Articles, and Presentations: William H. Andrews, Stanley P. Butchart, Donald L. Hughes, and Thomas R. Sisk, "Flight Tests Related to Jet Transport Upset and Turbulent-Air Penetration," *Conference on Aircraft Operating Problems*, NASA SP-83 (Washington, DC: NASA, 1965).

Jarvis J. Arthur, III, Randall E. Bailey, Lynda J. Kramer, R.M. Norman, Lawrence J. Prinzel, III, Kevin J. Shelton, and Steven P. Williams, "Synthetic Vision Enhanced Surface Operations With Head-Worn Display for Commercial Aircraft," *International Journal of Aviation Psychology*, vol. 19, no. 2 (Apr. 2009), pp. 158–181.

Stephen Augustine, Brian Capozzi, John DiFelici, Michael Graham, Raymond M.C. Miraflor, and Terry Thompson, "Environmental Impact Analysis with the Airspace Concept Evaluation System," *6th ATM Research and Development Seminar, Baltimore, June 27–30, 2005.*

"The Aviation System Monitoring and Modeling (ASMM) Project: A Documentation of its History and Accomplishments: 1999-2005," NASA TP-2007-214556 (2007).

R. Bach, C. Farrell, and H. Erzberger, "An Algorithm for Level-Aircraft Conflict Resolution," NASA CR-2009-214573 (2009).

Roger M. Bailey, Mark W. Frye, and Artie D. Jessup, "NASA-Langley Research Center's Aircraft Condition Analysis and Management System Implementation," NASA TM-2004-213276 (2004).

Battelle Memorial Institute, "NAOMS Reference Report: Concepts, Methods, and Development Roadmap" (2007).

Durand R. Begault and Marc T. Pittman, "Three Dimensional Audio Versus Head Down TCAS Displays," NASA CR-177636 (1994).

Christopher Bergh, Thomas J. Davis, and Ken J. Krzeczowski, "The Final Approach Spacing Tool," *IFAC Conference, Palo Alto, CA, Sept. 1994.*

Karl D. Bilimoria and Hilda Q. Lee, "Properties of Air Traffic Conflicts for Free and Structured Routing," *AIAA GN&C Conference, Montreal, Aug. 2001.*

C.E. Billings, "Human-Centered Aircraft Automation: A Concept and Guidelines," NASA TM-103885 (1991).

C.E. Billings, E.S. Cheaney, R. Hardy, and W.D. Reynard, "The Development of the NASA Aviation Safety Reporting System," NASA RP-1114 (1986).

Charles A. Bonsall, "NUSAT Update," *The 1986 Get Away Special Experimenter's Symposium* (Greenbelt, MD: NASA, 1987), p. 63.

A. Buige, "FAA Global Positioning System Program," *Global Positioning System for Gen. Aviation: Joint FAA–NASA Seminar, Washington, DC, 1978.*

T. Breen, R. Cassell, C. Evers, R. Hulstrom, and A. Smith, "System-Wide ADS-B Back Up and Validation," *Sixth Integrated Communications, Navigation and Surveillance Conference, Baltimore, May 1–3, 2006.*

Bruno Bruderer and Peter Steidinger, "Methods of Quantitative and Qualitative Analysis of Bird Migration with a Tracking Radar," *Animal Orientation and Navigation* (Washington, DC: NASA, 1972), pp. 151–167.

Brian Capozzi, Patrick Carlos, Vikram Manikonda, Larry Meyn, and Robert Windhorst, "The Airspace Concepts Evaluation System Architecture and System Plant," *AIAA Guidance, Navigation, and Control Conference, Keystone, CO, Aug. 21–24, 2006.*

Greg Carr and Frank Neuman, "A Fast-Time Study of Aircraft Reordering in Arrival Sequencing and Scheduling," *AIAA Guidance, Navigation and Control Conference, Boston, Aug. 10–12, 1998.*

W. Chan, R. Bach, and J. Walton, "Improving and Validating CTAS Performance Models," *AIAA Guidance, Navigation, and Control Conference, Denver, CO, Aug. 2000.*

Gano Chatterji, Kapil Sheth, and Banavar Sridhar, "Airspace Complexity and its Application in Air Traffic Management," *Second USA/Europe Air Traffic Management R & D Seminar, Dec. 1–4, 1998.*

Thomas R. Chidester, "Aviation Performance Measuring System," *Ames Research Center Research and Technology 2000* (Moffett Field: NASA, 2000).

Doreen A. Comerford, "Recommendations for a Cockpit Display that Integrates Weather Information with Traffic Information," NASA TM-2004-212830 (2004).

"Comprehensive Software Eases Air Traffic Management," *Spinoff 2007* (Washington, DC: NASA, 2007).

R.A. Coppenbarger, "Climb Trajectory Prediction Enhancement Using Airline Flight-Planning Information," AIAA-99-4147 (1999).

R.A. Coppenbarger, R. Lanier, D. Sweet, and S. Dorsky, "Design and Development of the En Route Descent Advisor (EDA) for Conflict-Free Arrival Metering," AIAA 2004-4875 (2004).

R.A. Coppenbarger, R.W. Mead, and D.N. Sweet, "Field Evaluation of the Tailored Arrivals Concept for Datalink-Enabled Continuous Descent Approach," AIAA 2007-7778 (2007).

Barry Crane, Everett Palmer, and Nancy Smith, "Simulator Evaluation of a New Cockpit Descent Procedure," *9th International Symposium on Aviation Psychology, Columbus, OH, Apr. 27–May 1, 1997.*

L. Credeur, W.R. Capron, G.W. Lohr, D.J. Crawford, D.A. Tang, and W.G. Rodgers, Jr., "Final-Approach Spacing Aids (FASA) Evaluation for Terminal-Area, Time-Based Air Traffic Control," NASA TP-3399 (1993).

T.J. Davis, H. Erzberger, and H. Bergeron, "Design of a Final Approach Spacing Tool for TRACON Air Traffic Control," NASA TM-102229 (1989).

T.J. Davis, H. Erzberger, and S.M. Green, "Simulator Evaluation of the Final Approach Spacing Tool," NASA TM-102807 (1990).

Thomas J. Davis and Steven M. Green, "Piloted Simulation of a Ground-Based Time-Control Concept for Air Traffic Control," NASA TM-10185 (1989).

Thomas J. Davis, Douglas R. Isaacson, Katharine K. Lee, and John E. Robinson, III, "Operational Test Results of the Final Approach Spacing Tool," *Transportation Systems 1997, Chania, Greece, June 16–18, 1997.*

Muna Demitri, Ian Harris, Byron Iijima, Ulf Lindqwister, Anthony Manucci, Xiaoqing Pi, and Brian Wilson, "Ionosphere Delay Calibration and Calibration Errors for Satellite Navigation of Aircraft," *Jet Propulsion Laboratory, Pasadena, CA, 2000.*

W. den Braven, "Design and Evaluation of an Advanced Data Link System for Air Traffic Control," NASA TM-103899 (1992).

Wim den Braven and John Schade, "Concept and Operation of the Performance Data Analysis and Reporting System (PDARS)," *SAE Conference, Montreal, 2003.*

Dallas G. Denery and Heinz Erzberger, "The Center-TRACON Automation System: Simulation and Field Testing," NASA TM-110366 (2006).

Dallas G. Denery and Heinz Erzberger, "The Future of Air Traffic Management," *NASA–ASEE Stanford University Seminars, Stanford, CA, 1998.*

D.G. Denery, H. Erzberger, T.J. Davis, S.M. Green, and B.D. McNally, "Challenges of Air Traffic Management Research: Analysis, Simulation, and Field Test," AIAA 97-3832 (1997).

E.B. Dobson, J.J. Hicks, and T.G. Konrad, "Radar Characteristics of Known, Single Birds in Flight," *Science*, vol. 159, no. 3812 (Jan. 19, 1968), pp. 274–280.

Philip Donely, "Safe Flight in Rough Air," NASA TM-X-51662 (Hampton, VA: NASA, 1964).

Sharon Doubek, Richard F. Haines, Stanton Harke, and Boris Rabin, "Information Presentation and Control in a Modern Air Traffic Control Tower Simulator," *7th International Conference on Human Computer Interface, San Francisco, Aug. 24–29, 1997.*

A.N. Engler, J.F. Nash, and J.D. Strange, "Applications Technology Satellite and Communications Technology/Satellite User Experiments for 1967-1980 Reference Book," NASA CR-165169-VOL-1 (1980).

Brian Evans, "MLS: Back to the Future?" *Aviation Today* (Apr. 1, 2003).

H. Erzberger, "Automation of On-Board Flight Management," NASA TM-84212 (1981).

H. Erzberger, "CTAS: Computer Intelligence for Air Traffic Control in the Terminal Area," NASA TM-103959 (1992).

H. Erzberger, "Transforming the NAS: The Next Generation Air Traffic Control System," NASA TP-2004-212828 (2004).

H. Erzberger, T.J. Davis, and S.M. Green, "Design of Center-TRACON Automation System," *AGARD Meeting on Machine Intelligence in Air Traffic Management, Berlin, Germany, May 11–14, 1993.*

H. Erzberger and L. Engle, "Conflict Detection Tool," NASA TM-102201 (1989).

H. Erzberger and W. Nedell, "Design of Automated System for Management of Arrival Traffic," NASA TM-102201 (1989).

Heinz Erzberger and William Nedell, "Design of Automation Tools for Management of Descent Traffic," NASA TM-101078 (1988).

H. Erzberger and L. Tobias, "A Time-Based Concept for Terminal-Area Traffic Management," NASA TM-88243 (1986).

Heinz Erzberger and Robert Windhorst, "Fast-time Simulation of an Automated Conflict Detection and Resolution Concept," *6th AIAA Aviation Technology, Integration and Operations Conference, Wichita, Sept. 25–27, 2006.*

FAA, Aeronautical Information Manual: Official Guide to Basic Flight Information and ATC Procedures (Washington, DC: Federal Aviation Administration, 2008).

FAA, "National Plan for Civil Aviation Human Factors: An Initiative for Research and Application" (Washington, DC: FAA, 1990).

FAA, Part 21 Aircraft Certification Procedures for Products and Parts, *Federal Aviation Regulations* (Washington, DC: Federal Aviation Administration, 2009).

FAA, Part 91 General Operating and Flight *Rules, Federal Aviation Regulations* (Washington, DC: Federal Aviation Administration, 2009).

T.C. Farley, J.D. Foster, T. Hoang, and K.K. Lee, "A Time-Based Approach to Metering Arrival Traffic to Philadelphia," AIAA 2001-5241 (2001).

T. Farley, M. Kupfer, and H. Erzberger, "Automated Conflict Resolution: A Simulation Evaluation Under High Demand Including Merging Arrivals," AIAA 2007-7736 (2007).

T.C. Farley, S.J. Landry, T. Hoang, M. Nickelson, K.M. Levin, D. Rowe, and J.D. Welch, "Multi-Center Traffic Management Advisor: Operational Test Results," AIAA 2005-7300 (2005).

Art Feinberg, Gary Lohr, Vikram Manikonda, and Michel Santos, "A Simulation Testbed for Airborne Merging and Spacing," *AIAA Atmospheric Flight Mechanics Conference, Honolulu, Aug. 18–21, 2008.*

Sanjay Garg, "NASA Glenn Research in Controls and Diagnostics for Intelligent Aerospace Propulsion Systems," *Integrated Condition Management 2006, Anaheim, Nov. 14–16, 2006.*

"The Glass Cockpit: Technology First Used in Military, Commercial Aircraft," FS-2000-06-43-LaRC (2000).

C. Gong and W.N. Chan, "Using Flight Manual Data to Derive Aero-Propulsive Models for Predicting Aircraft Trajectories," *AIAA Aircraft Technology, Integration, and Operations (ATIO) Conference, Los Angeles, CA, Oct. 1–3, 2002.*

Shon Grabbe and Banavar Sridhar, "Modeling and Evaluation of Miles-in-Trail Restrictions in the National Air Space," *AIAA Guidance, Navigation, and Control Conference, Austin, Aug. 11–14, 2003.*

S. Grabbe, B. Sridhar, and N. Cheng, "Central East Pacific Flight Routing," AIAA 2006-6773 (2006).

S. Grabbe, B. Sridhar, and A. Mukherjee, "Central East Pacific Flight Scheduling," AIAA 2007-6447 (2007).

Sara Graves, Mahabaleshwa Hegde, Ken Keiser, Christopher Lynnes, Manil Maskey, Long Pham, and Rahul Ramachandran, "Earth Science Mining Web Services," *American Geophysical Union Meeting, San Francisco, Dec. 15–19, 2008.*

S.M. Green, W. den Braven, D.H. Williams, "Development and Evaluation of a Profile Negotiation Process for Integrating Aircraft and Air Traffic Control Automation," NASA TM-4360 (1993).

S.M. Green, T. Goka, and D.H. Williams, "Enabling User Preferences Through Data Exchange," *AIAA Guidance, Navigation and Control Conference, New Orleans, LA, Aug. 1997.*

S.M. Green and R.A. Vivona, "En route Descent Advisor Concept for Arrival Metering," AIAA 2001-4114 (2001).

Steven Green and Robert Vivona, "En Route Descent Advisor Multi-Sector Planning Using Active and Provisional Controller Plans," AIAA Paper 2003-5572 (2003).

S.M. Green and R.A. Vivona, "Field Evaluation of Descent Advisor Trajectory Prediction Accuracy," AIAA 96-3764 (1996).

S.M. Green, R.A. Vivona, and B. Sanford, "Descent Advisor Preliminary Field Test," AIAA 95-3368 (1995).

Michael Griffin, "Letter from NASA Administrator Mike Griffin" (Washington, DC: NASA, 2008).

Gautam Gupta and Matthew Stephen Kistler, "Effect of Surface Traffic Count on Taxi Time at Dallas-Fort Worth International Airport," NASA ARC-E-DAA-TN286 (2008).

S. Hart, J.G. Kreifeldt, and L. Parkin, "Air Traffic Control by Distributed Management in a MLS Environment," *13th Conference on Manual Control, Cambridge, MA, 1977.*

K. Harwood and B. Sanford, "Denver TMA Assessment," NASA CR-4554 (1993).

K.R. Heere and R.E. Zelenka, "A Comparison of Center/TRACON Automation System and Airline Time of Arrival Predictions," NASA TM-2000-209584 (2000).

Susan T. Heers and Gregory M. Pisanich, "A Laboratory Glass-Cockpit Flight Simulator for Automation and Communications Research," *Human Factors Society Conference, San Diego, Oct. 9–13, 1995.*

R.A. Hess and Y. Zeyada, "A Methodology for Evaluating the Fidelity of Ground-Based Flight Simulators," AIAA Paper 99-4034 (1999).

M.S. Hirschbein, "Bird Impact Analysis Package for Turbine Engine Fan Blades," *23rd Structures, Structural Dynamics and Materials Conference, New Orleans, LA, May 10–12, 1982.*

T. Hoang and H. Swenson, "The Challenges of Field Testing the Traffic Management Advisor (TMA) in an Operational Air Traffic Control Facility," NASA TM-112211 (1997).

"Horns and Hollers," CALLBACK From NASA's Aviation Safety Reporting System, No. 359 (Nov. 2009), p. 2.

"Improving Airline Safety," *Spinoff* (Washington, DC: NASA, 1998), p. 62.

Dave Jara and Yoon C. Jung, "Development of the Surface Management System Integrated with CTAS Arrival Tools," *AIAA 5th Aviation Technology, Integration, and Operations Forum, Arlington, TX, Sept. 2005.*

F.W. Jefferson, "ATS-1 VHF Communications Experimentation," FAA 0444707 (1970).

Lisa E. Jones, "Overview of the NASA Systems Approach to Crashworthiness Program," *American Helicopter Society 58th Annual Forum, Montreal, Canada, June 11–13, 2002.*

Jon E. Jonsson and Leland G. Summers, "Crew Procedures and Workload of Retrofit Concepts for Microwave Landing System," NASA CR-181700 (1989).

Y.C. Jung and G.A. Monroe, "Development of Surface Management System Integrated with CTAS Arrival Tool," AIAA 2005-7334 (2005).

David B. Kaber and Lawrence J. Prinzel, III, "Adaptive and Adaptable Automation Design: A Critical Review of the Literature and Recommendations for Future Research," NASA TM-2006-214504 (2006).

A. Klein, P. Kopardekar, M. Rodgers, and H. Kaing, "Airspace Playbook: Dynamic Airspace Reallocation Coordinated with the National Severe Weather Playbook," AIAA 2007-7764 (2007).

P. Kopardekar, K. Bilimoria, and B. Sridhar, "Initial Concepts for Dynamic Airspace Configuration," AIAA 2007-7763 (2007).

K.J. Krzeczowski, T.J. Davis, H. Erzberger, I. Lev-Ram, and C.P. Bergh, "Knowledge-Based Scheduling of Arrival Aircraft in the Terminal Area," AIAA 95-3366 (1995).

Greg Kubat and Don Vandrei, "Airspace Concept Evaluation System, Concept Simulations using Communication, Navigation and Surveillance System Models," *Proceedings of the Sixth Integrated Communications, Navigation and Surveillance Conference & Workshop, Baltimore, May 1–3, 2006.*

I.V. Laudeman, C.L. Brasil, and P. Stassart, "An Evaluation and Redesign of the Conflict Prediction and Trial Planning Planview Graphical User Interface," NASA TM-1998-112227 (1998).

Katherine Lee, "CTAS and NASA Air Traffic Management Fact Sheets for En Route Descent Advisor and Surface Management System," *2004 NATCA Safety Conference, Fort Worth, Apr. 2004.*

K.K. Lee and T.J. Davis, "The Development of the Final Approach Spacing Tool (FAST): A Cooperative Controller-Engineer Design Approach," NASA TM-110359 (1995).

K.K. Lee and B.D. Sanford, "Human Factors Assessment: The Passive Final Approach Spacing Tool (pFAST) Operational Evaluation," NASA TM-208750 (1998).

K.K. Lee, C.M. Quinn, T. Hoang, and B.D. Sanford, "Human Factors Report: TMA Operational Evaluations 1996 & 1998," NASA TM-2000-209587 (2000).

H.Q. Lee and H. Erzberger, "Time-Controlled Descent Guidance Algorithm for Simulation of Advanced ATC Systems," NASA TM-84373 (1983).

H.Q. Lee, P.J. Obrien, L.L. Peach, L. Tobias, and F.M. Willett, Jr., "Helicopter IFR Approaches into Major Terminals Using RNAV, MLS and CDTI," *Journal of Aircraft*, vol. 20 (Aug. 1983).

K. Leiden, J. Kamienski, and P. Kopardekar, "Initial Implications of Automation on Dynamic Airspace Configuration," AIAA 2007-7886 (2007).

Micheline Maynard, "Bird Hazard is Persistent for Planes," *New York Times* (Jan. 19, 2009).

Jim McClenahen, "Virtual Planning at Work: A Tour of NASA Future Flight Central," NASA Tech Server Document ID: 7 (2000).

Michael W. McGreevy, "Searching the ASRS Database Using QUORUM Keyword Search, Phrase Search, Phrase Generation, and Phrase Discovery," NASA TM-2001-210913 (2001).

B.D. McNally, H. Erzberger, R.E. Bach, and W. Chan, "A Controller Tool for Transition Airspace," AIAA 99-4298 (1999).

B.D. McNally and C. Gong, "Concept and Laboratory Analysis of Trajectory-Based Automation for Separation Assurance," AIAA 2006-6600 (2006).

B.D. McNally and J. Walton, "A Holding Function for Conflict Probe Applications," AIAA 2004-4874 (2004).

Larry Meyn and Shannon Zelinski, "Validating the Airspace Concept Evaluation System for Different Weather Days," *AIAA Modeling and Simulation Technologies Conference, Keystone, CO, Aug. 21–24, 2006.*

R.G. Moore, "A Proof-of-Principle Getaway Special Free-Flying Satellite Demonstration," *2nd Symposium on Space Industrialization* (Huntsville, AL: NASA, 1984), p. 349.

Samuel A. Morello and Wendell R. Ricks, "Aviation Safety Issues Database," NASA TM-2009-215706 (2009).

D.G. Moss, P.F. Rieder, B.P. Stapleton, A.D. Thompson, and D.B. Walen, "MLS: Airplane System Modeling," NASA CR-165700 (1981).

E. Mueller, "Experimental Evaluation of an Integrated Datalink and Automation-Based Strategic Trajectory Concept," AIAA 2007-7777 (2007).

K.T. Mueller, R. Bortins, D.R. Schleicher, D. Sweet, and R. Coppenbarger, "Effect of Uncertainty on En Route Descent Advisor (EDA) Predictions," AIAA 2004-6347 (2004).

J. Murphy and J. Robinson, "Design of a Research Platform for En Route Conflict Detection and Resolution," AIAA 2007-7803 (2007).

NASA, "VHF Ranging and Position Fixing Experiment using ATS Satellites," NASA CR-125537 (1971).

"Navigating the Airways," *Spinoff* (Washington, DC: NASA, 1999), p. 50.

R. Nehl and J. Schade, "Update: Concept and Operation of the Performance Data Analysis and Reporting System (PDARS)," *2007 IEEE Aerospace Conference, Big Sky, MT, Mar. 3–10, 2007.*

John O'Neill and Roxana Wales, "Information Management for Airline Operations," Ames Research Center Research and Technology Report (Moffett Field: NASA, 1998).

R.A. Paielli and H. Erzberger, "Conflict Probability Estimation for Free Flight," NASA TM-110411 (1997).

M.T. Palmer, W.H. Rogers, H.N. Press, K.A. Latorella, and T.A. Abbott, "A Crew Centered Flight Deck Design Philosophy for High-Speed Civil Transport (HSCT) Aircraft," NASA TM-109171 (1995).

Sarah Stock Patterson, "Dynamic Flow Management Problems in Air Transportation," NASA CR-97-206395 (1997).

M.M. Poulose, "Microwave Landing System Modeling with Application to Air Traffic Control Automation," *Journal of Aircraft*, vol. 29, no. 3 (May–June 1992).

M.M. Poulose, "Terrain Modeling for Microwave Landing System," *IEEE Transactions on Aerospace and Electronic Systems*, vol. 27 (May 1991).

"Proceedings of the Second NASA Aviation Safety Program Weather Accident Review," NASA CP-2003-210964 (2003).

Doug Rohn and Rick Young, "Aircraft Aging and Durability Project: Technical Plan Summary" (Washington, DC: NASA, 2007).

Marianne Rudisill, "Crew/Automation Interaction in Space Transportation Systems: Lessons Learned from the Glass Cockpit," NASA Langley Research Center (2000).

Glenn Sakamoto, "Intelligent Data Mining Capabilities as Applied to Integrated Vehicle Health Management," *2007 Research and Engineering Annual Report* (Edwards, CA: NASA, 2008), p. 65.

John L. Seubert, "Activities of the FAA Inter-Agency Bird Hazard Committee" (Washington, DC: FAA, 1968).

Jaiwon Shin, "The NASA Aviation Safety Program: Overview," NASA TM-2000-209810 (2000).

J.N. Sivo, W.H. Robbins, and D.M. Stretchberry, "Trends in NASA Communications Satellites," NASA TM-X-68141 (1972).

R.A. Slattery and S.M. Green, "Conflict-Free Trajectory Planning for Air Traffic Control Automation," NASA TM-108790 (1994).

Banavar Sridhar, "Future Air Traffic Management Concepts Evaluation Tool," *Ames Research Center Research and Technology 2000* (Moffett Field: NASA, 2000), p. 5.

Irving Statler, "APMS: An Integrated Set of Tools for Measuring Safety," *ISASI Flight Recorder Working Group Workshop, Santa Monica, CA, Apr. 16–18, 1996.*

Irving Statler, "The Aviation Performance Measuring System (APMS): An Integrated Suite of Tools for Measuring Performance and Safety," *World Aviation Congress, Anaheim, CA, Sept. 28–30, 1998.*

Irving C. Statler, "The Aviation System Monitoring and Modeling (ASMM) Project: A Documentation of its History and Accomplishments: 1999–2005," NASA TP-2007-214556 (2007).

Harry N. Swenson and Danny Vincent, "Design and Operational Evaluation of the Traffic Management Advisor at the Ft. Worth Air Route Traffic Control Center," *United States/Europe Air Traffic Management Research and Development Seminar, Paris, June 16–19, 1997.*

R.G. Synnestvedt, H. Swenson, and H. Erzberger, "Scheduling Logic for Miles-In-Trail Traffic Management," NASA TM-4700 (1995).

J. Thipphavong and S.J. Landry, "The Effects of the Uncertainty of Departures on Multi-Center Traffic Management Advisor Scheduling," AIAA 2005-7301 (2005).

L. Tobias, U. Volckers, and H. Erzberger, "Controller Evaluations of the Descent Advisor Automation Aid," NASA TM-102197 (1989).

Luis Trevino, Deidre E. Paris, and Michael D. Watson, "Intelligent Vehicle Health Management," *41st AIAA–ASME–SAE–ASEE Joint Propulsion Conference and Exhibit, Tucson, July 10–13, 2005.*

R.A. Vivona, M.G. Ballin, S.M. Green, R.E. Bach, and B.D. McNally, "A System Concept for Facilitating User Preferences in En Route Airspace," NASA TM-4763 (1996).

Lane E. Wallace, "Airborne Trailblazer: Two Decades with NASA Langley's 737 Flying Laboratory," NASA SP-4216 (1994).

Earl L. Wiener, "Flight Training and Management for High-Technology Transport Aircraft," NASA CR-199670 (1995).

Earl L. Wiener, "Flight Training and Management for High-Technology Transport Aircraft," NASA CR-200816 (1996).

E.L. Wiener, "Human Factors of Advanced Technology Transport Aircraft," NASA CR-177528 (1989).

E.L. Wiener and R.E. Curry, "Flight-Deck Automation: Promises and Problems," NASA TM-81206 (1980).

D.H. Williams and S.M. Green, "Airborne Four-Dimensional Flight Management in a Time-Based Air Traffic Control Environment," NASA TM-4249 (1991).

D.H. Williams and S.M. Green, "Flight Evaluation of Center-TRACON Automation System Trajectory Prediction Process," NASA TP-1998-208439 (1998).

D.H. Williams and S.M. Green, "Piloted Simulation of an Air-Ground Profile Negotiation Process in a Time-Based Air Traffic Control Environment," NASA TM-107748 (1993).

G.L. Wong, "The Dynamic Planner: The Sequencer, Scheduler, and Runway Allocator for Air Traffic Control Automation," NASA TM-2000-209586 (2000).

Michael S. Wusk, "ARIES: NASA Langley's Airborne Research Facility," AIAA 2002-5822 (2002).

Books and Monographs:

Roger E. Bilstein, *Flight in America, 1900–1983* (Baltimore: Johns Hopkins University Press, 1984).

Robert Burkhardt, *CAB—The Civil Aeronautics Board* (Dulles International Airport, VA: Green Hills Publishing, Co., 1974).

Robert Burkhardt, *The Federal Aviation Administration* (NY: Frederick A. Praeger, 1967).

R.E.G. Davies, *Airlines of the United States Since 1914* (Washington, DC: Smithsonian Institution Press, 1972).

Glenn A. Gilbert, *Air Traffic Control: The Uncrowded Sky* (Washington, DC: Smithsonian Institution Press, 1973).

Najeeb E. Halaby, *Crosswinds: An Airman's Memoir* (Garden City: Doubleday, 1978).

T.A. Heppenheimer, *Turbulent Skies: The History of Commercial Aviation* (Hoboken, NJ: John Wiley & Sons, 1995).

V.D. Hopkin, *Human Factors in Air Traffic Control* (Bristol, PA: Taylor & Francis, 1995).

William E. Jackson, ed., *The Federal Airway System* (Institute of Electrical and Electronics Engineers, 1970).

Robert M. Kane and Allan D. Vose, *Air Transportation* (Dubuque, IA: Kendall/Hunt Publishing Company, 8th ed., 1982).

Richard J. Kent, *Safe, Separated, and Soaring: A History of Federal Civil Aviation Policy, 1961–1972* (Washington, DC: DOT–FAA, 1980).

A. Komons, *Bonfires to Beacons: Federal Civil Aviation Policy Under the Air Commerce Act, 1926–1938* (Washington, DC: DOT–FAA, 1978).

Nick A. Komons, *The Cutting Air Crash* (Washington, DC: DOT–FAA, 1984).

Nick A. Komons, *The Third Man: A History of the Airline Crew Complement Controversy, 1947–1981* (Washington, DC: DOT–FAA, 1987).

William M. Leary, ed., *Aviation's Golden Age: Portraits from the 1920s and 1930s* (Iowa City: University of Iowa Press, 1989).

William M. Leary, ed., *Encyclopedia of American Business History and Biography: The Airline Industry* (New York: Bruccoli Clark Layman and Facts on File, 1992).

NASA, *NASA Chronology on Science, Technology and Policy*, NASA SP-4005 (Washington, DC: NASA, 1965), p. 233.

National Air Traffic Controllers Association, *NATCA: A History of Air Traffic Control* (Washington, DC: National Air Traffic Controllers Association, 2009), p. 16.

Michael S. Nolan, *Fundamentals of Air Traffic Control,* (Pacific Grove, CA, Brooks/Cole Publishing, Co., 1999).

Michael Osborn and Joseph Riggs, eds., *"Mr. Mac:" William P. MacCracken, Jr., on Aviation, Law, Optometry* (Memphis: Southern College of Optometry, 1970).

Dominick Pisano, *To Fill the Skies with Pilots: The Civilian Pilot Training Program, 1939–1949* (Urbana: University of Illinois Press, 1993).

Edmund Preston, *FAA Historical Chronology, Civil Aviation and the Federal Government 1926–1996* (Washington, DC: Federal Aviation Administration, 1996).

Edmund Preston, *Troubled Passage: The Federal Aviation Administration During the Nixon-Ford Term, 1973–1977* (Washington, DC: DOT–FAA, 1987).

Bob Richards, *Secrets From the Tower* (Ithaca, NY: Ithaca Press, 2007).

Billy D. Robbins, *Air Cops: A Personal History of Air Traffic Control* (Lincoln, NE: iUniverse, 2006).

Stuart I. Rochester, *Takeoff at Mid-Century: Federal Civil Aviation Policy in the Eisenhower Years, 1953–1961* (Washington, DC: DOT–FAA, 1976).

Alex Roland, *Model Research: The National Advisory Committee for Aeronautics 1915–1958,* NASA SP-4103 (Washington, DC: NASA, 1985).

Laurence F. Schmeckebier, *The Aeronautics Branch, Department of Commerce: Its History, Activities and Organization* (Washington, DC: The Brookings Institution, 1930).

Page Shamburger, *Tracks Across the Sky* (New York: J.B. Lippincott Company, 1964).

Patricia Strickland, *The Putt-Putt Air Force: The Story of the Civilian Pilot Training Program and The War Training Service, 1939–1944* (DOT–FAA, Aviation Education Staff, 1971).

Scott A. Thompson, *Flight Check!: The Story of FAA Flight Inspection* (DOT–FAA, Office of Aviation System Standards, 1993).

Donald R. Whitnah, *Safer Airways: Federal Control of Aviation, 1926–1966* (IA: Iowa State University Press, 1966).

John R.M. Wilson, *Turbulence Aloft: The Civil Aeronautics Administration Amid Wars and Rumors of Wars, 1938–1953* (Washington, DC: DOT–FAA, 1979).

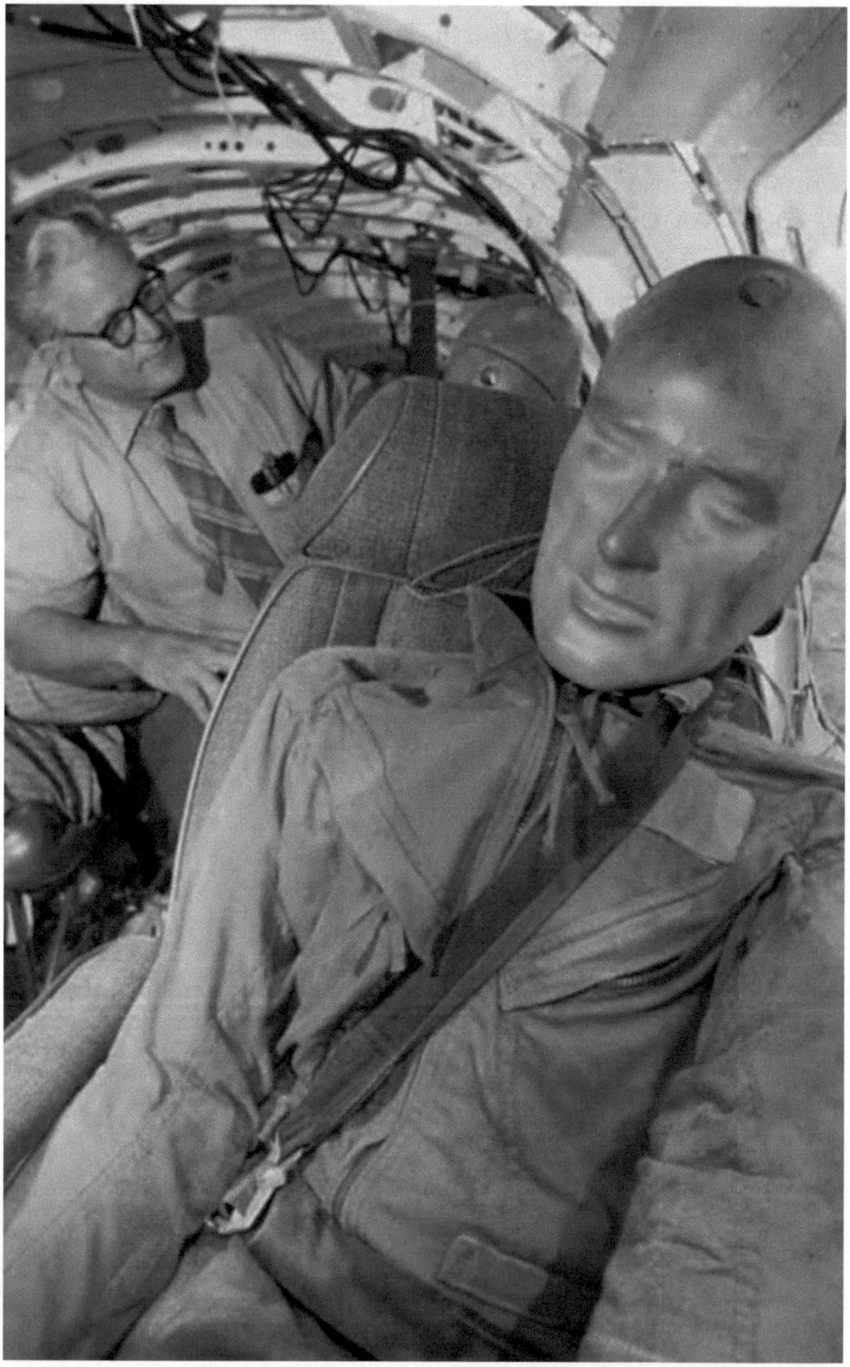

A Langley Research Center human factors research engineer inspects the interior of a light business aircraft after a simulated crash to assess the loads experienced during accidents and develop means of improving survivability. NASA.

CASE 4

Human Factors Research: Meshing Pilots with Planes

Steven A. Ruffin

The invention of flight exposed human limitations. Altitude effects endangered early aviators. As the capabilities of aircraft grew, so did the challenges for aeromedical and human factors researchers. Open cockpits gave way to pressurized cabins. Wicker seats perched on the leading edge of frail wood-and-fabric wings were replaced by robust metal seats and eventually sophisticated rocket-boosted ejection seats. The casual cloth work clothes and hats presaged increasingly complex suits.

AS MERCURY ASTRONAUT ALAN B. SHEPARD, JR., lay flat on his back, sealed in a metal capsule perched high atop a Redstone rocket on the morning of May 5, 1961, many thoughts probably crossed his mind: the pride he felt of becoming America's first man in space, or perhaps, the possibility that the powerful rocket beneath him would blow him sky high . . . in a bad way, or maybe even a greater fear he would "screw the pooch" by doing something to embarrass himself—or far worse—jeopardize the U.S. space program.

After lying there nearly 4 hours and suffering through several launch delays, however, Shepard was by his own admission not thinking about any of these things. Rather, he was consumed with an issue much more down to earth: his bladder was full, and he desperately needed to relieve himself. Because exiting the capsule was out of the question at this point, he literally had no place to go. The designers of his modified Goodrich U.S. Navy Mark IV pressure suit had provided for nearly every contingency imaginable, but not this; after all, the flight was only scheduled to last a few minutes.

Finally, Shepard was forced to make his need known to the controllers below. As he candidly described later, "You heard me, I've got to pee. I've been in here forever."[1] Despite the unequivocal reply of "No!" to

1. Alan Shepard and Deke Slayton, with Jay Barbree and Howard Benedict, *Moon Shot: The Inside Story of America's Race to the Moon* (Atlanta: Turner Publishers, Inc., 1994), p. 107.

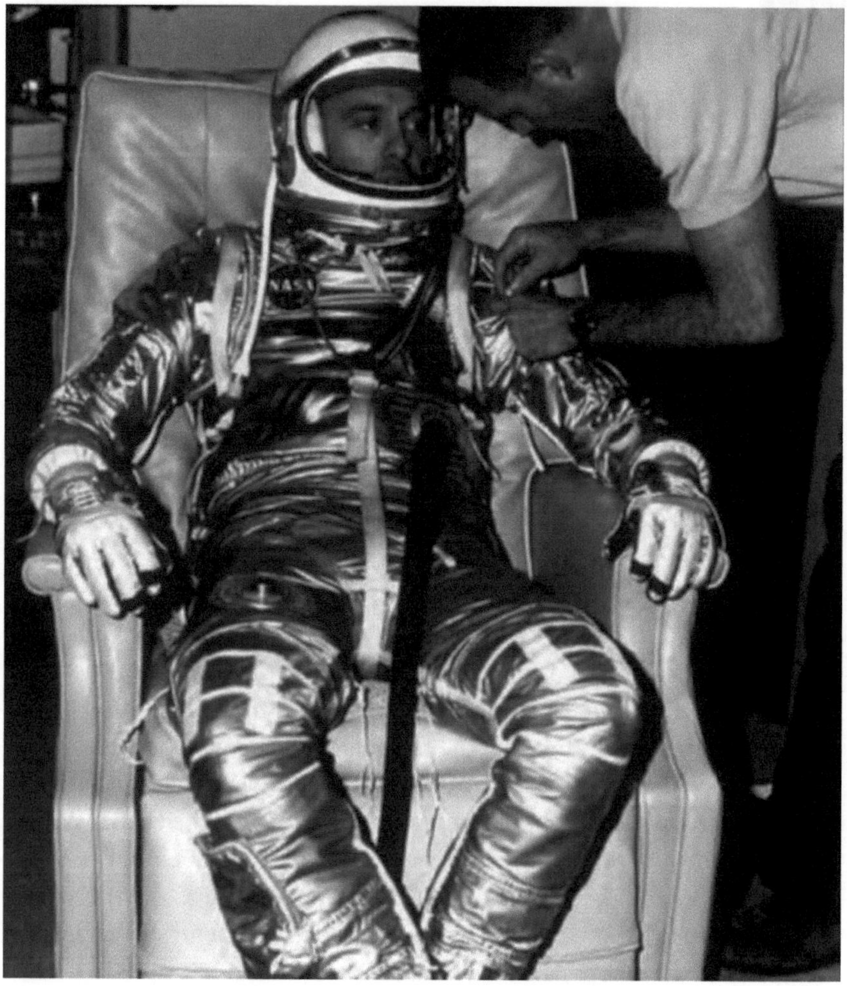

Mercury 7 astronaut Alan B. Shepard, Jr., preparing for his historic flight of May 5, 1961. His gleaming silver pressure suit had all the bells and whistles . . . except for one. NASA.

his request, Shepard's bladder gave him no alternative but to persist. Historic flight or not, he had to go—and now.

When the powers below finally accepted that they had no choice, they gave the suffering astronaut a reluctant thumbs up: so, "pee," he did . . . all over his sensor-laden body and inside his gleaming silver spacesuit. And then, while the world watched—unaware of this behind-the-scenes drama—Shepard rode his spaceship into history . . . drenched in his own urine.

This inauspicious moment should have been something of an epiphany for the human factors scientists who worked for the newly formed

National Aeronautics and Space Administration (NASA). It graphically pointed out the obvious: human requirements—even the most basic ones—are not optional; they are real, and accommodations must always be made to meet them. But NASA's piloted space program had advanced so far technologically in such a short time that this was only one of many lessons that the Agency's planners had learned the hard way. There would be many more in the years to come.

As described in the Tom Wolfe book and movie of the same name, *The Right Stuff*, the first astronauts were considered by many of their contemporary non-astronaut pilots—including the ace who first broke the sound barrier, U.S. Air Force test pilot Chuck Yeager—as little more than "spam in a can."[2] In fact, Yeager's commander in charge of all the test pilots at Edwards Air Force Base had made it known that he didn't particularly want his top pilots volunteering for the astronaut program; he considered it a "waste of talent."[3] After all, these new astronauts—more like lab animals than pilots—had little real function in the early flights, other than to survive, and sealed as they were in their tiny metal capsules with no realistic means of escape, the cynical "spam in a can" metaphor was not entirely inappropriate.

But all pilots appreciated the dangers faced by this new breed of American hero: based on the space program's much-publicized recent history of one spectacular experimental launch failure after another, it seemed like a morbidly fair bet to most observers that the brave astronauts, sitting helplessly astride 30 tons of unstable and highly explosive rocket fuel, had a realistic chance of becoming something akin to America's most famous canned meat dish. It was indeed a dangerous job, even for the 7 overqualified test-pilots-turned-astronauts who had been so carefully chosen from more than 500 actively serving military test pilots.[4] Clearly, piloted space flight had to become considerably more human-friendly if it were to become the way of the future.

NASA had existed less than 3 years before Shepard's flight. On July 19, 1958, President Dwight D. Eisenhower signed into law the National Aeronautics and Space Act of 1958, and chief among the provisions was the establishment of NASA. Expanding on this act's stated purpose of

2. Tom Wolfe, *The Right Stuff* (Toronto: McGraw-Hill Ryerson, Ltd., 1979), p. 78.

3. Ibid.

4. John A. Pitts, The *Human Factor: Biomedicine in the Manned Space Program to 1980*, NASA SP-4213 (Washington, DC: NASA, 1985), p. 18.

conducting research into the "problems of flight within and outside the earth's atmosphere" was an objective to develop vehicles capable of carrying—among other things—"living organisms" through space.[5]

Because this official directive clearly implied the intention of sending humans into space, NASA was from its inception charged with formulating a piloted space program. Consequently, within 3 years after it was created, the budding space agency managed to successfully launch its first human, Alan Shepard, into space. The astronaut completed NASA Mercury mission MR-3 to become America's first man in space. Encapsulated in his Freedom 7 spacecraft, he lifted off from Cape Canaveral, FL, and flew to an altitude of just over 116 miles before splashing down into the Atlantic Ocean 302 miles downrange.[6] It was only a 15-minute suborbital flight and, as related above, not without problems, but it accomplished its objective: America officially had a piloted space program.

This was no small accomplishment. Numerous major technological barriers had to be surmounted during this short time before even this most basic of piloted space flights was possible. Among these obstacles, none was more challenging than the problems associated with maintaining and supporting human life in the ultrahostile environment of space. Thus, from the beginning of the Nation's space program and continuing to the present, human factors research has been vital to NASA's comprehensive research program.

The Science of Human Factors

To be clear, however, NASA did not invent the science of human factors. Not only has the term been in use long before NASA ever existed, the concept it describes has existed since the beginning of mankind. Human factors research encompasses nearly all aspects of science and technology and therefore has been described with several different names. In simplest terms, human factors studies the interface between humans and the machines they operate. One of the pioneers of this science, Dr. Alphonse Chapanis, provided a more inclusive and descriptive definition:

5. "National Aeronautics and Space Act of 1958," Public Law No. 85-568, 72 Stat., 426-438, July 29, 1958, Record Group 255, National Archives and Records Administration, Washington, DC, Introduction and Sec. 102(d)(3).

6. Loyd S. Swenson, Jr., James M. Grimwood, and Charles C. Alexander, *This New Ocean: A History of Project Mercury* (Washington, DC: NASA, 1966), p. 341.

"Human factors discovers and applies information about human behavior, abilities, limitations, and other characteristics to the design of tools, machines, systems, tasks, jobs, and environments for productive, safe, comfortable, and effective human use."[7] The goal of human factors research, therefore, is to reduce error, while increasing productivity, safety, and comfort in the interaction between humans and the tools with which they work.[8]

As already suggested, the study of human factors involves a myriad of disciplines. These include medicine, physiology, applied psychology, engineering, sociology, anthropology, biology, and education.[9] These in turn interact with one another and with other technical and scientific fields, as they relate to behavior and usage of technology. Human factors issues are also described by many similar—though not necessarily synonymous—terms, such as human engineering, human factors engineering, human factors integration, human systems integration, ergonomics, usability, engineering psychology, applied experimental psychology, biomechanics, biotechnology, man-machine design (or integration), and human-centered design.[10]

The Changing Human Factors Dimension Over Time

The consideration of human factors in technology has existed since the first man shaped a wooden spear with a sharp rock to help him grasp it more firmly. It therefore stands to reason that the dimension of human factors has changed over time with advancing technology—a trend that has accelerated throughout the 20th century and into the current one.[11]

Man's earliest requirements for using his primitive tools and weapons gave way during the Industrial Revolution to more refined needs in operating more complicated tools and machines. During this period, the emergence of more complex machinery necessitated increased consideration of the needs of the humans who were to operate this machinery—even

7. Alphonse Chapanis, "Some reflections on progress," paper presented at the Proceedings of the *Human Factors Society 29th Annual Meeting* (Santa Monica, CA: Human Factors Society, 1985), pp. 1–8.
8. Christopher D. Wickens, Sallie E. Gordon, and Yili Liu, *An Introduction to Human Factors Engineering* (New York: Longman, 1998), p. 2.
9. Peggy Tillman and Barry Tillman, *Human Factors Essentials: An Ergonomics Guide for Designers, Engineers, Scientists, and Managers* (New York: McGraw-Hill, 1991), p. 4.
10. Ibid., p. 5.
11. Ibid., pp. 9–10.

if it was nothing more complicated than providing a place for the operator to sit, or a handle or step to help this person access instruments and controls. In the years after the Industrial Revolution, human factors concerns became increasingly important.[12]

The Altitude Problem

The interface between humans and technology was no less important for those early pioneers, who, for the first time in history, were starting to reach for the sky. Human factors research in aeronautics did not, however, begin with the Wright brothers' first powered flight in 1903; it began more than a century earlier.

Much of this early work dealt with the effects of high altitude on humans. At greater heights above the Earth, barometric pressure decreases. This allows the air to expand and become thinner. The net effect is diminished breathable oxygen at higher altitudes. In humans operating high above sea level without supplemental oxygen, this translates to a medical condition known as hypoxia. The untoward effects on humans of hypoxia, or altitude sickness, had been known for centuries—long before man ever took to the skies. It was a well-known entity to ancient explorers traversing high mountains, thus the still commonly used term mountain sickness.[13]

The world's first aeronauts—the early balloonists—soon noticed this phenomenon when ascending to higher altitudes; eventually, some of the early flying scientists began to study it. As early as 1784, American physician John Jeffries ascended to more than 9,000 feet over London with French balloonist Jean Pierre Blanchard.[14] During this flight, they recorded changes in temperature and barometric pressure and became perhaps the first to record an "aeromedical" problem, in the form of ear pain associated with altitude changes.[15] Another early flying doctor, British physician John Shelton, also wrote of the detrimental effects of high-altitude flight on humans.[16]

12. Ibid., pp. 9–10.

13. John B. West, *High Life: A History of High-Altitude Physiology and Medicine* (New York: Oxford University Press, 1998), pp. xi–xv.

14. Ibid., pp. 51–52.

15. Ibid., p. 52.

16. Eloise Engle and Arnold S. Lott, *Man in Flight: Biomedical Achievements in Aerospace* (Annapolis: Leeward, 1979), pp. 31–34.

During the 1870s—with mankind's first powered, winged human flight still decades in the future—French physiologist Paul Bert conducted important research on the manner in which high-altitude flight affects living organisms. Using the world's first pressure chamber, he studied the effects of varying barometric pressure and oxygen levels on dogs and later humans—himself included. He conducted 670 experiments at simulated altitudes of up to 36,000 feet. His findings clarified the effects of high-altitude conditions on humans and established the requirement for supplemental oxygen at higher altitudes.[17] Later studies by other researchers followed, so that by the time piloted flight in powered aircraft became a reality at Kitty Hawk, NC, on December 17, 1903, the scientific community already had a substantial amount of knowledge concerning the physiology of high-altitude flight. Even so, there was much more to be learned, and additional research in this important area would continue in the decades to come.

Early Flight and the Emergence of Human Factors Research

During the early years of 20th century aviation, it became apparent that the ability to maintaining human life and function at high altitude was only one of many human factors challenges associated with powered flight. Aviation received its first big technological boost during the World War I years of 1914–1918.[18] Accompanying this advancement was a new set of human-related problems associated with flight.[19] As a result of the massive, nearly overnight wartime buildup, there were suddenly tens of thousands of newly trained pilots worldwide, flying on a daily basis in aircraft far more advanced than anyone had ever imagined possible. In the latter stages of the war, aeronautical know-how had become so sophisticated that aircraft capabilities had surpassed that of their human operators. These Great War pilots, flying open-cockpit aircraft capable of altitudes occasionally exceeding 20,000 feet, began to routinely

17. West, *High Life*, pp. 62–73; Engle and Lott, *Man in Flight*, pp. 34–37.

18. Richard P. Hallion, *Rise of the Fighter Aircraft, 1914–1918* (Annapolis, MD: The Nautical & Aviation Publishing Company of America, 1984), pp. iii–iv.

19. Steven A. Ruffin, "Flying in the Great War: Rx for Misery; An Overview of the Medical and Physiological and Psychological Aspects of Combat Flying During the First World War," *Over the Front*, vol. 14, no. 2 (summer 1999), pp. 115–124, and vol. 17, no. 2 (summer 2002), pp. 117–136.

suffer from altitude sickness and frostbite.[20] They were also experiencing pressure-induced ear, sinus, and dental pain, as well as motion sickness and vertigo.[21] In addition, these early open-cockpit pilots endured the effects of ear-shattering noise, severe vibration, noxious engine fumes, extreme acceleration or gravitational g forces, and a constant hurricane-force wind blast to their faces.[22] And as if these physical challenges were not bad enough, these early pilots also suffered devastating injuries from crashes in aircraft unequipped with practically any basic safety features.[23] Less obvious, but still a very real human problem, these early high flyers were exhibiting an array of psychological problems, to which these stresses undoubtedly contributed.[24] Indeed, though proof of the human limitations in flying during this period was hardly needed, the British found early in the war that only 2 percent of aviation fatalities came at the hands of the enemy, while 90 percent were attributed to pilot deficiencies; the remainder came from structural and engine failure, and a variety of lesser causes.[25] By the end of World War I, it was painfully apparent to flight surgeons, psychologists, aircraft designers, and engineers that much additional work was needed to improve the human-machine interface associated with piloted flight.

Because of the many flight-related medical problems observed in airmen during the Great War, much of the human factors research accomplished during the following two decades leading to the Second World War focused largely on the aeromedical aspects of flight. Flight surgeons, physiologists, engineers, and other professionals of this period devoted themselves to developing better life-support equipment and other protective gear to improve safety and efficiency during flight operations. Great emphasis was also placed on improving pilot selection.[26]

---

20. Harry G. Armstrong, *Principles and Practice of Aviation Medicine* (Baltimore: Williams & Wilkins, 1939), pp. 38, 279; William H. Wilmer, *Aviation Medicine in the A.E.F.* (Washington, DC: Government Printing Office, 1920), pp. 61–62; Armstrong, *Principles and Practice of Aviation Medicine,* pp. 184–187.
21. Wilmer, *Aviation Medicine in the A.E.F.*, p. 58.
22. Ruffin, "Flying in the Great War"; Russell R. Burton, "G-Induced Loss of Consciousness: Definition, History, Current Status," *Aviation, Space, and Environmental Medicine,* Jan. 1988, p. 2.
23. Wilmer, *Aviation Medicine in the A.E.F.*, p. 217.
24. Armstrong, *Principles and Practice of Aviation Medicine,* p. 6.
25. U.S. Army Air Service, *Air Service Medical* (Washington, DC: Government Printing Office, 1919).
26. Harry W. Orlady and Linda M. Orlady, *Human Factors in Multi-Crew Flight Operations* (Brookfield, VT: Ashgate Publishing, Ltd., 1999), pp. 46–47.

Of particular note during the interwar period of the 1920s and 1930s were several piloted high-altitude balloon flights conducted to further investigate conditions in the upper part of the Earth's atmosphere known as the stratosphere. Perhaps the most ambitious and fruitful of these was the 1935 joint U.S. Army Air Corps/National Geographic Society flight that lifted off from a South Dakota Black Hills natural geological depression known as the "Stratobowl." The two Air Corps officers, riding in a sealed metal gondola—much like a future space capsule—with a virtual laboratory full of scientific monitoring equipment, traveled to a record altitude of 72,395 feet.[27] Little did they know it at the time, but the data they collected while aloft would be put to good use decades later by human factors scientists in the piloted space program. This included information about cosmic rays, the distribution of ozone in the upper atmosphere, and the spectra and brightness of sun and sky, as well as the chemical composition, electrical conductivity, and living spore content of the air at that altitude.[28]

Although the U.S. Army Air Corps and Navy conducted the bulk of the human factors research during this interwar period of the 1920s and 1930s, another important contributor was the National Advisory Committee for Aeronautics (NACA). Established in 1915, the NACA was actively engaged in a variety of aeronautical research for more than 40 years. Starting only with a miniscule $5,000 budget and an ambitious mission to "direct and conduct research and experimentation in aeronautics, with a view to their practical solution,"[29] the NACA became one of this country's leading aeronautical research agencies and remained so up until its replacement in 1958 by the newly established space agency NASA. The work that the NACA accomplished during this era in design engineering and life-support systems, in cooperation with the U.S. military and other agencies and institutions, contributed greatly to information and technology that would become vital to the piloted space program, still decades—and another World War—in the future.[30]

27. National Geographic Society, *The National Geographic Society-U.S. Army Air Corps Stratosphere Flight of 1935*, (Washington, DC: National Geographic Society, 1936); Steven A. Ruffin, "Explorer Over Dakota: From Stratobowl to Stratosphere," *Aviation History*, May 1996, pp. 22–28, 72.
28. *The National Geographic Society-U.S. Army Air Corps Stratosphere Flight of 1935*; Ruffin, "Explorer Over Dakota."
29. Public Law 271, 63rd Congress, 3rd session, Mar. 3, 1915 (38 Stat. 930).
30. Frank W. Anderson, Jr., *Orders of Magnitude: A History of NACA and NASA, 1915–1976*, NASA SP-4403 (Washington, DC: NASA, 1976).

World War II and the Birth of Human Factors Engineering

During World War II, human factors was pushed into even greater prominence as a science. During this wartime period of rapidly advancing military technology, greater demands were being placed on the users of this technology. Success or failure depended on such factors as the operators' attention span, hand-eye coordination, situational awareness, and decision-making skills. These demands made it increasingly challenging for operators of the latest military hardware—aircraft, tanks, ships, and other complex military machinery—to operate their equipment safely and efficiently.[31] Thus, the need for greater consideration of human factors issues in technological design became more obvious than ever before; as a consequence, the discipline of human engineering emerged.[32] This branch of human factors research is involved with finding ways of designing "machines, operations, and work environments so that they match human capacities and limitations." Or, to put it another way, it is the "engineering of machines for human use and the engineering of human tasks for operating machines."[33]

During World War II, no area of military technology had a more critical need for both human factors and human engineering considerations than did aviation.[34] Many of the biomedical problems afflicting airmen in the First World War had by this time been addressed, but new challenges had appeared. Most noticeable were the increased physiological strains for air crewmen who were now flying faster, higher, for longer periods of time, and—because of wartime demands—more aggressively than ever before. High-performance World War II aircraft were capable of cruising several times faster than they were in the previous war and were routinely approaching the speed of sound in steep dives. Because of these higher speeds, they were also exerting more than enough gravitational g forces during turns and pullouts to render pilots almost instantly unconscious. In addition, some of these advanced air-

31. David Meister, *The History of Human Factors and Ergonomics* (Mahwah, NJ: Lawrence Erlbaum Associates, 1999), pp. 151–153.

32. Alphonse Chapanis, *Research Techniques in Human Engineering* (Baltimore: The Johns Hopkins Press, 1958), p. vii; R.A. Behan and H.W. Wendhausen, *Some NASA Contributions to Human Factors Engineering*, NASA SP-5117 (Washington, DC: NASA, 1973), pp. 1–2.

33. Chapanis, *Research Techniques in Human Engineering*, p. vii.

34. Earl L. Wiener and David C. Nagel, *Human Factors in Aviation* (San Diego: Academic Press, Inc., 1988), p. 7.

craft could climb high into the stratosphere to altitudes exceeding 40,000 feet and were capable of more hours of flight-time endurance than their human operators possessed. Because of this phenomenal increase in aircraft technology, human factors research focused heavily on addressing the problems of high-performance flight.[35]

The other aspect of the human factors challenge coming into play involved human engineering concerns. Aircraft of this era were exhibiting a rapidly escalating degree of complexity that made flying them—particularly under combat conditions—nearly overwhelming. Because of this combination of challenges to the mortals charged with operating these aircraft, human engineering became an increasingly vital aspect of aircraft design.[36]

During these wartime years, high-performance military aircraft were still crashing at an alarmingly high rate, in spite of rigorous pilot training programs and structurally well-designed aircraft. It was eventually accepted that not all of these accidents could be adequately explained by the standard default excuse of "pilot error." Instead, it became apparent that many of these crashes were more a result of "designer error" than operator error.[37] Military aircraft designers had to do more to help the humans charged with operating these complex, high-performance aircraft. Thus, not only was there a need during these war years for greater human safety and life support in the increasingly hostile environment aloft, but the crews also needed better-designed cockpits to help them perform the complex tasks necessary to carry out their missions and safely return.[38]

In earlier aircraft of this era, design and placement of controls and gauges tended to be purely engineer-driven; that is, they were constructed to be as light as possible and located wherever designers could most conveniently place them, using the shortest connections and simplest attachments. Because the needs of the users were not always taken into account, cockpit designs tended not to be as user-friendly as they should have been. This also meant that there was no attempt to standardize

35. Jefferson M. Koonce, "A Historical Overview of Human Factors in Aviation," in Daniel J. Garland, John A. Wise, and V. David Hopkin, etc., *Handbook of Aviation Human Factors* (Mahwah, NJ: Lawrence Erlbaum Associates, 1999), pp. 3–13.

36. W.F. Moroney, "The Evolution of Human Engineering: A Selected Review," in J. Weimer, ed., *Research Techniques in Human Engineering* (Englewood Cliffs, NJ: Prentice-Hall, 1995), pp. 1–19; Chapanis, *Research Techniques in Human Engineering*, pp. 1–2.

37. Alphonse Chapanis, *The Chapanis Chronicles: 50 Years of Human Factors Research, Education, and Design* (Santa Barbara, CA: Aegean Publishing, Co., 1999), pp. 15–16.

38. Moroney, "The Evolution of Human Engineering: A Selected Review," pp. 1–19.

the cockpit layout between different types of aircraft. This contributed to longer and more difficult transitions to new aircraft with different instrument and control arrangements. This disregard for human needs in cockpit design resulted in decreased aircrew efficiency and performance, greater fatigue, and, ultimately, more mistakes.[39]

An example of this lack of human consideration in cockpit design was one that existed in an early model Boeing B-17 bomber. In this aircraft, the flap and landing gear handles were similar in appearance and proximity, and therefore easily confused. This unfortunate arrangement had already inducted several pilots into the dreaded "gear-up club," when, after landing, they inadvertently retracted the landing gear instead of the intended flaps. To address this problem, a young Air Corps physiologist and Yale psychology Ph.D. named Alphonse Chapanis proved that the incidence of such pilot errors could be greatly reduced by more logical control design and placement. His ingeniously simple solution of moving the controls apart from one another and attaching different shapes to the various handles allowed pilots to determine by touch alone which control to activate. This fix—though not exactly rocket science—was all that was needed to end a dangerous and costly problem.[40]

As a result of a host of human-operator problems, such as those described above, wartime aircraft design engineers began routinely working with industrial and engineering psychologists and flight surgeons to optimize human utilization of this technology. Thus was born in aviation the concept of human factors in engineering design, a discipline that would become increasingly crucial in the decades to come.[41]

The Jet Age: Man Reaches the Edge of Space

By the end of the Second World War, aviation was already well into the jet age, and man was flying yet higher and faster in his quest for space. During the years after the end of the war, human factors research continued to evolve in support of this movement. A multiplicity of human and animal studies were conducted during this period by military, civilian, and Government researchers to learn more about such problems as acceleration and deceleration, emergency egress from high-speed jet aircraft, explosive decompression, pressurization of suits and cockpits,

39. Wiener and Nagel, *Human Factors in Aviation*, pp. 7–9.

40. Chapanis, *The Chapanis Chronicles*, pp. 15–16.

41. Engle and Lott, *Man in Flight*, p. 79.

and the biological effects of various types of cosmic rays. In addition, a significant amount of work concentrated on instrument design and cockpit display.[42]

During the years leading up to America's space program, humans were already operating at the edge of space. This was made possible in large part by the cutting-edge performance of the NACA–NASA high-speed, high-altitude rocket "X-planes"—progressing from the Bell X-1, in which Chuck Yeager became the first person to officially break the sound barrier, on October 14, 1947, to the phenomenal hypersonic X-15 rocket plane, which introduced man to true space flight.[43]

These unique experimental rocket-propelled aircraft, developed and flown from 1946 through 1968, were instrumental in helping scientists understand how best to sustain human life during high-speed, high-altitude flight.[44] One of the more important human factors developments employed in the first of this series, the Bell X-1 rocket plane, was the T-1 partial pressure suit designed by Dr. James Henry of the University of Southern California and produced by the David Clark Company.[45] This suit proved its worth during an August 25, 1949, test flight, when X-1 pilot Maj. Frank K. "Pete" Everest lost cabin pressure at an altitude of more than 65,000 feet. His pressure suit automatically inflated, and though it constricted him almost to the point of incapacitation, it nevertheless kept him alive until he could descend. He thus became the first pilot saved by the emergency use of a pressure suit.[46]

During the 1950s and 1960s, the NACA and NASA tested several additional experimental rocket planes after the X-1 series; however, the most famous and accomplished of these by far was the North American X-15. During the 199 flights this phenomenal rocket plane made from 1959 to 1968, it carried its pilots to unprecedented hypersonic speeds of

42. Wiener and Nagel, *Human Factors in Aviation*, p. 9.
43. J. Miller, *The X-Planes* (Arlington, TX: Aerofax, Inc., 1988); Lane E. Wallace, *Flights of Discovery: 60 Years at the Dryden Flight Research Center*, NASA SP-2006-4318 (Washington, DC: NASA, 2006), pp. 33–72.
44. Ibid.
45. Kenneth S. Thomas and Harold J. McMann, *U.S. Spacesuits* (Chichester UK, Praxis Publishing, Ltd., 2006), p. 8; Lillian D. Kozloski, *U.S. Space Gear: Outfitting the Astronaut* (Washington, DC: Smithsonian Institution Press, 1994), pp. 26–31.
46. Richard P. Hallion, *Supersonic Flight: Breaking the Sound Barrier and Beyond* (London: Brassey's, 1997), pp. 130, 214.

nearly 7 times the speed of sound (4,520 mph) and as high as 67 miles above the Earth.[47] The wealth of information these flights continued to produce, nearly right up until the first piloted Moon flight, enabled technology vital to the success of the NASA piloted space program.

One of the X-15 program's more important challenges was how to keep its pilots alive and functioning in a craft traveling through space at hypersonic speeds. The solution was the development of a full-pressure suit capable of sustaining its occupant in the vacuum of space yet allowing him sufficient mobility to perform his duties. This innovation was an absolute must before human space flight could occur.

The MC-2 full-pressure suit provided by the David Clark Co. met these requirements, and more.[48] The suit in its later forms, the A/P-22S-2 and A/P-22S-6, not only provided life-sustaining atmospheric pressure, breathable oxygen, temperature control, and ventilation, but also a parachute harness, communications system, electrical leads for physiological monitoring, and an antifogging system for the visor. Even with all these features, the pilot still had enough mobility to function inside the aircraft. By combining the properties of this pressure suit with those of the X-15 ejection seat, the pilot at least had a chance for emergency escape from the aircraft. This suit was so successful that it was also adapted for use in high-altitude military aircraft, and it served as the template for the suit developed by B.F. Goodrich for the Mercury and Gemini piloted space programs.[49]

The development of a practical spacesuit was not the only human factors contribution of the X-15 program. Its pioneering emphasis on the physiological monitoring of the pilot also formed the basis of that used in the piloted space program. These in-flight measurements and later analysis were an important aspect of each X-15 flight. The aeromedical data collected included heart and respiratory rates, electrocardiograph, skin temperature, oxygen flow, suit pressure, and blood pressure.

---

47. Milton O. Thompson, *At the Edge of Space* (Washington, DC: Smithsonian Institution Press, 1992), pp. 281–355.

48. Dennis R. Jenkins, *Hypersonics Before the Shuttle: A Concise History of the X-15 Research Airplane,* NASA SP-2000-4518, *Monographs in Aerospace History,* No. 18 (Washington: GPO, June 2000), pp. 39, 70; Kozloski, *U.S. Space Gear*, pp. 31–40.

49. Wendell H. Stillwell, *X-15 Research Results* (Washington, DC: NASA, 1965), pp. 86–88; Jenkins, *Hypersonics Before the Shuttle,* p. 70; Thomas and McCann, *U.S. Spacesuits,* pp. 25–31; Kozloski, *U.S. Space Gear,* pp. 31–40.

The X-15 on lakebed with B-52 mother ship flying overhead. Lessons learned from this phenomenal rocket plane helped launch humans into space. NASA.

Through this information, researchers were able to better understand human adaptation to hypersonic high-altitude flight.[50]

The many lessons learned from these high-performance rocket planes were invaluable in transforming space flight into reality. From a human factors standpoint, these flights provided the necessary testbed for ushering humans into the deadly environment of high-altitude, high-speed flight—and ultimately, into space.

Another hazardous type of human research activity conducted after World War II that contributed to piloted space operations was the series of U.S. military piloted high-altitude balloon flights conducted in the 1950s and 1960s. Most significant among these were the U.S. Navy Strato-Lab flights and the Air Force Manhigh and Excelsior programs.[51]

50. Stillwell, *X-15 Research Results*, p. 89.

51. Craig Ryan, *The Pre-Astronauts: Manned Ballooning on the Threshold of Space* (Annapolis, MD: Naval Institute Press, 1995); Gregory P. Kennedy, *Touching Space: The Story of Project Manhigh* (Atglen, PA: Schiffer Military History, 2007); National Museum of the United States Air Force Fact Sheet, "Excelsior Gondola," *http://www.nationalmuseum.af.mil/factsheets/factsheet.asp?id=562*, accessed Oct. 7, 2009.

The information these flights provided paved the way for the design of space capsules and astronaut pressure suits, and they gained important biomedical and astronomical data.

The Excelsior program, in particular, studied the problem of emergency egress high in the stratosphere. During the flight of August 16, 1960, Air Force pilot Joseph Kittinger, Jr., ascended in Excelsior III to an altitude of 102,800 feet before parachuting to Earth. During this highest-ever jump, Kittinger went into a freefall for a record 4 minutes 36 seconds and attained a record speed for a falling human body outside of an aircraft of 614 mph.[52] Although, thankfully, no astronaut has had to repeat this performance, Kittinger showed how it could be done.

Yet another human research contribution from this period that proved to be of great value to the piloted space program was the series of impact deceleration tests conducted by U.S. Air Force physician Lt. Col. John P. Stapp. While strapped to a rocket-propelled research sled on a 3,500-foot track at Holloman Air Force Base (AFB), NM, Stapp made 29 sled rides during the years of 1947–1954. During these, he attained speeds of up to 632 mph, making him—at least in the eyes of the press—the fastest man on Earth, and he withstood impact deceleration forces of as high as 46 times the force of gravity. To say this work was hazardous would be an understatement. While conducting this research, Stapp suffered broken bones, concussions, bruises, retinal hemorrhages, and even temporary blindness. But the knowledge he gained about the effects of acceleration and deceleration forces was invaluable in delineating the human limitations that astronauts would have while exiting and reentering the Earth's atmosphere.[53]

All of these flying and research endeavors involved great danger for the humans directly involved in them. Injuries and fatalities did occur, but such was the dedication of pioneers such as Stapp and the pilots of these trailblazing aircraft. The knowledge they gained by putting their lives on the line—knowledge that could have been acquired in no other way—would be essential to the establishment of the piloted space program, looming just over the horizon.

52. Ibid.

53. David Bushnell, *History of Research in Space Biology and Biodynamics, 1946–58*, AF Missile Dev. Center, Holloman AFB, NM (1958); Engle and Lott, *Man in Flight*, pp. 210–215; Eugene M. Emme, *Aeronautics and Astronautics: An American Chronology of Science and Technology in the Exploration of Space, 1915–1960* (Washington, DC: NASA, 1961), pp. 62, 68.

NASA Arrives: Taking Human Factors Research to the Next Level

It is therefore abundantly evident that when the NACA handed over the keys of its research facilities to NASA on October 1, 1958, the Nation's new space agency began operations with a large database of information relating to the human factors and human engineering aspects of piloted flight. But though this mass of accumulated knowledge and technology was of inestimable value, the prospect of taking man to the next level, into the great unknown of outer space, was a different proposition from any ever before tackled by aviation research.[54] No one had yet comprehensively dealt with such human challenges as the effects of long-term weightlessness, exposure to ionizing radiation and extreme temperature changes, maintaining life in the vacuum of space, or withstanding prolonged impact deceleration forces encountered by humans violently reentering the Earth's atmosphere.[55]

NASA began operations in 1958 with a final parting report from the NACA's Special Committee on Space Technology. This report recommended several technical areas in which NASA should proceed with its human factors research. These included acceleration, high-intensity radiation in space, cosmic radiation, ionization effects, human information processing and communication, displays, closed-cycle living, space capsules, and crew selection and training.[56] This Committee's Working Group on Human Factors and Training further suggested that all experimentation consider crew selection, survival, safety, and efficiency.[57] With that, America's new space agency had its marching orders. It proceeded to assemble "the largest group of technicians and greatest body of knowledge ever used to define man's performance on the ground and in space environments."[58]

Thus, from NASA's earliest days, it has pioneered the way in human-centered aerospace research and technology. And also from its beginning—and extending to the present—it has shared the benefits of this research with the rest of the world, including the same industry that contributed so much to NASA during its earliest days—aeronautics. This 50-year storehouse of knowledge produced by NASA human factors research has been shared with all areas of the aviation community—

54. Behan and Wendhausen, *Some NASA Contributions to Human Factors Engineering*, p. 5.
55. Pitts, *The Human Factor*, pp. 8–10.
56. Engle and Lott, *Man in Flight*, p. 130.
57. Ibid., p. 131.
58. Behan and Wendhausen, *Some NASA Contributions to Human Factors Engineering*, p. 5.

4

both the Department of Defense (DOD) and all realms of civil aviation, including the Federal Aviation Administration (FAA), the National Transportation and Safety Board (NTSB), the airlines, general aviation, aircraft manufacturing companies, and producers of aviation-related hardware and software.

Bioastronautics, Bioengineering, and Some Hard-Learned Lessons

Over the past 50 years, NASA has indeed encountered many complex human factors issues. Each of these had to be resolved to make possible the space agency's many phenomenal accomplishments. Its initial goal of putting a man into space was quickly accomplished by 1961. But in the years to come, NASA progressed beyond that at warp speed—at least technologically speaking.[59] By 1973, it had put men into orbit around the Earth; sent them outside the relative safety of their orbiting craft to "walk" in space, with only their pressurized suit to protect them; sent them around the far side of the Moon and back; placed them into an orbiting space station, where they would live, function, and perform complex scientific experiments in weightlessness for months at a time; and, certainly most significantly, accomplished mankind's greatest technological feat by landing humans onto the surface of the Moon—not just once, but six times—and bringing them all safely back home to Mother Earth.[60]

NASA's magnificent accomplishments in its piloted space program during the 1960s and 1970s—nearly unfathomable only a few years before—thus occurred in large part as a result of years of dedicated human factors research. In the early years of the piloted space program, researchers from the NASA Environmental Physiology Branch focused on the biodynamics—or more accurately, the bioastronautics–of man in space. This discipline, which studies the biological and medical effects of space flight on man, evaluated such problems as noise, vibration, acceleration and deceleration, weightlessness, radiation, and the physiology, behavioral aspects, and performance of astronauts operating under confined and often stressful conditions.[61] These researchers thus focused on providing life support and ensuring the best possi-

59. Steven J. Dick, ed., *America in Space: NASA's First Fifty Years* (New York: Abrams, 2007).
60. Andrew Chaikin, *A Man on the Moon: The Voyages of the Apollo Astronauts* (New York: Viking, 1994).
61. George B. Smith, Siegfried J. Gerathewohl, and Bo E. Gernandt, *Bioastronautics*, NASA Publication No. SP-18 (Washington, DC: 1962), pp. 1–18.

Mercury astronauts experiencing weightlessness in a C-131 aircraft flying a "zero-g" trajectory. This was just one of many aspects of piloted space flight that had never before been addressed. NASA.

ble medical selection and maintenance of the humans who were to fly into space.

Also essential for this work to progress was the further development of the technology of biomedical telemetry. This involved monitoring and transmitting a multitude of vital signs from an astronaut in space on a real-time basis to medical personnel on the ground. The comprehensive data collected included such information as body temperature, heart rate and rhythm, blood and pulse pressure, blood oxygen content, respiratory and gastrointestinal functions, muscle size and activity, urinary functions, and varying types of central nervous system activity.[62] Although much work had already been done in this field, particularly in the X-15 program, NASA further perfected it during the Mercury program when the need to carefully monitor the physiological condition of astronauts in space became critical.[63]

62. Engle and Lott, *Man in Flight*, p. 180.

63. Stillwell, *X-15 Research Results*, p. 89; *Project Mercury Summary*, U.S. Manned Spacecraft Center, Houston, TX (Washington, DC: NASA, 1963), pp. 203–207; Stillwell, *X-15 Research Results*, p. 89.

Finally, this early era of NASA human factors research included an emphasis on the bioengineering aspects of piloted space flight, or the application of engineering principles in order to satisfy the physiological requirements of humans in space. This included the design and application of life-sustaining equipment to maintain atmospheric pressure, oxygen, and temperature; provide food and water; eliminate metabolic waste products; ensure proper restraint; and combat the many other stresses and hazards of space flight. This research also included finding the most expeditious way of arranging the multitude of dials, switches, knobs, and displays in the spacecraft so that the astronaut could efficiently monitor and operate them.[64]

In addition to the knowledge gained and applied while planning these early space flights was that gleaned from the flights themselves. The data gained and the lessons learned from each flight were essential to further success, and they were continually factored into future piloted space endeavors. Perhaps even more important, however, was the information gained from the failures of this period. They taught NASA researchers many painful but nonetheless important lessons about the cost of neglecting human factors considerations. Perhaps the most glaring example of this was the Apollo 1 fire of January 27, 1967, that killed NASA astronauts Virgil "Gus" Grissom, Roger Chaffee, and Edward White. While the men were sealed in their capsule conducting a launch pad test of the Apollo/Saturn space vehicle that was to be used for the first flight, a flash fire occurred. That such a fire could have happened in such a controlled environment was hard to explain, but the fact that there had been provided no effective means for the astronauts' rescue or escape in such an emergency was inexplicable.[65] This tragedy did, however, serve some purpose; it gave impetus to tangible safety and engineering improvements, including the creation of an escape hatch through which astronauts could more quickly open and egress during an emergency.[66] Perhaps more importantly, this tragedy caused NASA to step back and reevaluate all of its safety and human engineering procedures.

64. Richard S. Johnston, *Bioastronautics,* NASA SP-18 (1962), pp. 21–28; Pitts, *The Human Factor,* pp. 20–28; Engle and Lott, *Man in Flight,* p. 233.

65. Erik Bergaust, *Murder on Pad 34* (New York: G.P. Putnam's Sons, 1968).

66. G.E. Mueller, "Design, Construction and Procedure Changes in Apollo Following Fire of January 1967," *Astronautics and Aeronautics,* vol. 5, no. 8 (Aug. 1967), pp. 28–33.

Apollo 1 astronauts, left to right, Gus Grissom, Ed White, and Roger Chaffee. Their deaths in a January 27, 1967, capsule fire prompted vital changes in NASA's safety and human engineering policies. NASA.

A New Direction for NASA's Human Factors Research

By the end of the Apollo program, NASA, though still focused on the many initiatives of its space ventures, began to look in a new direction for its research activities. The impetus for this came from a 1968 Senate Committee on Aeronautical and Space Sciences report recommending that NASA and the recently created Department of Transportation jointly determine which areas of civil aviation might benefit from further research.[67] A subsequent study prompted the President's Office of Science and Technology to direct NASA to begin similar research. The resulting Terminal Configured Vehicle program led to a new focus in NASA human factors research. This included the all-important interface between not only the pilot and airplane, but also the pilot and the air traffic controller.[68]

67. Senate Committee on Aeronautical and Space Sciences, *Aeronautical Research and Development Policy Report,* 90th Congress, 2nd session, 1968, S. Rept. 957.

68. The name Terminal Configured Vehicle was changed in 1982 to Advanced Transport Operating Systems (ATOPS) to reflect additional emphasis on air transportation systems, as opposed to individual aircraft technologies.

The goal of this ambitious program was

> . . . to provide improvements in the airborne systems (avionics and air vehicle) and operational flight procedures for reducing approach and landing accidents, reducing weather minima, increasing air traffic controller productivity and airport and airway capacity, saving fuel by more efficient terminal area operations, and reducing noise by operational procedures.[69]

With this directive, NASA's human factors scientists were now officially involved with far more than "just" a piloted space program; they would now have to extend their efforts into the expansive world of aviation.

With these new aviation-oriented research responsibilities, NASA's human factors programs would continue to evolve and increase in complexity throughout the remaining decades of the 20th century and into the present one. This advancement in development was inevitable, given the growing technology, especially in the realm of computer science and complex computer-managed systems, as well as the changing space and aeronautical needs that arose throughout this period.

During NASA's first three decades, more and more of the increasingly complex aerospace operating systems it was developing for its space initiatives and the aviation industry were composed of multiple subsystems. For this reason, the need arose for a human systems integration (HSI) plan to help maximize their efficiency. HSI is a multidisciplinary approach that stresses human factors considerations, along with other such issues as health, safety, training, and manpower, in the early design of fully integrated systems.[70]

To better address the human factors research needs of the aviation community, NASA formed the Flight Management and Human Factors Division at Ames Research Center, Moffett Field, CA.[71] Its name was

---

69. NASA Langley Research Center, *Terminal Configured Vehicle Program Plan* (Hampton, VA: Dec. 1, 1973), p. 2.

70. D.J. Fitts and A. Sandor, "Human Systems Integration," *http://www.dsls.usra.edu/meetings/hrp2008/pdf/SHFH/1065DFitts.pdf*, accessed Oct. 7, 2009.

71. "National Plan for Civil Aviation Human Factors: An Initiative for Research and Application," 1st ed., Federal Aviation Administration (Feb. 3, 1995), *http://www.hf.faa.gov/docs/natplan.doc*, accessed Oct. 7, 2009.

later changed to the Human Factors Research & Technology Division; today, it is known as the Human Systems Integrations Division (HSID).[72]

For the past three decades, this division and its precursors have sponsored and participated in most of NASA's human factors research affecting both aviation and space flight. HSID describes its goal as "safe, efficient, and cost-effective operations, maintenance, and training, both in space, in flight, and on the ground," in order to "advance human-centered design and operations of complex aerospace systems through analysis, experimentation and modeling of human performance and human-automation interaction to make dramatic improvements in safety, efficiency and mission success."[73] To accomplish this goal, the division, in its own words,

- Studies how humans process information, make decisions, and collaborate with human and machine systems.
- Develops human-centered automation and interfaces, decision support tools, training, and team and organizational practices.
- Develops tools, technologies, and countermeasures for safe and effective space operations.[74]

More specifically, the Human Systems Integrations Division focuses on the following three areas:

- Human performance: This research strives to better define how people react and adapt to various types of technology and differing environments to which they are exposed. By analyzing such human reactions as visual, auditory, and tactile senses; eye movement; fatigue; attention; motor control; and such perceptual cognitive processes as memory, it is possible to better predict and ultimately improve human performance.

72. Personal communication with Jeffrey W. McCandless, Deputy Division Chief, Human Systems Integration Division, NASA Ames Research Center, May 8, 2009.
73. NASA Human Systems Integration Division Web site, *http://human-factors.arc.nasa.gov*, accessed Oct. 7, 2009.
74. "Human Systems Integration Division Overview," NASA Human Systems Integration Division Fact Sheet, *http://hsi.arc.nasa.gov/factsheets/TH_Division_Overview.pdf*, accessed Oct. 7, 2009.

- Technology interface design: This directly affects human performance, so technology design that is patterned to efficient human use is of utmost importance. Given the complexity and magnitude of modern pilot/aircrew cockpit responsibilities—in commercial, private, and military aircraft, as well as space vehicles—it is essential to simplify and maximize the efficiency of these tasks. Only with cockpit instruments and controls that are easy to operate can human safety and efficiency be maximized. Interface design might include, for example, the development of cockpit instrumentation displays and arrangement, using a graphical user interface.
- Human-computer interaction: This studies the "processes, dialogues, and actions" a person uses to interact with a computer in all types of environment. This interaction allows the user to communicate with the computer by inputting instructions and then receiving responses back from the computer via such mechanisms as conventional monitor displays or head monitor displays that allows the user to interact with a virtual environment. This interface must be properly adapted to the individual user, task, and environment.[75]

Some of the more important research challenges HSID is addressing and will continue to address are proactive risk management, human performance in virtual environments, distributed air traffic management, computational models of human-automation interaction, cognitive models of complex performance, and human performance in complex operations.[76]

Over the years, NASA's human factors research has covered an almost unbelievably wide array of topics. This work has involved—and benefitted—nearly every aspect of the aviation world, including the FAA, DOD, the airline industry, general aviation, and a multitude of nonaviation areas. To get some idea of the scope of the research with which NASA has been involved, one need only search the NASA Technical Report Server using the term "human factors," which produces more

75. NASA Human Systems Integration Division Web site.

76. "Human Systems Integration Division Overview," NASA Human Systems Integration Division Fact Sheet.

A full-scale aircraft drop test being conducted at the 240-foot-high NASA Langley Impact Dynamics Research Facility. The gantry previously served as the Lunar Landing Research Facility. NASA.

than 3,600 records.[77] It follows that no single paper or document—and this case study is no exception—could ever comprehensively describe NASA's human factors research. It is possible, however, to get some idea of the impact that NASA human factors research has had on aviation safety and technology by reviewing some of the major programs that have driven the Agency's human factors research over the past decades.

NASA's Human Factors Initiatives: A Boon to Aviation Safety

No aspect of NASA's human factors research has been of greater importance than that which has dealt with improving the safety of those humans who occupy all different types of aircraft—both as operators and as passengers. NASA human factors scientists have over the past several decades joined forces with the FAA, DOD, and nearly all members of the aviation industry to make flying safer for all parties. To understand the scope of the work that has helped accomplish this goal, one should review some of the major safety-oriented human factors programs in which NASA has participated.

77. NASA Technical Reports Server (NTRS), *http://ntrs.nasa.gov/search.jsp.*

4

A full-scale aircraft drop test being conducted at the Langley Impact Dynamics Research Facility. These NASA–FAA tests helped develop technology to improve crashworthiness and passenger survivability in general-aviation aircraft. NASA.

Landing Impact and Aircraft Crashworthiness/Survivability Research

Among NASA's earliest research conducted primarily in the interest of aviation safety was its Aircraft Crash Test program. Aircraft crash survivability has been a serious concern almost since the beginning of flight. On September 17, 1908, U.S. Army Lt. Thomas E. Selfridge became powered aviation's first fatality, after the aircraft in which he was a passenger crashed at Fort Myers, VA. His pilot, Orville Wright, survived the crash.[78] Since then, untold thousands of humans have perished in aviation accidents. To address this grim aspect of flight, NASA Langley Research Center began in the early 1970s to investigate ways to increase the human survivability of aircraft crashes. This important series of studies has been instrumental in the development of important safety improvements in commercial, general aviation, and military aircraft, as well as NASA space vehicles.[79]

78. A.J. Launay, *Historic Air Disasters* (London: Ian Allan, 1967), p. 13.

79. Karen E. Jackson, Richard L. Boitnott, Edwin L. Fasanella, Lisa Jones, and Karen H. Lyle "A Summary of DOD-Sponsored Research Performed at NASA Langley's Impact Dynamics Research Facility," NASA Langley Research Center and U.S. Army Research Laboratory, Hampton, VA, presented at the *American Helicopter Society 60th Annual Forum, Baltimore, MD, June 7–10, 2004.*

These unique experiments involved dropping various types and components of aircraft from a 240-foot-high gantry structure at NASA Langley. This towering structure had been built in the 1960s as the Lunar Landing Research Facility to provide a realistic setting for Apollo astronauts to train for lunar landings. At the end of the Apollo program in 1972, the gantry was converted for use as a full-scale crash test facility. The goal was to learn more about the effects of crash impact on aircraft structures and their occupants, and to evaluate seat and restraint systems. At this time, the gantry was renamed the Impact Dynamics Research Facility (IDRF).[80]

This aircraft test site was the only such testing facility in the country capable of slinging a full-scale aircraft into the ground, similar to the way it would impact during a real crash. To add to the realism, many of the aircraft dropped during these tests carried instrumented anthropomorphic test dummies to simulate passengers and crew. The gantry was able to support aircraft weighing up to 30,000 pounds and drop them from as high as 200 feet above the ground. Each crash was recorded and evaluated using both external and internal cameras, as well as an array of onboard scientific instrumentation.[81]

Since 1974, NASA has conducted crash tests on a variety of aircraft, including high and low wing, single- and twin-engine general-aviation aircraft and fuselage sections, military rotorcraft, and a variety of other aviation and space components. During the 30-year period after the first full-scale crash test in February 1974, this system was employed to conduct 41 crash/impact tests on full-sized general-aviation aircraft and 11 full-scale rotorcraft tests. It also provided for 48 Wire Strike Protection System (WSPS) Army helicopter qualification tests, 3 Boeing 707 fuselage section vertical drop tests, and at least 60 drop tests of the F-111 crew escape module.[82]

The massive amount of data collected in these tests has been used to determine what types of crashes are survivable. More specifically, this information has been used to establish guidelines for aircraft seat design that are still used by the FAA as its standard for certification. It has also contributed to new technologies, such as energy-absorbing seats, and to

80. V.L. Vaughan, Jr., and E. Alfaro-Bou, "Impact Dynamics Research Facility for Full-Scale Aircraft Crash Testing," NASA TN-D-8179 (Apr. 1976).

81. Jackson, et al., "A Summary of DOD-Sponsored Research."

82. Ibid.; Edwin L. Fasanella and Emilio Alfaro-Bou, "Vertical Drop Test of a Transport Fuselage Section Located Aft of the Wing," NASA TM-89025 (Sept. 1986).

improving the impact characteristics of new advanced composite materials, cabin floors, engine support fittings, and other aircraft components and equipment.[83] Indeed, much of today's aircraft safety technology can trace its roots to NASA's pioneering landing impact research.

Full-Scale Transport Controlled Impact Demonstration

This dramatic and elaborate crash test program of the early 1980s was one of the most ambitious and well-publicized experiments that NASA has conducted in its decades-long quest for increased aviation safety. In this 1980–1984 study, the NASA Dryden and Langley Research Centers joined with the FAA to quantitatively assess airline crashes. To do this, they set out to intentionally crash a remotely controlled Boeing 720 airliner into the ground. The objective was not simply to crash the airliner, but rather to achieve an "impact-survivable" crash, in which many passengers might be expected to survive.[84] This type of crash would allow a more meaningful evaluation of both the existing and experimental cabin safety features that were being observed. Much of the information used to determine just what was "impact-survivable" came from Boeing 707 fuselage drop tests conducted previously at Dryden's Impact Dynamics Research Facility and a similar but complete aircraft drop conducted by the FAA.[85]

The FAA's primary interest in the Controlled Impact Demonstration (CID, also sometimes jokingly referred to as "Crash in the Desert") was to test an anti-misting kerosene (AMK) fuel additive called FM-9. This high-molecular-weight polymer, when combined with Jet-A fuel, had shown promise during simulated impact tests in inhibiting the spontaneous combustion of fuel spilling from ruptured fuel tanks. The possible benefits of this test were highly significant: if the fireball that usually follows an aircraft crash could be eliminated or diminished, countless lives might be saved. The FAA was also interested, secondarily, in testing new safety-related design features. NASA's main interest in this study, on the other hand, was to measure airframe structural loads and collect crash dynamics data.[86]

83. Joseph R. Chambers, *Concept to Reality: Contributions of the NASA Langley Research Center to U.S. Civil Aircraft of the 1990s,* NASA SP-2003-4529 (2003).

84. Ibid.

85. Edwin L. Fasanella, Emilio Alfaro-Bou, and Robert J. Hayduk, "Impact Data from a Transport Aircraft During a Controlled Impact Demonstration," NASA TP-2589 (Sept. 2, 1986).

86. Chambers, *Concept to Reality.*

A remotely controlled Boeing 720 airliner explodes in flame on December 1, 1984, during the Controlled Impact Demonstration. Although the test sank hopes for a new anti-misting kerosene fuel, other information from the test helped increase airline safety. NASA.

With these objectives in mind, researchers from the two agencies filled the seats of the "doomed" passenger jet with anthropomorphic dummies instrumented to measure the transmission of impact loads. They also fitted the airliner with additional crash-survivability testing equipment, such as burn-resistant windows, fireproof cabin materials, experimental seat designs, flight data recorders, and galley and stowage-bin attachments.[87]

The series of tests included 15 remote-controlled flights, the first 14 of which included safety pilots onboard. The final flight took place on the morning of December 1, 1984. It started at Edwards AFB, NV, and ended with the intentional crash of the four-engine jet airliner onto the bed of Rogers Dry Lake. The designated target was a set of eight steel posts, or cutters, cemented into the lakebed to ensure that the jet's fuel tanks ruptured. During this flight, NASA Dryden's Remotely Controlled Vehicle Facility research pilot, Fitzhugh Fulton, controlled the aircraft from the ground.[88]

The crash was accomplished more or less as planned. As expected, the fuel tanks, containing 76,000 pounds of the anti-misting kerosene jet fuel, were successfully ruptured; unfortunately, the unexpectedly

87. Fasanella, et al., "Impact Data from a Transport Aircraft During a Controlled Impact Demonstration."
88. Ibid.

Instrumented test dummies installed in Boeing 720 airliner for the Controlled Impact Demonstration of December 1, 1984. NASA.

spectacular fireball that ensued—and that took an hour to extinguish—was a major disappointment to the FAA. Because of the dramatic failure of the anti-misting fuel, the FAA was forced to curtail its plan to require the use of this additive in airliners.[89]

In most other ways, however, the CID was a success. Of utmost importance were the lessons learned about crash survivability. New safety initiatives had been tested under realistic conditions, and the effects of a catastrophic crash on simulated humans were filmed inside the aircraft by multiple cameras and later visualized at the crash site. Analysis of these data showed, among many other things, that in a burning airliner, seat cushions with fire-blocking layers were indeed superior to conventional cushions. This finding resulted in FAA-mandated flammability standards requiring these safer seat cushions.[90] Another important safety finding that the crash-test data revealed was that the airliner's adhesive-fastened tritium aisle lights, which would be of utmost importance during postcrash emergency egress, became dislodged and

89. Ibid.

90. "Full-Scale Transport Controlled Impact Demonstration Program: Final Summary Report," NASA TM-89642 (Sept. 1987), p. 33.

nonfunctional during the crash. As a result, the FAA mandated that these lights be mechanically fastened, to maximize their time of usefulness after a crash.[91] These and other lessons from this unique research project have made commercial travel safer.

Aviation Safety Reporting System

NASA initiated and implemented this important human-based safety program in 1976 at the request of the FAA. Its importance can best be judged by the fact it is still in full operation—funded by the FAA and managed by NASA. The Aviation Safety Reporting System (ASRS) collects information voluntarily and confidentially submitted by pilots, controllers, and other aviation professionals. This information is used to identify deficiencies in the National Aviation System (NAS), some of which include those of the human participants themselves. The ASRS analyzes these data and refers them in the form of an "alerting message" to the appropriate agencies so that problems can be corrected. To date, nearly 5,000 alert messages have been issued.[92] The ASRS also educates through its operational issues bulletins, its newsletter *CALLBACK* and its journal *ASRS Directline*, as well as through the more than 60 research studies it has published.[93] The massive database that the ASRS maintains benefits not only NASA and the FAA, but also other agencies worldwide involved in the study and promotion of flight safety. Perhaps most importantly, this system serves to foster further aviation human factors safety research designed to prevent aviation accidents.[94] After more than 30 years in operation, the ASRS has been an unqualified success. During this period, pilots, air traffic controllers, and others have provided more than 800,000 reports.[95] The many types of ASRS responses to the data it has collected have triggered a variety of safety-oriented actions, including modifications to the Federal Aviation Regulations.[96]

---

91. Ibid., p. 39.

92. "ASRS Program Briefing," via personal communication with Linda Connell, ASRS Program Director, Sept. 25, 2009.

93. Corrie, "The US Aviation Safety Reporting System," pp. 1–7; "ASRS Program Briefing," via personal communication with Connell.

94. Ibid.

95. Amy Pritchett, "Aviation Safety Program," Integrated Intelligent Flight Deck Technologies presentation dated June 17, 2008, *http://www.jpdo.gov/library/20080618AllHands/04_20080618_Amy_Pritchett.pdf*, accessed Oct. 7, 2009; "ASRS Program Briefing."

96. Wiener and Nagel, *Human Factors in Aviation*, pp. 268–269.

It is impossible to quantify the number of lives saved by this important long-running human-based program, but there is little dispute that its wide-ranging effect on the spectrum of flight safety has benefitted all areas of aviation.

Fatigue Countermeasures Program

NASA Ames Research Center began the Fatigue Countermeasures program in the 1980s in response to a congressional request to determine if there existed a safety problem "due to transmeridian flying and a potential problem due to fatigue in association with various factors found in air transport operations."[97] Originally termed the NASA Ames Fatigue/Jet Lag program, this ongoing program, jointly funded by the FAA, was created to study such issues as fatigue, sleep, flight operations performance, and the biological clock—otherwise known as circadian rhythms. This research was focused on (1) determining the level of fatigue, sleep loss, and circadian rhythm disruption that exists during flight operations, (2) finding out how these factors affect crew performance, and (3) developing ways to counteract these factors to improve crew alertness and proficiency. Many of the findings from this series of field studies, which included such fatigue countermeasures as regular flightcrew naps, breaks, and better scheduling practices, were subsequently adopted by the airlines and the military.[98] This research also resulted in Federal Aviation Regulations that are still in effect, which specify the amount of rest flightcrews must have during a 24-hour period.[99]

97. Michael B. Mann, NASA Office of Aero-Space Technology, Hearing on Pilot Fatigue, Aviation Subcommittee of the Committee on Transportation and Infrastructure, U.S. House of Representatives, Aug. 3, 1999, *http://www.hq.nasa.gov/office/legaff/mann8-3.html,* accessed Oct. 7, 2009.
98. Ibid.; "Human Fatigue Countermeasures: Aviation," NASA Fact Sheet, *http://hsi.arc.nasa.gov/factsheets/Caldwell_fatigue_aero.pdf,* accessed Oct. 7, 2009; Mark R. Rosekind, et al., "Crew Factors in Flight Operations IX: Effects of Planned Cockpit Rest on Crew Performance and Alertness in Long-Haul Operations," NASA TM-108839 (July 1994); Rosekind, et. al, "Crew Factors in Flight Operations X: Alertness Management in Flight Operations," NASA TM-2001-211385, DOT/FAA/AR-01-01, NASA Ames Research Center (Nov. 2001); Rosekind, et al., "Crew Factors in Flight Operations XII: A Survey of Sleep Quantity and Quality in On-Board Crew Rest Facilities," NASA TM-2000-209611 (Sept. 2000); Rosekind, et al., "Crew Factors in Flight Operations XIV: Alertness Management in Regional Flight Operations," NASA TM-2002-211393 (Feb. 2002); Rosekind, et al., "Crew Factors in Flight Operations XV: Alertness Management in General Aviation," NASA TM-2002-211394 (Feb. 2002).
99. "Pilot Flight Time, Rest, and Fatigue," FAA Fact Sheet (June 10, 2009).

Crew Factors and Resource Management Program

After a series of airline accidents in the 1970s involving aircraft with no apparent problems, findings were presented at a 1979 NASA workshop indicating that most aviation accidents were indeed caused by human error, rather than mechanical malfunctions or weather. Specifically, there were communication, leadership, and decision-making failures within the cockpit that were causing accidents.[100] The concept of Cockpit Resource Management (now often referred to as Crew Resource Management, or CRM) was thus introduced. It describes the process of helping aircrews reduce errors in the cockpit by improving crew coordination and better utilizing all available resources on the flight deck, including information, equipment, and people.[101] Such training has been shown to improve the performance of aircrew members and thus increase efficiency and safety.[102] It is considered so successful in reducing accidents caused by human error that the aviation industry has almost universally adopted CRM training. Such training is now considered mandatory not only by NASA, but also the FAA, the airlines, the military, and even a variety of nonaviation fields, such as medicine and emergency services.[103] Most recently, measures have been taken to further expand mandatory CRM training to all U.S. Federal Aviation Regulations Part 135 operators, including commuter aircraft. Also included is Single-Pilot Resource Management (SRM) training for on-demand pilots who fly without additional crewmembers.[104]

---

100. G.E. Cooper, M.D. White, and J.K. Lauber, "Resource Management on the Flightdeck: Proceedings of a NASA/Industry Workshop," NASA CP-2120 (1980).

101. J.K. Lauber, "Cockpit Resource Management: Background and Overview," in H.W. Orlady and H.C. Foushee, eds., "Cockpit Resource Management Training: Proceedings of the NASA/Military Airlift Command Workshop," NASA CP-2455 (1987).

102. Robert L. Helmreich, John A. Wilhelm, Steven E. Gregorich, and Thomas R. Chidester, "Preliminary Results from the Evaluation of Cockpit Resource Management Training: Performance Ratings of Flightcrews," *Aviation, Space & Environmental Medicine*, vol. 61, no. 6 (June 1990), pp. 576–579.

103. Earl Wiener, Barbara Kanki, and Robert Helmreich, *Cockpit Resource Management* (San Diego: Academic Press, 1993), pp. 495–496; "Crew Resource Management and its Applications in Medicine," pp. 501–510; Steven K. Howard, David M. Gabe, Kevin J. Fish, George Yang, and Frank H. Sarnquist, "Anesthesia Crisis Resource Management Training: Teaching Anesthesiologists to Handle Critical Incidents," *Aviation, Space & Environmental Medicine*, vol. 63, no. 9 (Sept. 1992), pp. 763–770; "Crew Resource Management Training," FAA Advisory Circular AC No: 120-51E (Jan. 22, 2004).

104. Paul Lowe, "NATA urges mandate for single-pilot CRM," *Aviation International News Online* (Sept. 2, 2009); "Crew Resource Management Training for Crewmembers in Part 135 Operations," Docket No. FAA-2009-0023 (May 1, 2009).

Presently, the NASA Ames Human Systems Integration Division's Flight Cognition Laboratory is involved with the evaluation of the thought processes that determine the behavior of air crewmen, controllers, and others involved with flight operations. Among the areas they are studying are prospective memory, concurrent task management, stress, and visual search. As always, the Agency actively shares this information with other governmental and nongovernmental aviation organizations, with the goal of increasing flight safety.[105]

Workload, Strategic Behavior, and Decision-Making

It is well-known that more than half of aircraft incidents and accidents have occurred because of human error. These errors resulted from such factors as flightcrew distractions, interruptions, lapses of attention, and work overload.[106] For this reason, NASA researchers have long been interested in characterizing errors made by pilots and other crewmembers while performing the many concurrent flight deck tasks required during normal flight operations. Its Attention Management in the Cockpit program analyzes accident and incident reports, as well as questionnaires completed by experienced pilots, to set up appropriate laboratory experiments to examine the problem of concurrent task management and to develop methods and training programs to reduce errors. This research will help design simulated but realistic training scenarios, assist flightcrew members in understanding their susceptibility to errors caused by lapses in attention, and create ways to help them manage heavy workload demands. The intended result is increased flight safety.[107]

Likewise, safety in the air can be compromised by errors in judgment and decision making. To tackle this problem, NASA Ames Research

---

105. "Flight Cognition Laboratory," NASA Web site, *http://human-factors.arc.nasa.gov/ihs/flightcognition/index.html,* accessed Oct. 7, 2009.

106. Charles E. Billings, and William D. Reynard, "Human Factors in Aircraft Incidents: Results of a 7-Year Study," *Aviation, Space & Environmental Medicine,* vol. 55, no. 10 (Oct. 1992), pp. 960–965.

107. "Attention Management in the Cockpit," NASA Human Systems Integration Division Fact Sheet, *http://hsi.arc.nasa.gov/factsheets/Dismukes_attention_manage.pdf,* accessed Oct. 7, 2009; David F. Dinges, "Crew Alertness Management on the Flight Deck: Cognitive and Vigilance performance," summary of research Feb. 1, 1989, to Oct. 31, 1998, Grant No. NCC-2-599 (1998); M.R. Rosekind and P.H. Gander, "Alertness Management in Two-Person Long-Haul Flight Operations," NASA Ames Research Center, *Aerospace Medical Association 63rd Annual Scientific Meeting Program, May 14, 1992;* H.P. Ruffell-Smith, "A Simulator Study of the Interaction of Pilot Workload with Errors, Vigilance, and Decisions," NASA TM-78482 (1979), pp. 1–54.

Center joined with the University of Oregon to study how decisions are made and to develop techniques to decrease the likelihood of bad decision making.[108] Similarly, mission success has been shown to depend on the degree of cooperation between crewmembers. NASA research specifically studied such factors as building trust, sharing information, and managing resources in stressful situations. The findings of this research will be used as the basis for training crews to manage interpersonal problems on long missions.[109]

It can therefore be seen that NASA has indeed played a primary role in developing many of the human factors models in use, relating to aircrew efficiency and mental well-being. These models and the training programs that incorporate them have helped both military and civilian flightcrew members improve their management of resources in the cockpit and make better individual and team decisions in the air. This knowledge has also helped more clearly define and minimize the negative effects of crew fatigue and excessive workload demands in the cockpit. Further, NASA has played a key role in assisting both the aviation industry and DOD in setting up many of the training programs that are utilizing this new technology to improve flight safety.

Traffic Collision Avoidance System

By the 1980s, increasing airspace congestion had made the risk of catastrophic midair collision greater than ever before. Consequently, the 100th Congress passed Public Law 100-223, the Airport and Airway Safety and Capacity Expansion Improvement Act of 1987. This required, among other provisions, that passenger-carrying aircraft be equipped with a Traffic Collision Avoidance System (TCAS), independent of air traffic control, that would alert pilots of other aircraft flying in their surrounding airspace.[110]

In response to this mandate, NASA, the FAA, the Air Transport Association, the Air Line Pilots Association, and various aviation

108. "Affect & Aeronautical Decision-Making," NASA Human systems Integration Division Fact Sheet, *http://hsi.arc.nasa.gov/factsheets/Barshi_Dec_Making.pdf*, accessed Oct. 7, 2009; Judith M. Orasanu, Ute Fischer, and Richard J. Tarrel, "A Taxonomy of Decision Problems on the Flight Deck," *7th International Symposium on Aviation Psychology, Columbus, OH, Apr. 26–29, 1993*, vols. 1–2, pp. 226–232.
109. "Distributed Team Decision-Making," NASA Human Systems Integration Division Fact Sheet, *http://hsi.arc.nasa.gov/factsheets/Orasanu_dtdm.pdf*, accessed Oct. 7, 2009.
110. U.S. Congress, Office of Technology Assessment, *Safer Skies with TCAS: Traffic Alert and Collision Avoidance System—A Special Report*, OTA-SET-431 (Washington, DC: GPO, Feb. 1989), *http://www.fas.org/ota/reports/8929.pdf*, accessed Oct. 7, 2009.

technology industries teamed up to develop and evaluate such a system, TCAS I, which later evolved to the current TCAS II. From 1988 to 1992, NASA Ames Research Center played a pivotal role in this major collaborative effort by evaluating the human performance factors that came into play with the use of TCAS. By employing ground-based simulators operated by actual airline flightcrews, NASA showed that this system was practicable, at least from a human factors standpoint.[111] The crews were found to be able to accurately use the system. This research also led to improved displays and aircrew training procedures, as well as the validation of a set of pilot collision-evading performance parameters.[112] One example of the new technologies developed for incorporation into the TCAS system is the Advanced Air Traffic Management Display. This innovative system provides pilots with a three-dimensional air traffic virtual-visualization display that increases their situational awareness while decreasing their workload.[113] This visualization system has been incorporated into TCAS system displays and has become the industry standard for new designs.[114]

Automation Design

Automation technology is an important factor in helping aircrew members to perform more wide-ranging and complicated cockpit activities. NASA engineers and psychologists have long been actively engaged in developing automated cockpit displays and other technologies.[115] These

111. S.L. Chappell, C.E. Billings, B.C. Scott, R.J. Tuttell, M.C. Olsen, and T.E. Kozon, "Pilots' Use of a Traffic Alert and Collision-Avoidance System (TCAS II) in Simulated Air Carrier Operations," vol. 1: "Methodology, Summary and Conclusions," NASA TM-100094, Moffett Field, CA: NASA Ames Research Center.

112. B. Grandchamp, W.D. Burnside, and R.G. Rojas, "A study of the TCAS II Collision Avoidance System Mounted on a Boeing 737 Aircraft," NASA CR-182457 (1988); R.G. Rojas, P. Law, and W.D. Burnside, "Simulation of an Enhanced TCAS II System in Operation," NASA CR-181545 (1988); K.S. Sampath, R.G. Rojas, and W.D. Burnside, "Modeling and Performance Analysis of Four and Eight Element TCAS," NASA CR-187414 (1991).

113. Durand R. Begault and Marc T. Pittman, "3-D Audio Versus Head Down TCAS Displays," NASA CR-177636 (1994).

114. Durand R. Begault, "Head-Up Auditory Displays for Traffic Collision Avoidance System Advisories: A Preliminary Investigation," *Human Factors,* vol. 35, no. 4 (1993), pp. 707–717.

115. Allen C. Cogley, "Automation of Closed Environments in Space for Human Comfort and Safety: Report for Academic Year 1989–1990," Kansas State University College of Engineering, NASA CR-186834 (1990); John P. Dwyer, "Crew Aiding and Automation: A System Concept for Terminal Area Operations and Guidelines for Automation Design," NASA CR-4631 (1995); Yvette J. Tenney, William H. Rogers, and Richard W. Pew, "Pilot Opinions on High Level Flight Deck Automation Issues: Toward the Development of a Design Philosophy," NASA CR-4669 (1995).

will be essential to pilots in order for them to safely and effectively operate within a new air traffic system being developed by NASA and others, called Free Flight. This system will use technically advanced aircraft computer systems to reduce the need for air traffic controllers and allow pilots to choose their path and speed, while allowing the computers to ensure proper aircraft separation. It is anticipated that Free Flight will in the upcoming decades become incorporated into the Next Generation Air Transportation System.[116]

NASA Aviation Safety & Security Program

As is apparent from the foregoing discussions, a recurring theme in NASA's human factors research has been its dedication to improving aviation safety. The Agency's many human factors research initiatives have contributed to such safety issues as crash survival, weather knowledge and information, improved cockpit systems and displays, security, management of air traffic, and aircraft control.[117]

Though NASA's involvement with aviation safety has been an important focus of its research activities since its earliest days, this involvement was formalized in 1997. In response to a report by the White House Commission on Aviation Safety and Security, NASA created its Aviation Safety Program (AvSP).[118] As NASA's primary safety program, AvSP dedicated itself and $500 million to researching and developing technologies that would reduce the fatal aircraft accident rate 80 percent by 2007.[119]

In pursuit of this goal, NASA researchers at Langley, Ames, Dryden, and Glenn Research Centers teamed with the FAA, DOD, the aviation industry, and various aviation employee groups—including the Air Line Pilots Association (ALPA), Allied Pilots Association (APA), Air Transport Association (ATA), and National Air Traffic Controllers Association

116. Robert Jacobsen, "NASA's Free Flight Air Traffic Management Research," *NASA Free Flight/ DAGATM Workshop, 2000, http://www.asc.nasa.gov/aatt/wspdfs/Jacobsen_Overview.pdf,* accessed Oct. 7, 2009.

117. "NASA's Aviation Safety Accomplishments," NASA Fact Sheet; Chambers, *Concept to Reality: Contributions of the NASA Langley Research Center to U.S. Civil Aircraft of the 1990s.*

118. Al Gore, White House Commission on Aviation Safety and Security: Final Report to President Clinton (Washington, DC: Executive Office of the President, Feb. 12, 1997).

119. "NASA Aviation Safety Program," NASA Facts Online, FS-2000-02-47-LaRC, *http://oea.larc.nasa.gov/PAIS/AvSP-factsheet.html,* accessed Oct. 7, 2009; Chambers, *Innovation in Flight: Research of the NASA Langley Research Center on Revolutionary Advanced Concepts for Aeronautics,* NASA SP-2005-4539 (2005), p. 97.

4

(NATCA)—to form the Commercial Aviation Safety Team (CAST) in 1998. The purpose of this all-inclusive consortium was to develop an integrated and data-driven strategy to make commercial aviation safer.[120]

As highlighted by the White House Commission report, statistics had shown that the overwhelming majority of the aviation accidents and fatalities in previous years had been caused by human error—specifically, loss of control in flight and so-called controlled flight into terrain (CFIT).[121] NASA—along with the FAA, DOD, the aviation industry, and human factors experts—had previously formed a National Aviation Human Factors Plan to develop strategies to decrease these human-caused mishaps.[122] Consequently, NASA joined with the FAA and DOD to further develop a human performance research plan, based on the NASA–FAA publication Toward a Safer 21st Century—Aviation Safety Research Baseline and Future Challenges.[123] The new AvSP thus incorporated many of the existing human factors initiatives, such as crew fatigue, resource management, and training. Human factors concerns were also emphasized by the program's focus on developing more sophisticated human-assisting aviation technology.

To accomplish its goals, AvSP focused not only on preventing accidents, but also minimizing injuries and loss of life when they did occur. The program also emphasized collection of data to find and address problems. The comprehensive nature of AvSP is beyond the scope of this case study, but some aspects of the program (which, in 2005, became the Aviation Safety & Security Program, or AvSSP) with the greatest human factors implications include accident mitigation, synthetic vision systems, system wide accident prevention, and aviation system monitoring and modeling.[124]

- Accident mitigation: The goal of this research is to find ways to make accidents more survivable to aircraft

120. "CAST: The Commercial Aviation Safety Team," *http://www.cast-safety.org*, accessed Oct. 7, 2009.

121. Gore, White House Commission Final Report to the President.

122. *The National Plan for Aviation Human Factors*, Federal Aviation Administration (Washington, DC, 1990).

123. *Toward a Safer 21st Century Aviation—Safety Research Baseline and Future Challenges*, NASA NP-1997-12-2321-HQ (1997).

124. "NASA Aviation Safety Program Initiative Will Reduce Aviation Fatalities," NASA Facts Online, FS-2000-02-47-LaRC, *http://oea.larc.nasa.gov/PAIS/AvSP-factsheet.html*, accessed Oct. 7, 2009.

occupants. This includes a range of activities, some of which have been discussed, to include impact tests, in-flight and postimpact fire prevention studies, improved restraint systems, and the creation of airframes better able to withstand crashes.

- Synthetic vision systems: Unrestricted vision is vital for a pilot's situational awareness and essential for him to control his aircraft safely. Limited visibility contributes to more fatal air accidents than any other single factor; since 1990, more than 1,750 deaths have been attributed to CFIT—crashing into the ground—not to mention numerous runway incursion accidents that have taken even more lives.[125]
- The traditional approach to this problem has been the development of sensor-based enhanced vision systems to improve pilot awareness. In 2000, however, NASA Langley researchers initiated a different approach. They began developing cockpit displays, termed Synthetic Vision Systems, which incorporate such technologies as Global Positioning System (GPS) and photo-realistic terrain databases to allow pilots to "see" a synthetically derived 3-D digital reproduction of what is outside the cockpit, regardless of the meteorological visibility. Even in zero visibility, these systems allow pilots to synthetically visualize runways and ground obstacles in their path. At the same time, this reduces their workload and decreases the disorientation they experience during low-visibility flying. Such systems would be useful in avoiding CFIT crashes, loss of aircraft control, and approach and landing errors that can occur amid low visibility.[126]
- Such technology could also be of use in decreasing the risk of runway incursions. For example, the Taxiway

125. Chambers, *Innovation in Flight,* p. 93.

126. Randall E. Bailey, Russell V. Parrish, Lynda J. Kramer, Steve Harrah, and J.J. Arthur, III, "Technical Challenges in the Development of a NASA Synthetic Vision System Concept," NASA Langley Research Center, AIAA Paper 2002-5188, *11th AIAA/AAAF International Space Planes and Hypersonic Systems and Technologies Conference, Sept. 29–Oct. 4, 2002, Orleans, France* (2002); Chambers, *Innovation in Flight,* p. 98.

Navigation and Situation Awareness System (T-NASA) was developed to help pilots taxiing in conditions of decreased visibility to "see" what is in front of them. This system allows them to visualize the runway by presenting them with a head-up display (HUD) of a computer-generated representation of the taxi route ahead of them.[127]

- One of the most important synthetic vision systems initiatives arose from the Advanced General Aviation Transport Experiments (AGATE) program, which NASA formed in the mid-1990s to help revitalize the lagging general-aviation industry. NASA joined with the FAA and some 80 industry members, in part to develop an affordable Highway in the Sky (HITS) cockpit display that would enhance safety and pilot situational awareness. In 2000, such a system was installed and demonstrated in a small production aircraft.[128] Today, nearly every aviation manufacturer has a Synthetic Vision System either in use or in the planning stages.[129]
- System wide accident prevention: This research, which focuses on the human causes of accidents, is involved with improving the training of aviation professionals and in developing models that would help predict human error before it occurs. Many of the programs addressing this issue were discussed earlier in greater detail.[130]
- Aviation system monitoring and modeling (ASMM) project: This program, which was in existence from 1999 to 2005, involved helping personnel in the aviation indus-

127. David C. Foyle, "HSCL Research: Taxiway Navigation and Situation Awareness System (T-NASA) Overview," Human Factors Research & Technology Division, NASA Ames Research Center, *http://human-factors.arc.nasa.gov/ihi/hcsl.inactive/T-NASA.html,* accessed Oct. 7, 2009.
128. Chambers, *Innovation in Flight,* p. 100.
129. Ibid., p. 121.
130. Stephen Darr, "NASA Aviation Safety & Security Program (AvSSP) Concept of Operation (CONOPS) for Health Monitoring and Maintenance Systems Products," National Institute of Aerospace NIA Report No. 2006-04 (2006); "Aviation Safety Program," NASA Fact Sheet, *http://www.nasa.gov/centers/langley/news/factsheets/AvSP-factsheet.html,* accessed Oct. 7, 2009; "NASA's Aviation Safety Accomplishments," NASA Fact Sheet, *http://www.nasa.gov/centers/langley/news/factsheets/AvSP-Accom.html,* accessed Oct. 7, 2009.

try to preemptively identify aviation system risk. This included using data collection and improved monitoring of equipment to predict problems before they occur.[131] One important element of the ASMM project is the Aviation Performance Measuring System (APMS).[132] In 1995, NASA and the FAA coordinated with the airlines to develop this program, which utilizes large amounts of information taken from flight data recorders to improve flight safety. The techniques developed are designed to use the data collected to formulate a situational awareness feedback process that improves flight performance and safety.[133]

- Yet another spinoff of ASMM is the National Aviation Operational Monitoring Service (NAOMS). This systemwide survey mechanism serves to quantitatively assess the safety of the National Airspace System and evaluate the effects of technologies and procedures introduced into the system. It uses input from pilots, controllers, mechanics, technicians, and flight attendants. NAOMS therefore serves to assess flight safety risks and the effectiveness of initiatives to decrease these risks.[134] APMS impacts air carrier operations by making routine monitoring of flight data possible, which in turn can allow evaluators to identify risks and develop changes that will improve quality and safety of air operations.[135]
- A similar program originating from ASMM is the Performance Data Analysis and Report and System (PDARS). This joint FAA–NASA initiative provides a

131. Irving C. Statler, ed., "The Aviation System Monitoring and Modeling (ASMM) Project: A Documentation of its History and Accomplishments: 1999–2005," NASA TP-2007-214556 (June 2007).
132. Ibid., pp. 15–16.
133. Ibid., pp. 15–16; Griff Jay, Gary Prothero, Timothy Romanowski, Robert Lynch, Robert Lawrence, and Loren Rosenthal, "APMS 3.0 Flight Analyst Guide: Aviation Performance Measuring System," NASA CR-2004-212840 (Oct. 2004).
134. Irving C. Statler, ed., "The Aviation System Monitoring and Modeling (ASMM) Project," pp. 16–17; "National Aviation Operational Monitoring Service (NAOMS)," NASA Human Systems Integration Division Fact Sheet, *http://hsi.arc.nasa.gov/factsheets/Connors_naoms.pdf,* accessed Oct. 7, 2009.
135. "Aviation Performance Measuring System (APMS)," NASA Human Systems Integration Division Fact Sheet, *http://hsi.arc.nasa.gov/factsheets/Statler_apms.pdf,* accessed Oct. 7, 2009.

> way to monitor daily operations in the NAS and to evaluate the effectiveness of air traffic control (ATC) services. This innovative system, which provides daily analysis of huge volumes of real-time information, including radar flight tracks, has been instituted throughout the continental U.S.[136]

The highly successful AvSP ended in 2005, when it became the Aviation Safety & Security Program. AvSSP exceeded its target goal of reducing aircraft fatalities 80 percent by 2007. In 2008, NASA shared with the other members of CAST the prestigious Robert J. Collier Trophy for its role in helping produce "the safest commercial aviation system in the world."[137] AvSSP continues to move forward with its goal of identifying and developing by 2016 "tools, methods, and technologies for improving overall aircraft safety of new and legacy vehicles operating in the Next Generation Air Transportation System."[138] NASA estimates that the combined efforts of the ongoing safety-oriented programs it has initiated or in which it has participated will decrease general-aviation fatalities by as much as another 90 percent from today's levels over the next 10–15 years.[139]

Taking Human Factors Technology into the 21st Century

From the foregoing, it is clear that NASA's human factors research has over the past decades specifically focused on aviation safety. This work, however, has also maintained an equally strong focus on improving the human-machine interface of aviation professionals, both in the air and on the ground. NASA has accomplished this through its many highly developed programs that have emphasized human-centered considerations in the design and engineering of increasingly complex flight systems.

These human factors considerations in systems design and integration have directly translated to increased human performance and efficiency and, indirectly, to greater flight safety. The scope of these contributions is

136. "Performance Data Analysis and Reporting System," Human Systems Integration Division Fact Sheet, *http://hsi.arc.nasa.gov/factsheets/Statler_pdars.pdf,* accessed Oct. 7, 2009.
137. "NASA Shares Collier Trophy Award for Aviation Safety Technologies," NASA Press Release 09-112, May 21, 2009.
138. Amy Pritchett, "Aviation Safety Program."
139. "NASA's Aviation Safety Accomplishments," NASA Fact Sheet, *http://www.nasa.gov/centers/langley/news/factsheets/AvSP-Accom.html,* accessed Oct. 7, 2009.

best illustrated by briefly discussing a representative sampling of NASA programs that have benefitted aviation in various ways, including the Man-Machine Integration Design and Analysis System (MIDAS), Controller-Pilot Data Link Communications (CPDLC), NASA's High-Speed Research (HSR) program, the Advanced Air Transportation Technologies (AATT) program, and the Agency's Vision Science and Technology effort.

Man-Machine Integration Design and Analysis System

NASA jointly initiated this research program in 1980 with the U.S. Army, San Jose State University, and Sterling Software/QSS/Perot Systems, Inc. This ongoing, work-station–based simulation system, which was designed to further develop human performance modeling, links a "virtual human" of a certain physical anthropometric description to a cognitive (visual, auditory, and memory) structure that is representative of human abilities and limitations. MIDAS then uses these human performance models to assess a system's procedures, displays, and controls. Using these models, procedural and equipment problems can be identified and human-system performance measures established before more expensive testing using human subjects.[140] The aim of MIDAS is to "reduce design cycle time, support quantitative predictions of human-system effectiveness, and improve the design of crew stations and their associated operating procedures."[141] These models thus demonstrate the behavior that might be expected of human operators working with a given automated system without the risk and cost of subjecting humans to these conditions. An important aspect of MIDAS is that it can be applied to any human-machine domain once adapted to the particular requirements of that system. It has in fact been employed in the development of such varied functions as establishing baseline performance measures for U.S. Army crews flying Longbow Apache helicopters with and without chemical warfare gear, evaluating crew performance/workload issues for steep noise abatement approaches into a vertiport, developing an advanced

140. Carolyn Banda, et al., "Army-NASA Aircrew/Aircraft Integration Program: Phase IV A³I Man-Machine Integration Design and Analysis System (MIDAS)," NASA CR-177593 (1991); Banda, et al., "Army-NASA Aircrew/Aircraft Integration Program: Phase V A³I Man-Machine Integration Design and Analysis System (MIDAS)," NASA CR-177596 (1992); Lowell Staveland, "Man-machine Integration Design and Analysis System (MIDAS), Task Loading Model (TLM)," NASA CR-177640 (1994).
141. "Man-machine Integration Design and Analysis System (MIDAS)," NASA Web site, *http://humansystems.arc.nasa.gov/groups/midas/index.html*, accessed Oct. 7, 2009.

NASA Shuttle orbiter cockpit with an improved display/control design, and upgrading emergency 911 dispatch facility and procedures.[142]

Controller-Pilot Data Link Communications

Research for this program, conducted by NASA's Advanced Transport Operating System (ATOPS), was initiated in the early 1980s to improve the quality of communication between aircrew and air traffic control personnel.[143] With increased aircraft congestion, radio frequency overload had become a potential safety issue. With so many pilots trying to communicate with ATC at the same time on the same radio frequency, the potential for miscommunication, errors, and even missed transmissions had become increasingly great.

One solution to this problem was a two-way data link system. This allows communications between aircrew and controllers to be displayed on computer screens both in the cockpit and at the controller's station on the ground. Here they can be read, verified, and stored for future reference. Additionally, flightcrew personnel flying in remote locations, well out of radio range, can communicate in real time with ground personnel via computers hooked up to a satellite network. The system also allows such enhanced capabilities as the transfer of weather data, charts, and other important information to aircraft flying at nearly any location in the world.[144]

Yet another aspect of this system allows computers in aircraft and on the ground to "talk" to one another directly. Controllers can thus arrange closer spacing and more direct routing for incoming and outgoing aircraft. This important feature has been calculated to save an estimated 3,000–6,000 pounds of fuel and up to 8 minutes of flight time on a typical transpacific flight.[145] Digitized voice communications have even been

142. Sandra G. Hart, Brian F. Gore, and Peter A. Jarvis, "The Man-Machine Integration Design & Analysis System (MIDAS): Recent Improvements," NASA Ames Research Center, *http://humansystems.arc.nasa.gov/groups/midas/documents/MIDAS(HFS%2010-04).ppt*, accessed Oct. 7, 2009; Kevin Corker and Christian Neukom, "Man-Machine Integrated Design and Analysis System (MIDAS): Functional Overview," Ames Research Center (Dec. 1998).

143. Marvin C. Waller and Gary W. Lohr, "A Piloted Simulation Study of Data Link ATC Message Exchange," NASA TP-2859 (1989); Charles E. Knox and Charles H. Scanlon, "Flight Tests with a Data Link Used for Air Traffic Control Information Exchange," NASA TP-3135 (1991).

144. Lane E. Wallace, *Airborne Trailblazer*, ch. 7-3, "Data Link," NASA SP-4216 (Washington, DC: 1994).

145. Ibid.

NASA's Future Flight Central, which opened at NASA Ames Research Center in 1999, was the first full-scale virtual control tower. Such synthetic vision systems can be used by both aircraft and controllers to visualize clearly what is taking place around them in any conditions. NASA.

added to decrease the amount of aircrew "head-down" time spent reading messages on the screen. This system has gained support from both pilots and the FAA, especially after NASA investigations showed that the system decreased communication errors, aircrew workload, and the need to repeat ATC messages.[146]

High-Speed Research Program

NASA and a group of U.S. aerospace corporations began research for this ambitious program in 1990. Their goal was to develop a jet capable of transporting up to 300 passengers at more than twice the speed of sound. An important human factors–related spinoff of the so-called High-Speed Civil Transport (HSCT) was an External Visibility System. This system replaced forward cockpit windows with displays of video images with computer-generated graphics. This system would have allowed better performance and safety than unaided human vision while

146. Ibid.

4

NASA's Boeing 737 test aircraft in 1974. Note the numerous confusing and hard-to-read conventional analog dials and gauges. NASA.

eliminating the need for the "droop nose" that the supersonic Concorde required for low-speed operations. Although this program was phased out in fiscal year (FY) 1999 for budgetary reasons, the successful vision technology produced was handed over to the previously discussed AvSP–AvSSP's Synthetic Vision Systems element for further development.[147]

Advanced Air Transportation Technologies Program

NASA established this project in 1996 to increase the capability of the Nation's air transport activities. This program's specific goal was to develop a set of "decision support tools" that would help air traffic service providers, aircrew members, and airline operations centers in streamlining gate-to-gate operations throughout the NAS.[148] Project personnel were tasked with researching and developing advanced

147. "NASA's High Speed Research Program: Developing Tomorrow's Supersonic Passenger Jet," NASA Facts Online, *http://oea.larc.nasa.gov/PAIS/HSR-Overview2.html*, accessed Oct. 7, 2009; Chambers, *Innovation in Flight*, p. 100; Ibid., p. 102.

148. Bruce Kaplan and David Lee, "Key Metrics and Goals for NASA's Advanced Air Transportation Technologies Program," NASA CR-1998-207678 (1998).

NASA's Boeing 737 in 1987 after significant cockpit upgrades. Note its much more user-friendly "glass cockpit" display, featuring eight 8- by 8-inch color monitors. NASA.

concepts within the air traffic management system to the point where the FAA and the air transport industry could develop a preproduction prototype. The program ended in 2004, but implementation of these tools into the NAS addressed such air traffic management challenges as complex airspace operations and assigning air and ground responsibilities for aircraft separation. Several of the technologies developed by this program received "Turning Goals into Reality" awards, and some of these—for example, the traffic management adviser and the collaborative arrival planner—are in use by ATC and the airlines.[149]

Vision Science and Technology

Scientists at NASA Ames Research Center have for many years been heavily involved with conducting research on visual technology for humans. The major areas explored include vision science, image

149. Advanced Air Transportation Technologies (AATT) project, NASA Web site, *http://www.nasa.gov/centers/ames/research/lifeonearth/lifeonearth-aatt.html*, accessed Oct. 7, 2009; "Advanced Air Transportation Technologies Overview," *http://www.asc.nasa.gov/aatt/overview.html*, accessed Oct. 7, 2009.

compression, imaging and displays, and visual human factors. Specific projects have investigated such issues as eye-tracking accuracy, image enhancement, metrics for measuring image quality, and methods to measure and improve the visibility of in-flight and air traffic control monitor displays.[150]

The information gained from this and other NASA-conducted research has played an important role in the development of such important and innovative human-assisting technologies as virtual reality goggles, helmet-mounted displays, and so-called glass cockpits.[151]

The latter concept, which NASA pioneered in the 1970s, refers to the replacement of conventional cockpit analog dials and gauges with a system of cathode ray tubes (CRT) or liquid crystal display (LCD) flatpanels that display the same information in a more readable and usable form.[152] Conventional instruments can be difficult to accurately read and monitor, and they are capable of providing only one level of information. Computerized "glass" instrumentation, on the other hand, can display both numerical and graphic color-coded readouts in 3-D format; furthermore, because each display can present several layers of information, fewer are needed. This provides the pilot larger and more readable displays. This technology, which is now used in nearly all airliners, business jets, and an increasing number of general-aviation aircraft, has improved flight safety and aircrew efficiency by decreasing workload, fatigue, and instrument interpretation errors.[153]

A related vision technology that NASA researchers helped develop is the head-up display.[154] This transparent display allows a pilot to view flight data while looking outside the aircraft. This is especially useful during approaches for landing, when the pilot's attention needs to be focused on events outside the cockpit. This concept was originally developed for the Space Shuttle and military aircraft but has since been

150. "NASA Vision Group," NASA Ames Research Center, *http://vision.arc.nasa.gov/publications.php*, accessed Oct. 7, 2009.

151. Andries van Dam, "Three Dimensional User Interfaces for Immersive Virtual Reality: Final Report," NASA CR-204997 (1997); Joseph W. Clark, "Integrated Helmet Mounted Display Concepts for Air Combat," NASA CR-198207 (1995); Earl L. Wiener, "Human Factors of Advanced Technology ('Glass Cockpit') Transport Aircraft," NASA CR-177528 (1989).

152. Ibid.

153. Wallace, *Airborne Trailblazer.*

154. Richard L. Newman, *Head-up Displays: Designing the Way Ahead* (Brookfield, VT: Ashgate, 1995).

adapted to commercial and civil aircraft, air traffic control towers, and even automobiles.[155]

Into the Future

The preceding discussion can serve only as a brief introduction to NASA's massive research contribution to aviation in the realm of human factors. Hopefully, however, it has clearly made the following point: NASA, since its creation in 1958, has been an equally contributing partner with the aeronautical industry in the sharing of new technology and information resulting from their respective human factors research activities.

Because aerospace is but an extension of aeronautics, it is difficult to envision how NASA could have put its first human into space without the knowledge and technology provided by the aeronautical human factors research and development that occurred in the decades leading up to the establishment of NASA and its piloted space program. In return, however, today's high-tech aviation industry is immeasurably more advanced than it would have been without the past half century of dedicated scientific human factors research conducted and shared by the various components of NASA.

Without the thousands of NASA human factors–related research initiatives during this period, many—if not most—of the technologies that are a normal part of today's flight, air traffic control, and aircraft maintenance operations, would not exist. The high cost, high risk, and lack of tangible cost effectiveness the research and development these advances entailed rendered this kind of research too expensive and speculative for funding by commercial concerns forced to abide by "bottom-line" considerations. As a result of NASA research and the many safety programs and technological innovations it has sponsored for the benefit of all, countless additional lives and dollars were saved as many accidents and losses of efficiency were undoubtedly prevented.

It is clear that NASA is going to remain in the business of improving aviation safety and technology for the long haul. NASA's Aeronautics Research Mission Directorate (ARMD), one of the Agency's four major directorates, will continue improving the safety and efficiency of aviation

155. E. Fisher, R.F. Haines, and T.A. Price, "Cognitive Issues in Head-up Displays," NASA TP-1711 (1980); J.K. Lauber, R.S. Bray, R.L. Harrison, J.C. Hemingway, and B.C. Scott, "An Operational Evaluation of Head-up Displays for Civil Transport Operations," NASA TP-1815 (1982).

with its aviation safety, fundamental aeronautics, airspace systems, and aeronautics test programs. Needless to say, a major aspect of these programs will involve human factors research, as it pertains to aeronautics.[156]

It is impossible to predict precisely in which direction NASA's human factors research will go in the decades to come; however, based on the Agency's remarkably unique 50-year history, it seems safe to assume it will continue to contribute to an ever-safer and more efficient world of aviation.

156. Lisa Porter and ARMD Program Directors, "NASA's New Aeronautics Research Program," presented at the *45th AIAA Aerospace Sciences Meeting & Exhibit, Jan. 11, 2007, http://www.aeronautics.nasa.gov/pdf/armd_overview_reno_4.pdf,* accessed Oct. 7, 2009.

Recommended Additional Readings

Reports, Papers, Articles, and Presentations:

Randall E. Bailey, Russell V. Parrish, Lynda J. Kramer, Steve Harrah, and J.J. Arthur, III, "Technical Challenges in the Development of a NASA Synthetic Vision System Concept," NASA Langley Research Center, AIAA Paper 2002-5188, *11th AIAA/AAAF International Space Planes and Hypersonic Systems and Technologies Conference, Sept. 29–Oct. 4, 2002, Orleans, France.*

Carolyn Banda, et al., "Army-NASA Aircrew/Aircraft Integration Program: Phase IV A³I Man-Machine Integration Design and Analysis System (MIDAS)," NASA CR-177593 (1991).

Carolyn Banda, et al., "Army-NASA Aircrew/Aircraft Integration Program: Phase V A³I Man-Machine Integration Design and Analysis System (MIDAS)," NASA CR-177596 (1992).

Durand R. Begault, "Head-Up Auditory Displays for Traffic Collision Avoidance System Advisories: A Preliminary Investigation," *Human Factors*, vol. 35, no. 4 (1993), pp. 707–717.

Durand R. Begault and Marc T. Pittman, "3-D Audio Versus Head Down TCAS Displays," NASA CR-177636 (1994).

Russell R. Burton, "G-Induced Loss of Consciousness: Definition, History, Current Status," *Aviation, Space, and Environmental Medicine* (Jan. 1988).

Alphonse Chapanis, "Some Reflections on Progress," in *Proceedings of the Human Factors Society 29th Annual Meeting* (Santa Monica, CA: Human Factors Society, 1985), pp. 1–8.

S.L. Chappell, C.E. Billings, B.C. Scott, R.J. Tuttell, M.C. Olsen, and T.E. Kozon, "Pilots' Use of a Traffic Alert and Collision-Avoidance System (TCAS II) in Simulated Air Carrier Operations," vol. 1: "Methodology, Summary and Conclusions," NASA TM-100094 (1989).

Joseph W. Clark, "Integrated Helmet Mounted Display Concepts for Air Combat," NASA CR-198207 (1995).

Allen C. Cogley, "Automation of Closed Environments in Space for Human Comfort and Safety: Report for Academic Year 1989-1990," Kansas State University College of Engineering, NASA CR-186834 (1990).

G.E. Cooper, M.D. White, and J.K. Lauber, "Resource Management on the Flightdeck: Proceedings of a NASA/Industry Workshop," NASA CP-2120 (1980).

Kevin Corker and Christian Neukom, "Man-Machine Integrated Design and Analysis System (MIDAS): Functional Overview" (Sunnyvale, CA: NASA Ames Research Center, 1998).

John P. Dwyer, "Crew Aiding and Automation: A System Concept for Terminal Area Operations and Guidelines for Automation Design," NASA CR-4631 (1995).

Edwin L. Fasanella, Emilio Alfaro-Bou, and Robert J. Hayduk, "Impact Data from a Transport Aircraft During a Controlled Impact Demonstration," NASA TP-2589 (1986).

Federal Aviation Administration, The National Plan for Aviation Human Factors (Washington, DC: FAA, 1990).

E. Fisher, R.F. Haines, and T.A. Price, "Cognitive Issues in Head-up Displays," NASA TP-1711 (1980).

Vice President Albert Gore, White House Commission on Aviation Safety and Security: Final Report to President Clinton (Washington, DC: Executive Office of the President, Feb. 12, 1997).

B. Grandchamp, W.D. Burnside, and R.G. Rojas, "A study of the TCAS II Collision Avoidance System Mounted on a Boeing 737 Aircraft," NASA CR-182457 (1988).

Robert L. Helmreich, John A. Wilhelm, Steven E. Gregorich, and Thomas R. Chidester, "Preliminary Results from the Evaluation of Cockpit Resource Management Training: Performance Ratings of Flightcrews," *Aviation, Space & Environmental Medicine*, vol. 61, no. 6 (June 1990), pp. 576–579.

Steven K. Howard, David M. Gabe, Kevin J. Fish, George Yang, and Frank H. Sarnquist, "Anesthesia Crisis Resource Management Training: Teaching Anesthesiologists to Handle Critical Incidents," *Aviation, Space & Environmental Medicine*, vol. 63, no. 9 (Sept. 1992), pp. 763–770.

K.E. Jackson, R.L. Boitnott, E.L. Fasanella, Lisa Jones, and Karen H. Lyle, "A Summary of DOD-Sponsored Research Performed at NASA Langley's Impact Dynamics Research Facility," presented at the *American Helicopter Society 60th Annual Forum, Baltimore, MD, June 7–10, 2004.*

Robert Jacobsen, "NASA's Free Flight Air Traffic Management Research," Free Flight DAGATM Workshop, 2000, at *http://www.asc.nasa.gov/aatt/wspdfs/Jacobsen_Overview.pdf*, accessed Oct. 7, 2009.

Griff Jay, Gary Prothero, Timothy Romanowski, Robert Lynch, Robert Lawrence, and Loren Rosenthal, "APMS 3.0 Flight Analyst Guide: Aviation Performance Measuring System," NASA CR-2004-212840 (2004).

Bruce Kaplan and David Lee, "Key Metrics and Goals for NASA's Advanced Air Transportation Technologies Program," NASA CR-1998-207678 (1998).

Charles E. Knox and Charles H. Scanlon, "Flight Tests with a Data Link Used for Air Traffic Control Information Exchange," NASA TP-3135 (1991).

J.K. Lauber, R.S. Bray, R.L. Harrison, J.C. Hemingway, and B.C. Scott, "An Operational Evaluation of Head-up Displays for Civil Transport Operations," NASA TP-1815 (1982).

Michael B. Mann, NASA Office of Aero-Space Technology, Hearing on Pilot Fatigue, Aviation Subcommittee of the Committee on Transportation and Infrastructure, U.S. House of Representatives, Aug. 3, 1999, at *http://www.hq.nasa.gov/office/legaff/mann8-3.html,* accessed Oct. 7, 2009.

George E. Mueller, "Design, Construction and Procedure Changes in Apollo Following Fire of January 1967," *Astronautics and Aeronautics*, vol. 5, no. 8 (Aug. 1967), pp. 28–33.

NASA, *Toward a Safer 21st Century Aviation—Safety Research Baseline and Future Challenges,* NASA NP-1997-12-2321-HQ (1997).

NASA Langley Research Center, Terminal Configured Vehicle Program Plan (Hampton, VA, Dec. 1, 1973).

Judith M. Orasanu, Ute Fischer, and Richard J. Tarrel, "A Taxonomy of Decision Problems on the Flight Deck," paper presented at the *7th International Symposium on Aviation Psychology, Columbus, OH, Apr. 26–29, 1993.*

H.W. Orlady and H.C. Foushee, eds., "Cockpit Resource Management Training: Proceedings of the NASA/Military Airlift Command Workshop," NASA CP-2455 (1987).

Harry W. Orlady and Linda M. Orlady, *Human Factors in Multi-Crew Flight Operations* (Brookfield, VT: Ashgate Publishing, Ltd., 1999).

Lisa Porter and ARMD Program Directors, "NASA's New Aeronautics Research Program," presented at the *45th AIAA Aerospace Sciences Meeting & Exhibit, Jan. 11, 2007, http://www.aeronautics.nasa.gov/pdf/armd_overview_reno_4.pdf,* accessed Oct. 7, 2009.

R.G. Rojas, P. Law, and W.D. Burnside, "Simulation of an Enhanced TCAS II System in Operation," NASA CR-181545 (1988).

Mark R. Rosekind, Elizabeth L. Co, David F. Neri, Raymond L. Oyung, and Melissa M. Mallis, "Crew Factors in Flight Operations XIV: Alertness Management in Regional Flight Operations," NASA TM-2002-211393 (2002).

Mark R. Rosekind, Elizabeth L. Co, Raymond L. Oyung, and Melissa M. Mallis, "Crew Factors in Flight Operations XV: Alertness Management in General Aviation," NASA TM-2002-211394 (2002).

Mark R. Rosekind and Philippa H. Gander, "Alertness Management in Two-Person Long-Haul Flight Operations," paper presented at the NASA Ames Research Center, *Aerospace Medical Association 63rd Annual Scientific Meeting Program, May 14, 1992.*

Mark R. Rosekind, Philippa H. Gander, Linda J. Connell, and Elizabeth L. Co, "Crew Factors in Flight Operations X: Alertness Management in Flight Operations," NASA TM-2001-211385, DOT/FAA/AR-01-01 (2001).

Mark R. Rosekind, R. Curtis Graeber, David Dinges, Linda J. Connell, Michael S. Rountree, Cheryl L. Spinweber, and Kelly A. Gillen, "Crew Factors in Flight Operations IX: Effects of Planned Cockpit Rest on Crew Performance and Alertness in Long-Haul Operations," NASA TM-108839 (July 1994).

Mark R. Rosekind, Kevin B. Gregory, Elizabeth L. Co, Donna L. Miller, and David F. Dinges, "Crew Factors in Flight Operations XII: A Survey of Sleep Quantity and Quality in On-Board Crew Rest Facilities," NASA TM-2000-209611 (Sept. 2000).

Steven A. Ruffin, "Explorer Over Dakota: From Stratobowl to Stratosphere," *Aviation History,* May 1996, pp. 22–28, 72.

Steven A. Ruffin, "Flying in the Great War: Rx for Misery; An Overview of the Medical and Physiological and Psychological Aspects of Combat Flying During the First World War," *Over the Front,* vol. 14, no. 2 (summer 1999), pp. 115–124, and vol. 17, no. 2 (summer 2002), pp. 117–136.

H.P. Ruffell-Smith, "A Simulator Study of the Interaction of Pilot Workload with Errors, Vigilance, and Decisions," NASA TM-78482 (1979).

K.S. Sampath, R.G. Rojas, and W.D. Burnside, "Modeling and Performance Analysis of Four and Eight Element TCAS," NASA CR-187414 (1991).

Lowell Staveland, "Man-machine Integration Design and Analysis System (MIDAS), Task Loading Model (TLM)," NASA CR-177640 (1994).

Yvette J. Tenney, William H. Rogers, and Richard W. Pew, "Pilot Opinions on High Level Flight Deck Automation Issues: Toward the Development of a Design Philosophy," NASA CR-4669 (1995).

U.S. Congress, Senate Committee on Aeronautical and Space Sciences, Aeronautical Research and Development Policy Report, 90th Congress, 2nd session, S. Rept. 957 (Washington, DC: GPO, 1968).

U.S. Congress, Office of Technology Assessment, *Safer Skies with TCAS: Traffic Alert and Collision Avoidance System—A Special Report,* OTA-SET-431 (Washington, DC: GPO, 1989).

Andries van Dam, "Three Dimensional User Interfaces for Immersive Virtual Reality: Final Report," NASA CR-204997 (1997).

V.L. Vaughan, Jr., and E. Alfaro-Bou, "Impact Dynamics Research Facility for Full-Scale Aircraft Crash Testing," NASA TN-D-8179 (1976).

Marvin C. Waller and Gary W. Lohr, "A Piloted Simulation Study of Data Link ATC Message Exchange," NASA TP-2859 (1989).

Earl L. Wiener, "Human Factors of Advanced Technology ('Glass Cockpit') Transport Aircraft," NASA CR-177528 (1989).

News Releases and Fact Sheets:

"Advanced Air Transportation Technologies (AATT) project," NASA Web site, *http://www.nasa.gov/centers/ames/research/lifeonearth/lifeonearth-aatt.html,* accessed Oct. 7, 2009.

"Affect & Aeronautical Decision-Making," NASA Human Systems Integration Division Fact Sheet, *http://hsi.arc.nasa.gov/factsheets/Barshi_Dec_Making.pdf,* accessed Oct. 7, 2009.

"Aviation Performance Measuring System," NASA Web site, *http://www.nasa.gov/centers/ames/research/technology-onepagers/aviation-performance.html,* accessed Oct. 7, 2009.

"Aviation Safety Program," NASA Fact Sheet, *http://www.nasa.gov/centers/langley/news/factsheets/AvSP-factsheet.html,* accessed Oct. 7, 2009.

"Distributed Team Decision-Making," NASA Human Systems Integration Division Fact Sheet, *http://hsi.arc.nasa.gov/factsheets/Orasanu_dtdm.pdf,* accessed Oct. 7, 2009.

D.J. Fitts and A. Sandor, "Human Systems Integration," *http://www.dsls.usra.edu/meetings/hrp2008/pdf/SHFH/1065DFitts.pdf,* accessed Oct. 7, 2009.

Sandra G. Hart, Brian F. Gore, and Peter A. Jarvis, "The Man-Machine Integration Design & Analysis System (MIDAS): Recent Improvements," NASA Ames Research Center, *http://humansystems.arc.nasa.gov/groups/midas/documents/MIDAS(HFS%2010-04).ppt,* accessed Oct. 7, 2009.

"Human Systems Integration Division Overview," NASA Human Systems Integration Division Fact Sheet, *http://hsi.arc.nasa.gov/factsheets/TH_Division_Overview.pdf,* accessed Oct. 7, 2009.

"Man-machine Integration Design and Analysis System (MIDAS)," NASA Web site, *http://humansystems.arc.nasa.gov/groups/midas/index.html,* accessed Oct. 7, 2009.

"NASA Aviation Safety Program," NASA Facts Online, FS-2000-02-47-LaRC, *http://oea.larc.nasa.gov/PAIS/AvSP-factsheet.html,* accessed Oct. 7, 2009.

"NASA Aviation Safety Program Initiative Will Reduce Aviation Fatalities," NASA Facts Online, FS-2000-02-47-LaRC, *http://oea.larc.nasa.gov/PAIS/AvSP-factsheet.html,* accessed Oct. 7, 2009.

NASA Human Systems Integration Division Web site, *http://human-factors.arc.nasa.gov,* accessed Oct. 7, 2009.

"NASA Vision Group," NASA Ames Research Center, *http://vision.arc.nasa.gov/publications.php,* accessed Oct. 7, 2009.

"NASA's Aviation Safety Accomplishments," NASA Fact Sheet, *http://www.nasa.gov/centers/langley/news/factsheets/AvSP-Accom.html,* accessed Oct. 7, 2009.

"NASA's High Speed Research Program: Developing Tomorrow's Supersonic Passenger Jet," NASA Facts Online, *http://oea.larc.nasa.gov/PAIS/HSR-Overview2.html,* accessed Oct. 7, 2009.

"National Plan for Civil Aviation Human Factors: An Initiative for Research and Application," 1st ed., Federal Aviation Administration (Feb. 3, 1995), *http://www.hf.faa.gov/docs/natplan.doc,* accessed Oct. 7, 2009.

"Pilot Flight Time, Rest, and Fatigue," *FAA Fact Sheet* (June 10, 2009).

Books and Monographs:

Frank W. Anderson, Jr., *Orders of Magnitude: A History of NACA and NASA, 1915–1976,* NASA SP-4403 (Washington, DC: NASA, 1976).

Harry G. Armstrong, *Principles and Practice of Aviation Medicine* (Baltimore: Williams & Wilkins, 1939).

R.A. Behan and H.W. Wendhausen, *Some NASA Contributions to Human Factors Engineering*, NASA SP-5117 (Washington, DC: NASA, 1973).

Charles E. Billings and William D. Reynard, "Human Factors in Aircraft Incidents: Results of a 7-Year Study," *Aviation, Space & Environmental Medicine*, vol. 55, no. 10 (Oct. 1992), pp. 960–965.

David Bushnell, *History of Research in Space Biology and Biodynamics, 1946–58* (Holloman AFB, NM: AF Missile Development Center, 1958).

Joseph R. Chambers, *Concept to Reality: Contributions of the NASA Langley Research Center to U.S. Civil Aircraft of the 1990s*, NASA SP-2003-4529 (Washington, DC: GPO, 2003).

Joseph R. Chambers, *Innovation in Flight: Research of the NASA Langley Research Center on Revolutionary Advanced Concepts for Aeronautics*, NASA SP-2005-4539 (Washington, DC: GPO, 2005).

Alphonse Chapanis, *Research Techniques in Human Engineering* (Baltimore: The Johns Hopkins Press, 1958).

Alphonse Chapanis, *The Chapanis Chronicles: 50 Years of Human Factors Research, Education, and Design* (Santa Barbara, CA: Aegean Publishing, Co., 1999).

Stephen Darr, "NASA Aviation Safety & Security Program (AvSSP) Concept of Operation (CONOPS) for Health Monitoring and Maintenance Systems Products," National Institute of Aerospace NIA Report No. 2006-04 (2006).

Eugene M. Emme, *Aeronautics and Astronautics: An American Chronology of Science and Technology in the Exploration of Space, 1915–1960* (Washington, DC: NASA, 1961).

Eloise Engle and Arnold S. Lott, *Man in Flight: Biomedical Achievements in Aerospace* (Annapolis: Leeward, 1979).

Daniel J. Garland, John A. Wise, and V. David Hopkin, eds., *Handbook of Aviation Human Factors* (Mahway, NJ: Lawrence Erlbaum Associates, 1999).

Gregory P. Kennedy, *Touching Space: The Story of Project Manhigh* (Atglen, PA: Schiffer Military History, 2007).

Lillian D. Kozloski, *U.S. Space Gear: Outfitting the Astronaut* (Washington, DC: Smithsonian Institution Press, 1994).

David Meister, *The History of Human Factors and Ergonomics* (Mahwah, NJ: Lawrence Erlbaum Associates, 1999).

Richard L. Newman, *Head-up Displays: Designing the Way Ahead* (Brookfield, VT: Ashgate, 1995).

John A. Pitts, The *Human Factor: Biomedicine in the Manned Space Program to 1980,* NASA SP-4213 (Washington, DC: NASA, 1985).

Craig Ryan, *The Pre-Astronauts: Manned Ballooning on the Threshold of Space* (Annapolis, MD: Naval Institute Press, 1995).

Alan Shepard and Donald K. "Deke" Slayton, with Jay Barbree and Howard Benedict, *Moon Shot: The Inside Story of America's Race to the Moon* (Atlanta: Turner Publishers, Inc., 1994).

George B. Smith, Siegfried J. Gerathewohl, and Bo E. Gernandt, *Bioastronautics,* NASA SP-18 (Washington, DC, 1962).

Irving C. Statler, ed., *The Aviation System Monitoring and Modeling (ASMM) Project: A Documentation of its History and Accomplishments: 1999–2005,* NASA TP-2007-214556 (2007).

Loyd S. Swenson, Jr., James M. Grimwood, and Charles C. Alexander, *This New Ocean: A History of Project Mercury* (Washington, DC: NASA, 1966).

Kenneth S. Thomas and Harold J. McMann, *U.S. Spacesuits* (Chichester, U.K.: Praxis Publishing Ltd., 2006).

Milton O. Thompson, *At the Edge of Space* (Washington, DC: Smithsonian Institution Press, 1992), pp. 281–355.

Peggy Tillman and Barry Tillman, *Human Factors Essentials: An Ergonomics Guide for Designers, Engineers, Scientists, and Managers* (New York: McGraw-Hill, 1991).

Lane E. Wallace, *Airborne Trailblazer,* NASA SP-4216 (Washington, DC: NASA, 1994).

Lane E. Wallace, *Flights of Discovery: 60 Years at the Dryden Flight Research Center,* NASA SP-2006-4318 (Washington, DC: NASA, 2006).

J. Weimer, ed., *Research Techniques in Human Engineering* (Englewood Cliffs, NJ: Prentice-Hall, 1995).

John B. West, *High Life: A History of High-Altitude Physiology and Medicine* (New York: Oxford University Press, 1998).

Christopher D. Wickens, Sallie E. Gordon, and Yili Liu, *An Introduction to Human Factors Engineering* (New York: Longman, 1998).

Earl Wiener, Barbara Kanki, and Robert Helmreich, *Cockpit Resource Management* (San Diego: Academic Press, 1993).

Earl L. Wiener and David C. Nagel, *Human Factors in Aviation* (San Diego: Academic Press, Inc., 1988).

William H. Wilmer, *Aviation Medicine in the A.E.F.* (Washington: GPO, 1920).

Tom Wolfe, *The Right Stuff* (Toronto: McGraw-Hill Ryerson, Ltd., 1979).

Hovering flight test of a free-flight model of the Hawker P.1127 V/STOL fighter underway in the return passage of the Full-Scale Tunnel. Flying-model demonstrations of the ease of transition to and from forward flight were key in obtaining the British government's support. NASA.

CASE 5

Dynamically Scaled Free-Flight Models

Joseph R. Chambers

The earliest flying machines were small models and concept demonstrators, and they dramatically influenced the invention of flight. Since the invention of the airplane, free-flight atmospheric model testing—and tests of "flying" models in wind tunnel and ground research facilities—has been a means of undertaking flight research critical to ensuring that designs meet mission objectives. Much of this testing has helped identify problems and solutions while reducing risk.

ON A HOT, MUGGY DAY IN SUMMER 1959, Joe Walker, the crusty old head of the wind tunnel technicians at the legendary NASA Langley Full-Scale Tunnel, couldn't believe what he saw in the test section of his beloved wind tunnel. Just a few decades earlier, Walker had led his technician staff during wind tunnel test operations of some of the most famous U.S. aircraft of World War II in its gigantic 30- by 60-foot test section. With names like Buffalo, Airacobra, Warhawk, Lightning, Mustang, Wildcat, Hellcat, Avenger, Thunderbolt, Helldiver, and Corsair, the test subjects were big, powerful fighters that carried the day for the United States and its allies during the war. Early versions of these aircraft had been flown to Langley Field and installed in the tunnel for exhaustive studies of how to improve their aerodynamic performance, engine cooling, and stability and control characteristics.

On this day, however, Walker was witnessing a type of test that would markedly change the research agenda at the Full-Scale Tunnel for many years to come. With the creation of the new National Aeronautics and Space Administration (NASA) in 1958 and its focus on human space flight, massive transfers of the old tunnel's National Advisory Committee for Aeronautics (NACA) personnel to new space flight priorities such as Project Mercury at other facilities had resulted in significant reductions in the tunnel's staff, test schedule, and workload. The situation had not, however, gone unnoticed by a group of brilliant engineers that had pioneered the use of remotely controlled free-flying model airplanes for predictions of the flying behavior of full-scale aircraft using a unique testing technique that had been developed and applied in a much smaller tunnel known as the Langley 12-Foot Free Flight Tunnel. The engineers' activities would benefit tremendously by use of the gigantic test section of the Full-Scale Tunnel, which would provide a tremendous increase in flying space and allow for a significant increase in the size of models used in their experiments. In view of the operational changes occurring at the tunnel, they began a strong advocacy to move their free-flight studies to the larger facility. The decision to transfer the free-flight model testing to the Full-Scale Tunnel was made in 1959 by Langley's management, and the model flight-testing was underway.

Joe Walker was observing a critical NASA free-flight model test that had been requested under joint sponsorship between NASA, industry, and the Department of Defense (DOD) to determine the flying characteristics of a 7-foot-long model of the North American X-15 research aircraft. As Walker watched the model maneuvering across the test section, he lamented the radical change of test subjects in the tunnel with several profanities and a proclamation that the testing had "gone from big-iron hardware to a bunch of damn butterflies."[1] What Walker didn't appreciate was that the revolutionary efforts of the NACA and NASA to develop tools, facilities, and testing techniques based on the use of subscale flying models were rapidly maturing and being sought by military and civil aircraft designers—not only in the Full-Scale Tunnel, but in several other unique NASA testing facilities.

For over 80 years, thousands of flight tests of "butterflies" in NACA and NASA wind tunnel facilities and outdoor test ranges have contributed valuable predictions, data, and risk reduction for the Nation's high-priority aircraft programs, space flight vehicles, and instrumented planetary probes. Free-flight models have been used in a myriad of studies as far ranging as aerodynamic drag reduction, loads caused by atmospheric gusts and landing impacts, ditching, aeroelasticity and flutter, and dynamic stability and control. The models used in the studies have been flown at conditions ranging from hovering flight to hypersonic speeds. Even a brief description of the wide variety of free-flight model applications is far beyond the intent of this essay; therefore, the following discussion is limited to activities in flight dynamics, which

1. Interview of Joseph Walker by author, NASA Langley Research Center, July 3, 1962.

includes dynamic stability and control, flight at high angles of attack, spin entry, and spinning.

Birthing the Testing Techniques

The development and use of free-flying model techniques within the NACA originated in the 1920s at the Langley Memorial Aeronautical Laboratory at Hampton, VA. The early efforts had been stimulated by concerns over a critical lack of understanding and design criteria for methods to improve aircraft spin behavior.[2] Although early aviation pioneers had been frequently using flying models to demonstrate concepts for flying machines, many of the applications had not adhered to the proper scaling procedures required for realistic simulation of full-scale aircraft motions. The NACA researchers were very aware that certain model features other than geometrical shape required application of scaling factors to ensure that the flight motions of the model would replicate those of the aircraft during flight. In particular, the requirements to scale the mass and the distribution of mass within the model were very specific.[3] The fundamental theories and derivation of scaling factors for free-flight models are based on the science known as dimensional analysis. Briefly, dynamic free-flight models are constructed so that the linear and angular motions and rates of the model can be readily scaled to full-scale values. For example, a dynamically scaled 1/9-scale model will have a wingspan 1/9 that of the airplane and it will have a weight of 1/729 that of the airplane. Of more importance is the fact that the scaled model will exhibit angular velocities that are three times faster than those of the airplane, creating a potential challenge for a remotely located human pilot to control its rapid motions.

Initial NACA testing of dynamically scaled models consisted of spin tests of biplane models that were hand-launched by a researcher or catapulted from a platform about 100 feet above the ground in an airship hangar at Langley Field.[4] As the unpowered model spun toward the ground, its path was tracked and followed by a pair of researchers holding a retrieval net similar to those used in fire rescues. To an observer,

2. Max Scherberg and R.V. Rhode, "Mass Distribution and Performance of Free Flight Models," NACA TN-268 (1927).
3. Ibid.
4. C.H. Zimmerman, "Preliminary Tests in the N.A.C.A. Free-Spinning Wind Tunnel," NACA TR-557 (1935).

the testing technique contained all the elements of an old silent movie, including the dash for the falling object. The information provided by this free-spin test technique was valuable and provided confidence (or lack thereof) in the ability of the model to predict full-scale behavior, but the briefness of the test and the inevitable delays caused by damage to the model left much to be desired.

The free-flight model testing at Langley was accompanied by other forms of analysis, including a 5-foot vertical wind tunnel in which the aerodynamic characteristics of the models could be measured during simulated spinning motions while attached to a motor-driven spinning apparatus. The aerodynamic data gathered in the Langley 5-Foot Vertical Tunnel were used for analyses of spin modes, the effects of various airplane components in spins, and the impact of configuration changes. The airstream in the tunnel was directed downward, therefore free-spinning tests could not be conducted.[5]

Meanwhile, in England, the Royal Aircraft Establishment (RAE) was aware of the NACA's airship hangar free-spinning technique and had been inspired to explore the use of similar catapulted model spin tests in a large building. The RAE experience led to the same unsatisfactory conclusions and redirected its interest to experiments with a novel 2-foot-diameter vertical free-spinning tunnel. The positive results of tests of very small models (wingspans of a few inches) in the apparatus led the British to construct a 12-foot vertical spin tunnel that became operational in 1932.[6] Tests in the facility were conducted with the model launched into a vertically rising airstream, with the model's weight being supported by its aerodynamic drag in the rising airstream. The model's vertical position in the test section could be reasonably maintained within the view of an observer by precise and rapid control of the tunnel speed, and the resulting test time could be much longer than that obtained with catapulted models. The advantages of this technique were very apparent to the international research community, and the facility features of the RAE tunnel have influenced the design of all other vertical spin tunnels to this day.

5. C. Wenzinger and T. Harris, "The Vertical Wind Tunnel of the National Advisory Committee for Aeronautics," NACA TR-387 (1931). The tunnel's vertical orientation was to minimize cyclical gravitational loads on the spinning model and apparatus as would have occurred in a horizontal tunnel.
6. H.E. Wimperis, "New Methods of Research in Aeronautics," *Journal of the Royal Aeronautical Society* (Dec. 1932), p. 985.

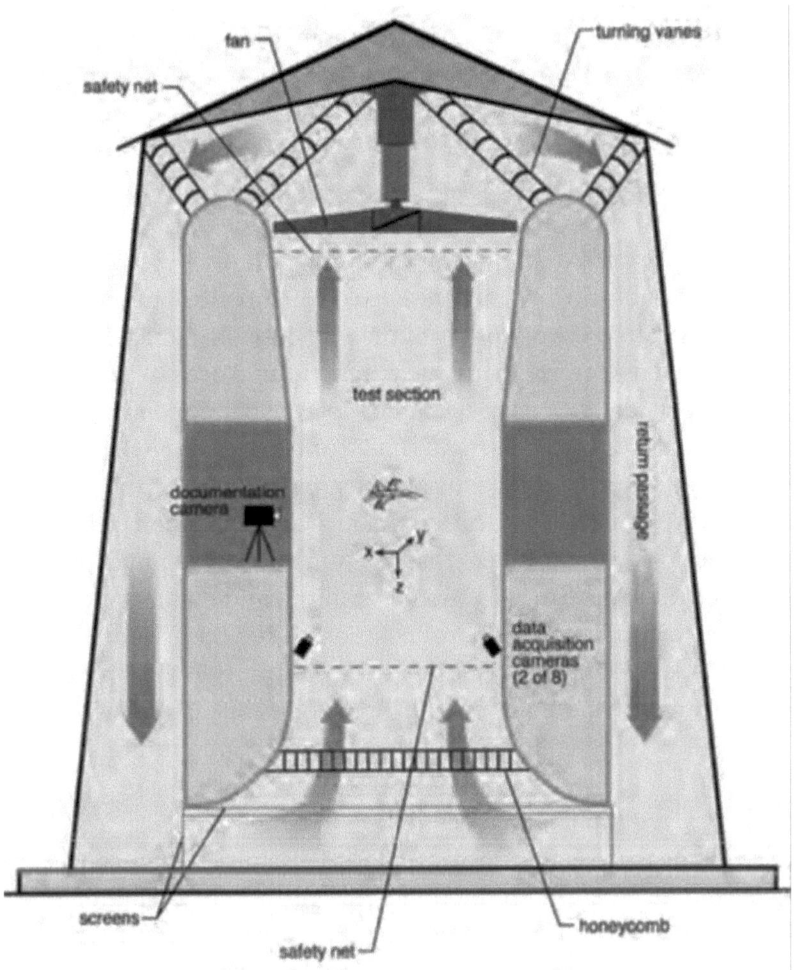

This cross-sectional view of the Langley 20-Foot Vertical Spin Tunnel shows the closed-return tunnel configuration, the location of the drive fan at the top of the facility, and the locations of safety nets above and below the test section to restrain and retrieve models. NASA.

When the NACA learned of the new British tunnel, Charles H. Zimmerman of the Langley staff led the design of a similar tunnel known as the Langley 15-Foot Free-Spinning Wind Tunnel, which became operational in 1935.[7] The use of clockwork delayed-action mechanisms to move the control surfaces of the model during the spin enabled the researchers

7. Zimmerman, "Preliminary Tests in the N.A.C.A. Free-Spinning Wind Tunnel." Zimmerman was a brilliant engineer with a notable career involving the design of dynamic wind tunnels, advanced aircraft configurations, and flying platforms, and he served NASA as a member of aerospace panels.

to evaluate the effectiveness of various combinations of spin recovery techniques. The tunnel was immediately used to accumulate design data for satisfactory spin characteristics, and its workload increased dramatically.

Langley replaced its 15-Foot Free-Spinning Wind Tunnel in 1941 with a 20-foot spin tunnel that produced higher test speeds to support scaled models of the heavier aircraft emerging at the time. Control inputs for spin recovery were actuated at the command of a researcher rather than the preset clockwork mechanisms of the previous tunnel. Copper coils placed around the periphery of the tunnel set up a magnetic field in the tunnel when energized, and the magnetic field actuated a magnetic device in the model to operate the model's aerodynamic control surfaces.[8]

The Langley 20-Foot Vertical Spin Tunnel has since continued to serve the Nation as the most active facility for spinning experiments and other studies requiring a vertical airstream. Data acquisition is based on a model space positioning system that uses retro-reflective targets attached on the model for determining model position, and results include spin rate, model attitudes, and control positions.[9] The Spin Tunnel has supported the development of nearly all U.S. military fighter and attack aircraft, trainers, and bombers during its 68-year history, with nearly 600 projects conducted for different aerospace configurations to date.

Wind Tunnel Free-Flight Techniques

Charles Zimmerman energetically continued his interest in free-flight models after the successful introduction of his 15-foot free-spinning tunnel. His next ambition was to provide a capability of investigating the dynamic stability and control of aircraft in conventional flight. His approach to this goal was to simulate the unpowered gliding flight of a model airplane in still air but to accomplish this goal in a wind tunnel with the model within view of the tunnel operators. Without power, the model would be in equilibrium in descending flight, so the tunnel airstream had to be at an inclined angle relative to the horizon. Zimmerman designed a 5-foot-diameter wind tunnel that was mounted in a yoke-like support structure such that the tunnel could be pivoted and its airstream could

8. Anshal I. Neihouse, Walter J. Klinar, and Stanley H. Scher, "Status of Spin Research for Recent Airplane Designs" NASA TR-R-57 (1962).

9. D. Bruce Owens, Jay M. Brandon, Mark A. Croom, Charles M. Fremaux, Eugene H. Heim, and Dan D. Vicroy, "Overview of Dynamic Test Techniques for Flight Dynamics Research at NASA LaRC," AIAA Paper 2006-3146 (2006).

The Langley 5-Foot Free-Flight Tunnel was mounted in a yoke assembly that permitted the test section to be tilted down for simulation of gliding flight. Its inventor, Charles Zimmerman, is on the left controlling the model, while the tunnel operator is behind the test section. NASA.

simulate various descent angles. Known as the Langley 5-Foot Free-Flight Tunnel, this exploratory apparatus was operated by two researchers—a tunnel operator, who controlled the airspeed and tilt angle of the tunnel, and a pilot, who controlled the model and assessed its behavior via a control box with a fine wire connection to the model's control actuators.[10]

Very positive results obtained in this proof-of-concept apparatus led to the design and construction of a larger 12-Foot Free-Flight Tunnel in 1939. Housed in a 60-foot-diameter sphere that permitted the tunnel to tilt upward and downward, the Langley 12-Foot Free-Flight Tunnel was designed for free-flight testing of powered as well as unpowered models. A three-person crew was used in the testing, including a tunnel airspeed controller, a tunnel tilt-angle operator, and an evaluation pilot.

The tunnel operated as the premier NACA low-speed free-flight facility for over 20 years, supporting advances in fundamental dynamic

10. Joseph R. Chambers and Mark A. Chambers, *Radical Wings and Wind Tunnels* (Specialty Press, 2008). Zimmerman was a very proficient model pilot and flew most of the tests in the apparatus.

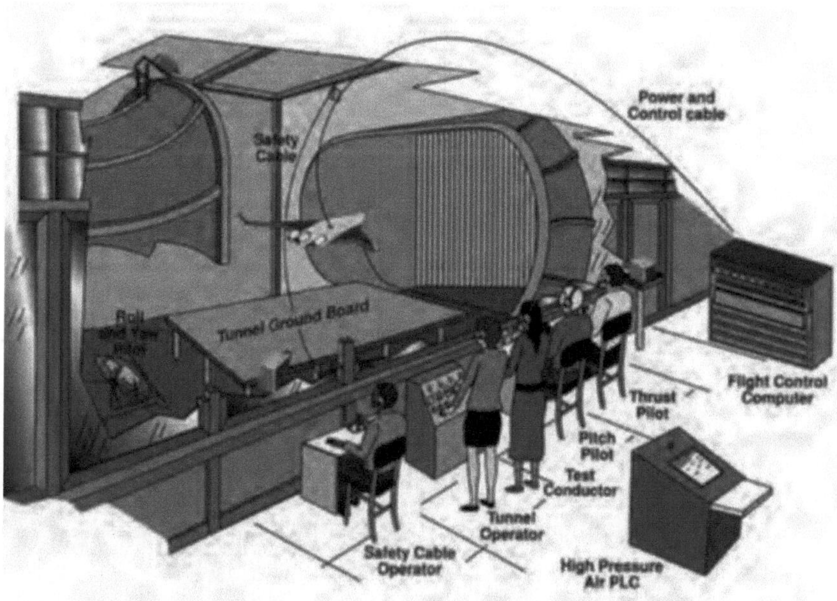

Test setup for free-flight studies at Langley. The pitch pilot is in a balcony at the side of the test section. The pilot who controls the rolling and yawing motions is at the rear of the tunnel. NASA.

stability and control theory as well as specific airplane development programs. After the 1959 decision to transfer the free-flight activities to the Full-Scale Tunnel, the tunnel pivot was fixed in a horizontal position, and the facility has continued to operate as a NASA low-cost laboratory-type tunnel for exploratory testing of advanced concepts.

Relocation of the free-flight testing to the Full-Scale Tunnel made that tunnel the focal point of free-flight applications at Langley for the next 50 years.[11] The move required updates to the test technique and the free-flight models. The test crew increased to four or more individuals responsible for piloting duties, thrust control, tunnel operations, and model retrieval and was located at two sites within the wind tunnel building. One group of researchers was in a balcony at one side of the open-throat test section, while a pilot who controlled the rolling and yawing motions of the model was in an enclosure at the rear of the test section within the structure of the tunnel exit-flow collector. Models of jet aircraft were typically powered by compressed air, and the level of

11. John P. Campbell, Jr., was head of the organization at the time of the move. Campbell was one of the youngest research heads ever employed at Langley. In addition to being an expert in flight dynamics, he later became recognized for his expertise in V/STOL aircraft technology.

thrust was controlled by a thrust pilot in the balcony. Next to the thrust pilot was a pitch pilot who controlled the longitudinal motions of the model and conducted assessments of dynamic longitudinal stability and control during flight tests. Other key members of the test crew in the balcony included the test conductor and the tunnel airspeed operator.

A light, flexible cable attached to the model supplied the model with the compressed air, electric power for control actuators, and transmission of signals for the controls and sensors carried within the model. A portion of the cable was made up of steel cable that passed through a pulley above the test section and was used to retrieve the model when the test was terminated or when an uncontrollable motion occurred. The flight cable was kept slack during the flight tests by a safety-cable operator in the balcony who accomplished the job with a high-speed winch.[12]

Free-flight models in the Full-Scale Tunnel typically had model wingspans of about 6 feet and weighed about 100 pounds. Propulsion was provided by compressed air ejectors, miniature turbofans, and high thrust/weight propeller motors. The materials used to fabricate models changed from the simple balsa free-flight construction used in the 12-Foot Free-Flight Tunnel to high-strength, lightweight composite materials. The control systems used by the free-flight models simulated the complex feedback and stabilization logic used in flight control systems for contemporary aircraft. The control signals from the pilot stations were transmitted to a digital computer in the balcony, and a special software program computed the control surface deflections required in response to pilot inputs, sensor feedbacks, and other control system inputs. Typical sensor packages included control-position indicators, linear accelerometers, and angular-rate gyros. Many models used nose-boom–mounted vanes for feedback of angle of attack and angle of sideslip, similar to systems used on full-scale aircraft. Data obtained from the flights included optical and digital recordings of model motions and pilot comments as well as analysis of the model's response characteristics.

The NACA and NASA also developed wind tunnel free-flight testing techniques to determine high-speed aerodynamic characteristics, dynamic stability of aircraft, Earth atmosphere entry configurations, planetary probes, and aerobraking concepts. The NASA Ames Research Center led the development of such facilities starting in the 1940s with the Ames

12. Owens, et al., "Overview of Dynamic Test Techniques," AIAA Paper 2006-3146.

Supersonic Free-Flight Tunnel (SFFT).[13] The SFFT, which was similar in many respects to ballistic range facilities used for testing munitions, was designed for aerodynamic and dynamic stability research at high supersonic Mach numbers (Mach numbers in excess of 10). In the SFFT, the model was fired at high speeds upstream into a supersonic airstream (typically Mach 2.0). Windows for shadowgraph photography were along the top and sides of the test section.

Data obtained from motion time histories and measurements of the model's attitudes during the brief flights were used to obtain aerodynamic and dynamic stability characteristics. The small research models had to be extremely strong to withstand high accelerations during the launch (up to 100,000 g's), yet light enough to meet requirements for dynamic mass scaling (moments of inertia). Launching the models without angular disturbances or damage was challenging and required extensive development and experience. The SFFT was completed in late 1949 and became operational in the early 1950s.

Ames later brought online its most advanced aeroballistic testing capability, the Ames Hypervelocity Free-Flight Aerodynamic Facility (HFFAF), in 1964. This facility was initially developed in support of the Apollo program and utilized both light-gas gun and shock tube technology to produce lunar return and atmospheric entry. At one end of the test section, a family of light-gas gun was used to launch specimens into the test section, while at the opposite end, a large shock tube could be simultaneously used to produce a counterflowing airstream (the result being Mach numbers of about 30). This counterflow mode of operation proved to be very challenging and was used for only a brief time from 1968 to 1971. Throughout much of the 1970s and 1980s, this versatile facility was operated as a traditional aeroballistic range, using the guns to launch models into quiescent air (or some other test gas), or as a hypervelocity impact test facility. From 1989 through 1995, the facility was operated as a shock tube–driven wind tunnel for scramjet propulsion testing. In 1997, the HFFAF underwent a major refurbishment and was returned to an aeroballistic mode of operation. It continues to operate in this mode and is NASA's only remaining aeroballistic test facility.[14]

13. Alvin Seiff, Carlton S. James, Thomas N. Canning, and Alfred G. Boissevain, "The Ames Supersonic Free-Flight Wind Tunnel," NACA RM-A52A24 (1952).

14. Charles J. Cornelison, "Status Report for the Hypervelocity Free-Flight Aerodynamic Facility," *48th Aero Ballistic Range Association Meeting, Austin, TX, Nov. 1997.*

Outdoor Free-Flight Facilities and Test Ranges

Wind tunnel free-flight testing facilities provide unique and very valuable information regarding the flying characteristics of advanced aerospace vehicles. However, they are inherently limited or unsuitable for certain types of investigations in flight dynamics. For example, vehicle motions involving large maneuvers at elevated g's, out-of-control conditions, and poststall gyrations result in significant changes in flight trajectories and altitude, which can only be studied in the expanded spaces provided by outdoor facilities. In addition, critical studies associated with high-speed flight could not be conducted in Langley's low-speed wind tunnels. Outdoor testing of dynamically scaled powered and unpowered free-flight models was therefore developed and applied in many research activities. Although outdoor test techniques are more expensive than wind tunnel free-flight tests, are subject to limitations because of weather conditions, and have inherently slower turnaround time than tunnel tests, the results obtained are unique and especially valuable for certain types of flight dynamics studies.

One of the most important outdoor free-flight test techniques developed by NASA is used in the study of aircraft spin entry motions, which includes investigations of spin resistance, poststall gyrations, and recovery controls. A significant void of information exists between the prestall and stall-departure results produced by the wind tunnel free-flight test technique in the Full-Scale Tunnel discussed earlier and the results of fully developed spin evaluations obtained in the Spin Tunnel. The lack of information in this area can be critically misleading for some aircraft designs. For example, some free-flight models exhibit severe instabilities in pitch, yaw, or roll at stall during wind tunnel free-flight tests, and they may also exhibit potentially dangerous spins from which recovery is impossible during spin tunnel tests. However, a combination of aerodynamic, control, and inertial properties can result in this same configuration exhibiting a high degree of resistance to enter the dangerous spin following a departure, despite forced spin entry attempts by a pilot. On the other hand, some configurations easily enter developed spins despite recovery controls applied by the pilot.

To evaluate the resistance of aircraft to spins, in 1950 Langley revisited the catapult techniques of the 1930s and experimented with an indoor catapult-launching technique.[15] Once again, however, the catapult technique proved to be unsatisfactory, and other approaches to study spin entry were pursued.[16] Disappointed by the inherent limitations of the catapult-launched technique, the Langley researchers began to explore the feasibility of an outdoor drop-model technique in which unpowered models would be launched from a helicopter at higher altitudes, permitting more time to study the spin entry and the effects of recovery controls. The technique would use much larger models than those used in the Spin Tunnel, resulting in a desirable increase in the test Reynolds number. After encouraging feasibility experiments were conducted at Langley Air Force Base, a search was conducted to locate a test site for research operations. A suitable low-traffic airport was identified near West Point, VA, about 40 miles from Langley, and research operations began in 1958.[17]

As testing progressed at West Point, the technique evolved into an operation consisting of launching the unpowered model at an altitude of about 2,000 feet and evaluating its spin resistance with separately located, ground-based pilots who attempted to promote spins by various combinations of control inputs and maneuvers. At the end of the test, an onboard recovery parachute was deployed and used to recover the model and lower it to a ground landing. This approach proved to be the prototype of the extremely successful drop-model testing technique that was continually updated and applied by NASA for over 50 years.

Initially, two separate tracking units consisting of modified power-driven antiaircraft gun trailer mounts were used by two pilots and two tracking operators to track and control the model. One pilot and tracker were to the side of the model's flight path, where they could control the longitudinal motions following launch, while the other pilot and tracker were about 1,000 feet away, behind the model, to control lateral-directional motions. However, as the technique was refined in later

15. Ralph W. Stone, Jr., William G. Garner, and Lawrence J. Gale, "Study of Motion of Model of Personal-Owner or Liaison Airplane Through the Stall and into the Incipient Spin by Means of a Free-Flight Testing Technique," NACA TN-2923 (1953).
16. NASA has, however, used catapulted models for spin entry studies on occasion. See James S. Bowman, Jr., "Spin-Entry Characteristics of a Delta-Wing Airplane as Determined by a Dynamic Model," NASA TN-D-2656 (1965).
17. Charles E. Libby and Sanger M. Burk, Jr., "A Technique Utilizing Free-Flying Radio-Controlled Models to Study the Incipient-and Developed-Spin Characteristics of Airplanes," NASA Memo 2-6-59L (1959).

F/A-18A drop model mounted on its launch rig on a NASA helicopter in preparation for spin entry investigations at the Langley Plum Tree test site. NASA.

years, both pilots used a single dual gun mount arrangement with a single tracker operator.

Researchers continued their search for a test site nearer to Langley, and in 1959, Langley requested and was granted approval by the Air Force to conduct drop tests at the abandoned Plum Tree bombing range near Poquoson, VA, about 5 miles from Langley. The marshy area under consideration had been cleared by the Air Force of depleted bombs and munitions left from the First and Second World War eras. A temporary building and concrete landing pad for the launch helicopter were added for operations at Plum Tree, and a surge of request jobs for U.S. high-performance military aircraft in the mid- to-late 1960s (F-14, F-15, B-1, F/A-18, etc.) brought a flurry of test activities that continued until the early 1990s.[18]

During operations at Plum Tree, the sophistication of the drop-model technique dramatically increased.[19] High-resolution video cameras were

18. In addition to specific requests from DOD, Langley conducted fundamental research on spin entry, such as the impact of automatic spin prevention.

19. David J. Fratello, Mark A. Croom, Luat T. Nguyen, and Christopher S. Domack, "Use of the Updated NASA Langley Radio-Controlled Drop-Model Technique for High-Alpha Studies of the X-29A Configuration," AIAA Paper 1987-2559 (1987).

used for tracking the model, and graphic displays were presented to a remote pilot control station, including images of the model in flight and the model's location within the range. A high-resolution video image of the model was centrally located in front of a pilot station within a building. In addition, digital displays of parameters such as angle of attack, angle of sideslip, altitude, yaw rate, and normal acceleration were also in the pilot's view. The centerpiece of operational capability was a digital flight control computer programmed with variable research flight control laws and a flight operations computer with telemetry downlinks and uplinks within the temporary building. NASA operations at Plum Tree lasted about 30 years and included a broad scope of free-flight model investigations of military aircraft, general aviation aircraft, parawings, gliding parachutes, and reentry vehicles. In the early 1990s, however, several issues regarding environmental protection forced NASA to close its research activities at Plum Tree and remove all its facilities. After considerable searching and consideration of several candidate sites, the NASA Wallops Flight Facility was chosen for Langley's drop-model activities.

The last NASA drop-model tests of a military fighter for poststall studies began in 1996 and ended in 2000.[20] This project, which evaluated the spin resistance of a 22-percent-scale model of the U.S. Navy F/A-18E Super Hornet, was the final evolution of drop-model technology for Langley. Launched from a helicopter at an altitude of about 15,000 feet in the vicinity of Wallops, the Super Hornet model weighed about 1,000 pounds. Recovery of the model at the end of the flight test was again initiated with the deployment of onboard parachutes. The model used a flotation bag after water impact and was retrieved from the Atlantic Ocean by a recovery boat.

Outdoor free-flight model testing has also flourished at NASA Dryden Flight Research Center. Dryden's primary advocate and highly successful user of free-flight models for low-speed research on advanced aerospace vehicles was the late Robert Dale Reed. An avid model builder, pilot, and researcher, Reed was inspired by his perceived need for a subscale free-flight model demonstrator of an emerging lifting body reentry configuration created by NASA Ames in 1962.[21] After initial testing of gliders of the Ames M2-F1 lifting body concept, he progressed into

20. Mark A. Croom, Holly M. Kenney, and Daniel G. Murri, "Research on the F/A-18E/F Using a 22%-Dynamically-Scaled Drop Model," AIAA Paper 2000-3913 (2000).
21. R. Dale Reed, *Wingless Flight: The Lifting Body Story*, NASA SP-4220 (1997).

Dryden free-flight research models of reentry lifting bodies. Dale Reed, second from left, and his test team pose with the mother ship and models of the M2-F2 and the Hyper III configurations. NASA.

the technique of using radio-controlled model tow planes to tow and release M2-F1 models. In the late 1960s, the launching technique for the unpowered models evolved with a powered radio-controlled mother ship, and by 1968, Reed's mother ship had conducted over 120 launches. Dale Reed's innovation and approach to using radio-controlled mother ships for launching drop models of radical configurations have endured to this day as the preferred method for small-scale free-flight activities at Dryden.

In the early 1970s, Reed's work at Dryden expanded into a series of flight tests of powered and unpowered remotely piloted research vehicles (RPRVs). These activities, which included remote-control evaluations of subscale and full-scale test subjects, used a ground-based cockpit equipped with flight instruments and sensors typical of a representative full-scale airplane. These projects included the Hyper III lifting body and a three-eighths-scale dynamically scaled model of the F-15. The technique used for the F-15 model consisted of air launches of the test article from a B-52 and control by a pilot in a ground cockpit outfitted with a sophisticated control system.[22] The setup featured a digital uplink capability, a ground computer, a television monitor, and a telemetry system. Initially, the F-15 model was recovered on its parachute in flight by helicopter midair snatch, but in later flights, it was landed on skids by the evaluation pilot.

NASA Ames also conducted and sponsored outdoor free-flight powered model testing in the 1970s as a result of interest in the oblique wing concept championed by Robert T. Jones. The progression of sophistication in these studies started with simple unpowered catapult-launched models at Ames, followed by cooperative powered model tests at Dryden in the 1970s and piloted flight tests of the AD-1 oblique wing demonstrator aircraft in the 1980s.[23] In the 1990s, Ames and Stanford University collaborated on potential designs for oblique wing supersonic transport designs, which led to flight tests of two free-flight models by Stanford.

Yet another historic high-speed outdoor free-flight facility was spun off Langley's interests. In 1945, a proposal was made to develop a new NACA high-speed test range known as the Pilotless Aircraft Research Station, which would use rocket-boosted models to explore the transonic and supersonic flight regimes. The facility ultimately became known as the NACA Wallops Island Flight Test Range.[24] From 1945 through 1959, Wallops served as a rocket-model "flying wind tunnel" for researchers in Langley's Pilotless Aircraft Research Division (PARD), which conducted vital investigations for the Nation's emerging supersonic aircraft, especially the Century series of advanced fighters in the 1950s. Rocket-boosted models were used by the Pilotless Aircraft Research Division of the NACA's Langley Laboratory in flight tests at Wallops to obtain valuable information on aerodynamic drag, dynamic stability, and control effectiveness at transonic conditions.

---

22. Euclid C. Holleman, "Summary of Flight Tests to Determine the Spin and Controllability Characteristics of a Remotely Piloted, Large-Scale (3/8) Fighter Airplane Model," NASA TN-D-8052 (1976).
23. Michael J. Hirschberg and David M. Hart, "A Summary of a Half-Century of Oblique Wing Research," AIAA Paper 2007-150 (2007).
24. Joseph A. Shortal, *A New Dimension. Wallops Island Flight Test Range: The First Fifteen Years*, NASA RP-1028 (1978).

Applications

Free-flight models are complementary to other tools used in aeronautical engineering. In the absence of adverse scale effects, the aerodynamic characteristics of the models have been found to agree very well with data obtained from other types of wind tunnel tests and theoretical analyses. By providing insight into the impact of aerodynamics on vehicle dynamics, the free-flight results help build the necessary understanding of critical aerodynamic parameters and the impact of modifications to resolve problems. The ability to conduct free-flight tests and aerodynamic measurements with the same model is a powerful advantage for the testing technique. When coupled with more sophisticated static wind tunnel tests, computational fluid dynamics methods, and piloted simulator technology, these tests are extremely informative. Finally, even the very visual results of free-flight tests are impressive, whether they demonstrate to critics and naysayers that radical and unconventional designs can be flown or identify a critical flight problem and potential solutions for a new configuration.

The most appropriate applications of free-flight models involve evaluations of unconventional designs for which no experience base exists and the analysis of aircraft behavior for flight conditions that are not easily studied with other methods because of complex aerodynamic phenomena that cannot be modeled at the present time.[25] Examples include flight in which separated flows, nonlinear aerodynamic behavior, and large dynamic motions are typically encountered.

The following discussion presents a brief overview of the historical applications and technological impacts of the use of free-flight models for studies of flight dynamics by the NACA and NASA in selected areas.

The most important applications have been in

- Dynamic stability and control.
- Flight at high angles of attack.[26]
- Spinning and spin recovery.
- Spin entry and poststall motions.

25. Campbell, "Free and Semi-Free Model Flight-Testing Techniques Used in Low-Speed Studies of Dynamic Stability and Control," NATO Advisory Group for Aeronautical Research and Development AGARDograph 76 (1963).
26. This topic is discussed for military applications in another case study in this volume by the same author.

Dynamic Stability: Early Applications and a Lesson Learned

When Langley began operations of its 12-Foot Free-Flight Tunnel in 1939, it placed a high priority on establishing correlation with full-scale flight results. Immediately, requests came from the Army and Navy for correlation of model tests with flight results for the North American BT-9, Brewster XF2A-1, Vought-Sikorsky V-173, Naval Aircraft Factory SBN-1, and Vought Sikorsky XF4U-1. Meanwhile, the NACA used a powered model of the Curtiss P-36 fighter for an in-house calibration of the free-flight process.[27]

The results of the P-36 study were, in general, in fair agreement with airplane flight results, but the dynamic longitudinal stability of the model was found to be greater (more damped) than that of the airplane, and the effectiveness of the model's ailerons was less than that for the airplane. Both discrepancies were attributed to aerodynamic deficiencies of the model caused by the low Reynolds number of the tunnel test and led to one of the first significant lessons learned with the free-flight technique. Using the wing airfoil shape (NACA 2210) of the full-scale P-36 for the model resulted in poor wing aerodynamic performance at the low Reynolds number of the model flight tests. The maximum lift of the model and the angle of attack for maximum lift were both decreased because of scale effects. As a result, the stall occurred at a slightly lower angle of attack for the model. After this experience, researchers conducted an exhaustive investigation of other airfoils that might have more satisfactory performance at low Reynolds numbers. In planning for subsequent tests, the researchers were trained to anticipate the potential existence of scale effects for certain airfoils, even at relatively low angles of attack. As a result of this experience, the wing airfoils of free-flight tunnel models were sometimes modified to airfoil shapes that provided better results at low Reynolds number.[28]

Progress and Design Data

In the 1920s and 1930s, researchers in several wind tunnel and full-scale aircraft flight groups at Langley conducted analytical and experimental investigations to develop design guidelines to ensure satisfactory stability

27. Joseph A. Shortal and Clayton J. Osterhout, "Preliminary Stability and Control Tests in the NACA Free-Flight Tunnel and Correlation with Flight Tests," NACA TN-810 (1941).
28. Charles L. Seacord, Jr., and Herman O. Ankenbruck, "Determination of the Stability and Control Characteristics of a Straight-Wing, Tailless Fighter-Airplane Model in the Langley Free-Flight Tunnel," NACA Wartime Report ACR L5K05 (1946).

and control behavior.[29] Such studies sought to develop methods to reliably predict the inherent flight characteristics of aircraft as affected by design variables such as the wing dihedral angle, sizes and locations of the vertical and horizontal tails, wing planform shape, engine power, mass distribution, and control surface geometry. The staff of the Free-Flight Tunnel joined in these efforts with several studies that correlated the qualitative behavior of free-flight models with analytical predictions of dynamic stability and control characteristics. Coupled with the results from other facilities and analytical groups, the free-flight results accelerated the maturity of design tools for future aircraft from a qualitative basis to a quantitative methodology, and many of the methods and design data derived from these studies became classic textbook material.[30]

By combining free-flight testing with theory, the researchers were able to quantify desirable design features, such as the amount of wing-dihedral angle and the relative size of vertical tail required for satisfactory behavior. With these data in hand, methods were also developed to theoretically solve the dynamic equations of motion of aircraft and determine dynamic stability characteristics such as the frequency of inherent oscillations and the damping of motions following inputs by pilots or turbulence.

During the final days of model flight projects in the Free-Flight Tunnel in the mid-1950s, various Langley organizations teamed to quantify the effects of aerodynamic dynamic stability parameters on flying characteristics. These efforts included correlation of experimentally determined aerodynamic stability derivatives with theoretical predictions and comparisons of the results of qualitative free-flight tests with theoretical predictions of dynamic stability characteristics. In some cases, rate gyroscopes and servos were used to artificially vary the magnitudes of dynamic aerodynamic stability parameters such as yawing moment because of rolling.[31] In these studies, the free-flight model result served as a critical test of the validity of theory.

29. M.O. McKinney, "Experimental Determination of the Effects of Dihedral, Vertical Tail Area, and Lift Coefficient on Lateral Stability and Control Characteristics," NACA TN-1094 (1946).
30. Campbell and Seacord, "The Effect of Mass Distribution on the Lateral Stability and Control Characteristics of an Airplane as Determined by Tests of a Model in the Free-Flight Tunnel," NACA TR-769 (1943).
31. Robert O. Schade and James L. Hassell, Jr., "The Effects on Dynamic Lateral Stability and Control of Large Artificial Variations in the Rotary Stability Derivatives," NACA TN-2781 (1953).

High-Speed Investigations

High-speed studies of dynamic stability were very active at Wallops. The scope and contributions of the Wallops rocket-boosted model research programs for aircraft configurations, missiles, and airframe components covered an astounding number of technical areas, including aerodynamic performance, flutter, stability and control, heat transfer, automatic controls, boundary-layer control, inlet performance, ramjets, and separation behavior of aircraft components and stores. As an example of test productivity, in just 3 years beginning in 1947, over 386 models were launched at Wallops to evaluate a single topic: roll control effectiveness at transonic conditions. These tests included generic configurations and models with wings representative of the historic Douglas D-558-2 Skyrocket, Douglas X-3 Stiletto, and Bell X-2 research aircraft.[32] Fundamental studies of dynamic stability and control were also conducted with generic research models to study basic phenomena such as longitudinal trim changes, dynamic longitudinal stability, control-hinge moments, and aerodynamic damping in roll.[33] Studies with models of the D-558-2 also detected unexpected coupling of longitudinal and lateral oscillations, a problem that would subsequently prove to be common for configurations with long fuselages and relatively small wings.[34] Similar coupled motions caused great concern in the X-3 and F-100 aircraft development programs and spurred on numerous studies of the phenomenon known as inertial coupling.

More than 20 specific aircraft configurations were evaluated during the Wallops studies, including early models of such well-known aircraft as the Douglas F4D Skyray, the McDonnell F3H Demon, the Convair B-58 Hustler, the North American F-100 Super Sabre, the Chance Vought F8U Crusader, the Convair F-102 Delta Dagger, the Grumman F11F Tiger, and the McDonnell F-4 Phantom II.

32. Carl A. Sandahl, "Free-Flight Investigation at Transonic and Supersonic Speeds of a Wing-Aileron Configuration Simulating the D558-2 Airplane," NACA RM-L8E28 (1948); and Sandahl, "Free-Flight Investigation at Transonic and Supersonic Speeds of the Rolling Effectiveness for a 42.7° Sweptback Wing Having Partial-Span Ailerons," NACA RM-L8E25 (1948).

33. Examples include James H. Parks and Jesse L. Mitchell, "Longitudinal Trim and Drag Characteristics of Rocket-Propelled Models Representing Two Airplane Configurations," NACA RM-L9L22 (1949); and James L. Edmondson and E. Claude Sanders, Jr., "A Free-Flight Technique for Measuring Damping in Roll by Use of Rocket-Powered Models and Some Initial Results for Rectangular Wings," NACA RM-L9101 (1949).

34. Parks, "Experimental Evidence of Sustained Coupled Longitudinal and Lateral Oscillations From Rocket-Propelled Model of a 35° Swept-Wing Airplane Configuration," NACA RM-L54D15 (1954).

Shadowgraph of X-15 model in free flight during high-speed tests in the Ames SFFT facility. Shock wave patterns emanating from various airframe components are visible. NASA.

High-speed dynamic stability testing techniques at the Ames SFFT included studies of the static and dynamic stability of blunt-nose reentry shapes, including analyses of boundary-layer separation.[35] This work included studies of the supersonic dynamic stability characteristics of the Mercury capsule. Noting the experimental observation of nonlinear variations of pitching moment with angle of attack typically exhibited by blunt bodies, Ames researchers contributed a mathematical method for including such nonlinearities in theoretical analyses and predictions of capsule dynamic stability at supersonic speeds. During the X-15 program, Ames conducted free-flight testing in the SFFT to define stability, control, and flow-field characteristics of the configuration at high supersonic speeds.[36]

Out of the Box: V/STOL Configurations

International interest in Vertical Take-Off and Landing (VTOL) and Vertical/Short Take-Off and Landing (V/STOL) configurations escalated during the 1950s and persisted through the mid-1960s with a huge number of radical propulsion/aircraft combinations proposed and evaluated

35. Maurice L. Rasmussen, "Determination of Nonlinear Pitching-Moment Characteristics of Axially Symmetric Models From Free-Flight Data," NASA TN-D-144 (1960).

36. Alfred G. Boissevain and Peter F. Intrieri, "Determination of Stability Derivatives from Ballistic Range Tests of Rolling Aircraft Models," NASA TM-X-399 (1961).

5

throughout industry, DOD, the NACA, and NASA. The configurations included an amazing variety of propulsion concepts to achieve hovering flight and the conversion to and from conventional forward flight. However, all these aircraft concepts were plagued with common issues regarding stability, control, and handling qualities.[37]

The first VTOL nonhelicopter concept to capture the interests of the U.S. military was the vertical-attitude tail-sitter concept. In 1947, the Air Force and Navy initiated an activity known as Project Hummingbird, which requested design approaches for VTOL aircraft. At Langley, discussions with Navy managers led to exploratory NACA free-flight studies in 1949 of simplified tail-sitter models to evaluate stability and control during hovering flight. Conducted in a large open area within a building, powered-model testing enabled researchers to explore the dynamic stability and control of such configurations.[38] The test results provided valuable information on the relative severity of unstable oscillations encountered during hovering flight. The instabilities in roll and pitch were caused by aerodynamic interactions of the propeller during forward or sideward translation, but the period of the growing oscillations was sufficiently long to permit relatively easy control. The model flight tests also provided guidance regarding the level of control power required for satisfactory maneuvering during hovering flight.

Navy interest in the tail-sitter concept led to contracts for the development of the Consolidated-Vultee (later Convair) XFY-1 "Pogo" and the Lockheed XFV-1 "Salmon" tail-sitter aircraft in 1951. The Navy asked Langley to conduct dynamic stability and control investigations of both configurations using its free-flight model test techniques. In 1952, hovering flights of the Pogo were conducted within the huge return passage of the Langley Full-Scale Tunnel, followed by transition flights from hovering to forward flight in the tunnel test section during a brief break in the tunnel's busy test schedule.[39] Observed by Convair

37. Chambers, *Radical Wings and Wind Tunnels*.

38. William R. Bates, Powell M. Lovell, Jr., and Charles C. Smith, Jr., "Dynamic Stability and Control Characteristics of a Vertically Rising Airplane Model in Hovering Flight," NACA RM-L50J16 (1951).

39. Hovering and transition tests included: Lovell, Smith, and R.H. Kirby, "Stability and Control Flight Tests of a 0.13-Scale Model of the Consolidated Vultee XFY-1 Airplane in Take-Offs, Landings, and Hovering Flight," NACA RM-SL52I26 (1952); and Lovell, Smith, and Kirby, "Flight Investigation of the Stability and Control Characteristics of a 0.13-Scale Model of the Convair XFY-1 Vertically Rising Airplane During Constant-Altitude Transitions," NACA RM-SL53E18 (1953).

personnel (including the XFY-1 test pilot), the flight tests provided encouragement and confidence to the visitors and the Navy.

Without doubt, the most successful NASA application of free-flight models for VTOL research was in support of the British P.1127 vectored-thrust fighter program. As the British Hawker Aircraft Company matured its design of the revolutionary P.1127 in the late 1950s, Langley's senior manager, John P. Stack, became a staunch supporter of the activity and directed that tests in the 16-Foot Transonic Tunnel and free-flight research activities in the Full-Scale Tunnel be used for cooperative development work.[40]

In response to the directive, a one-sixth-scale free-flight model was flown in the Full-Scale Tunnel to examine the hovering and transition behavior of the design. Results of the free-flight tests were witnessed by Hawker staff members, including the test pilot slated to conduct the first transition flights, were very impressive. The NASA researchers regarded the P.1127 model as the most docile V/STOL configuration ever flown during their extensive experiences with free-flight VTOL designs. As was the case for many free-flight model projects, the motion-picture segments showing successful transitions from hovering to conventional flight in the Full-Scale Tunnel were a powerful influence in convincing critics that the concept was feasible. In this case, the model flight demonstrations helped sway a doubtful British government to fund the project. Refined versions of the P.1127 design were subsequently developed into today's British Harrier and Boeing AV-8 fighter/attack aircraft.

The NACA and NASA also conducted pioneering free-flight model research on tilt wing aircraft for V/STOL missions. In the early 1950s, several generic free-flight propeller-powered models were flown to evaluate some of the stability and control issues that were anticipated to limit the feasibility of the concept.[41] The fundamental principle used by the tilt wing concept to convert from hovering to forward flight involves reorienting the wing from a vertical position for takeoff to a conventional position for forward flight. However, this simple conversion of the wing angle relative to the fuselage brings major challenges. For example, the

40. Smith, "Flight Tests of a 1/6-Scale Model of the Hawker P.1127 Jet VTOL Airplane," NASA TM-SX-531 (1961).

41. Lovell and Lysle P. Parlett, "Hovering-Flight Tests of a Model of a Transport Vertical Take-Off Airplane with Tilting Wing and Propellers," NACA TN-3630 (1956); Lovell and Parlett, "Flight Tests of a Model of a High-Wing Transport Vertical-Take-Off Airplane With Tilting Wing and Propellers and With Jet Controls at the Rear of the Fuselage for Pitch and Yaw Control," NACA TN-3912 (1957).

wing experiences large changes in its angle of attack relative to the flight path during the transition, and areas of wing stall may be encountered during the maneuver. The asymmetric loss of wing lift during stall can result in wing-dropping, wallowing motions and uncommanded transient maneuvers. Therefore, the wing must be carefully designed to minimize or eliminate flow separation that would otherwise result in degraded or unsatisfactory stability and control characteristics. Extensive wind tunnel and flight research on many generic NACA and NASA models, as well as the Hiller X-18, Vertol VZ-2, and Ling-Temco-Vought XC-142A tilt wing configurations at Langley, included a series of free-flight model tests in the Full-Scale Tunnel.[42]

Coordinated closely with full-scale flight tests, the model testing initially focused on providing early information on dynamic stability and the adequacy of control power in hovering and transition flight for the configurations. However, all projects quickly encountered the anticipated problem of wing stall, especially in reduced-power descending flight maneuvers. Tilt wing aircraft depend on the high-energy slipstream of large propellers to prevent local wing stall by reducing the effective angle of attack across the wingspan. For reduced-power conditions, which are required for steep descents to accomplish short-field missions, the energy of the slipstream is severely reduced, and wing stall is experienced. Large uncontrolled dynamic motions may be exhibited by the configuration for such conditions, and the undesirable motions can limit the descent capability (or safety) of the airplane. Flying model tests provided valuable information on the acceptability of uncontrolled motions such as wing dropping and lateral-directional wallowing during descent, and the test technique was used to evaluate the effectiveness of aircraft modifications such as wing flaps or slats, which were ultimately adapted by full-scale aircraft such as the XC-142A.

As the 1960s drew to a close, the worldwide engineering community began to appreciate that the weight and complexity required for VTOL missions presented significant penalties in aircraft design. It therefore

42. Louis P. Tosti, "Flight Investigation of Stability and Control Characteristics of a 1/8-Scale Model of a Tilt-Wing Vertical-Take-Off-And-Landing Airplane," NASA TN-D-45 (1960); Tosti, "Longitudinal Stability and Control of a Tilt-Wing VTOL Aircraft Model with Rigid and Flapping Propeller Blades," NASA TN-D-1365 (1962); William A. Newsom and Robert H. Kirby, "Flight Investigation of Stability and Control Characteristics of a 1/9-Scale Model of a Four-Propeller Tilt-Wing V/STOL Transport," NASA TN-D-2443 (1964).

turned its attention to the possibility of providing less demanding STOL capability with fewer penalties, particularly for large military transport aircraft. Langley researchers had begun to explore methods of using propeller or jet exhaust flows to induce additional lift on wing surfaces in the 1950s, and although the magnitude of lift augmentation was relatively high, practical propulsion limitations stymied the application of most concepts.

A particularly promising concept known as the externally blown flap (EBF) used the redirected jet engine exhausts from conventional pod-mounted engines to induce additional circulation lift at low speeds for takeoff and landing.[43] However, the relatively hot exhaust temperatures of turbojets of the 1950s were much too high for structural integrity and feasible applications. Nonetheless, Langley continued to explore and mature such ideas, known as powered-lift concepts. These research studies embodied conventional powered model tests in several wind tunnels, including free-flight investigations of the dynamic stability and control of multiengine EBF configurations in the Full-Scale Tunnel, with emphasis on providing satisfactory lateral control and lateral-directional trim after the failure of an engine. Other powered-lift concepts were also explored, including the upper-surface-blowing (USB) configuration, in which the engine exhaust is directed over the upper surface of the wing to induce additional circulation and lift.[44] Advantages of this approach included potential noise shielding and flow-turning efficiency.

While Langley continued its fundamental research on EBF and USB configurations, in the early 1970s, an enabling technology leap occurred with the introduction of turbofan engines, which inherently produce relatively cool exhaust fan flows.[45] The turbofan was the perfect match for these STOL concepts, and industry's awareness and participation in the basic NASA research program matured the state of the art for design data for powered-lift aircraft. The free-flight model results, coupled with NASA piloted simulator studies of full-scale aircraft STOL missions, helped provide the fundamental knowledge and data required to reduce

43. Campbell and Joseph L. Johnson, Jr., "Wind-Tunnel Investigation of an External-Flow Jet-Augmented Slotted Flap Suitable for Applications to Airplanes with Pod-Mounted Jet Engines," NACA TN-3898 (1956).

44. Parlett, "Free-Flight Wind-Tunnel Investigation of a Four-Engine Sweptwing Upper-Surface Blown Transport Configuration," NASA TM-X-71932 (1974).

45. Parlett, "Free-Flight Investigation of the Stability and Control Characteristics of a STOL Model with an Externally Blown Jet Flap," NASA TN-D-7411 (1974); Chambers, *Radical Wings and Wind Tunnels*.

John P. Campbell, Jr., left, inventor of the externally blown flap, and Gerald G. Kayten of NASA Headquarters pose with a free-flight model of an STOL configuration at the Full-Scale Tunnel. Slotted trailing-edge flaps were used to deflect the exhaust flows of turbofan engines. NASA.

risk in development programs. Ultimately applied to the McDonnell-Douglas YC-15 and Boeing YC-14 prototype transports in the 1970s and to today's Boeing C-17, the EBF and USB concepts were the result of over 30 years of NASA research and development, including many valuable studies of free-flight models in the Full-Scale Tunnel.[46]

Breakthrough: Variable Sweep

Spurred on by postwar interests in the variable-wing-sweep concept as a means to optimize mission performance at both low and high speeds, the NACA at Langley initiated a broad research program to identify the potential benefits and problems associated with the concept. The disappointing experiences of the Bell X-5 research aircraft, which used a single wing pivot to achieve variable sweep in the early 1950s, had clearly identified the unacceptable weight penalties associated with the concept of translating the wing along the fuselage centerline to maintain satisfactory levels of longitudinal stability while the wing sweep angle was varied from forward to aft sweep. After the X-5 experience, military interest in variable sweep quickly diminished while aerodynamicists at

46. Campbell originally conceived the EBF concept and was awarded a patent for his invention.

Langley continued to explore alternate concepts that might permit variations in wing sweep without moving the wing pivot location and without serious degradation in longitudinal stability and control.

After years of intense research and wind tunnel testing, Langley researchers conceived a promising concept known as the outboard pivot.[47] The basic principle involved in the NASA solution was to pivot the movable wing panels at two outboard pivot locations on a fixed inner wing and share the lift between the fixed portion of the wing and the movable outer wing panel, thereby minimizing the longitudinal movement of the aerodynamic center of lift for various flight speeds. As the concept was matured in configuration studies and supporting tests, refined designs were continually submitted to intense evaluations in tunnels across the speed range from supersonic cruise conditions to subsonic takeoff and landing.[48]

The use of dynamically scaled free-flight models to evaluate the stability and control characteristics of variable-sweep configurations was an ideal application of the testing technique. Since variable-sweep designs are capable of an infinite number of wing sweep angles between the forward and aft positions, the number of conventional wind tunnel force tests required to completely document stability and control variations with wing sweep for every sweep angle could quickly become unacceptable. In contrast, a free-flight model with continually variable wing sweep angles could be used to quickly examine qualitative characteristics as its geometry changed, resulting in rapid identification of significant problems. Free-flight model investigations of a configuration based on a proposed Navy combat air patrol (CAP) mission in the Full-Scale Tunnel provided a convincing demonstration that the outboard pivot was ready for applications.

The oblique wing concept (sometimes referred to as the "switchblade wing" or "skewed wing") had originated in the German design studies of the Blohm & Voss P202 jet aircraft during World War II and was pursued at Langley by R.T. Jones. Oblique wing designs use a single-pivot, all-moving wing to achieve variable sweep in an asymmetrical fashion. The wing is positioned in the conventional unswept position for takeoff and landings, and it is rotated about its single pivot point for high-speed flight. As part of a general research effort that included

47. Chambers, *Radical Wings and Wind Tunnels*. Langley researchers Polhamus and William J. Alford were awarded a patent for the outboard pivot concept.

48. Polhamus and Thomas A. Toll, "Research Related to Variable Sweep Aircraft Development," NASA TM-83121 (1981).

theoretical aerodynamic studies and conventional wind tunnel tests, a free-flight investigation of the dynamic stability and control of a simplified model was conducted in the Free-Flight Tunnel in 1946.[49] This research on the asymmetric swept wing actually predated NACA wind tunnel research on symmetrical variable sweep concepts with a research model of the Bell X-1.[50] The test objectives were to determine whether such a radical aircraft configuration would exhibit satisfactory stability characteristics and remain controllable in the swept wing asymmetric state at low-speed flight conditions. The results of the flight tests, which were the first U.S. flight studies of oblique wings ever conducted, showed that the wing could be swept as much as 40 degrees without significant degradation in behavior. However, when the sweep angle was increased to 60 degrees, an unacceptable longitudinal trim change was experienced, and a severe reduction in lateral control occurred at moderate and high angles of attack. Nonetheless, the results obtained with the simple free-flight model provided optimism that the unconventional oblique wing concept might be feasible from a perspective of stability and control.

R.T. Jones transferred to the NACA Ames Aeronautical Laboratory in 1947 and continued his brilliant career there, which included his continuing interest in the application of oblique wing technology. In the early 1970s, the scope of NASA studies on potential civil supersonic transport configurations included an effort by an Ames team headed by Jones that examined a possible oblique wing version of the supersonic transport. Although wind tunnel testing was conducted at Ames, the demise and cancellation of the American SST program in the early 1970s terminated this activity. Wind tunnel and computational studies of oblique wing designs continued at Ames throughout the 1970s for subsonic, transonic, and supersonic flight applications.[51] Jones stimulated and participated in flight tests of several oblique wing radio-controlled models, and a joint Ames-Dryden project was initiated to use a remotely piloted research aircraft known as the Oblique Wing Research Aircraft (OWRA) for studies of the aerodynamic characteristics and control requirements to achieve satisfactory handling qualities.

49. Campbell and Hubert M. Drake, "Investigation of Stability and Control Characteristics of an Airplane Model with Skewed Wing in the Langley Free-Flight Tunnel," NACA TN-1208 (1947).
50. Polhamus and Toll, "Variable Sweep Aircraft Development," NASA TM-83121.
51. Michael J. Hirschberg and David M. Hart, "A Summary of a Half-Century of Oblique Wing Research," AIAA Paper 2007-150 (2007).

Growing interest in the oblique wing and the success of the OWRA remotely piloted vehicle project led to the design and low-speed flight demonstrations of a full-scale research aircraft known as the AD-1 in the late 1970s. Designed as a low-cost demonstrator, the radical AD-1 proved to be a showstopper during air shows and generated considerable public interest.[52] The flight characteristics of the AD-1 were quite satisfactory for wing-sweep angles of less than about 45 degrees, but the handling qualities degraded for higher values of sweep, in agreement with the earlier Langley exploratory free-flight model study.

After his retirement, Jones continued his interest in supersonic oblique wing transport configurations. When the NASA High-Speed Research program to develop technologies necessary for a viable supersonic transport began in the 1990s, several industry teams revisited the oblique wing for potential applications. Ames sponsored free-flight radio-controlled model studies of oblique wing configurations at Stanford University in the early 1990s. As a result of free-flight model contributions from Langley, Ames, Dryden, and academia, major issues regarding potential dynamic stability and control problems for oblique wing configurations have been addressed for low-speed conditions. Unfortunately, funding for transonic and supersonic model flight studies has not been forthcoming, and high-speed studies have not yet been accomplished.

Safe Return: Space Capsules

The selection of blunt capsule designs for the Mercury, Gemini, and Apollo programs resulted in numerous investigations of the dynamic stability and recovery of such shapes. Nonlinear, unstable variations of aerodynamic forces and moments with angle of attack and sideslip were known to exist for these configurations, and extensive conventional force tests, dynamic free-flight model tests, and analytical studies were conducted to define the nature of potential problems that might be encountered during atmospheric reentry. At Ames, the supersonic and hypersonic free-flight aerodynamic facilities have been used to observe dynamic stability characteristics, extract aerodynamic data from flight tests, provide stabilizing concepts, and develop mathematical models for flight simulation at hypersonic and supersonic speeds.

52. Weneth D. Painter, "AD-1 Oblique Wing Research Aircraft Pilot Evaluation Program," AIAA Paper 1983-2509 (1983).

Meanwhile, at Langley, researchers in the Spin Tunnel were conducting dynamic stability investigations of the Mercury, Gemini, and Apollo capsules in vertically descending subsonic flight.[53]

Results of these studies dramatically illustrated potential dynamic stability issues during the spacecraft recovery procedure. For example, the Gemini capsule model was very unstable; it would at various times oscillate, tumble, or spin about a vertical axis with its symmetrical axis tilted as much as 90 degrees from the vertical. However, the deployment of a drogue parachute during any spinning or tumbling motions quickly terminated these unstable motions at subsonic speeds. Extensive tests of various drogue-parachute configurations resulted in definitions of acceptable parachute bridle-line lengths and attachment points. Spin Tunnel results for the Apollo command module configuration were even more dramatic. The Apollo capsule with blunt end forward was dynamically unstable and displayed violent gyrations, including large oscillations, tumbling, and spinning motions. With the apex end forward, the capsule was dynamically stable and would trim at an angle of attack of about 40 degrees and glide in large circles. Once again, the use of a drogue parachute stabilized the capsule, and the researchers also found that retention of the launch escape system, with either a drogue parachute or canard surfaces attached to it, would prevent an unacceptable apex-forward trim condition during launch abort.

Following the Apollo program, NASA conducted a considerable effort on unpiloted space probes and planetary exploration. In the Langley Spin Tunnel, several planetary-entry capsule configurations were tested to evaluate their dynamic stability during descent, with a priority in simulating descent in the Martian atmosphere.[54] Studies also included assessments of the Pioneer Venus probe in the 1970s. These tests provided considerable design information on the dynamic stability of a variety of potential planetary exploration capsule shapes. Additional studies

53. James S. Bowman, Jr., "Dynamic Model Tests at Low Subsonic Speeds of Project Mercury Capsule Configurations With and Without Drogue Parachutes," NASA TM-X-459 (1961); Henry A. Lee, Peter S. Costigan, and Bowman, "Dynamic Model Investigation of a 1/20-Scale Gemini Spacecraft in the Langley Spin Tunnel," NASA TN-D-2191 (1964); Henry A. Lee and Sanger M. Burk, "Low-Speed Dynamic Model Investigation of Apollo Command Module Configuration in the Langley Spin Tunnel," NASA TN-D-3888 (1967).

54. Costigan, "Dynamic-Model Study of Planetary-Entry Configurations in the Langley Spin Tunnel," NASA TN-D-3499 (1966).

Photograph of a free-flight model of the Project Mercury capsule in vertical descent in the Spin Tunnel with drogue parachute deployed. Tests to improve the dynamic stability characteristics of capsules have continued to this day. NASA.

of the stability characteristics of blunt, large-angle capsules were conducted in the late 1990s in the Spin Tunnel.

As the new millennium began, NASA's interests in piloted and unpiloted planetary exploration resulted in additional studies of dynamic stability in the Spin Tunnel. Currently, the tunnel and its dynamic model testing techniques are supporting NASA's Constellation program for lunar exploration. Included in the dynamic stability testing are the Orion launch abort vehicle, the crew module, and alternate launch abort systems.[55]

5

A Larger Footprint: Reentry Vehicles and Lifting Bodies

The NACA and military visionaries initiated early efforts for the X-15 hypersonic research aircraft, in-house design studies for hypersonic vehicles were started at Langley and Ames, and the Air Force began its X-20 Dyna-Soar space plane program. The evolution of long, slender configurations and others with highly swept lifting surfaces was yet another perturbation of new and unusual vehicles with unconventional aerodynamic, stability, and control characteristics requiring the use of free-flight models for assessments of flight dynamics.

In addition to the high-speed studies of the X-15 in the Ames supersonic free-flight facility previously discussed, the X-15 program sponsored low-speed investigations of free-flight models at Langley in the Full-Scale Tunnel, the Spin Tunnel, and an outdoor helicopter drop model.[56] The most significant contribution of the NASA free-flight tests of the X-15 was confirmation of the effectiveness of the differential tail for control. North American had followed pioneering research at Langley on the use of the tail for roll control. It had used such a design in its YF-107A aircraft and opted to use the concept for the X-15 to avoid ailerons that would have complicated wing design for the hypersonic aircraft. Nonetheless, skepticism existed over the potential effectiveness of the application until the free-flight tests at Langley provided a dramatic demonstration of its success.[57]

In the late 1950s, scientists at NASA Ames conducted in-depth studies of the aerodynamic and aerothermal challenges of hypersonic reentry and concluded that blunted half-cone shapes could provide adequate thermal protection for vehicle structures while also producing

55. David E. Hahne and Charles M. Fremaux, "Low-Speed Dynamic Tests and Analysis of the Orion Crew Module Drogue Parachute System," AIAA Paper 2008-09-05 (2008).
56. Peter C. Boisseau, "Investigation of the Low-Speed Stability and Control Characteristics of a 1/7-Scale Model of the North American X-15 Airplane," NACA RM-L57D09 (1957); Donald E. Hewes and James L. Hassell, Jr., "Subsonic Flight Tests of a 1/7-Scale Radio-Controlled Model of the North American X-15 Airplane With Particular Reference to High Angle-of-Attack Conditions," NASA TM-X-283 (1960).
57. Dennis R. Jenkins and Tony R. Landis, *Hypersonic-The Story of the North American X-15* (Specialty Press, 2008).

a significant expansion in operational range and landing options. As interest in the concept intensified following a major conference in 1958, a series of half-cone free-flight models provided convincing proof that such vehicles exhibited satisfactory flight behavior.

The most famous free-flight model activity in support of lifting body development was stimulated by the advocacy and leadership of Dale Reed of the Dryden Flight Research Center. In 1962, Reed became fascinated with the lifting body concept and proposed that a piloted research vehicle be used to validate the potential of lifting bodies.[58] He was particularly interested in the flight characteristics of a second-generation Ames lifting body design known as the M2-F1 concept. After Reed's convincing flights of radio-controlled models of the M2-F1 ranging from kite-like tows to launches from a larger radio-controlled mother ship demonstrated its satisfactory flight characteristics, Reed obtained approval for the construction and flight-testing of his vision of a low-cost piloted unpowered glider. The impact of motion-picture films of Reed's free-flight model flight tests on skeptics was overwhelming, and management's support led to an entire decade of highly successful lifting body flight research at Dryden.

At Langley, support for the M2-F1 flight program included free-flight tow tests of a model in the Full-Scale Tunnel, and the emergence of Langley's own lifting body design known as the HL-10 resulted in wind tunnel tests in virtually every facility at Langley. Free-flight testing of a dynamic model of the HL-10 in the Full-Scale Tunnel demonstrated outstanding dynamic stability and control to angles of attack as high as 45 degrees, and rolling oscillations that had been exhibited by the earlier highly swept reentry bodies were completely damped for the HL-10 with three vertical fins.[59]

In the early 1970s, a new class of lifting body emerged, dubbed "racehorses" by Dale Reed.[60] Characterized by high fineness ratios, long pointed noses, and flat bottoms, these configurations were much more efficient at hypersonic speeds than the earlier "flying bathtubs." One Langley-developed configuration, known as the Hyper III, was evaluated at Dryden by Reed and his team using free-flight models and the

58. Reed, *Wingless Flight,* NASA SP-4220.

59. George M. Ware, "Investigation of the Flight Characteristics of a Model of the HL-10 Manned Lifting Entry Vehicle," NASA TM-X-1307 (1967).

60. Reed, *Wingless Flight.*

mother ship test technique. Although the Hyper III was efficient at high speeds, it exhibited a very low lift-to-drag ratio at low speeds requiring some form of variable geometry such as a pivot wing, flexible wing, or gliding parachute.

Reed successfully advocated for a low-cost, 32-foot-long helicopter-launched demonstration vehicle of the Hyper III with a pop-out wing, which made its first flight in 1969. Flown from a ground-based cockpit, the Hyper III flight was launched from a helicopter at an altitude of 10,000 feet. After being flown in research maneuvers by a research pilot using instruments, the vehicle was handed off to a safety pilot, who safely landed it. Unfortunately, funding for a low-cost piloted project similar to the earlier M2-F1 activity was not forthcoming for the Hyper III.

Avoiding Catastrophe: Vehicle/Store Separation

One of the more complex and challenging areas in aerospace technology is the prediction of paths of aircraft components following the release of items such as external stores, canopies, crew modules, or vehicles dropped from mother ships. Aerodynamic interference phenomena between vehicles can cause major safety-of-flight issues, resulting in catastrophic impact of the components with the airplane. Unexpected pressures and shock waves can dramatically change the expected trajectory of stores. Conventional wind tunnel tests used to obtain aerodynamic inputs for calculations of separation trajectories must cover a wide range of test parameters, and the requirement for dynamic aerodynamic information further complicates the task. Measurement of aerodynamic pressures, forces, and moments on vehicles in proximity to one another in wind tunnels is a highly challenging technical procedure. The use of dynamically scaled free-flight models can quickly provide a qualitative indication of separation dynamics, thereby providing guidance for wind tunnel test planning and early identification of potentially critical flight conditions.

Separation testing for military aircraft components using dynamic models at Langley evolved into a specialty at the Langley 300-mph 7- by 10-Foot Tunnel, where subsonic separation studies included assessments of the trajectories taken by released cockpit capsules, stores, and canopies. In addition, bomb releases were simulated for several bomb-bay configurations, and the trajectories of model rockets fired from the wingtips of models were also evaluated. As requests for specific separation studies mounted, the staff rapidly accumulated unique expertise in testing techniques for separation clearance.[61] One of the more important separation studies conducted in the Langley tunnel was an assessment of the launch dynamics of the X-15/B-52 combination for launches of the X-15. Prior to the X-15, launches of research aircraft from carrier aircraft had only been made from the fuselage centerline location of the mother ship. In view of the asymmetrical location of the X-15 under the right wing of the B-52, concern arose as to the aerodynamic loads encountered during separation and the safety of the launching procedure. Separation studies were therefore conducted in the Langley 300-mph 7- by 10-Foot Tunnel and the Langley High-Speed 7- by 10-Foot Tunnel.[62]

Detailed measurements of the aerodynamic loads on the X-15 in proximity to the B-52 under its right wing were made during conventional force tests in the high-speed tunnel, while the trajectory of a dynamically scaled X-15 model was observed during a separate investigation in the low-speed tunnel. The test set up for the low-speed drop tests used a dynamically scaled X-15 model under the left wing of the B-52 model to accommodate viewing stations in the tunnel. Initial trim settings for the X-15 were determined to avoid contact with the B-52, and the drop tests showed that the resulting trajectory motions provided adequate clearance for all conditions investigated.

During successful subsonic separation events, a bomb or external store is released, and gravity typically pulls it away safely. At supersonic speeds, however, aerodynamic forces are appreciably higher relative to the store weight, shock waves may cause unexpected pressures that severely influence the store trajectory or bomb guidance system, and aerodynamic interference effects may cause catastrophic collisions after launch. Under some conditions, bombs released from within a fuselage bomb bay at supersonic speeds have encountered adverse flow fields, to the extent that the bombs have reentered the bomb bay. In the early 1950s, the NACA advisory committees strongly recommended that focused efforts be initiated by the Agency in store separation, especially for supersonic flight conditions. Researchers within Langley's Pilotless Aircraft Research Division used their Preflight Jet facility at Wallops to conduct research on supersonic separation characteristics for several

61. Linwood W. McKinney and Polhamus, "A Summary of NASA Data Relative to External-Store Separation Characteristics," NASA TN-D-3582 (1966).
62. Alford and Robert T. Taylor, "Aerodynamic Characteristics of the X-15/B-52 Combination," NASA Memo-8-59L (1958).

Langley researcher William J. Alford, Jr., observes a free-flight drop model of the X-15 research aircraft as it undergoes separation testing beneath a B-52 model in a Langley tunnel. NASA.

high-priority military programs.[63] The Preflight Jet facility was designed to check out ramjet engines prior to rocket launches, consisting of a "blow down"–type tunnel powered by compressed air exhausted through a supersonic nozzle. Test Mach number capability was from 1.4 to 2.25. With an open throat and no danger to a downstream facility drive system, the facility proved to be ideal for dynamic studies of bombs or stores following supersonic releases.

One of the more crucial tests conducted in the Wallops Preflight Jet facility was support for the development of the Republic F-105 fighter-bomber, which was specifically designed with forcible ejection of bombs from within the bomb bay to avoid the issues associated with external releases at supersonic speeds. For the test program, a half-fuselage model (with bomb bay) was mounted to the top of the nozzle, and the ejection sequence included extension of folding fins on the store after release. A piston and rod assembly from the open bomb bay forcefully ejected the

63. Shortal, *A New Dimension.*

store, and high-speed photography documented the motion of the store and its trajectory. The F-105 program expanded to include numerous specific and generic bomb and store shapes requiring almost 2 years of tests in the facility. Numerous generic and specific aircraft separation studies in the Preflight Jet facility from 1954 to 1959 included F-105 pilot escape, F-104 wing drop-tank separations, F-106 store releases from an internal bomb bay, and B-58 pod drops.

Glimpse of the Future: Advanced Civil Aircraft

Most of the free-flight model research conducted by NASA to evaluate dynamic stability and control within the flight envelope has focused on military configurations and a few radical civil aviation designs. This situation resulted from advances in the state of the art for design methods for conventional subsonic configurations over the years and many experiences correlating results of model and airplane tests. As a result, transport design teams have collected massive data and experience bases for transports that serve as the corporate knowledge base for derivative aircraft. For example, companies now have considerable experience with the accuracy of their conventional static wind tunnel model tests for the prediction of full-scale aircraft characteristics, including the effects of Reynolds number. Consequently, testing techniques such as free-flight tests do not have high technical priority for such organizations.

The radical Blended Wing-Body (BWB) flying wing configuration has been a notable exception to the foregoing trend. Initiated with NASA sponsorship at McDonnell-Douglas (now Boeing) in 1993, the subsonic BWB concept carries passengers or payload within its wing structure to minimize drag and maximize aerodynamic efficiency.[64] Over the past 16 years, wind tunnel research and computational studies of various BWB configurations have been conducted by NASA–Boeing teams to assess cruise conditions at high subsonic speeds, takeoff and landing characteristics, spinning and tumbling tendencies, emergency spin/tumble recovery parachute systems, and dynamic stability and control.

By 2005, the BWB team had conducted static and dynamic force tests of models in the 12-Foot Low-Speed Tunnel and the 14- by 22-Foot Tunnel to define aerodynamic data used to develop control laws and control limits, as well as trade studies of various control effectors available

64. Chambers, *Radical Wings and Wind Tunnels*; Chambers, *Innovation in Flight: Research of the Langley Research Center on Revolutionary Advanced Concepts for Aeronautics*, NASA SP-4539 (2005).

on the trailing edge of the wing. Free-flight testing then occurred in the Full-Scale Tunnel with a 12-foot-span model.[65] Results of the flight test indicated satisfactory flight behavior, including assessments of engine-out asymmetric thrust conditions.

In 2002, Boeing contracted with Cranfield Aerospace, Ltd., for the design and production of a pair of 21-foot-span remotely piloted models of BWB vehicles known as the X-48B configuration. After conventional wind tunnel tests of the first X-48B vehicle in the Langley Full-Scale Tunnel in 2006, the second X-48B underwent its first flight in July 2007 at the Dryden Flight Research Center. The BWB flight-test team is a cooperative venture between NASA, Boeing Phantom Works, and the Air Force Research Laboratory. The first 11 flight tests of the 8.5-percent-scale vehicle in 2007 focused on low-speed dynamic stability and control with wing leading-edge slats deployed. In a second series of flights, which began in April 2008, the slats were retracted, and higher speed studies were conducted. Powered by three model aircraft turbojet engines, the 500-pound X-48B is expected to have a top speed of about 140 mph. A sequence of flight phases is scheduled for the X-48B with various objectives within each study directed at the technology issues facing the implementation of the innovative concept.

Final Maturity: Concept Demonstrators

The efforts of the NACA and NASA in developing and applying dynamically scaled free-flight model testing techniques have progressed through a truly impressive maturation process. Although the scaling relationships have remained constant since the inception of free-flight testing, the facilities and test attributes have become dramatically more sophisticated. The size and construction of models have changed from unpowered balsa models weighing a few ounces with wingspans of less than 2 feet to very large powered composite models with weights of over 1,000 pounds. Control systems have changed from simple solenoid bang-bang controls operated by a pilot with visual cues provided by model motions to hydraulic systems with digital flight controls and full feedbacks from an array of sensors and adaptive control systems. The level of sophistication integrated into the model testing techniques has now given rise

65. Dan D. Vicroy, "Blended-Wing-Body Low-Speed Flight Dynamics: Summary of Ground Tests and Sample Results," AIAA Invited Paper presented at the *47th AIAA Aerospace Sciences Meeting and Exhibit, Jan. 2009.*

The Boeing X-48B Blended Wing Body flying model in flight at NASA Dryden. The configuration has undergone almost 15 years of research, including free-flight testing at Langley and Dryden. NASA.

to a new class of free-flight models that are considered to be integrated concept demonstrators rather than specific technology tools. Thus, the lines between free-flight models and more complex remotely piloted vehicles have become blurred, with a noticeable degree of refinement in the concept demonstrators.

Research activities at the NASA Dryden Flight Research Center vividly illustrate how far free-flight testing has come. Since the 1970s, Dryden has continually conducted a broad program of demonstrator applications with emphasis on integrations of advanced technology. In 1997, another milestone was achieved at Dryden in remotely piloted research vehicle technology, when an X-36 vehicle demonstrated the feasibility of using advanced technologies to ensure satisfactory flying qualities for radical tailless fighter designs. The X-36 was designed as a joint effort between the NASA Ames Research Center and the Boeing Phantom Works (previously McDonnell-Douglas) as a 0.28-scale powered free-flight model of an advanced fighter without vertical or horizontal tails to enhance survivability. Powered by a F112 turbofan engine and weighing about 1,200 pounds, the 18-foot-long configuration used a canard, split aileron surfaces, wing leading- and trailing-edge flaps, and a thrust-vectoring nozzle for control. A single-channel digital fly-by-wire system provided artificial stability for the configuration, which was inherently unstable about the pitch and yaw axes.[66]

Spinning

Qualitatively, recovery from the various spin modes is dependent on the type of spins exhibited, the mass distribution of the aircraft, and the sequence of controls applied. Recovering from the steep steady spin tends to be relatively easy because the nose-down orientation of the aircraft control surfaces to the free stream enables at least a portion of the control effectiveness to be retained. In contrast, during a flat spin, the fuselage may be almost horizontal, and the control surfaces are oriented so as to provide little recovery moment, especially a rudder on a conventional vertical tail. In addition to the ineffectiveness of controls for recovery from the flat spin, the rotation of the aircraft about a near-vertical axis near its center of gravity results in extremely high centrifugal forces at the cockpit for configurations with long fuselages. In many cases, the negative ("eyeballs out") g-loads may be so high as to incapacitate the crewmembers and prevent them from escaping from the aircraft.

Establishing Creditability: The Early Days

Following the operational readiness of the Langley 15-Foot Free-Spinning Tunnel in 1935, initial testing centered on establishing correlation with full-scale flight-test results of spinning behavior for the XN2Y-1 and F4B-2 biplanes.[67] Critical comparisons of earlier results obtained on small-scale models from the Langley 5-Foot Vertical Tunnel and full-scale flight tests indicated considerable scale effects on aerodynamic characteristics; therefore, calibration tests in the new tunnel were deemed imperative. The results of the tests for the two biplane models were very encouraging in terms of the nature of recovery characteristics and served to inspire confidence in the testing technique and promote future tests. During those prewar years, the NACA staff was afforded time to conduct fundamental research studies and to make general conclusions for emerging monoplane designs. Systematic series of investigations were conducted in which, for example, models were tested for combinations

66. Laurence A. Walker, "Flight Testing the X-36-The Test Pilot's Perspective," NASA CR-198058 (1997).
67. Zimmerman, "N.A.C.A. Free-Spinning Wind Tunnel," NACA TR-557.

of eight different wings and three different tails.[68] Other investigations of tunnel-to-flight correlations occurred, including comparison of results for the BT-9 monoplane trainer.

As experience with spin tunnel testing increased, researchers began to observe more troublesome differences between results obtained in flight and in the tunnel. The effects of Reynolds number, model accuracies, control-surface rigging of full-scale aircraft, propeller slipstream effects not present during unpowered model tests, and other factors became appreciated to the point that a general philosophy began to emerge for which model tests were viewed as good predictors of full-scale characteristics but also examples of poor correlation that required even more correlation studies and a conservative interpretation of model results. Critics of small-scale model testing did not accept a growing philosophy that spin predictions were an "art" based on extensive testing to determine the relative sensitivity of results to configuration variables, model damage, and testing technique. Nonetheless, pressure mounted to arrive at design guidelines for satisfactory spin recovery characteristics.

Quest for Guidelines: Tail Damping Power Factor

An empirical criterion based on the projected side area and mass distribution of the airplane was derived in England, and the Langley staff proposed a design criterion in 1939 based solely on the geometry of aircraft tail surfaces. Known as the tail-damping power factor (TDPF), it was touted as a rapid estimation method for determining whether a new design was likely to comply with the minimum requirements for safety in spinning.[69]

The beginning of World War II and the introduction of a new Langley 20-Foot Spin Tunnel in 1941 resulted in a tremendous demand for spinning tests of high-priority military aircraft. The workload of the staff increased dramatically, and a tremendous amount of data was gathered for a large number of different configurations. Military requests for spin tunnel tests filled all available tunnel test times, leaving no time for general research. At the same time, configurations were tested with

68. Oscar Seidman and Anshal I. Neihouse, "Free-Spinning Wind-Tunnel Tests on a Low-Wing Monoplane with Systematic Changes in Wings and Tails III. Mass Distributed Along the Wings," NACA TN-664 (1938).
69. Seidman and Charles J. Donlan, "An Approximate Spin Design Criteria for Monoplanes," NACA TN-711 (1939).

radical differences in geometry and mass distribution. Tailless aircraft with their masses distributed in a primarily spanwise direction were introduced, along with twin-engine bombers and other unconventional designs with moderately swept wings and canards.

In the 1950s, advances in aircraft performance provided by the introduction of jet propulsion resulted in radical changes in aircraft configurations, creating new challenges for spin technology. Military fighters no longer resembled the aircraft of World War II, as the introduction of swept wings and long, pointed fuselages became commonplace. Suddenly, certain factors, such as mass distribution, became even more important, and airflow around the unconventional, long fuselage shapes during spins dominated the spin behavior of some configurations. At the same time, fighter aircraft became larger and heavier, resulting in much higher masses relative to the atmospheric density, especially during flight at high altitudes.

Effect of Reynolds Number

In the mid-1950s, the NACA encountered an unexpected aerodynamic scale effect related to the long fuselage forebodies being introduced at the time. This experience led to one of the more important and lasting lessons learned in the use of free-spinning models for spin predictions. One particular project stands out as a key experience regarding this topic. As part of the ongoing military requests for NACA support of new aircraft development programs, the Navy requested Langley to conduct spin tunnel tests of a model of its new Chance Vought XF8U-1 Crusader fighter in 1955. The results of spin tunnel tests of a 1/25-scale model indicated that the airplane would exhibit two spin modes.[70] The first mode would be a potentially dangerous fast, flat spin at an angle of attack of approximately 87 degrees, from which recoveries were unsatisfactory or unobtainable. The second spin was much steeper, with a lower rate of rotation, and recoveries would probably be satisfactory.

As the spin tunnel results were analyzed, Chance Vought engineers directed their focus to identifying factors that were responsible for the flat spin exhibited by the model. The scope of activities stimulated by the XF8U-1 spin tunnel results included, in addition to extended spin tunnel tests, one-degree-of-freedom autorotation tests of a model of the

70. Walter J. Klinar, Henry A. Lee, and L. Faye Wilkes, "Free-Spinning-Tunnel Investigation of a 1/25-Scale Model of the Chance Vought XF8U-1 Airplane," NACA RM-SL56L31b (1956).

XF8U-1 configuration in the Chance Vought Low Speed Tunnel and a NACA wind tunnel research program that measured the aerodynamic sensitivity of a wide range of two-dimensional, noncircular cylinders to Reynolds number.[71] The wind tunnel tests were designed and conducted to include variations in Reynolds number from the low values associated with spin tunnel testing to much higher values more representative of flight.

With results from the static and autorotation wind tunnel studies in hand, researchers were able to identify an adverse effect of Reynolds number on the forward fuselage shape of the XF8U-1 such that, at the relatively low values of Reynolds number of the spin tunnel tests (about 90,000 based on fuselage-forebody depth), the spin model exhibited a powerful pro-spin aerodynamic yawing moment dominated by forces produced on the forebody. The pro-spin moment caused an autorotative spinning tendency, resulting in the fast flat spin observed in the spin tunnel tests. As the Reynolds number in the tunnel tests was increased to values approaching 300,000, however, the moments produced by the forward fuselage reversed direction and became antispin, remaining so for higher values of Reynolds number. Fundamentally, the researchers had clearly identified the importance of cross-sectional shapes of modern aircraft—particularly those with long forebodies—on spin characteristics and the possibility of erroneous spin tunnel predictions because of the low test Reynolds number. When the full-scale spin tests were conducted, the XF8U-1 airplane exhibited only the steeper spin mode and the fast, flat spin predicted by the spin model that had caused such concern was never encountered.

During and after the XF8U-1 project, Langley's spin tunnel personnel developed expertise in the anticipation of potential Reynolds number effects on the forebody, and in the art of developing methods to geometrically modify models to minimize unrealistic spin predictions, caused by the phenomenon. In this approach, cross-sectional shapes of aircraft are examined before models are constructed, and if the forebody cross section is similar to those known to exhibit scale effects at low Reynolds number, static tests at other wind tunnels are

71. M.H. Clarkson, "Autorotation of Fuselages," *Aeronautical Engineering Review*, vol. 17 (Feb. 1958); Polhamus, "Effect of Flow Incidence and Reynolds Number on Low-Speed Aerodynamic Characteristics of Several Noncircular Cylinders with Applications to Directional Stability and Spinning," NACA TN-4176 (1958).

conducted for a range of Reynolds number to determine if artificial devices, such as nose-mounted strakes at specific locations, can be used to artificially alter the flow separation on the nose at low Reynolds number and cause it to more accurately simulate full-scale conditions.[72]

In addition to the XF8U-1, it was necessary to apply scale-correction fuselage strakes to the spin tunnel models of the Northrop F-5A and F-5E fighters, the Northrop YF-17 lightweight fighter prototype, and the Fairchild A-10 attack aircraft to avoid erroneous predictions because of fuselage forebody effects. In the case of the X-29, a specific study of the effects of forebody devices for correcting low Reynolds number effects was conducted in detail.[73]

Effect of External Stores

External stores have been found to have large effects on spin and recovery, especially for asymmetric loadings in which stores are located asymmetrically along the wing, resulting in a lateral displacement of the center of gravity of the configuration. For example, some aircraft may not spin in the direction of the "heavy" wing but will spin fast and flat into the "light" wing. In most cases, model tests in which the shapes of the external stores were replaced with equivalent weight ballast indicated that the effects of asymmetric loadings were primarily due to a mass effect, with little or no aerodynamic effect detected. However, very large stores such as fuel tanks were found, on occasion, to have unexpected effects because of aerodynamic characteristics of the component. During the aircraft development phase, spin characteristics of high-performance military aircraft must be assessed for all loadings proposed, including symmetric and asymmetric configurations. Spin tunnel tests can therefore be extensive for some aircraft, especially those with variable-sweep wing capabilities. Testing

72. D.N. Petroff, S.H. Scher, and L.E. Cohen, "Low Speed Aerodynamic Characteristics of an 0.075-Scale F-15 Airplane Model at High Angles of Attack and Sideslip," NASA TM-X-62360 (1974); Petroff, Scher, and C.E. Sutton, "Low-Speed Aerodynamic Characteristics of a 0.08-Scale YF-17 Airplane Model at High Angles of Attack and Sideslip," NASA TM-78438 (1978); Raymond D. Whipple and J.L. Ricket, "Low-Speed Aerodynamic Characteristics of a 1/8-scale X-29A Airplane Model at High Angles of Attack and Sideslip," NASA TM-87722 (1986).

73. Stanley H. Scher and William L. White, "Spin-Tunnel Investigation of the Northrop F-5E Airplane," NASA TM-SX-3556 (1977); C. Michael Fremaux, "Wind-Tunnel Parametric Investigation of Forebody Devices for Correcting Low Reynolds Number Aerodynamic Characteristics at Spinning Attitudes," NASA CR-198321 (1996).

of the General Dynamics F-111, for example, required several months of test time to determine spin and recovery characteristics for all potential conditions of wing-sweep angles, center-of-gravity positions, and symmetric and asymmetric store loadings.[74]

Parachute Technology

The use of tail-mounted parachutes for emergency spin recovery has been common practice from the earliest days of flight to the present day. Properly designed and deployed parachutes have proven to be relatively reliable spin recovery device, always providing an antispin moment, regardless of the orientation of the aircraft or the disorientation or confusion of the pilot. Almost every military aircraft spin program conducted in the Spin Tunnel includes a parachute investigation. Free-spinning model tests are used to determine the critical geometric variables for parachute systems. Paramount among these variables is the minimum size of parachute required for recovery from the most dangerous spin modes. As would be expected, the size of the parachute is constrained by issues regarding system weight and the opening shock loads transmitted to the rear of the aircraft. In addition to parachute size, the length of parachute riser (attachment) lines and the attachment point location on the rear of the aircraft are also critical design parameters.

The importance of parachute riser line length can be especially critical to the inflation and effectiveness of the parachute for spin recovery. Results of free-spin tests of hundreds of models in the Spin Tunnel has shown that if the riser length is too short, the parachute will be immersed in the low-energy wake of the spinning airplane and will not inflate. On the other hand, if the towline length is too long, the parachute will inflate but will drift inward and align itself with the axis of rotation, thereby providing no antispin contribution. The design and operational implementation of emergency spin recovery parachutes are a stringent process that begins with spin tunnel tests and proceeds through the design and qualification of the parachute system, including the deployment and release mechanisms. By participation in each of these segments of the process, Langley researchers have amassed tremendous amount of knowledge regarding parachute systems and are called upon frequently by the aviation community for consultation

74. A discussion of the powerful effects of asymmetric mass loadings for the F-15 fighter is presented in an accompanying case study in this volume by the same author.

before designing and fabricating parachute systems for spin tests of full-scale aircraft.[75]

General-Aviation Spin Technology

The dramatic changes in aircraft configurations after World War II required almost complete commitment of the Spin Tunnel to development programs for the military, resulting in stagnation of any research for light personal-owner–type aircraft. In subsequent years, designers had to rely on the database and design guidelines that had been developed based on experiences during the war. Unfortunately, stall/spin accidents in the early 1970s in the general aviation community increased at an alarming rate. Even more troublesome, on several occasions aircraft that had been designed according to the NACA tail-damping power factor criterion had exhibited unsatisfactory recovery characteristics, and the introduction of features such as advanced general aviation airfoils resulted in concern over the technical adequacy and state of the database for general aviation configurations.

Finally, in the early 1970s, the pressure of new military aircraft development programs eased, permitting NASA to embark on new studies related to spin technology for general aviation aircraft. A NASA General Aviation Spin Research program was initiated at Langley that focused on the use of radio-control and spin tunnel models to assess the impact of design features on spin and recovery characteristics, and to develop testing techniques that could be used by the industry. The program also included the acquisition of several full-scale aircraft that were modified for spin tests to produce data for correlation with model results.[76]

One of the key objectives of the program was to evaluate the impact of tail geometry on spin characteristics. The approach taken was to design alternate tail configurations so as to produce variability in the TDPF parameter by changing the vertical and horizontal locations of the

75. Scher, "Wind-Tunnel Investigation of the Behavior of Parachutes in Close Proximity to One Another," NACA RM-L53G07 (1953); Scher and John W. Draper, "The Effects of Stability of Spin-Recovery Tail Parachutes on the Behavior of Airplanes in Gliding Flight and in Spins," NACA TN-2098 (1950); Sanger M. Burk, Jr., "Summary of Design Considerations for Airplane Spin-Recovery Parachute Systems," NASA TN-D-6866 (1972); H. Paul Stough, III, "A Summary of Spin-Recovery Parachute Experience on Light Airplanes," AIAA Paper 90-1317 (1990).

76. James S. Bowman, Jr., and Burk, "Stall/Spin Studies Relating to Light General-Aviation Aircraft," SAE Paper presented at the *Society of Automotive Engineers Business Aircraft Meeting, Wichita, KS, Apr. 1973.*

Involved in a study of spinning characteristics of general-aviation configurations in the 1970s were Langley test pilot Jim Patton, center, and researchers Jim Bowman, left, and Todd Burk. NASA.

horizontal tail. A spin tunnel model of a representative low wing configuration was constructed with four interchangeable tails, and results for the individual tail configurations were compared with predictions based on the tail design criteria. The range of tails tested included conventional cruciform-tail configurations, low horizontal tail locations, and a T-tail configuration.

As expected, results of the spin tunnel testing indicated that tail configuration had a large influence on spin and recovery characteristics, but many other geometric features also influenced the characteristics, including fuselage cross-sectional shape. In addition, seemingly small configuration features such as wing fillets at the wing trailing-edge juncture with the fuselage had large effects. Importantly, the existing TDPF criterion for light airplanes did not correctly predict the spin recovery characteristics of models for some conditions, especially for those in which ailerons were deflected. NASA's report to the industry following the tests stressed that, based on these results, TDPF should not be used to predict spin recovery characteristics. However, the recommendation did provide a recommended "best practice" approach to overall design of the tail of the airplane for spin behavior.[77]

As part of its General Aviation Spin Research program, NASA continued to provide information on the design of emergency spin recovery parachute systems.[78] Parachute diameters and riser line lengths were sized based on free-spinning model results for high and low wing configurations and a variety of tail configurations. Additionally, guidelines for the design and implementation of the mechanical systems required for parachute deployment (such as mechanical jaws and pyrotechnic deployment) and release of the parachute were documented.

NASA also encouraged industry to use its spin tunnel facility on a fee-paying basis. Several industry teams proceeded to use the opportunity to conduct proprietary tests for configurations in the tunnel. For example, the Beech Aircraft Corporation sponsored the first fee-paid test in the Langley Spin Tunnel for free-spinning model tests of its Model 77 "Skipper" trainer.[79] In such proprietary tests, the industry provided models and personnel for joint participation in the testing experience.

Spin Entry

The helicopter drop-model technique has been used since the early 1950s to evaluate the spin entry behavior of relatively large unpowered models of military aircraft. The objective of these tests has been to evaluate the relative spin resistance of configurations following various combinations of control inputs, and the effects of timing of recovery control inputs following departures. A related testing technique used to evaluate spin resistance of spin entry evaluations of general aviation configurations employs remotely controlled powered models that take off from ground runways and fly to the test condition.

In the late 1950s, industry had become concerned over potential scale effects on long pointed fuselage shapes as a result of the XF8U-1

77. Burk, Bowman, and White, "Spin-Tunnel Investigation of the Spinning Characteristics of Typical Single-Engine General Aviation Airplane Designs: Part I-Low-Wing Model A.: Effects of Tail Configurations," NASA TP-1009 (1977).

78. Stough, "A Summary of Spin-Recovery Parachute Experience on Light Airplanes," AIAA Paper 90-1317 (1990).

79. M.L. Holcomb, "The Beech Model 77 'Skipper' Spin Program," AIAA Paper 79-1835 (1979).

experiences in the Spin Tunnel, as discussed earlier. Thus, interest was growing over the possible use of much larger models than those used in spin tunnel tests, to eliminate or minimize undesirable scale effects. Finally, a major concern arose for some airplane designs over the launching technique used in the Spin Tunnel. Because the spin tunnel model was launched by hand in a very flat attitude with forced rotation, it would quickly seek the developed spin modes—a very valuable output—but the full-scale airplane might not easily enter the spin because of control limitations, poststall motions, or other factors.

One of the first configurations tested, in 1958, to establish the credibility of the drop-model program was a 6.3-foot-long, 90-pound model of the XF8U-1 configuration.[80] With previously conducted spin tunnel results in hand, the choice of this design permitted correlation with the earlier tunnel and aircraft flight-test results. As has been discussed, wind tunnel testing of the XF8U-1 fuselage forebody shape had indicated that pro-spin yawing moments would be produced by the fuselage for values of Reynolds number below about 400,000, based on the average depth of the fuselage forebody. The Reynolds number for the drop-model tests ranged from 420,000 to 505,000, at which the fuselage contribution became antispin and the spins and recovery characteristics of the drop model were found to be very similar to the full-scale results. In particular, the drop model did not exhibit a flat-spin mode predicted by the smaller spin tunnel model, and results were in agreement with results of the aircraft flight tests, demonstrating the value of larger models from a Reynolds number perspective.

Success in applications of the drop-model technique for studies of spin entry led to the beginning of many military requests for evaluations of emerging fighter aircraft. In 1959, the Navy requested an evaluation of the McDonnell F4H-1 Phantom II airplane using the drop technique.[81] Earlier spin tunnel tests of the configuration indicated the possibility of two types of spins: one of which was steep and oscillatory, from which recoveries were satisfactory, and the other was fast and flat, from which recovery was difficult or impossible. As mentioned previously, the spin tunnel launching technique had led to questions regarding whether the airplane would exhibit a tendency toward the steeper spin or the more

80. Libby, "A Technique Utilizing Free-Flying Radio-Controlled Models," NASA Memo 2-6-59L.

81. Burk and Libby, "Large-Angle Motion Tests, Including Spins, of a Free-Flying Radio-Controlled 0.13-Scale Model of a Twin-Jet Swept-Wing Fighter Airplane," NASA TM-SX-445 (1960).

dangerous flat spin. The objective of the drop tests was to determine if it was likely, or even possible, for the F4H-1 to enter the flat spin.

In the F4H-1 investigation, an additional launching technique was used in an attempt to obtain a developed spin more readily and to possibly obtain the flat spin to verify its existence. This technique consisted of prespinning the model on the helicopter launch rig before it was released in a flat attitude with the helicopter in a hovering condition. To achieve even higher initial rotation rates than could be achieved on the launch rig, a detachable flat metal plate was attached to one wingtip of the model to propel it to spin even faster. After the model appeared to be rotating sufficiently fast after release, the vane was jettisoned by the ground-based pilot, who, at the same time, moved the ailerons against the direction of rotation to help promote the spin. The model was then allowed to spin for several turns, after which recovery controls were applied. In some aspects, this approach to testing replicated the spin tunnel launch technique but at a larger scale.

Results of the drop-model investigation for the F4H-1 are especially notable because it established the value of the testing technique to predict spin tendencies as verified by subsequent full-scale results. A total of 35 flights were made, with the model launched 15 times in the prerotated condition and 20 times in forward flight. During these 35 flights, poststall gyrations were obtained on 21 occasions, steep spins were obtained on 10 flights, and only 4 flat spins were obtained. No recoveries were possible from the flat spins, but only one flat spin was obtained without prerotation. The conclusions of the tests stated that the aircraft was more susceptible to poststall gyrations than spins; that the steeper, more oscillatory spin would be more readily obtainable and recovery could be made by the NASA-recommended control technique; and that the likelihood of encountering a fast, flat spin was relatively remote. Ultimately, these general characteristics of the airplane were replicated at full-scale test conditions during spin evaluations by the Navy and Air Force.

The Pace Quickens

Beginning in the early 1960s, a flurry of new military aircraft development programs resulted in an unprecedented workload for the drop-model personnel. Support was requested by the military services for the General Dynamics F-111, Grumman F-14, McDonnell-Douglas F-15, Rockwell B-1A, and McDonnell-Douglas F/A-18 development programs. In addition, drop-model tests were conducted in support of the Grumman

X-29 and the X-31—sponsored by the Defense Advanced Research Projects Agency (DARPA)—research aircraft programs, which were scheduled for high-angle-of-attack full-scale flight tests at the Dryden flight facility. The specific objectives and test programs conducted with the drop models were considerably different for each configuration. Overviews of the results of the military programs are given in this volume, in another case study by this author.

General-Aviation Configurations

As part of its General Aviation Spin Research program in the 1970s, Langley included the development of a testing technique using powered radio-controlled models to study spin resistance, spin entry, and spin recovery during the incipient phase of the spin.[82] Equally important was a focus on developing a reliable, low-cost model testing technique that could be used by the industry for spin predictions in early design stages. The dynamically scaled models, which were about 1/5-scale (wingspan of about 4–5 feet), were powered and flown with hobby equipment.

Although resembling conventional radio-control models flown by hobbyists, the scaling process discussed earlier resulted in models that were much heavier (about 15–20 pounds) than conventional hobby models (about 6–8 pounds).

The radio-controlled model activities in the Langley program consisted of three distinct phases. Initially, model testing and analysis was directed at producing timely data for correlation with spin tunnel and full-scale flight results to establish the accuracy of the model results in predicting spin and recovery characteristics, and to gain experience with the testing technique. The second phase of the radio-controlled model program involved assessments of the effectiveness of NASA-developed wing leading-edge modifications to enhance the spin resistance of several general-aviation configurations. The focus of this research was a concept consisting of a drooped leading edge on the outboard wing panel with a sharp discontinuity at the inboard edge of the droop. The third phase of radio-controlled model testing involved cooperative studies of specific general-aviation designs with industry. In this segment of the program, studies centered on industry's assessment of the radio-controlled model technique.

82. Bowman and Burk, "Stall/Spin Studies Relating to Light General-Aviation Aircraft," *Society of Automotive Engineers Business Aircraft Meeting, Wichita, KS.*

Direct correlation of results for radio-controlled model tests and full-scale airplane results for a low wing NASA configuration was very good, especially with regard to susceptibility of the design to enter a fast, flat spin with poor or no recovery.[83] In addition, the effects of various control input strategies agreed very well. For example, with normal pro-spin controls and any use of ailerons, the radio-controlled model and the airplane were both reluctant to enter the flat spin mode that had been predicted by spin tunnel tests; they only exhibited steeper spins from which recovery could still be accomplished. Subsequently, the test pilot and flight-test engineers of the full-scale airplane developed a unique control scheme during spin tests that would aggravate the steeper spin and propel the airplane into a flat spin requiring the emergency parachute for recovery. When a similar control technique was used on the radio-controlled model, it also would enter the flat spin, also requiring its parachute for recovery.

Some of the more impressive results of the radio-controlled model program for the low wing configuration related to the ability of the model to demonstrate effects of the discontinuous leading-edge droop concept that had been developed by Langley for improved spin resistance.[84] Several wing-leading-edge droop configurations had been derived in wind tunnel tests with the objective to delay wing autorotation and spin entry to high angles of attack. Tests with the radio-controlled model when modified with a full-span droop indicated better stall characteristics than the basic configuration did, but the resistance of the model to enter the unrecoverable flat spin was significantly degraded. The flat spin could be obtained on virtually every flight if pro-spin controls were maintained beyond about three turns after stall.

In contrast to this result, when the discontinuous droop was applied to the outer wing, the model would enter a very steep spin from which recovery could be obtained by simply neutralizing controls. When the discontinuity on the inboard edge of the droop was faired over, the model reverted to the same characteristics that had been displayed with the full-span droop and could easily be flown into the flat spin. Correlation between the radio-controlled model and aircraft results in this phase of the project was outstanding. The agreement was particularly noteworthy

83. Bowman, Stough, Burk, and Patton, "Correlation of Model and Airplane Spin Characteristics for a Low-Wing General Aviation Research Airplane," AIAA Paper 78-1477 (1978).

84. Staff of the Langley Research Center, "Exploratory Study of the Effects of Wing-Leading-Edge Modifications" NASA TP-1589 (1979).

in view of the large differences between the model and full-scale flight Reynolds numbers. All of the important stall/spin characteristics displayed by the low wing, radio-controlled model with the full-span droop configuration and the outboard droop configuration (with and without the fairing on the discontinuous juncture) were nearly identical to those exhibited by the full-scale aircraft, including stall characteristics, spin modes, spin resistance, and recovery characteristics.[85]

While researchers were conducting the technical objectives of the radio-controlled model program, an effort was directed at developing test techniques that might be used by industry for relatively low-cost testing. Innovative instrumentation techniques were developed that used relatively inexpensive hobby-type onboard sensors to measure control positions, angle of attack, airspeed, angular rates, and other variables. Data output from the sensors was transmitted to a low-cost ground-based data acquisition station by modifying a conventional seven-channel radio-control model transmitter. The ground station consisted of separate receivers for monitoring angle of attack, angle of sideslip, and control commands. The receivers operated servos to drive potentiometers, whose signals were recorded on an oscillograph recorder. Tracking equipment and cameras were also developed. Other facets of the test technique development included the design and operational deployment of emergency spin recovery parachutes for the models.

One particularly innovative testing technique demonstrated by NASA in the radio-controlled model flight programs was the use of miniature auxiliary rockets mounted on the wingtips of models to artificially promote flat spins. This approach was particularly useful in determining the potential existence of dangerous flat spins that were difficult to enter from conventional flight. In this application, the pilot remotely ignited one of the rockets during a spin entry, resulting in extremely high spin rates and a transition to very high angles of attack and flat-spin attitudes. After the "spin up" maneuver was complete, the rocket thrust subsided, and the model either remained in a stable flat spin or pitched down to a steeper spin mode. Beech Aircraft used this technique in its subsequent applications to radio-controlled models.

85. The impressive results of NASA's full-scale and model flight-testing, together with evaluations of the droop concept by FAA pilots, led to the creation of a new spin certification category known as "spin resistant design." See Chambers, *Concept to Reality: Contributions of the Langley Research Center to U.S. Civil Aircraft of the 1990s*, NASA SP-4529 (2003).

General-aviation manufacturers maintained a close liaison with Langley researchers during the NASA stall/spin program, absorbing data produced by the coordinated testing of models and full-scale aircraft. The radio-controlled testing technique was of great interest, and following frequent interactions with Langley's test team, industry conducted its own evaluations of radio-controlled models for spin testing. In the mid-1970s, Beech Aircraft conducted radio-controlled testing of its T-34 trainer aircraft, the Model 77 Skipper trainer, and the twin-engine Model 76 Duchess.[86] Piper Aircraft also conducted radio-controlled model testing to explore the spin entry, developed spin, and recovery techniques of a light twin-engine configuration.[87] Later in the 1980s, a joint program was conducted with the DeVore Aviation Corporation to evaluate the spin resistance of a model of a high wing trainer design that incorporated the NASA-developed leading-edge droop concept.[88]

As a result of these cooperative ventures, industry obtained valuable experience in model construction techniques, spin recovery parachute system technology, methods of measuring moments of inertia and scaling engine thrust, the cost and time required to conduct such programs, and correlation with full-scale flight-test results.

The Future of Dynamic Model Testing

Efforts by the NACA and NASA over the last 80 years with applications of free-flying dynamic model test techniques have resulted in significant contributions to the civil and military aerospace communities. The results of the investigations have documented the testing techniques and lessons learned, and they have been especially valuable in defining critical characteristics of radical new configurations. With the passing of each decade, the free-flight techniques have become more sophisticated, and the accumulation of correlation between model and full-scale results has rapidly increased. In view of this technical progress, it

86. M.L. Holcomb and R.R. Tumlinson, "Evaluation of a Radio-Control Model for Spin Simulation," SAE Paper 77-0482 (1977); Holcomb, "The Beech Model 77 "Skipper" Spin Program," AIAA 1979-1835; Tumlinson, Holcomb and V.D. Gregg, "Spin Research on a Twin-Engine Aircraft," AIAA Paper 1981-1667 (1981).
87. Burk and Calvin F. Wilson, "Radio-Controlled Model Design and Testing Techniques for Stall/Spin Evaluation of General-Aviation Aircraft," NASA TM-80510 (1975).
88. Yip, et al., "Model Flight Test of a Spin-Resistant Trainer," AIAA 88-2146.

Langley researchers Long Yip, left, and David Robelen with a radio-controlled model used in a program on spin resistance with the DeVore Aviation Corporation. The model was equipped with NASA-developed discontinuous outboard droops and was extremely spin resistant. NASA.

is appropriate to reflect on the state of the art in free-flight technology and the challenges and opportunities of the future.

Forcing Factors

One of the more impressive advances in aerospace capability in the last few years has been the acceptance and accelerated development of remotely piloted unmanned aerial vehicles (UAVs) by the military. The progress in innovative hardware and software products to support this focus has truly been impressive and warrants a consideration that properly scaled free-flight models have reached the appropriate limits of development. In comparison to today's capabilities, the past equipment used by the NACA and NASA seems primitive. It is difficult to anticipate hardware breakthroughs in free-flight model technologies beyond those currently employed, but NASA's most valuable contributions have come from the applications of the models to specific aerospace issues—especially those that require years of difficult research and participation in model-to-flight correlation studies.

Changes in the world situation are now having an impact on aeronautics, with a trickle-down effect on technical areas such as free-flight testing. The end of the Cold War and industrial mergers have resulted in a dramatic reduction in new aircraft designs, especially for unconventional configurations that would benefit from free-flight testing. Reductions in research budgets for industry and NASA have further aggravated the situation.

These factors have led to a slowdown in requirements for the ongoing NASA capabilities in free-flight testing at a time when rollover changes in the NASA workforce is resulting in the retirements of specialists in this and other technologies without adequate transfer of knowledge and mentoring to the new research staffs. In addition, planned closures of key NASA facilities will challenge new generations of researchers to reinvent the free-flight capabilities discussed herein. For example, the planned demolition of the Langley Full-Scale Tunnel in 2009 will terminate that historic 78-year-old facility's role in providing free-flight testing capability, and although exploratory free-flight tests have been conducted in the much smaller test section of the Langley 14- by 22-Foot Tunnel, it remains to be seen if the technique will continue as a testing capability. Based on the foregoing observations, NASA will be challenged to provide the facilities and expertise required to continue to provide the Nation with contributions from free-flight models.

Remaining Technical Challenges

Without doubt, the most important technical issues in the application of dynamically scaled free-flight models are the effects of Reynolds number. Although a few research agencies have attempted to minimize these effects by the use of pressurized wind tunnels, a practical approach to free-flight testing without concern for Reynolds number effects has not been identified.

In the author's opinion, the challenge of eliminating Reynolds number effects in spin studies is worthy of an investigation. In particular, the research community should seriously examine the possibilities of combining recent advances in cryogenic wind tunnel technology, magnetic suspension systems, and other relevant fields in a feasibility study of free-spinning tests at full-scale values of Reynolds number. The obvious issues of cost, operational efficiencies, and value added versus today's testing would be critical factors in the study, although one would hope that the operational experiences gained in the U.S. and Europe with cryogenic tunnels in recent years might provide some optimism for success.

Other approaches to analyzing and correcting for Reynolds number effects might involve the application of computational fluid dynamics (CFD) methods. Although applications of CFD methods to dynamic stability and control issues are in their infancy, one can visualize their use in evaluating the impact of Reynolds number on critical phenomena such as the effect of fuselage cross-sectional shape on spin damping.

In summary, the next major breakthroughs in dynamic free-flight model technology should come in the area of improving the prediction of Reynolds number effects. However, to make advances toward this goal will require programmatic commitments similar to the ones made during the past 80 years for the continued support of model testing in the specialty areas discussed herein.

Recommended Additional Reading

Reports, Papers, Articles, and Presentations:

5

William J. Alford, Jr., and Robert T. Taylor, "Aerodynamic Characteristics of the X-15/B-52 Combination," NASA Memo-8-59L (1958).

Ernie L. Anglin, James S. Bowman, Jr., and Joseph R. Chambers, "Effects of a Pointed Nose on Spin Characteristics of a Fighter Airplane Model Including Correlation with Theoretical Calculations," NASA TN D-5921 (1970).

H.O. Ankenbruck, "Determination of the Stability and Control Characteristics of a 1/10-Scale Model of the MCD-387-A Swept-Forward Wing, Tailless Fighter Airplane in the Langley Free-Flight Tunnel," NACA MR-L6G02 (1946).

R. Bates, Powell M. Lovell, Jr., and Charles C. Smith, Jr., "Dynamic Stability and Control Characteristics of a Vertically Rising Airplane Model in Hovering Flight," NACA RM-L50J16 (1951).

Peter C. Boisseau, "Investigation in the Langley Free-Flight Tunnel of the Low-Speed Stability and Control Characteristics of a 1/10-Scale Model Simulating the Convair F-102A Airplane," NACA RM-SL55B21 (1955).

Peter C. Boisseau, "Investigation of the Low-Speed Stability and Control Characteristics of a 1/7-Scale Model of the North American X-15 Airplane," NACA RM-L57D09 (1957).

Alfred G. Boissevain and Peter F. Intrieri, "Determination of Stability Derivatives from Ballistic Range Tests of Rolling Aircraft Models," NASA TM-X-399 (1961).

James S. Bowman, Jr., "Dynamic Model Tests at Low Subsonic Speeds of Project Mercury Capsule Configurations With and Without Drogue Parachutes," NASA TM-X-459 (1961).

James S. Bowman, Jr., "Free-Spinning-Tunnel Investigation of Gyroscopic Effects of Jet-Engine Rotating Parts (or of Rotating Propellers) on Spin and Spin Recovery," NACA TN-3480 (1955).

James S. Bowman, Jr., "Spin-Entry Characteristics of a Delta-Wing Airplane as Determined by a Dynamic Model," NASA TN-D-2656 (1965).

James S. Bowman, Jr., and Sanger M. Burk, Jr., "Stall/Spin Studies Relating to Light General-Aviation Aircraft," SAE Paper Presented at the *Society of Automotive Engineers Business Aircraft Meeting, Wichita, KS, Apr. 1973.*

James S. Bowman, Jr., H. Paul Stough, III, Sanger M. Burk, Jr., and James S. Patton, Jr., "Correlation of Model and Airplane Spin Characteristics for a Low-Wing General Aviation Research Airplane," AIAA Paper 78-1477 (1978).

Jay M. Brandon, Frank L. Jordan, Jr., Robert A. Stuever, and Catherine W. Buttrill, "Application of Wind Tunnel Free-Flight Technique for Wake Vortex Encounters," NASA TP-3672 (1997).

Sanger M. Burk, Jr., "Free-Flight Investigation of the Deployment, Dynamic Stability, and Control Characteristics of a 1/12-Scale Dynamic Radio-Controlled Model of a Large Booster and Parawing," NASA TN-D-1932 (1963).

Sanger M. Burk, Jr., "Summary of Design Considerations for Airplane Spin-Recovery Parachute Systems," NASA TN-D-6866 (1972).

Sanger M. Burk, Jr., James S. Bowman, Jr., and William L. White, "Spin-Tunnel Investigation of the Spinning Characteristics of Typical Single-Engine General Aviation Airplane Designs: Part I-Low-Wing Model A.: Effects of Tail Configurations," NASA TP-1009 (1977).

Sanger M. Burk, Jr., and Charles E. Libby, "Large-Angle Motion Tests, Including Spins, of a Free-Flying Radio-Controlled 0.13-Scale Model of a Twin-Jet Swept-Wing Fighter Airplane," NASA TM-SX-445 (1960).

Sanger M. Burk, Jr., and Calvin F. Wilson, "Radio-Controlled Model Design and Testing Techniques for Stall/Spin Evaluation of General-Aviation Aircraft," NASA TM-80510 (1975).

John P. Campbell, Jr., "Free and Semi-Free Model Flight-Testing Techniques Used in Low-Speed Studies of Dynamic Stability and Control," North Atlantic Treaty Organization Advisory Group for Aeronautical Research and Development, NATO *AGARDograph* 76 (1963).

John P. Campbell, Jr., and Hubert M. Drake, "Investigation of Stability and Control Characteristics of an Airplane Model with Skewed Wing in the Langley Free-Flight Tunnel," NACA TN-1208 (1947).

John P. Campbell, Jr., and Joseph L. Johnson, Jr., "Wind-Tunnel Investigation of an External-Flow Jet-Augmented Slotted Flap Suitable for Applications to Airplanes with Pod-Mounted Jet Engines," NACA TN-3898 (1956).

John P. Campbell, Jr., and Charles L. Seacord, Jr., "The Effect of Mass Distribution on the Lateral Stability and Control Characteristics of an Airplane as Determined by Tests of a Model in the Free-Flight Tunnel," NACA TR-769 (1943).

Joseph R. Chambers and Sue B. Grafton, "Aerodynamic Characteristics of Airplanes at High Angles of Attack," NASA TM-74097 (1977).

William L. Clarke and R.L. Maltby, "The Vertical Spinning Tunnel at the National Aeronautical Establishment, Bedford," Royal Aircraft Establishment [U.K.] Technical Note Aero 2339 (1954).

M.H. Clarkson, "Autorotation of Fuselages," *Aeronautical Engineering Review*, vol. 17 (Feb. 1958).

Charles J. Cornelison, "Status Report for the Hypervelocity Free-Flight Aerodynamic Facility," *48th Aero Ballistic Range Association Meeting, Austin, TX, Nov. 1997.*

Peter S. Costigan, "Dynamic-Model Study of Planetary-Entry Configurations in the Langley Spin Tunnel," NASA TN-D-3499 (1966).

Mark A. Croom, Holly M. Kenney, and Daniel G. Murri, "Research on the F/A-18E/F Using a 22%-Dynamically-Scaled Drop Model," AIAA Paper 2000-3913 (2000).

James L. Edmondson and E. Claude Sanders, Jr., "A Free-Flight Technique for Measuring Damping in Roll by Use of Rocket-Powered Models and Some Initial Results for Rectangular Wings," NACA RM-L9101 (1949).

David J. Fratello, Mark A. Croom, Luat T. Nguyen, and Christopher S. Domack, "Use of the Updated NASA Langley Radio-Controlled Drop-Model Technique for High-Alpha Studies of the X-29A Configuration," AIAA Paper 1987-2559 (1987).

Delma C. Freeman, Jr., "Low Subsonic Flight and Force Investigation of a Supersonic Transport Model with a Highly Swept Arrow Wing," NASA TN-D-3887 (1967).

Delma C. Freeman, Jr., "Low Subsonic Flight and Force Investigation of a Supersonic Transport Model with a Variable-Sweep Wing," NASA TN-D-4726 (1968).

C. Michael Fremaux, "Wind-Tunnel Parametric Investigation of Forebody Devices for Correcting Low Reynolds Number Aerodynamic Characteristics at Spinning Attitudes," NASA CR-198321 (1996).

C.M. Fremaux, D.M. Vairo, and R.D. Whipple, "Effect of Geometry and Mass Distribution on Tumbling Characteristics of Flying Wings," NASA TM-111858 (1993).

David E. Hahne and Charles M. Fremaux, "Low-Speed Dynamic Tests and Analysis of the Orion Crew Module Drogue Parachute System," AIAA Paper 2008-09-05 (2008).

Donald E. Hewes, "Free-Flight Investigation of Radio-Controlled Models with Parawings," NASA TN-D-927 (1961).

Donald E. Hewes and James L. Hassell, Jr., "Subsonic Flight Tests of a 1/7-Scale Radio-Controlled Model of the North American X-15 Airplane With Particular Reference to High Angle-of-Attack Conditions," NASA TM-X-283 (1960).

Michael J. Hirschberg and David M. Hart, "A Summary of a Half-Century of Oblique Wing Research," AIAA Paper 2007-150 (2007).

M.L. Holcomb, "The Beech Model 77 'Skipper' Spin Program," AIAA Paper 79-1835 (1979).

M.L. Holcomb and R.R. Tumlinson, "Evaluation of a Radio-Control Model for Spin Simulation," Society of Automotive Engineers [SAE] Paper 77-0482 (1977).

Euclid C. Holleman, "Summary of Flight Tests to Determine the Spin and Controllability Characteristics of a Remotely Piloted, Large-Scale (3/8) Fighter Airplane Model," NASA TN-D-8052 (1976).

Joseph L. Johnson, Jr., "Investigation of the Low-Speed Stability and Control Characteristics of a 1/10-Scale Model of the Douglas XF4D-1 Airplane in the Langley Free-Flight Tunnel," NACA RM-SL51J22 (1951).

Joseph L. Johnson, Jr., "Low-Speed Wind-Tunnel Investigation to Determine the Flight Characteristics of a Model of a Parawing Utility Vehicle," NASA TN-D-1255 (1962).

Joseph L. Johnson, Jr., "Stability and Control Characteristics of a 1/10-Scale Model of the McDonnell XP-85 Airplane While Attached to the Trapeze," NACA RM-L7J16 (1947).

Robert W. Kamm and Philip W. Pepoon, "Spin-Tunnel Tests of a 1/57.33-Scale Model of the Northrop XB-35 Airplane," NACA Wartime Report L-739 (1944).

Mark W. Kelly and Lewis H. Smaus, "Flight Characteristics of a 1/4-Scale Model of the XFV-1 Airplane," NACA RM-SA52J15 (1952).

P. Sean Kenney and Mark A. Croom, "Simulating the ARES Aircraft in the Mars Environment," AIAA Paper 2003-6579 (2003).

David D. Kershner, "Miniature Flow-Direction and Airspeed Sensor for Airplanes and Radio-Controlled Models in Spin Studies," NASA TP-1467 (1979).

Robert H. Kirby, "Flight Investigation of the Stability and Control Characteristics of a Vertically Rising Airplane with Swept or Unswept Wings and X- or +- Tails," NACA TN-3812 (1956).

Walter J. Klinar, Henry A. Lee, and L. Faye Wilkes, "Free-Spinning-Tunnel Investigation of a 1/25-Scale Model of the Chance Vought XF8U-1 Airplane," NACA RM-SL56L31b (1956).

Henry A. Lee and Sanger M. Burk, "Low-Speed Dynamic Model Investigation of Apollo Command Module Configuration in the Langley Spin Tunnel," NASA TN-D-3888 (1967).

Henry A. Lee, Peter S. Costigan, and James S. Bowman, Jr., "Dynamic Model Investigation of a 1/20-Scale Gemini Spacecraft in the Langley Spin Tunnel," NASA TN-D-2191 (1964).

Charles E. Libby, "Free-Flight Investigation of the Deployment of a Parawing Recovery Device for a Radio-Controlled 1/5-Scale Dynamic Model Spacecraft," NASA TN-D-2044 (1963).

Charles E. Libby and Sanger M. Burk, Jr., "A Technique Utilizing Free-Flying Radio-Controlled Models to Study the Incipient-and Developed-Spin Characteristics of Airplanes," NASA Memo 2-6-59L (1959).

Charles E. Libby and Sanger M. Burk, Jr., "Large-Angle Motion Tests, Including Spins, of a Free-Flying Dynamically Scaled Radio-Controlled 1/9-Scale Model of an Attack Airplane," NASA TM-X-551 (1961).

Charles E. Libby and Joseph L. Johnson, Jr., "Stalling and Tumbling of a Radio-Controlled Parawing Airplane Model," NASA TN-D-2291 (1964).

Powell M. Lovell, Jr., and Lysle P. Parlett, "Flight Tests of a Model of a High-Wing Transport Vertical-Take-Off Airplane With Tilting Wing and Propellers and With Jet Controls at the Rear of the Fuselage for Pitch and Yaw Control," NACA TN 3912 (1957).

Powell M. Lovell, Jr., and Lysle P. Parlett, "Hovering-Flight Tests of a Model of a Transport Vertical Take-Off Airplane with Tilting Wing and Propellers," NACA TN-3630 (1956).

P.M. Lovell, Jr., C.C. Smith, and R.H. Kirby, "Flight Investigation of the Stability and Control Characteristics of a 0.13-Scale Model of the Convair XFY-1 Vertically Rising Airplane During Constant-Altitude Transitions," NACA RM-SL53E18 (1953).

P.M. Lovell, Jr., C.C. Smith, and R.H. Kirby, "Stability and Control Flight Tests of a 0.13-Scale Model of the Consolidated Vultee XFY-1 Airplane in Take-Offs, Landings, and Hovering Flight," NACA RM-SL52I26 (1952).

B. Maggin and C.V. Bennett, "Tow Tests of a 1/17-Scale Model of the P-80A Airplane in the Langley Free-Flight Tunnel," NACA MR-L5K06 (1945).

B. Maggin and A.H. LeShane, "Tow Tests of a 1/17.8-Scale Model of the XFG-1 Glider in the Langley Free-Flight Tunnel," NACA MR-L5H21 (1945).

Linwood W. McKinney and Edward C. Polhamus, "A Summary of NASA Data Relative to External-Store Separation Characteristics," NASA TN-D-3582 (1966).

Marion O. McKinney, Jr., "Experimental Determination of the Effects of Dihedral, Vertical Tail Area, and Lift Coefficient on Lateral Stability and Control Characteristics," NACA TN-1094 (1946).

Marion O. McKinney, Jr., and Hubert M. Drake, "Flight Characteristics at Low Speed of Delta-Wing Models," NACA RM-L7K07 (1948).

James E. Murray, Joseph W. Pahle, Stephen P. Thornton, Shannon Vogus, Tony Frackowiak, Joe Mello, and Brook Norton, "Ground and Flight Evaluation of a Small-Scale Inflatable-Winged Aircraft," NASA TM-2002-210721 (2002).

NASA, Langley Research Center staff, "Exploratory Study of the Effects of Wing-Leading-Edge Modifications" NASA TP-1589 (1979).

Anshal I. Neihouse, Walter J. Klinar, and Stanley H. Scher, "Status of Spin Research for Recent Airplane Designs" NASA TR-R-57 (1962).

Anshal I. Neihouse and Philip W. Pepoon, "Dynamic Similitude Between a Model and a Full-Scale Body for Model Investigation at Full-Scale Mach Number," NACA TN-2062 (1950).

W.A. Newsom, Jr., and E.L. Anglin, "Free-Flight Model Investigation of a Vertical Attitude VTOL Fighter," NASA TN-D-8054 (1975).

W.A. Newsom, Jr., and S.B. Grafton, "Free-Flight Investigation of Effects of Slats on Lateral-Directional Stability of a 0.13-Scale Model of the F-4E Airplane" NASA TM-SX-2337 (1971).

William A. Newsom, Jr., and Robert H. Kirby, "Flight Investigation of Stability and Control Characteristics of a 1/9-Scale Model of a Four-Propeller Tilt-Wing V/STOL Transport," NASA TN-D-2443 (1964).

D. Bruce Owens, Jay M. Brandon, Mark A. Croom, Charles M. Fremaux, Eugene H. Heim, and Dan D. Vicroy, "Overview of Dynamic Test Techniques for Flight Dynamics Research at NASA LaRC," AIAA Paper 2006-3146 (2006).

Weneth D. Painter, "AD-1 Oblique Wing Research Aircraft Pilot Evaluation Program," AIAA Paper 83-2509 (1983).

James H. Parks, "Experimental Evidence of Sustained Coupled Longitudinal and Lateral Oscillations From Rocket-Propelled Model of a 35° Swept-Wing Airplane Configuration," NACA RM-L54D15 (1954).

James H. Parks and Jesse L. Mitchell, "Longitudinal Trim and Drag Characteristics of Rocket-Propelled Models Representing Two Airplane Configurations," NACA RM-L9L22 (1949).

Lysle P. Parlett, "Free-Flight Investigation of the Stability and Control Characteristics of a STOL Model with an Externally Blown Jet Flap," NASA TN-D-7411 (1974).

Lysle P. Parlett, "Free-Flight Wind-Tunnel Investigation of a Four-Engine Sweptwing Upper-Surface Blown Transport Configuration," NASA TM-X-71932 (1974).

D.N. Petroff, S.H. Scher, and L.E. Cohen, "Low Speed Aerodynamic Characteristics of an 0.075-Scale F-15 Airplane Model at High Angles of Attack and Sideslip," NASA TM-X-62360 (1974).

D.N. Petroff, S.H. Scher, and C.E. Sutton, "Low-Speed Aerodynamic Characteristics of a 0.08-Scale YF-17 Airplane Model at High Angles of Attack and Sideslip," NASA TM-78438 (1978).

Edward C. Polhamus, "Effect of Flow Incidence and Reynolds Number on Low-Speed Aerodynamic Characteristics of Several Noncircular Cylinders with Applications to Directional Stability and Spinning," NACA TN-4176 (1958).

Edward C. Polhamus and Thomas A. Toll, "Research Related to Variable Sweep Aircraft Development," NASA TM-83121 (1981).

Robert W. Rainey, "A Wind-Tunnel Investigation of Bomb Release at a Mach Number of 1.62," NACA RM-L53L29 (1954).

Maurice L. Rasmussen, "Determination of Nonlinear Pitching-Moment Characteristics of Axially Symmetric Models From Free-Flight Data," NASA TN-D-144 (1960).

Carl A. Sandahl, "Free-Flight Investigation at Transonic and Supersonic Speeds of a Wing-Aileron Configuration Simulating the D558-2 Airplane," NACA RM-L8E28 (1948).

Carl A. Sandahl, "Free-Flight Investigation at Transonic and Supersonic Speeds of the Rolling Effectiveness for a 42.7° Sweptback Wing Having Partial-Span Ailerons," NACA RM-L8E25 (1948).

Carl A. Sandahl and Maxime A. Faget, "Similitude Relations for Free-Model Wind-Tunnel Studies of Store-Dropping Problems," NACA TN-3907 (1957).

Dale R. Satran, "Wind-Tunnel Investigation of the Flight Characteristics of a Canard General-Aviation Airplane Configuration," NASA TP-2623 (1986).

Robert O. Schade, "Flight Test Investigation on the Langley Control-Line Facility of a Model of a Propeller-Driven Tail-Sitter-Type Vertical-Take-off Airplane with Delta Wing during Rapid Transitions," NACA TN-4070 (1957).

Robert O. Schade and James L. Hassell, Jr., "The Effects on Dynamic Lateral Stability and Control of Large Artificial Variations in the Rotary Stability Derivatives," NACA TN-2781 (1953).

Robert E. Shanks, "Investigation of the Dynamic Stability of Two Towed Models of a Flat-Bottom Lifting Reentry Configuration," NASA TM-X-1150 (1966).

Robert E. Shanks and George M. Ware, "Investigation of the Flight Characteristics of a 1/5-Scale Model of a Dyna-Soar Glider Configuration at Low Subsonic Speeds," NASA TM-X-683 (1962).

Stanley H. Scher, "Wind-Tunnel Investigation of the Behavior of Parachutes in Close Proximity to One Another," NACA RM-L53G07 (1953).

Stanley H. Scher and John W. Draper, "The Effects of Stability of Spin-Recovery Tail Parachutes on the Behavior of Airplanes in Gliding Flight and in Spins," NACA TN-2098 (1950).

Stanley H. Scher and William L. White, "Spin-Tunnel Investigation of the Northrop F-5E Airplane," NASA TM-SX-3556 (1977).

Max Scherberg and R.V. Rhode, "Mass Distribution and Performance of Free Flight Models," NACA TN-268 (1927).

Charles L. Seacord, Jr., and Herman O. Ankenbruck, "Determination of the Stability and Control Characteristics of a Straight-Wing, Tailless Fighter-Airplane Model in the Langley Free-Flight Tunnel," NACA Wartime Report ACR No. L5K05 (1946).

Oscar Seidman and Charles J. Donlan, "An Approximate Spin Design Criteria for Monoplanes," NACA TN-711 (1939).

Oscar Seidman and Anshal I. Neihouse, "Free-Spinning Wind-Tunnel Tests on a Low-Wing Monoplane with Systematic Changes in Wings and Tails III. Mass Distributed Along the Wings," NACA TN-664 (1938).

Alvin Seiff, Carlton S. James, Thomas N. Canning, and Alfred G. Boissevain, "The Ames Supersonic Free-Flight Wind Tunnel," NACA RM-A52A24 (1952).

Robert E. Shanks, "Aerodynamic and Hydrodynamic Characteristics of Models of Some Aircraft-Towed Mine-Sweeping Devices," NACA RM-SL55K21 (1955).

Robert E. Shanks, "Experimental Investigation of the Dynamic Stability of a Towed Parawing Glider Model," NASA TN-D-1614 (1963).

Robert E. Shanks, "Investigation of the Dynamic Stability and Controllability of a Towed Model of a Modified Half-Cone Reentry Vehicle," NASA TN-D-2517 (1965).

Joseph A. Shortal and Clayton J. Osterhout, "Preliminary Stability and Control Tests in the NACA Free-Flight Tunnel and Correlation with Flight Tests," NACA TN-810 (1941).

Charles C. Smith, Jr., "Flight Tests of a 1/6-Scale Model of the Hawker P.1127 Jet VTOL Airplane," NASA TM-SX-531 (1961).

Ralph W. Stone, Jr., William G. Garner, and Lawrence J. Gale, "Study of Motion of Model of Personal-Owner or Liaison Airplane Through the Stall and into the Incipient Spin by Means of a Free-Flight Testing Technique," NACA TN-2923 (1953).

Ralph W. Stone, Jr., and Walter J. Klinar, "The Influence of Very Heavy Fuselage Mass Loadings and Long Nose Lengths Upon Oscillations in the Spin," NACA TN-1510 (1948).

H. Paul Stough, III, "A Summary of Spin-Recovery Parachute Experience on Light Airplanes," AIAA Paper 90-1317 (1990).

Tim Tam, Stephen Ruffin, Leslie Yates, John Morgenstern, Peter Gage, and David Bogdanoff, "Sonic Boom Testing of Artificially Blunted Leading Edge (ABLE) Concepts in the NASA Ames Aeroballistic Range," AIAA Paper 2000-1011 (2000).

Louis P. Tosti, "Flight Investigation of Stability and Control Characteristics of a 1/8-Scale Model of a Tilt-Wing Vertical-Take-Off-And-Landing Airplane," NASA TN-D-45 (1960).

Louis P. Tosti, "Longitudinal Stability and Control of a Tilt-Wing VTOL Aircraft Model with Rigid and Flapping Propeller Blades," NASA TN-D-1365 (1962).

Louis P. Tosti, "Rapid-Transition Tests of a ¼-Scale Model of the VZ-2 Tilt-Wing Aircraft," NASA TN-D-946 (1961).

R.R. Tumlinson, M.L. Holcomb, and V.D. Gregg, "Spin Research on a Twin-Engine Aircraft," AIAA Paper 81-1667 (1981).

Dan D. Vicroy, "Blended-Wing-Body Low-Speed Flight Dynamics: Summary of Ground Tests and Sample Results," AIAA Invited Paper Presented at the *47th AIAA Aerospace Sciences Meeting and Exhibit, Jan. 2009.*

Laurence A. Walker, "Flight Testing the X-36-The Test Pilot's Perspective," NASA CR-198058 (1997).

George M. Ware, "Investigation of the Flight Characteristics of a Model of the HL-10 Manned Lifting Entry Vehicle," NASA TM-X-1307 (1967).

C. Wenzinger and T. Harris, "The Vertical Wind Tunnel of the National Advisory Committee for Aeronautics," NACA TR-387 (1931).

Raymond D. Whipple and J.L. Ricket, "Low-Speed Aerodynamic Characteristics of a 1/8-scale X-29A Airplane Model at High Angles of Attack and Sideslip," NASA TM-87722 (1986).

H.E. Wimperis, "New Methods of Research in Aeronautics," *Journal of the Royal Aeronautical Society* (Dec. 1932), p. 985.

C.H. Wolowicz, J.S. Bowman, Jr., and W.P. Gilbert, "Similitude Requirements and Scaling Relationships as Applied to Model Testing," NASA TP-1435 (1979).

C.H. Zimmerman, "Preliminary Tests in the N.A.C.A. Free-Spinning Wind Tunnel," NACA TR-557 (1935).

Books and Monographs:

Donald D. Baals and William R. Corliss, *Wind Tunnels of NASA*, NASA SP-440 (Washington, DC: GPO, 1981).

John V. Becker, *The High Speed Frontier*, NASA SP-445 (Washington, DC: GPO, 2005).

Joseph R. Chambers, *Concept to Reality: Contributions of the Langley Research Center to U.S. Civil Aircraft of the 1990s*, NASA 2003-SP-4529 (Washington, DC: GPO, 2003).

Joseph R. Chambers, *Innovation in Flight: Research of the Langley Research Center on Revolutionary Advanced Concepts for Aeronautics*, NASA 2005-SP-4539 (Washington, DC: GPO, 2005).

Joseph R. Chambers, *Partners in Freedom: Contributions of the Langley Research Center to U.S. Military Aircraft of the 1990s,* NASA SP-2000-4519 (Washington, DC: GPO, 2000).

Joseph R. Chambers and Mark A. Chambers, *Radical Wings and Wind Tunnels* (North Branch, MN: Specialty Press, 2008).

James R. Hansen, *Engineer in Charge*, NASA SP-4305 (Washington, DC: GPO, 1987).

Edwin P. Hartman, *Adventures in Research: A History of Ames Research Center 1940–1965*, NASA SP-4302 (Washington, DC: GPO, 1970).

Dennis R. Jenkins and Tony R. Landis, *Hypersonic-The Story of the North American X-15* (North Branch, MN: Specialty Press, 2008).

R. Dale Reed, *Wingless Flight: The Lifting Body Story*, NASA SP-4220 (Washington, DC: GPO, 1997).

Joseph A. Shortal, *A New Dimension. Wallops Island Flight Test Range: The First Fifteen Years*, NASA RP-1028 (Washington, DC: GPO, 1978).

Blended Wing-Body free-flight model in the Langley Full-Scale Tunnel. NASA.

CASE 6

NASA and the Evolution of the Wind Tunnel

Jeremy Kinney

Even before the invention of the airplane, wind tunnels have been key in undertaking fundamental research in aerodynamics and evaluating design concepts and configurations. Wind tunnels are essential for aeronautical research, whether for subsonic, transonic, supersonic, or hypersonic flight. The swept wing, delta wing, blended wing body shapes, lifting bodies, hypersonic boost-gliders, and other flight concepts have been evaluated and refined in NACA and NASA tunnels.

IN NOVEMBER 2004, the small X-43A scramjet hypersonic research vehicle achieved Mach 9.8, roughly 6,600 mph, the fastest speed ever attained by an air-breathing engine. During the course of the vehicle's 10-second engine burn over the Pacific Ocean, the National Aeronautics and Space Administration (NASA) offered the promise of a new revolution in aviation, that of high-speed global travel and cost-effective entry into space. Randy Voland, project engineer at Langley Research Center, exclaimed that the flight "looked really, really good" and that "in fact, it looked like one of our simulations."[1] In the early 21st century, the public's awareness of modern aeronautical research recognized advanced computer simulations and dramatic flight tests, such as the launching of the X-43A mounted to the front of a Pegasus rocket booster from NASA's venerable B-52 platform. A key element in the success of the X-43A was a technology as old as the airplane itself: the wind tunnel, a fundamental research tool that also has evolved over the past century of flight.

NASA and its predecessor, the National Advisory Committee for Aeronautics (NACA), have been at the forefront of aerospace research since the early 20th century and on into the 21st. NASA made fundamental contributions to the development and refinement of aircraft and spacecraft—from commercial airliners to the Space Shuttle—for

1. Warren E. Leary, "NASA Jet Sets Record For Speed," *New York Times*, Nov. 17, 2004, p. A24.

operation at various speeds. The core of this success has been NASA's innovation, development, and use of wind tunnels. At crucial moments in the history of the United States, the NACA–NASA introduced state-of-the-art testing technologies as the aerospace community needed them, placing the organization onto the world stage.

The Anatomy of a Wind Tunnel

The design of an efficient aircraft or spacecraft involves the use of the wind tunnel. These tools simulate flight conditions, including Mach number and scale effects, in a controlled environment. Over the late 19th, 20th, and early 21st centuries, wind tunnels evolved greatly, but they all incorporate five basic features, often in radically different forms. The main components are a drive system, a controlled fluid flow, a test section, a model, and instrumentation. The drive system creates a fluid flow that replicates flight conditions in the test section. That flow can move at subsonic (up to Mach 1), transonic (Mach 0.75 to 1.25), supersonic (up to Mach 5), or hypersonic (above Mach 5) speeds. The placement of a scale model of an aircraft or spacecraft in the test section via balances allows the measurement of the physical forces acting upon that model with test instrumentation. The specific characteristics of each of these components vary from tunnel to tunnel and reflect the myriad of needs for this testing technology and the times in which experimenters designed them.[2]

Wind tunnels allow researchers to focus on isolating and gathering data about particular design challenges rooted in the four main systems of aircraft: aerodynamics, control, structures, and propulsion. Wind tunnels measure primarily forces such as lift, drag, and pitching moment, but they also gauge air pressure, flow, density, and temperature. Engineers convert those measurements into aerodynamic data to evaluate performance and design and to verify performance predictions. The data represent design factors such as structural loading and strength, stability and control, the design of wings and other elements, and, most importantly, overall vehicle performance.[3]

Most NACA and NASA wind tunnels are identified by their location, the size of their test section, the speed of the fluid flow, and the main design characteristic. For example, the Langley 0.3-Meter Transonic

2. Donald D. Baals and William R. Corliss, *Wind Tunnels of NASA*, NASA SP-440 (Washington, DC: GPO, 1981), p. 2.

3. NASA Ames Applied Aerodynamics Branch, "The Unitary Plan Wind Tunnels" (July 1994), pp. 10–11.

Cryogenic Tunnel evaluates scale models in its 0.3-meter test section between speeds of Mach 0.2 to 1.25 in a fluid flow of nitrogen gas. A specific application, 9- by 6-Foot Thermal Structures Tunnel, or the exact nature of the test medium, 8-Foot Transonic Pressure Tunnel, can be other characterizing factors for the name of a wind tunnel.

The Prehistory of the Wind Tunnel to 1958

The growing interest in and institutionalization of aeronautics in the late 19th century led to the creation of the wind tunnel.[4] English scientists and engineers formed the Royal Aeronautical Society in 1866. The group organized lectures, technical meetings, and public exhibitions, published the influential *Annual Report of the Aeronautical Society,* and funded research to spread the idea of powered flight. One of the more influential members was Francis Herbert Wenham. Wenham, a professional engineer with a variety of interests, found his experiments with a whirling arm to be unsatisfactory. Funded by a grant from the Royal Aeronautical Society, he created the world's first operating wind tunnel in 1870–1872. Wenham and his colleagues conducted rudimentary lift and drag studies and investigated wing designs with their new research tool.[5]

Wenham's wing models were not full-scale wings. In England, University of Manchester researcher Osborne Reynolds recognized in 1883 that the airflow pattern over a scale model would be the same for its full-scale version if a certain flow parameter were the same in both cases. This basic parameter, attributed to its discoverer as the Reynolds number, is a measure of the relative effects of the inertia and viscosity of air flowing over an aircraft. The Reynolds number is used to describe all types of fluid flow, including the shape of flow, heat transfer, and the start of turbulence.[6]

While Wenham invented the wind tunnel and Reynolds created the basic parameter for understanding its application to full-scale aircraft, Wilbur and Orville Wright were the first to use a wind tunnel in the systematic way that later aeronautical engineers would use it. The brothers, not aware of Wenham's work, saw their "invention" of the wind tunnel become part of their revolutionary program to create a practical heavier-than-air flying machine from 1896 to 1903. Frustrated by the

4. For a detailed history of wind tunnel development before World War II, see J. Lawrence Lee, "Into the Wind: A History of the American Wind Tunnel, 1896–1941," dissertation, Auburn University, 2001.
5. Baals and Corliss, *Wind Tunnels of NASA*, p. 3.
6. Ibid.

poor performance of their 1900 and 1901 gliders on the sandy dunes of the Outer Banks—they did not generate enough lift and were uncontrollable—the Wright brothers began to reevaluate their aerodynamic calculations. They discovered that Smeaton's coefficient, one of the early contributions to aeronautics, and Otto Lilienthal's groundbreaking airfoil data were wrong. They found the discrepancy through the use of their wind tunnel, a 6-foot-long box with a fan at one end to generate air that would flow over small metal models of airfoils mounted on balances, which they had created in their bicycle workshop. The lift and drag data they compiled in their notebooks would be the key to the design of wings and propellers during the rest of their experimental program, which culminated in the first controlled, heavier-than-air flight December 17, 1903.[7]

Over the early flight and World War I eras, aeronautical enthusiasts, universities, aircraft manufacturers, military services, and national governments in Europe and the United States built 20 wind tunnels. The United States built the most at 9, with 4 rapidly appearing during American involvement during the Great War. Of the European countries, Great Britain built 4, but the tunnels in France (2) and Germany (3) proved to be the most innovative. Gustav Eiffel's 1912 tunnel at Auteiul, France, became a practical tool for the French aviation industry to develop high-performance aircraft for the Great War. At the University of Göttingen in Germany, aerodynamics pioneer Ludwig Prandtl designed what would become the model for all "modern" wind tunnels in 1916. The tunnel featured a closed circuit; a contraction cone, or nozzle, just before the test section that created uniform air velocity and reduced turbulence in the test section; and a chamber upstream of the test section that stilled any remaining turbulent air further.[8]

The NACA and the Wind Tunnel

For the United States, the Great War highlighted the need to achieve parity with Europe in aeronautical development. Part of that effort was the creation of the Government civilian research agency, the NACA, in March 1915. The committee established its first facility, Langley Memorial Aeronautical Laboratory—named in honor of aeronautical experimenter and Smithsonian Secretary Samuel P. Langley—2 years

7. Peter Jakab, *Visions of a Flying Machine* (Washington, DC: Smithsonian Institution Press, 1990), p. 155.
8. Baals and Corliss, *Wind Tunnels of NASA*, pp. 9–12.

NACA Wind Tunnel No. 1 with a model of a Curtiss JN-4D Trainer in the test section. NASA.

later near Hampton, VA, on the Chesapeake Bay. In June 1920, NACA Wind Tunnel No. 1 became operational. A close copy of a design built at the British National Physical Laboratory a decade earlier, the tunnel produced no data directly applicable to aircraft design.[9]

9. Ibid., pp. 13–15.

One of the major obstacles facing the effective use of a wind tunnel was scale effects, meaning the Reynolds number of model did not match the full-scale airplane. Prandtl protege Max Munk proposed the construction of a high-pressure tunnel to solve the problem. His Variable Density Tunnel (VDT) could be used to test a 1/20th-scale model in an airflow pressurized to 20 atmospheres, which would generate identical Reynolds numbers to full-scale aircraft. Built in the Newport News shipyards, the VDT was radical in design with its boilerplate and rivets. More importantly, it proved to be a point of departure from previous tunnels with the data that it produced.[10]

The VDT became an indispensable tool to airfoil development that effectively reshaped the subsequent direction of American airfoil research and development after it became operational in 1923. Munk's successor in the VDT, Eastman Jacobs, and his colleagues in the VDT pioneered airfoil design methods with the pivotal Technical Report 460, which influenced aircraft design for decades after its publication in 1933.[11] Of the 101 distinct airfoil sections employed on modern Army, Navy, and commercial airplanes by 1937, 66 were NACA designs. Those aircraft included the venerable Douglas DC-3 airliner, considered by many to be the first truly "modern" airplane, and the highly successful Boeing B-17 Flying Fortress of World War II.[12]

The NACA also addressed the fundamental problem of incorporating a radial engine into aircraft design in the pioneering Propeller Research Tunnel (PRT). Lightweight, powerful, and considered a revolutionary aeronautical innovation, a radial engine featured a flat frontal configuration that created a lot of drag. Engineer Fred E. Weick and his colleagues tested full-size aircraft structures in the tunnel's 20-foot opening. Their solution, called the NACA cowling, arrived at the right moment to increase the performance of new aircraft. Spectacular demonstrations—such as Frank Hawks flying the Texaco Lockheed Air Express, with a NACA cowling installed, from Los Angeles to New York nonstop in a record time of 18 hours 13 minutes in February 1929—led to the organization's first Collier Trophy, in 1929.

10. Ibid., pp. 15–17.

11. Eastman N. Jacobs, Kenneth E. Ward, and Robert M. Pinkerton, "The Characteristics of 78 Related Airfoil Sections from Tests in the Variable-Density Wind Tunnel," NACA TR-460 (1933).

12. R.C. Platt, memorandum for Dr. Lewis, "Airfoil sections employed for wings of modern airplanes," Sept. 2, 1937, RA file 290, Langley Research Center Historical Archives; Baals and Corliss, *Wind Tunnels of NASA*, pp. 15–17 (quote).

With the basic formula for the modern airplane in place, the aeronautical community began to push the limits of conventional aircraft design. The NACA built upon its success with the cowling research in the PRT and concentrated on the aerodynamic testing of full-scale aircraft in wind tunnels. The Full-Scale Tunnel (FST) featured a 30- by 60-foot test section and opened at Langley in 1931. The building was a massive structure at 434 feet long, over 200 feet wide, and 9 stories high. The first aircraft to be tested in the FST was a Navy Vought O3U-1 Corsair observation airplane. Testing in the late 1930s focused on removing as much drag from an airplane in flight as possible. NACA engineers—through an extensive program involving the Navy's first monoplane fighter, Brewster XF2A-1 Buffalo—showed that attention to details such as air intakes, exhaust pipes, and gun ports effectively reduced drag.

In the mid- to late 1920s, the first generation of university-trained American aeronautical engineers began to enter work with industry, the Government, and academia. The philanthropic Daniel Guggenheim Fund for the Promotion of Aeronautics created aeronautical engineering schools, complete with wind tunnels, at the California Institute of Technology, Georgia Institute of Technology, Massachusetts Institute of Technology, University of Michigan, New York University, Stanford University, and University of Washington. The creation of these dedicated academic programs ensured that aeronautics would be an institutionalized profession. The university wind tunnels quickly made their mark. The prototype Douglas DC airliner, the DC-1, flew in July 1933. In every sense of the word, it was a streamline airplane because of the extensive amount of wind tunnel testing at Guggenheim Aeronautical Laboratory at the California Institute of Technology used in its design.

By the mid-1930s, it was obvious that the sophisticated wind tunnel research program undertaken by the NACA had contributed to a new level of American aeronautical capability. Each of the major American manufacturers built wind tunnels or relied upon a growing number of university facilities to keep up with the rapid pace of innovation. Despite those additions, it was clear in the minds of the editors at the influential trade journal *Aviation* that the NACA led the field with the grace, style, and coordinated virtuosity of a symphonic orchestra.[13]

13. Edward P. Warner, "Research to the Fore," *Aviation*, vol. 33, no. 6 (June 1934), p. 186; "Research Symphony: The Langley Philharmonic in Opus No. 10," *Aviation*, vol. 34, no. 6 (June 1935), pp. 15–18.

6

World War II stimulated the need for sophisticated aerodynamic testing, and new wind tunnels met the need. Langley's 20-Foot Vertical Spin Tunnel (VST) became operational in March 1941. The major difference between the VST and those that came before was its vertical closed-throat, annular return. A variable-speed three-blade, fixed-pitch fan provided vertical airflow at an approximate velocity of 85 feet per second at atmospheric conditions. Researchers threw dynamically scaled, free-flying aircraft models into the tunnel to evaluate their stability as they spun and tumbled out of control. The installation of remotely actuated control surfaces allowed the study of spin recovery characteristics. The NACA solution to spin problems for aircraft was to enlarge the vertical tail, raise the horizontal tail, and extend the length of the ventral fin.[14]

The NACA founded the Ames Aeronautical Laboratory on December 20, 1939, in anticipation of the need for expanded research and flight-test facilities for the West Coast aviation industry. The NACA leadership wanted to reach parity with European aeronautical research based on the belief that the United States would be entering World War II. The cornerstone facility at Ames was the 40 by 80 Tunnel capable of generating airflow of 265 mph for even larger full-scale aircraft when it opened in 1944. Building upon the revolutionary drag reduction studies pioneered in the FST, Ames researchers continued to modify existing aircraft with fillets and innovated dive recovery flaps to offset a new problem encountered when aircraft entered high-speed dives called compressibility.[15]

The NACA also desired a dedicated research facility that specialized in aircraft propulsion systems. Construction of the Aircraft Engine Research Laboratory (AERL) began at Cleveland, OH, in January 1941, with the facility becoming operational in May 1943.[16] The cornerstone

---

14. George W. Gray, *Frontiers of Flight: The Story of NACA Research* (New York: A.A. Knopf, 1948), p. 156; James R. Hansen, *Engineer in Charge: A History of the Langley Aeronautical Laboratory, 1917–1958*, NASA SP-4305 (Washington, DC: GPO, 1987), pp. 462–463; NASA, "Wind Tunnels at NASA Langley Research Center," FS-2001-04-64-LaRC, 2001, *http://www.nasa.gov/centers/langley/news/factsheets/windtunnels.html*, accessed May 28, 2009.

15. Glenn E. Bugos, *Atmosphere of Freedom: Sixty Years at the NASA Ames Research Center*, NASA SP-4314 (Washington, DC: GPO, 2000), pp. 6–13.

16. The NACA renamed the AERL the Propulsion Research Laboratory in 1947 and changed the name of the facility once again to the Lewis Flight Propulsion Laboratory a year later in honor of George W. Lewis, the committee's first Director of Aeronautical Research. Virginia P. Dawson, *Engines and Innovation: Lewis Laboratory and American Propulsion Technology*, NASA SP-4306 (Washington, DC: GPO, 1991), pp. 2–14, 36.

facility was the Altitude Wind Tunnel (AWT), which became operational in 1944. The AWT was the only wind tunnel in the world capable of evaluating full-scale aircraft engines in realistic flight conditions that simulated altitudes up to 50,000 feet and speeds up to 500 mph. AERL researchers began first with large radial engines and propellers and continued with the new jet technology on through the postwar decades.[17]

The AERL soon became the center of the NACA's work on alleviating aircraft icing. The Army Air Forces lost over 100 military transports along with their crews and cargoes over the "Hump," or the Himalayas, as it tried to supply China by air. The problem was the buildup of ice on wings and control surfaces that degraded the aerodynamic integrity and overloaded the aircraft. The challenge was developing de-icing systems that removed or prevented the ice buildup. The Icing Research Tunnel (IRT) was the largest of its kind when it opened in 1944. It featured a 6- by 9-foot test section, a 160-horsepower electric motor capable of generating a 300 mph airstream, and a 2,100-ton refrigeration system that cooled the airflow down to -40 degrees Fahrenheit (°F).[18] The tunnel worked well during the war and the following two decades, before NASA closed it. However, a new generation of icing problems for jet aircraft, rotary wing, and Vertical/Short Take-Off and Landing (V/STOL) aircraft resulted in the reopening of the IRT in 1978.[19]

During World War II, airplanes ventured into a new aerodynamic regime, the so-called "transonic barrier." American propeller-driven aircraft suffered from aerodynamic problems caused by high-speed flight. Flight-testing of the P-38 Lightning revealed compressibility problems that resulted in the death of a test pilot in November 1941. As the Lightning dove from 30,000 feet, shock waves formed over the wings and hit the tail, causing violent vibration, which caused the airplane to plummet into a vertical, and unrecoverable, dive. At speeds approaching Mach 1, aircraft experienced sudden changes in stability and control,

17. Ernest G. Whitney, "Altitude Tunnel at AERL," Lecture 22, June 23, 1943, *http://awt.grc.nasa.gov/resources/Research_Documents/Altitude_Wind_Tunnel_at_AERL.pdf*, accessed Oct. 12, 2009.
18. William M. Leary, *"We Freeze to Please": A History of NASA's Icing Research Tunnel and the Quest for Flight Safety*, NASA SP-2002-4226 (Washington, DC: NASA, 2002), pp. 19–37; Baals and Corliss, *Wind Tunnels of NASA*, pp. 45–46; NASA Langley, "NASA's Wind Tunnels," IS-1992-05-002-LaRC, May 1992, *http://oea.larc.nasa.gov/PAIS/WindTunnel.html*, accessed May 26, 2009.
19. In 1987, the American Society of Mechanical Engineers designated the IRT an International Historic Mechanical Engineering Landmark.

extreme buffeting, and, most importantly, a dramatic increase in drag, which created challenges for the aeronautical community involving propulsion, research facilities, and aerodynamics. Bridging the gap between subsonic and supersonic speeds was a major aerodynamic challenge.[20]

The transonic regime was unknown territory in the 1940s. Four approaches—putting full-size aircraft into terminal velocity dives, dropping models from aircraft, installing miniature wings mounted on flying aircraft, and launching models mounted on rockets—were used in lieu of an available wind tunnel in the 1940s for transonic research. Aeronautical engineers faced a daunting challenge rooted in developing tools and concepts because no known wind tunnel was able to operate and generate data at transonic speeds.

NACA Manager John Stack took the lead in American work in transonic development. As the central NACA researcher in the development of the first research airplane, the Bell X-1, he was well-qualified for high-speed research. His part in the first supersonic flight resulted in a joint award of the 1947 Collier Trophy. He ordered the conversion of the 8- and 16-Foot High-Speed Tunnels in spring 1948 to a slotted throat to enable research in the transonic regime. Slots in the tunnels' test sections, or throats, enabled smooth operation at high subsonic speeds and low supersonic speeds. The initial conversion was not satisfactory. Physicist Ray Wright and engineers Virgil S. Ritchie and Richard T. Whitcomb hand-shaped the slots based on their visualization of smooth transonic flow. Working directly with Langley woodworkers, they designed and fabricated a channel at the downstream end of the test section that reintroduced air that traveled through the slots. Their painstaking work led to the inauguration of operations in the newly christened 8-Foot Transonic Tunnel (TT) 7 months later, on October 6, 1950.[21]

Rumors had been circulating throughout the aeronautical community about the NACA's new transonic tunnels: the 8-Foot TT and the 16-Foot TT. The NACA wanted knowledge of their existence to remain confidential among the military and industry. Concerns over secrecy were

20. John Becker, *The High Speed Frontier: Case Histories of Four NACA Programs 1920–1950*, NASA SP-445 (Washington, DC: GPO, 1980), p. 61.

21. Hansen, *Engineer in Charge*, pp. 327–328, 454; Steven T. Corneliussen, "The Transonic Wind Tunnel and the NACA Technical Culture," in Pamela E. Mack, ed., *From Engineering Science to Big Science: The NACA and NASA Collier Trophy Research Project Winners*, NASA SP-4219 (Washington, DC: GPO, 1998), p. 133.

deemed less important than the acknowledgement of the development of the slotted-throat tunnel, for which John Stack and 19 of his colleagues received a Collier Trophy in 1951. The award specifically recognized the importance of a research tool, which was a first in the 40-year history of the award. When used with already available wind tunnel components and techniques, the tunnel balance, pressure orifice, tuft surveys, and schlieren photographs, slotted-throat tunnels resulted in a new theoretical understanding of transonic drag. The NACA claimed that its slotted-throat transonic tunnels gave the United States a 2-year lead in the design of supersonic military aircraft.[22] John Stack's leadership affected the NACA's development of state-of-the-art wind tunnel technology. The researchers inspired by or working under him developed a generation of wind tunnels that, according to Joseph R. Chambers, became "national treasures."[23]

The Transition to NASA

In the wake of the launch of Sputnik I in October 1957, the National Air and Space Act of 1958 combined the NACA's research facilities at Langley, Ames, Lewis, Wallops Island, and Edwards with the Army and Navy rocket programs and the California Institute of Technology's Jet Propulsion Laboratory to form NASA. Suddenly, the NACA's scope of American civilian research in aeronautics expanded to include the challenges of space flight driven by the Cold War competition between the United States and the Soviet Union and the unprecedented growth of American commercial aviation on the world stage.

NASA inherited an impressive inventory of facilities from the NACA. The wind tunnels at Langley, Ames, and Lewis were the start of the art and reflected the rich four-decade legacy of the NACA and the ever-evolving need for specialized tunnels. Over the next five decades of NASA history, the work of the wind tunnels reflected equally in the first "A" and the "S" in the administration's acronym.

The Unitary Plan Tunnels

In the aftermath of World War II and the early days of the Cold War, the Air Force, Army, Navy, and the NACA evaluated what the aeronautical

22. Ibid., p. 91; Hansen, *Engineer in Charge*, pp. 329, 330–331.

23. Joseph R. Chambers, *Innovation in Flight: Research of the NASA Langley Research Center on Revolutionary Advanced Concepts for Aeronautics*, NASA SP-2005-4539 (Washington, DC: GPO, 2005), pp. 18–19.

industry needed to continue leadership and innovation in aircraft and missile development. Specifically, the United States needed more transonic and supersonic tunnels. The joint evaluation resulted in proposal called the Unitary Plan. President Harry S. Truman's Air Policy Commission urged the passage of the Unitary Plan in January 1948. The draft plan, distributed to the press at the White House, proposed the installation of the 16 wind tunnels "as quickly as possible," with the remainder to quickly follow.[24]

Congress passed the Unitary Wind Tunnel Plan Act, and President Truman signed it October 27, 1949. The act authorized the construction of a group of wind tunnels at U.S. Air Force and NACA installations for the testing of supersonic aircraft and missiles and for the high-speed and high-altitude evaluation of engines. The wind tunnel system was to benefit industry, the military, and other Government agencies.[25]

The portion of the Unitary Plan assigned to the U.S. Air Force led to the creation of the Arnold Engineering Development Center (AEDC) at Tullahoma, TN. Dedicated in June 1951, the AEDC took advantage of abundant hydroelectric power provided by the nearby Tennessee Valley Authority. The Air Force erected facilities, such as the Propulsion Wind Tunnel and two individual 16-Foot wind tunnels that covered the range of Mach 0.2 to Mach 4.75, for the evaluation of full-scale jet and rocket engines in simulated aircraft and missile applications. Starting with 2 wind tunnels and an engine test facility, the research equipment at the AEDC expanded to 58 aerodynamic and propulsion wind tunnels.[26] The Aeropropulsion Systems Test Facility, operational in 1985, was the finishing touch, which made the AEDC, in the words of one observer, "the world's most complete aerospace ground test complex."[27]

The sole focus of the AEDC on military aeronautics led the NACA to focus on commercial aeronautics. The Unitary Plan provided two benefits for the NACA. First, it upgraded and repowered the NACA's existing wind tunnel facilities. Second, and more importantly, the Unitary

24. "Report of President's Air Policy Commission Calling for a Greatly Enlarged Defense Force," *New York Times*, Jan. 14, 1948, p. 21.

25. NASA Ames Applied Aerodynamics Branch, "The Unitary Plan Wind Tunnels" (July 1994), pp. 3–4; Hansen, *Engineer in Charge*, pp. 474–475.

26. Arnold Air Force Base, "Arnold Engineering Development Center," 2007, *http://www.arnold.af.mil/library/factsheets/factsheet.asp?id=12977*, accessed July 30, 2009.

27. Baals and Corliss, *Wind Tunnels of NASA*, pp. 65–66.

Plan and provided for three new tunnels at each of the three NACA laboratories at the cost of $75 million. Overall, those three tunnels represented, to one observer, "a landmark in wind tunnel design by any criterion—size, cost, performance, or complexity."[28]

The NACA provided a manual for users of the Unitary Plan Wind Tunnel system in 1956, after the facilities became operational. The document allowed aircraft manufacturers, the military, and other Government agencies to plan development testing. Two general classes of work could be conducted in the Unitary Plan wind tunnels: company or Government projects. Industrial clients were responsible for renting the facility, which amounted to between $25,000 and $35,000 per week (approximately $190,000 to $265,000 in modern currency), depending on the tunnel, the utility costs required to power the facility, and the labor, materials, and overhead related to the creation of the basic test report. The test report consisted of plotted curves, tabulated data, and a description of the methods and procedures that allowed the company to properly interpret the data. The NACA kept the original report in a secure file for 2 years to protect the interests of the company. There were no fees for work initiated by Government agencies.[29]

The Langley Unitary Plan Wind Tunnel began operations in 1955. NACA researcher Herbert Wilson led a design team that created a closed-circuit, continual flow, variable density supersonic tunnel with two test sections. The test sections, each measuring 4 by 4 feet and 7 feet long, covered the range between low Mach (1.5 to 2.9) and high Mach (2.3 to 4.6). Tests in the Langley Unitary Plan Tunnel included force and moment, surface pressure measurements and distribution, visualization of on- and off-surface airflow patterns, and heat transfer. The tunnel operated at 150 °F, with the capability of generating 300–400 °F in short bursts for heat transfer studies. Built at an initial cost of $15.4 million, the Langley facility was the cheapest of the three NACA Unitary Plan wind tunnels.[30]

The original intention of the Langley Unitary Plan tunnel was missile development. A long series of missile tests addressed high-speed

28. "Manual for Users of the Unitary Plan Wind Tunnel Facilities of the National Advisory Committee for Aeronautics," NACA TM-80998 (1956), p. 1; Baals and Corliss, *Wind Tunnels of NASA*, pp. 66, 71 (quote).

29. "Manual for Users of the Unitary Plan Wind Tunnel Facilities," pp. i, 1, 5–9.

30. William T. Shaefer, Jr., "Characteristics of Major Active Wind Tunnels at the Langley Research Center," NASA TM-X-1130 (July 1965), p. 32; Hansen, *Engineer in Charge*, pp. 474–475.

A model of the Apollo Launch Escape System in the Unitary Wind Tunnel at NASA Ames. NASA.

performance, stability and control, maneuverability, jet-exhaust effects, and other factors. NACA researchers quickly placed models of the McDonnell-Douglas F-4 Phantom II in the tunnel in 1956, and soon after, various models of the North American X-15, the General Dynamics F-111 Aardvark, proposed supersonic transport configurations, and spacecraft appeared in the tunnel.[31]

The Ames Unitary Plan Wind Tunnel opened in 1956. It featured three test sections: an 11- by 11-foot transonic section (Mach 0.3 to 1.5) and two supersonic sections that measured 9 by 7 feet (Mach 1.5 to 2.6) and 8 by 7 feet (Mach 2.5 to 3.5). Tunnel personnel could adjust the airflow to simulate flying conditions at various altitudes in each section.[32]

The power and magnitude of the tunnel facility called for unprecedented design and construction. The 11-stage axial-flow compressor featured a 20-foot diameter and was capable of moving air at 3.2 million cubic feet per minute. The complete assembly, which included over 2,000 rotor and stator blades, weighed 445 tons. The flow diversion valve allowed the compressor to drive either the 9- by 7-foot or 8- by 7-foot

31. Baals and Corliss, *Wind Tunnels of NASA*, pp. 68–69.

32. NASA Ames Applied Aerodynamics Branch, "The Unitary Plan Wind Tunnels" (July 1994), p. 9; NASA Langley, "NASA's Wind Tunnels," IS-1992-05-002-LaRC, May 1992, *http://oea.larc.nasa.gov/PAIS/WindTunnel.html*, accessed May 26, 2009.

supersonic wind tunnels. At 24 feet in diameter, the compressor was the largest of its kind in the world in 1956 but took only 3.5 minutes to switch between the two wind tunnels. Four main drive rotors, weighing 150 tons each, powered the facility. They could generate 180,000 horsepower on a continual basis and 216,000 horsepower at 1-hour intervals. Crews used 10,000 cubic yards of concrete for the foundation and 7,500 tons of steel plate for the major structural components. Workers expended 100 tons of welding rods during construction. When the facility began operations in 1956, the project had cost the NACA $35 million.[33]

The personnel of the Ames Unitary Plan Wind Tunnel evaluated every major craft in the American aerospace industry from the late 1950s to the late 20th century. In aeronautics, models of nearly every commercial transport and military fighter underwent testing. For the space program, the Unitary Plan Wind Tunnel was crucial to the design of the landmark Mercury, Gemini, and Apollo spacecraft, and the Space Shuttle. That record led NASA to assert that the facility was a "unique national asset of vital importance to the nation's defense and its competitive position in the world aerospace market." It also reflected the fact that the Unitary Plan facility was NASA's most heavily used wind tunnel, with over 1,000 test programs conducted during 60,000 hours of operation by 1994.[34]

SAMPLE AEROSPACE VEHICLES EVALUATED IN THE UNITARY PLAN WIND TUNNEL		
MILITARY	COMMERCIAL	SPACE
Convair B-58	McDonnell-Douglas DC-8	Mercury spacecraft
Lockheed A-12/YF-12/SR-71	McDonnell-Douglas DC-10	Gemini spacecraft
Lockheed F-104	Boeing 727	Apollo Command Module
North American XB-70	Boeing 767	Space Shuttle orbiter
Rockwell International B-1		
General Dynamics F-111		
McDonnell-Douglas F/A-18		
Northrop/McDonnell-Douglas YF-23		

33. NASA Ames Applied Aerodynamics Branch, "The Unitary Plan Wind Tunnels" (July 1994), pp. 3–7.
34. Ibid., pp. 1, 3, 12–14.

The National Park Service designated the Ames Unitary Plan Wind Tunnel Facility a national historic landmark in 1985. The Unitary Plan Wind Tunnel represented "the logical crossover point from NACA to NASA" and "contributed equally to both the development of advanced American aircraft and manned spacecraft."[35]

The Unitary Plan facility at Lewis Research Center allowed the observation and development of full-scale jet and rocket engines in a 10- by 10-foot supersonic wind tunnel that cost $24.6 million. Designed by Abe Silverstein and Eugene Wasliewski, the test section featured a flexible wall made up of 10-foot-wide polished stainless steel plates, almost 1.5 inches thick and 76 feet long. Hydraulic jacks changed the shape of the plates to simulate nozzle shapes covering the range of Mach 2 to Mach 3.5. Silverstein and Wasliewski also incorporated both open and closed operation. For propulsion tests, air entered the tunnel and exited on the other side of the test section continually. In the aerodynamic mode, the same air circulated repeatedly to maintain a higher atmospheric pressure, desired temperature, or moisture content. The Lewis Unitary Plan Wind Tunnel contributed to the development of the General Electric F110 and Pratt & Whitney TF30 jet engines intended for the Grumman F-14 Tomcat and the liquid-fueled rocket engines destined for the Space Shuttle.[36]

Many NACA tunnels found long-term use with NASA. After NASA made modifications in the 1950s, the 20-Foot VST allowed the study of spacecraft and recovery devices in vertical descent. In the early 21st century, researchers used the 20-Foot VST to test the free-fall and dynamic stability characteristics of spacecraft models. It remains one of only two operation spin tunnels in the world.[37]

---

35. National Park Service, "National Advisory Committee for Aeronautics Wind Tunnels: Unitary Plan Wind Tunnel," Jan. 2001, *http://www.nps.gov/history/history/online_books/butowsky4/space4.htm*, accessed May 28, 2009; and Unitary Plan Wind Tunnel," n.d., *http://www.nps.gov/history/nr/travel/aviation/uni.htm*, accessed July 31, 2009.

36. Baals and Corliss, *Wind Tunnels of NASA*, pp. 69–70; NASA Langley, "NASA's Wind Tunnels," IS-1992-05-002-LaRC, May 1992, *http://oea.larc.nasa.gov/PAIS/WindTunnel.html*, accessed May 26, 2009.

37. Hansen, *Engineer in Charge*, pp. 462–463; NASA, "Wind Tunnels at NASA Langley Research Center," FS-2001-04-64-LaRC, 2001, *http://www.nasa.gov/centers/langley/news/factsheets/windtunnels.html*, accessed May 28, 2009; Langley Research Center, "Research and Test Facilities," NASA TM-1096859 (1993), p. 17; Rachel C. Samples, "A New Spin on the Constellation Program," Aug. 2007, *http://www.nasa.gov/mission_pages/constellation/orion/orion-spintunnel.html*, accessed Sept. 14, 2009.

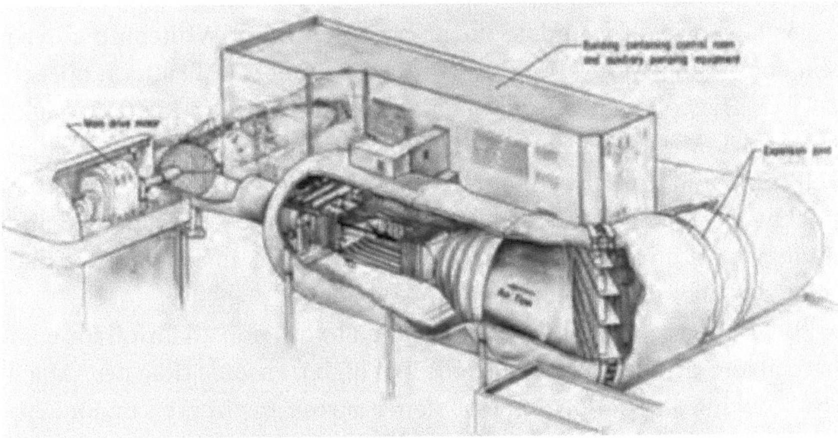

The 8-Foot Transonic Pressure Tunnel (TPT). NASA.

Tunnel Visions: Dick Whitcomb's Creative Forays

The slotted-throat transonic tunnels pioneered by John Stack and his associates at Langley proved valuable, especially in the hands of one of the Center's more creative minds, Richard. T. Whitcomb. In the 8-Foot TT, he investigated the transonic regime. Gaining a better understanding of aircraft speeds between Mach 0.75 and 1.25 was one of the major aerodynamic challenges of the 1950s and a matter of national security during the Cold War. The Air Force's Convair YF-102 Delta Dagger interceptor was unable to reach supersonic speeds during its first flights in 1953. Tests in the 8-Foot TT revealed that the increase in drag as an airplane approached supersonic speeds was not the result of shock waves forming at the nose but of those forming just behind the wings. Whitcomb created a rule of thumb that decreased transonic drag by narrowing, or pinching, the fuselage where it met the wings.[38] The improved YF-102A, with its new "area rule" fuselage, achieved supersonic flight in December 1954. The area rule fuselage increased the YF-102A's top speed by 25 percent. Embraced by the aviation industry, Whitcomb's revolutionary idea enabled a generation of military aircraft to achieve supersonic speeds.[39]

38. Richard T. Whitcomb, "A Study of the Zero-Lift Drag-Rise Characteristics of Wing-Body Combinations Near the Speed of Sound," NACA RM-L52H08 (Sept. 3, 1952).

39. Richard T. Whitcomb and Thomas L. Fischetti, "Development of a Supersonic Area Rule and an Application to the Design of a Wing-Body Combination Having High Lift-to-Drag Ratios," NACA RM-L53H31A (Aug. 18, 1953); Richard T. Whitcomb, "Some Considerations Regarding the Application of the Supersonic Area Rule to the Design of Airplane Fuselages," NACA RM-L56E23a (July 3, 1956).

As he worked to validate the area rule concept, Whitcomb moved next door to the 8-Foot Transonic Pressure Tunnel (TPT) after it opened in 1953. His colleagues John Stack, Eugene C. Draley, Ray H. Wright, and Axel T. Mattson designed the facility from the outset as a slotted-wall transonic tunnel with a maximum speed of Mach 1.2.[40] In what quickly became known as "Dick Whitcomb's tunnel," he validated and made two additional aerodynamic contributions in the decades that followed—the supercritical wing and winglets.

Beginning in 1964, Whitcomb wanted to develop an airfoil for commercial aircraft that delayed the onset of high transonic drag near Mach 1 by reducing air friction and turbulence across an aircraft's major aerodynamic surface, the wing. Whitcomb went intuitively against conventional airfoil design by envisioning a smoother flow of air by turning a conventional airfoil upside down. Whitcomb's airfoil was flat on top with a downward curved rear section. The blunt leading edge facilitated better takeoff, landing, and maneuvering performance as the airfoil slowed airflow, which lessened drag and buffeting and improved stability. Spending days at a time in the 8-Foot TPT, he validated his concept with a model he made with his own hands. He called his innovation a "supercritical wing," combining "super" (meaning "beyond") with "critical" Mach number, which is the speed supersonic flow revealed itself above the wing.[41] After a successful flight program was conducted at NASA Dryden from 1971 to 1973, the aviation industry incorporated the supercritical wing into a new generation of aircraft, including subsonic transports, business jets, Short Take-Off and Landing (STOL) aircraft, and unmanned aerial vehicles (UAVs).[42]

Whitcomb's continual quest to improve subsonic aircraft led him to investigate the wingtip vortex, the turbulent air found at the end of an airplane wing that created induced drag, as part of the Aircraft Energy Efficiency (ACEE) program. His solution was the winglet, a vertical wing-like surface that extended above and sometimes below the tip of each

40. Hansen, *Engineer in Charge*, p. 474.

41. Richard T. Whitcomb and Larry L. Clark, "An Airfoil Shape for Efficient Flight at Supercritical Mach Numbers," NASA TM-X-1109 (Apr. 20, 1965); Michael Gorn, *Expanding the Envelope: Flight Research at NACA and NASA* (Lexington: University Press of Kentucky, 2001), p. 331.

42. Thomas C. McMurtry, Neil W. Matheny, and Donald H. Gatlin, "Piloting and Operational Aspects of the F-8 Supercritical Wing Airplane," in *Supercritical Wing Technology—A Progress Report on Flight Evaluations*, NASA SP-301 (1972), p. 102; Gorn, *Expanding the Envelope*, pp. 335, 337.

wing. Whitcomb and his research team in the 8-Foot TPT investigated the drag-reducing properties of winglets for a first-generation, narrow-body subsonic jet transport from 1974 to 1976.[43] Whitcomb found that winglets reduced drag by approximately 20 percent and doubled the improvement in the lift-to-drag (L/D) ratio, to 9 percent, which boosted performance by enabling higher cruise speeds. The first jet-powered airplane to enter production with winglets was the Learjet Model 28 in 1977. The first large U.S. commercial transport to incorporate winglets, the Boeing 747-400, followed in 1985.[44]

Unlocking the Mysteries of Flutter: Langley's Transonic Dynamics Tunnel

The example of the Langley Transonic Dynamics Tunnel (TDT) illustrates how the NACA and NASA took an unsatisfactory tunnel and converted it into one capable of contributing to longstanding aerospace research. The Transonic Dynamics Tunnel began operations as the 19-Foot Pressure Tunnel in June 1939. The NACA design team, which included Smith J. DeFrance and John F. Parsons, wanted to address continued problems with scale effects. Their solution resulted in the first large-scale high-pressure tunnel. Primarily, the tunnel was to evaluate propellers and wings at high Reynolds numbers. Researchers were to use it to study the stability and control characteristics of aircraft models as well. Only able to generate a speed of 330 mph in the closed-throat test section, the NACA shifted the high-speed propeller work to another new facility, the 500 mph 16-Foot High-Speed Tunnel. The slower 19-Foot Pressure Tunnel pressed on in the utilitarian work of testing models at high Reynolds numbers.[45]

Dissatisfied with the performance of the 19-Foot Pressure Tunnel, the NACA converted it into a closed-circuit, continual flow, variable pressure Mach 1.2 wind tunnel to evaluate such dynamic flight characteristics as aeroelasticity, flutter, buffeting, vortex shedding, and gust loads. From 1955 to 1959, the conversion involved the installation of new components, including a slotted test section, mounts, a quick-stop drive system, an airflow oscillator (or "gust maker"), and a system that

43. Richard T. Whitcomb, "A Design Approach and Selected Wind-Tunnel Results at High Subsonic Speeds for Wing-Tip Mounted Winglets," NASA TN-D-8260 (July 1976), p. 1; Chambers, *Concept to Reality*, p. 35.
44. Ibid., pp. 38, 41, 43.
45. Hansen, *Engineer in Charge*, pp. 459, 462.

generated natural air or a refrigerant (Freon-12 and later R-134a) test medium. The use of gas improved full-scale aircraft simulation.[46] It produced higher Reynolds numbers, eased fabrication of scaled models, reduced tunnel power requirements, and, in the case of rotary wing models, reduced model power requirements.[47]

After 8 years of design, calibration, and conversion, the TDT became the world's first aeroelastic testing tunnel, becoming operational in 1960. The tunnel was ready for its first challenge: the mysterious crashes of the first American turboprop airliner, the Lockheed L-188 Electra II. The Electra entered commercial service with American Airlines in December 1958. Powered by 4 Allison 501 turboprop engines, the $2.4-million Electra carried approximately 100 passengers while cruising at 400 mph. On September 29, 1959, Braniff Airways Flight 542 crashed near Buffalo, TX, with the loss of all 34 people aboard the new Electra airliner. A witness saw what appeared to be lightning followed by a ball of fire and a shrieking explosion. The 2.5- by 1-mile debris field included the left wing, which settled over a mile away from the main wreckage. The initial Civil Aeronautics Board crash investigation revealed that failure of the left wing about a foot from the fuselage in flight led to the destruction of the airplane.[48]

There was no indication of the exact cause of the wing failure. The prevailing theories were sabotage or pilot and crew error. The crash of a Northwest Orient Airlines Electra near Tell City, IN, on March 17, 1960, with a loss of 63 people provided an important clue. The right wing landed 2 miles from the crash site. Federal and Lockheed investigators believed that violent flutter ripped the wings off both Electras, but they did not know the specific cause.[49]

46. E. Carson Yates, Jr., Norman S. Land, and Jerome T. Foughner, Jr., "Measured and Calculated Subsonic and Transonic Flutter Characteristics of a 45 Degree Sweptback Wing Planform in Air and Freon-12 in the Langley Transonic Dynamics Tunnel," NASA TN-D-1616 (Mar. 1963), p. 13.
47. Hansen, *Engineer in Charge*, 459, 462; Baals and Corliss, *Wind Tunnels of NASA*, pp. 78–80; NASA, "Wind Tunnels at NASA Langley Research Center," FS-2001-04-64-LaRC, 2001, *http://www.nasa.gov/centers/langley/news/factsheets/windtunnels.html*, accessed May 28, 2009.
48. William E. Giles, "Air Crash Aftermath," *Wall Street Journal*, Oct. 13, 1959, p. 1; "Airliner Lost Wing," *New York Times*, Oct. 28, 1959, p. 75.
49. Wayne Thomis, "What Air Crash Probers Seek," *Chicago Daily Tribune*, Mar. 18, 1960, p. 3; "Wild Flutter Split Wings Of Electras," *Washington Post, Times Herald*, May 13, 1960, p. D6; Baals and Corliss, *Wind Tunnels of NASA*, p. 80.

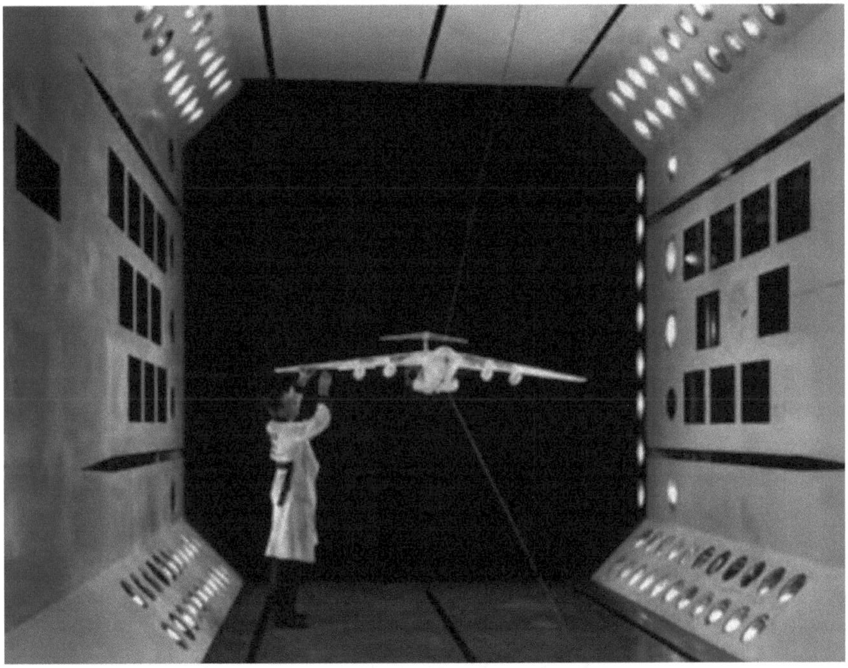

A Lockheed C-141 model undergoing evaluation in the Transonic Dynamics Tunnel (TDT). NASA.

The future of the new American jet airliner fleet was a stake. While the tragic story of the Electra unfolded, the Langley Transonic Dynamics Tunnel became operational in early 1960. NASA quickly prepared a one-eighth-scale model of an Electra that featured rotating propellers, simulated fuel load changes, and different engine-mount structural configurations. Those features would be important to the wind tunnel tests because a Lockheed engineer believed that the Electra experienced propeller-whirl flutter, a phenomenon stimulated by engine gyroscopic torques, propeller forces and moments, and the aerodynamic loads acting on the wings. Basically, a design flaw, weakened engine mounts, allowed the engine nacelles and the wings to oscillate at the same frequency, which led to catastrophic failure. Reinforced engine mounts ensured that the Electra continued operations through the 1960s and 1970s.[50]

Flutter has been a consistent problem for aircraft since the 1960s, and the Transonic Dynamics Tunnel contributed to the refinement of many aircraft, including frontline military transports and fighters.

50. Ibid., p. 80.

The Lockheed C-141 Starlifter transport experienced tail flutter in its original configuration. The horizontal tail of the McDonnell-Douglas F-15 Eagle all-weather air superiority fighter-bomber fluttered.[51] The inclusion of air-to-air and air-to-ground missiles, bombs, electronic countermeasures pods, and fuel tanks produced wing flutter on the General Dynamics F-16 Fighting Falcon lightweight fighter. NASA and General Dynamics underwent a combined computational, wind tunnel, and flight program from June 1975 to March 1977. The TDT tests sought to minimize expensive flight-testing. They verified analytical methods in determining flutter and determined practical operational methods in which portions of fuel tanks needed to be emptied first to delay the onset of flutter.[52]

The TDT offered versatility beyond the investigation of flutter on fixed wing aircraft. Tunnel personnel also conducted performance, load, and stability tests of helicopter and tilt rotor configurations. Researchers in the space program used the tunnel to determine the effects of ground-wind loads on launch vehicles. Whether it is for a fixed or rotary wing airplane or a spacecraft, the TDT was used to evaluate the effect of wind gusts on flying vehicles.[53]

The Cold War and the Space Age

In 1958, NASA was on a firm foundation for hypersonic and space research. Throughout the 1950s, NACA researchers first addressed the challenge of atmospheric reentry with their work on intercontinental ballistic missiles (ICBMs) for the military. The same fundamental design problems existed for ICBMs, spacecraft, interplanetary probes, and hypersonic aircraft. Each of the NASA Centers specialized in a specific aspect of hypersonic and hypervelocity research that resulted from their heritage as NACA laboratories. Langley's emphasis was in the creation of facilities applicable to hypersonic cruise aircraft and reentry vehicles—including winged reentry. Ames explored the extreme temperatures and the design shapes that could withstand them as vehicles

51. Ibid., p. 155.

52. Jerome T. Foughner, Jr., and Charles T. Bensinger, "F-16 Flutter Model Studies with External Wing Stores," NASA TM-74078 (Oct. 1977), pp. 1, 7, 14.

53. NASA, "Wind Tunnels at NASA Langley Research Center," FS-2001-04-64-LaRC, 2001, *http://www.nasa.gov/centers/langley/news/factsheets/windtunnels.html*, accessed May 28, 2009.

John Becker with his 11-Inch Hypersonic Tunnel of 1947. NASA.

returned to Earth from space. Researchers at Lewis focused on propulsion systems for these new craft. With the impetus of the space race, each Center worked with a growing collection of hypersonic and hypervelocity wind tunnels that ranged from conventional aerodynamic facilities to radically different configurations such as shock tubes, arc-jets, and new tunnels designed for the evaluation of aerodynamic heating on spacecraft structures.[54]

The Advent of Hypersonic Tunnel and Aeroballistic Facilities

John V. Becker at Langley led the way in the development of conventional hypersonic wind tunnels. He built America's first hypersonic wind tunnel in 1947, with an 11-inch test section and the capability of Mach 6.9 flow. To T.A. Heppenheimer, it is "a major advance in hypersonics," because Becker had built the discipline's first research instrument.[55] Becker and Eugene S. Love followed that success with their design of the 20-Inch Hypersonic Tunnel in 1958. Becker, Love, and their colleagues used the tunnel for the investigation of heat transfer, pressure,

54. Baals and Corliss, *Wind Tunnels of NASA*, pp. 86, 101; T.A. Heppenheimer, *Facing the Heat Barrier: A History of Hypersonics*, NASA SP-2007-4232 (Washington, DC: GPO, 2007), p. 42.
55. Ibid., pp. xi, 2.

and forces acting on inlets and complete models at Mach 6. The facility featured an induction drive system that ran for approximately 15 minutes in a nonreturn circuit operating at 220–550 psia (pounds-force per square inch absolute).[56]

The need for higher Mach numbers led to tunnels that did not rely upon the creation of a flow of air by fans. A counterflow tunnel featured a gun that fired a model into a continual onrushing stream of gas or air, which was an effective tool for supersonic and hypersonic testing. An impulse wind tunnel created high temperature and pressure in a test gas through an explosive release of energy. That expanded gas burst through a nozzle at hypersonic speeds and over a model in the test section in milliseconds. The two types of impulse tunnels—hotshot and shock—introduced the test gas differently and were important steps in reaching ever-higher speeds, but NASA required even faster tunnels.[57]

The companion to a hotshot tunnel was an arc-jet facility, which was capable of evaluating spacecraft heat shield materials under the extreme heat of planetary reentry. An electric arc preheated the test gas in the stilling chamber upstream of the nozzle to temperatures of 10,000–20,000 °F. Injected under pressure into the nozzle, the heated gas created a flow that was sustainable for several minutes at low-density numbers and supersonic Mach numbers. The electric arc required over 100,000 kilowatts of power. Unlike the hotshot, the arc-jet could operate continually.[58]

NASA combined these different types of nontraditional tunnels into the Ames Hypersonic Ballistic Range Complex in the 1960s.[59] The Ames Vertical Gun Range (1964) simulated planetary impact with various model-launching guns. Ames researchers used the Hypervelocity Free-Flight Aerodynamic Facility (1965) to examine the aerodynamic characteristics of atmospheric entry and hypervelocity vehicle configurations. The research programs investigated Earth atmosphere entry (Mercury, Gemini, Apollo,

56. James C. Emery, "Appendix: Description and Calibration of the Langley 20-inch Mach 6 Tunnel," in Theodore J. Goldberg and Jerry N. Hefner, "Starting Phenomena for Hypersonic Inlets with Thick Turbulent Boundary Layers at Mach 6," NASA TN-D-6280 (Aug. 1971), pp. 13–15; Hansen, *Engineer in Charge*, p. 478.

57. Baals and Corliss, *Wind Tunnels of NASA*, pp. 84–85, 90.

58. Ibid., p. 85.

59. For more information on ballistic ranges, see Alvin Seiff and Thomas N. Canning, "Modern Ballistic Ranges and Their Uses," NASA TM-X-66530 (Aug. 1970).

and Shuttle), planetary entry (Viking, Pioneer-Venus, Galileo, and Mars Science Lab), supersonic and hypersonic flight (X-15), aerobraking configurations, and scramjet propulsion studies. The Electric Arc Shock Tube (1966) enabled the investigation of the effects of radiation and ionization that occurred during high-velocity atmospheric entries. The shock tube fired a gaseous bullet at a light-gas gun, which fired a small model into the onrushing gas.[60]

The NACA also investigated the use of test gases other than air. Designed by Antonio Ferri, Macon C. Ellis, and Clinton E. Brown, the Gas Dynamics Laboratory at Langley became operational in 1951. One facility was a high-pressure shock tube consisting of a constant area tube 3.75 inches in diameter, a 20-inch test section, a 14-foot-long high-pressure chamber, and 70-foot-long low-pressure section. The induction drive system consisted of a central 300-psi tank farm that provided heated fluid flow at a maximum speed of Mach 8 in a nonreturn circuit at a pressure of 20 atmospheres. Langley researchers investigated aerodynamic heating and fluid mechanical problems at speeds above the capability of conventional supersonic wind tunnels to simulate hypersonic and space-reentry conditions. For the space program, NASA used pure nitrogen and helium instead of heated air as the test medium to simulate reentry speeds.[61]

NASA built the similar Ames Thermal Protection Laboratory in the early 1960s to solve reentry materials problems for a new generation of craft, whether designed for Earth reentry or the penetration of the atmospheres of the outer planets. A central bank of 10 test cells provided the pressurized flow. Specifically, the Thermal Protection Laboratory found solutions for many vexing heat shield problems associated with the Space Shuttle, interplanetary probes, and intercontinental ballistic missiles.

Called the "suicidal wind tunnel" by Donald D. Baals and William R. Corliss because it was self-destructive, the Ames Voitenko Compressor was the only method for replicating the extreme velocities required for the design of interplanetary space probes. It was based on the Voitenko

60. Heppenheimer, *Facing the Heat Barrier*, pp. 32, 40–42; NASA Ames Research Center, "Thermophsyics Facilities Branch Range Complex," n.d., *http://thermo-physics.arc.nasa.gov/fact_sheets/Range%20Fact%20Sheet.pdf*, accessed Oct. 13, 2009.

61. Jim J. Jones, "Resume of Experiments Conducted in the High-Pressure Shock Tube of the Gas Dynamics Tube at NASA," NASA TM-X-56214 (Mar. 1959), p. 1; Hansen, *Engineer in Charge*, pp. 473–474.

The Continuous Flow Hypersonic Tunnel at Langley in 1961. NASA.

concept from 1965 that a high-velocity explosive, or shaped, charge developed for military use be used for the acceleration of shock waves. Voitenko's compressor consisted of a shaped charge, a malleable steel plate, and the test gas. At detonation, the shaped charge exerts pressure on the steel plate to drive it and the test gas forward. Researchers at the Ames Laboratory adapted the Voitenko compressor concept to a self-destroying shock tube comprised of a 66-pound shaped charge and a glass-walled tube 1.25 inches in diameter and 6.5 feet long. Observation of the tunnel in action revealed that the shock wave traveled well ahead of the rapidly disintegrating tube. The velocities generated upward of 220,000 feet per second could not be reached by any other method.[62]

Langley, building upon a rich history of research in high-speed flight, started work on two tunnels at the moment of transition from the NACA

62. Baals and Corliss, *Wind Tunnels of NASA*, p. 92.

to NASA. Eugene Love designed the Continuous Flow Hypersonic Tunnel for nonstop operation at Mach 10. A series of compressors pushed high-speed air through a 1.25-inch square nozzle into the 31-inch square test section. A 13,000-kilowatt electric resistance heater raised the air temperature to 1,450 °F in the settling chamber, while large water coolers and channels kept the tunnel walls cool. The tunnel became operational in 1962 and became instrumental in study of the aerodynamic performance and heat transfer on winged reentry vehicles such as the Space Shuttle.[63]

The 8-Foot High-Temperature Structures Tunnel, opened in 1967, permitted full-scale testing of hypersonic and spacecraft components. By burning methane in air at high pressure and through a hypersonic nozzle in the tunnel, Langley researchers could test structures at Mach 7 speeds and at temperatures of 3,000 °F. Too late for the 1960s space program, the tunnel was instrumental in the testing of the insulating tiles used on the Space Shuttle.[64]

NASA researchers Richard R. Heldenfels and E. Barton Geer developed the 9- by 6-Foot Thermal Structures Tunnel to test aircraft and missile structural components operating under the combined effects of aerodynamic heating and loading. The tunnel became operational in 1957 and featured a Mach 3 drive system consisting of 600-psia air stored in a tank farm filled by a high-capacity compressor. The spent air simply exhausted to the atmosphere. Modifications included additional air storage (1957), a high-speed digital data system (1959), a subsonic diffuser (1960), a Topping compressor (1961), and a boost heater system that generated 2,000 °F of heat (1963). NASA closed the 9- by 6-Foot Thermal Structures Tunnel in September 1971. Metal fatigue in the air storage field led to an explosion that destroyed part of the facility and nearby buildings.[65]

NASA's wind tunnels contributed to the growing refinement of spacecraft technology. The multiple design changes made during the transition from the Mercury program to the Gemini program and the need for more information on the effects of angle of attack, heat transfer, and surface pressure resulted in a new wind tunnel and flight-test program. Wind tunnel tests of the Gemini spacecraft were conducted in the range

63. Ibid., p. 95.
64. Ibid., pp. 94–97.
65. Hansen, *Engineer in Charge*, pp. 475–478.

of Mach 3.51 to 16.8 at the Langley Unitary Plan and tunnels at AEDC and Cornell University. The flight-test program gathered data from the first four launches and reentries of Gemini spacecraft.[66] Correlation revealed that both independent sets of data were in agreement.[67]

Applying Hypersonic Test Facilities to Hypersonic Vehicle Design

One of NASA's first flight research studies was the X-15 program (1959–1968). The program investigated flight at five or more times the speed of sound at altitudes reaching the fringes of space. Launched from the wing of NASA's venerable Boeing B-52 mother ship, the North American X-15 was a true "aerospace" plane, with performance that went well beyond the capabilities of existing aircraft within and beyond the atmosphere. Long, black, rocket-powered, and distinctive with its cruciform tail, the X-15 became the highest-flying airplane in history. In one flight, the X-15 flew to 67 miles (354,200 feet) above the Earth at a speed of Mach 6.7, or 4,534 mph. At those speeds and altitudes, the X-15 pilots, made up of the leading military and civilian aviators, had to wear pressure suits, and many of them earned astronaut's wings. North American used titanium as the primary structural material and covered it with a new high-temperature nickel alloy called Inconel-X. The X-15 relied upon conventional controls in the atmosphere but used reaction-control jets to maneuver in space. The 199 flights of X-15 program generated important data on high-speed flight and provided valuable lessons for NASA's space program.

The air traveling over the X-15 at hypersonic speeds generated enough friction and heat that the outside surface of the airplane reached 1,200 °F. A dozen Langley and Ames wind tunnels contributed to the X-15 program. The sole source of aerodynamic data for the X-15 came from tests generated in the pioneering Mach 6.8 11-Inch Hypersonic Tunnel developed by John Becker at Langley in the late 1940s. Fifty percent of the work conducted in the tunnel was for the X-15 program, which focused on aerodynamic heating, stability and control, and load

66. Gemini I (Apr. 1964) and II (Jan. 1965) were unpiloted missions. Gemini III (Mar. 1965) and IV (June 1965) included astronaut crews. NASA Kennedy Space Center, "Gemini Missions," Nov. 9, 2000, *http://www-pao.ksc.nasa.gov/history/gemini/gemini-manned.htm*, accessed Aug. 14, 2009.
67. Richard M. Raper, "Heat-Transfer and Pressure Measurements Obtained During Launch and Reentry of the First Four Gemini-Titan Missions and Some Comparisons with Wind-Tunnel Data," NASA TM-X-1407 (Aug. 1967), pp. 1, 14.

Part of the Project Fire study included the simulation of reentry heating on high-temperature materials in the 9- by 6-Foot Thermal Structures Tunnel. NASA.

distribution studies. The stability and control investigations contributed to the research airplane's distinctive cruciform tail. The 7- by 10-Foot High-Speed Wind Tunnel enabled the study of the X-15's separation from the B-52 at subsonic speeds, a crucial phase in the test flight. At Ames, gun-launched models fired into the free-flight tunnels obtained shadowgraphs of the shock wave patterns between Mach 3.5 and 6, the performance regime for the X-15. The Unitary Plan Supersonic Tunnel generated data on aerodynamic forces and heat transfer. The Lewis Research Center facilities provided additional data on supersonic jet-plumes and rocket-nozzle studies.[68]

There was a concern that wind tunnel tests would not provide correct data for the program. First, the cramped size of the tunnel test sections did not facilitate more accurate full-scale testing. Second, none of NASA's tunnels was capable of replicating the extreme heat generated by hypersonic flight, which was believed to be a major factor in flying at those speeds. The flights of the X-15 validated the wind tunnel

68. Baals and Corliss, *Wind Tunnels of NASA*, p. 107; Heppenheimer, *Facing the Heat Barrier*, p. 55.

testing and revealed that lift, drag, and stability values were in agreement with one another at speeds up to Mach 10.[69]

The wind tunnels of NASA continued to reflect the Agency's flexibility in the development of craft that operated in and out of the Earth's atmosphere. Specific components evaluated in the 9- by 6-Foot Thermal Structures Tunnel included the X-15 vertical tail, the heat shields for the Centaur launch vehicle and Project Fire entry vehicle, and components of the Hawk, Falcon, Sam-D, and Minuteman missiles. Researchers also subjected humans, equipment, and structures such as the Mercury Spacecraft to the 162-decibel, high-intensity noise at the tunnel exit. As part of Project Fire, in the early 1960s, personnel in the tunnel evaluated the effects of reentry heating on spacecraft materials.[70]

The Air Force's failed X-20 Dyna-Soar project attempted to develop a winged spacecraft. The X-20 never flew, primarily because of bureaucratic entanglements. NASA researchers H. Julian Allen and Alfred J. Eggers, Jr., working on ballistic missiles, found that a blunt shape made reentry possible.[71] NASA developed a series of "lifting bodies"—capable of reentry and then being controlled in the atmosphere—to test unconventional blunt configurations. The blunt nose and wing-leading edge of the Space Shuttles that are launched into space and then glide to a landing after reentry, starting with Columbia in April 1981, owe their success to the lifting body tests flown by NASA in the 1960s and 1970s.

The knowledge gained in these programs contributed to the Space Shuttle of the 1980s. Analyses of the Shuttle reflected the tradition dating back to the Wright brothers of correlating ground, or wind tunnel, data with flight data. Langley researchers conducted an extended aerodynamic and aerothermodynamic comparison of hypersonic flight- and ground-test results for the program. The research team asserted that the "survival of the vehicle is a tribute to the overall design philosophy, including ground test predictions, and to the designers of the Space Shuttle."[72]

---

69. Ibid., p. 82.

70. Hansen, *Engineer in Charge*, pp. 475–478.

71. H. Julian Allen and Alfred J. Eggers, Jr., "A Study of the Motion and Aerodynamic Heating of Ballistic Missiles Entering the Earth's Atmosphere at High Supersonic Speeds," NACA TR-1381 (1958).

72. Kenneth W. Iliff and Mary F. Shafer, "Space Shuttle Hypersonic Aerodynamic and Aerothermodynamic Flight Research and the Comparison to Ground Test Results," NASA TM-4499 (1993), p. 3.

H. Julian Allen used the 8- by 7-foot test section of the NACA Ames Unitary Plan Wind Tunnel during the development of his blunt-body theory. NASA.

The latest NASA research program, called Hyper-X, investigated hypersonic flight with a new type of aircraft engine, the X-43A scramjet, or supersonic combustion ramjet. The previous flights of the X-15, the lifting bodies, and the Space Shuttle relied upon rocket power for hypersonic propulsion. A conventional air-breathing jet engine, which relies upon the mixture of air and atomized fuel for combustion, can only propel aircraft to speeds approaching Mach 4. A scramjet can operate well past Mach 5 because the process of combustion takes place at supersonic speeds. Launch-mounted to the front of rocket booster from a B-52 at 40,000 feet, the 12-foot-long, 2,700-pound X-43A first flew in March 2004. During the 11-second flight, the little engine reached Mach 6.8 and demonstrated the first successful operation of a scramjet. In November 2004, a second flight achieved Mach 9.8, the fastest speed ever attained by an air-breathing engine. Much like Frank Whittle and Hans von Ohain's turbojets and the Wrights' invention of the airplane, the X-43A offered the promise of a new revolution in aviation, that of high-speed global travel and a cheaper means to access space.

The diminutive X-43A allowed for realistic testing at NASA Langley. First, it was at full-scale for the specific scramjet tests. Moreover, it served as a scale model for the hypersonic engines intended for future aerospace craft. The majority of the testing for the Hyper-X program occurred in the Arc-Heated Scramjet Test Facility, which was the primary Mach 7 scramjet test facility. Introduced in the late 1970s, the Langley facility generated the appropriate flows at 3,500 °F. Additional transonic and supersonic tests of 30-inch X-43A models took place in the 16-Foot Transonic Tunnel and the Unitary Plan Wind Tunnel.[73]

Researchers in the Langley Aerothermodynamics Branch worked on a critical phase of the flight: the separation of the X-43A from the Pegasus booster. The complete Hyper-X Launch Vehicle stack, consisting of the scramjet and booster, climbed to 20,000 feet under the wing of NASA's Boeing B-52 Stratofortress in captive/carry flight. Clean separation between the two within less than a second ensured the success of the flight. The X-43A, with its asymmetrical shape, did not facilitate that clean separation. The Langley team required a better aerodynamic understanding of multiple configurations: the combined stack, the X-43A and the Pegasus in close proximity, and each vehicle in open, free flight. The Langley 20-Inch Mach 6 and 31-Inch Mach 10 blow-down tunnels were used for launch, postlaunch, and free-flyer hypersonic testing.[74]

Matching the Tunnel to the Supercomputer

The use of sophisticated wind tunnels and their accompanying complex mathematical equations led observers early on to call aerodynamics the

73. Heppenheimer, *Facing the Heat Barrier*, pp. 208, 271, 273.

74. William C. Woods, Scott D. Holland, and Michael DiFulvio, "Hyper-X Stage Separation Wind-Tunnel Test Program," *Journal of Spacecraft and Rockets*, vol. 38 (Nov.–Dec. 2001), p. 811.

A model of the X-43A and the Pegasus Launch Vehicle in the Langley 31-Inch Mach 10 Tunnel. NASA.

"science" of flight. There were three major methods of evaluating an aircraft or spacecraft: theoretical analysis, the wind tunnel, and full-flight testing. The specific order of use was ambiguous. Ideally, researchers originated a theoretical goal and began their work in a wind tunnel, with the final confirmation of results occurring during full-flight testing. Researchers at Langley sometimes addressed a challenge first by studying it in flight, then moving to the wind tunnel for more extreme testing, such as dangerous and unpredictable high speeds, and then following up with the creation of a theoretical framework. The lack of knowledge of the effect of Reynolds number was at the root of the inability to trust wind tunnel data. Moreover, tunnel structures such as walls, struts, and supports affected the performance of a model in ways that were hard to quantify.[75]

From the early days of the NACA and other aeronautical research facilities, an essential component of the science was the "computer." Human computers, primarily women, worked laboriously to finish the myriad of calculations needed to interpret the data generated in wind

75. Baals and Corliss, *Wind Tunnels of NASA*, p. 136; Hansen, *Engineer in Charge*, p. 479.

tunnel tests. Data acquisition became increasingly sophisticated as the NACA grew in the 1940s. The Langley Unitary Plan Wind Tunnel possessed the capability of remote and automatic collection of pressure, force, temperature data from 85 locations at 64 measurements a second, which was undoubtedly faster than manual collection. Computers processed the data and delivered it via monitors or automated plotters to researchers during the course of the test. The near-instantaneous availability of test data was a leap from the manual (and visual) inspection of industrial scales during testing.[76]

Computers beginning in the 1970s were capable of mathematically calculating the nature of fluid flows quickly and cheaply, which contributed to the idea of what Baals and Corliss called the "electronic wind tunnel."[77] No longer were computers only a tool to collect and interpret data faster. With the ability to perform billions of calculations in seconds to mathematically simulate conditions, the new supercomputers potentially could perform the job of the wind tunnel. The Royal Aeronautical Society published *The Future of Flight* in 1970, which included an article on computers in aerodynamic design by Bryan Thwaites, a professor of theoretical aerodynamics at the University of London. His essay would be a clarion call for the rise of computational fluid dynamics (CFD) in the late 20th century.[78] Moreover, improvements in computers and algorithms drove down the operating time and cost of computational experiments. At the same time, the time and cost of operating wind tunnels increased dramatically by 1980. The fundamental limitations of wind tunnels centered on the age-old problems related to model size and Reynolds number, temperature, wall interference, model support ("sting") interference, unrealistic aeroelastic model distortions under load, stream nonuniformity, and unrealistic turbulence levels. Problematic results from the use of test gases were a concern for the design of vehicles for flight in the atmospheres of other planets.[79]

76. Baals and Corliss, *Wind Tunnels of NASA*, p. 71.

77. Ibid., p. 136.

78. Reference in James R. Hansen, *The Bird is on the Wing: Aerodynamics and the Progress of the American Airplane* (College Station: Texas A & M University Press, 2004), p. 221.

79. Victor L. Peterson and William F. Ballhaus, Jr., "History of the Numerical Aerodynamic Simulation Program," in Paul Kutler and Helen Yee, *Supercomputing in Aerospace: Proceedings of a Symposium Held at the NASA Ames Research Center, Moffett Field, CA, Mar. 10–12, 1987*, NASA CP-2454 (1987), pp. 1, 3.

The control panels of the Langley Unitary Wind Tunnel in 1956. NASA.

The work of researchers at NASA Ames influenced Thwaites's assertions about the potential of CFD to benefit aeronautical research. Ames researcher Dean Chapman highlighted the new capabilities of supercomputers in his Dryden Lecture in Research for 1979 at the American Institute of Aeronautics and Astronautics Aerospace Sciences Meeting in New Orleans, LA, in January 1979. To Chapman, innovations in computer speed and memory led to an "extraordinary cost reduction trend in computational aerodynamics," while the cost of wind tunnel experiments had been "increasing with time." He brought to the audience's attention that a meager $1,000 and 30 minutes computer time allowed the numerical simulation of flow over an airfoil. The same task in 1959 would have cost $10 million and would have been completed 30 years later. Chapman made it clear that computers could cure the "many ills of wind-tunnel and turbomachinery experiments" while providing "important new technical capabilities for the aerospace industry."[80]

80. Dean R. Chapman, "Computational Aerodynamics Development and Outlook," Dryden Lecture in Research for 1979, American Institute of Aeronautics and Astronautics, *Aerospace Sciences Meeting, New Orleans, LA, Jan. 15–17, 1979*, AIAA-1979-129 (1979), p. 1; Baals and Corliss, *Wind Tunnels of NASA*, p. 137.

The crowning achievement of the Ames work was the establishment of the Numerical Aerodynamic Simulation (NAS) Facility, which began operations in 1987. The facility's Cray-2 supercomputer was capable of 250 million computations a second and 1.72 billion per second for short periods, with the possibility of expanding capacity to 1 billion computations per second. That capability reduced the time and cost of developing aircraft designs and enabled engineers to experiment with new designs without resorting to the expense of building a model and testing it in a wind tunnel. Ames researcher Victor L. Peterson said the new facility, and those like it, would allow engineers "to explore more combinations of the design variables than would be practical in the wind tunnel."[81]

The impetus for the NAS program arose from several factors. First, its creation recognized that computational aerodynamics offered new capabilities in aeronautical research and development. Primarily, that meant the use of computers as a complement to wind tunnel testing, which, because of the relative youth of the discipline, also placed heavy demands on those computer systems. The NAS Facility represented the committed role of the Federal Government in the development and use of large-scale scientific computing systems dating back to the use of the ENIAC for hydrogen bomb and ballistic missile calculations in the late 1940s.[82]

It was clear to NASA that supercomputers were part of the Agency's future in the late 1980s. Futuristic projects that involved NASA supercomputers included the National Aero-Space Plane (NASP), which had an anticipated speed of Mach 25; new main engines and a crew escape system for the Space Shuttle; and refined rotors for helicopters. Most importantly from the perspective of supplanting the wind tunnel, a supercomputer generated data and converted them into pictures that captured flow phenomena that had been previously unable to be simulated.[83] In other words, the "mind's eye" of the wind tunnel engineer could be captured on film.

Nevertheless, computer simulations were not to replace the wind tunnel. At a meeting sponsored by Advisory Group for Aerospace

81. John Noble Wilford, "Advanced Supercomputer Begins Operation," *New York Times*, Mar. 10, 1987, p. C3.

82. Peterson and Ballhaus, "History of the Numerical Aerodynamic Simulation Program," pp. 1, 9.

83. John Noble Wilford, "Advanced Supercomputer Begins Operation," *New York Times*, Mar. 10, 1987, p. C3.

Research & Development (AGARD) on the Integration of Computers and Wind Testing in September 1980, Joseph G. Marvin, the chief of the Experimental Fluid Dynamics Branch at Ames, asserted CFD was an "attractive means of providing that necessary bridge between wind-tunnel simulation and flight." Before that could happen, a careful and critical program of comparison with wind tunnel experiments had to take place. In other words, the wind tunnel was the tool to verify the accuracy of CFD.[84] Dr. Seymour M. Bogdonoff of Princeton University commented in 1988 that "computers can't do anything unless you know what data to put in them." The aerospace community still had to discover and document the key phenomena to realize the "future of flight" in the hypersonic and interplanetary regimes. The next step was inputting the data into the supercomputers.[85]

Researchers Victor L. Peterson and William F. Ballhaus, Jr., who worked in the NAS Facility, recognized the "complementary nature of computation and wind tunnel testing," where the "combined use" of each captured the "strengths of each tool." Wind tunnels and computers brought different strengths to the research. The wind tunnel was best for providing detailed performance data once a final configuration was selected, especially for investigations involving complex aerodynamic phenomena. Computers facilitated the arrival and analysis of that final configuration through several steps. They allowed development of design concepts such as the forward-swept wing or jet flap for lift augmentation and offered a more efficient process of choosing the most promising designs to evaluate in the wind tunnel. Computers also made the instrumentation of test models easier and corrected wind tunnel data for scaling and interference errors.[86]

The Future of the Tunnel in the Era of CFD

A longstanding flaw with wind tunnels was the aerodynamic interference caused by the "sting," or the connection between the model and the test instrumentation. Researchers around the world experimented with magnetic suspension systems beginning in the late 1950s. Langley,

84. Joseph G. Marvin, "Advancing Computational Aerodynamics through Wind-Tunnel Experimentation," NASA TM-109742 (Sept. 1980), p. 4.
85. William J. Broad, "In the Space Age, the Old Wind Tunnel Is Being Left Behind," *New York Times*, Jan. 5, 1988, p. C4.
86. Peterson and Ballhaus, "History of the Numerical Aerodynamic Simulation Program," p. 3.

6

in collaboration with the AEDC, constructed the 13-Inch Magnetic Suspension and Balance System (MSBS). The transparent test section measured about 12.6 inches high and 10.7 inches wide. Five powerful electromagnets installed in the test section suspended the model and provided lift, drag, side forces, and pitching and yaw moments. Control of the iron-cored model over these five axes removed the need for a model support. The lift force of the system enabled the suspension of a 6-pound iron-cored model. The rest of the tunnel was conventional: a continual-flow, closed-throat, open-circuit design capable of speeds up to Mach 0.5.[87]

When the 13-Inch MSBS became operational in 1965, NASA used the tunnel for wake studies and general research. Persistent problems with the system led to its closing in 1970. New technology and renewed interest revived the tunnel in 1979, and it ran until the early 1990s.[88]

NASA's work on magnetic suspension and balance systems led to a newfound interest in a wind tunnel capable of generating cryogenic test temperatures in 1971. Testing a model at below -150 °F permitted theoretically an increase in Reynolds number. There was a precedent for a cryogenic wind tunnel. R. Smelt at the Royal Aircraft Establishment at Farnborough conducted an investigation into the use of airflow at cryogenic temperatures in a wind tunnel. His work revealed that a cryogenic wind tunnel could be reduced in size and required less power as compared with a similar ambient temperature wind tunnel operated at the same pressure, Mach number, and Reynolds number.[89]

The state of the art in cooling techniques and structural materials required to build a cryogenic tunnel did not exist in the 1940s. American and European interest in the development of a transonic tunnel that generated high Reynolds numbers, combined with advances in cryogenics and structures in the 1960s, revived interest in Smelt's findings. A team of Langley researchers led by Robert A. Kilgore initiated a study of the viability of a cryogenic wind tunnel. The first experiment with a low-

87. R.P. Boyden, "A Review of Magnetic Suspension and Balance Systems," AIAA Paper 88-2008 (May 1988); NASA Langley, "NASA's Wind Tunnels," IS-1992-05-002-LaRC, May 1992, *http://oea.larc.nasa.gov/PAIS/WindTunnel.html*, accessed May 26, 2009.

88. Marie H. Tuttle, Deborah L. Moore, and Robert A. Kilgore, "Magnetic Suspension and Balance Systems: A Comprehensive, Annotated Bibliography," NASA TM-4318 (1991), p. iv; Langley Research Center, "Research and Test Facilities," p. 9.

89. R. Smelt, "Power Economy in High Speed Wind Tunnels by Choice of Working Fluid and Temperature," Report No. Aero. c081, Royal Aircraft Establishment, Farnborough, England, Aug. 1945.

speed tunnel during summer 1972 resulted in an extension of the program into the transonic regime. Kilgore and his team began design of the tunnel in December 1972, and the Langley Pilot Transonic Cryogenic Tunnel became operational in September 1973.[90]

The pilot tunnel was a continual-flow, fan-driven tunnel with a slotted octagonal test section, 0.3 meters (1 foot) across the flats, and was constructed almost entirely out of aluminum alloy. The normal test medium was gaseous nitrogen, but air could be used at ambient temperatures. The experimental tunnel provided true simulation of full-scale transonic Reynolds numbers (up to 100 x 10^6 per foot) from Mach 0.1 to 0.9 and was a departure from conventional wind tunnel design. The key was decreasing air temperature, which increased the density and decreased the viscosity factor in the denominator of the Reynolds number. The result was the simulation of full-scale flight conditions at transonic speeds with great accuracy.[91]

Kilgore and his team's work generated fundamental conclusions about cryogenic tunnels. First, cooling with liquid nitrogen was practical at the power levels required for transonic testing. It was also simple to operate. Researchers could predict accurately the amount of time required to cool the tunnel, a basic operational parameter, and the amount of liquid nitrogen needed for testing. Through the use of a simple liquid nitrogen injection system, tunnel personnel could control and evenly distribute the temperature. Finally, the cryogenic tunnel was quieter than was an identical tunnel operating at ambient temperature. The experiment was such a success and generated such promising results that NASA reclassified the temporary tunnel as a "permanent" facility and renamed it the 0.3-Meter Transonic Cryogenic Tunnel (TCT).[92]

---

90. Robert A. Kilgore, "Design Features and Operational Characteristics of the Langley Pilot Transonic Cryogenic Tunnel," NASA TM-X-72012 (Sept. 1974), pp. 1, 3, 4.

91. Ibid., p. 6; Edward J. Ray, Charles L. Ladson, Jerry B. Adcock, Pierce L. Lawing, and Robert M. Hall, "Review of Design and Operational Characteristics of the 0.3-Meter Transonic Cryogenic Tunnel," NASA TM-80123 (Sept. 1979), pp. 1, 4; Baals and Corliss, *Wind Tunnels of NASA*, p. 133; NASA, "Wind Tunnels at NASA Langley Research Center," FS-2001-04-64-LaRC, 2001, *http://www.nasa.gov/centers/langley/news/factsheets/windtunnels.html*, accessed May 28, 2009; NASA Langley, "NASA's Wind Tunnels," IS-1992-05-002-LaRC, May 1992, *http://oea.larc.nasa.gov/PAIS/WindTunnel.html*, accessed May 26, 2009.

92. Robert A. Kilgore, "Design Features and Operational Characteristics of the Langley 0.3-meter Transonic Cryogenic Tunnel," NASA TN-D-8304 (Dec. 1976), pp. 1, 3, 19; Baals and Corliss, *Wind Tunnels of NASA*, pp. 106–107.

The 0.3-Meter Transonic Cryogenic Tunnel. NASA.

After 6 years of operation, NASA researchers shared their experiences at the First International Symposium on Cryogenic Wind Tunnels at the University of Southampton, England, in 1979. Their operation of the 0.3-Meter TCT demonstrated that there were no insurmountable problems associated with a variety of aerodynamic tests with gaseous nitrogen at transonic Mach numbers and high Reynolds numbers. The team found that the injection of liquid nitrogen into the tunnel circuit to induce cryogenic cooling caused no problems with temperature distribution or dynamic response characteristics. Not everything, however, was known about cryogenic tunnels. There would be a significant learning process, which included the challenges of tunnel control, run logic, economics, instrumentation, and model technology.[93]

Developments in computer technology in the mid-1980s allowed continual improvement in transonic data collection in the 0.3-Meter TCT, which alleviated a long-term problem with all wind tunnels. The walls, floor, and ceiling of all tunnels provided artificial constraints on flight simulation. The installation of computer-controlled adaptive, or "smart," tunnel walls in March 1986 lessened airflow disturbances, because they allowed the addition or expulsion of air through the expansion and contraction along the length, width, and height of the tunnel walls. The result was a more realistic simulation of an aircraft flying in the open atmosphere. The 0.3-Meter TCT's computer system also automatically tailored Mach number, pressure, temperature, and angle of attack to a specific test program and monitored the drive, electrical, lubrication, hydraulic, cooling, and pneumatic systems for dangerous leaks and failures. The success of the 0.3-Meter TCT led to further investigation of smart walls at Langley and Lewis.[94]

NASA's success with the 0.3-Meter Transonic Cryogenic Tunnel led to the creation of the National Transonic Facility (NTF) at Langley. Both NASA and the Air Force were considering the construction of a large transonic wind tunnel. NASA proposed a larger cryogenic tunnel, and the Air Force wanted a Ludweig-tube tunnel. The Federal Government decided in 1974 to fund a facility to meet commercial, military, and scientific needs based on NASA's pioneering operation of the cryogenic tunnel. Contractors built the tunnel on the site of the 4-Foot Supersonic Pressure Tunnel and incorporated the old tunnel's drive motors, support buildings, and cooling towers.[95]

93. Ray, et.al, "Review of Design and Operational Characteristics of the 0.3-Meter Transonic Cryogenic Tunnel," p. 1.

94. NASA Langley, "NASA's Wind Tunnels," IS-1992-05-002-LaRC, May 1992, *http://oea.larc.nasa.gov/PAIS/WindTunnel.html*, accessed May 26, 2009; NASA, "Wind Tunnels at NASA Langley Research Center," FS-2001-04-64-LaRC, 2001, *http://www.nasa.gov/centers/langley/news/factsheets/windtunnels.html*, accessed May 28, 2009.

95. Baals and Corliss, *Wind Tunnels of NASA*, p. 133.

Becoming operational in 1983, the NTF was a high-pressure, cryogenic, closed-circuit wind tunnel with a Mach number range from 0.1 to 1.2 and a Reynolds number range of 4 x 10^6 to 145 x 10^6 per foot. It featured a 2.5-meter test section with 12 slots and 14 reentry flaps in the ceiling and floor. Langley personnel designed a drive system to include a fan with variable inlet guide vanes for precise Mach number control. Injected as super-cold liquid and evaporated into a gas, nitrogen is the primary test medium. Air is the test gas in the ambient temperature mode, while a heat exchanger maintains the tunnel temperature. Thermal insulation of the tunnel's pressure shell ensured minimal energy consumption. The NTF continues to be one of Langley's more advanced facilities as researchers evaluate the stability and control, cruise performance, stall buffet onset, and aerodynamic configurations of model aircraft and airfoil sections.[96]

The movement toward the establishment of national aeronautical facilities led NASA to expand the operational flexibility of the highly successful subsonic 40- by 80-foot wind tunnel at Ames Research Center. A major renovation project added an additional 80- by 120-foot test section capable of testing a full-size Boeing 737 airliner, making it the world's largest wind tunnel. A central drive system that featured fans almost 4 stories tall and electric motors capable of generating 135,000 horsepower created the airflow for both sections through movable vanes that directed air through either section. The 40- by 80-foot test section acted as a closed circuit up to 345 mph. The air driven through the 80- by 120-foot test section traveled up to 115 mph before exhausting into the atmosphere. Each section incorporated a range of model supports to facilitate a variety of experiments. The two sections became operational in 1987 (40- by 80-foot) and 1988 (80- by 120-foot). NASA christened the tunnel the National Full-Scale Aerodynamics Complex (NFAC) at Ames Research Center.[97]

---

96. Marie H. Tuttle, Robert A. Kilgore, and Deborah L. Moore, "Cryogenic Wind Tunnels: A Comprehensive, Annotated Bibliography," NASA TM-4273 (1991), p. iv; NASA Langley, "NASA's Wind Tunnels," IS-1992-05-002-LaRC, May 1992, *http://oea.larc.nasa.gov/PAIS/WindTunnel.html*, accessed May 26, 2009; NASA, "Wind Tunnels at NASA Langley Research Center," FS-2001-04-64-LaRC, 2001, *http://www.nasa.gov/centers/langley/news/factsheets/windtunnels.html*, accessed May 28, 2009.

97. H. Kipling Edenborough, "Research at NASA's NFAC Wind Tunnels," NASA TM-102827 (June 1990), pp. 1–6; NASA Langley, "NASA's Wind Tunnels," IS-1992-05-002-LaRC, May 1992, *http://oea.larc.nasa.gov/PAIS/WindTunnel.html*, accessed May 26, 2009.

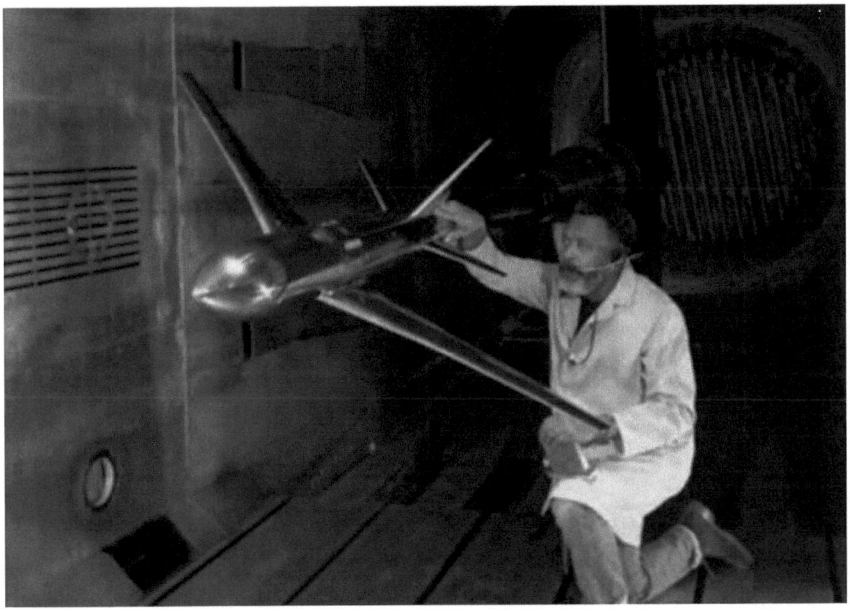

A Pathfinder I advanced transport model being prepared for a test in the super-cold nitrogen and high-pressure environment of the National Transonic Facility (NTF) in 1986. NASA.

Bringing the Tunnel to Industry and Academia

NASA has always justified its existence by making itself available for outside research. In an effort to advertise the services and capabilities of Langley's wind tunnels, NASA published the technical memorandum, "Characteristics of Major Active Wind Tunnels at the Langley Research Center," by William T. Shaefer, Jr., in July 1965. Unlike the NACA's goal of assisting industry through the use of its pioneering wind tunnels at a time when there were few facilities to rely upon, NASA's wind tunnels first and foremost met the needs of the Agency's fundamental research and development. Secondary to that priority were projects that were important to other Government agencies. Two specific committees handled U.S. Army, Navy, and Air Force requests concerning aircraft and missiles and propulsion projects. Finally, the aerospace industry had access to NASA facilities, primarily the Unitary Plan Wind Tunnels, on a fee basis for the evaluation of proprietary designs. No NASA wind tunnel was to be used for testing that could be done at a commercial facility, and all projects had to be "clearly in the national interest."[98]

98. Shaefer, "Characteristics of Major Active Wind Tunnels at the Langley Research Center," p. 2.

NASA continued to "sell" its tunnels on through the following decades. In 1992, the Agency confidently announced:

> NASA's wind tunnels are a national technological resource. They have provided vast knowledge that has contributed to the development and advancement of the nation's aviation industry, space program, economy and the national security. Amid today's increasingly fierce international, commercial and technological competition, NASA's wind tunnels are crucial tools for helping the United States retain its global leadership in aviation and space flight.[99]

According to this rhetoric, NASA's wind tunnels were central to the continued leadership of the United States in aerospace.

As part of the selling of the tunnels, NASA initiated the Technology Opportunities Showcase (TOPS) in the early 1990s. The program distributed to the aerospace industry a catalog of available facilities similar to a real estate sampler. A prospective user could check a box marked "Please Send More Information" or "Would Like To Discuss Facility Usage" as part of the process. NASA wind tunnels were used on a space-available basis. If the research was of interest to NASA, there would be no facility charge, and the Agency would publish the results. If a manufacturing concern had a proprietary interest and the client did not want the test results to be public, then it had to bear all costs, primarily the use of the facility.[100]

The TOPS evolved into the NASA Aeronautics Test Program (ATP) in the early 21st century to include all four Research Centers at Langley, Ames, Glenn, and Dryden.[101] The ATP offered Government, corporations, and institutions the opportunity to contract 14 facilities, which included a "nationwide team of highly trained and certified staff, whose backgrounds and education encompass every aspect of aerospace testing and engineering," for a "wide range" of experimental test services that reflected "sixty years of unmatched aerospace test history." The ATP

99. NASA Langley, "NASA's Wind Tunnels," IS-1992-05-002-LaRC, May 1992, *http://oea.larc.nasa.gov/PAIS/WindTunnel.html*, accessed May 26, 2009.

100. Langley Research Center, "Research and Test Facilities," p. 12.

101. NASA changed the name of the Lewis Research Center to the John H. Glenn Research Center at Lewis Field in 1999 to recognize the achievements of the astronaut and Ohio Senator.

and, by extension, NASA maintained that they could provide clients test results of "unparalleled superiority."[102]

THE NASA AERONAUTICS TEST PROGRAM WIND TUNNELS, 2009		
WIND TUNNEL	SPEED	LOCATION
9- by 15-Foot Low-Speed Wind Tunnel	Mach 0 to 0.2	Glenn
14- by 22-Foot Subsonic Tunnel	Mach 0 to 0.3	Langley
20-Foot Vertical Spin Tunnel	Mach 0 to 0.08	Langley
Icing Research Tunnel	Mach 0.06 to 0.56	Glenn
11-Foot Transonic Unitary Plan Facility	Mach 0.2 to 1.45	Ames
National Transonic Facility	Mach 0.1 to 1.2	Langley
Transonic Dynamics Tunnel	Mach 0.1 to 1.2	Langley
10- by 10-Foot Supersonic Wind Tunnel	Mach 0 to 0.4/2.0 to 3.5	Glenn
8- by 6-Foot Supersonic Wind Tunnel	Mach 0.25 to 2.0/0.0 to 0.1	Glenn
4-Foot Supersonic Unitary Plan Wind Tunnel	Mach 1.5 to 2.9/2.3 to 4.6	Langley
9- by 7-Foot Supersonic Wind Tunnel	Mach 1.55 to 2.55	Ames
Propulsion Systems Laboratory	Mach 4	Glenn
8-Foot High-Temperature Tunnel	Mach 3, 4, 5, 7	Langley
Aerothermodynamics Laboratory	Mach 6, 10	Langley

The Wind Tunnel's Future

Is the wind tunnel obsolete? In a word, no. But the value and merit of the tunnel in the early 21st century must be evaluated in the light of manifold other techniques that researchers can now employ. The range of these new techniques, particularly CFD, coupled with the seeming maturity of the airplane, has led some observers to conclude that there is little need for extensive investment in research, development, and infrastructure.[103] That facile assumption has been carried over into the question of whether there is a continued need for wind tunnels. It brings into question the role of the wind tunnel in contemporary aerospace research and development.

A 1988 *New York Times* article titled "In the Space Age, the Old Wind Tunnel Is Being Left Behind" proclaimed "aerospace engineers have hit

102. NASA, "NASA's Aeronautics Test Program: The Right Facility at the Right Time," B–1240 (Oct. 2006); NASA, "Aeronautics Test Program," NF-2009-03-486-HQ (n.d. [2009]).
103. Hansen, *The Bird is on the Wing*, p. 212.

a dead end in conventional efforts to test designs for the next generation of spaceships, planetary probes and other futuristic flying machines." The technology for the anticipated next generation in spacecraft technology that would appear in the 21st century included speeds in the escape velocity range and the ability to maneuver in and out of planetary atmospheres rather than the now-familiar single direction and uncontrolled descents of today. At the core of the problem was getting realistic flight data from a "nineteenth century invention used by the Wright brothers," the wind tunnel. William I. Scallion of NASA Langley asserted, "We've pushed beyond the capacity of most of our ground facilities." NASA, the Air Force, and various national universities began work on methods to simulate the speeds, temperatures, stress, forces, and vibration challenging the success of these new craft. The proposed solutions were improved wind tunnels capable of higher speeds, the firing of small-scale models atop rockets into the atmosphere, and the dropping of small test vehicles from the Space Shuttle while in orbit.[104]

The need for new testing methods and facilities reflected the changing nature of aerospace craft missions and design. Several programs perceived to be pathways to the future in the 1980s exemplified the need for new testing facilities. Proponents of the X-30 aerospace plane believed it would be able to take off and fly directly into space by reaching Mach 25, or 17,000 mph, while being powered by air-breathing engines. In 1988, wind tunnels could only simulate speeds up to Mach 12.5. NASA intended the Aeromanuevering Orbit Transfer Vehicle to be a low-cost "space tug" that could move payloads between high- and low-Earth orbits beginning in the late 1990s. The vehicle slowed itself in orbit by grazing the Earth's outer atmosphere with an aerobrake, or a lightweight shield, rather than relying upon heavy retrorockets, a technique that was impossible to replicate in a wind tunnel. NASA planned to launch small models from the Space Shuttle for evaluation. The final program concerned new interplanetary probes destined for Mars; Jupiter; Saturn's moon, Titan; and their atmospheres, which were much unlike Earth's. They no longer just dropped back into Earth's or another planet's atmosphere from space. The craft required maneuverability and flexibility as incorporated into the Space Shuttle for better economy.[105]

---

104. William J. Broad, "In the Space Age, the Old Wind Tunnel Is Being Left Behind," *New York Times*, Jan. 5, 1988, p. C1.
105. Ibid., p. C4.

NASA allocated funds for the demolition of unused facilities for the first time in the long history of the Agency in 2003. The process required that each of the Research Centers submit listings of target facilities.[106] NASA's Assistant Inspector General for Auditing conducted a survey of the utilization of NASA's wind tunnels at three Centers in 2003 and reported the findings to the directors of Langley, Ames, and Lewis and to the Associate Administrator for Aerospace Technology. Private industry and the Department of Defense spent approximately 28,000 hours in NASA tunnels in 2002. The number dwindled to 10,000 hours in 2003, dipping to about 2,500 hours in 2008. NASA managers acknowledged there was a direct correlation between a higher user fee schedule introduced in 2002 and the decline in usage. The audit also included the first complete list of tunnel closures for the Agency. Of the 19 closed facilities, NASA classified 5 as having been "mothballed," with the remaining 14 being "abandoned."[107]

Budget pressures also forced NASA to close running facilities. Unfortunately, NASA's operation of the NFAC was short-lived when the Agency closed the facility in 2003. Recognizing the need for full-scale testing of rotorcraft and powered-lift V/STOL aircraft, the Air Force leased the facility in 2006 for use by the AEDC. The NFAC became operational again in 2008. Besides aircraft, the schedule at the NFAC accommodated nontraditional test subjects, including wind turbines, parachutes, and trucks.[108]

In 2005, NASA announced its plan to reduce its aeronautics budget by 20 percent over the following 5 years. The budget cuts included the closing of wind tunnels and other research facilities and the elimination of hundreds of jobs. NASA had spread thin what was left of the aeronautics budget (down $54 million to $852 million) over too many programs. NASA did receive a small increase in its overall budget to cover the costs of the new Moon-Mars initiative, which meant cuts in aviation-related research. In a hearing before the House Science Subcommittee

106. Glenn Research Center, "Mitigation: Altitude Wind Tunnel and Space Power Chambers," Apr. 24, 2008, *http://awt.grc.nasa.gov/mitigation_demolition.aspx*, accessed Oct. 12, 2009.
107. David M. Cushing, "Final Management Letter on Audit of Wind Tunnel Utilization," Sept. 26, 2003, *oig.nasa.gov/audits/reports/FY03/pdfs/ig-03-027.pdf*, accessed June 17, 2009; Cory Nealon, "Winds of Change at NASA," *Newport News Daily Press*, Aug. 25, 2009.
108. Arnold Air Force Base, "The National Full-Scale Aerodynamics Complex," 2009, *http://www.arnold.af.mil/library/factsheets/factsheet.asp?id=13107*, accessed July 30, 2009.

on Space and Aeronautics to discuss the budget cuts, aerospace industry experts and politicians commented on the future of fundamental aeronautics research in the United States. Dr. John M. Klineberg, a former NASA official and industry executive, asserted that the NASA aeronautics program was "on its way to becoming irrelevant to the future of aeronautics in this country and in the world." Representative Dennis Kucinich, whose district included Cleveland, the home of NASA Glenn, warned that the United States was "going to take the 'A' out" of NASA and that the new Agency was "just going to be the National Space Administration."[109]

Philip S. Antón, Director of the RAND Corporation's Acquisition and Technology Policy Center, spoke before the Committee. RAND concluded a 3-year investigation that revealed that only 2 of NASA's 31 wind tunnels warranted closure.[110] As to the lingering question of the supremacy of CFD, Antón asserted that NASA should pursue wind tunnel facility, CFD, and flight-testing to meet national testing needs. RAND recommended a veritable laundry list of suggested improvements that ranged from the practical—the establishment of a minimum set of facilities that could serve national needs and the financial support to keep them running—to the visionary—continued investment in CFD and focus on the challenge of hypersonic air-breathing research.

RAND analysts had concluded in 2004 that NASA's wind tunnel facilities continued to be important to continued American competitiveness in the military, commercial, and space sectors of the world aerospace industry while "management issues" were "creating real risks." NASA needed a clear aeronautics test technology vision based on the idea of a national test facility plan that identified and maintained a minimum set of facilities.

For RAND, the bottom line was the establishment of shared financial support that kept NASA's underutilized but essential facilities from crumbling into ruin.[111] Antón found the alternative—the use of foreign tunnels, a practice many of the leading

109. Warren E. Leary, "NASA Plan to Cut Aviation Research 20% Dismays Experts," *New York Times*, Mar. 17, 2005, p. A24.

110. Philip S. Antón, "Testimony: Roles and Issues of NASA's Wind Tunnel and Propulsion Test Facilities for American Aeronautics," RAND Publication CT-239 (Mar. 2005), p. 8, 17.

111. "Do NASA's Wind Tunnel and Propulsion Test Facilities Serve National Needs?" RB-9066-NASA/OSD (2004).

aerospace manufacturers embraced—problematic because of the myriad of security, access, and availability challenges.[112]

NASA's wind tunnel heritage and the Agency's viability in the international aerospace community came to a head in 2009. Those issues centered on the planned demolition of the most famous, recognizable, and oldest operating research facility at Langley, the 30- by 60-Foot Tunnel, in 2009 or 2010. Better known by its NACA name, the Full-Scale Tunnel was, according to many, "old, inefficient and not designed for the computer age" in 2009.[113] The Deputy of NASA's Aeronautics Test Program, Tim Marshall, explained that the Agency decided "to focus its abilities on things that are strategically more important to the nation." NASA's focus was supersonic and hypersonic research that required smaller, faster tunnels for experiments on new technologies such as scramjets, not subsonic testing. In the case of the last operator of the FST, Old Dominion University, it had an important mission, refining the aerodynamics of motor trucks at a time of high fuel prices. It was told that economics, NASA's strategic mission, and the desire of the Agency's landlord, the U.S. Air Force, to regain the land, even if only for a parking lot in a flood zone, overrode its desire to continue using the FST for landlocked aerodynamic research.[114]

In conclusion, wind tunnels have been a central element in the success of NACA and NASA research throughout the century of flight. They are the physical representation of the rich and dynamic legacy of the organization. Their evolution, shaped by the innovative minds at Langley, Ames, and Glenn, paralleled the continual development of aircraft and spacecraft as national, economic, and technological missions shaped both. As newer, smaller, and cheaper digital technologies emerged in the late 20th century, wind tunnels and the testing methodologies pioneered in them still retained a place in the aerospace engineer's toolbox, no matter how low-tech they appeared. What resulted was a richer fabric of opportunities and modes of research that continued to contribute to the future of flight.

112. Antón, "Testimony," p. 15.

113. Cory Nealon, "Winds of Change at NASA," *Newport News Daily Press*, Aug. 25, 2009.

114. Barry Newman, "Shutting This Wind Tunnel Should Be a Breeze, But Its Fans Won't Be Silent," *The Wall Street Journal*, Aug. 26, 2009, p. A1.

Recommended Additional Readings

Reports, Papers, Articles, and Presentations:

H. Julian Allen and Alfred J. Eggers, Jr., "A Study of the Motion and Aerodynamic Heating of Ballistic Missiles Entering the Earth's Atmosphere at High Supersonic Speeds," NACA TR-1381 (1958).

Philip S. Antón, "Testimony: Roles and Issues of NASA's Wind Tunnel and Propulsion Test Facilities for American Aeronautics," RAND Publication CT-239 (2005).

R.P. Boyden, "A Review of Magnetic Suspension and Balance Systems," AIAA Paper 88-2008 (1988).

Dean R. Chapman, "Computational Aerodynamics Development and Outlook," Dryden Lecture in Research for 1979, American Institute of Aeronautics and Astronautics, *Aerospace Sciences Meeting, New Orleans, LA, Jan. 15–17, 1979*, AIAA-1979-129 (1979).

H. Kipling Edenborough, "Research at NASA's NFAC Wind Tunnels," NASA TM-102827 (1990).

Jerome T. Foughner, Jr., and Charles T. Bensinger, "F-16 Flutter Model Studies with External Wing Stores," NASA TM-74078 (1977).

Theodore J. Goldberg and Jerry N. Hefner, "Starting Phenomena for Hypersonic Inlets with Thick Turbulent Boundary Layers at Mach 6," NASA TN-D-6280 (1971).

Kenneth W. Iliff and Mary F. Shafer, "Space Shuttle Hypersonic Aerodynamic and Aerothermodynamic Flight Research and the Comparison to Ground Test Results," NASA TM-4499 (1993).

Eastman N. Jacobs, Kenneth E. Ward, and Robert M. Pinkerton, "The Characteristics of 78 Related Airfoil Sections from Tests in the Variable-Density Wind Tunnel," NACA TR-460 (1933).

Jim J. Jones, "Resume of Experiments Conducted in the High-Pressure Shock Tube of the Gas Dynamics Tube at NASA," NASA TM-X-56214 (1959).

Robert A. Kilgore, "Design Features and Operational Characteristics of the Langley 0.3-meter Transonic Cryogenic Tunnel," NASA TN-D-8304 (1976).

Robert A. Kilgore, "Design Features and Operational Characteristics of the Langley Pilot Transonic Cryogenic Tunnel," NASA TM-X-72012 (1974).

Paul Kutler and Helen Yee, *Supercomputing in Aerospace: Proceedings of a Symposium Held at the NASA Ames Research Center, Moffett Field, CA, Mar. 10–12, 1987*, NASA CP-2454 (1987).

Langley Research Center, "Research and Test Facilities," NASA TM-1096859 (1993).

"Manual for Users of the Unitary Plan Wind Tunnel Facilities of the National Advisory Committee for Aeronautics," NACA TM-80998 (1956).

Joseph G. Marvin, "Advancing Computational Aerodynamics through Wind-Tunnel Experimentation," NASA TM-109742 (1980).

Richard M. Raper, "Heat-Transfer and Pressure Measurements Obtained During Launch and Reentry of the First Four Gemini-Titan Missions and Some Comparisons with Wind-Tunnel Data," NASA TM-X-1407 (1967).

Edward J. Ray, Charles L. Ladson, Jerry B. Adcock, Pierce L. Lawing, and Robert M. Hall, "Review of Design and Operational Characteristics of the 0.3-Meter Transonic Cryogenic Tunnel," NASA TM-80123 (1979).

Alvin Seiff and Thomas N. Canning, "Modern Ballistic Ranges and Their Uses," NASA TM-X-66530 (1970).

William T. Shaefer, Jr., "Characteristics of Major Active Wind Tunnels at the Langley Research Center," NASA TM-X-1130 (1965).

Supercritical Wing Technology—A Progress Report on Flight Evaluations, NASA SP-301 (1972).

Marie H. Tuttle, Robert A. Kilgore, and Deborah L. Moore, "Cryogenic Wind Tunnels: A Comprehensive, Annotated Bibliography," NASA TM-4273 (1991).

Marie H. Tuttle, Deborah L. Moore, and Robert A. Kilgore, "Magnetic Suspension and Balance Systems: A Comprehensive, Annotated Bibliography," NASA TM-4318 (1991).

Edward P. Warner, "Research Symphony: The Langley Philharmonic in Opus No. 10," *Aviation*, vol. 34, no. 6 (1935).

Edward P. Warner, "Research to the Fore," *Aviation*, vol. 33, no. 6 (1934).

Richard T. Whitcomb, "A Design Approach and Selected Wind-Tunnel Results at High Subsonic Speeds for Wing-Tip Mounted Winglets," NASA TN-D-8260 (1976).

Richard T. Whitcomb, "A Study of the Zero-Lift Drag-Rise Characteristics of Wing-Body Combinations Near the Speed of Sound," NACA RM-L52H08 (1952).

Richard T. Whitcomb, "Some Considerations Regarding the Application of the Supersonic Area Rule to the Design of Airplane Fuselages," NACA RM-L56E23a (1956).

Richard T. Whitcomb and Larry L. Clark, "An Airfoil Shape for Efficient Flight at Supercritical Mach Numbers," NASA TM-X-1109 (1965).

Richard T. Whitcomb and Thomas L. Fischetti, "Development of a Supersonic Area Rule and an Application to the Design of a Wing-Body Combination Having High Lift-to-Drag Ratios," NACA RM-L53H31A (1953).

William C. Woods, Scott D. Holland, and Michael DiFulvio, "Hyper-X Stage Separation Wind-Tunnel Test Program," *Journal of Spacecraft and Rockets*, vol. 38 (2001).

E. Carson Yates, Jr., Norman S. Land, and Jerome T. Foughner, Jr., "Measured and Calculated Subsonic and Transonic Flutter Characteristics of a 45 Degree Sweptback Wing Planform in Air and Freon-12 in the Langley Transonic Dynamics Tunnel," NASA TN-D-1616 (1963).

Books and Monographs:

Donald D. Baals and William R. Corliss, *Wind Tunnels of NASA*, NASA SP-440 (Washington, DC: GPO, 1981).

John Becker, *The High Speed Frontier: Case Histories of Four NACA Programs 1920–1950*, NASA SP-445 (Washington, DC: GPO, 1980).

Glenn E. Bugos, *Atmosphere of Freedom: Sixty Years at the NASA Ames Research Center*, NASA SP-4314 (Washington, DC: GPO, 2000).

Joseph R. Chambers, *Innovation in Flight: Research of the Langley Research Center on Revolutionary Advanced Concepts for Aeronautics*, NASA 2005-SP-4539 (Washington, DC: GPO, 2005).

Virginia P. Dawson, *Engines and Innovation: Lewis Laboratory and American Propulsion Technology*, NASA SP-4306 (Washington, DC: GPO, 1991).

Michael Gorn, *Expanding the Envelope: Flight Research at NACA and NASA* (Lexington: University Press of Kentucky, 2001).

George W. Gray, *Frontiers of Flight: The Story of NACA Research* (New York: A.A. Knopf, 1948).

James R. Hansen, *The Bird is on the Wing: Aerodynamics and the Progress of the American Airplane* (College Station: Texas A & M University Press, 2004).

James R. Hansen, *Engineer in Charge: A History of the Langley Aeronautical Laboratory, 1917–1958*, NASA SP-4305 (Washington, DC: GPO, 1987).

T. A. Heppenheimer, *Facing the Heat Barrier: A History of Hypersonics*, NASA SP-2007-4232 (Washington, DC: GPO, 2007).

Peter Jakab, *Visions of a Flying Machine: The Wright Brothers and the Process of Invention* (Washington, DC: Smithsonian Institution Press, 1990).

William M. Leary, *"We Freeze to Please": A History of NASA's Icing Research Tunnel and the Quest for Flight Safety*, NASA SP-2002-4226 (Washington, DC: GPO, 2002).

J. Lawrence Lee, "Into the Wind: A History of the American Wind Tunnel, 1896–1941," dissertation, Auburn University, 2001.

Pamela E. Mack, ed., *From Engineering Science to Big Science: The NACA and NASA Collier Trophy Research Project Winners*, NASA SP-4219 (Washington, DC: GPO, 1998).

THIS PAGE INTENTIONALLY BLANK

The Micarta Controllable Pitch Propeller, pictured second from left, at the National Museum of the U.S. Air Force. Designed by McCook Field (now Wright-Patterson Air Force Base) engineers in 1922, this 9-foot propeller changed pitch in flight. U.S. Air Force.

CASE 7

Evolving the Modern Composite Airplane

Stephen Trimble

7

Structures and structural materials have undergone progressive refinement. Originally, aircraft were fabricated much like ships and complex wooden musical instruments: of wood, wire, and cloth. Then, metal gradually supplanted these materials. Now, high-strength composite materials have become the next generation, allowing for synthetic structures with even better structural properties for much less weight. NASA has assiduously pursued development of composite structures.

WHEN THE LOCKHEED MARTIN X-55 advanced composite cargo aircraft (ACCA) took flight early on the morning of June 2, 2009,[1] it marked a watershed moment in a century-long quest to marry the high-strength yet lightweight properties of plastics with the structure required to support a heavily loaded flying vehicle. As the X-55, a greatly modified Dornier 328Jet, headed east from the runway at the U.S. Air Force's Plant 42 outside Palmdale, CA, it gave the appearance of a conventional cargo aircraft. But the X-55's fuselage structure aft of the fuselage represented perhaps the promising breakthrough in four decades of composite technology development.

The single barrel, measuring 55 feet long by 9 feet wide,[2] revolutionizes expectations for structural performance at the same time that it proposes to dramatically reduce manufacturing costs. In the long history of applying composites to aircraft structures, the former seemed always to come at the expense of the latter, or vice versa. Yet the X-55 defies experience, with both aluminum skins and traditional composites. To distinguish it from the aluminum skin of the 328Jet, Lockheed used fewer than 4,000 fasteners to assemble the aircraft with the single-

1. "Cargo X-Plane Shows Benefits of Advanced Composites," *Aviation Week & Space Technology,* June 8, 2009, p. 18.
2. Stephen Trimble, "Skunk Works nears flight for new breed of all-composite aircraft," *Flight International,* June 5, 2009.

piece fuselage barrel. The metal 328Jet requires nearly 30,000 fasteners for all the pieces to fit together.[3] Unlike traditional composites, the X-55 did not require hours of time baking in a complex and costly industrial oven called an autoclave. Neither was the X-55 skin fashioned from textile preforms with resins requiring a strictly controlled climate that can be manipulated only within a precise window of time. Instead, Lockheed relied on an advanced composite resin called MTM45-1, an "out-of-autoclave" material flexible enough to assemble on a production line yet strong enough to support the X-55's normal aerodynamic loads and payload of three 463L-standard cargo pallets.[4]

Lockheed attributed the program's success to the fruits of a 10-year program sponsored by the Air Force Research Laboratory called the composites affordability initiative.[5] In truth, the X-55 bears the legacy of nearly a century's effort to make plastic suitable in terms of both performance and cost for serving as a load-bearing structure for large military and commercial aircraft.

It was an effort that began almost as soon as a method to mass-produce plastic became viable within 4 years after the Wright brothers' first flight in 1903. In aviation's formative years, plastics spread from cockpit dials to propellers to the laminated wood that formed the fuselage structure for small aircraft. Several decades would pass, however, before the properties of all but the most advanced plastics could be considered for mainstream aerospace applications. The spike in fuel prices of the early 1970s accelerated the search for a basic construction material for aircraft more efficient than aluminum, and composites finally moved to the forefront. Just as the National Advisory Committee for Aeronautics (NACA) fueled the industry's transition from spruce to metal in the early 1930s, the National Aeronautics and Space Administration (NASA) would pioneer the progression from all-metal airframes to all-composite material over four decades.

The first flight of the X-55 moved the progression of composite technology one step further. As a reward, the Air Force Research Laboratory announced 4 months later that it would continue to support the X-55

3. "USAF Advanced Composite Cargo Aircraft Makes First Flight," U.S. Air Force Aeronautical Systems Center press release, June 3, 2009.

4. Trimble, "Skunk Works nears flight for new breed of all-composite aircraft."

5. "Lockheed Martin Conducts Successful Flight of AFRL's Advanced Composite Cargo Aircraft," Lockheed Martin Corporation press release, June 3, 2009.

program, injecting more funding to continue a series of flight tests.[6] Where the X-55 technology goes from here can only be guessed.

Composites and the Airplane: Birth Through the 1930s

The history of composite development reveals at least as many false starts and technological blind alleys as genuine progress. Leo Baekeland, an American inventor of Dutch descent, started a revolution in materials science in 1907. Forming a new polymer of phenol and formaldehyde, Baekeland had succeeded in inventing the first thermosetting plastic, called Bakelite. Although various types of plastic had been developed in previous decades, Bakelite was the first commercial success. Baekeland's true breakthrough was inventing a process that allowed the mass production of a thermosetting plastic to be done cheaply enough to serve the mechanical and fiscal needs of a huge cross section of products, from industrial equipment to consumer goods.

It is no small irony that powered flight and thermosetting plastics were invented within a few years of each other. William F. Durand, the first Chairman of the NACA, the forerunner of NASA, in 1918 summarized the key structural issue facing any aircraft designer. Delivering the sixth Wilbur Wright Memorial Lecture to the Royal Aeronautical Society, the former naval officer and mechanical engineer said, "Broadly speaking, the fundamental problem in all airplane construction is adequate strength or function on minimum weight."[7] A second major structural concern, which NACA officials would soon come to fully appreciate, was the effect of corrosion on first wood, then metal, structures. Thermosetting plastics, one of two major forms of composite materials, present a tantalizing solution to both problems. The challenge has been to develop composite matrices and production processes that can mass-produce materials strong enough to replace wood and metal, yet affordable enough to meet commercial interests.

While Baekeland's grand innovation in 1907 immediately made strides in other sectors, aviation would be slow to realize the benefit of thermosetting plastics.

6. Guy Norris, "Advanced Composite Cargo Aircraft Gets Green Light For Phase III," *Aerospace Daily & Defense Report,* Oct. 2, 2009, p. 2.

7. William F. Durand, "Some Outstanding Problems in Aeronautics," in NACA, Fourth Annual Report (Washington, DC: GPO, 1920).

The substance was too brittle and too week in tensional strength to be used immediately in contemporary aircraft structures. But Bakelite eventually found its place by 1912, when some aircraft manufacturers started using the substance as a less corrosive glue to bind the joints between wooden structures.[8] The material shortages of World War I, however, would force the Government and its fledgling NACA organization to start considering alternative sources to wood for primary structures. In 1917, in fact, the NACA began what would become a decades-long effort to investigate and develop alternatives to wood, beginning with metal. As a very young bureaucracy with few resources for staffing or research, the NACA would not gain its own facilities to conduct research until the Langley laboratory in Virginia was opened in 1920. Instead, the NACA committee formed to investigate potential solutions to materials problems, such as a shortage of wood for war production of aircraft, and recommended that the Army and the Bureau of Standards study commercially available aluminum alloys and steels for their suitability as wing spars.[9]

Even by this time, Bakelite could be found inside cockpits for instruments and other surfaces, but it was not yet considered as a primary or secondary load-bearing structure, even for the relatively lightweight aircraft of this age. Perhaps the first evidence that Bakelite could serve as an instrumental component in aircraft came in 1924. With funding provided by the NACA, two early aircraft materials scientists—Frank W. Caldwell and N.S. Clay—ran tests on propellers made of Micarta material. The material was a generational improvement upon the phenolic resin introduced by Baekeland. Micarta is a laminated fabric—in this case cotton duck, or canvas—impregnated with the Bakelite resin.[10] Caldwell was the Government's chief propeller engineer through 1928 and later served as chief engineer for Hamilton Standard. Caldwell is credited with the invention of variable pitch propellers during the interwar period, which would eventually enable the Boeing Model 247 to achieve altitudes greater than 6,000 feet, thus clearing the Rocky Mountains and becoming a truly intercontinental aircraft. Micarta had already served

8. Eric Schatzberg, *Wings of Wood, Wings of Metal: Culture and Technical Choice in American Airplane Materials 1914–1945* (Princeton: Princeton University Press, 1999), p. 176.

9. Ibid., p. 32.

10. Meyer Fishbein, "Physical Properties of Synthetic Resin Materials," NACA Technical Note No. 694 (1939), p. 2.

as a material for fixed-pitch blades in World War I engines, including the Liberty and the 300-horsepower Wright.[11] Fixed-pitch blades were optimized neither for takeoff or cruise. Caldwell wanted to allow the pilot to change the pitch of the blade as the airplane climbed, allowing the pitch to remain efficient in all phases of flight. Using the same technique, the pilot could also reverse the pitch of the blade after landing. The propeller blades now functioned as a brake, allowing the aircraft to operate on shorter runways. Finding the right material to use for the blades was foremost among the challenges for Caldwell and Clay. It had to be strong enough to survive the stronger aerodynamic forces as the blade changed its pitch. The extra strength had to be balanced with the weight of the material, and metal alloys had not yet advanced far enough in the early 1920s. However, Caldwell and Clay found that Micarta was suitable. In an NACA technical report, they concluded: "The reversible and adjustable propeller with micarta blades . . . is one of the most practical devices yet worked out for this purpose. It is quite strong in all details, weighs very little more than the fixed pitch propeller and operates so easily that the pitch may be adjusted with two fingers on the control level when the engine is running." The authors had performed flight tests comparing the same aircraft and engine using both Micarta and wooden propeller blades. The former exceeded the top speed of the wooden propeller by 2 miles per hour (mph), while turning the engine at about 120 fewer revolutions per minute (rpm) and maintaining a similar rate of climb. The Micarta propeller was not only faster, it was also 7 percent more fuel efficient.[12]

The propeller work on Micarta showed that even if full-up plastics remained too weak for load-bearing applications, laminating wood with plastic glues provided a suitable alternative for that era's demands for structural strength in aircraft designs. While American developers continued to make advances, critical research also was occurring overseas. By the late 1920s, Otto Kraemer—a research scientist at Deutsche Versuchsanstalt fur Luftfahrt (DVL), the NACA's equivalent body in Germany—had started combining phenolic resins with paper or cloth. When this fiber-reinforced resin failed to yield a material with a structural stiffness superior to wood, Kraemer in 1933 started to investigate

11. Frank W. Caldwell and N.S. Clay, "Micarta Propellers III: General Description of the Design," NACA Technical Note No. 200 (1924), p. 3.
12. Ibid., pp. 7–9.

birch veneers instead as a filler. Thin sheets of birch veneer impregnated with the phenolic resin were laminated into a stack 1 centimeter thick. The material proved stronger than wood and offered the capability of being molded into complex shapes, finally making plastic a viable option for aircraft production.[13] Kraemer also got the aviation industry's attention by testing the durability of fiber-reinforced plastic resins. He exposed 1-millimeter-thick sheets of the material to outdoor exposure for 15 months. His results showed that although the material frayed at the edges, its strength had eroded by only 14 percent. In comparison to other contemporary materials, these results were observed as "practically no loss of strength."[14] In the late 1930s, European designers also fabricated propellers using a wood veneer impregnated with a resin varnish.[15]

A critical date in aircraft structural history is March 31, 1931, the day a Fokker F-10A Trimotor crashed in Kansas, with Notre Dame football coach Knute Rockne among the eight passengers killed. Crash investigators determined that the glues joining the wing strut to the F-10A's fuselage had been seriously deteriorated by exposure to moisture. The cumulative weakening of the joint caused the wing to break off in flight. The crash triggered a surge of nationwide negative publicity about the weaknesses of wood materials used in aircraft structures. This caused the aviation industry and passengers to embrace the transition from wood to metal for airplane materials, even as progress in synthetic materials, especially involving wood impregnated with phenolic resins, had started to develop in earnest.[16]

In his landmark text on the aviation industry's transition from wood to metal, Eric Schatzberg sharply criticizes the ambivalence of the NACA's leadership toward nonmetal alternatives as shortsightedness. For example, "In the case of the NACA, this neglect involved more than passive ignorance," Schatzberg argues, "but rather an active rejection of research on the new adhesives." However, with the military, airlines, and the traveling public all "voting with their feet," or, more precisely, their bank accounts, in favor of the metal option, it is not difficult to understand the NACA leadership's reluctance to invest scarce resources to develop

13. Schatzberg, *Wings of Wood, Wings of Metal*, p. 180.

14. G.M. Kline, "Plastics as Structural Materials for Aircraft," NACA Technical Note No. 628 (1937), p. 10.

15. Fishbein, "Physical Properties of Synthetic Resin Materials," p. 2.

16. Schatzberg, *Wings of Wood, Wings of Metal*, p. 133.

wood-based synthetic aircraft materials. The specimens developed during this period clearly lacked the popular support devoted to metal. Indeed, given the dominant role that metal structures were to play in aircraft and aerospace technology for most of the next 70 years, the priority placed on metal by the NACA's experts could be viewed as strategically prescient.

That is not to say that synthetic materials, such as plastic resins, were ignored by the aerospace industry in the 1930s. The technology of phenol- and formaldehyde-based resins had already grown beyond functioning as an adhesive with superior properties for resisting corrosion. The next step was using these highly moisture-resistant mixtures to form plywood and other laminated wood parts.[17] Ultimately, the same resins could be used as an impregnant that could be reinforced by wood,[18] essentially a carbon-based material. These early researchers had discovered the building blocks for what would become the carbon-fiber-reinforced plastic material that dominates the composite structures market for aircraft. Of course, there were also plenty of early applications, albeit with few commercial successes. A host of early attempts to bypass the era of metal aircraft, with its armies of riveters and concerns over corrosion and metal fatigue, would begin in the mid-1930s.

Clarence Chamberlin, who missed his chance by a few weeks to beat Charles Lindbergh across the Atlantic in 1927, flew an all-composite airplane. Called the Airmobile, it was designed by Harry Atwood, once a pupil of the Wright brothers, who flew from Boston to Washington, DC, in 1910, landing on the White House lawn.[19] Unfortunately, the full story of the Airmobile would expose Atwood as a charlatan and fraud. However, even if Atwood's dubious financing schemes ultimately hurt his reputation, his design for the Airmobile was legitimate; for its day, it was a major achievement. With a 22-foot wingspan and a 16-foot-long cabin, the Airmobile weighed only 800 pounds. Its low weight was achieved by constructing the wings, fuselage, tail surfaces, and ailerons with a new material called Duply, a thin veneer from a birch tree impregnated with a cellulose acetate.[20]

17. Arthur R. von Hippel and A.G.H. Dietz, "Curing of Resin-Wood Combinations By High-Frequency Heating," NACA Technical Note No. 874 (1938), pp. 1–3.
18. Ibid.
19. Howard Mansfield, *Skylark: The Life, Lies and Inventions of Harry Atwood* (Lebanon, NH: University Press of New England, 1999).
20. Fishbein, "Physical Properties of Synthetic Resin Materials."

Writing a technical note for the NACA in 1937, G.M. Kline, working for the Bureau of Standards, described the Airmobile's construction: "The wings and fuselage were each molded in one piece of extremely thin films of wood and cellulose acetate."[21] To raise money and attract public attention, however, Atwood oversold his ability to manufacture the aircraft cheaply and reliably. According to his farfetched publicity claims, 10 workers starting at 8 a.m. could build a new Airmobile from a single, 6-inch-diameter birch tree and have the airplane flying by dinner.

After a 12-minute first flight before 2,000 gawkers at the Nashua, NH, airport, Chamberlin complained that the aircraft was "nose heavy" but otherwise flew well. But any chance of pursuing full-scale manufacturing of the Airmobile would be short-lived. To develop the Airmobile, Atwood had accumulated more than 200 impatient creditors and a staggering debt greater than $100,000. The Airmobile's manufacturing process needed a long time to mature, and the Duply material was not nearly as easy to fabricate as advertised. The Airmobile idea was dropped as Atwood's converted furniture factory fell into insolvency.[22]

Also in the late 1930s, two early aviation legends—Eugene Vidal and Virginius Clark—pursued separate paths to manufacture an aircraft made of a laminated wood. Despite the military's focus on developing and buying all-metal aircraft, Vidal secured a contract in 1938 to provide a wing assembly molded from a thermoplastic resin. Vidal also received a small contract to deliver a static test model for a basic trainer designated the BT-11. Schatzberg writes: "A significant innovation in the Vidal process was the molding of stiffeners and the skin in a single step." Clark, meanwhile, partnered with Fairchild and Haskelite to build the F-46, the first airliner type made of all-synthetic materials. Haskelite reported that only nine men built the first half-shell of the fuselage within 2 hours. The F-46 first flew in 1937 and generated a great amount of interest. However, the estimated costs to develop the molds necessary to build Clark's proposed production system (greater than $230,000) exceeded the amount private or military investors were willing to spend. Clark's duramold technology was later acquired by Howard Hughes and put to use on the HK-1 flying boat (famously nicknamed—inaccurately—the "Spruce Goose").[23]

21. Kline, "Plastics as Structural Materials for Aircraft."

22. Howard Mansfield, *Skylark: The Life, Lies and Inventions of Harry Atwood.*

23. Schatzberg, *Wings of Wood, Wings of Metal,* pp. 182–191.

The February 16, 1939, issue of the U.K.-based *Flight* magazine offers a fascinating contemporary account of Clark's progress:

> Recent reports from America paint in glowing terms a new process said to have been invented by Col Virginius Clark (of Clark Y wing section fame) by which aeroplane fuselages and wings can, it is claimed, be built of plastic materials in two hours by nine men. . . . There is little doubt that Col Clark and his associates of the Bakelite Corporation and the Haskelite Manufacturing Corporation have evolved a method of production which is rapid and cheap. Exactly how rapid and how cheap time will show. In the meantime, it is well to remember that we are not standing still in this country. Dr. Norman de Bruyne has been doing excellent work on plastics at Duxford, and the Airscrew Company of Weybridge is doing some very interesting and promising experimental and development work with reinforced wood.[24]

The NACA first moved to undertake research in plastics for aircraft in 1936, tasking Kline to conduct a review of the technical research already completed.[25] Kline conducted a survey of "reinforced phenol-formaldehyde resin" as a structural material for aircraft. The survey was made with the "cooperation and financial support" of the NACA. Kline also summarized the industry's dilemma in an NACA technical note:

> In the fabrication of aircraft today the labor costs are high relative to the costs of tools. If large sections could be molded in one piece, the labor costs would be reduced but the cost of the molds and presses would be very high. Such a change in type construction would be economically practicable excepting the mass production of aircraft of a standard design. Langley suggests, therefore, that progress in the utilization of plastics in aircraft construction will be made by the gradual introduction of these materials into an otherwise orthodox

24. "Towards an Ideal," *Flight*, Feb. 16, 1939.
25. Schatzberg, *Wings of Wood, Wings of Metal*, p. 181.

> structure, and that the early stages of this development will involve the molding of such small units as fins and rudders and the fabrication of the larger units from reinforced sheets and molded sections by conventional methods of jointing.[26]

Kline essentially was predicting the focus of a massive NASA research program that would not get started for nearly four more decades. The subsequent effort was conducted along the lines that Kline prescribed and will be discussed later in this essay. Kline also seemed to understand how far ahead the age of composite structure would be for the aviation industry, especially as aircraft would quickly grow larger and more capable than he probably imagined. "It is very difficult to outline specific problems on this subject," Kline wrote, "because the exploration of the potential applications of reinforced plastics to aircraft construction is in its infancy, and is still uncharted."[27]

In 1939, an NACA technical report noted that synthetic materials had already started making an impact in aircraft construction of that era. The technology was still unsuited for supporting the weight of the aircraft in flight or on the ground, but the relative lightness and durability of synthetics made them popular for a range of accessories. Inside a wood or metal cockpit, a pilot scanned instruments with dials and casings made of synthetics and looked out a synthetic windshield. Synthetics also were employed for cabin soundproofing, lights encasings, pulleys, and the streamlined housings around loop antennas. The 1939 NACA paper concludes: "It is realized, at present, that the use of synthetic resin materials in the aircraft industry have been limited to miscellaneous accessories. The future is promising, however, for with continued development, resin materials suitable for aircraft structures will be produced."[28]

The Second World War Impetus

One man's vision for the possibilities of new synthetic adhesives had a powerful impact on history. Before World War II, Geoffrey de Havilland had designed the recordbreaking Comet racer and Albatross airliner, both

26. Kline, "Plastics as Structural Materials for Aircraft."

27. Ibid.

28. Fishbein, "Physical Properties of Synthetic Resin Materials," pp. 1, 16–17.

made of wood.[29] Delivering a speech at the Royal Aeronautical Society in London in April 1935, however, de Havilland seemed to have already written off wooden construction. "Few will doubt, however," he said, "that metal or possibly synthetic material will eventually be used universally, because it is in this direction we must look for lighter construction."[30] Yet de Havilland would introduce 6 years later the immortal D.H. 98 Mosquito, a lightweight, speedy, multirole aircraft mass-produced for the Royal Air Force (RAF).

De Havilland's decision to offer the RAF an essentially all-wooden aircraft might seem to be based more on logistical pragmatism than aerodynamic performance. After all, the British Empire's metal stocks were already committed to building the heavy Lancaster bombers and Spitfire fighters. Wooden materials were all that were left, not to mention the thousands of untapped and experienced woodworkers.[31] But the Mosquito, designed as a lightweight bomber, became a success because it could outperform opposing fighters. Lacking guns for self-defense, the Merlin-powered Mosquito survived by outracing its all-metal opponents.[32] Unlike metal airplanes, which obtain rigidity by using stringers to connect a series of bulkheads,[33] the Mosquito employed a plywood fuselage that was built in two halves and glued together.[34] De Havilland used a new resin called Aerolite as the glue, replacing the casein-type resins that had proved so susceptible to corrosion.[35] The Mosquito's construction technique anticipated the simplicity and strength of one-piece fuselage structures, not seen again until the first flight of Lockheed's X-55 ACCA, nearly six decades later.

For most of the 1940s, both the Government and industry focused on keeping up with wartime demand for vast fleets of all-metal aircraft. Howard Hughes pushed the boundaries of conventional flight at the

29. Ian Thirsk, *de Havilland Mosquito: An Illustrated History*, v. 2 (Manchester, U.K.: Crecy Publishing, Ltd., 2006), p. 39.

30. Kline, "Plastics as Structural Materials for Aircraft."

31. William S. Friedman, "Flying Plywood With a Sting," *Popular Science*, No. 6, Dec. 1943, pp. 100–103.

32. Robert L. O'Connell, *Soul of the Sword: An Illustrated History of Weaponry and Warfare from Prehistory to the Present* (New York: The Free Press, 2002), pp. 289–290.

33. Friedman, "Flying Plywood With a Sting," pp. 100–103.

34. Thirsk, *de Havilland Mosquito*, p. 39.

35. Phillippe Cognard, ed., *Adhesives and Sealants: Basic Concepts and High-Tech Bonding* (Amsterdam: Elsevier, 2005).

time with the first—and ultimately singular—flight of the Spruce Goose, which adopted a fuselage structure developed from the same Haskelite material pioneered by Clark in the late 1930s.

Pioneering work on plastic structures continued, with researchers focusing on the basic foundations of the processes that would later gain wide application. For example, the NACA funded a study by the Laboratory for Insulation Research at the Massachusetts Institute of Technology (MIT) that would explore problems later solved by autoclaves. The goal of the MIT researchers was to address a difficulty in the curing process for thermoset plastics based on heating a wood-resin composite between hot plates. Because wood and resin were poor heat conductors, it would take several hours to raise the center of the material to the curing temperature. In the process, temperatures at the surface could rise above desired levels, potentially damaging the material even as it was being cured. The NACA-funded study looked for new ways to rapidly heat the material uniformly on the surface and at the center. The particular method involved inserting the material into a high-frequency electrical field, attempting to heat the material from the inside using the "dielectric loss of the material."[36] This was an ambitious objective, anticipating and appropriating the same principles used in microwave ovens for building aircraft structures. Not surprisingly, the study's authors hoped to manage expectations. As they were not attempting to arrive at a final solution, the authors of the final report said their contribution was to "lay the groundwork for further development." Their final conclusion: "The problem of treating complicated shapes remains to be solved."[37]

Meanwhile, a Douglas Aircraft engineer hired shortly before World War II began would soon have a profound impact on the plastic composite industry. Brandt Goldsworthy served as a plastics engineer at Douglas during the war, where he was among the first to combine fiberglass and phenolic resin to produce laminated tooling.[38] The invention did not spark radical progress in the aviation industry, although the

36. Von Hippel and Dietz, "Curing of Resin-Wood Combinations By High-Frequency Heating," pp. 1–3.

37. Ibid.

38. "Brandt Goldsworthy: Composites Visionary," *High Performance Composites*, May 1, 2003, *http://www.compositesworld.com/articles/brandt-goldsworthy-composites-visionary.aspx*, accessed Oct. 3, 2009.

material was used to design ammunition chutes used to channel machine gun cartridges from storage boxes and into aircraft machine guns.[39] More noteworthy, after leaving Douglas in 1945 to start his own company, Goldsworthy would pioneer the automation of the manufacturing process for composite materials. Goldsworthy's invention of the pultrusion process in the 1950s would make durable and high-strength composites affordable for a range of applications, from cars to aircraft parts to fishing rods.[40]

As plastic composites continued to mature, the U.S. Army Air Corps began an ambitious series of experiments in the early 1940s on new composite material made from fiberglass-polyester blends. In the next two decades, the material would prove useful on aircraft as nose radomes and as both helicopter and propeller blades.[41] The combination of fiberglass and polyester also proved tempting to the military as a potential new load-bearing structural material for aircraft. In 1943, researchers at Wright-Patterson Air Force Base fabricated an aft fuselage for the Vultee BT-15 basic trainer using fiberglass and a polyester material called Plaskon, with balsa used as a sandwich core material.[42] The Wright Field experiments also included the development of an outer wing panel made of cloth and cellulose acetate for a North American AT-6C.[43] The BT-15 experiment proved unsuccessful, but the plastic wing of the AT-6C was more promising, showing only minor wing cracks after 245 flight hours.[44]

Into the Jet Age

Materials used in aircraft construction changed little from the early 1950s to the late 1970s. Aluminum alloyed with zinc metals, first introduced in 1943,[45] grew steadily in sophistication, leading to the introduction of a new line of even lighter-weight aluminum-lithium alloys in 1957.

39. Ibid.

40. Ibid.

41. Allen M. Shibley, Adolph E. Slobodzinski, John Nardone, and Martin Cutler, "Special Report: Structural Plastics in Aircraft," Plastics Technical Evaluation Center, U.S. Army Picatinny Arsenal (Mar. 1965), p. 7.

42. Ibid., p. 9.

43. Ibid.

44. Ibid.

45. John M. Swihart, "Commercial Jet Transportation Structures and Materials Evolution," Apr. 1985, presented at *AIAA Evolution of Aircraft/Aerospace Structures and Materials Symposium, Dayton, OH, Apr. 24–25, 1985*, p. 5-10.

Composite structure remained mostly a novelty item in aerospace construction. Progress continued to be made with developing composites, but demand was driven mainly by unique performance requirements, such as for high-speed atmospheric flight or exo-atmospheric travel.

A few exceptions emerged in the general-aviation market. The Federal Aviation Agency (FAA) certified the Taylorcraft Model 20 in 1955, which was based on a steel substructure but incorporated fiberglass for the skins and cowlings.[46] Even more progress was made by Piper Aircraft, which launched the PA-29 "plastic plane" project a few years later.[47] The PA-29 was essentially a commercial X-plane, experimenting with materials that could replace aluminum alloy for light aircraft.[48] The PA-29's all-fiberglass structure demonstrated the potential strength properties of composite material. Piper's engineers reported that the wing survived to 200 percent of ultimate load in static tests; the fuselage cracked at 180 percent because of a weakened bolt hole near the cockpit.[49] Piper concluded that it "is not only possible but also quite practical to build primary aircraft structures of fiberglass reinforced plastic."[50]

Commercial airliners built in the early 1950s relied almost exclusively upon aluminum and steel for structures. Boeing selected 2024 aluminum alloy for the fuselage skin and lower wing cover of the four-engine 707.[51] It was not until Boeing started designing the 747 jumbo airliner in 1966 that it paid serious attention to composites. Composites were used on the 747's rudder and elevators. Fiberglass, however, was in even greater demand on the 747, used as the structure for variable-camber leading-edge flaps.[52]

In 1972, NASA started a program with Boeing to redesign the 737's aluminum spoilers with skins made of graphite-epoxy composite and an aluminum honeycomb core, while the rest of the spoiler structure—the hinges and spar—remained unchanged. Each of the four spoilers on the 737 measures roughly 24 inches wide by 52 inches long. The composite

46. Ibid., p. 20.

47. F.S. Snyder and R.E. Drake, "Experience with Reinforced Plastic Primary Aircraft Structures," presented at the *Society of Automotive Engineers' Automotive Engineering Congress in Detroit, MI, Jan. 14–18, 1963*, p. 1.

48. Ibid.

49. Ibid., p. 4.

50. Ibid., p. 4.

51. Swihart, "Commercial Jet Transportation Structures and Materials Evolution," p. 5-3.

52. Ibid., pp. 5–6.

material comprised about 35 percent of the weight of the new structure of each spoiler, which measured about 13 pounds, or 17 percent less than an all-metal structure.[53] The composite spoilers initiated flight operations on 27 737s owned by the airlines Aloha, Lufthansa, New Zealand National, Piedmont, PSA, and VASP. Five years later, Boeing reported no problems with durability and projected a long service life for the components.[54]

The impact of the 1973 oil embargo finally forced airlines to start reexamining their fuel-burn rates. After annual fuel price increases of 5 percent before the embargo, the gas bill for airlines jumped by 10 cents to 28 cents per gallon almost overnight.[55] Most immediately, airframers looked to the potential of the recently developed high-bypass turbofan engine, as typified by the General Electric TF39/CF6 engine family, to gain rapid improvements in fuel efficiency for airliners. But against the backdrop of the oil embargo, the potential of composites to drive another revolution in airframe efficiency could not be ignored. Graphite-epoxy composite weighed 25 percent less than comparable aluminum structure, potentially boosting fuel efficiency by 15 percent.[56]

The stage was set for launching the most significant change in aircraft structural technology since the rapid transition to aluminum in the early 1930s. However, it would be no easy transition. In the early 1970s, composite design for airframes was still in its infancy, despite its many advances in military service. Recalling this period, a Boeing executive would later remember the words of caution from one of his mentors in 1975: "One of Boeing most senior employees said, when composites were first introduced in 1975, that he had lived through the transition from spruce and fabric to aluminum. It took three airplane generations before the younger designers were able to put aluminum to its best use, and he thought that we would have to be very clever to avoid that with composites."[57] The anonymous commentary would prove eerily prescient. From 1975, Boeing would advance through two generations of aircraft—beginning with the 757/767 and progressing with the 777 and

53. Richard A. Pride, "Composite Fibres and Composites," NASA CP-2074 (1979).

54. Ibid.

55. Robert L. James and Dal V. Maddalon, "The Drive for Aircraft Energy Efficiency," *Aerospace America*, Feb. 1984, p. 54.

56. Ibid.

57. Swihart, "Commercial Jet Transportation Structures and Materials Evolution."

Next Generation 737—before mastering the manufacturing and design requirements to mass-produce an all-composite fuselage barrel, one of the key design features of the 787, launched in 2003.

By the early 1970s, the transition to composites was a commercial imperative, but it took projects and studies launched by NASA and the military to start building momentum. Unlike the transition from spruce to metal structures four decades before, the industry's leading aircraft makers now postured conservatively. The maturing air travel industry presented manufacturers with a new set of regulatory and legal barriers to embracing innovative ideas. In this new era, passengers would not be the unwitting guinea pigs as engineers worked out the problems of a new construction material. Conservatism in design would especially apply to load-bearing primary structures. "Today's climate of government regulatory nervousness and aircraft/airline industry liability concerns demand that any new structural material system be equally reliable," Boeing executive G.L. Brower commented in 1978.[58]

The Path to the Modern Era

A strategy began forming in 1972 with the launch of the Air Force–NASA Long Range Planning Study for Composites (RECAST), which focused priorities for the research projects that would soon begin.[59] That was prelude to what NASA research Marvin Dow would later call the "golden age of composites research,"[60] a period stretching from roughly 1975 until funding priorities shifted in 1986. As airlines looked to airframers for help, military aircraft were already making great strides with composite structure. The Grumman F-14 Tomcat, then the McDonnell-Douglas F-15 Eagle, incorporated boron-epoxy composites into the empennage skin, a primary structure.[61] With the first flight of the McDonnell-Douglas AV-8B Harrier in 1978, composite usage had drifted to the wing as well. In all,

58. Richard G. O'Lone, "Industry Tackles Composites Challenge," *Aviation Week & Space Technology*, Sept. 15, 1980, p. 22.
59. Marvin B. Dow, "The ACEE Program and Basic Composites Research at Langley Research Centre (1975 to 1986): Summary and Bibliography," NASA RP-1177 (1987), p. 1.
60. Ibid., p. 14
61. Ravi B. Deo, James H. Starnes, Jr., and Richard C. Holzwarth, "Low-Cost Composite Materials and Structures for Aircraft Applications," presented at the *NATO Research and Technology Agency Applied Vehicle Technical Panel Specialists' Meeting on Low-Cost Composite Structures, Loen, Norway, May 7–8, 2001*, p. 1-1.

Air Force engineer Norris Krone prompted NASA to develop the X-29 to prove that high-strength composites were capable of supporting forward-swept wings. NASA.

about one-fourth of the AV-8B's weight,[62] including 75 percent in the weight of the wing alone,[63] was made of composite material. Meanwhile, composite materials studies by top Grumman engineer Norris Krone opened the door to experimenting with forward-swept wings. NASA responded to Krone's papers in 1976 by launching the X-29 technology demonstrator, which incorporated an all-composite wing.[64]

Composites also found a fertile atmosphere for innovation in the rotorcraft industry during this period. As NASA pushed the commercial aircraft industry forward in the use of composites, the U.S. Army spurred progress among its helicopter suppliers. In 1981, the Army selected Bell Helicopter Textron and Sikorsky to design all-composite airframes under the advanced composite airframe program (ACAP).[65]

62. Ibid., p. SM 1-2.

63. O'Lone, "Industry Tackles Composites Challenge," p. 22.

64. Richard N. Hadcock, "X-29 Composite Wing," presented at the *AIAA Evolution of Aircraft/ Aerospace Structures and Materials Symposium, Dayton, OH, Apr. 24–25, 1985*, p. 7–1.

65. Joseph J. Klumpp, "Parametric Cost Estimation Applied to Composite Helicopter Airframes" (Monterey, CA: Naval Postgraduate School master's thesis, 1994).

Perhaps already eyeing the need for a new light airframe to replace the Bell OH-58 Kiowa scout helicopter, the Army tasked the contractors to design a new utility helicopter under 10,000 pounds that could fly for up to 2 hours 20 minutes.[66] Bell first flew the D-292 in 1984, and Sikorsky flew the S-75 ACAP in 1985.[67] Boeing complemented their efforts by designing the Model 360, an all-composite helicopter airframe with a gross weight of 30,500 pounds.[68] Each of these projects provided the steppingstones needed for all three contractors to fulfill the design goals for both the now-canceled Sikorsky–Boeing RAH-66 Comanche and the Bell–Boeing V-22 Osprey tilt rotor. The latter also drove developments in automated fiber placement technology, relieving the need to lay up by hand about 50 percent of the airframe's weight.[69]

In the midst of this rapid progress, the makers of executive and "general" aircraft required neither the encouragement nor the financial assistance of the Government to move wholesale into composite airframe manufacturing. While Boeing dabbled with composite spoilers, ailerons, and wing covers on its new 767, William P. Lear, founder of LearAvia, was developing the Lear Fan 2100—a twin-engine, nine-seat aircraft powered by a pusher-propeller with a 3,650-pound airframe made almost entirely from a graphite-epoxy composite.[70] About a decade later, Beechcraft unveiled the popular and stylish Starship 1, an 8- to 10-passenger twin turboprop weighing 7,644 pounds empty.[71] Composite materials—mainly using graphite-epoxy and NOMEX sandwich panels—accounted for 72 percent of the airframe's weight.[72]

Actual performance fell far short of the original expectations during this period. Dow's NASA colleagues in 1975 had outlined a strategy that should have led to full-scale tests of an all-composite fuselage and wing box for a civil airliner by the late 1980s. Although the dream was delayed by more than a decade, it is true that state of knowledge and

66. Ibid.

67. Ibid., p. 68.

68. D.A. Reed and R. Gable, "Ground Shake Test of the Boeing Model 360 Helicopter Airframe," NASA CR-181766 (1989), p. 6.

69. Deo, Starnes, and Holzwarth, "Low-Cost Composite Materials and Structures for Aircraft Applications."

70. "Lightweight Composites Are Displacing Metals," *Business Week*, July 30, 1979, p. 36D.

71. E.H. Hooper, "Starship 1," presented at the *AIAA Evolution of Aircraft/Aerospace Structures and Materials Symposium, Dayton, OH, Apr. 24–25, 1985*, p. 6–1.

72. Ibid.

understanding of composite materials leaped dramatically during this period. The three major U.S. commercial airframers of the era—Boeing, Lockheed, and McDonnell-Douglas—each made contributions. However, the agenda was led by NASA's $435-million investment in the Aircraft Energy Efficiency (ACEE) program. ACEE's top goal, in terms of funding priority, was to develop an energy-efficient engine. The program also invested greatly to improve how airframers control for laminar flow. But a major pillar of ACEE was to drive the civil industry to fundamentally change its approach to aircraft structures and shift from metal to the new breed of composites then emerging from laboratories. As of 1979, NASA had budgeted $75 million toward achieving that goal,[73] with the manufacturers responsible for providing a 10-percent match.

ACEE proposed a gradual development strategy. The first step was to install a graphite-epoxy composite material called Narmco T300/5208[74] on lightly loaded secondary structures of existing commercial aircraft in operational service. For their parts, Boeing selected the 727 elevator, Lockheed chose the L-1011 inboard aileron, and Douglas opted to change the DC-10 upper aft rudder.[75] From this starting point, NASA engaged the manufacturers to move on to medium-primary components, which became the 737 horizontal stabilizer, the L-1011 vertical fin, and the DC-10 vertical stabilizer.[76] The weight savings for each of the medium primary components was estimated to be 23 percent, 30 percent, and 22 percent, respectively.[77]

The leap from secondary to medium-primary components yielded some immediate lessons for what not to do in composite structural design. All three components failed before experiencing ultimate loads in initial ground tests.[78] The problems showed how different composite material could be from the familiar characteristics of metal. Compared to aluminum, an equal amount of composite material can support a heavier load. But, as experience revealed, this was not true in every condition experienced by an aircraft in normal flight. Metals are known to

73. "Energy Efficiency Funding Detailed," *Aviation Week & Space Technology*, Nov. 12, 1979, p. 122.

74. Dow, "The ACEE Program and Basic Composites Research at Langley Research Centre (1975 to 1986): Summary and Bibliography," p. 3.

75. Ibid.

76. Ibid., pp. 4–5.

77. "Composite Programs Pushed by NASA," *Aviation Week & Space Technology*, Nov. 12, 1979, p. 203.

78. James and Maddalon, "The Drive for Aircraft Energy Efficiency," p. 54.

distribute stresses and loads to surrounding structures. In simple terms, they bend more than they break. Composite material does the opposite. It is brittle, stiff, and unyielding to the point of breaking.

Boeing's horizontal stabilizer and Douglas's vertical stabilizer both failed before the predicted ultimate load for similar reasons. The brittle composite structure did not redistribute loads as expected. In the case of the 737 component, Boeing had intentionally removed one lug pin to simulate a fail-safe mode. The structure under the point of stress buckled rather than redistributed the load. Douglas had inadvertently drilled too big of a hole for a fastener where the web cover for the rear spar met a cutout for an access hole.[79] It was an error by Douglas's machinists but a tolerable one if the same structure were designed with metal. Lockheed faced a different kind of problem with the failure of the L-1011 vertical fin during similar ground tests. In this case, a secondary interlaminar stress developed after the fin's aerodynamic cover buckled at the attachment point with the front spar cap. NASA later noted: "Such secondary forces are routinely ignored in current metals design."[80] The design for each of these components was later modified to overcome these unfamiliar weaknesses of composite materials.

In the late 1970s, all three manufacturers began working on the basic technology for the ultimate goal of the ACEE program: designing full-scale, composite-only wing and fuselage. Control surfaces and empennage structures provided important steppingstones, but it was expected that expanding the use of composites to large sections of the fuselage and wing could improve efficiency by an order of magnitude.[81] More specifically, Boeing's design studies estimated a weight savings of 25–30 percent if the 757 fuselage was converted to an all-composite design.[82] Further, an all-composite wing designed with a metal-like allowable strain could reduce weight by as much as 40 percent for a large commercial aircraft, according to NASA's design analysis.[83] Each manufacturer was assigned a different task, with all three collaborating on their results to gain maximum results. Lockheed explored

79. Herman L. Bohon and John G. Davis, Jr., "Composites for Large Transports—Facing the Challenge," *Aerospace America*, June 1984, p. 58.

80. Ibid.

81. James and Maddalon, "The Drive for Aircraft Energy Efficiency," p. 54.

82. Bohon and Davis, "Composites for Large Transports—Facing the Challenge," p. 58.

83. Ibid.

design techniques for a wet wing that could contain fuel and survive lightning strikes.[84] Boeing worked on creating a system for defining degrees of damage tolerance for structures[85] and designed wing panes strong enough to endure postimpact compression of 50,000 pounds per square inch (psi) at strains of 0.006.[86] Meanwhile, Douglas concentrated on methods for designing multibolted joints.[87] By 1984, NASA and Lockheed had launched the advanced composite center wing project, aimed at designing an all-composite center wing box for an "advanced" C-130 airlifter. This project, which included fabricating two 35-foot-long structures for static and durability tests, would seek to reduce the weight of the C-130's center wing box by 35 percent and reduce manufacturing costs by 10 percent compared with aluminum structure.[88] Meanwhile, Boeing started work in 1984 to design, fabricate, and test full-scale fuselage panels.[89]

Within a 10-year period, the U.S. commercial aircraft industry had come very far. From the near exclusion of composite structure in the early 1970s, composites had entered the production flow as both secondary and medium-primary components by the mid-1980s. This record of achievement, however, was eclipsed by even greater progress in commercial aircraft technology in Europe, where the then-upstart DASA Airbus consortium had pushed composites technology even further.

While U.S. commercial programs continued to conduct demonstrations, the A300 and A310 production lines introduced an all-composite rudder in 1983 and achieved a vertical tailfin in 1985. The latter vividly demonstrated the manufacturing efficiencies promised by composite designs. While a metal vertical tail contained more than 2,000 parts, Airbus designed a new structure with a carbon fiber epoxy-honeycomb core sandwich that required fewer than 100 parts, reducing both the weight of the structure and the cost of assembly.[90] A few years later, Airbus unveiled the A320 narrow body with 28 percent of its structural weight filled by

84. James and Maddalon, "The Drive for Aircraft Energy Efficiency," p. 54.
85. Ibid.
86. Dow, "The ACEE Program and Basic Composites Research at Langley Research Centre (1975 to 1986): Summary and Bibliography," p. 6.
87. Bohon and Davis, "Composites for Large Transports—Facing the Challenge," p. 58.
88. Ibid.
89. Dow, "The ACEE Program and Basic Composites Research at Langley Research Centre (1975 to 1986): Summary and Bibliography," p. 6.
90. Deo, Starnes, and Holzwarth, "Low-Cost Composite Materials and Structures for Aircraft Applications," p. 6–7.

composite materials, including the entire tail structure, fuselage belly skins, trailing-edge flaps, spoilers, ailerons, and nacelles.[91] It would be another decade before a U.S. manufacturer eclipsed Airbus's lead, with the introduction of the Boeing 777 in 1995. Consolidating experience gained as a major structural supplier for the Northrop B-2A bomber program, Boeing designed the 777, with an all-composite empennage one-tenth of the weight.[92] By this time, the percentage of composites integrated into a commercial airliner's weight had become a measure of the manufacturer's progress in gaining a competitive edge over a rival, a trend that continues to this day with the emerging Airbus A350/Boeing 787 competition.

As European manufacturers assumed a technical lead over U.S. rivals for composite technology in the 1980s, the U.S. still retained a huge lead with military aircraft technology. With fewer operational concerns about damage tolerance, crash survivability, and manufacturing cost, military aircraft exploited the performance advantages of composite material, particularly for its weight savings. The V-22 Osprey tilt rotor employed composites for 70 percent of its structural weight.[93] Meanwhile, Northrop and Boeing used composites extensively on the B-2 stealth bomber, which is 37-percent composite material by weight.

Steady progress on the military side, however, was not enough to sustain momentum for NASA's commercial-oriented technology. The ACEE program folded after 1985, following several years of real progress but before it had achieved all of its goals. The full-scale wing and fuselage test program, which had received a $92-million, 6-year budget from NASA in fiscal year 1984,[94] was deleted from the Agency's spending plans a year later.[95] By 1985, funding available to carry out the goals of the ACEE program had been steadily eroding for several years. The Reagan Administration took office in 1981 with a distinctly different view on the responsibility of Government to support the validation of commercial technologies.[96]

91. Ibid.

92. Richard Piellisch, "Composites Roll Sevens," *Aerospace America*, Oct. 1992, p. 26.

93. Mark T. Wright, et al., "Composite Materials in Aircraft Mishaps Involving Fire: A Literature Review," Naval Air Warfare Center Weapons Division TP-8552, June 2003, p. 13.

94. Bohon and Davis, "Composites for Large Transports—Facing the Challenge," p. 58.

95. Dow, "The ACEE Program and Basic Composites Research at Langley Research Centre (1975 to 1986): Summary and Bibliography," p. 6.

96. Jay C. Lowndes, "Keeping a Sharp Technology Edge," *Aerospace America*, Feb. 1988, p. 24.

In constant 1988 dollars, ACEE funding dropped from a peak $300 million in 1980 to $80 million in 1988, with funding for validating high-strength composite materials in flight wiped out entirely.[97] The shift in technology policy corresponded with priority disagreements between aeronautics and space supporters in industry, with the latter favoring boosting support for electronics over pure aeronautics research.[98]

In its 10-year run, the composite structural element of the ACEE program had overcome numerous technical issues. The most serious issue erupted in 1979 and caused NASA to briefly halt further studies until it could be fully analyzed. The story, always expressed in general terms, has become an urban myth for the aircraft composites community. Precise details of the incident appear lost to history, but the consequences of its impact were very real at the time. The legend goes that in the late 1970s, waste fibers from composite materials were dumped into an incinerator. Afterward, whether by cause or coincidence, a nearby electric substation shorted out.[99] Carbon fibers set loose by the incinerator fire were blamed for the malfunction at the substation.

The incident prompted widespread concerns among aviation engineers at a time when NASA was poised to spend hundreds of millions of dollars to transition composite materials from mainly space and military vehicles to large commercial transports. In 1979, NASA halted work on the ACEE program to analyze the risk that future crashes of increasingly composite-laden aircraft would spew blackout-causing fibers onto the Nation's electrical grid.[100]

Few seriously question the potential benefits that composite materials offer society. By the mid-1970s, it was clear that composites dramatically raise the efficiency of aircraft. The cost of manufacturing the materials was higher, but the life-cycle cost of maintaining noncorroding composite structures offered a compelling offset. Concerns about the economic and health risks poised by such a dramatic transition to a different structural material have also been very real.

---

97. Ibid.

98. Ibid.

99. Wright, et al., "Composite Materials in Aircraft Mishaps Involving Fire: A Literature Review," pp. 8–9.

100. "Carbon Fire Hazard Concerns NASA," *Aviation Week & Space Technology*, Mar. 5, 1979, p. 47.

7

It was up to the aviation industry, with Government support, to answer these vital questions before composite technology could move further.

With the ACEE program suspended to study concerns about the risks to electrical equipment, both NASA and the U.S. Air Force by 1978 had launched separate efforts to overcome these concerns. In a typical aircraft fire after a crash, the fuel-driven blaze can reach temperatures between 1,800 to 3,600 degrees Fahrenheit (°F). At temperatures higher than 750 °F, the matrix material in a composite structure will burn off, which creates two potential hazards. As the matrix polymer transforms into fumes, the underlying chemistry creates a toxic mixture called pyrolysis product, which if inhaled can be harmful. Secondly, after the matrix material burns away, the carbon fibers are released into the atmosphere.[101]

These liberated fibers, which as natural conductors have the power to short circuit a power line, could be dispersed over wide areas by wind. This led to concerns that the fibers would could come into contact with local power cables or, even worse, exposed power substations, leading to widespread power blackouts as the fibers short circuit the electrical equipment.[102] In the late 1970s, the U.S. Air Force started a program to study aircraft crashes that involved early-generation composite materials.

Another incident in 1997 was typical of different type of concern about the growing use of composite materials for aircraft structures. A U.S. Air Force F-117 flying a routine at the Baltimore airshow crashed when a wing-strut failed. Emergency crews who rushed to the scene extinguished fires that destroyed and damaged several dwellings, blanketing the area with a "wax-like" substance that contained carbon fibers embedded in the F-117's structures that could have otherwise been released into the atmosphere. Despite these precautions, the same firefighters and paramedics who rushed to the scene later reported becoming "ill from the fumes emitted by the fire. It was believed that some of these fumes resulted from the burning of the resin in the composite materials," according a U.S. Navy technical paper published in 2003.[103]

101. Lt. John M. Olson, USAF, "Mishap Risk Control for Advanced Aerospace/Composite Materials" (Wright-Patterson AFB: USAF Advanced Composites Program Office, 1995), p. 4.

102. "Carbon Fibre Risk Analysis," NASA CP-2074 (1979), p. iii.

103. Wright, et al., "Composite Materials in Aircraft Mishaps Involving Fire: A Literature Review," pp. 8–9.

Yet another issue has sapped the public's confidence in composite materials for aircraft structures for several decades. As late as 2007, the risk presented by lightning striking a composite section of an aircraft fuselage was the subject of a primetime investigation by Dan Rather, who extensively quoted a retired Boeing Space Shuttle engineer. The question is repeatedly asked: If the aluminum structure of a previous generation of airliners created a natural Faraday cage, how would composite materials with weaker properties for conductivity respond when struck by lightning?

Technical hazards were not the only threat to the acceptance of composite materials. To be sure, proving that composite material would be safe to operate in commercial service constituted an important endorsement of the technology for subsequent application, as the ACEE projects showed. But the aerospace industry also faced the challenge of establishing a new industrial infrastructure from the ground up that would supply vast quantities of composite materials. NASA officials anticipated the magnitude of the infrastructure issue. The shift from wood to metal in the 1930s occurred in an era when airframers acted almost recklessly by today's standards. Making a similar transition in the regulatory and business climate of the late 1970s would be another challenge entirely. Perhaps with an eye on the rapid progress being made by European competitors in commercial aircraft, NASA addressed the issue head-on. In 1980, NASA Deputy Administrator Alan M. Lovelace urged industry to "anticipate this change," adding that he realized "this will take considerable capital, but I do worry that if this is not done then might we not, a decade from now, find ourselves in a position similar to that in which the automobile industry is at the present time?"[104]

Of course, demand drives supply, and the availability of the raw material for making composite aerospace parts grew precipitously throughout the 1980s. For example, 2 years before Lovelace issued his warning to industry, U.S. manufacturers consumed 500,000 pounds of composites every 12 months, with the aerospace industry accounting for half of that amount.[105] Meanwhile, a single supplier for graphite fiber, Union Carbide, had already announced plans to increase annual output to 800,000 pounds by the end of 1981.[106] U.S. consumption would soon be driven by the automobile industry, which was also struggling

104. O'Lone, "Industry Tackles Composites Challenge," p. 22.
105. "Lightweight Composites Are Displacing Metals."
106. Ibid.

to keep up with the innovations of foreign competition, as much as by the aerospace industry throughout the 1980s.

Challenges and Opportunities

If composites were to receive wide application, the cost of the materials would have to dramatically decline from their mid-1980s levels. ACEE succeeded in making plastic composites commonplace not just in fairings and hatches for large airliners but also on control surfaces, such as the ailerons, flaps, and rudder. On these secondary structures, cash-strapped airlines achieved the weight savings that prompted the shift to composites in the first place. The program did not, however, result in the immediate transition to widespread production of plastic composites for primary structures. Until the industry could make that transition, it would be impossible to justify the investment required to create the infrastructure that Lovelace described to produce composites at rates equivalent to yearly aluminum output.

To the contrary, tooling costs for composites remained high, as did the labor costs required to fabricate the composite parts.[107] A major issue driving costs up under the ACEE program was the need to improve the damage tolerance of the composite parts, especially as the program transitioned from secondary components to heavily loaded primary structures. Composite plastics were still easy to damage and costly to replace. McDonnell-Douglas once calculated that the MD-11 trijet contained about 14,000 pounds of composite structure, which the company estimated saved airlines about $44,000 in yearly fuel costs per plane.[108] But a single incident of "ramp rash" requiring the airline to replace one of the plastic components could wipe away the yearly return on investment provided by all 14,000 pounds of composite structure.[109]

The method that manufacturers devised in the early 1980s involved using toughened resins, but these required more intensive labor to fabricate, which aggravated the cost problem.[110] From the early 1980s, NASA

107. O'Lone, "Industry Tackles Composites Challenge," p. 22.

108. "NASA Langley Reorienting Advanced Composites Technology Program," *Aerospace Daily*, Aug. 29, 1994, p. 328.

109. Ibid.

110. H. Benson Dexter, "Development of Textile Reinforced Composites for Aircraft Structures," presented at the *4th International Symposium for Textile Composites in Kyoto, Japan, Oct. 12–14, 1998*, p. 1.

worked to solve this dilemma by investigation new manufacturing methods. One research program sponsored by the Agency considered whether textile-reinforced composites could be a cost-effective way to build damage-tolerant primary structures for aircraft.[111] Composite laminates are not strong so much as they are stiff, particularly in the direction of the aligned fibers. Loads coming from different directions have a tendency to damage the structure unless it is properly reinforced, usually in the form of increased thickness or other supports. Another poor characteristic of laminated composites is how the material reacts to damage. Instead of buckling like aluminum, which helps absorb some of the energy caused by the impact, the stiff composite material tends to shatter.

Some feared that such materials could prove too much for cash-strapped airlines of the early 1990s to accept. If laminated composites were the problem, some believed the solution was to continue investigating textile composites. That meant shifting to a new process in which carbon fibers could be stitched or woven into place, then infused with a plastic resin matrix. This method seemed to offer the opportunity to solve both the damage tolerance and the manufacturing problems simultaneously. Textile fibers could be woven in a manner that made the material strong against loads coming from several directions, not just one. Moreover, some envisioned the deployment of giant textile composite sewing machines to mass-produce the stronger material, dramatically lowering the cost of manufacture in a single stroke.

The reality, of course, would prove far more complex and challenging than the visionaries of textile composites had imagined. To be sure, the concept faced many skeptics within the conservative aerospace industry even as it gained force in the early 1990s. Indeed, there have been many false starts in the composite business. The *Aerospace America* journal in 1990 proposed that thermoplastics, a comparatively little-used form of composites, could soon eclipse thermoset composites to become the "material of the '90s." The article wisely contained a cautionary note from a wry Lockheed executive, who recalled a quote by a former boss in the structures business: "The first thing I hear about a new material is the best thing I ever hear about it. Then reality sinks in, and it's a matter of slow and steady improvements until you achieve the properties you want."[112] The visionaries of textile composite in the late

111. Ibid.

112. Alan S. Brown, "Material of the '90s?" *Aerospace America*, Jan. 1990, p. 28.

1980s could not foresee it, but they would contend with more than the normal challenges of introducing any technology for widespread production. A series of industry forces were about to transform the competitive landscape of the aerospace industry over the next decade, with a wave of mergers wreaking particular havoc on NASA's best-laid plans.

It was in this environment when NASA began the plunge into developing ever-more-advanced forms of composites. The timeframe came in the immediate aftermath of the ACEE program's demise. In 1988, the Agency launched an ambitious effort called the Advanced Composites Technology (ACT) program. It was aimed at developing hardware for composite wing and fuselage structures. The goals were to reduce structural weight for large commercial aircraft by 30–50 percent and reduce acquisition costs by 20–25 percent.[113] NASA awarded 15 contracts under the ACT banner a year later, signing up teams of large original equipment manufacturers, universities, and composite materials suppliers to work together to build an all-composite fuselage mated to an all-composite wing by the end of the century.[114]

During Phase A, from 1989 to 1991, the program focused on manufacturing technologies and structural concepts, with stitched textile preform and automated tow placement identified as the most promising new production methods.[115] "At that point in time, textile reinforced composites moved from being a laboratory curiosity to large scale aircraft hardware development," a NASA researcher noted.[116] Phase B, from 1992 to 1995, focused on testing subscale components.

Within the ACT banner, NASA sponsored projects of wide-ranging scope and significance. Sikorsky, for example, which was selected after 1991 to lead development and production of the RAH-66 Comanche, worked on a new process using flowable silicone powder to simplify the process of vacuum-bagging composites before being heated in an autoclave.[117] Meanwhile, McDonnell-Douglas Helicopter investigated 3-D

113. Joseph R. Chambers, *Concept to Reality: Contributions of the Langley Research Center to U.S. Civil Aircraft of the 1990s*, NASA SP-2005-4539 (Washington, DC: GPO, 2005), p. 80.

114. John G. Davis, "Overview of the ACT program," NASA Langley Research Center, NTIS Report N95-28463 (1995), p. 577.

115. Chambers, *Concept to Reality.*

116. Dexter, "Development of Textile Reinforced Composites for Aircraft Structures."

117. Alan Dobyns, "Aerospace 1991: The Year in Review—Structures," *Aerospace America*, Dec. 1991, p. 38.

finite element models to discover how combined loads create stresses through the thickness of composite parts during the design process.

The focus of ACT, however, would be aimed at developing the technologies that would finally commercialize composites for heavily loaded structures. The three major commercial airliner firms that dominated activity under the ACEE remained active in the new program despite huge changes in the commercial landscape.

Lockheed already had decided not to build any more commercial airliners after ceasing production of the L-1011 Tristar in 1984 but pursued ACT contracts to support a new strategy—also later dropped—to become a structures supplier for the commercial market.[118] Lockheed's role involved evaluating textile composite preforms for a wide variety of applications on aircraft.

It was still 8 years before Boeing and McDonnell-Douglas agreed to their fateful merge in 1997, but ACT set each on a path for developing new composites that would converge around the same time as their corporate identities. NASA set Douglas engineers to work on producing an all-composite wing. Part of Boeing's role under ACT involved constructing several massive components, such as a composite fuselage barrel; a window belt, introducing the complexity of material cutouts; and a full wing box, allowing a position to mate the Douglas wing and the Boeing fuselage. As ambitious as this roughly 10-year plan was, it did not overpromise. NASA did not intend to validate the airworthiness of the technologies. That role would be assigned to industry, as a private investment. Rather, the ACT program sought to merely prove that such structures could be built and that the materials were sound in their manufactured configuration. Thus, pressure tests would be performed on the completed structures to verify the analytical predictions of engineers.

Such aims presupposed some level of intense collaboration between the two future partners, Boeing and McDonnell-Douglas, but NASA may have been disappointed about the results before the merger of 1997. Although the former ACEE program had achieved a level of unique collaboration between the highly competitive commercial aircraft prime contractors, that spirit appeared to have eroded under the intense market pressures of the early 1990s airline industry. One unnamed industry source explained to an *Aerospace Daily* reporter in 1994: "Each company

118. Piellisch, "Materials Notebook: Weaving an Aircraft," *Aerospace America,* Feb. 1992, p. 54.

wants to do its own work. McDonnell doesn't want to put its [composite] wing on a Boeing [composite] fuselage and Boeing doesn't trust its composite fuselage mated to a McDonnell composite wing."[119]

NASA, facing funding shortages after 1993, ultimately scaled back the goal of ACT to mating an all-composite wing made by either McDonnell-Douglas or Boeing to an "advanced aluminum" fuselage section.[120] Boeing's work on completing an all-composite fuselage would continue, but it would transition to a private investment, leveraging the extensive experiences provided by the NASA and military composite development programs.

In 1995, McDonnell-Douglas was selected to enter Phase C of the ACT program with the goal to construct the all-composite wing, but industry developments intervened. After McDonnell-Douglas was absorbed into Boeing's brand, speculation swirled about the fate of the former's active all-composite wing program. In 1997, McDonnell-Douglas had plans to eventually incorporate the new wing technology on the legacy MD-90 narrow body.[121] (Boeing later renamed MD-90 by filling a gap created when the manufacturer skipped from the 707 to the 727 airliners, having internally designated the U.S. Air Force KC-135 refueler the 717.[122]) One postmerger speculative report suggested that Boeing might even consider adopting McDonnell-Douglas's all-composite wing for the Next Generation 737 or a future variant of the 757. Boeing, however, would eventually drop the all-composite wing concept, even closing 717 production in 2006.

The ACT program produced an impressive legacy of innovation. Amid the drive under ACT to finally build full-scale hardware, NASA also pushed industry to radically depart from building composite structures through the laborious process of laying up laminates. This process not only drove up costs by requiring exorbitant touch labor; it also produced material that was easy to damage without adding bulk—and weight—to the structure in the form of thicker laminates and extra stiffeners and doublers.

The ACT formed three teams that combined one major airframer each, with several firms that represented part of a growing and

119. "NASA Langley Reorienting Advanced Composites Technology Program."

120. Ibid.

121. Norris, "Boeing Studies Composite Primary-Wing Technology," *Flight International*, Aug. 27, 1997.

122. Swihart, "Commercial Jet Transportation Structures and Materials Evolution."

increasingly sophisticated network of composite materials suppliers to the aerospace industry. A Boeing/Hercules team focused on a promising new method called automated tow placement. McDonnell-Douglas was paired with Dow Chemical to develop a process that could stitch the fibers roughly into the shape of the finished parts, then introduce the resin matrix through the resin transfer molding (RTM) process.[123] That process is known as "stitched/RTM."[124] Lockheed, meanwhile, was tasked with BASF Structural Materials to work on textile preforms.

NASA and the ACT contractors had turned to textiles full bore to both reduce manufacturing costs and enhance performance. Preimpregnating fibers aligned unidirectionally into layers of laminate laid up by hand and cured in an autoclave had been the predominant production method throughout the 1980s. However, layers arranged in this manner have a tendency to delaminate when damaged.[125] The solution proposed under the ACT program was to develop a method to sew or weave the composites three-dimensionally roughly into their final configuration, then infuse the "preform" mold with resin through resin transfer molding or vacuum-assisted resin transfer molding.[126] It would require the invention of a giant sewing machine large and flexible enough to stitch a carbon fabric as large as an MD-90 wing.

McDonnell-Douglas began the process with the goal of building a wing stub box test article measuring 8 feet by 12 feet. Pathe Technologies, Inc., built a single-needle sewing machine. Its sewing head was computer controlled and could move by a gantry-type mechanism in the x- and y-axes to sew materials up to 1 inch in thickness. The machine stitched prefabricated stringers and intercostal clips to the wing skins.[127] The wings skins had been prestitched using a separate multineedle machine.[128] Both belonged to a first generation of sewing machines that accomplished their purpose, which was to provide valuable data and experience. The single-needle head, however, would prove far too limited. It moved only 90 degrees in the vertical and horizontal planes,

123. Piellisch, "Materials Notebook: Weaving an Aircraft," p. 54.
124. Davis, "Overview of the ACT program," p. 583.
125. Piellisch, "Materials Notebook: Weaving an Aircraft," p. 54.
126. Dexter, "Development of Textile Reinforced Composites for Aircraft Structures," p. 2.
127. M. Karal, "AST Composite Wing Program—Executive Summary," NASA CR-2001-210650 (2001).
128. Ibid.

The Advanced Composite Cargo Aircraft is a modified Dornier 328Jet aircraft. The fuselage aft of the crew station and the vertical tail were removed and replaced with new structural designs made of advanced composite materials fabricated using out-of-autoclave curing. It was developed by the Air Force Research Laboratory and Lockheed Martin. Lockheed Martin.

meaning it was limited to stitching only panels with a flat outer mold line. The machine also could not stitch materials deeply enough to meet the requirement for a full-scale wing.[129]

NASA and McDonnell-Douglas recognized that a high-speed multineedle machine, combined with an improved process for multiaxial warp knitting, would achieve affordable full-scale wing structures. This so-called advanced stitching machine would have to handle "cover panel preforms that were 3.0m wide by 15.2m long by 38.1mm thick at speeds up to 800 stitches per minute. The multiaxial warp knitting machine had to be capable of producing 2.5m wide carbon fabric with an areal weight of 1,425g/m^2."[130] Multiaxial warp knitting automates the process of producing multilayer broad goods. NASA and Boeing selected the resin film infusion (RFI) process to develop a wing cost-effectively.

Boeing's advanced stitching machine remains in use today, quietly producing landing gear doors for the C-17 airlifter. The thrust of

129. Ibid.

130. Chambers, *Concept to Reality*.

innovation in composite manufacturing technology, however, has shifted to other places. Lockheed's ACCA program spotlighted the emergence of a third generation of out-of-autoclave materials. Small civil aircraft had been fashioned out of previous generations of this type of material, but it was not nearly strong enough to support loads required for larger aircraft such as, of course, a 328Jet. In the future, manufacturers hope to build all-composite aircraft on a conventional production line, with localized ovens to cure specific parts. Parts or sections will no longer need to be diverted to cure several hours inside an autoclave to obtain their strength properties. Lockheed's move with the X-55 ACCA jet represents a critical first attempt, but others are likely to soon follow. For its part, Boeing has revealed two major leaps in composite technology development on the military side, from the revelation of the 1990s-era Bird of Prey demonstrator, which included a single-piece composite structure, to the co-bonded, all-composite wing section for the X-45C demonstrator (now revived and expected to resume flight-testing as the Phantom Ray).

The key features of new out-of-autoclave materials are measured by curing temperature and a statistic vital for determining crashworthiness called compression after impact strength. Third-generation resins now making an appearance in both Lockheed and Boeing demonstration programs represent major leaps in both categories. In terms of raw strength, Boeing states that third-generation materials can resist impact loads up to 25,000 pounds per square inch (psi), compared to 18,000 psi for the previous generation. That remains below the FAA standard for measuring crashworthiness of large commercial aircraft but may fit the standard for a new generation of military cargo aircraft that will eventually replace the C-130 and C-17 after 2020. In September 2009, the U.S. Air Force awarded Boeing a nearly $10-million contract to demonstrate such a nonautoclave manufacturing technology.

Toward the Future

NASA remains active in the pursuit of new materials that will support fresh objectives for enabling a step change in efficiency for commercial aircraft of the next few decades. A key element of NASA's strategy is to promote the transition from conventional, fuselage-and-wing designs for large commercial aircraft to flying wing designs, with the Boeing X-48 Blended Wing-Body subscale demonstrator as the model. The concept assumes many changes in current approaches to

NASA's Langley Research Center started experimenting with this stitching machine in the early 1990s. The machine stitches carbon, Kevlar, and fiberglass composite preforms before they are infused with plastic epoxy through the resin transfer molding process. The machine was limited to stitching only small and nearly flat panels. NASA.

flight controls, propulsion, and, indeed, expectations for the passenger experience. Among the many innovations to maximize efficiency, such flying wing airliners also must be supported by a radical new look at how composite materials are produced and incorporated in aircraft design.

To support the structural technology for the BWB, Boeing faces the challenge of manufacturing an aircraft with a flat bottom, no constant section, and a diversity of shapes across the outer mold line.[131] To meet these challenges, Boeing is returning to the stitching method, although with a different concept. Boeing's concept is called pultruded rod stitched efficient unitized structure (PRSEUS). *Aviation Week & Space Technology* described the idea: "This stitches the composite frames and stringers to the skin to produce a fail-safe structure. The frames and stringers provide continuous load paths and the nylon stitching stops cracks. The design allows the use of minimum-gauge-post-buckled-skins, and Boeing

131. Graham Warwick, "Shaping the Future," *Aviation Week & Space Technology*, Feb. 2, 2009, p. 50.

estimates a PRSEUS pressure vessel will be 28% lighter than a composite sandwich structure."[132]

Under a NASA contract, Boeing is building a 4-foot by 8-foot pressure box with multiple frames and a 30-foot-wide test article of the double-deck BWB airframe. The manufacturing process resembles past experience with the advanced stitching machine. Structure laid up by dry fabric is stitched before a machine pulls carbon fiber rods through pickets in the stringers. The process locks the structure and stringers into a preform without the need for a mold-line tool. The parts are cured in an oven, not an autoclave.[133]

The dream of designing a commercially viable, large transport aircraft made entirely out of plastic may finally soon be realized. The all-composite fuselage of the Boeing 787 and the proposed Airbus A350 are only the latest markers in progress toward this objective. But the next generation of both commercial and military transports will be the first to benefit from composite materials that may be produced and assembled nearly as efficiently as are aluminum and steel.

132. Ibid.
133. Ibid.

Recommended Additional Readings

Reports, Papers, Articles, and Presentations:

"Brandt Goldsworthy: Composites Visionary," *High Performance Composites*, May 1, 2003, *http://www.compositesworld.com/articles/brandt-goldsworthy-composites-visionary.aspx*, accessed Oct. 3, 2009.

Herman L. Bohon and John G. Davis, Jr., "Composites for Large Transports—Facing the Challenge," *Aerospace America*, June 1984.

Alan S. Brown, "Material of the '90s?" *Aerospace America*, Jan. 1990.

Frank W. Caldwell and N.S. Clay, "Micarta Propellers III: General Description of the Design," NACA Technical Note No. 200 (1924).

"Carbon Fibre Risk Analysis," NASA CP-2074 (1979).

"Carbon Fire Hazard Concerns NASA," *Aviation Week & Space Technology*, Mar. 5, 1979.

"Cargo X-Plane Shows Benefits of Advanced Composites," *Aviation Week & Space Technology*, June 8, 2009.

"Composite Programs Pushed by NASA," *Aviation Week & Space Technology*, Nov. 12, 1979.

John G. Davis, "Overview of the ACT program," NASA Langley Research Center, NTIS Report N95-28463 (1995).

Ravi B. Deo, James H. Starnes, Jr., and Richard C. Holzwarth, "Low-Cost Composite Materials and Structures for Aircraft Applications," presented at the *NATO Research and Technology Agency Applied Vehicle Technical Panel Specialists' Meeting on Low-Cost Composite Structures, Loen, Norway, May 7–8, 2001.*

H. Benson Dexter, "Development of Textile Reinforced Composites for Aircraft Structures," presented at the *4th International Symposium for Textile Composites in Kyoto, Japan, Oct. 12–14, 1998.*

Alan Dobyns, "Aerospace 1991: The Year in Review—Structures," *Aerospace America*, Dec. 1991.

Marvin B. Dow, "The ACEE Program and Basic Composites Research at Langley Research Centre (1975 to 1986): Summary and Bibliography," NASA RP-1177 (1987).

William F. Durand, "Some Outstanding Problems in Aeronautics," in NACA, *Fourth Annual Report* (Washington, DC: GPO, 1920).

"Energy Efficiency Funding Detailed," *Aviation Week & Space Technology*, Nov. 12, 1979.

Meyer Fishbein, "Physical Properties of Synthetic Resin Materials," NACA Technical Note No. 694 (1939).

William S. Friedman, "Flying Plywood With a Sting," *Popular Science*, No. 6, Dec. 1943, pp. 100–103.

Richard N. Hadcock, "X-29 Composite Wing," presented at the *AIAA Evolution of Aircraft/Aerospace Structures and Materials Symposium, Dayton, OH, Apr. 24–25, 1985.*

E.H. Hooper, "Starship 1," presented at the *AIAA Evolution of Aircraft/Aerospace Structures and Materials Symposium, Dayton, OH, Apr. 24–25, 1985.*

Robert L. James and Dal V. Maddalon, "The Drive for Aircraft Energy Efficiency," *Aerospace America*, Feb. 1984.

D.C. Jegley, H.G. Bush, and A.E. Lovejoy, "Structural Response and Failure of a Full-Scale Stitched Graphite-Epoxy Wing," AIAA Paper 2001-1334-CP (2001).

N.J. Johnston, T.W. Towell, J.M. Marchello, and R.W. Grenoble, "Automated Fabrication of High-Performance Composites: An Overview of Research at the Langley Research Center," presented at the *11th International Conference on Composite Materials, Gold Coast, Australia, July 1997.*

M. Karal, "AST Composite Wing Program—Executive Summary," NASA CR-2001-210650 (2001).

G.M. Kline, "Plastics as Structural Materials for Aircraft," NACA Technical Note No. 628 (1937).

Joseph J. Klumpp, "Parametric Cost Estimation Applied to Composite Helicopter Airframes" (Monterey, CA: Naval Postgraduate School master's thesis, 1994).

"Lightweight Composites Are Displacing Metals," *Business Week,* July 30, 1979.

"Lockheed Martin Conducts Successful Flight of AFRL's Advanced Composite Cargo Aircraft," Lockheed Martin Corporation press release, June 3, 2009.

Jay C. Lowndes, "Keeping a Sharp Technology Edge," *Aerospace America*, Feb. 1988.

"NASA Langley Reorienting Advanced Composites Technology Program," *Aerospace Daily*, Aug. 29, 1994.

Guy Norris, "Advanced Composite Cargo Aircraft Gets Green Light For Phase III," *Aerospace Daily & Defense Report,* Oct. 2, 2009.

Guy Norris, "Boeing Studies Composite Primary-Wing Technology," *Flight International*, Aug. 27, 1997.

Richard G. O'Lone, "Industry Tackles Composites Challenge," *Aviation Week & Space Technology*, Sept. 15, 1980, p. 22.

Lt. John M. Olson, USAF, "Mishap Risk Control for Advanced Aerospace/Composite Materials" (Wright-Patterson AFB: USAF Advanced Composites Program Office, 1995).

Richard Piellisch, "Composites Roll Sevens," *Aerospace America*, Oct. 1992.

Richard Piellisch, "Materials Notebook: Weaving an Aircraft," *Aerospace America,* Feb. 1992.

Richard A. Pride, "Composite Fibres and Composites," NASA CP-2074 (1979).

D.A. Reed and R. Gable, "Ground Shake Test of the Boeing Model 360 Helicopter Airframe," NASA CR-181766 (1989).

Allen M. Shibley, Adolph E. Slobodzinski, John Nardone, and Martin Cutler, "Special Report: Structural Plastics in Aircraft," Plastics Technical Evaluation Center, U.S. Army Picatinny Arsenal (Mar. 1965).

F.S. Snyder and R.E. Drake, "Experience with Reinforced Plastic Primary Aircraft Structures," presented at the *Society of Automotive Engineers' Automotive Engineering Congress in Detroit, MI, Jan. 14–18, 1963.*

James H. Starnes, Jr., H. Benson Dexter, Norman J. Johnston, Damodar R. Ambur, and Roberto J. Cano, "Composite Structures and Materials Research at NASA Langley Research Center," Paper MP-69-P-17, presented at the *NATO Research and Technology Agency Applied Vehicle Technical Panel Specialists' Meeting on Low-Cost Composite Structures, Loen, Norway, May 7–8, 2001.*

John M. Swihart, "Commercial Jet Transportation Structures and Materials Evolution," Apr. 1985, presented at *AIAA Evolution of Aircraft/Aerospace Structures and Materials Symposium, Dayton, OH, Apr. 24–25, 1985.*

"Towards an Ideal," *Flight,* Feb. 16, 1939.

Stephen Trimble, "Skunk Works nears flight for new breed of all-composite aircraft," *Flight International,* June 5, 2009.

"USAF Advanced Composite Cargo Aircraft Makes First Flight," U.S. Air Force Aeronautical Systems Center press release, June 3, 2009.

Arthur R. von Hippel and A.G.H. Dietz, "Curing of Resin-Wood Combinations By High-Frequency Heating," NACA Technical Note No. 874 (1938).

Graham Warwick, "Shaping the Future," *Aviation Week & Space Technology*, Feb. 2, 2009.

Mark T. Wright, et al., "Composite Materials in Aircraft Mishaps Involving Fire: A Literature Review," Naval Air Warfare Center Weapons Division TP-8552, June 2003.

Books and Monographs:

Joseph R. Chambers, *Concept to Reality: Contributions of the Langley Research Center to U.S. Civil Aircraft of the 1990s,* NASA SP-2005-4539 (Washington, DC: GPO, 2005).

Joseph R. Chambers, *Innovations in Flight: Research of the NASA Langley Flight Research Center on Revolutionary Concepts for Advanced Aeronautics*, NASA SP-2005-4539 (Washington, DC: GPO 2005), pp. 39–48.

Joseph R. Chambers, *Partners in Freedom: Contributions of the Langley Research Center to U.S. Military Aircraft of the 1990s,* NASA SP-2000-4519 (Washington, DC: GPO, 2000).

Phillippe Cognard, ed., *Adhesives and Sealants: Basic Concepts and High-Tech Bonding* (Amsterdam: Elsevier, 2005).

Phillippe Cognard, ed., *Handbook of Adhesives and Sealants*, v. 2: *General Knowledge, Application of Adhesives, New Curing Techniques* (Amsterdam: Elsevier, 2006).

Howard Mansfield, *Skylark: The Life, Lies and Inventions of Harry Atwood, (Lebanon, NH:* University Press of New England, 1999).

Robert L. O'Connell, *Soul of the Sword: An Illustrated History of Weaponry and Warfare from Prehistory to the Present* (New York: The Free Press, 2002).

Eric Schatzberg, *Wings of Wood, Wings of Metal: Culture and Technical Choice in American Airplane Materials 1914–1945* (Princeton: Princeton University Press, 1999).

Ian Thirsk, *de Havilland Mosquito: An Illustrated History*, v. 2 (Manchester, U.K.: Crecy Publishing, Ltd., 2006).

NASA Beech King Air general aviation aircraft over the Dryden Flight Research Center. NASA.

CASE 8

NACA-NASA's Contribution to General Aviation

By Weneth D. Painter

8

General Aviation has always been an essential element of American aeronautics. The NACA and NASA have contributed greatly to its efficiency, safety, and reliability via research across many technical disciplines. The mutually beneficial bonds linking research in civil and military aeronautics have resulted in such developments as the supercritical wing, electronic flight controls, turbofan propulsion, composite structures, and advanced displays and instrumentation systems.

THOUGH COMMONLY ASSOCIATED IN THE PUBLIC MIND with small private aircraft seen buzzing around local airports and air parks, the term "General Aviation" (hereafter GA) is primarily a definition of aircraft utilization rather than a classification per se of aircraft physical characteristics or performance. GA encompasses flying machines ranging from light personal aircraft to Mach 0.9+ business jets, comprising those elements of U.S. civil aviation which are neither certified nor supplemental air carriers: kit planes and other home-built aircraft, personal pleasure aircraft, commuter airlines, corporate air transports, aircraft manufacturers, unscheduled air taxi operations, and fixed-base operators and operations.

Overall, NACA-NASA's research has profoundly influenced all of this, contributing notably to the safety and efficiency of GA worldwide. Since the creation of the NACA in 1915, and continuing after establishment of NASA in 1958, Agency engineers have extensively investigated design concepts for GA, GA aircraft themselves, and the operating environment and related areas of inquiry affecting the GA community. In particular, they have made great contributions by documenting the results of various wind tunnel and flight tests of GA aircraft. These results have strengthened both industrial practice within the GA industry itself and the educational training of America's science, technology, engineering, and mathematics workforce, helping buttress and advance America's stature as an aerospace nation. This study discusses the advancements in GA through a review of selected applications of flight disciplines and aerospace technology.

8

The Early Evolution of General Aviation

The National Advisory Committee for Aeronautics (NACA) was formed on March 3, 1915, to provide advice and carry out much of cutting-edge research in aeronautics in the United States. This organization was modeled on the British Advisory Committee for Aeronautics. President Woodrow Wilson created the advisory committee in an effort to organize American aeronautical research and raise it to the level of European aviation. Its charter and $5,000 initial appropriation (low even in 1915) were appended to a naval appropriations bill and passed with little fanfare. The committee's mission was "to supervise and direct the scientific study of the problems of flight, with a view to their practical solution," and to "direct and conduct research and experiment in aeronautics."[1] Thus, from its outset, it was far more than simply a bureaucratic panel distanced from design-shop, laboratory, and flight line.

The NACA soon involved itself across the field of American aeronautics, advising the Government and industry on a wide range of issues including establishing the national air mail service, along with its night mail operations, and brokering a solution—the cross-licensing of aeronautics patents—to the enervating Wright-Curtiss patent feud that had hampered American aviation development in the pre-World War I era and that continued to do so even as American forces were fighting overseas. The NACA proposed establishing a Bureau of Aeronautics in the Commerce Department, granting funds to the Weather Bureau to promote safety in aerial navigation, licensing of pilots, aircraft inspections, and expanding airmail. It also made recommendations in 1925 to President Calvin Coolidge's Morrow Board that led to passage of the Air Commerce Act of 1926, the first Federal legislation regulating civil aeronautics. It continued to provide policy recommendations on the Nation's aviation until its incorporation in the National Aeronautics and Space Administration (NASA) in 1958.[2]

1. NACA enabling legislation, March 3, 1915; see George W. Gray, *Frontiers of Flight: The Story of NACA Research* (New York: Alfred A. Knopf, 1948), pp. 9–13.

2. Roger E. Bilstein, *The American Aerospace Industry* (New York: Twayne Publishers, 1996), pp. 14–15, 29–30.

The NACA started working in the field of GA almost as soon as it was established. Its first research airplane programs, undertaken primarily by F.H. Norton, involved studying the flight performance, stability and control, and handling qualities of Curtiss JN-4H, America's iconic "Jenny" of the "Great War" time period, and one that became first great American GA airplane as well.[3] The initial aerodynamic and performance studies of Dr. Max M. Munk, a towering figure in the history of fluid mechanics, profoundly influenced the Agency's subsequent approach to aerodynamic research. Munk, the inventor of the variable-density wind tunnel (which put NACA aerodynamics research at the forefront of the world standard) and architect of American aerodynamic research methodology, dramatically transformed the Agency's approach to airfoil design by introducing the methods of the "Prandtl school" at Göttingen and by designing and supervising the construction of a radical new form of wind tunnel, the so-called "variable density tunnel." His GA influence began with a detailed study of the airflow around and through a biplane wing cellule (the upper and lower wings, connected with struts and wires, considered as a single design element). He produced a report in which the variation of the section, chord, gap, stagger, and decalage (the angle of incidence of the respective chords of the upper and lower wings) and their influence upon the available wing cell space for engines, cockpits, passenger, and luggage, were investigated with a great number of calculated examples in which all of the numerical results were given in tables. Munk's report was in some respects a prototypical example of subsequent NACA-NASA research reports that, over the years, would prove beneficial to the development of GA by investigating a number of areas of particular concern, such as aircraft aerodynamic design, flight safety, spin prevention and recoveries, and handling qualities.[4] Arguably these reports that conveyed Agency research results to a public audience were the most influential product

3. Edward P. Warner and F.H. Norton, "Preliminary Report on Free Flight Tests," NACA TR-70 (1920); F.H. Norton and E.T. Allen, "Accelerations in Flight," NACA TR-99 (1921); F.H. Norton, "A Preliminary Study of Airplane Performance," NACA TN-120 (1922); F.H. Norton, "Practical Stability and Controllability of Airplanes," NACA TR-120 (1923); F.H. Norton and W.G. Brown, "Controllability and Maneuverability of Airplanes," NACA TR-153 (1923); F.H. Norton, "A Study of Longitudinal Dynamic Stability in Flight," NASA TR-170 (1924); F.H. Norton, "The Measurement of the Damping in Roll of a JN-4H in Flight," NACA TR-167 (1924).

4. Max M. Munk, "General Biplane Theory," NACA TR-151 (1922).

of NACA-NASA research. They influenced not only the practice of engineering within the various aircraft manufacturers, but provided the latest information incorporated in many aeronautical engineering textbooks used in engineering schools.

Though light aircraft are often seen as the by-product of the air transport revolution, in fact, they led, not followed, the expansion of commercial aviation, particularly in the United States. The interwar years saw an explosive growth in American aeronautics, particularly private flying and GA. It is fair to state that the roots of the American air transport revolution were nurtured by individual entrepreneurs manufacturing light aircraft and beginning air mail and air transport services, rather than (as in Europe) largely by "top-down" government direction. As early as 1923, American fixed-base operators "carried 80,888 passengers and 208,302 pounds of freight."[5] In 1926, there were a total of 41 private airplanes registered with the Federal Government. Just three years later, there were 1,454. The Depression severely curtailed private ownership, but although the number of private airplanes plummeted to 241 in 1932, it rose steadily thereafter to 1,473 in 1938, with Wichita, KS, emerging as the Nation's center of GA production, a distinction it still holds.[6]

Two of the many notable NACA-NASA engineers who were influenced by their exposure to Max Munk and had a special interest in GA, and who in turn greatly influenced subsequent aircraft design, were Fred E. Weick and Robert T. Jones. Weick arrived at NACA Langley Field, VA, in the 1920s after first working for the U.S. Navy's Bureau of Aeronautics.[7] Weick subsequently conceived the NACA cowling that became a feature of radial-piston-engine civil and military aircraft design. The cowling both improved the cooling of such engines and streamlined the engine installation, reducing drag and enabling aircraft to fly higher and faster.

5. Roger E. Bilstein, *Flight Patterns: Trends of Aeronautical Development in the United States, 1918–1929* (Athens: The University of Georgia Press, 1983), p. 63.

6. Donald M. Pattillo, *A History in the Making: 80 Turbulent Years in the American General Aviation Industry* (New York: McGraw-Hill, 1998), pp. 5–44; and Tom D. Crouch, "General Aviation: The Search for a Market, 1910–1976," in Eugene M. Emme, *Two Hundred Years of Flight in America: A Bicentennial Survey* (San Diego: American Astronautical Society and Univelt, 1977), Table 2, p. 129. For Wichita, see Jay M. Price and the AIAA Wichita Section, *Wichita's Legacy of Flight* (Charleston, SC: Arcadia Publishing, 2003).

7. Fred E. Weick and James R. Hansen, *From the Ground Up: The Autobiography of an Aeronautical Engineer* (Washington, DC: Smithsonian Institution Press, 1988).

This Curtiss AT-5A validated Weick's NACA Cowling. The cowling increased its speed by 19 miles per hour, equivalent to adding 83 horsepower. Afterwards it became a standard design feature on radial-engine airplanes worldwide. NASA.

In late fall of 1934, Robert T. Jones, then 23 years old, started a temporary, 9-month job at Langley as a scientific aide. He would remain with the Agency and NASA afterwards for the next half-century, being particularly known for having independently discovered the benefits of wing sweep for transonic and supersonic flight. Despite his youth, Jones already had greater mathematical ability than any other of his coworkers, who soon sought his expertise for various theoretical analyses. Jones was a former Capitol Hill elevator operator and had previously been a designer for the Nicholas Beazley Company in Marshall, MO. The Great Depression collapsed the company and forced him to seek other employment. His work as an elevator operator allowed him to hone his mathematical abilities gaining him the patronage of senior officials who arranged for his employment by the NACA.[8]

Jones and Weick formed a fruitful collaboration, exemplified by a joint report they prepared on the status of NACA lateral control research. Two things were considered of primary importance in judging the effectiveness of different control devices: the calculated banking and yawing motion of a typical small airplane caused by control deflection, and the stick force required to produce this control deflection. The report included a table in which a number of different lateral control devices

8. Jones's seminal paper was his "Properties of Low-Aspect-Ratio Pointed Wings at Speeds Below and Above the Speed of Sound," NACA TN-1032 (1946).

8

were compared.[9] Unlike Jones, Weick eventually left the NACA to continue his work in the GA field, producing a succession of designs emphasizing inherent stability and stall resistance. His research mirrored Federal interest in developing cheap, yet safe, GA aircraft, an effort that resulted in a well-publicized design competition by the Department of Commerce that was won by the innovative Stearman-Hammond Model Y of 1936. Weick had designed a contender himself, the W-1, and though he did not win, his continued research led him to soon develop one of the most distinctive and iconic "safe" aircraft of all time, his twin-fin and single-engine Ercoupe. It is perhaps a telling comment that Jones, one of aeronautics' most profound scientists, himself maintained and flew an Ercoupe into the 1980s.[10]

The Weick W-1 was an early example of attempting to build a cheap yet safe General Aviation airplane. NASA.

The NACA-NASA contributions to GA have come from research, development, test, and evaluation within the classic disciplines of aerodynamics, structures, propulsion, and controls but have also involved functional areas such as aircraft handling qualities and aircrew

9. Fred E. Weick and Robert T. Jones, "Response and Analysis of NACA. Lateral Control Research," TR 605 (1937).

10. Weick and Hansen, *From the Ground Up*, pp. 137–140. Jones kept his Ercoupe at Half Moon Bay Airport, CA; recollection of R.P. Hallion, who knew Jones.

Weick's Ercoupe is one of the most distinctive and classic General Aviation aircraft of all time. RPH.

performance, aviation safety, aviation meteorology, air traffic control, and education and training. The following are selected examples of such work, and how it has influenced and been adapted, applied, and exploited by the GA community.

Airfoil Evolution and Its Application to General Aviation

In the early 1930s, largely thanks to the work of Munk, the NACA had risen to world prominence in airfoil design, such status evident when, in 1933, the Agency released a report cataloging its airfoil research and presenting a definitive guide to the performance and characteristics of a wide range of airfoil shapes and concepts. Prepared by Eastman N. Jacobs, Kenneth E. Ward, and Robert M. Pinkerton, this document, TR-460, became a standard industry reference both in America and abroad.[11] The Agency, of course, continued its airfoil research in the 1930s, making notable advances in the development of high-speed airfoil sections and low-drag and laminar sections as well. By 1945, as valuable as TR-460 had been, it was now outdated. And so, one of the

11. Eastman N. Jacobs; Kenneth E. Ward; and Robert M. Pinkerton, "The Characteristics of 78 Related Airfoil Sections from Tests in the Variable-Density Wind Tunnel," NACA TR-460 (1933); see also Ira H. Abbott and Albert E. von Doenhoff, *Theory of Wing Sections, Including a Summary of Airfoil Data* (New York: McGraw-Hill, 1949), p. 112.

most useful of all NACA reports, and one that likewise became a standard reference for use by designers and other aeronautical engineers in airplane airfoil/wing design, was its effective replacement prepared in 1945 by Ira H. Abbott, Albert E. von Doenhoff, and Louis S. Stivers, Jr. This study, TR-824, was likewise effectively a catalog of NACA airfoil research, its authors noting (with justifiable pride) that

> Recent information of the aerodynamic characteristics of NACA airfoils is presented. The historical development of NACA airfoils is briefly reviewed. New data are presented that permit the rapid of the approximate pressure distribution for the older NACA four-digital and five-digit airfoils, by the same methods used for the NACA 6-series airfoils. The general methods used to derive the basic thickness forms for NACA 6 and 7 series airfoils together with their corresponding pressure distributions are presented. Detailed data necessary for the application of the airfoils to wing design are presented in supplementary figures placed at the end of the paper. This report includes an analysis of the lift, drag, pitching moment, and critical-speed characteristics of the airfoils, together with a discussion of the effects of surface conditions available data on high-lift devices. Problems associated with the later-control devices, leading edge air intakes, and interference is briefly discussed, together with aerodynamic problems of application.[12]

While much of this is best remembered because of its association with the advanced high-speed aircraft of the transonic and supersonic era, much was as well applicable to new, more capable civil transport and GA designs produced after the war.

Two key contributions to the jet-age expansion of GA were the supercritical wing and the wingtip winglet, both developments conceived by Richard Travis Whitcomb, a legendary NACA-NASA Langley aerodynamicist who was, overall, the finest aeronautical scientist of the post-Second World War era. More comfortable working in the wind tunnel

12. Ira H. Abbott, Albert E. von Doenhoff, and Louis S. Stivers, Jr., "Summary of Airfoil Data," NACA TR-824 (1945), p. 1.

than sitting at a desk, Whitcomb first gained fame by experimentally investigating the zero lift drag of wing-body combinations through the transonic flow regime based on analyses by W.D. Hayes.[13] His resulting "Area Rule" for transonic flow represented a significant contribution to the aerodynamics of high-speed aircraft, first manifested by its application to the so-called "Century series" of Air Force jet fighters.[14] Whitcomb followed area rule a decade later in the 1960s and derived the supercritical wing. It delayed the sharp drag rise associated with shock wave formation by having a flattened top with pronounced curvature towards its trailing edge. First tested on a modified T-2C jet trainer, and then on a modified transonic F-8 jet fighter, the supercritical wing proved in actual flight that Whitcomb's concept was sound. This distinctive profile would become a key design element for both jet transports and high-speed GA aircraft in the 1980s and 1990s, offering a beneficial combination of lower drag, better fuel economy, greater range, and higher cruise speed exemplified by its application on GA aircraft such as the Cessna Citation X, the world's first business jet to routinely fly faster than Mach 0.90.[15]

The application of Whitcomb's supercritical wing to General Aviation began with the GA community itself, whose representatives approached Whitcomb after a Langley briefing, enthusiastically endorsing his concept. In response, Whitcomb launched a new Langley program, the Low-and-Medium-Speed Airfoil Program, in 1972. This effort, blending 2-D computer analysis and tests in the Langley Low-Turbulence Pressure Tunnel, led to development of the GA(W)-1 airfoil.[16] The GA(W)-1 employed a

13. W.D. Hayes, "Linearized Supersonic Flow," North American Aviation, Inc., Report AL-222 (18 Jun. 1947).

14. Richard T. Whitcomb, "A Study of the Zero-Lift Drag-Rise Characteristics of Wing Body Combinations Near the Speed of Sound," NACA Report 1237 (1956).

15. Whitcomb's work is covered in detail in other essays in these volumes. For the technological climate at the time of his work, see Albert L. Braslow and Theodore G. Ayers, "Application of Advanced Aerodynamics to Future Transport Aircraft," in Donely et al., *NASA Aircraft Safety and Operating Problems*, v. 1; re the Citation X, see Mark O. Schlegel, "Citation X: Development and Certification of a Mach 0.9+ Business Jet," in Society of Experimental Test Pilots, *1997 Report to the Aerospace Profession* (Lancaster, CA: Society of Experimental Test Pilots, 1997), pp. 349–368.

16. Robert J. McGhee, William D. Beasley, and Richard T. Whitcomb, "NASA Low- and Medium-Speed Airfoil Development," in NASA Langley Research Center Staff, *Advanced Technology Airfoil Research*, v. 2, NASA CP-2046 (Washington, DC: NASA Scientific and Technical Information Office, 1979), pp. 1–23.

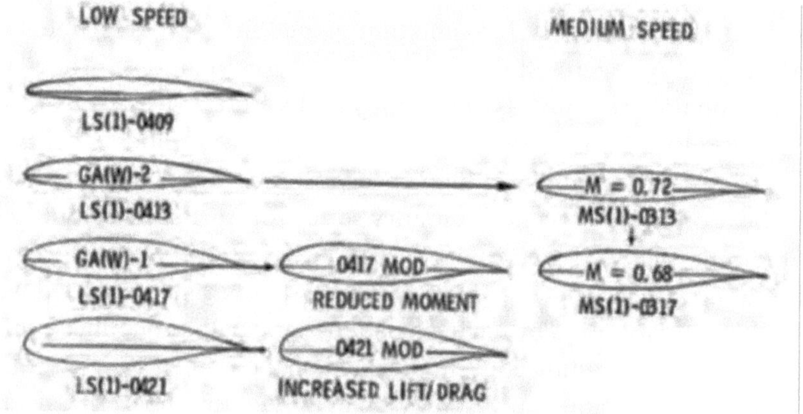

Low-and-Medium-Speed variants of the GA(W)-1 and -2 airfoil family. From NASA CP-2046 (1979).

17-percent-thickness-chord ratio low-speed airfoil, offering a beneficial mix of low cruise drag, high lift-to-drag ratios during climbs, high maximum lift properties, and docile stall behavior.[17] Whitcomb's team generated thinner and thicker variations of the GA(W)-1 that underwent its initial flight test validation in 1974 on NASA Langley's Advanced

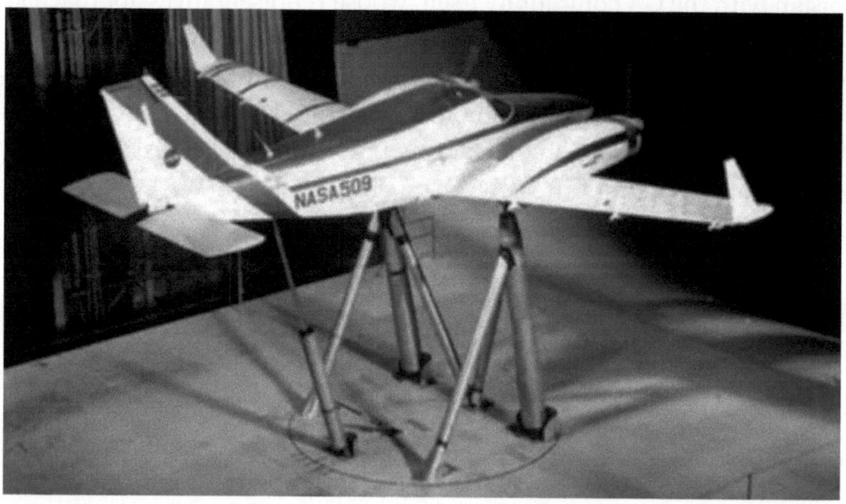

The Advanced Technology Light Twin-Engine airplane undergoing tests in the Langley 30 ft x 60 ft Full Scale Tunnel. NASA.

17. Joseph R. Chambers, *Concept to Reality: Contributions of the NASA Langley Research Center to U.S. Civil Aircraft of the 1990s*, NASA SP-2003-4529 (Washington, DC: GPO, 2003), pp. 22–25.

Technology Light Twin (ATLIT) engine airplane, a Piper PA-34 Seneca twin-engine aircraft modified to employ a high-aspect-ratio wing with a GA(W)-1 airfoil with winglets. Testing on ATLIT proved the practical advantages of the design, as did subsequent follow-on ground tests of the ATLIT in the Langley 30 ft x 60 ft Full-Scale-Tunnel.[18]

Subsequently, the NASA-sponsored General Aviation Airfoil Design and Analysis Center (GA/ADAC) at the Ohio State University, led by Dr. Gerald M. Gregorek, modified a single-engine Beech Sundowner light aircraft to undertake a further series of tests of a thinner variant, the GA(W)-2. GA/ADAC flight tests of the Sundowner from 1976–1977 confirmed that the Langley results were not merely fortuitous, paving the way for derivatives of the GA(W) family to be applied to a range of new aircraft designs starting with the Beech Skipper, the Piper Tomahawk, and the Rutan VariEze.[19]

Following on the derivation of the GA(W) family, NASA Langley researchers, in concert with industry and academic partners, continued refinement of airfoil development, exploring natural laminar flow (NLF) airfoils, previously largely restricted to exotic, smoothly finished sailplanes, but now possible thanks to the revolutionary development of smooth composite structures with easily manufactured complex shapes tailored to the specific aerodynamic needs of the aircraft under development.[20] Langley researchers subsequently blended their own conceptual and tunnel research with a computational design code developed at the University of Stuttgart to generate a new natural laminar flow

18. Bruce J. Holmes, "Flight Evaluation of an Advanced Technology Light Twin-Engine Airplane (ATLIT)," NASA CR-2832 (1977).

19. G.M. Gregorek and M.J. Hoffman, "An Investigation of the Aerodynamic Characteristics of a New General Aviation Airfoil in Flight," NASA CR-169477 (1982). For the GA/ADAC, see G.M. Gregorek, K.D. Korkan, and R.J. Freuler, "The General Aviation Airfoil Design and Analysis Service—A Progress Report," in NASA LRC Staff, *Advanced Technology Airfoil Research*, v. 2, NASA CP-2046, pp. 99–104.

20. Chambers, *Concept to Reality*, pp. 27–30; Roy V. Harris, Jr., and Jerry N. Hefner, "NASA Laminar-Flow Program—Past, Present, Future"; Bruce E. Peterman, "Laminar Flow: The Cessna Perspective"; J.K. Viken et al., "Design of the Low-Speed NLF (1)-0414F and the High-Speed HSNLF(1)-0213 Airfoils with High-Lift Systems"; Daniel G. Murri et al., "Wind Tunnel Results of the Low-Speed NLF(1)-0414F Airfoil"; and William G. Sewall et al., "Wind Tunnel Results of the High-Speed NLF(1)-0213 Airfoil," all in NASA Langley Research Center Staff, *Research in Natural Laminar Flow and Laminar-Flow Control*, Pts. 1-3, NASA CP-2487 (Hampton, VA: Langley Research Center, 1987).

airfoil section, the NLF(1).[21] Like the GA(W) before it, it served as the basis for various derivative sections. After flight testing on various test-beds, it was transitioned into mainstream GA design beginning with a derivative of the Cessna Citation II in 1990. Thereafter, it has become a standard feature of many subsequent aircraft.[22]

8

The second Whitcomb-rooted development that offered great promise in the 1970s was the so-called winglet.[23] The winglet promised to dramatically reduce energy consumption and reduce drag by minimizing the wasteful tip losses caused by vortex flow off the wingtip of the aircraft. Though reminiscent of tip plates, which had long been tried over the years without much success, the winglet was a more refined and

The Gates Learjet 28 Longhorn, which pioneered the application of Whitcomb winglets to a General Aviation aircraft. NASA.

21. Richard Eppler and Dan M. Somers, "A Computer Program for the Design and Analysis of Low-Speed Airfoils," NASA TM-80210 (1980); B.J. Holmes, C.C. Cronin, E.C. Hastings, Jr., C.J. Obara, and C.P. Vandam, "Flight Research on Natural Laminar Flow," NTRS ID 88N14950 (1986).
22. Michael S. Selig, Mark D. Maughmer, and Dan M. Somers, "An Airfoil for General Aviation Applications," in AIAA, *Proceedings of the 1990 AIAA/FAA Joint Symposium on General Aviation Systems* (Hampton, VA: NASA LRC, 1990), pp. 280–291, NTRS ID N91-12572 (1990); and Wayne A. Doty, "Flight Test Investigation of Certification Issues Pertaining to General-Aviation-Type Aircraft With Natural Laminar Flow," NASA CR-181967 (1990).
23. Richard T. Whitcomb, "A Design Approach and Selected High-Speed Wind Tunnel Results at High Subsonic Speeds for Wing-Tip Mounted Winglets," NASA TN D-8260 (1976); and Stuart G. Flechner, Peter F. Jacobs, and Richard T. Whitcomb, "A High Subsonic Speed Wind-Tunnel Investigation of Winglets on a Representative Second-Generation Jet Transport Wing," NASA TN D-8264 (1976). See also NASA Dryden Flight Research Center, "Winglets," TF-2004-15 (2004).

better-thought-out concept, which could actually take advantage of the strong flow-field at the wingtip to generate a small forward lift component, much as a sail does. Primarily, however, it altered the span-wise distribution of circulation along the wing, reducing the magnitude and energy of the trailing tip vortex. First to use it was the Gates Learjet Model 28, aptly named the "Longhorn," which completed its first flight in August 1977. The Longhorn had 6 to 8 percent better range than previous Lears.[24]

The winglet was experimentally verified for large aircraft application by being mounted on the wing tips of a first-generation jet transport, the Boeing KC-135 Stratotanker, progenitor of the civil 707 jetliner, and tested at Dryden from 1979–1980. The winglets, designed with a general-purpose airfoil that retained the same airfoil cross-section from root to tip, could be adjusted to seven different cant and incidence angles to enable a variety of research options and configurations. Tests revealed the winglets increased the KC-135's range by 6.5 percent—a measure of both aerodynamic and fuel efficiency—better than the 6 percent projected by Langley wind tunnel studies and consistent with results obtained with the Learjet Longhorn. With this experience in hand, the winglet was swiftly applied to GA aircraft and airliners, and today, most airliners, and many GA aircraft, use them.[25]

24. See Neil A. Armstrong and Peter T. Reynolds, "The Learjet Longhorn Series: The First Jets with Winglets," in Society of Experimental Test Pilots, *1978 Report to the Aerospace Profession* (Lancaster: SETP, 1978), pp. 57–66.

25. Richard T. Whitcomb, "A High Subsonic Speed Wind-Tunnel Investigation of Winglets on a Representative Second-Generation Jet Transport Wing," NASA TN D-8264 (1976); and NASA Dryden Flight Research Center, "Winglets," *NASA Technology Facts*, TF 2004-15 (2004), pp. 1–4. Another interesting project in this time period was the NASA AD-1 Oblique Wing, whose flight test was conducted at Dryden. The oblique wing concept originated with Ames's Robert T. Jones. The NASA Project Engineer was Weneth "Wen" Painter and the Project Pilot was Tom McMurtry. The team successfully demonstrated an aircraft wing could be pivoted obliquely from 0 to 60 degrees during flight. The aircraft was flown 79 times during the research program, which evaluated the basic pivot-wing concept and gathered information on handling qualities and aerodynamics at various speeds and degrees of pivot. The supersonic concept would have been design with a more complex control system, such as fly-by-wire. The AD-1 aircraft was flown by 19 pilots: 2 USAF pilots; 2 Navy pilots; and 15 NASA Dryden, Langley, and Ames research pilots. The final flights of the AD-1 occurred at the 1982 Experimental Aircraft Association's (EAA) annual exhibition at Oshkosh, WI, where it flew eight times to demonstrate it unique configuration, a swan song watched over by Jones and his old colleague Weick.

8

The Propulsion Perspective

Aerodynamics always constituted an important facet of NACA-NASA GA research, but no less significant is flight propulsion, for the aircraft engine is often termed the "heart" of an airplane. In the 1920s and 1930s, NACA research by Fred Weick, Eastman Jacobs, John Stack, and others had profoundly influenced the efficiency of the piston engine-propeller-cowling combination.[26] Agency work in the early jet age had been no less influential upon improving the performance of turbojet, turboshaft, and turbofan engines, producing data judged "essential to industry designers."[27]

The rapid proliferation of turbofan-powered GA aircraft—over 2,100 of which were in service by 1978, with 250 more being added each year—stimulated even greater attention.[28] NASA swiftly supported development of a specialized computer-based program for assessing engine performance and efficiency. In 1977, for example, Ames Research Center funded development of GASP, the General Aviation Synthesis Program, by the Aerophysics Research Corporation, to compute propulsion system performance for engine sizing and studies of overall aircraft performance. GASP consisted of an overall program routine, ENGSZ, to determine appropriate fanjet engine size, with specialized subroutines such as ENGDT and NACDG assessing engine data and nacelle drag. Additional subroutines treated performance for propeller powerplants, including PWEPLT for piston engines, TURBEG for turboprops, ENGDAT and PERFM for propeller characteristics and performance, GEARBX for gearbox cost and weight, and PNOYS for propeller and engine noise.[29]

Such study efforts reflected the increasing numbers of noisy turbine-powered aircraft operating into over 14,500 airports and airfields

26. For engine-and-cowling, see James R. Hansen, "Engineering Science and the Development of the NACA Low-Drag Engine Cowling," in Pamela E. Mack, ed., *From Engineering Science to Big Science: The NACA and NASA Collier Trophy Research Project Winners*, NASA SP-4219 (Washington, DC: NASA, 1998), pp. 1–27; for propellers, see John V. Becker, *The High-Speed Frontier: Case Histories of Four NACA Programs, 1920–1950*, NASA SP-445 (Washington, DC: NASA Scientific and Technical Information Branch, 1980), pp. 119–138.

27. Virginia P. Dawson, *Engines and Innovation: Lewis Laboratory and American Propulsion Technology*, NASA SP-4306 (Washington: NASA, 1991), p. 140.

28. Gilbert K. Sievers, "Overview of NASA QCGAT Program," in NASA Lewis Research Center Staff, *General Aviation Propulsion*, NASA CP-2126 (Cleveland, OH: NASA Lewis Research Center, 1980), p. 1; and Pattillo, *A History in the Making*, pp. 122–126.

29. Aerophysics Research Corporation, "GASP—General Aviation Synthesis Program," NASA CR-152303 (1978).

in the United States, most in suburban areas, as well as the growing cost of aviation fuel and the consequent quest for greater engine efficiency. NASA had long been interested in reducing jet engine noise, and the Agency's first efforts to find means of suppressing jet noise dated to the late NACA in 1957. The needs of the space program had necessarily focused Lewis research primarily on space, but it returned vigorously to air-breathing propulsion at the conclusion of the Apollo program, spurred by the widespread introduction of turbofan engines for military and civil purposes and the onset of the first oil crisis in the wake of the 1973 Arab-Israeli War.

Out of this came a variety of cooperative research efforts and programs, including the congressionally mandated ACEE program (for Aircraft Engine Efficiency, launched in 1975), the NASA-industry QCSEE (for Quiet Clean STOL Experimental Engine) study effort, and the QCGAT (Quiet Clean General Aviation Turbofan) program. All benefited future propulsion studies, the latter two particularly so.[30]

QCGAT, launched in 1975, involved awarding initial study contracts to Garrett AiResearch, General Electric, and Avco Lycoming to explore applying large turbofan technology to GA needs. Next, AiResearch and Avco were selected to build a small turbofan demonstrator engine suitable for GA applications that could meet stringent noise, emissions, and fuel consumption standards using an existing gas-generating engine core. AiResearch and Avco took different approaches, the former with a high-thrust engine suitable for long-range high-speed and high altitude GA aircraft (using as a baseline a stretched Lear 35), and the latter with a lower-thrust engine for a lower, slower, intermediate-range design (based upon a Cessna Citation I). Subsequent testing indicated that each company did an excellent job in meeting the QCGAT program goals, each having various strengths. The Avco engine was quieter, and both engines bettered the QCQAT emissions goals for carbon monoxide and unburned hydrocarbons. While the Avco engine was "right at the goal" for nitrous oxide emissions, the AiResearch engine was higher, though much better than the baseline TFE-731-2 turbofan used for comparative purposes. While the AiResearch engine met sea-level takeoff and design cruise thrust goals, the Avco engine missed

30. And are treated in other case studies. For Lewis and NASA aero-propulsion work in this period, see Dawson, *Engines and Innovation*, pp. 203–205; and Jeffrey L. Ethell, *Fuel Economy in Aviation*, NASA SP-462 (NASA Scientific and Technical Information Branch, 1983), passim.

both, though its measured numbers were nevertheless "quite respectable." Overall, NASA considered that the QCGAT program, executed on schedule and within budget, constituted "a very successful NASA joint effort with industry," concluding that it had "demonstrated that noise need not be a major constraint on the future growth of the GA turbofan fleet."[31] Subsequently, NASA launched GATE (General Aviation Turbine Engines) to explore other opportunities for the application of small turbine technology to GA, awarding study contracts to AiResearch, Detroit Diesel Allison, Teledyne CAE, and Williams Research.[32] GA propulsion study efforts gained renewed impetus through the Advanced General Aviation Transport Experiment (AGATE) program launched in 1994, which is discussed later in this study.

Understanding GA Aircraft Behavior and Handling Qualities

As noted earlier, the NACA research on aircraft performance began at the onset of the Agency. The steady progression of aircraft technology was matched by an equivalent progression in the understanding and comprehension of aircraft motions, beginning with extensive studies of the loads, stability, control, and handling qualities fighter biplanes encountered during steady and maneuvering flight.[33] At the end of the interwar period, NACA Langley researchers undertook a major evaluation of the flying qualities of American GA aircraft, though the results of that investigation were not disseminated because of the outbreak of the Second World War and the need for the Agency to focus its attention on military, not civil, needs. Langley test pilots flew five representative aircraft, and the test results, on the whole, were generally satisfactory. Control effectiveness was, on the overall, good, and the aircraft demonstrated a desirable degree of longitudinal (pitch) inherent stability, though two of the designs had degraded longitudinal stability at low speeds. Lateral

31. Gilbert K. Sievers, "Summary of NASA QCGAT Program," in NASA Lewis RC, *General Aviation Propulsion,* NASA CP-2126, pp. 189–190; see also his "Overview of NASA QCGAT Program" in the same volume, pp. 2–4.

32. See William C. Strack, "New Opportunities for Future, Small, General-Aviation Turbine Engines (GATE)," in NASA Lewis RC, *General Aviation Propulsion,* NASA CP-2126, pp. 195–197.

33. For example, James H. Doolittle, "Accelerations in Flight," NACA TR-203 (1925); Richard V. Rhode, "The Pressure Distribution Over the Horizontal and Vertical Tail Surfaces of the F6C-4 Pursuit Airplane in Violent Maneuvers," NACA TR-307 (1929); and Richard V. Rhode, "The Pressure Distribution Over the Wings and Tail Surfaces of a PW-9 Pursuit Airplane in Flight," NACA TR-364 (1931).

(roll) stability was likewise satisfactory, but "wide variations" were found in directional stability, though rudder inputs on each were sufficient to trim the aircraft for straight flight. Stall warning (exemplified by progressively more violent airframe buffeting) was good, and each aircraft possessed adequate stall recovery behavior, though departures from controlled flight during stalls in turns proved more violent (the airplane rolling in the direction of the downward wing) than stalls made from wings-level flight. In all cases, aileron power was inadequate to maintain lateral control. Stall recovery was "easily made" in every case simply by pushing forward on the elevator. Overall, if some performance deficiencies existed—for example, the tendency to spiral instability or the lack of lateral control effectiveness at the staff—such limitations were small compared with the dramatic handling qualities deficiencies of many early aircraft just two decades previously, at the end of the First World War. This survey demonstrated that by 1940 America had mastered the design of the practical, useful GA airplane. Indeed, such aircraft, built by the thousands, would play a critical role in initiating many young Americans into wartime service as combat and combat support pilots.[34]

The Aeronca Super Chief shown here was evaluated at Langley as part of a prewar survey of General Aviation aircraft handling and flying qualities. NASA.

During the Second World War, the NACA generated a new series of so-called Wartime Reports, complementing its prewar series of Technical

34. Paul A. Hunter, "Flight Measurements of the Flying Qualities of Five Light Airplanes," NACA TN-1573 (1948), pp. 1–2, 8–9, 19–20.

Reports (TR), Technical Memoranda (TM), and Technical Notes (TN). They subsequently had great influence upon aircraft design and engineering practice, particularly after the war, when applied to high-performance GA aircraft. The NACA studied various ways to improve aircraft performance through drag reduction of single-engine military fighter type aircraft and other designs resulting in improved handling qualities and increased airspeeds. The first Wartime Report was published in October 1940 by NACA engineers C.H. Dearborn and Abe Silverstein. This report described the test results that investigated methods for increasing the high speed for 11 single-engine military aircraft for the Army Air Corps. Their tests found inefficient design features on many of these airplanes indicating the desirability of analyzing and combining all of the results into a single paper for distribution to the designers. It highlighted one of the major problems afflicting aircraft design and performance analysis: understanding the interrelationship of design, performance, and handling qualities.[35]

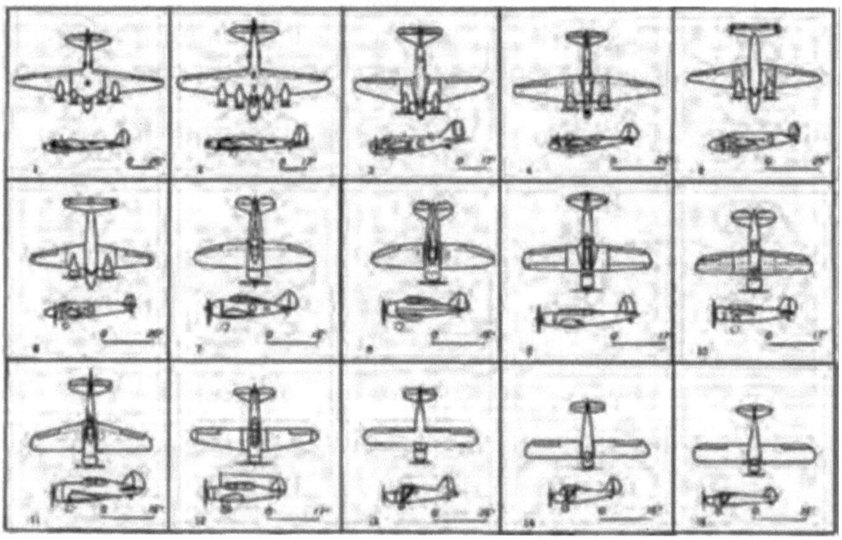

The fifteen different types of aircraft evaluated as part of a landmark study on longitudinal stability represented various configurations and design layouts, both single and multiengine, and from light general aviation designs to experimental heavy bombers. From NACA TR-711 (1941).

The NACA had long recognized "the need for quantitative design criterions for describing those qualities of an airplane that make up satis-

35. C.H. Dearborn and Abe Silverstein, "Drag Analysis of Single-Engine Military Airplanes Tested in the NACA Full-Scale Wind Tunnel," NACA WR-489 (1940).

factory controllability, stability, and handling characteristics," and the individual who, more than any other, spurred Agency development of them was Robert R. Gilruth, later a towering figure in the development of America's manned spaceflight program.[36] Gilruth's work built upon earlier preliminary efforts by two fellow Langley researchers, Hartley A. Soulé (later chairman of the NACA Research Airplane Projects Panel that oversaw the postwar X-series transonic and supersonic research airplane programs) and chief Agency test pilot Melvin N. "Mel" Gough, though it went considerably beyond.[37] In 1941, Gilruth and M.D. White assessed the longitudinal stability characteristics of 15 different airplanes (including bombers, fighters, transports, trainers, and GA sport aircraft).[38] Gilruth followed this with another study, in partnership with W.N. Turner, on the lateral control required for satisfactory flying qualities, again based on flight tests of numerous airplanes.[39] Gilruth capped his research with a landmark report establishing the requirements for satisfactory handling qualities in airplanes, issued first as an Advanced Confidential Report in April 1941, then as a Wartime Report, and, finally, in 1943, as one of the Agency's Technical Reports, TR-755. Based on "real-world" flight-test results, TR-755 defined what measured characteristics were significant in the definition of satisfactory flying qualities, what were reasonable to require from an airplane (and thus to establish as design requirements), and what influence various design features had upon the flying qualities of the aircraft once it entered flight testing.[40] Together, this trio profoundly influenced the field of flying qualities assessment.

But what was equally needed was a means of establishing a standard measure for pilot assessment of aircraft handling qualities.

36. R.R. Gilruth, "Requirements for Satisfactory Flying Qualities of Airplanes," NACA TR-755 (1943) [previously issued as Wartime Report L-276 in 1941—ed.], p. 49.

37. For Soulé, see his "Flight Measurements of the Dynamic Longitudinal Stability of Several Airplanes and a Correlation of the Measurements with Pilots' Observations of Handling Characteristics," NACA TR-578 (1936); and "Preliminary Investigation of the Flying Qualities of Airplanes," TR-700 (1940). For Gough, see his note, with A.P. Beard, "Limitations of the Pilot in Applying Forces to Airplane Controls," NACA TN-550 (1936).

38. R.R. Gilruth and M.D. White, "Analysis and Prediction of Longitudinal Stability of Airplanes," NACA TR-711 (1941).

39. R.R. Gilruth and W.N. Turner, "Lateral Control Required For Satisfactory Flying Qualities Based on Flight Tests of Numerous Airplanes," NACA TR-715 (1941).

40. R.R. Gilruth, "Requirements for Satisfactory Flying Qualities of Airplanes," NACA TR-755 (1943).

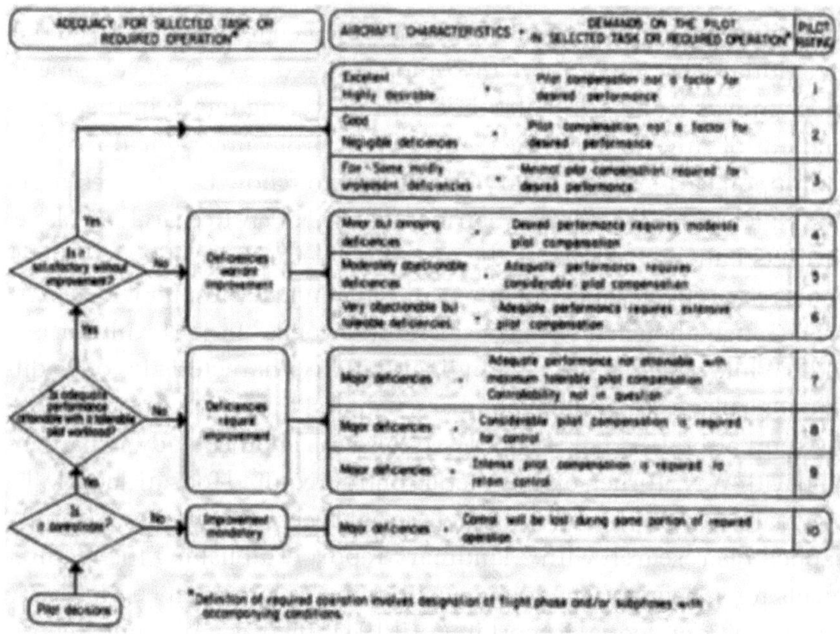

The Cooper-Harper Rating Scale. From NASA TN D-5153 (1969).

This proved surprisingly difficult to achieve and took a number of years of effort. Indeed, developing such measures took on such urgency and constituted such a clear requirement that it was one of the compelling reasons underlying the establishment of professional test pilot training schools, beginning with Britain's Empire Test Pilots' School established in 1943.[41] The measure was finally derived by two American test pilots, NASA's George Cooper and the Cornell Aeronautical Laboratory's Robert Harper, Jr., thereby establishing one of the essential tools of flight testing and flight research, the Cooper-Harper rating scale, issued in 1969 in a seminal report.[42] This evaluation tool quickly replaced earlier scales and measures and won international acceptance, influencing the flight-test evaluation of virtually all flying craft, from light GA aircraft through hypersonic lifting reentry vehicles and rotorcraft. The combination of the work undertaken by Gilruth, Cooper, and

41. Alan H. Wheeler, ". . . *That Nothing Failed Them:" Testing Aeroplanes in Wartime* (London: G.T. Foulis and Co., Ltd., 1963), pp. 2–3. Wheeler, a distinguished British test pilot, notably relates some of the problems caused the lack of established standards of measurement in this fine memoir.
42. George E. Cooper and Robert P. Harper, Jr., "The Use Of Pilot Rating In The Evaluation Of Aircraft Handling Qualities," NASA TN D 5153 (1969).

their associates dramatically improved flight safety and flight efficiency, and must therefore be considered one of the NACA-NASA's major contributions to aviation.[43]

The Cessna C-190 shown here was evaluated at Langley as part of an early postwar assessment of General Aviation aircraft performance. NASA.

Despite the demands of wartime research, the NACA and its research staff continued to maintain a keen interest in the GA field, particularly as expectations (subsequently frustrated by postwar economics) anticipated massive sales of GA aircraft as soon as conflict ended. While this was true in 1946—when 35,000 were sold in a single year!—the postwar market swiftly contracted by half, and then fell again, to just 3,000 in 1952, a "boom-bust" cycle the field would, alas, all too frequently repeat over the next half-century.[44] Despite this, hundreds of NACA general-aviation-focused reports, notes, and memoranda were produced—many reflecting flight tests of new and interesting GA designs—but, as well, some already-classic machines such as the Douglas DC-3, which underwent a flying qualities evaluation at Langley in 1950 as an exercise to calculate its stability derivatives, and, as well, update and refine the then-existing Air Force and Navy handling qualities specifications guidebooks. Not surprisingly, the project pilot concluded, "the DC-3

43. For background of this rating, see George Cooper, Robert Harper, and Roy Martin, "Pilot Rating Scales," in Society of Experimental Test Pilots, *2004 Report to the Aerospace Profession* (Lancaster, CA: Society of Experimental Test Pilots, 2004), pp. 319–337.
44. Crouch, "General Aviation: The Search for a Market," p. 126.

is a very comfortable airplane to fly through all normal flight regimes, despite fairly high control forces about all three axes."[45]

On October 4, 1957, Sputnik rocketed into orbit, heralding the onset of the "Space Age" and the consequent transformation of the NACA into the National Aeronautics and Space Administration (NASA). But despite the new national focus on space, NASA maintained a broad program of aeronautical research—the lasting legacy of the NACA—even in the shadow of Apollo and the Kennedy-mandated drive to Tranquility Base.

The Beech Debonair, one of many General Aviation aircraft types evaluated at the NASA Flight Research Center (now the NASA Dryden Flight Research Center). NASA.

This included, in particular, the field of GA flying and handling qualities. The first report written in 1960 under NASA presented the status of spin research—a traditional area of concern, particularly as it was a killer of low-flying-time pilots—from recent airplane design as interpreted at the NASA Langley Research Center, Langley, VA.[46] Sporadically, NASA researchers flight-tested new GA designs to assess their handling qualities, performance, and flight safety, their flight test reports frankly

45. John A. Harper, "DC-3 Handling Qualities Flight Tests: NACA—1950," in Society of Experimental Test Pilots, *1991 Report to the Aerospace Profession* (Lancaster, CA: SETP, 1991), pp. 264–265; also Arthur Assadourian and John A. Harper, "Determination of the Flying Qualities of the Douglas DC-3 Airplane," NACA TN-3088 (1953).

46. Anshal L. Neilhous, Walter L. Klinar, and Stanley H. Scher, "Status of Spin Research for Recent Airplane Designs," NASA TR-R-57 (1960).

detailing both strengths and deficiencies. In December 1964, for example, NASA Flight Research Center test pilot William Dana (one of the Agency's X-15 pilots) evaluated a Beech Debonair, a conventional-tailed derivative of the V-tail Beech Bonanza. Dana found the sleek Debonair a satisfactory aircraft overall. It had excellent longitudinal, spiral, and speed stability, with good roll damping and "honest" stall behavior in "clean" (landing gear retracted) configuration. But he faulted it for lack of rudder trim that hurt its climb performance, lack of "much warning, either by stick or airframe buffet" of impending stalls, and poor gear-down stall performance manifested by an abrupt left wing drop that hindered recovery. Finally, the plane's tendency to promote pilot-induced oscillations (PIO) during its landing flare earned it a pilot-rating grade of "C" for landings.[47]

The growing recognition that GA technology had advanced far beyond the state of GA that had existed at the time of the NACA's first qualitative examination of light aircraft handling qualities triggered one of the most significant of NASA's GA assessment programs. In 1966, at the height of the Apollo program, pilots and engineers at the Flight Research Center performed an evaluation of the handling qualities of seven GA aircraft, expanding upon this study subsequently to include the handling qualities of other light aircraft and advanced control systems and displays. The aircraft for the 1966 study were a mix of popular single-and twin-engine, high-and low-wing types. Project pilot was Fred W. Haise (subsequently an Apollo 13 astronaut); Marvin R. Barber, Charles K. Jones, and Thomas R. Sisk were project engineers.[48]

As a group, the seven aircraft all exhibited generally satisfactory stability and control characteristics. However, these characteristics, as researchers noted,

> Degraded with decreasing airspeed, increasing aft center of gravity, increasing power, and extension of gear and flaps.

47. William H. Dana, "Pilot's Flight Notes, Test of Debonair #430T," 30 Dec. 1964, in file L1-8-2A-13 "Beech Model 33 Debonair" file, NASA Dryden Flight Research Center Archives, Edwards, CA. The "C" rating, it might be noted, reflected pilot rating standards at that time, prior to the issuance and widespread adaptation of the Cooper-Harper rating.
48. Marvin R. Barber, Charles K. Jones, and Thomas R. Sisk, "An Evaluation Of The Handling Qualities of Seven General-Aviation Aircraft ," NASA TN D 3726 (1966).

8

> The qualitative portion of the program showed the handling qualities were generally satisfactory during visual and instrument flight in smooth air. However, atmosphere turbulence degraded these handling qualities, with the greatest degradation noted during instrument landing system approaches. Such factors as excessive control-system friction, low levels of static stability, high adverse yaw, poor Dutch roll characteristics, and control-surface float combined to make precise instrument tracking tasks, in the present of turbulence difficult even for experienced instrument pilots.
>
> The program revealed three characteristics of specific airplanes that were considered unacceptable if encountered by inexperienced or unsuspecting pilots: (1) A violent elevator force reversal or reduced load factors in the landing configuration, (2) power-on stall characteristics that culminate in rapid roll offs and/or spins, and (3) neutral-to-unstable static longitudinal stability at aft center gravity.
>
> A review indicated that existing criteria had not kept pace with aircraft development in areas of Dutch roll, adverse yaw, effective dihedral, and allowable trim changes with gear, flap and power. This study indicated that criteria should be specified for control-system friction and control-surface float.
>
> This program suggested a method of quantitative evaluating and handling qualities of aircraft by the use of pilot-work-load factor.[49]

As well, all of the aircraft tested had "undesirable and inconsistent placement of both primary flight instruments and navigational displays," increasing pilot workload, a matter of critical concern during precision instrument landing approaches.[50] Further, they all lacked good

49. Barber et al., "An Evaluation of the Handling Qualities of Seven General-Aviation Aircraft," p. 1.
50. Barber et al., "An Evaluation of the Handling Qualities of Seven General-Aviation Aircraft," p. 16.

stall warning (defined as progressively strong airframe buffet prior to stall onset). Two had "unacceptable" stall characteristics, one entering an "uncontrollable" left roll/yaw and altitude-consuming spin, and the other having "a rapid left rolloff in the power-on accelerated stall with landing flaps extended."[51]

The 1966 survey stimulated more frequent evaluations of GA designs by NASA research pilots and engineers, both out of curiosity and sometimes after accounts surfaced of marginal or questionable behavior. NASA test pilots and engineers found that while various GA designs had "generally satisfactory" handling qualities for flight in smooth air and under visual conditions, they had far different qualities in turbulent flight and with degraded visibility. Control system friction, longitudinal and spiral instability, adverse yaw, combined lateral-directional "Dutch roll" characteristics, abrupt trim changes when deploying landing gear flaps, and adding or subtracting power all inhibited effective precision instrument tracking. Thus, instrument landing approaches quickly taxed a pilot, markedly increasing pilot workload. The FRC team explored applying advanced control systems and displays, modifying a light twin-engine

The workhorse Piper PA-30 on final approach for a lakebed landing at the Dryden Flight Research Center. NASA.

Piper PA-30 Twin Comanche business aircraft as a GA testbed with a flight-director display and an attitude-command control system. The

51. Barber et al., "An Evaluation of the Handling Qualities of Seven General-Aviation Aircraft," p. 18.

result, demonstrated in 72 flight tests and over 120 hours of operation, was "a flying machine that borders on being perfect from a handling qualities standpoint during ILS approaches in turbulent air." The team presented their findings at a seminal NASA conference on aircraft safety and operating problems held at the Langley Research Center in May 1971.[52]

The little PA-30 proved a workhorse, employed for a variety of research studies including exploring remotely piloted vehicle technology.[53] During the time period of 1969–1972, NASA researchers Chester Wolowicz and Roxanah Yancey undertook wind tunnel and flight tests on it to investigate and assess its longitudinal and lateral static and dynamic stability characteristics.[54] These tests documented representative state-of-the-art analytical procedures and design data for predicting the subsonic longitudinal static and dynamic stability and control characteristics of a light, propeller-driven airplane.[55] But the tests also confirmed, as one survey undertaken by North Carolina State University researchers for NASA concluded, that much work remained to be done to define and properly quantify the desirable handling qualities of GA aircraft.[56]

Fortunately, a key tool was rapidly maturing that made such analysis far more attainable than it would have been just a few years previously: the computer. Given a properly written analytical program, it had the ability to rapidly extract relevant performance parameters from

52. Paul C. Loschke, Marvin R. Barber, Calvin R. Jarvis, and Einar K. Enevoldson, "Handling Qualities of Light Aircraft with Advanced Control Systems and Displays," in Philip Donely et al., *NASA Aircraft Safety and Operating Problems*, v. 1, NASA SP-270 (Washington, DC: NASA Scientific and Technical Information Office, 1971), p. 189. NASA has continued its research on applying sophisticated avionics to civil and military aircraft for flight safety purposes, as examined by Robert Rivers in a case on synthetic vision systems in this volume.

53. Discussed in a companion case study in this series by Peter Merlin.

54. Marvin P. Fink and Delma C. Freeman, Jr., "Full-Scale Wind-Tunnel Investigation of Static Longitudinal and Lateral Characteristics of a Light Twin-Engine Aircraft," NASA TN D-4983 (1969); Chester H. Wolowicz and Roxanah B. Yancey, "Longitudinal Aerodynamic Characteristics of Light Twin-Engine, Propeller-Driven Airplanes," NASA TN D-6800 (1972); and Chester H. Wolowicz and Roxanah B. Yancey, "Lateral-Directional Aerodynamic Characteristics of Light, Twin-Engine Propeller-Driven Airplanes," NASA TN D-6946 (1972).

55. Afterwards, Wolowicz and Yancey expanded their research to include experimental determination of airplane mass and inertial characteristics. See Chester H. Wolowicz and Roxanah B. Yancey, "Experimental Determination of Airplane Mass and Inertial Characteristics," NASA TR R-433 (1974).

56. Frederick O. Smetana, Delbert C. Summey, and W. Donald Johnson, "Riding and Handling Qualities of Light Aircraft—A Review and Analysis," NASA CR-1975 (1972).

flight-test data. Over several decades, estimating stability and control parameters from flight-test data had progressed through simple analog matching methodologies, time vector analysis, and regression analysis.[57] A joint program between the NASA Langley Research Center and the Aeronautical Laboratory of Princeton University using a Ryan Navion demonstrated that an iterative "maximum-likelihood minimum variance" parameter estimation procedure could be used to extract key aerodynamic parameters based on flight test results, but also showed that caution was warranted. Unanticipated relations between the various parameters had made it difficult to sort out individual values and indicated that prior to such studies, researchers should have a reliable mathematical model of the aircraft.[58] At the Flight Research Center, Richard E. Maine and Kenneth W. Iliff extended such work by applying IBM's FORTRAN programming language to ease determination of aircraft stability and control derivatives from flight data. Their resulting program, a maximum likelihood estimation method supported by two associated programs for routine data handling, was validated by successful analysis of 1,500 maneuvers executed by 20 different aircraft and was made available for use by the aviation community via a NASA Technical Note issued in April 1975.[59] Afterwards, NASA, the Beech Aircraft Corporation, and the Flight Research Laboratory at the University of Kansas collaborated on a joint flight test of a loaned Beech 99 twin-engine commuter aircraft, extracting longitudinal and lateral-directional stability derivatives during a variety of maneuvers at assorted angles of attack

57. William F. Milliken, Jr., "Progress is Dynamic Stability and Control Research," *Journal of the Aeronautical Sciences*, v. 14, no. 9 (Sep. 1947), pp. 493–519; Harry Greenberg, "A Survey of Methods for Determining Stability Parameters of an Airplane from Dynamic Flight Measurements," NACA TM-2340 (1951); Marvin Shinbrot, "On the Analysis of Linear and Nonlinear Dynamical Systems From Transient-Response Data," NACA TN-3288 (1954); Randall D. Grove, Roland L. Bowles, and Stanley C. Mayhew, "A Procedure for Estimating Stability and Control Parameters from Flight Test Data by Using Maximum Likelihood Methods Employing a Real-Time Digital System," NASA TN D-6735 (1972); and William T. Suit and Robert L. Cannaday, "Comparison of Stability and Control Parameters for a Light, Single-Engine, High-Winged Aircraft Using Different Flight Test and Parameter Estimation Techniques," NASA TM-80163 (1979).

58. William T. Suit, "Aerodynamic Parameters of the Navion Airplane Extracted from Flight Data," NASA TN D-6643 (1972).

59. Richard E. Maine and Kenneth W. Iliff, "A FORTRAN Program for Determining Aircraft Stability and Control Derivatives from Flight Data," NASA TN D-7831 (1975); and Kenneth W. Iliff and Richard E. Maine, "Practical Aspects of a Maximum Likelihood Estimation Method to Extract Stability and Control Derivatives from Flight Data," NASA TN D-8209 (1976).

and in clean and flaps-down condition. "In general," researchers concluded, "derivative estimates from flight data for the Beech 99 airplane were quite consistent with the manufacturer's predictions."[60] Another analytical tool was thus available for undertaking flying and handling qualities analysis.

Enhancing General Aviation Safety

Flying and handling qualities are, per se, an important aspect of operational safety. But many other issues affect safety as well. The GA airplane of the postwar era was very different from its prewar predecessor—gone was fabric and wood or steel tube, with some small engine and a two-bladed fixed-pitch propeller. Instead, many were sleek all-metal monoplanes with retractable landing gears, near-or-over-200-mph cruising speeds, and, as noted in the previous section, often challenging and demanding flying and handling qualities. In November 1971, NASA sponsored a meeting at the Langley Research Center to discuss technologies that might be applied to future civil aviation in the 1970s and beyond. Among the many papers presented was a survey of GA by Jack Fischel and Marvin Barber of the Flight Research Center.[61] Barber and Fischel offered an incisive survey and synthesis of applicable technologies, including the then-new concept of the supercritical wing, which was of course applicable to propeller design as well. They addressed opportunities to employ new structural design concepts and materials advances (as were then beginning to be explored for military aircraft). Boron and graphite composites, which could be laid up and injection molded, promised to reduce both weight and labor costs, offering higher strength-to-weight ratios than conventional aluminum and steel construction. They noted the potentiality of increasingly reliable and cheap gas turbine engines (and the then-fashionable rotary combustion engine as well), and improved avionics could provide greater utility and safety for pilots of lower flight experience. Barber and Fischel concluded that,

> On the basis of current and projected near-future technology, it is believed that the main technology effort in the next decade will be devoted to improving the

60. Russel R. Tanner and Terry D. Montgomery, "Stability and Control Derivative Estimates Obtained from Flight Data for the Beech 99 Aircraft," NASA TM-72863 (1979).
61. Barber and Fischel, "General Aviation: The Seventies and Beyond," pp. 317–332.

economy, performance, utility, and safety of General Aviation aircraft.[62]

Of these, the greatest challenges involved safety. By the early 1970s, the fatality rate for GA was 10 times higher per passenger miles than that of automobiles.[63] Many accidents were caused by pilots exceeding their flying abilities, leading one manufacturing executive to ruefully remark at a NASA conference, "If we don't soon find ways to improve the safety of our airplanes, we are going to be putting placards on the airplanes which say 'Flying airplanes may be hazardous to your health.'"[64] Alarmed, NASA set an aviation safety goal to reduce fatality rates by 80 percent by the mid-1980s.[65] While basic changes in pilot training and practices could accomplish a great deal of good, so, too, could better understanding of GA safety challenges to create aircraft that were easier and more tolerant of pilot error, together with sub-systems such as advanced avionics and flight controls that could further enhance flight safety. Underpinning all of this was a continuing need for the highest quality information and analysis that NASA research could furnish. The following examples offer an appreciation of some of the contributions NACA-NASA researchers made confronting some of the major challenges to GA safety.

Spin Research

One of the areas of greatest interest has been that of spin behavior. When an airplane stalls, it may enter a spin, typically following a steeply

62. Barber and Fischel, "General Aviation: The Seventies and Beyond," p. 325.

63. NASA Flight Research Center, Flight Programs Review Committee, "NASA Flight Research Center: Current and Proposed Research Programs" (Jan. 1973), "Development of Flight Systems for General Aviation" slide and attached briefing notes, NASA DFRC Archives.

64. The executive was Piper's chief of aerodynamics, flight test, and structures, Calvin F. Wilson; readers should note he was speaking of GA aircraft generically, not just Piper's. See Statement of Calvin F. Wilson, in NASA Langley Research Center Staff, *Vehicle Technology for Civil Aviation: The Seventies and Beyond, Panel Discussion,* Supplement to NASA SP-292 (Washington: NASA Scientific and Technical Information Office, 1972), p. 9.

65. "General Aviation Technology Program" NASA TM X-73051 (1976). Indeed, fatalities in civil aviation subsequently did fall a remarkable 91 percent between 1976 and 1986, though for various reasons and not exclusively through NASA activities. See National Transportation Safety Board, *Annual Review of Aircraft Accident Data, U.S. Air Carrier Operations,* 1976, NTSB-ARC-78-1 (1978), p. 3; and National Transportation Safety Board, *Annual Review of Aircraft Accident Data, U.S. Air Carrier Operations, 1986,* NTSB-ARC-89-01 (1989), p. 3.

descending flightpath accompanied by a rotational motion (sometimes accompanied by other rolling and pitching motions) that is highly disorientating to a pilot. Depending on the dynamics of the entry and the design of the aircraft, it may be easily recoverable, difficult to recover from, or irrecoverable. Spins were a killer in the early days of aviation, when their onset and recovery phenomena were imperfectly understood, but have remained a dangerous problem since, as well.[66] Using specialized vertical spin tunnels, the NACA, and later NASA, undertook extensive research on aircraft spin performance, looking at the dynamics of spins, the inertial characteristics of aircraft, the influence of aircraft design (such as tail placement and volume), corrective control input, and the like.[67]

As noted, spins have remained an area of concern as aviation has progressed, because of the strong influence of aircraft configuration upon spin behavior. During the early jet age, for example, the coupled motion dynamics of high-performance low-aspect-ratio and high-fineness-ratio jet fighters triggered intense interest in their departure and spin characteristics, which differed significantly from earlier aircraft because their mass was now primarily distributed along the longitudinal, not lateral, axis of the aircraft.[68] Because spins were not a normal part of GA flying operations, GA pilots often lacked the skills to recognize and cope with spin-onset, and GA aircraft themselves were often inadequately designed to deal with out-of-balance or out-of-trim conditions that might force a spin entry. If encountered at low altitude, such as approach to landing, the consequences could be disastrous. Indeed, landing accidents composed more than half of all GA accidents, and of these, as one NASA document noted, "the largest single factor in General Aviation fatal accidents is the stall/spin."[69]

The Flight Research Center's 1966 study of comparative handling qualities and behavior of a range of GA aircraft had underscored

66. For a historical perspective on spins, drawn from pilot accounts, see Dunstan Hadley, *One Second to Live: Pilots' Tales of the Stall and Spin* (Shrewsbury, U.K.: Airlife Publishing Ltd., 1997).
67. Anshal I. Neihouse, Walter J. Klinar, and Stanley H. Scher, "Status of Spin Research for Recent Airplane Designs," NASA TR R-57 (1960).
68. For example, see NACA High-Speed Flight Station, "Flight Experience with Two High-Speed Airplanes Having Violent Lateral-Longitudinal Coupling in Aileron Rolls," NACA RM-H55A13 (1955).
69. NASA Scientific and Technical Information Program, "General Aviation Technology Program," Release No. 76–51, NASA TM X-73051 (1976), p. 2; Barber and Fischel, "General Aviation: The Seventies and Beyond," p. 323.

the continuing need to study stall-spin behavior. Accordingly, in the 1970s, NASA devoted particular attention to studying GA spins (and continued studying the spins of high-performance aircraft as well), marking "the most progressive era of NASA stall/spin research for general aviation configurations."[70] Langley researchers James S. Bowman, Jr.; James M. Patton, Jr.; and Sanger M. Burk oversaw a broad program of stall/spin research. They and other investigators evaluated tail location and its influence upon spin recovery behavior using both spin-tunnel models,[71] and free-flight tests of radio-controlled models and actual aircraft at the Wallops Flight Center, on the Virginia coast of the Delmarva Peninsula.[72] Between 1977 and 1989, NASA instrumented and modified four aircraft of differing configuration for spin research: an experimental low-wing Piper design with a T-tail, a Grumman American AA-1 Yankee modified so that researchers could evaluate three different horizontal tail positions, a low-wing Beech Sundowner equipped with wingtip rockets to aid in stopping spin rotation, and a high-wing Cessna C-172. Overall, the tests revealed the critical importance of designers ensuring that the vertical fin and rudder of their new GA aircraft be in active airflow during a spin, so as to ensure their effectiveness in spin recovery. To do that, the horizontal tail needed to be located in such a position on the aft fuselage or fin so as not to shield the vertical fin and rudder from active flow. The program was not without danger and incident. Mission planners prudently equipped the four aircraft with an emergency 10.5-ft-diameter spin-recovery parachute. Over that time, the 'chute had to be deployed on 29 occasions when the test aircraft entered unrecoverable

8

70. Chambers, *Concept to Reality*, p. 122; James S. Bowman, Jr., "Summary of Spin Technology as Related to Light General-Aviation Airplanes," NASA TN D-6575 (1971).

71. William Bihrle, Jr., Randy S. Hultberg, and William Mulcay, "Rotary Balance Data for a Typical Single-Engine General Aviation Design for an Angle-of-Attack range of 30 deg. to 90 deg.," NASA CR-2972 (1978); William Bihrle, Jr., and Randy S. Hultberg, "Rotary Balance Data for a Typical Single-Engine General Aviation Design for an Angle-of-Attack range of 8 deg. to 90 deg.," NASA CR-3097 (1979); William J. Mulcay and Robert A. Rose, "Rotary Balance Data for a Typical Single-Engine General Aviation Design for an Angle of Attack Range of 8 deg. to 90 deg.," NASA CP-3200 (1980); and Mark G. Ballin, "An Experimental Study of the Effect of Tail Configuration on the Spinning Characteristics of General Aviation Aircraft," NASA CR-168578 (1982).

72. W.S. Blanchard, Jr., "A Flight Investigation of the Ultra-Deep-Stall Descent and Spin Recovery Characteristics of a 1/6 scale Radio-controlled Model of the Piper PA-38 Tomahawk," NASA CR-156871 (1981); J.R. Chambers and H.P. Stough III, "Summary of NASA Stall/Spin Research for General Aviation Configurations," AIAA Paper 86-2597 (1986).

spins; each of the four aircraft deployed the 'chute at least twice, a measure of the risk inherent in stall-spin testing.[73]

NASA's work in stall-spin research has continued, but at a lower level of effort than in the heyday of the late 1970s and 1980s, reflecting changes in the Agency's research priorities, but also that NASA's work had materially aided the understanding of spins, and hence had influenced the data and experience base available to designers shaping the GA aircraft of the future. As well, the widespread advent of electronic flight controls and computer-aided flight has dramatically improved spin behavior. Newer designs exhibit a degree of flying ease and safety unknown to earlier generations of GA aircraft. This does not mean that the spin is a danger of the past—only that it is under control. In the present and future, as in the past, ensuring GA aircraft have safe stall/spin behavior will continue to require high-order analysis, engineering, and test.

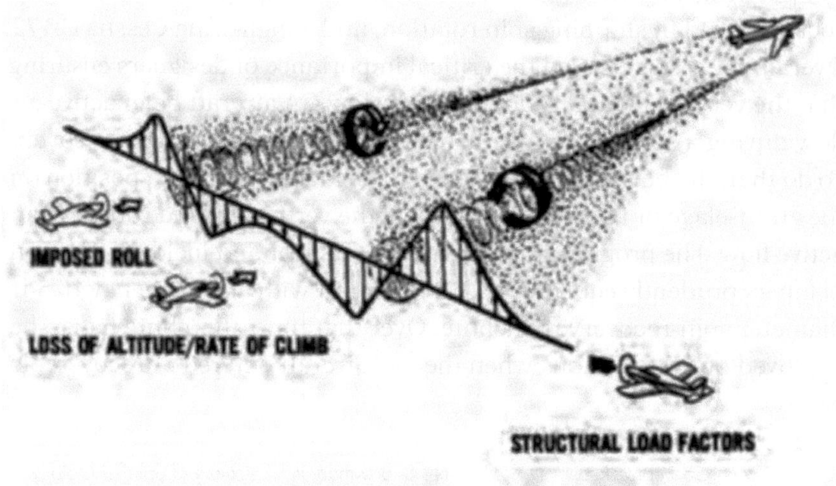

Aircraft entering wake vortex flow encountered a series of dangers, ranging from upset to structural failure, depending on their approach to the turbulent flow. From NASA SP-409 (1977).

Wake Vortex Research

The 1970s inauguration of widebody jumbo jets posed special problems for smaller aircraft because of the powerful streaming wake vortices generated by aircraft such as the Boeing 747, Douglas DC-10, and Lockheed L-1011. After several unexplained accidents caused by aircraft

73. H. Paul Stough III, "A Summary of Spin-Recovery Parachute Experience on Light Airplanes," AIAA Paper 90-1317 (1990).

upset, and urged by organizations such as the Flight Safety Foundation and the Aircraft Owners and Pilots Association, the Federal Aviation Administration (FAA) asked NASA and the U.S. Air Force to initiate a flight-test program to evaluate the effect of the wingtip vortex wake generated by large jet transport airplanes on a variety of smaller airplanes. The program began in December 1969 and, though initially ended in April 1970, was subsequently expanded and continued over the next decade. Operations were performed at Edwards Air Force Base, CA, under the supervision of the NASA Flight Research Center in cooperation with the Ames Research Center and the U.S. Air Force, using a range of research aircraft including 747, 727, and L-1011 airliners, and smaller test subjects such as the T-37 trainer and QF-86 drones, supported by extensive wind tunnel and water channel research.[74]

The Boeing 747 subsequently modified as carrier aircraft for the Space Shuttle Orbiter furnished NASA the opportunity to undertake vortex upset using the Lear Jet and Cessna T-37 trainer shown here flying formation on the larger aircraft. NASA.

Subsequently, in 1972, NASA intensified its wake vortex research to seek reducing vortex formation via aerodynamic modification and addition of wind devices. By the beginning of 1974, Alfred Gessow, the Chief of Fluid and Flight Dynamics at NASA Headquarters, announced the Agency was optimistic that wake vortex could be eliminated "as a constraint to airport operations by new aerodynamic designs or by ret-

74. For example, M.R. Barber and Joseph J. Tymczyszyn, "Wake Vortex Attenuation Flight Tests: A Status Report," in Joseph W. Stickle, ed., *1980 Aircraft Safety and Operating Problems*, Pt. 2 (Washington, DC: NASA Scientific and Technical Information Office, 1981), pp. 387–408.

rofit modifications to large transport aircraft."[75] Overall, the tests, and ones that followed, had clearly demonstrated the power of wake vortices to constrain the operations GA aircraft; light jet trainers and business aircraft such as the Lear Jet were buffeted and rolled, and researchers found that the vortices maintained significant strength up to 10 miles behind a widebody. As a result of NASA's studies, the FAA introduced a requirement for wake turbulence awareness training for all pilots, increased separation distances between aircraft, and mandated verbal warnings to pilots during the landing approach at control-towered airports when appropriate. NASA has continued its wake turbulence studies since that time, adding further to the understanding of this fascinating, if potentially dangerous, phenomenon.[76]

Crash Impact Research

In support of the Apollo lunar landing program, engineers at the Langley Research Center had constructed a huge steel A-frame gantry structure, the Lunar Landing Research Facility (LLRF). Longer than a football field and nearly half as high as the Washington Monument, this facility proved less useful for its intended purposes than free-flight jet-and-rocket powered training vehicles tested and flown at Edwards and Houston. In serendipitous fashion, however, it proved of tremendous value for aviation safety after having been resurrected as a crash-impact test facility, the Impact Dynamics Research Facility (IDRF) in 1974, coincident with the conclusion of the Apollo program.[77]

---

75. Alfred Gessow et al., *Wake Vortex Minimization*, NASA SP-409 (Washington, DC: NASA Scientific and Technical Information Office, 1977), p. iv. See, for example, Delwin R. Croom, Raymond D. Vogler, and John A. Thelander, "Low-Speed Wind-Tunnel Investigation of Flight Spoilers as Trailing-Vortex-Alleviation Devices on an Extended-Range Wide-Body Tri-Jet Airplane Model," NASA TN D-8373 (1976).

76. See for example, Harriet J. Smith, "A Flight Test Investigation of the Rolling Moments Induced on a T-37B Airplane in the Wake of a B-747 Airplane," NASA TM X-56031 (1975); and S.C. Crow and E.R. Bate, Jr., "Lifespan of Trailing Vortices in a Turbulent Atmosphere," *AIAA Journal of Aircraft*, v. 13, no. 7 (Jul. 1976), pp. 476–482.

77. The history of the facility is well covered in Karen E. Jackson, Richard L. Boitnott, Edwin L. Fasanella, Lisa Jones, and Karen H. Lyle, "A History of Full-Scale Aircraft and Rotorcraft Crash Testing and Simulation at NASA Langley Research Center," Paper presented at the 4th Triennial International Aircraft and Cabin Safety Research Conference, 15–18 Nov. 2004, Lisbon, Portugal, NTRS, CASI ID 20040191337.

Test Director Victor Vaughan studies the results of one 1974 crash impact test at the Langley Impact Dynamics Research Facility. NASA.

Over its first three decades, the IDRF was used to conduct 41 full-scale crash tests of GA aircraft and approximately 125 other impact tests of helicopters and aircraft components. The IDRF could pendulum-sling aircraft and components into the ground at precise impact angles and velocities, simulating the dynamic conditions of a full-scale accident or impact.[78] In the first 10 years of its existence, the IDRF served as the focal point for a joint NASA-FAA-GA industry study to improve the crashworthiness of light aircraft. It was a case of making the best of a bad situation: a flood had rendered a sizeable portion of Piper's single-and-twin-engine GA production at its Lock Haven, PA, plant unfit for sale and service.[79] Rather than simply scrap the aircraft, NASA and Piper worked together to turn them to the benefit of the GA industry and user communities. A variety of Piper Aztecs, Cherokees, and Navajos, and later some Cessna 172s, some adorned with colorful names like "Born to Lose," were instrumented, suspended from cable harnesses, and then "crashed" at various impact angles, attitudes, velocities, and sink-rates, and against hard and soft surfaces. To gain greater fidelity, some were accelerated during their drop by small solid-fuel rockets installed in their engine nacelles.[80]

Later tests, undertaken in 1995 as part of the Advanced General Aviation Transport Experiment (AGATE) study effort (discussed subsequently), tested Beech Starship, Cirrus SR-20, Lear Fan 2100, and Lancair aircraft.[81] The rapid maturation of computerized analysis programs led to its swift adoption for crash impact research. In partnership with NASA, researchers at the Grumman Corporation Research Center developed DYCAST (DYnamic Crash Analysis of STructures) to analyze structural response during crashes. DYCAST, a finite element program, was qualified during extensive NASA testing for light aircraft component testing, including seat and fuselage section analysis, and then made available for broader aviation community use in 1987.[82] Application of computa-

78. Victor. L. Vaughan, Jr., and Emilio Alfaro-Bou, "Impact Dynamics Research Facility for Full-Scale Aircraft Crash Testing," NASA TND-8179 (1976).

79. David A. Anderton, *NASA Aeronautics*, EP-85 (Washington, DC: NASA, 1982), p. 23.

80. For example, see Victor L. Vaughan, Jr., and Robert J. Hayduk, "Crash Tests of Four Identical High-Wing Single Engine Airplanes," NASA TP-1699 (1980); M. Susan Williams and Edwin L. Fasanella, "Results from Tests of Three Prototype General Aviation Seats," NASA TM-84533 (1982); M. Susan Williams and Edwin L. Fasanella, "Crash Tests of Four Low-Wing Twin-Engine Airplanes with Truss-Reinforced Fuselage Structure," NASA TP-2070 (1982); and Claude B. Castle and Emilio Alfaro-Bou, "Crash Tests of Three Identical Low-Wing Single-Engine Airplanes," NASA TP-2190 (1983); and Huey D. Carden, "Full-Scale Crash-Test Evaluation of Two Load-Limiting Subfloors for General Aviation Airframes," NASA TP-2380 (1984).

81. Jackson et al., "A History of Full-Scale Aircraft and Rotorcraft Crash Testing."

82. Allan B. Pifko, Robert Winter, and Patricia L. Ogilvie, "DYCAST—A Finite Element Program for the Crash Analysis of Structures," NACA CR-4040 (1987).

tional methodologies to crash impact research expanded so greatly that by the early 1990s, NASA, in partnership with the University of Virginia Center for Computational Structures Technology, held a seminal workshop on advances in the field.[83] Out of all of this testing came better understanding of the dynamics of an accident and the behavior of aircraft at and after impact, quantitative data applicable to the design of new and more survivable aircraft structures, better seats and restraint systems, comparative data on the relative merits of conventional versus composite construction, and computational methodologies for evermore precise and informed analysis of crashworthiness.

Avionics and Cockpit Research for Safer General Aviation Operations

Aircraft instrumentation has always been intrinsically related to flight safety. The challenge of blind and bad-weather flying in the 1920s led to development of both radio navigation equipment and techniques, and specialized blind-flying instrumentation, typified by the gyro-stabilized artificial horizon, which, like radar later, was one of the few truly transforming instruments developed in the history of flight, for it made possible instrument-only (IFR) flight. Taken together with advances in the Federal airway system, the development of lightweight airborne radars, digital electronics, sophisticated communications, and radar-based and later satellite navigation, as well as access to up-to-date weather information, revolutionized civil and military air operations. Ironically, accident rates remained high, particularly among GA pilots flying single-pilot (SP) aircraft under IFR conditions. By the early 1980s, the National Transportation Safety Board was reporting that "SPIFR" accidents accounted for 79 percent of all IFR-related accidents, with half of these occurring during high-workload landing approaches, totaling more than 100 serious accidents attributable to pilot error per year.[84] Analysis revealed five major problem areas: controller judgment and response, pilot judgment and response, Air Traffic Control (ATC) intrafacility and interfacility conflict, ATC-pilot communication, and IFR-VFR (instrument flight rules-visual flight rules) conflicts. Common to

83. Ahmed K. Noor and Huey D. Carden, "Computational Methods for Crashworthiness," NACA CP-3223 (1993).

84. John D. Shaughnessy, "Single-Pilot IFR Program Overview and Status," in NASA Langley Research Center Staff, *Controls, Displays, and Information Transfer for General Aviation IFR Operations*, NASA CP 2279 (Hampton, VA: NASA Langley Research Center, 1983), p. 3.

all of these were a mix of human error, communications deficiencies, conflicting or complex procedures and rules, and excessive workload. In particular, NASA researchers concluded that "methods, techniques, and systems for reducing work load are drastically needed."[85]

In the mid-1970s, NASA aeronautics planners had identified "design[ing] avionic systems to more effectively integrate the light airplane with the air-space system" as a priority, with researchers at Ames Research Center evaluating integration avionic functions with the goal of producing a single system concept.[86] In 1978, faced with the challenge of rising SPIFR accidents, NASA Langley Research Center launched a SPIFR program, holding a workshop in August 1983 at Langley to review and evaluate the progress to date on SPIFR studies and to disseminate it to an industry, academic, and governmental audience. The SPIFR program studied in depth the interface of the pilot and airplane, looking at a variety of issues ranging from the tradeoffs between complex autopilots and their potential benefits to simulator utility. Overall, researchers found that "[b]ecause of the increase in air traffic and the more sophisticated and complex ground control systems handling this traffic, IFR flight has become extremely demanding, frequently taxing the pilot to his limits. It is rapidly becoming imperative that all the pilot's sensory and manipulative skills be optimized in managing the aircraft systems"; hopefully, they reasoned, the rapid growth in computer capabilities could "enhance single-crewman effectiveness in future aircraft operations and automated ATC systems."[87] Encouragingly, in part because of NASA research, a remarkable 41-percent decrease in overall GA accidents did occur from the mid-1980s to the late 1990s.[88]

However, all was not well. Indeed, a key goad stimulating NASA's pursuit of avionics technology to enhance flight safety (particularly weather safety) was the decline of American General Aviation. In the late 1970s, America's GA aircraft industry reached the peak of its power: in 1978,

85. Hugh P. Bergeron, "Analysis of General Aviation Single-Pilot IFR Incident Data Obtained from the NASA Aviation Safety Reporting System," NASA TM-80206 (1980), p. 8.

86. NASA, *General Aviation Technology Program,* p. 13.

87. Shaughnessy, "Single-Pilot IFR Program Overview and Status," p. 4.

88. From 3,233 in 1982 to 1,989 in 1998; see Nanette Scarpellini Metz, *Partnership and the Revitalization of Aviation: A Study of the Advanced General Aviation Transport Experiments Program, 1994-2001,* UNOAI Report 02-5 (Omaha: University of Nebraska at Omaha Aviation Institute, 2002), p. 6.

manufacturers shipped 17,817 aircraft, and the next year, 1979, the top three manufacturers—Cessna, Beech, and Gates Learjet—had combined sales over $1.8 billion. It seemed poised for even greater success over the next decade. In fact, such did not occur, thanks largely to rapidly rising insurance costs added to aircraft purchase prices, a by-product of a "rash of product liability lawsuits against manufacturers stemming from aircraft accidents," some frivolously alleging inherent design flaws in aircraft that had flown safely for previous decades. Rising aircraft prices cooled any ardor for new aircraft purchases, particularly of single-engine light aircraft (business aircraft sales were affected, but more slowly). Other factors also contributed, including a global recession in the early 1980s, an increase in aircraft leasing and charter aircraft operations (lessening the need for personal ownership), and mergers within the aircraft industry that eliminated some production programs. The number of students taking flight instruction fell by over a third, from 150,000 in 1980 to 96,000 in 1994. That year, GA manufacturers produced just 928 aircraft, representing a production decline of almost 95 percent since the heady days of the late 1970s.[89]

The year 1994 witnessed both the near-extinction of American General Aviation and its fortuitous revival. At the nadir of its fortunes, relief, fortunately, was in hand, thanks to two initiatives launched by Congress and NASA. The first was the General Aviation Revitalization Act (GARA) of 1994, passed by Congress and signed into law in August that year by President William Jefferson Clinton.[90] GARA banned product liability claims against manufacturers later than 18 years after an aircraft or component first flew. By 1998, the 18-year provision could be applied to the large numbers of aircraft produced in the 1970s, bringing relief at last to manufacturers who had been so plagued by legal action that many had actually taken aircraft—including old classics such as the Cessna C-172—out of production.[91] It is not too strong to state that GARA saved the American GA industry from utter extinction, for it brought much needed stability and restored sanity to a litigation

89. Pattillo, *A History in the Making*, p. 127; see also John H. Winant, *Keep Business Flying: A History of The National Business Aircraft Association, Inc., 1946–1986* (Washington: The National Business Aircraft Association, 1989), pp. 151–152, 157, and 186–187; and Metz, *Partnership and the Revitalization of Aviation*, p. 7.
90. The General Aviation Revitalization Act of 1994, Public Law No. 103–298, 103 Stat. 1552.
91. Pattillo, *A History in the Making*, Table 7-2, p. 129, and pp. 169–170.

process that had gotten out of hand. Thus it constitutes the most significant piece of American aviation legislation passed in the modern era.

But important as well was a second initiative, the establishment by NASA of the AGATE program, a joint NASA-industry-FAA partnership. AGATE existed thanks to the persistency of Bruce Holmes, the Agency's Assistant Director of Aeronautics, who had vigorously championed it. Functionally organized within NASA's Advanced Subsonic Technology Project Office, AGATE dovetailed nicely with GARA. It sought to revitalize GA by focusing on innovative cockpit technologies that could achieve goals of safety, affordability, and ease of use, chief of which was the "Highway in the Sky" (HITS) initiative, which aimed to replace the dial-and-gauge legacy instrument technology of the 1920s with advanced computer-based graphical presentations. As well, it supported crashworthiness research. It served as well as single focal point to bring together NASA, industry, Government, and GA community representatives.

AGATE ran from 1994 through 2001, and a key aspect of its success was that it operated under a NASA-unique process, the Joint Sponsored Research Agreement (JSRA), a management process that streamlined research and internal management processes, while accelerating the results of technology development into the private sector. AGATE suffered in its early years from "learning problems" with internal communication, with building trust and openness among industry partners more used to seeing themselves as competitors, and with managerial oversight of its activities. Some participants were disappointed that AGATE never achieved its most ambitious objective, a fully automated aircraft. Others were bothered by the uncertainty of steady Federal support, a characteristic aspect of Federal management of research and development. But if not perfect—and no program ever is—AGATE proved vital to restoring GA, and as an end-of-project study concluded inelegantly if bluntly, "[a]ccording to participants from all parts of the program, AGATE revitalized an industry that had gone into the toilet."[92]

The legacy of AGATE is evident in much of NASA's subsequent avionics and cockpit presentation research, which, building upon earlier research, has involved improving a pilot's situational awareness. Since weather-related accidents account for one-third of all aviation accidents and over one-quarter of all GA accidents, a particular concern is present-

92. Metz, *Partnership and the Revitalization of Aviation,* p. 18.

ing timely and informative weather information, for example, graphics overlaid on navigational and geographical cockpit displays.[93] Another area of acute interest is improving pilot controllability via advanced flight control technology to close the gap between an automobile-like 2-D control system and the traditionally more complex 3-D aircraft system and generating a HITS-like synthetic vision capability to enhance flight safety. This, too, is a longstanding concern, related to the handling qualities and flight control capabilities of aircraft so that the pilot can concentrate more on what is going on around the aircraft than having to concentrate on flying it.[94]

Towards Tomorrow: Transforming the General Aviation Aircraft

In the mid-1970s, coincident with the beginning of the fuel and litigation crises that would nearly destroy GA, production of homebuilt and kit-built aircraft greatly accelerated, reflecting the maturity of light aircraft design technology, the widespread availability of quality engineering and technical education, and the frustration of would-be aircraft owners with rising aircraft prices. Indeed, by the early 1990s, kit sales would outnumber sales of production GA aircraft by more than four to one.[95] Today, in a far-different post-GARA era, kit sales remain strong. As well, new manufacturers appeared, some wedded to particular ideas or concepts, but many also showing a broader (and thus generally more successful) approach to light aircraft design.

Exemplifying this resurgence of individual creativity and insight was Burt Rutan of Mojave, CA. An accomplished engineer and flight-tester, Rutan designed a small two-seat canard light aircraft, the VariEze,

93. For example, see Shashi Seth, "Cockpit Weather Graphics Using Mobile Satellite Communications," in NASA Jet Propulsion Laboratory, *Proceedings of the Third International Mobile Satellite Conference* (Pasadena: NASA JPL, 1993), pp. 231–233; H. Paul Stough III, Daniel B. Shafer, Philip R. Schaffner, and Konstantinos S. Martzaklis, "Reducing Aviation Weather-Related Accidents Through High-Fidelity Weather Information Distribution and Presentation," Paper Presented at the International Council of the Aeronautical Sciences, 27 Aug.–1 Sep. 2000. Harrowgate, U.K.; and H. Paul Stough III, James F. Watson, Jr., Taumi S. Daniels, Konstantinos S. Martzaklis, Michael A. Jarrell, and Rodney K. Bougue, "New Technologies for Weather Accident Prevention," AIAA Paper 2005-7451 (2005).

94. "Highway-in-the-Sky" synthetic vision systems are the subject of a separate case study.

95. Pattillo, *A History in the Making*, Tables 7-2 and 8-1, pp. 129 and 172; the exact figures for 1994 were 928 aircraft produced and 4,085 kits sold (this does not include plans sales, which in 1994 numbered 2,831).

8

powered by a 100-hp Continental engine. Futuristic in look, the VariEze embodied very advanced thinking, including a GA(W)-1 wing section and Whitcomb winglets. The implications of applying the configuration to other civil and military aircraft of far greater performance were obvious, and NASA studied his work both in the tunnel and via flight tests of the VariEze itself.[96] Rutan's influence upon advanced general aviation aircraft thinking was immediate. Beech adopted a canard configuration for a proposed King Air replacement, the Starship, and Rutan built a subscale demonstrator of the aircraft.[97] Rutan subsequently expanded his range of work, becoming a noted designer of remarkable flying machines capable of performance—such as flying nonstop around the world or rocketing into the upper atmosphere—many would have held impossible to attain.

NASA followed Rutan's work with interest, for the canard configuration was one that had great applicability across the range of aircraft design, from light aircraft to supersonic military and civil designs. Langley tunnel tests in 1984 confirmed that with a forward center of gravity location, the canard configuration was extremely stall-resistant. Conversely, at an aft center of gravity location, and with high power, the canard had reduced longitudinal stability and a tendency to enter a high-angle-attack, deep-stall trim condition.[98] NASA researchers undertook a second series of tests, comparing the canard with other wing planforms including closely coupled dual wings, swept forward-swept rearward wings, joined wings, and conventional wing-tail configurations, evaluating their application to a hypothetical 350-mph, 1,500-mile-range 6- or 12-passenger aircraft operating at 30,000 to 40,000 feet. In these tests, the dual wing configuration prevailed, due to greater structural weight efficiencies than other approaches.[99]

Seeking optimal structural efficiency has always been an important aspect of aircraft design, and the balance between configuration choice

96. Burt Rutan, "Development of a Small High-Aspect-Ratio Canard Aircraft," in Society of Experimental Test Pilots, *1976 Report to the Aerospace Profession* (Lancaster, CA: SETP, 1976), pp. 93–101; Philip W. Brown and James M. Patton, Jr., "Pilots' Flight Evaluation of VariEze N4EZ," NASA TM-103457 (1978).

97. Subsequently, for reasons unrelated to the basic canard concept, the Starship did not prove a great success.

98. Joseph R. Chambers, Long P. Yip, and Thomas M. Moul, "Wind Tunnel Investigation of an Advanced General Aviation Canard Configuration," NASA TM-85760 (1984).

99. B.P. Selberg and D.L. Cronin, "Aerodynamic Structural Study of Canard Wing, Dual Wing, and Conventional Wing Systems for General Aviation Applications," NASA CR-172529 (1985).

and structural design is a fine one. The advent of composite structures enabled a revolution in structural and aerodynamic design fully as significant as that at the time of the transformation of the airplane from wood to metal. As designers then had initially simply replaced wooden components with metal ones, so, too, in the earliest stage of the composite revolution, designers had initially simply replaced metal components with composite ones. In many of their own GA proposals and studies, NASA researchers repeatedly stressed the importance of getting away from such a "metal replacement" approach and, instead, adopting composite structures for their own inherent merit.[100]

The blend of research strains coming from NASA's diverse work in structures, propulsion, controls, and aerodynamics, joined to the creative impact of outside sources in industry and academia—not least of which were student study projects, many reflecting an insight and expertise belying the relative inexperience of their creators—informed NASA's next steps beyond AGATE. Student design competitions offered a valuable means of both "growing" a knowledgeable future aerospace workforce and seeking fresh approaches and insight. Beginning in 1994, NASA joined with the FAA and the Air Force Research Laboratory to sponsor a yearly National General Aviation Design Competition establishing design baselines for single-pilot, 2- to 6-passenger vehicles, turbine or piston-powered, capable of 150 to 400 knots airspeed, and with a range of 800 to 1,000 miles. The Virginia Space Grant Consortium at Old Dominion University Peninsula Center, near Langley Research Center, coordinated the competition. Competing teams had to address "design challenges" in such technical areas as integrated cockpit systems; propulsion, noise, and emissions; integrated design and manufacturing; aerodynamics; operating infrastructure; and unconventional designs (such as roadable aircraft).[101] In cascading fashion, other opportunities existed for teams to take their designs to ever-more-advanced levels, even, ultimately, to building and test-flying them. Through these

100. For example, Robert F. Stengel, "It's Time to Reinvent the General Aviation Airplane," in Frederick R. Morrell, ed., *Joint University Program for Air Transportation Research-1986*, NASA CP-2502 (Washington, DC: NASA Scientific and Technical Information Office, 1988), pp. 81–105; J. Roskam and E. Wenninger, "A Revolutionary Approach to Composite Construction and Flight Management Systems for Small, General Aviation Airplanes," NTRS ID 94N25714 (1992).
101. NASA-FAA-AFRL, "National General Aviation Design Competition Guidelines, 1999–2000 Academic Year."

competitions, study teams explored integrating such diverse technical elements as advanced fiber optic flight control systems, laminar flow design, swept-forward wings, HITS cockpit technology, coupled with advanced Heads-up Displays (HUD) and sidestick flight control, and advanced composite materials to achieve increased efficiencies in performance and economic advantage over existing designs.[102]

Succeeding AGATE was SATS—the NASA Small Aircraft Transportation System Project. SATS (another Holmes initiative) sought to take the integrated products of this diverse research and form from it a distributed public airport network, with small aircraft flying on demand as users saw fit, thereby taking advantage of the ramp space capacity at over 5,000 public airports located around the country.[103] SATS would benefit as well by a Glenn Research Center initiative, the GAP (General Aviation Propulsion) program, seeking new propulsive efficiencies beyond those already obtained by previous NASA research.[104] In 2005, SATS concluded with a 3-day "Transformation of Air Travel" held at Danville Airport, VA, showcasing new aviation technologies with six aircraft equipped with advanced cockpit displays enabling them to operate from airports lacking radar or air traffic control services. Complementing SATS and GAP was PAV—a Langley initiative for Personal Air Vehicles, a reincarnation of an old dream of flight dating to the small ultralight aircraft and airships found at the dawn of flight, such as Alberto Santos-Dumont's little one-person dirigibles and his Demoiselle light aircraft. Like many such studies through the years, PAV studies in the 2002–2005 period generated many innovative and imaginative concepts, but the

102. See, for example, NASA LRC, "NASA and FAA Announce Design Competition Winners," Release No. 00-060, 29 Jul. 2000; University of Kansas, Department of Aerospace Engineering, "Preliminary Design Studies of an Advanced General Aviation Aircraft," in Universities Space Research Association, *Proceedings of the Seventh annual Summer Conference, NASA-USRA Advanced Design Program,* NTRS ID 93N29717 (Houston, TX: NASA-USRA, 1991), pp. 45–56.
103. Scott E. Tarry, Brent D. Bowen, and Jocelyn S. Nickerson, "The Small Aircraft Transportation System (SATS): Research Collaborations with the NASA Langley Research Center," in Brent D. Bowen et al., *The Aeronautics Education, Research, and Industry Alliance (AERIAL) 2002 Report,* UNOAI Report 02-7 (Omaha, NE: University of Nebraska at Omaha Aviation Institute, 2002), pp. 45–57; and Patrick D. O'Neil and Scott E. Tarry, *Annotated Bibliography of Enabling Technologies for the Small Aircraft Transportation System,* UNOAI Report 02-3 (Omaha, NE: University of Nebraska at Omaha Aviation Institute, 2002).
104. Williams International, "The General Aviation Propulsion (GAP) Program," NASA CR-2008-215266 (2008).

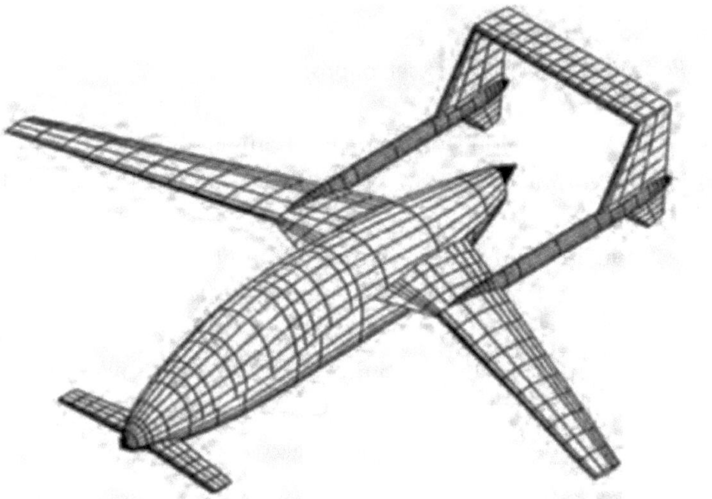

A computer-aided-design model of a six-passenger single-pilot Advanced Personal Transport concept developed as a University of Kansas-NASA-Universities Space Research Association student research project in 1991. NASA.

Agency did not support such studies afterwards, turning instead towards good stewardship and environmental responsibility, seeking to reduce emissions, noise, and improve economic efficiencies by reducing airport delays and fuel consumption. These are not innocuous challenges: in 2005, airspace system capacity limitations generated fully $5.9 billion in economic impact through airline delays, and the next year, fuel consumption constituted a full 26 percent of airline operating costs.[105]

The history of the NACA-NASA support of General Aviation is one of mutual endeavor and benefit. Examining that history reveals a surprising interdependency between the technologies of air transport, military, and general aviation. Developments such as the supercritical wing, electronic flight controls, turbofan propulsion, composite structures, synthetic vision systems, and heads-up displays that were first exploited for one have migrated and diffused more broadly across the entire aeronautical field. Once again, the lesson is clear: the many streams of NASA research form a rich and broad confluence that nourishes and invigorates the entire American aeronautical enterprise, ever renewing our nature as an aerospace nation.

105. Jaiwon Shin, "NASA Aeronautics Research Then and Now," a PowerPoint presentation at the 48th AIAA Aerospace Sciences Meeting, Orlando, FL, 4 January 2010, Slide 2; Chambers, *Innovation in Flight*, pp. 306–312.

Recommended Additional Readings

Reports, Papers, Articles, and Presentations:

Ira H. Abbott, Albert E. von Doenhoff, and Louis S. Stivers, Jr., "Summary of Airfoil Data," NACA TR 824 [formerly NACA ACR L5C05] (1945).

Arthur Assadourian and John A. Harper, "Determination of the Flying Qualities of the Douglas DC-3 Airplane," NACA TN-3088 (1953).

Mark G. Ballin, "An Experimental Study of the Effect of Tail Configuration on the Spinning Characteristics of General Aviation Aircraft," NASA CR-168578 (1982).

Marvin R. Barber, Charles K. Jones, and Thomas R. Sisk, "An Evaluation of the Handling Qualities of Seven General-Aviation Aircraft," NASA TN D-3726 (1966).

Marvin R. Barber and Jack Fischel, "General Aviation: The Seventies and Beyond," in NASA Langley Research Center staff, ed., *Vehicle Technology for Civil Aviation: the Seventies and Beyond,* NASA SP-292 (Washington, DC: NASA Scientific and Technical Information Office, 1971), pp. 317–332.

Marvin R. Barber and Joseph J. Tymczyszyn, "Wake Vortex Attenuation Flight Tests: A Status Report," in Joseph W. Stickle, ed., *1980 Aircraft Safety and Operating Problems*, Pt. 2 (Washington: NASA Scientific and Technical Information Office, 1981), pp. 387–408.

Hugh P. Bergeron, "Analysis of General Aviation Single-Pilot IFR Incident Data Obtained from the NASA Aviation Safety Reporting System," NASA TM-80206 (1980).

William Bihrle, Jr., Randy S. Hultberg, and William Mulcay, "Rotary Balance Data for a Typical Single-Engine General Aviation Design for an Angle-of-Attack range of 30 deg. to 90 deg.," NASA CR-2972 (1978).

William Bihrle, Jr., and Randy S. Hultberg, "Rotary Balance Data for a Typical Single-Engine General Aviation Design for an Angle-of-Attack range of 8 deg. to 90 deg.," NASA CR-3097 (1979).

F.W. Bircham, Jr., C. Gordon Fullerton, Glenn B. Gilyard, Thomas D. Wolf, and James F. Stewart, "A Preliminary Investigation of the Use of Throttles for Emergency Flight Control," NASA TM-4320 (1991).

Elaine M. Blount and Victoria I. Chung, "General Aviation Data Framework," AIAA Paper 2006-6478 (2006).

James S. Bowman, Jr., "Summary of Spin Technology as Related to Light General-Aviation Airplanes," NASA TN D-6575 (1971).

Albert L. Braslow and Theodore G. Ayers, "Application of Advanced Aerodynamics to Future Transport Aircraft," in NASA Langley Research Center staff, ed., *NASA Aircraft Safety and Operating Problems,* v. 1, NASA SP-270 (Washington, DC: NASA Scientific and Technical Information Office, 1971), pp. 165–176.

R.K. Breeze and G.M. Clark, "Light Transport and General Aviation Aircraft Icing Research Requirements," NASA CR-165290 (1981).

Carey S. Buttrill, ed., *NASA LaRC Workshop on Guidance, Navigation, Controls, and Dynamics for Atmospheric Flight, 1993*, NASA CP-10127 (Hampton, VA: NASA Langley Research Center, 1993).

Robert L. Cannaday, "Estimation of Moments and Products of Inertia of Light Aircraft Under Various Loading Conditions," NASA Langley Working Paper LWP-1135 (1973).

Huey D. Carden, "Full-Scale Crash-Test Evaluation of Two Load-Limiting Subfloors for General Aviation Airframes," NASA TP-2380 (1984).

Claude B. Castle and Emilio Alfaro-Bou, "Crash Tests of Three Identical Low-Wing Single-Engine Airplanes," NASA TP-2190 (1983).

Joseph R. Chambers, Long P. Yip, and Thomas M. Moul, "Wind Tunnel Investigation of an Advanced General Aviation Canard Configuration," NASA TM-85760 (1984).

J.R. Chambers and H.P. Stough III, "Summary of NASA Stall/Spin Research for General Aviation Configurations," AIAA Paper 86-2597 (1986).

Paul L. Coe, Jr., "Review of Drag Cleanup Tests in Langley Full-Scale Tunnel (From 1935 to 1945) Applicable to Current General Aviation Airplanes," NASA TN D-8206 (1976).

Jennifer H. Cole, Bruce R. Cogan, C. Gordon Fullerton, John J. Burken, Michael W. Venti, and Frank W. Burcham, "Emergency Flight Control of a Twin-Jet Commercial Aircraft Using Manual Throttle Manipulation," NASA DFRC-650 (2007).

George E. Cooper and Robert P. Harper, Jr., "The Use Of Pilot Rating in the Evaluation of Aircraft Handling Qualities," NASA TN D-5153 (1969).

George Cooper, Robert Harper, and Roy Martin, "Pilot Rating Scales," in Society of Experimental Test Pilots, *2004 Report to the Aerospace Profession* (Lancaster, CA: Society of Experimental Test Pilots, 2004), pp. 319–337.

Delwin R. Croom, Raymond D. Vogler, and John A. Thelander, "Low-Speed Wind-Tunnel Investigation of Flight Spoilers as Trailing-Vortex-Alleviation Devices on an Extended-Range Wide-Body Tri-Jet Airplane Model," NASA TN D-8373 (1976).

S.C. Crow and E.R. Bate, Jr., "Lifespan of Trailing Vortices in a Turbulent Atmosphere," *AIAA Journal of Aircraft,* v. 13, no. 7 (Jul. 1976), pp. 476–482.

C.H. Dearborn and Abe Silverstein, "Drag Analysis of Single-Engine Military Airplanes Tested in the NACA Full-Scale Wind Tunnel," NACA WR-489 (October 1940).

James H. Doolittle, "Accelerations in Flight," NACA TR-203 (1925).

Wayne A. Doty, "Flight Test Investigation of Certification Issues Pertaining to General-Aviation-Type Aircraft With Natural Laminar Flow," NASA CR-181967 (1990).

Richard Eppler and Dan M. Somers, "A Computer Program for the Design and Analysis of Low-Speed Airfoils," NASA TM-80210 (1980).

Marvin P. Fink and Delma C. Freeman, Jr., "Full-Scale Wind-Tunnel Investigation of Static Longitudinal and Lateral Characteristics of a Light Twin-Engine Aircraft," NASA TN D-4983 (Jan. 1969).

Stuart G. Flechner, Peter F. Jacobs, and Richard T. Whitcomb, "A High Subsonic Speed Wind-Tunnel Investigation of Winglets on a Representative Second-Generation Jet Transport Wing," NASA TN D-8264 (1976).

Alfred Gessow et al., *Wake Vortex Minimization*, NASA SP-409 (Washington, DC: NASA Scientific and Technical Information Office, 1977).

R.R. Gilruth, "Requirements for Satisfactory Flying Qualities of Airplanes," NACA TR-755 (1943) [previously issued as Wartime Report L-276 in 1941—ed.]

R.R. Gilruth and M.D. White, "Analysis and Prediction of Longitudinal Stability of Airplanes," NACA TR 711 (1941).

R.R. Gilruth and W.N. Turner, "Lateral Control Required For Satisfactory Flying Qualities Based on Flight Tests of Numerous Airplanes," NACA TR 715 (1941).

Melvin N. Gough and A.P. Beard, "Limitations of the Pilot in Applying Forces to Airplane Controls," NACA TN-550 (1936).

Harry Greenberg, "A Survey of Methods for Determining Stability Parameters of an Airplane from Dynamic Flight Measurements," NACA TM-2340 (1951).

G.M. Gregorek, K.D. Korkan, and R.J. Freuler, "The General Aviation Airfoil Design and Analysis Service—A Progress Report," in NASA LRC Staff, *Advanced Technology Airfoil Research*, v. 2, NASA CP-2046 (Washington, DC: NASA Scientific and Technical Information Office, 1979), pp. 99–104.

G.M. Gregorek and M.J. Hoffmann, "An Investigation of the Aerodynamic Characteristics of a New General Aviation Airfoil in Flight: Final Report," NASA CR-169477 (1982).

Randall D. Grove, Roland L. Bowles, and Stanley C. Mayhew, "A Procedure for Estimating Stability and Control Parameters from Flight Test Data by Using Maximum Likelihood Methods Employing a Real-Time Digital System," NASA TN D-6735 (1972).

John A. Harper, "DC-3 Handling Qualities Flight Tests: NACA—1950," in Society of Experimental Test Pilots, *1991 Report to the Aerospace Profession* (Lancaster, CA: Society of Experimental Test Pilots, 1991), pp. 264–265.

Charles E. Harris, ed., *FAA/NASA International Symposium on Advanced Structural Integrity Methods for Airframe Durability and Damage Tolerance*, Pts. 1–2, NASA CP-3274 (Hampton, VA: NASA Langley Research Center, 1994).

Charles E. Harris, ed., *The Second Joint NASA/FAA/DoD Conference on Aging Aircraft: Proceedings*, Pts. 1–2, NASA CP-1999-208982 (Hampton, VA: NASA Langley Research Center, 1999).

Roy V. Harris, Jr., and Jerry N. Hefner, "NASA Laminar-Flow Program—Past, Present, Future," in NASA Langley Research Center Staff, *Research in Natural Laminar Flow and Laminar-Flow Control,* Pts. 1–3, NASA CP-2487 (Hampton, VA: NASA Langley Research Center, 1987).

W.D. Hayes, "Linearized Supersonic Flow," North American Aviation, Inc., Report AL-222 (1947).

Jerry N. Hefner and Frances E. Sabo, eds., *Research in Natural Laminar Flow and Laminar-Flow Control,* NASA CP-2487, Pts. 1–3 (Washington, DC: NASA Scientific and Technical Information Division, 1987).

William C. Hoffman and Walter M. Hollister, "Forecast of the General Aviation Air Traffic Control Environment for the 1980's," NASA CR-137909 (1976).

Bruce J. Holmes, "Flight Evaluation of an Advanced Technology Light Twin-Engine Airplane (ATLIT)," NASA CR-2832 (1977).

B.J. Holmes, C.C. Cronin, E.C. Hastings, Jr., C.J. Obara, and C.P. Vandam, "Flight Research on Natural Laminar Flow," NTRS ID 88N14950 (1986).

Bruce J. Holmes, "U.S. General Aviation: The Ingredients for a Renaissance—A Vision and Technology Strategy for U.S. Industry, NASA, FAA, Universities," in Carey S. Buttrill, ed., *NASA LaRC Workshop on Guidance, Navigation, Controls, and Dynamics for Atmospheric Flight, 1993,* NASA CP-10127 (Hampton, VA: NASA Langley Research Center, 1993), pp. 1–53.

Lawrence D. Huebner, Scott C. Asbury, John E. Lamar, Robert E. McKinley, Jr., Robert C. Scott, William J. Small, and Abel O. Torres, *Transportation Beyond 2000: Technologies Needed for Engineering Design: Proceedings,* NASA CP-10184, Pts. 1–2 (Hampton, VA: NASA Langley Research Center, 1996).

Paul A. Hunter, "Flight Measurements of the Flying Qualities of Five Light Airplanes," NACA TN-1573 (1948).

Kenneth W. Iliff and Richard E. Maine, "Practical Aspects of a Maximum Likelihood Estimation Method to Extract Stability and Control Derivatives from Flight Data," NASA TN D-8209 (1976).

8

Karen E. Jackson, Richard L. Boitnott, Edwin L. Fasanella, Lisa Jones, and Karen H. Lyle, "A History of Full-Scale Aircraft and Rotorcraft Crash Testing and Simulation at NASA Langley Research Center," Paper presented at the 4th Triennial International Aircraft and Cabin Safety Research Conference, 15–18 Nov. 2004, Lisbon, Portugal, NTRS, CASI ID 20040191337.

B.V. Keppel, H. Eysink, J. Hammer, K. Hawley, P. Meredith, and J. Roskam, "A Study of Commuter Airplane Design Optimization," NASA CR-157210 (1978).

Vladislav Klein, "Determination of Stability and Control Parameters of a Light Airplane from Flight Data Using Two Estimation Methods," NASA TP-1306 (1979).

Paul C. Loschke, Marvin R. Barber, Calvin R. Jarvis, and Einar K. Enevoldson, "Handling Qualities of Light Aircraft with Advanced Control Systems and Displays," in Philip Donely et al., *NASA Aircraft Safety and Operating Problems*, v. 1, NASA SP-270 (Washington, DC: NASA Scientific and Technical Information Office, 1971), pp. 189–206.

Richard E. Maine and Kenneth W. Iliff, "A FORTRAN Program for Determining Aircraft Stability and Control Derivatives from Flight Data," NASA TN D-7831 (1975).

Robert J. McGhee, William D. Beasley, and Richard T. Whitcomb, "NASA Low- and Medium-Speed Airfoil Development," in NASA Langley Research Center Staff, *Advanced Technology Airfoil Research*, v. 2, NASA CP-2046, (Washington, DC: NASA Scientific and Technical Information Office, 1979), pp. 1–23.

G.L. Merrill, "Study of Small Turbofan Engines Applicable to Single-Engine Light Airplanes," NASA CR-137944 (1976).

S.R. Mettu, V. Shivakumar, J.M. Beek, F. Yeh, L.C. Williams, R.G. Forman, J.J. McMahon, and J.C. Newman, Jr., "NASGRO 3.0: A Software For Analyzing Aging Aircraft," in Charles E. Harris, ed., *The Second Joint NASA/FAA/DoD Conference on Aging Aircraft: Proceedings*, Pt. 2, NASA CP-1999-208982 (Hampton, VA: NASA Langley Research Center, 1999), pp. 792–801.

William F. Milliken, Jr., "Progress is Dynamic Stability and Control Research," *Journal of the Aeronautical Sciences*, v. 14, no. 9 (Sep. 1947), pp. 493–519.

Richard V. Rhode, "The Pressure Distribution Over the Horizontal and Vertical Tail Surfaces of the F6C-4 Pursuit Airplane in Violent Maneuvers," NACA TR-307 (1929).

Richard V. Rhode, "The Pressure Distribution Over the Wings and Tail Surfaces of a PW-9 Pursuit Airplane in Flight," NACA TR-364 (1931).

Nanette Scarpellini Metz, *Partnership and the Revitalization of Aviation: A Study of the Advanced General Aviation Transport Experiments Program, 1994–2001,* UNOAI Report 02-5 (Omaha, NE: University of Nebraska at Omaha Aviation Institute, 2002).

William J. Mulcay and Robert A. Rose, "Rotary Balance Data for a Typical Single-Engine General Aviation Design for an Angle of Attack Range of 8 deg. to 90 deg.," NASA CP-3200 (1980).

Max M. Munk, "General Biplane Theory," NACA TR 151 (1922).

Daniel G. Murri et al., "Wind Tunnel Results of the Low-Speed NLF(1)-0414F Airfoil," in NASA Langley Research Center Staff, *Research in Natural Laminar Flow and Laminar-Flow Control,* Pt. 3, NASA CP-2487 (Hampton, VA: NASA Langley Research Center, 1987).

NACA High-Speed Flight Station, "Flight Experience with Two High-Speed Airplanes Having Violent Lateral-Longitudinal Coupling in Aileron Rolls," NACA RM-H55A13 (1955).

NASA Dryden Flight Research Center, "Winglets," *NASA Technology Facts*, TF 2004-15 (2004), pp. 1–4.

NASA Langley Research Center Staff, *Vehicle Technology for Civil Aviation: The Seventies and Beyond*, NASA SP-292 (Washington, DC: NASA Scientific and Technical Information Office, 1972).

NASA Langley Research Center Staff, *Vehicle Technology for Civil Aviation: The Seventies and Beyond, Panel Discussion,* Supplement to NASA SP-292 (Washington, DC: NASA Scientific and Technical Information Office, 1972).

NASA Langley Research Center Staff, *NASA Aircraft Safety and Operating Problems, v. 1,* NASA SP-270 (Washington, DC: NASA Scientific and Technical Information Office, 1972).

NASA Langley Research Center Staff, *Advanced Technology Airfoil Research*, 2 vols., NASA CP-2046 (Washington, DC: NASA Scientific and Technical Information Office, 1979).

NASA Langley Research Center Staff, *Joint University Program for Air Transportation Research-1980,* NASA CP-2176 (Washington, DC: NASA Scientific and Technical Information Office, 1981).

NASA Langley Research Center Staff, *Joint University Program for Air Transportation Research-1981,* NASA CP-2224 (Washington, DC: NASA Scientific and Technical Information Office, 1982).

NASA Langley Research Center Staff, *Joint University Program for Air Transportation Research-1982,* NASA CP-2285 (Washington, DC: NASA Scientific and Technical Information Office, 1983).

NASA Langley Research Center, Frederick R. Morrell, ed., *Joint University Program for Air Transportation Research-1983,* NASA CP-2451 (Washington, DC: NASA Scientific and Technical Information Office, 1987).

NASA Langley Research Center, Frederick R. Morrell, ed., *Joint University Program for Air Transportation Research-1984,* NASA CP-2452 (Washington, DC: NASA Scientific and Technical Information Office, 1987).

NASA Langley Research Center, Frederick R. Morrell, ed., *Joint University Program for Air Transportation Research-1985,* NASA CP-2453 (Washington, DC: NASA Scientific and Technical Information Office, 1987).

NASA Langley Research Center, Frederick R. Morrell, ed., *Joint University Program for Air Transportation Research-1986,* NASA CP-2502 (Washington, DC: NASA Scientific and Technical Information Office, 1988).

NASA Langley Research Center, Frederick R. Morrell, ed., *Joint University Program for Air Transportation Research-1987,* NASA CP-3028 (Washington, DC: NASA Scientific and Technical Information Office, 1988).

NASA Langley Research Center, Frederick R. Morrell, ed., *Joint University Program for Air Transportation Research-1988–1989,* NASA CP-3063 (Washington, DC: NASA Scientific and Technical Information Office, 1990).

NASA Langley Research Center, Frederick R. Morrell, ed., *Joint University Program for Air Transportation Research-1989–1990,* NASA CP-3095 (Washington, DC: NASA Scientific and Technical Information Office, 1990).

NASA Langley Research Center, Frederick R. Morrell, ed., *Joint University Program for Air Transportation Research-1990–1991,* NASA CP-3131 (Washington, DC: NASA Scientific and Technical Information Office, 1991).

NASA Langley Research Center, Frederick R. Morrell, ed., *Joint University Program for Air Transportation Research-1991–1992,* NASA CP-3193 (Washington, DC: NASA Scientific and Technical Information Office, 1993).

NASA Langley Research Center, Frederick R. Morrell, ed., *FAA/NASA Joint University Program for Air Transportation Research-1992–1993,* NASA CP-3246, DOT/FAA/CT-94/03 (Hampton, VA: NASA Langley Research Center, 1994).

NASA Langley Research Center, Richard M. Hueschen, ed., *FAA/NASA Joint University Program for Air Transportation Research-1993–1994,* NASA CP-3305, DOT/FAA/CT-95/39 (Hampton, VA: NASA Langley Research Center, 1995).

NASA Langley Research Center, J.H. Remer, ed., *FAA/NASA Joint University Program for Air Transportation Research-1994–1995,* DOT/FAA/AR-98/3 (Hampton, VA: NASA Langley Research Center, 1998).

NASA Langley Research Center Staff, *Controls, Displays, and Information Transfer for General Aviation IFR Operations,* NASA CP 2279 (Hampton, VA: NASA Langley Research Center, 1983).

NASA Lewis Research Center Staff, *General Aviation Propulsion,* NASA CP-2126 (Cleveland, OH: NASA Lewis Research Center, 1980).

NASA Scientific and Technical Information Program, "General Aviation Technology Program," Release No. 76-51, NASA TM X-73051 (1976).

Anshal L. Neilhous, Walter L. Klinar, and Stanley H. Scher, "Status of Spin Research for Recent Airplane Designs," NASA TR R-57 (1960).

Ahmed K. Noor and Huey D. Carden, "Computational Methods for Crashworthiness," NACA CP-3223 (1993).

F.H. Norton, "A Preliminary Study of Airplane Performance," NACA TN-120 (1922).

F.H. Norton, "Practical Stability and Controllability of Airplanes," NACA TR-120 (1923).

F.H. Norton, "A Study of Longitudinal Dynamic Stability in Flight," NACA TR-170 (1924).

F.H. Norton, "The Measurement of the Damping in Roll of a JN-4H in Flight," NACA TR-167 (1924).

F.H. Norton and E.T. Allen, "Accelerations in Flight," NACA TR-99 (1921).

F.H. Norton and W.G. Brown, "Controllability and Maneuverability of Airplanes," NACA TR-153 (1923).

J.W. Olcott, E. Sackel, and D.R. Ellis, "Analysis and Flight Evaluation of a Small, Fixed-Wing Aircraft Equipped with Hinged Plate Spoilers," NASA CR-166247 (1981).

Patrick D. O'Neil and Scott E. Tarry, *Annotated Bibliography of Enabling Technologies for the Small Aircraft Transportation System,* UNOAI Report 02-3 (Omaha, NE: University of Nebraska at Omaha Aviation Institute, 2002).

Ladislao Pazmany et al., "Potential Structural Materials and Design Concepts for Light Aircraft," NASA CR-1285 (1969).

Bruce E. Peterman, "Laminar Flow: The Cessna Perspective," in NASA Langley Research Center Staff, *Research in Natural Laminar Flow and Laminar-Flow Control,* Pt. 1, NASA CP-2487 (Hampton, VA: NASA Langley Research Center, 1987).

Allan B. Pifko, Robert Winter, and Patricia L. Ogilvie, "DYCAST—A Finite Element Program for the Crash Analysis of Structures," NACA CR-4040 (1987).

R.H. Rickets, "Structural Testing for Static Failure, Flutter and Other Scary Things," NASA TM 84606 (1983).

Jan Roskam, ed., *Proceedings of the NASA Industry University General Aviation Drag Reduction Workshop*, NASA CR-145627 (Lawrence, KS: The University of Kansas Space Technology Center, 1976).

J. Roskam and M. Williams, "Comparison of Selected Lift and Sideslip Characteristics of the Ayres Thrush S2R-800, Winglets Off and Winglets On, to Full-Scale Wind Tunnel Data," NASA CR-165710 (1981).

J. Roskam and E. Wenninger, "A Revolutionary Approach to Composite Construction and Flight Management Systems for Small, General Aviation Airplanes," NTRS ID 94N25714 (1992).

R.A. Rozendaal, "Natural Laminar Flow Flight Experiments on a Swept Wing Business Jet-Boundary Layer Stability Analyses," NASA CR-3975 (1986).

8

Burt Rutan, "Development of a Small High-Aspect-Ratio Canard Aircraft," in Society of Experimental Test Pilots, *1976 Report to the Aerospace Profession* (Lancaster, CA: Society of Experimental Test Pilots, 1976), pp. 93–101.

Mark O. Schlegel, "Citation X: Development and Certification of a Mach 0.9+ Business Jet," in Society of Experimental Test Pilots, *1997 Report to the Aerospace Profession* (Lancaster, CA: Society of Experimental Test Pilots, 1997).

Michael S. Selig, Mark D. Maughmer, and Dan M. Somers, "An Airfoil for General Aviation Applications," in AIAA, *Proceedings of the 1990 AIAA/FAA Joint Symposium on General Aviation Systems* (Hampton, VA: NASA LRC, 1990), pp. 280–291, NTRS ID N91-12572 (1990).

B.P. Selberg and D.L. Cronin, "Aerodynamic Structural Study of Canard Wing, Dual Wing, and Conventional Wing Systems for General Aviation Applications," NASA CR-172529 (1985).

Shashi Seth, "Cockpit Weather Graphics Using Mobile Satellite Communications," in NASA Jet Propulsion Laboratory, *Proceedings of the Third International Mobile Satellite Conference* (Pasadena: NASA JPL, 1993), pp. 231–233.

William G. Sewall, Robert J. McGhee, David E. Hahne, and Frank L. Jordan, Jr., "Wind Tunnel Results of the High-Speed NLF(1)-0213 Airfoil," in Jerry N. Hefner and Frances E. Sabo, eds., *Research in Natural Laminar Flow and Laminar-Flow Control,* NASA CP-2487, Pt. 3 (Washington, DC: NASA Scientific and Technical Information Division, 1987).

Marvin Shinbrot, "On the Analysis of Linear and Nonlinear Dynamical Systems From Transient-Response Data," NACA TN-3288 (1954).

Michael Z. Sincoff et al., "General Aviation and Community Development," NASA CR-145776 (1975).

Frederick O. Smetana, Delbert C. Summey, and W. Donald Johnson, "Riding and Handling Qualities of Light Aircraft—A Review and Analysis," NASA CR-1975 (1972).

Harriet J. Smith, "A Flight Test Investigation of the Rolling Moments Induced on a T-37B Airplane in the Wake of a B-747 Airplane," NASA TM X-56031 (1975).

Hartley A. Soulé, "Flight Measurements of the Dynamic Longitudinal Stability of Several Airplanes and a Correlation of the Measurements with Pilots' Observations of Handling Characteristics," NACA TR-578 (1936).

Hartley A. Soulé, "Preliminary Investigation of the Flying Qualities of Airplanes," TR-700 (1940).

Robert F. Stengel, "It's Time to Reinvent the General Aviation Airplane," in Frederick R. Morrell, ed., *Joint University Program for Air Transportation Research-1986,* NASA CP-2502 (Washington, DC: NASA Scientific and Technical Information Office, 1988), pp. 81–105.

H. Paul Stough III, "A Summary of Spin-Recovery Parachute Experience on Light Airplanes," AIAA Paper 90-1317 (1990).

H. Paul Stough III, Daniel B. Shafer, Philip R. Schaffner, and Konstantinos S. Martzaklis, "Reducing Aviation Weather-Related Accidents Through High-Fidelity Weather Information Distribution and Presentation," Paper presented at the International Council of the Aeronautical Sciences, 27 Aug.–1 Sep. 2000, Harrowgate, U.K.

H. Paul Stough III, James F. Watson, Jr., Taumi S. Daniels, Konstantinos S. Martzaklis, Michael A. Jarrell, and Rodney K. Bougue, "New Technologies for Weather Accident Prevention," AIAA Paper 2005-7451 (2005).

William T. Suit, "Aerodynamic Parameters of the Navion Airplane Extracted from Flight Data," NASA TN D-6643 (1972).

William T. Suit and Robert L. Cannaday, "Comparison of Stability and Control Parameters for a Light, Single-Engine, High-Winged Aircraft Using Different Flight Test and Parameter Estimation Techniques," NASA TM-80163 (1979). T. Swift, "Wide-Spread Fatigue Damage Monitoring: Issues and Concerns," in Charles E. Harris, ed., *FAA/NASA International Symposium on Advanced Structural Integrity Methods for Airframe Durability and Damage Tolerance*, Pt. 2, NASA CP-3274 (Hampton, VA: NASA Langley Research Center, 1994), pp. 829–870.

Russell R. Tanner and Terry D. Montgomery, "Stability and Control Derivative Estimates Obtained from Flight Data for the Beech 99 Aircraft," NASA TM-72863 (1979).

Scott E. Tarry, Brent D. Bowen, and Jocelyn S. Nickerson, "The Small Aircraft Transportation System (SATS): Research Collaborations with the NASA Langley Research Center," in Brent D. Bowen et al., *The Aeronautics Education, Research, and Industry Alliance (AERIAL) 2002 Report,* UNOAI Report 02-7 (Omaha, NE: University of Nebraska at Omaha Aviation Institute, 2002), pp. 45–57.

University of Kansas, Department of Aerospace Engineering, "An Investigation of Separate Stability Augmentation Systems for General Aviation Aircraft," NASA CR-138120 (1974).

University of Kansas, Department of Aerospace Engineering, "Preliminary Design Studies of an Advanced General Aviation Aircraft," in Universities Space Research Association, *Proceedings of the Seventh Annual Summer Conference, NASA-USRA Advanced Design Program*, NTRS ID 93N29717 (Houston, TX: NASA-USRA, 1991), pp. 45–56.

Victor. L. Vaughan, Jr., and Emilio Alfaro-Bou, "Impact Dynamics Research Facility for Full-Scale Aircraft Crash Testing," NASA TND-8179 (1976).

Victor L. Vaughan, Jr., and Robert J. Hayduk, "Crash Tests of Four Identical High-Wing Single Engine Airplanes," NASA TP-1699 (1980).

M. Susan Williams and Edwin L. Fasanella, "Crash Tests of Four Low-Wing Twin-Engine Airplanes with Truss-Reinforced Fuselage Structure," NASA TP-2070 (1982).

J.K. Viken et al., "Design of the Low-Speed NLF (1)-0414F and the High-Speed HSNLF(1)-0213 Airfoils with High-Lift Systems," in NASA Langley Research Center Staff, *Research in Natural Laminar Flow and Laminar-Flow Control,* Pt. 3, NASA CP-2487 (Hampton, VA: NASA Langley Research Center, 1987).

Edward P. Warner and F.H. Norton, "Preliminary Report on Free Flight Tests," NACA TR-70 (1920).

Fred E. Weick and Robert T. Jones, "Response and Analysis of NACA Lateral Control Research" TR-605 (1937).

Richard T. Whitcomb, "A Study of the Zero-Lift Drag-Rise Characteristics of Wing Body Combinations Near the Speed of Sound," NACA Report 1237 (1956).

Richard T. Whitcomb, "A Design Approach and Selected Wind-Tunnel Results at High Subsonic Speeds for Wing-Tip Mounted Winglets," NASA TN D-8260 (1976).

M. Susan Williams and Edwin L. Fasanella, "Results from Tests of Three Prototype General Aviation Seats," NASA TM-84533 (1982).

M. Susan Williams and Edwin L. Fasanella, "Crash Tests of Four Low-Wing Twin-Engine Airplanes with Truss-Reinforced Fuselage Structure," NASA TP-2070 (1982).

Williams International, "The General Aviation Propulsion (GAP) Program," NASA CR-2008-215266 (2008).

Chester H. Wolowicz and Roxanah B. Yancey, "Longitudinal Aerodynamic Characteristics of Light Twin-Engine, Propeller-Driven Airplanes," NASA TN D-6800 (1972).

Chester H. Wolowicz and Roxanah B. Yancey, "Lateral-Directional Aerodynamic Characteristics of Light, Twin-Engine Propeller-Driven Airplanes," NASA TN D-6946 (1972).

Chester H. Wolowicz and Roxanah B. Yancey, "Experimental Determination of Airplane Mass and Inertial Characteristics," NASA TR R-433 (1974).

Books and Monographs:

David A. Anderton, *NASA Aeronautics*, NASA SP-85 (Washington, DC: National Aeronautics and Space Administration, 1982).

John V. Becker, *The High-Speed Frontier: Case Histories of Four NACA Programs, 1920–1950,* NASA SP-445 (Washington, DC: NASA Scientific and Technical Information Branch, 1980).

Roger E. Bilstein, *Flight Patterns: Trends of Aeronautical Development in the United States, 1918–1929* (Athens, GA: The University of Georgia Press, 1983).

Roger E. Bilstein, *The American Aerospace Industry* (New York: Twayne Publishers, 1996).

Paul F. Borchers, James A. Franklin, and Jay W. Fletcher, *Flight Research at Ames: Fifty-Seven Years of Development and Validation of Aeronautical Technology,* NASA SP-1998-3300 (Moffett Field, CA: Ames Research Center, 1998).

Joseph R. Chambers, *Partners in Freedom: Contributions of the Langley Research Center to U.S. Military Aircraft of the 1990s,* NASA SP-2000-4519 (Washington, DC: GPO, 2000).

Joseph R. Chambers, *Concept to Reality: Contributions of the NASA Langley Research Center to U.S. Civil Aircraft of the 1990s,* NASA SP-2003-4529 (Washington, DC: GPO, 2003).

Joseph R. Chambers, *Innovation in Flight: Research of the NASA Langley Research Center on Revolutionary Advanced Concepts for Aeronautics,* NASA SP-2005-4539 (Washington, DC: GPO, 2005).

Virginia P. Dawson, *Engines and Innovation: Lewis Laboratory and American Propulsion Technology,* NASA SP-4306 (Washington, DC: NASA, 1991).

Eugene M. Emme, ed., *Two Hundred Years of Flight in America: A Bicentennial Survey* (San Diego, CA: American Astronautical Society and Univelt, 1977).

Jeffrey L. Ethell, *Fuel Economy in Aviation,* NASA SP-462 (Washington, DC: NASA Scientific and Technical Information Branch, 1983).

Michael H. Gorn, *Expanding the Envelope: Flight Research at NACA and NASA* (Lexington, KY: The University Press of Kentucky, 2001).

George W. Gray, *Frontiers of Flight: The Story of NACA Research* (New York: Alfred A. Knopf, 1948).

Dunstan Hadley, *One Second to Live: Pilots' Tales of the Stall and Spin* (Shrewsbury, U.K.: Airlife Publishing Ltd., 1997).

Richard P. Hallion, *On the Frontier: Flight Research at Dryden, 1946–1981,* NASA SP-4303 (Washington, DC: GPO, 1984).

James R. Hansen, *Engineer in Charge: A History of the Langley Aeronautical Laboratory, 1917–1958,* NASA SP-4305 (Washington, DC: GPO, 1987).

Edwin P. Hartman, *Adventures in Research: A History of Ames Research Center 1940–1965,* NASA SP-4302 (Washington, DC: GPO, 1970).

Laurence K. Loftin, Jr., *Quest for Performance: The Evolution of Modern Aircraft,* NASA SP-468 (Washington, DC: GPO, 1965).

Pamela E. Mack, ed., *From Engineering Science to Big Science: The NACA and NASA Collier Trophy Research Project Winners,* NASA SP-4219 (Washington, DC: NASA, 1998).

Donald L. Mallick with Peter W. Merlin, *The Smell of Kerosene: A Test Pilot's Odyssey,* NASA SP-4108 (Washington, DC: GPO, 2003).

Kenneth Munson, *Private Aircraft, Business and General Purpose Since 1946* (London: Blandford Press, 1967).

Donald M. Pattillo, *A History in the Making: 80 Turbulent Years in the American General Aviation Industry* (New York: McGraw-Hill, 1998).

Jay M. Price and the AIAA Wichita Section, *Wichita's Legacy of Flight* (Charleston, SC: Arcadia Publishing, 2003).

Fred E. Weick and James R. Hansen, *From the Ground Up: The Autobiography of an Aeronautical Engineer* (Washington, DC: Smithsonian Institution Press, 1988).

Alan H. Wheeler, *". . . That Nothing Failed Them:" Testing Aeroplanes in Wartime* (London: G.T. Foulis and Co., Ltd., 1963).

John H. Winant, *Keep Business Flying: A History of The National Business Aircraft Association, Inc., 1946–1986* (Washington, DC: The National Business Aircraft Association, 1989).

THIS PAGE INTENTIONALLY BLANK

One of two small APV-3 aircraft flown in the joint Ames-Dryen Networked UAV Teaming Experiment flares for landing on a roadway on a remote area of Edwards Air Force Base. NASA.

CASE 9

The Evolution of Remotely Piloted Research Vehicles

Peter W. Merlin

9

For over a half century, NASA researchers have worked to make remotely piloted research vehicles to complement piloted aircraft, in the forms of furnishing cheap "quick look" design validations, undertaking testing too hazardous for piloted aircraft, and furnishing new research capabilities such as high-altitude solar-powered environmental monitoring. The RPRV has evolved to sophisticated fly-by-wire inherently unstable vehicles with composite structures and integrated propulsion.

SINCE THE MID-1990S, researchers at the National Aeronautics and Space Administration (NASA) have increasingly relied on unmanned aerial vehicles (UAVs) to fill roles traditionally defined by piloted aircraft. Instead of strapping themselves into the cockpit and taking off into the unknown, test pilots more often fly remotely piloted research vehicles (RPRVs) from the safety of a ground-based control station. Such craft are ideally suited to serve as aerodynamic and systems testbeds, airborne science platforms, and launch aircraft, or to explore unorthodox flight modes. NASA scientists began exploring the RPRV concept at Dryden Flight Research Center, Edwards, CA, in the 1960s. Since then, NASA RPRV development has contributed significantly to such technological innovations as autopilot systems, data links, and inertial navigation systems, among others. By the beginning of the 21st century, use of the once-novel RPRV concept had become standard practice.

There is no substitute—wind tunnel and computer modeling notwithstanding—for actual flight data. The RPRV provides real-world results while providing the ground pilot with precisely the same responsibilities and tasks as if he were sitting in a cockpit onboard a research airplane. As in piloted flight-testing, the remote pilot is responsible for performing data maneuvers, evaluating vehicle and systems performance, and reacting to emergency situations.

A ground pilot may, in fact, be considered the most versatile element of an RPRV system. Since experimental vehicles are designed to venture into unexplored engineering territory, the remote pilot may be called upon to repeat or abort a test point, or execute additional tasks not included in the original flight plan. Not all unmanned research vehicles require a pilot in the loop, but having one adds flexibility and provides an additional level of safety when performing hazardous maneuvers.[1]

Reducing the High Cost of Flight Research

Research aircraft are designed to explore advanced technologies and new fight regimes. Consequently, they are often relatively expensive to build and operate, and inherently risky to fly. Flight research from the earliest days of aviation well into the mid-20th century resulted in a staggering loss of life and valuable, often one-of-a-kind, aircraft.

This was tragically illustrated during experimental testing of advanced aircraft concepts, early jet-powered aircraft, and supersonic rocket planes of the 1940s and 1950s at Muroc Army Air Field in the Mojave Desert. Between 1943 and 1959, more than two-dozen research airplanes and prototypes were lost in accidents, more than half of them fatal. Among these were several of Northrop's flying wing designs, including the N9M-1, XP-56, and both YB-49 prototypes. Early variants of Lockheed P-80 and F-104 jet fighters were lost, along with the two Martin XB-51 bomber prototypes. A rocket-powered Bell X-1 and its second-generation stablemates, the X-1A and X-1D, were lost to explosions—all fortunately nonfatal—and Capt. Milburn Apt died in the Bell X-2 after becoming the first human to fly more than three times the speed of sound.

By the 1960s, researchers began to recognize the value of using remotely piloted vehicles (RPVs) to mitigate the risks associated with flight-testing. During World War I and World War II, remotely controlled aircraft had been developed as weapons. In the postwar era, drones served as targets for missile tests and for such tasks as flying through clouds of radioactive fallout from nuclear explosions to collect particulate samples without endangering aircrews. By the 1950s, cruise-missile prototypes, such as the Regulus and X-10, were taking off and landing under radio control. Several of these vehicles crashed, but without a crew on board, there was no risk of losing a valuable test pilot.[2] Over the

1. R. Dale Reed, "Flight Research Techniques Utilizing Remotely Piloted Research Vehicles," *UAV Flight Test Lessons Learned Workshop, NASA Dryden Flight Research Center, Edwards, CA, Dec. 18, 1996.*
2. Peter W. Merlin and Tony Moore, *X-Plane Crashes: Exploring Experimental, Rocket Plane, and Spycraft Incidents, Accidents and Crash Sites* (North Branch: Specialty Press, 2008), pp. 130–149.

years, advances in electronics greatly increased the reliability of control systems, rendering development of RPRVs more practical. Early efforts focused on guidance and navigation, stabilization, and remote control. Eventually, designers worked to improve technologies to support these capabilities through the integration of improved avionics, microprocessors, and computers. The RPRV concept was attractive to researchers because it built confidence in new technology through demonstration under actual flight conditions, at relatively low cost, in quick response to demand, and at no risk to the pilot.

Taking the pilot out of the airplane provided additional savings in terms of development and fabrication. The cost and complexity of robotic and remotely piloted vehicles are generally less than those of comparable aircraft that require an onboard crew, because there is no need for life-support systems, escape and survival equipment, or hygiene facilities. Hazardous testing can be accomplished with a vehicle that may be considered expendable or semiexpendable.

Quick response to customer requirements and reduced program costs resulted from the elimination of redundant systems (usually added for crew safety) and man-rating tests, and through the use of less complex structures and systems. Subscale test vehicles generally cost less than full-size airplanes while providing usable aerodynamic and systems data. The use of programmable ground-based control systems provides additional flexibility and eliminates downtime resulting from the need for extensive aircraft modifications.[3]

Modeling the Future: Radio-Controlled Lifting Bodies

Robert Dale Reed, an engineer at NASA's Flight Research Center (later renamed NASA Dryden Flight Research Center) at Edwards Air Force Base and an avid radio-controlled (R/C) model airplane hobbyist, was one of the first to recognize the RPRV potential. Previous drone aircraft had been used for reconnaissance or strike missions, flying a restricted number of maneuvers with the help of an autopilot or radio signals from a ground station. The RPRV, on the other hand, offered a versatile platform for operating in what Reed called "unexplored engineering territory."[4] In 1962, when astronauts returned from space

3. Reed, "Flight Research Techniques Utilizing Remotely Piloted Research Vehicles."

4. Richard P. Hallion and Michael H. Gorn, *On the Frontier: Experimental Flight at NASA Dryden* (Washington, DC: Smithsonian Books, 2002), p. 207.

in capsules that splashed down in the ocean, NASA and Air Force engineers were discussing a revolutionary concept for spacecraft reentry vehicles. Wingless lifting bodies—half-cone-shaped vehicles capable of controlled flight using the craft's fuselage shape to produce stability and lift—could be controlled from atmospheric entry to gliding touchdown on a conventional runway. Skeptics believed such craft would require deployable wings and possibly even pop-out jet engines.

Reed believed the basic lifting body concept was sound and set out to convince his peers. His first modest efforts at flight demonstration were confined to hand-launching small paper models in the hallways of the Flight Research Center. His next step involved construction, from balsa wood, of a 24-inch-long free-flight model.

The vehicle's shape was a half-cone design with twin vertical-stabilizer fins with rudders and a bump representing a cockpit canopy. Elevons provided longitudinal trim and turning control. Spring-wired tricycle wheels served as landing gear. Reed adjusted the craft's center of gravity until he was satisfied and began a series of hand-launched flight tests. He began at ground level and finally moved to the top of the NASA Administration building, gradually expanding the performance envelope. Reed found the model had a steep gliding angle but remained upright and landed on its gear.

He soon embarked on a path that presaged eventual testing of a full-scale, piloted vehicle. He attached a thread to the upper part of the nose gear and ran to tow the lifting body aloft, as one would launch a kite. Reed then turned to one of his favorite hobbies: radio-controlled, gas-powered model airplanes. He had previously used R/C models to tow free flight model gliders with great success. By attaching the towline to the top of the R/C model's fuselage, just at the trailing edge of the wing, he ensured minimum effect on the tow plane from the motions of the lifting body model behind it.

Reed conducted his flight tests at Sterk's Ranch in nearby Lancaster while his wife, Donna, documented the demonstrations with an 8-millimeter motion picture camera. When the R/C tow plane reached a sufficient altitude for extended gliding flight, a vacuum timer released the lifting body model from the towline. The lifting body demonstrated stable flight and landing characteristics, inspiring Reed and other researchers to pursue development of a full-scale,

Radio-controlled mother ship and models of Hyper III and M2-F2 on lakebed with research staff. Left to right: Richard C. Eldredge, Dale Reed, James O. Newman, and Bob McDonald. NASA.

piloted lifting body, dubbed the M2-F1.[5] Reed's R/C model experiments provided a low-cost demonstration capability for a revolutionary concept. Success with the model built confidence in proposals for a full-scale lifting body. Essentially, the model was scaled up to a length of 20 feet, with a span of 14.167 feet. A tubular steel framework provided internal support for the cockpit and landing gear. The outer

5. R. Dale Reed, with Darlene Lister, *Wingless Flight: The Lifting Body Story*, NASA SP-4220 (Washington, DC: NASA, 1997), pp. 8–23.

shell was comprised of mahogany ribs and spars covered with plywood and doped cloth skin. As with the small model, the full-scale M2-F1 was towed into the air—first behind a Pontiac convertible and later behind a C-47 transport for extended glide flights. Just as the models paved the way for full-scale, piloted testing, the M2-F1 served as a pathfinder for a series of air-launched heavyweight lifting body vehicles—flown between 1966 and 1975—that provided data eventually used in development of the Space Shuttle and other aerospace vehicles.[6]

By 1969, Reed had teamed with Dick Eldredge, one of the original engineers from the M2-F1 project, for a series of studies involving modeling spacecraft-landing techniques. Still seeking alternatives to splashdown, the pair experimented with deployable wings and paraglider concepts. Reed discussed his ideas with Max Faget, director of engineering at the Manned Spacecraft Center (now NASA Johnson Space Center) in Houston, TX. Faget, who had played a major role in designing the Mercury, Gemini, and Apollo spacecraft, had proposed a Gemini-derived vehicle capable of carrying 12 astronauts. Known as the "Big G," it was to be flown to a landing beneath a gliding parachute canopy.

Reed proposed a single-pilot test vehicle to demonstrate paraglider-landing techniques similar to those used with his models. The Parawing demonstrator would be launched from a helicopter and glide to a landing beneath a Rogallo wing, as used in typical hang glider designs. Spacecraft-type viewports would provide visibility for realistic simulation of Big G design characteristics.[7] Faget offered to lend a borrowed Navy SH-3A helicopter—one being used to support the Apollo program—to the Flight Research Center and provide enough money for several Rogallo parafoils. Hugh Jackson was selected as project pilot, but for safety reasons, Reed suggested that the test vehicle initially be flown by radio control with a dummy on board.

Eldredge designed the Parawing vehicle, incorporating a generic ogival lifting body shape with an aluminum internal support structure, Gemini-style viewing ports, a pilot's seat mounted on surplus shock struts from Apollo crew couches, and landing skids. A general-aviation autopilot servo was used to actuate the parachute control lines. A side stick controller was installed to control the servo. On planned piloted flights, it would be hand-actuated, but in the test configuration, model airplane

6. Ibid.

7. Reed and Lister, *Wingless Flight*, pp. 158–161.

servos were used to move the side stick. For realism, engineers placed an anthropomorphic dummy in the pilot's seat and tied the dummy's hands in its lap to prevent interference with the controls. The dummy and airframe were instrumented to record accelerations, decelerations, and shock loads as the parachute opened.

The Parawing test vehicle was then mounted on the side of the helicopter using a pneumatic hook release borrowed from the M2-F2 lifting body launch adapter. Donald Mallick and Bruce Peterson flew the SH-3A to an altitude of approximately 10,000 feet and released the Parawing test vehicle above Rosamond Dry Lake. Using his R/C model controls, Reed guided the craft to a safe landing. He and Eldredge conducted 30 successful radio-controlled test flights between February and October 1969. Shortly before the first scheduled piloted tests were to take place, however, officials at the Manned Spacecraft Center canceled the project. The next planned piloted spacecraft, the Space Shuttle orbiter, would be designed to land on a runway like a conventional airplane does. There was no need to pursue a paraglider system.[8] This, however, did not spell the end of Reed's paraglider research. A few decades later, he would again find himself involved with paraglider recovery systems for the Spacecraft Autoland Project and the X-38 Crew Return Vehicle technology demonstration.

Hyper III: The First True RPRV

In support of the lifting body program, Dale Reed had built a small fleet of models, including variations on the M2-F2 and FDL-7 concepts. The M2-F2 was a half cone with twin stabilizer fins like the M2-F1 but with the cockpit bulge moved forward from midfuselage to the nose. The full-scale heavyweight M2-F2 suffered some stability problems and eventually crashed, although it was later rebuilt as the M2-F3 with an additional vertical stabilizer. The FDL-7 had a sleek shape (somewhat resembling a flatiron) with four stabilizer fins, two horizontal, and two that were canted outward. Engineers at the Air Force Flight Dynamics Laboratory at Wright-Patterson Air Force Base, OH, designed it with hypersonic-flight characteristics in mind. Variants included wingless versions as well as those equipped with fixed or pop-out wings for extended gliding.[9] Reed launched his creations from a twin-engine R/C model

8. Ibid.

9. Hallion and Gorn, *On the Frontier*, p. 207.

The Hyper III, with its ground cockpit visible at upper left, was a full-scale lifting body remotely piloted research vehicle. NASA.

plane he dubbed "Mother," since it served as a mother ship for his lifting body models. With a 10.5-foot wingspan, Mother was capable of lofting models of various sizes to useful altitudes for extended glide flights. By the end of 1968, Reed's mother ship had successfully made 120 drops from an altitude of around 1,000 feet.

One day, Reed asked research pilot Milton O. Thompson if he thought he would be able to control a research airplane from the ground using an attitude-indicator instrument as a reference. Thompson thought this was possible and agreed to try it using Reed's mother ship. Within a month, at a cost of $500, Mother was modified, and Thompson had successfully demonstrated the ability to fly the craft from the ground using the instrument reference.[10] Next, Reed wanted to explore the possibility of flying a full-scale research airplane from a ground cockpit. Because of his interest in lifting bodies, he selected a simplified variant of the FDL-7 configuration based on research accomplished at NASA Langley Research Center. Known as Hyper III—because the shape would have a lift-to-drag (L/D) ratio of 3.0 at hypersonic speeds—the test vehicle had a 32-foot-long fuselage with a narrow delta planform and trapezoidal

10. Ibid.

cross-section, stabilizer fins, and fixed straight wings spanning 18.5 feet to simulate pop-out airfoils that could be used to improve the low-speed glide ratio of a reentry vehicle. The Hyper III RPRV weighed about 1,000 pounds.[11]

Reed recruited numerous volunteers for his low-budget, low-priority project. Dick Fischer, a designer of R/C models as well as full-scale homebuilt aircraft, joined the team as operations engineer and designed the vehicle's structure. With previous control-system engineering experience on the X-15, Bill "Pete" Peterson designed a control system for the Hyper III. Reed also recruited aircraft inspector Ed Browne, painter Billy Schuler, crew chief Herman Dorr, and mechanics Willard Dives, Bill Mersereau, and Herb Scott.

The craft was built in the Flight Research Center's fabrication shops. Frank McDonald and Howard Curtis assembled the fuselage, consisting of a Dacron-covered, steel-tube frame with a molded fiberglass nose assembly. LaVern Kelly constructed the stabilizer fins from sheet aluminum. Daniel Garrabrant borrowed and assembled aluminum wings from an HP-11 sailplane kit. The vehicle was built at a cost of just $6,500 and without interfering with the Center's other, higher-priority projects.[12] The team managed to scrounge and recycle a variety of items for the vehicle's control system. These included a Kraft uplink from a model airplane radio-control system and miniature hydraulic pumps from the Air Force's Precision Recovery Including Maneuvering Entry (PRIME) lifting body program. Peterson designed the Hyper III control system to work from either of two Kraft receivers, mounted on the top and bottom of the vehicle, depending on signal strength. If either malfunctioned or suffered interference, an electronic circuit switched control signals to the operating receiver to actuate the elevons. Keith Anderson modified the PRIME hydraulic actuator system for use on the Hyper III.

The team also developed an emergency-recovery parachute system in case control of the vehicle was lost. Dave Gold, of Northrop, who had helped design the Apollo spacecraft parachute system, and John Rifenberry, of the Flight Research Center life-support shop, designed a system that included a drogue chute and three main parachutes that would safely lower the vehicle to the ground onto its landing skids. Pyrotechnics expert Chester Bergener assumed responsibility for the

11. Ibid.

12. Reed and Lister, *Wingless Flight*, pp. 161–165.

drogue's firing system.[13] To test the recovery system, technicians mounted the Hyper III on a flatbed truck and fired the drogue-extraction system while racing across the dry lakebed, but weak radio signals kept the three main chutes from deploying. To test the clustered main parachutes, the team dropped a weight equivalent to the vehicle from a helicopter.

Tom McAlister assembled a ground cockpit with instruments identical to those in a fixed-base flight simulator. An attitude indicator displayed roll, pitch, heading, and sideslip. Other instruments showed airspeed, altitude, angle of attack, and control-surface position. Don Yount and Chuck Bailey installed a 12-channel downlink telemetry system to record data and drive the cockpit instruments. The ground cockpit station was designed to be transported to the landing area on a two-wheeled trailer.[14] On December 12, 1969, Bruce Peterson piloted the SH-3A helicopter that towed the Hyper III to an altitude of 10,000 feet above the lakebed. Hanging at the end of a 400-foot cable, the nose of the Hyper III had a disturbing tendency to drift to one side or another. Reed realized later that he should have added a small drag chute to stabilize the craft's heading prior to launch. Peterson started and stopped forward flight several times until the Hyper III stabilized in a forward climb attitude, downwind with a northerly heading.

As soon as Peterson released the hook, Thompson took control of the lifting body. He flew the vehicle north for 3 miles, then reversed course and steered toward the landing site, covering another 3 miles. During each straight course, Thompson performed pitch doublets and oscillations in order to collect aerodynamic data. Since the Hyper III was not equipped with an onboard video camera, Thompson was forced to fly on instruments alone. Gary Layton, in the Flight Research Center control room, watched the radar data showing the vehicle's position and relayed information to Thompson via radio.

Dick Fischer stood beside Thompson to take control of the Hyper III just before the landing flare, using the model airplane radio-control box. Several miles away, the Hyper III was invisible in the hazy sky as it descended toward the lakebed. Thompson called out altitude readings as Fischer strained to see the vehicle. Suddenly, he spotted the lifting body, when it was on final approach just 1,000 feet above the ground. Thompson relinquished control, and Fischer commanded

13. Ibid.

14. Ibid.

a slight left roll to confirm he had established radio contact. He then leveled the aircraft and executed a landing flare, bringing the Hyper III down softly on its skids.

Thompson found the experience of flying the RPRV exciting and challenging. After the 3-minute flight, he was as physically and emotionally drained as he had been after piloting first flights in piloted research aircraft. Worries that lack of motion and visual cues might hurt his piloting performance proved unfounded. It seemed as natural to control the Hyper III on gauges as it did any other airplane or simulator, responding solely to instrument readings. Twice during the flight, he used his experience to compensate for departures from predicted aerodynamic characteristics when the lift-to-drag ratio proved lower than expected, thus demonstrating the value of having a research pilot at the controls.[15]

The Next, More Ambitious Step: The Piper PA-30

Encouraged by the results of the Hyper III experiment, Reed and his team decided to convert a full-scale production airplane into a RPRV. They selected the Flight Research Center's modified Piper PA-30 Twin Comanche, a light, twin-engine propeller plane that was equipped with both conventional and fly-by-wire control systems. Technicians installed uplink/downlink telemetry equipment to transmit radio commands and data. A television camera, mounted above the cockpit windscreen, transmitted images to the ground pilot to provide a visual reference—a significant improvement over the Hyper III cockpit. To provide the pilot with physical cues, as well, the team developed a harness with small electronic motors connected to straps surrounding the pilot's torso. During maneuvers such as sideslips and stalls, the straps exerted forces to simulate lateral accelerations in accordance with data telemetered from the RPRV, thus providing the pilot with a more natural "feel."[16] The original control system of pulleys and cables was left intact, but a few minor modifications were incorporated. The right-hand, or safety pilot's, controls were connected directly to the flight control surfaces via conventional control cables and to the nose gear steering system via pushrods. The left-hand control wheel and rudder pedals were completely independent of the control cables, instead operating the control surfaces via hydraulic actuators through an electronic stability-augmentation system.

15. Ibid.

16. Hallion and Gorn, *On the Frontier*, pp. 208–209.

Bungees were installed to give the left-hand controls an artificial "feel." A friction control was added to provide free movement of the throttles while still providing friction control on the propellers when the remote throttle was in operation.

When flown in RPRV configuration, the left-hand cockpit controls were disabled, and signals from a remote control receiver fed directly into the control system electronics. Control of the airplane from the ground cockpit was functionally identical to control from the pilot's seat. A safety trip channel was added to disengage the control system whenever the airborne remote control system failed to receive intelligible commands. In such a situation, the safety pilot would immediately take control.[17] Flight trials began in October 1971, with research pilot Einar Enevoldson flying the PA-30 from the ground while Thomas C. McMurtry rode on board as safety pilot, ready to take control if problems developed. Following a series of incremental buildup flights, Enevoldson eventually flew the airplane unassisted from takeoff to landing, demonstrating precise instrument landing system approaches, stall recovery, and other maneuvers.[18] By February 1973, the project was nearly complete. The research team had successfully developed and demonstrated basic RPRV hardware and operating techniques quickly and at relatively low cost. These achievements were critical to follow-on programs that would rely on the use of remotely piloted vehicles to reduce the cost of flight research while maintaining or expanding data return.[19]

Extending the Vision: The Evolution of Mini-Sniffer

The Mini-Sniffer program was initiated in 1975 to develop a small, unpiloted, propeller-driven aircraft with which to conduct research on turbulence, natural particulates, and manmade pollutants in the upper atmosphere. Unencumbered and flying at speeds of around 45 mph, the craft was designed to reach a maximum altitude of 90,000 feet. The Mini-Sniffer was capable of carrying a 25-pound instrument package to 70,000 feet and cruising there for about 1 hour within a 200-mile range.

17. Operations Fact Sheet OFS-808-77-1, Model PA-30, prepared by W. Albrecht, Sept. 20, 1977, DFRC Historical Reference Collection, NASA Dryden Flight Research Center, Edwards, CA.
18. Hallion and Gorn, *On the Frontier*, pp. 208–209.
19. OAST Flight Research Operations Review Committee minutes, Jan. 31–Feb. 1, 1973, DFRC Historical Reference Collection, NASA Dryden Flight Research Center, Edwards, CA.

The Aircraft Propulsion Division of NASA's Office of Aeronautics and Space Technology sponsored the project and a team at the Flight Research Center, led by R. Dale Reed, was charged with designing and testing the airplane. Researchers at Johnson Space Center developed a hydrazine-fueled engine for use at high altitudes, where oxygen is scarce. To avoid delays while waiting for the revolutionary new engine, Reed's team built two Mini-Sniffer aircraft powered by conventional gasoline engines. These were used for validating the airplane's structure, aerodynamics, handling qualities, guidance and control systems, and operational techniques.[20] As Reed worked on the airframe design, he built small, hand-launched balsa wood gliders for qualitative evaluation of different configurations. He decided from the outset that the Mini-Sniffer should have a pusher engine to leave the nose-mounted payload free to collect air samples without disruption or contamination from the engine. Climb performance was given priority over cruise performance.

Eventually, Reed's team constructed three configurations. The first two—using the same airframe—were powered by a single two-stroke, gasoline-fueled go-cart engine driving a 22-inch-diameter propeller. The third was powered by a hydrazine-fueled engine developed by James W. Akkerman, a propulsion engineer at Johnson Space Center. Thirty-three flights were completed with the three airplanes, each of which provided experimental research results. Thanks to the use of a six-degree-of-freedom simulator, none of the Mini-Sniffer flights had to be devoted to training. Simulation also proved useful for designing the control system and, when compared with flight results, proved an accurate representation of the vehicle's flight characteristics.

The Mini-Sniffer I featured an 18-foot-span, aft-mounted wing, and a nose-mounted canard. Initially, it was flown via a model airplane radio-control box. Dual-redundant batteries supplied power, and fail-safe units were provided to put the airplane into a gliding turn for landing descent in the event of a transmitter failure. After 12 test flights, Reed abandoned the flying-wing canard configuration for one with substantially greater stability.[21] The Mini-Sniffer II design had a 22-foot wingspan with twin tail booms supporting a horizontal stabilizer. This configuration was less susceptible to flat spin, encountered with the Mini-Sniffer I on its

20. R. Dale Reed, "High-Flying Mini-Sniffer RPV: Mars Bound?" *Astronautics and Aeronautics* (June 1978), pp. 26–39.
21. Ibid.

final flight when the ground pilot's timing between right and left yaw pulses coupled the adverse yaw characteristics of the ailerons with the vehicle's Dutch roll motions. The ensuing unrecoverable spin resulted in only minor damage to the airplane, as the landing gear absorbed most of the impact forces. It took 3 weeks to restore the airframe to flying condition and convert it to the Mini-Sniffer II configuration. Dihedral wingtips provided additional roll control.

The modified craft was flown 20 times, including 10 flights using wing-mounted ailerons to evaluate their effectiveness in controlling the aircraft. Simulations showed that summing a yaw-rate gyro and pilot inputs to the rudders gave automatic wings leveling at all altitudes and yaw damping at altitudes above 60,000 feet. Subsequently, the ailerons were locked and a turn-rate command system introduced in which the ground controller needed only to turn a knob to achieve desired turning radius. Flight-testing indicated that the Mini-Sniffer II had a high static-stability margin, making the aircraft very easy to trim and minimizing the effects of altering nose shapes and sizes or adding pods of various shapes and sizes under the fuselage to accommodate instrumentation. A highly damped short-period longitudinal oscillation resulted in rapid recovery from turbulence or upset. When an inadvertent hardover rudder command rolled the airplane inverted, the ground pilot simply turned the yaw damper on and the vehicle recovered automatically, losing just 200 feet of altitude.[22] The Mini-Sniffer III was a completely new airframe, similar in configuration to the Mini-Sniffer II but with a lengthened forward fuselage. An 18-inch nose extension provided better balance and greater payload capacity—up to 50 pounds plus telemetry equipment, radar transponder, radio-control gear, instrumentation, and sensors for stability and control investigations. Technicians at a sailplane repair company constructed the fuselage and wings from fiberglass and plastic foam, and they built tail surfaces from Kevlar and carbon fiber. Metal workers at Dryden fashioned an aluminum tail assembly, while a manufacturer of mini-RPVs designed and constructed an aluminum hydrazine tank to be integral with the fuselage. The Mini-Sniffer III was assembled at Dryden and integrated with Akkerman's engine.

The 15-horsepower, hydrazine-fueled piston engine drove a 38-inch-diameter, 4-bladed propeller. Plans called for eventually using

22. Ibid.

Ground crew for the Mini-Sniffer III wore self-contained suits and oxygen tanks because the engine was fueled with hydrazine. NASA.

a 6-foot-diameter, 2-bladed propeller for high-altitude flights. A slightly pressurized tank fed liquid hydrazine into a fuel pump, where it became pressurized to 850 pounds per square inch (psi). A fuel valve then routed some of the pressurized hydrazine to a gas generator, where liquid fuel was converted to hot gas at 1,700 degrees Fahrenheit (°F). Expansion of the hot gas drove the piston.[23] Since hydrazine doesn't need to be mixed with oxygen for combustion, it is highly suited to use in the thin upper atmosphere. This led to a proposal to send a hydrazine-powered aircraft, based on the Mini-Sniffer concept, to Mars, where it would be flown in the thin Martian atmosphere while collecting data and transmitting it back to scientists on Earth. Regrettably, such a vehicle has yet to be built.

During a 1-hour shakedown flight on November 23, 1976, the Mini-Sniffer III reached an altitude of 20,000 feet. Power fluctuations prevented the airplane from attaining the planned altitude of 40,000 feet, but otherwise, the engine performed well. About 34 minutes into the flight, fuel tank pressure was near zero, so the ground pilot closed the throttle and initiated a gliding descent. Some 30 minutes later, the Mini-Sniffer III touched down on the dry lakebed. The retrieval crew, wearing

23. Ibid.

protective garments to prevent contact with toxic and highly flammable fuels, found that there had been a hydrazine leak. This in itself did not account for the power reduction, however. Investigators suggested a possible fuel line blockage or valve malfunction might have been to blame.[24] Although the mission successfully demonstrated the operational characteristics of a hydrazine-fueled, non–air-breathing aircraft, the Mini-Sniffer III never flew again. Funding for tests with a variable-pitch propeller needed for flights at higher altitudes was not forthcoming, although interest in a Mars exploration airplane resurfaced from time to time over the next few decades.[25] The Mini-Sniffer project yielded a great deal of useful information for application to future RPRV efforts. One area of interest concerned procedures for controlling the vehicle. On the first flights of Mini-Sniffer I, ordinary model radio-control gear was used. This was later replaced with a custom-made, multichannel radio-control system for greater range and equipped with built-in fail-safe circuits to retain control when more than one transmitter was used. The onboard receiver was designed to respond only to the strongest signal. To demonstrate this feature, one of the vehicles was flown over two operating transmitter units located 50 feet apart on the ground. As the Mini-Sniffer passed overhead, the controller of the transmitter nearest the airplane took command from the other controller, with both transmitters broadcasting on the same frequency. With typical model radio-control gear, interference from two simultaneously operating transmitters usually results in loss of control regardless of relative signal strength.[26] A chase truck was used during developmental flights to collect early data on control issues. A controller, called the visual pilot, operated the airplane from the truck bed while observing its response to commands. Speed and trim curves were plotted based on the truck's speed and a recording of the pilot's inputs. During later flights, a remote pilot controlled the Mini-Sniffer from a chase helicopter. Technicians installed a telemetering system and radar transponder in the airplane so that it could be controlled at altitude from the NASA Mission Control Room at Dryden. Plot boards at the control station displayed position and altitude, airspeed, turn rate, elevator trim, and engine data. A miniature

24. J.W. Akkerman, "Hydrazine Monopropellant Reciprocating Engine Development," *Journal of Engineering for Industry*, vol. 101, no. 4 (Nov. 1979), pp. 456–462.

25. Reed, "Flight Research Techniques Utilizing Remotely Piloted Research Vehicles."

26. Reed, "High-Flying Mini-Sniffer RPV: Mars Bound?"

television camera provided a visual reference for the pilot. In most cases, a visual pilot took control for landing while directly observing the airplane from a vantage point adjacent to the landing area. Reed, however, also demonstrated a solo flight, which he controlled unassisted from takeoff to landing.

"I got a bigger thrill from doing this than from my first flight in a light plane as a teenager," he said, "probably because I felt more was at stake."[27]

The RPV Comes of Age as RDT&E Asset: The F-15 RPRV/SRV

NASA's work with the RPV concept came of age when the agency applied RPV technology to support the Research, Development, Test, and Evaluation (RDT&E) of a new Air Force fighter, the McDonnell-Douglas (subsequently Boeing) F-15 Eagle. In 1969, the Air Force selected McDonnell-Douglas Aircraft Corporation to build the F-15, a Mach-2–capable air superiority fighter airplane designed using lessons learned during aerial combat over Vietnam. The prototype first flew in July 1972. In the months leading up to that event, Maj. Gen. Benjamin Bellis, chief of the F-15 System Program Office at Wright-Patterson Air Force Base, OH, requested NASA assistance in testing a three-eights-scale model F-15 RPRV to explore aerodynamic and control system characteristics of the F-15 configuration in spins and high-angle-of-attack flight. Such maneuvers can be extremely hazardous. Rather than risk harm to a valuable test pilot and prototype, a ground pilot would develop stall/spin recovery techniques with the RPRV and pass lessons learned to test pilots flying the actual airplanes.

In April 1972, NASA awarded McDonnell-Douglas a $762,000 contract to build three F-15 RPRV models. Other contractors provided electronic components and parachute-recovery equipment. NASA technicians installed avionics, hydraulics, and other subsystems. The F-15 RPRV was 23.5 feet long, was made primarily of fiberglass and wood, and weighed 2,500 pounds. It had no propulsion system and was designed for midair recovery using a helicopter. Each model cost a little over $250,000, compared with $6.8 million for a full-scale F-15 aircraft.[28] Every effort was made to use off-the-shelf components and equipment readily available at the Flight Research Center, including

27. Ibid.

28. Hallion and Gorn, *On the Frontier*, p. 210.

hydraulic components, gyros, and telemetry systems from the lifting body research programs. A proportional uplink, then being used for instrument-landing system experiments, was acquired for the RPRV Ground Control Station (GCS). The ground cockpit itself was fashioned from a general-purpose simulator that had been used for stability-and-control studies. Data-processing computers were adapted for use in a programmable ground-based control system. A television camera provided forward visibility. The midair recovery system (MARS) parachute mechanism was taken from a Firebee drone.[29] The first F-15 RPRV arrived at the Flight Research Center in December 1972 but wasn't flown until October 12, 1973. The model was carried to an altitude of about 45,000 feet beneath the wing of a modified B-52 Stratofortress known as the NB-52B. Following release from the launch pylon at a speed of 175 knots, ground pilot Einar Enevoldson guided the craft through a flawless 9-minute flight, during which he explored the vehicle's basic handling qualities. At 15,000 feet altitude, a 12-foot spin-recovery parachute deployed to stabilize the descent. An 18-foot engagement chute and a 79-foot-diameter main chute then deployed so that the RPRV could be snagged in flight by a hook and cable beneath a helicopter, and set down gently on an inflated bag.[30] Enevoldson found the task of flying the RPRV very challenging, both physically and psychologically. The lack of physical cues left him feeling remote from the essential reassuring sensations of flight that provide a pilot with situational feedback. Lacking sensory input, he found that his workload increased and that subjective time seemed to speed up. Afterward, he reenacted the mission in a simulator at 1.5 times actual time and found that the pace seemed the same as it had during the flight.

Researchers had monitored his heart rate during the flight to see if it would register the 70 to 80 beats per minute typical for a piloted test flight. They were surprised to see the readings indicate 130 to 140 beats per minute as the pilot's stress level increased. Enevoldson found flying the F-15 RPRV less pleasant or satisfying than he normally did a difficult or demanding test mission.[31] "The results were gratifying," he wrote in his postflight report, "and some satisfaction is gained from the

29. Reed, "Flight Research Techniques Utilizing Remotely Piloted Research Vehicles."

30. Hallion and Gorn, *On the Frontier*, pp. 210–211.

31. F-15 Drone Flight Report, Flight No. D-1-3, Oct. 12, 1973, DFRC Historical Reference Collection, NASA Dryden Flight Research Center, Edwards, CA.

NASA's three-eighths-scale F-15 remotely piloted research vehicle landing on Rogers Dry Lake at Edwards Air Force Base, CA. NASA.

success of the technical and organizational achievement—but it wasn't fun."[32] In subsequent tests, Enevoldson and other research pilots explored the vehicle's stability and control characteristics. Spin testing confirmed the RPRV's capabilities for returning useful data, encouraging officials at the F-15 Joint Test Force to proceed with piloted spin trials in the preproduction prototypes at Edwards.[33] William H. "Bill" Dana piloted the fourth F-15 RPRV flight, on December 21, 1973. He collected about 100 seconds of data at angles of attack exceeding 30 degrees and 90 seconds of control-response data. Dana had a little more difficulty controlling the RPRV in flight than he had in the simulator but otherwise felt everything went well. At Enevoldson's suggestion, the simulator flights had been sped up to 1.4 times actual speed, and Dana later acknowledged that this had provided a more realistic experience.

During a postflight debriefing, Dana was asked how he liked flying the RPRV. He responded that it was quite different from sitting in the cockpit of an actual research vehicle, where he generally worried and

32. Ibid.

33. Hallion and Gorn, *On the Frontier*, p. 211.

fretted until just before launch. Then he could settle down and just fly the airplane. With the RPRV, he said, he was calm and cool until launch and then felt keyed up through the recovery.[34] The first of several incidents involving the MARS parachute gear occurred during the ninth flight. The recovery helicopter failed to engage the chute, and the RPRV descended to the ground, where it was dragged upside down for about a quarter mile. Fortunately, damage was limited to the vertical tails, canopy bulge, and nose boom. The RPRV was severely damaged at the end of the 14th flight, when the main parachute did not deploy because of failure of the MARS disconnect fitting.

Rather than repair the vehicle, it was replaced with the second F-15 RPRV. During the craft's second flight, on January 16, 1975, research pilot Thomas C. McMurtry successfully completed a series of planned maneuvers and then deployed the recovery parachute. During MARS retrieval, with the RPRV about 3,000 feet above the ground, the towline separated. McMurtry quickly assumed control and executed an emergency landing on the Edwards Precision Impact Range Area (PIRA). As a result of this success and previous parachute-recovery difficulties, further use of MARS was discontinued. The RPRV was modified with landing skids, and all flights thereafter ended with horizontal touchdowns on the lakebed.[35] The F-15 RPRV project came to a halt December 17, 1975, following the 26th flight, but this did not spell the end of the vehicle's career. In November 1977, flights resumed under the Spin Research Vehicle (SRV) project. Researchers were interested in evaluating the effect of nose shape on the spin susceptibility of modern high-performance fighters. Flight-testing with the F-15 model would augment previous wind tunnel experiments and analytical studies. Baseline work with the SRV consisted of an evaluation of the basic nose shape with and without two vortex strips installed. In November 1978, following nine baseline-data flights, the SRV was placed in inactive status pending the start of testing with various nose configurations for spin-mode determination, forebody pressure-distribution studies, and nose-mounted spin-recovery parachute evaluation. Flights resumed in February 1981.[36]

34. F-15 Drone Flight Report, Flight No. D-4-6, Dec. 21, 1973, DFRC Historical Reference Collection, NASA Dryden Flight Research Center, Edwards, CA.
35. Project Document OPD 80-67, Spin Research Vehicle Nose Shape Project, Feb. 5, 1980, DFRC Historical Reference Collection, NASA Dryden Flight Research Center, Edwards, CA.
36. Ibid.

When the SRV program ended in July 1981, the F-15 models had been carried aloft 72 times: 41 times for the RPRV flights and 31 times for the SRV. A total of 52 research missions were flown with the two aircraft: 26 free flights with each one. There had been only 2 ground aborts, 1 aborted planned-captive flight, and 15 air aborts prior to launch. Of 16 MARS recoveries, 13 were successful. Five landings occurred on the PIRA and 34 on the lakebed.[37] Flight data were correlated with wind tunnel and mathematical modeling results and presented in various technical papers. Tests of the subscale F-15 models clearly demonstrated the value of the RPRV concept for making bold, rapid advances in free-flight testing of experimental aircraft with minimal risk and maximum return on investment. R. Dale Reed wrote that, "If information obtained from this program avoids the loss of just one full-scale F-15, then the program will have been a tremendous bargain."[38]

Indeed it was: spin test results of the F-15 model identified a potentially dangerous "yaw-trip" problem with the full-scale F-15 if it had an offset airspeed boom. Such a configuration, the F-15 RPRV showed, might exhibit abrupt departure characteristics in turning flight as angle of attack increased. Subsequently, during early testing of F-15C aircraft equipped with fuselage-hugging conformal fuel tanks (like those subsequently employed on the F-15E Strike Eagle) and an offset nose boom, Air Force test pilot John Hoffman experienced just such a departure. Review of the F-15 RPRV research results swiftly pinpointed the problem and alleviated fears that the F-15 suffered from some inherent and major flaw that would force a costly and extensive redesign. This one "save" likely more than paid for the entire NASA F-15 RPRV effort.[39]

Skewed Logic: The RPRV Explores Jones's Oblique Wing

In the early 1970s—a time when fuel prices were soaring—scientists at NASA Ames Research Center and NASA Dryden began investigating an aircraft concept featuring a wing that could be rotated about a

37. F-15 RPRV/SRV Flight Log, compiled by Peter W. Merlin, July 2001, DFRC Historical Reference Collection, NASA Dryden Flight Research Center, Edwards, CA.

38. R. Dale Reed, "RPRVs—The First and Future Flights," *Astronautics and Aeronautics* (Apr. 1974), pp. 26–42.

39. Recollection of Hallion, who, as AFFTC historian, was present at the postflight briefing to the commander, AFFTC; see James O. Young, *History of the Air Force Flight Test Center, Jan. 1, 1982–Dec. 31, 1982,* vol. 1 (Edwards AFB, CA: AFFTC History Office, 1984), pp. 348–352.

central pivot. For low-speed flight, the planform would present a conventional straight wing, perpendicular to the fuselage. At higher speeds, the wing would be skewed to an oblique angle, with one side swept forward and the other aft to enhance transonic cruise efficiency by reducing drag. Dr. Robert T. Jones, a senior scientist at Ames (and, early in his career, the American father of the swept wing), proposed the single-pivot oblique wing concept for a future supersonic transport. Studies indicated that such a plane flying at 1,000 mph would achieve twice the fuel economy of supersonic transports then operational, including the Concorde and Tu-144.

Jones built a 5.5-foot wingspan, radio-controlled model to test the configuration's basic handling qualities. The wing, mounted atop the fuselage, pivoted so that the left side moved forward and the right side moved aft to take advantage of propeller torque to cancel rolling moment. Burnett L. Gadberg controlled the model during flight tests at wing angles up to 45 degrees and speeds between 50 and 100 mph. He found that the model remained stable at high sweep angles and could be controlled with decoupled aerodynamic control surfaces.[40] In order to further investigate the aerodynamic characteristics of an oblique wing and develop control laws necessary to achieve acceptable handling qualities, a $200,000 contract was awarded for design and development of a subsonic, remotely piloted Oblique Wing Research Aircraft (OWRA). Rod Bailey at Ames led the design effort, originally conceiving an all-wing vehicle. Because of stability and control issues, however, a tail assembly was eventually added.

Built by Developmental Sciences, Inc., of City of Industry, CA, the OWRA had a narrow cylindrical fuselage tipped with a glass dome—like a cyclopean eye—containing a television camera. Power was provided by a McCullough 90-horsepower, 4-cylinder, air-cooled, reciprocating engine mounted in the center of a 22-foot-span, oval planform wing. The engine drove a pusher propeller, shrouded in a 50-inch-diameter duct to reduce risk of crash damage. To further ensure survivability and ease of repair, key structural components were constructed of fiberglass epoxy composites. A two-axis, gyro-controlled autopilot provided stabilization for pitch, roll, and altitude hold, but the vacuum-tube-based

40. Michael J. Hirschberg, David M. Hart, and Thomas J. Beutner, "A Summary of a Half-Century of Oblique Wing Research," paper No. AIAA-2007-150, 45th *AIAA Aerospace Sciences Meeting*, *Reno, NV, Jan. 2007.*

sensors resulted in a significant weight penalty.[41] By December 1975, following 3 years of development with minimal resources, construction of the OWRA was essentially complete. Engineers evaluated the vehicle in two rounds of wind tunnel testing to collect preliminary data. Tests in a 7- by 10-foot tunnel helped designers refine the basic layout of the aircraft and confirmed trends noted with the original subscale model.

Milton O. Thompson, chief engineer at Dryden, recommended flying the vehicle from a remote site such as Bicycle Lake, at nearby U.S. Army Fort Irwin, or Mud Lake, NV, in order to minimize any adverse publicity should an incident occur. Based on his recommendation, Bicycle Lake was selected for taxi testing.[42] During these preliminary trials, engineers discovered that the OWRA—designed to have a top speed of 146 knots—was considerably underpowered. Additionally, the aircraft was damaged when it flipped over on the lakebed following loss of signal from the control transmitter. After being rebuilt, the OWRA was tested in a 40- by 80-foot Ames wind tunnel in order to evaluate three different tail configurations and determine static aerodynamic characteristics at varying wing-sweep angles. The results of these tests provided data required for ground simulation and training for pilot Jim Martin.[43] In April 1976, the OWRA was delivered to Dryden for testing. Technicians spent the next several months installing avionics and instrumentation, conducting systems checkouts, and developing a flight plan through detailed simulations. Taxi testing took place August 3, and the first flight was accomplished 3 days later at Rogers Dry Lake.

The results of the 24-minute flight indicated insufficient longitudinal stability because of a center of gravity located too far aft. Subsequently, the aircraft was modified with a 33-percent-larger vertical stabilizer, which was also moved back 3 feet, and a redesigned flight control system, which alleviated trim and stability problems. During a second flight, on September 16, stability and control data were collected to wing skew angles up to 30 degrees. Although severe radio-control system problems were encountered throughout the flight, all mission objectives were accomplished. A third and final flight was made October 20. Despite some control difficulties, researchers

41. Ibid.

42. Milton O. Thompson's correspondence and project files, 1975, DFRC Historical Reference Collection, NASA Dryden Flight Research Center, Edwards, CA.

43. Hirschberg, Hart, and Beutner, "A Summary of a Half-Century of Oblique Wing Research."

were able to obtain data at wing-skew angles up to 45 degrees, boosting confidence in plans for development of piloted oblique wing aircraft designs such as the Ames-Dryden AD-1 research airplane that was successfully flown in the early 1980s.[44]

Exploring the Torsionally Free Wing

9

Aeronautical researchers have long known that low wing loading contributes to poor ride quality in turbulence. This problem is compounded by the fact that lightweight aircraft, such as general aviation airplanes, spend a great deal of their flight time at lower altitudes, where measurable turbulence is most likely to occur. One way to improve gust alleviation is through the use of a torsionally free wing, also known as a free wing.

The free-wing concept involves unconventional attachment of a wing to an airplane's fuselage in such a way that the airfoil is free to pivot about its spanwise axis, subject to aerodynamic pitching moments but otherwise unrestricted by mechanical constraints. To provide static pitch stability, the axis of rotation is located forward of the chordwise aerodynamic center of the wing panel. Angle-of-attack equilibrium is established through the use of a trimming control surface and natural torque from lift and drag. Gust alleviation, and thus improved ride quality, results from the fact that a stable lifting surface tends to maintain a prescribed lift coefficient by responding to natural pitching moments that accompany changes in airflow direction.[45] Use of a free wing offers other advantages as well. Use of full-span flaps permits operation at a higher lift coefficient, thus allowing lower minimum-speed capability. A free stabilizer helps eliminate stalls. Use of differentially movable wings instead of ailerons permits improved roll control at low speeds. During takeoff, the wing rotates for lift-off, eliminating pitching movements caused by landing-gear geometry issues. Lift changes are accommodated without body-axis rotation. Because of independent attitude control, fuselage pitch can be trimmed for optimum visibility during landing approach. Negative lift can be applied to increase deceleration during

44. Oblique Wing Research Aircraft (OWRA) Flight Log, compiled by Peter W. Merlin, Jan. 2002, DFRC Historical Reference Collection, NASA Dryden Flight Research Center, Edwards, CA.
45. Richard F. Porter, David W. Hall, Joe H. Brown, Jr., and Gerald M. Gregorek, "Analytical Study of a Free-Wing/Free-Trimmer Concept," Battelle Columbus Laboratories, Columbus, OH, Dec. 7, 1977, DFRC Historical Reference Collection, NASA Dryden Flight Research Center, Edwards, CA.

Dick Eldredge, left, and Dan Garrabrant prepare the Free-Wing RPRV for flight. NASA.

landing roll. Fuselage drag can be reduced through attitude trim. Finally, large changes in the center of gravity do not result in changes to longitudinal static stability.[46] To explore this concept, researchers at NASA Dryden, led by Shu Gee, proposed testing a radio-controlled model airplane with a free-wing/free-canard configuration. Quantitative and qualitative flight-test data would provide proof of the free-wing concept and allow comparison with analytical models. The research team included engineers Gee and Chester Wolowicz of Dryden. Dr. Joe H. Brown, Jr., served as principal investigator for Battelle Columbus Laboratories of Columbus, OH. Professor Gerald Gregorek of Ohio State University's Aeronautical Engineering Department, along with Battelle's Richard F. Porter and Richard G. Ollila, calculated aerodynamics and equations of motion. Battelle's Professor David W. Hall, formerly of Iowa State University, assisted with vehicle layout and sizing.[47] Technicians at Dryden modified a radio-controlled airplane with a 6-foot wingspan to the test configuration. A small free-wing airfoil was rigidly mounted on twin booms forward of the primary flying surface. The ground pilot could change wing lift by actuating a flap on the free wing for longitudinal

46. Shu Gee, "Preliminary Research Proposal—Free-Wing Concept (First Draft)," Dec. 2, 1976, DFRC Historical Reference Collection, NASA Dryden Flight Research Center, Edwards, CA.
47. "Proposed Research Program (Technical Proposal) on Analytical Study of a Free-Wing, Free-Stabilizer Concept," Battelle Columbus Laboratories, Columbus, OH, Oct. 20, 1976, DFRC Historical Reference Collection, NASA Dryden Flight Research Center, Edwards, CA.

control. Elevators provided pitch attitude control, while full-span ailerons were used for roll control.

For data acquisition, the Free-Wing RPRV was flown at low altitude in a pacing formation with a ground vehicle. Observers noted the positions of protractors on the sides of the aircraft to indicate wing and canard position relative to the fuselage. Instrumentation in the vehicle, along with motion picture film, allowed researchers to record wing angle, control-surface positions, velocity, and fuselage angle relative to the ground. Another airplane model with a standard wing configuration was flown under similar conditions to collect baseline data for comparison with the Free-Wing RPRV performance.[48]

Researchers conducted eight flights at Dryden during spring 1977. They found that the test vehicle exhibited normal stability and control characteristics throughout the flight envelope for all maneuvers performed. Pitch response appeared to be faster than that of a conventional airplane, apparently because the inertia of the free-wing assembly was lower than that of the complete airplane. Handling qualities appeared to be as good or better than those of the baseline fixed-wing airplane. The investigators noted that separate control of the decoupled fuselage enhanced vehicle performance by acting as pseudo-thrust vectoring. The Free-Wing RPRV had excellent stall/spin characteristics, and the pilot was able to control the aircraft easily under gusty conditions. As predicted, center of gravity changes had little or no effect on longitudinal stability.[49] Some unique and unexpected problems were also encountered. When the canard encountered a mechanical trailing-edge position limit, it became aerodynamically locked, resulting in an irreversible stall and hard landing. Increased deflection limits for the free canard eliminated this problem. Researchers had difficulty matching the wing-hinge margin (the distance from the wing's aerodynamic center to the pivot) and canard control effectiveness. Designers improved handling qualities by increasing the wing hinge margin, the canard area aft of the pivot, and the canard flap area. Canard pivot friction caused some destabilizing effects during taxi, but these abated during takeoff. The ground pilot experienced control difficulty

48. Shu W. Gee and Samuel R. Brown, "Flight Tests of a Radio-Controlled Airplane Model with a Free-Wing, Free-Canard Configuration," NASA TM-72853, Mar. 1978, NASA Dryden Flight Research Center, Edwards, CA.
49. Ibid.

A research pilot controls the DAST vehicle from a ground cockpit. NASA.

because wing-fuselage decoupling made it difficult to visually judge approach and landing speeds, but it was concluded that this would not be a problem for a pilot flying a full-scale airplane equipped with conventional flight instruments.[50]

DAST: Exploring the Limits of Aeroelastic Structural Design

In the early 1970s, researchers at Dryden and NASA Langley Research Center sought to expand the use of RPRVs into the transonic realm. The Drones for Aerodynamic and Structural Testing (DAST) program was conceived as a means of conducting high-risk flight experiments using specially modified Teledyne-Ryan BQM-34E/F Firebee II supersonic target drones to test theoretical data under actual flight conditions. Described by NASA engineers as a "wind-tunnel in the sky," the DAST program merged advances in electronic remote-control systems with advanced airplane-design techniques. The drones were relatively inexpensive and easy to modify for research purposes

50. Ibid.

and, moreover, were readily available from an existing stock of Navy target drones.[51] The unmodified Firebee II had a maximum speed of Mach 1.1 at sea level and almost Mach 1.8 at 45,000 feet, and was capable of 5 g turns. Firebee II drones in the basic configuration provided baseline data. Researchers modified two vehicles, DAST-1 and DAST-2, to test several wing configurations during maneuvers at transonic speeds in order to compare flight results with theoretical and wind tunnel findings. For captive and free flights, the drones were carried aloft beneath a DC-130A or the NB-52B. The DAST vehicles were equipped with remotely augmented digital flight control systems, research instrumentation, an auxiliary fuel tank for extended range, and a MARS recovery system. On the ground, a pilot controlled the DAST vehicle from a remote cockpit while researchers examined flight data transmitted via pulse-mode telemetry. In the event of a ground computer failure, the DAST vehicle could also be flown using a backup control system in the rear cockpit of a Lockheed F-104B chase plane.[52]

The primary flight control system for DAST was remotely augmented. In this configuration, control laws for augmenting the airplane's flying characteristics were programmed into a general-purpose computer on the ground. Closed-loop operation was achieved through a telemetry uplink/downlink between the ground cockpit and the vehicle. This technique had previously been tested using the F-15 RPRV.[53] Baseline testing was conducted between November 1975 and June 1977, using an unmodified BQM-34F drone. It was carried aloft three times for captive flights, twice by a DC-130A and once by the NB-52B. These flights gave ground pilot William H. Dana a chance to check out the RPRV systems and practice prelaunch procedures. Finally, on July 28, 1977, the Firebee II was launched from the NB-52B for the first time. Dana flew the vehicle using an unaugmented control mode called Babcock-direct. He found the Firebee less controllable in roll than had been indicated in simulations, but overall performance was higher.

51. H.N Murrow and C.V. Eckstrom, "Drones for Aerodynamic and Structural Testing (DAST)—A Status Report," *AIAA Aircraft Systems and Technology Conference, Los Angeles, CA, Aug. 21–23, 1978.*

52. Ibid.

53. David L. Grose, "The Development of the DAST I Remotely Piloted Research Vehicle for Flight Testing an Active Flutter Suppression Control System," NASA CR-144881, University of Kansas, Feb. 1979.

Dana successfully transferred control of the drone to Vic Horton in the rear seat of an F-104B chase plane. Horton flew the Firebee through the autopilot to evaluate controllability before transferring control back to Dana just prior to recovery.

Technicians then installed instrumented standard wings, known as the Blue Streak configuration. Thomas C. McMurtry flew a mission March 9, 1979, to evaluate onboard systems such as the autopilot and RAV system. Results were generally good, with some minor issues to be addressed prior to flying the DAST-1 vehicle.[54] The DAST researchers were most interested in correlating theoretical predictions and experimental flight results of aeroelastic effects in the transonic speed range. Such tests, particularly those involving wing flutter, would be extremely hazardous with a piloted aircraft.

One modified Firebee airframe, which came to be known as DAST-1, was fitted with a set of swept supercritical wings of a shape optimized for a transport-type aircraft capable of Mach 0.98 at 45,000 feet. The ARW-1 aeroelastic research wing, designed and built by Boeing in Wichita, KS, was equipped with an active flutter-suppression system (FSS). Research goals included validation of active controls technology for flutter suppression, enhancement, and verification of transonic flutter prediction techniques, and providing a database for aerodynamic-loads prediction techniques for elastic structures.[55] The basic Firebee drone was controlled through collective and differential horizontal stabilizer and rudder deflections because it had no wing control surfaces. The DAST-1 retained this control system, leaving the ailerons free to perform the flutter suppression function. During fabrication of the wings, it became apparent that torsional stiffness was higher than predicted. To ensure that the flutter boundary remained at an acceptable Mach number, 2-pound ballast weights were added to each wingtip. These weights consisted of containers of lead shot that could be jettisoned to aid recovery from inadvertent large-amplitude wing oscillations. Researchers planned to intentionally fly the DAST-1 beyond its flutter boundary to demonstrate the effectiveness of the FSS.[56] Along with the remote cockpit, there were two other

54. DAST flight logs and mission reports, 1975–1983, DFRC Historical Reference Collection, NASA Dryden Flight Research Center, Edwards, CA.
55. John W. Edwards, "Flight Test Results of an Active Flutter Suppression System Installed on a Remotely Piloted Vehicle," NASA TM-83132, May 1981, NASA Langley Research Center, Hampton, VA.
56. Ibid.

ground-based facilities for monitoring and controlling the progress of DAST flight tests. Dryden's Control Room contained radar plot boards for monitoring the flight path, strip charts indicating vehicle rigid-body stability and control and operational functions, and communications equipment for coordinating test activities. A research pilot stationed in the Control Room served as flight director. Engineers monitoring the flutter tests were located in the Structural Analysis Facility (SAF). The SAF accommodated six people, one serving as test director to oversee monitoring of the experiments and communicate directly with the ground pilot.[57] The DAST-1 was launched for the first time October 2, 1979. Following release from the NB-52B, Tom McMurtry guided the vehicle through FSS checkout maneuvers and a subcritical-flutter investigation. An uplink receiver failure resulted in an unplanned MARS recovery about 8 minutes after launch. The second flight was delayed until March 1980. Again only subcritical-flutter data were obtained, this time because of an unexplained oscillation in the left FSS aileron.[58] During the third flight, unknown to test engineers, the FSS was operating at one-half nominal gain. Misleading instrument indications concealed a trend toward violent flutter conditions at speeds beyond Mach 0.8. As the DAST-1 accelerated to Mach 0.825, rapidly divergent oscillations saturated the FSS ailerons. The pilot jettisoned the wingtip masses, but this failed to arrest the flutter. Less than 6 seconds after the oscillations began, the right wing broke apart, and the vehicle crashed near Cuddeback Dry Lake, CA.

Investigators concluded that erroneous gain settings were the primary cause. The error resulted in a configuration that caused the wing to be unstable at lower Mach numbers than anticipated, causing the vehicle to experience closed-loop flutter. The ARW-1 wing was rebuilt as the ARW-1R and installed in a second DAST vehicle in order to continue the research program.[59] The DAST-2 underwent a captive systems-checkout flight beneath the wing of the NB-52B on October 29, 1982, followed by a subcritical-flutter envelope expansion flight 5 days later. Unfortunately, the flight had to be aborted early because of unexplained wing structural

57. Ibid.

58. Merlin, DAST flight logs, Sept. 1999, DFRC Historical Reference Collection, NASA Dryden Flight Research Center, Edwards, CA.

59. Edwards, "Flight Test Results of an Active Flutter Suppression System Installed on a Remotely Piloted Vehicle."

vibrations and control-system problems. The next three flight attempts were also aborted—the first because of a drone engine temperature warning, the second because of loss of telemetry, and a third time for unspecified reasons prior to taxi.[60] Further testing of the DAST-2 vehicle was conducted using a Navy DC-130A launch aircraft. Following two planned captive flights for systems checkout, the vehicle was ready to fly.

On June 1, 1983, the DC-130A departed Edwards as the crew executed a climbing turn over Mojave and California City. Rogers Smith flew the TF-104G with backup pilot Ray Young, while Einar Enevoldson began preflight preparations from the ground cockpit. The airplanes passed abeam of Cuddeback Dry Lake, passed north of Barstow, and turned west. The launch occurred a few minutes later over Harper Dry Lake. Immediately after separation from the launch pylon, the drone's recovery-system drag chute deployed, but the main parachute was jettisoned while still packed in its canister.[61] The drone plummeted to the ground in the middle of a farm field west of the lakebed. It was completely destroyed, but other than loss of a small patch of alfalfa at the impact site, there was no property damage. Much later, when it was possible to joke about such things, a few wags referred to this event as the "alfalfa impact study."[62] An investigation board found that a combination of several improbable anomalies—a design flaw, a procedural error, and a hardware failure—simultaneously contributed to loss of the vehicle. These included an uncommanded recovery signal produced by an electrical spike, failure to reset a drag chute timer, and improper grounding of an electrical relay. Another section of the investigation focused on project management issues. Criticism of Dryden's DAST program management was hotly debated, and several dissenting opinions were filed along with the main report.[63] Throughout its history, the DAST program was plagued by difficulties. Between December 1973 and November 1983, five different project managers oversaw the program. As early as December 1978, Dryden's Center Director, Isaac T. Gillam, had requested

60. Merlin, DAST flight logs.

61. Paul C. Loschke, Garrison P. Layton, George H. Kidwell, William P. Albrecht, William H. Dana, and Eugene L. Kelsey, "Report of DAST-1R Test Failure Investigation Board," Dec. 30, 1983, DFRC Historical Reference Collection, NASA Dryden Flight Research Center, Edwards, CA.

62. This event is recorded with other DAST mission markings painted on the side of the NB-52B even though the modified Stratofortress was not the launch aircraft for the ill-fated mission.

63. Loschke, et al., "Report of DAST-1R Test Failure Investigation Board."

chief engineer Milton O. Thompson and chief counsel John C. Mathews to investigate management problems associated with the project. This resulted from the project team's failure to meet an October 1978 flight date for the Blue Streak wing, Langley managers' concern that Dryden was not properly discharging its project obligations, repeated requests by the project manager for schedule slips, and various other indications that the project was in a general state of confusion. The resulting report indicated that problems had been caused by a lack of effective planning at Dryden, exacerbated by poor internal communication among project personnel.[64] Only 7 flights were achieved in 10 years. Several flights were aborted for various reasons, and two vehicles crashed, problems that drove up testing costs. Meanwhile, flight experiments with higher-profile, better-funded remotely piloted research vehicles took priority over DAST missions at Dryden. Organizational upheaval also took a toll, as Dryden was consolidated with Ames Research Center in 1981 and responsibility for projects was transferred to the Flight Operations Directorate in 1983.

Exceptionally good test data had been obtained through the DAST program but not in an efficient and timely manner. Initially, the Firebee drone was selected for use in the DAST project in the belief that it offered a quick and reasonably inexpensive option for conducting a task too hazardous for a piloted vehicle. Experience proved, however, that using off-the-shelf hardware did not guarantee expected results. Just getting the vehicle to fly was far more difficult and far less successful than originally anticipated.[65] Hardware delays created additional difficulties. The Blue Streak wing was not delivered until mid-1978. The ARW-1 wing arrived in April 1979, 1½ years behind schedule, and was not flown until 6 months later. Following the loss of the DAST-1 vehicle, the program was delayed nearly 2 years until delivery of the ARW-1R wing. After the 1983 crash, the program was terminated.[66]

Pursuing Highly Maneuverable Aircraft Technology

In 1973, NASA and Air Force officials began exploring a project to develop technologies for advanced fighter aircraft. Several aerospace

64. Milton O. Thompson and John C. Mathews, "Investigation of DAST Project," Jan. 22, 1979, DFRC Historical Reference Collection, NASA Dryden Flight Research Center, Edwards, CA.
65. Loschke, et al., "Report of DAST-1R Test Failure Investigation Board."
66. DAST briefing material, Thompson collection, April 1987, DFRC Historical Reference Collection, NASA Dryden Flight Research Center, Edwards, CA.

The HiMAT research vehicle demonstrated advanced technologies for use in high-performance military aircraft. NASA.

contractors submitted designs for a baseline advanced-fighter concept with performance goals of a 300-nautical-mile mission radius, sustained 8 g maneuvering capability at Mach 0.9, and a maximum speed of Mach 1.6 at 30,000 feet altitude. The Los Angeles Division of Rockwell International was selected to build a 44-percent-scale, remotely piloted model for a project known as Highly Maneuverable Aircraft Technology (HiMAT). Testing took place at Dryden, initially under the leadership of Project Manager Paul C. Loschke and later under Henry Arnaiz.[67] The scale factor for the RPRV was determined by cost considerations, payload requirements, test-data fidelity, close matching of thrust-to-weight ratio and wing loading between the model and the full-scale design, and availability of off-the-shelf hardware. The overall geometry of the design was faithfully scaled with the exception of fuselage diameter and inlet-capture area, which were necessarily over-scale in order to accommodate a 5,000-pound-thrust General Electric J85-21 afterburning turbojet engine.

Advanced technology features included maximum use of lightweight, high-strength composite materials to minimize airframe weight; aeroelastic tailoring to provide aerodynamic benefits from the airplane's

67. L.E. Brown, Jr., M.H. Roe, and R.A. Quam, "HiMAT Systems Development Results and Projections," *Society of Automotive Engineers Aerospace Congress and Exposition, Los Angeles, CA, Oct. 13–16, 1980.*

structural-flexibility characteristics; relaxed static stability, to provide favorable drag effects because of trimming; digital fly-by-wire controls; a digital integrated propulsion-control system; and such advanced aerodynamic features as close-coupled canards, winglets, variable-camber leading edges, and supercritical wings. Composite materials, mostly graphite/epoxy, comprised about 95 percent of the exterior surfaces and approximately 29 percent of the total structural weight of the airplane. Researchers were interested in studying the interaction of the various new technologies.[68] To keep development costs low and allow for maximum flexibility for proposed follow-on programs, the HiMAT vehicle was modular for easy reconfiguration of external geometry and propulsion systems. Follow-on research proposals included forward-swept wings, a two-dimensional exhaust nozzle, alternate canard configurations, active flutter suppression, and various control-system modifications. These options, however, were never pursued.[69] Rockwell built two HiMAT air vehicles, known as AV-1 and AV-2, at a cost of $17.3 million. Each was 22.5 feet long, spanned 15.56 feet, and weighed 3,370 pounds. The vehicle was carried to a launch altitude of about 40,000 to 45,000 feet beneath the wing of the NB-52B. Following release from the wing pylon at a speed of about Mach 0.7, the HiMAT dropped for 3 seconds in a preprogrammed maneuver before transitioning to control of the ground pilot. Research flight-test maneuvers were restricted to within a 50-nautical-mile radius of Edwards and ended with landing on Rogers Dry Lake. The HiMAT was equipped with steel skid landing gear. Maximum flight duration varied from about 15 to 80 minutes, depending on thrust requirements, with an average planned flight duration of about 30 minutes.

As delivered, the vehicles were equipped with a 227-channel data collection and recording system. Each RPRV was instrumented with 128 surface-pressure orifices with 85 transducers, 48 structural load and hinge-moment strain gauges, 6 buffet accelerometers, 7 propulsion system parameters, 10 control-surface-position indicators, and 15 airplane motion and air data parameters. NASA technicians later added more transducers for a surface-pressure survey.[70] The HiMAT project represented a shift in focus by researchers at Dryden. Through the Vietnam

68. Ibid.

69. Ibid.

70. L.E. Brown, M. Roe, and C.D. Wiler, "The HiMAT RPRV System," AIAA Paper 78-1457, *AIAA Aircraft Systems and Technology Conference, Los Angeles, CA, Aug. 21–23, 1978.*

era, the focal point of fighter research had been speed. In the 1970s, driven by a national energy crisis, new digital technology, and a changing combat environment, researchers sought to develop efficient research models for experiments into the extremes of fighter maneuverability. As a result, the quest for speed, long considered the key component of successful air combat, became secondary.

HiMAT program goals included a 100-percent increase in aerodynamic efficiency over 1973 technology and maneuverability that would allow a sustained 8 g turn at Mach 0.9 and an altitude of 25,000 feet. Engineers designed the HiMAT aircraft's rear-mounted swept wings, digital flight-control system, and forward-mounted controllable canards to give the plane a turn radius twice as tight as that of conventional fighter planes. At near-sonic speeds and at an altitude of 25,000 feet, the HiMAT aircraft could perform an 8 g turn, nearly twice the capability of an F-16 under the same conditions.[71] Flying the HiMAT from the ground-based cockpit using the digital fly-by-wire system required control techniques similar to those used in conventional aircraft, although design of the vehicle's control laws had proved extremely challenging. The HiMAT was equipped with a flight-test-maneuver autopilot based on a design developed by Teledyne Ryan Aeronautical Company, which also developed the aircraft's backup flight control system (with modifications made by Dryden engineers). The autopilot system provided precise, repeatable control of the vehicle during prescribed maneuvers so that large quantities of reliable test data could be recorded in a comparatively short period of flight time. Dryden engineers and pilots tested the control laws for the system in simulations and in flight, making any necessary adjustments based on experience. Once adjusted, the autopilot was a valuable tool for obtaining high-quality, precise data that would not have been obtainable using standard piloting methods. The autopilot enabled the pilot to control multiple parameters simultaneously and to do so within demanding, repeatable tolerances. As such, the flight-test-maneuver autopilot showed itself to be a broadly applicable technique for flight research with potential benefit to any flight program.[72]

71. Henry H. Arnaiz and Paul C. Loschke, "Current Overview of the Joint NASA/USAF HiMAT Program," NASA CP-2162 (1980), pp. 91–121.

72. E.L. Duke, F.P. Jones, and R.B. Roncoli, "Development of a Flight Test Maneuver Autopilot for a Highly Maneuverable Aircraft," AIAA Paper 83-0061, *AIAA 21st Aerospace Sciences Meeting, Reno, NV, Jan. 10–13, 1983.*

The maiden flight of HiMAT AV-1 took place July 27, 1979, with Bill Dana at the controls. All objectives were met despite some minor difficulty with the telemetry receiver. Subsequent flights resulted in acquisition of significant data and cleared the HiMAT to a maximum speed of Mach 0.9 and an altitude of 40,000 feet, as well as demonstrating a 4 g turning capability. By the end of October 1980, the HiMAT had been flown to Mach 0.925 and performed a sustained 7 g turn. The ground pilot was occasionally challenged to respond to unexpected events, including an emergency engine restart during flight and a gear-up landing.

AV-2 was flown for the first time July 24, 1981. The following week, Stephen Ishmael joined the project as a ground pilot. After several airspeed calibration flights, researcher began collecting data with AV-2.

On February 3, 1982, AV-1 was flown to demonstrate the 8 g maneuver capabilities that had been predicted for the vehicle. A little over 3 months later, researchers obtained the first supersonic data with the HiMAT, achieving speeds of Mach 1.2 and Mach 1.45. Research with both air vehicles continued through January 1983. Fourteen flights were completed with AV-1 and 12 with AV-2, for a total of 26 over 3½ years.[73] The HiMAT research successfully demonstrated a synergistic approach to accelerating development of an advanced high-performance aircraft. Many high-risk technologies were incorporated into a single, low-cost vehicle and tested—at no risk to the pilot—to study interaction among systems, advanced materials, and control software. Design requirements dictated that no single failure should result in loss of the vehicle. Consequently, redundant systems were incorporated throughout the aircraft, including computer microprocessors, hydraulic and electrical systems, servo-actuators, and data uplink/downlink equipment.[74] The HiMAT program resulted in several important contributions to flight technology. The foremost of these was the use of new composite materials in structural design. HiMAT engineers used materials such as fiberglass and graphite epoxy composites to strengthen the airframe and allow it to withstand high g conditions during maneuverability tests. Knowledge gained in composite construction of the HiMAT vehicle strongly influenced other advanced research projects, and such materials are now used extensively on commercial and military aircraft.

73. HiMAT Flight Reports, 1979–1983, DFRC Historical Reference Collection, NASA Dryden Flight Research Center, Edwards, CA.

74. Reed, "Flight Research Techniques Utilizing Remotely Piloted Research Vehicles."

Designers of the X-29 employed many design concepts developed for HiMAT, including the successful use of a forward canard and the rear-mounted swept wing constructed from lightweight composite materials. Although the X-29's wings swept forward rather than to the rear, the principle was the same. HiMAT research also brought about far-reaching advances in digital flight control systems, which can monitor and automatically correct potential flight hazards.[75]

On TARGIT: Civil Aviation Crash Testing in the Desert

On December 1, 1984, a Boeing 720B airliner crashed near the east shore of Rogers Dry Lake. Although none of the 73 passengers walked away from the flaming wreckage, there were no fatalities. The occupants were plastic, anthropomorphic dummies, some of them instrumented to collect research data. There was no flight crew on board; the pilot was seated in a ground-based cockpit 6 miles away at NASA Dryden.

As early as 1980, Federal Aviation Administration (FAA) and NASA officials had been planning a full-scale transport aircraft crash demonstration to study impact dynamics and new safety technologies to improve aircraft crashworthiness. Initially dubbed the Transport Crash Test, the project was later renamed Transport Aircraft Remotely Piloted Ground Impact Test (TARGIT). In August 1983, planners settled on the name Controlled Impact Demonstration (CID). Some wags immediately twisted the acronym to stand for "Crash in the Desert" or "Cremating Innocent Dummies."[76] In point of fact, no fireball was expected. One of the primary test objectives included demonstration of anti-misting kerosene (AMK) fuel, which was designed to prevent formation of a postimpact fireball. While many airplane crashes are survivable, most victims perish in postcrash fire resulting from the release of fuel from shattered tanks in the wings and fuselage. In 1977, FAA officials looked into the possibility of using an additive called Avgard FM-9 to reduce the volatility of kerosene fuel released during catastrophic crash events. Ground-impact studies using surplus Lockheed SP-2H airplanes

75. HiMAT fact sheet (FS-2002-06-025), NASA Dryden Flight Research Center, Edwards, CA, June 2002.

76. Controlled Impact Demonstration project files, 1978–1984, and personal diary of Timothy W. Horton, 1980–1986, DFRC Historical Reference Collection, NASA Dryden Flight Research Center, Edwards, CA.

showed great promise, because the FM-9 prevented the kerosene from forming a highly volatile mist as the airframe broke apart.[77] As a result of these early successes, the FAA planned to implement the requirement that airlines add FM-9 to their fuel. Estimates made calculated that the impact of adopting AMK would have included a one-time cost to airlines of $25,000–$35,000 for retrofitting each high-bypass turbine engine and a 3- to 6-percent increase in fuel costs, which would drive ticket prices up by $2–$4 each. In order to definitively prove the effectiveness of AMK, officials from the FAA and NASA signed a Memorandum of Agreement in 1980 for a full-scale impact demonstration. The FAA was responsible for program management and providing a test aircraft, while NASA scientists designed the experiments, provided instrumentation, arranged for data retrieval, and integrated systems.[78] The FAA supplied the Boeing 720B, a typical intermediate-range passenger transport that entered airline service in the mid-1960s. It was selected for the test because its construction and design features were common to most contemporary U.S. and foreign airliners. It was powered by four Pratt & Whitney JT3C-7 turbine engines and carried 12,000 gallons of fuel. With a length of 136 feet, a 130-foot wingspan, and maximum takeoff weight of 202,000 pounds, it was the world's largest RPRV. FAA Program Manager John Reed headed overall CID project development and coordination with all participating researchers and support organizations.

Researchers at NASA Langley were responsible for characterizing airframe structural loads during impact and developing a data-acquisition system for the entire aircraft. Impact forces during the demonstration were characterized as being survivable for planning purposes, with the primary danger to be from postimpact fire. Study data to be gathered included measurements of structural, seat, and occupant response to impact loads, to corroborate analytical models developed at Langley, as well as data to be used in developing a crashworthy seat and restraint system. Robert J. Hayduk managed NASA crashworthiness and cabin-instrumentation requirements.[79] Dryden personnel,

77. Julian Moxon, "Crash for Safety," *Flight International*, May 12, 1984, pp. 1270–1274.

78. Richard DeMeis, "What Really Happened in Safe Fuel Test," *Aerospace America*, July 1985, pp. 47–49. Additional information from Moxon, "Crash for Safety."

79. Les Reinertson, "NASA/FAA Full-Scale Transport Controlled Impact Demonstration Fact Sheet," Release No. 84-26, 1984, NASA Ames Research Center, Dryden Flight Research Facility, Edwards, CA.

under the direction of Marvin R. "Russ" Barber, were responsible for overall flight research management, systems integration, and flight operations. These included RPRV control and simulation, aircraft/ground interface, test and systems hardware integration, impact-site preparation, and flight-test operations.

The Boeing 720B was equipped to receive uplinked commands from the ground cockpit. Commands providing direct flight path control were routed through the autopilot, while other functions were fed directly to appropriate systems. Information on engine performance, navigation, attitude, altitude, and airspeed was downlinked to the ground pilot.[80] Commands from the ground cockpit were conditioned in control-law computers, encoded, and transmitted to the aircraft from either a primary or backup antenna. Two antennas on the top and bottom of the Boeing 720B provided omnidirectional telemetry coverage, each feeding a separate receiver. The output from the two receivers was then combined into a single input to a decoder that processed uplink data and generated commands to the controls. Additionally, the flight engineer could select redundant uplink transmission antennas at the ground station. There were three pulse-code modulation systems for downlink telemetry, two for experimental data, and one to provide aircraft control and performance data.

The airplane was equipped with two forward-facing television cameras—a primary color system and a black-and-white backup—to give the ground pilot sufficient visibility for situational awareness. Ten high-speed motion picture cameras photographed the interior of the passenger cabin to provide researchers with footage of seat and occupant motion during the impact sequence.[81] Prior to the final CID mission, 14 test flights were made with a safety crew on board. During these flights, 10 remote takeoffs, 13 remote landings (the initial landing was made by the safety pilot), and 69 CID approaches were accomplished. All remote takeoffs were flown from the Edwards Air Force Base main runway. Remote landings took place on the emergency recovery runway (lakebed Runway 25).

Research pilots for the project included Edward T. Schneider, Fitzhugh L. Fulton, Thomas C. McMurtry, and Donald L. Mallick.

80. Ibid.

81. Paul F. Harney, James B. Craft, Jr., and Richard G. Johnson, "Remote Control of an Impact Demonstration Vehicle," NASA TM-85925, Apr. 1985, NASA Ames Research Center, Dryden Flight Research Facility, Edwards, CA.

William R. "Ray" Young, Victor W. Horton, and Dale Dennis served as flight engineers. The first flight, a functional checkout, took place March 7, 1984. Schneider served as ground pilot for the first three flights, while two of the other pilots and one or two engineers acted as safety crew. These missions allowed researchers to test the uplink/downlink systems and autopilot, as well as to conduct airspeed calibration and collect ground-effects data. Fulton took over as ground pilot for the remaining flight tests, practicing the CID flight profile while researchers qualified the AMK system (the fire retardant AMK had to pass through a degrader to convert it into a form that could be burned by the engines) and tested data-acquisition equipment. The final pre-CID flight was completed November 26. The stage was set for the controlled impact test.[82] The CID crash scenario called for a symmetric impact prior to encountering obstructions as if the airliner were involved in a gear-up landing short of the runway or an aborted takeoff. The remote pilot was to slide the airplane through a corridor of heavy steel structures designed to slice open the wings, spilling fuel at a rate of 20 to 100 gallons per second. A specially prepared surface consisting of a rectangular grid of crushed rock peppered with powered electric landing lights provided ignition sources on the ground, while two jet-fueled flame generators in the airplane's tail cone provided onboard ignition sources.

On December 1, 1984, the Boeing 720B was prepared for its final flight. The airplane had a gross takeoff weight of 200,455 pounds, including 76,058 gallons of AMK fuel. Fitz Fulton initiated takeoff from the remote cockpit and guided the Boeing 720B into the sky for the last time.[83]At an altitude of 200 feet, Fulton lined up on final approach to the impact site. He noticed that the airplane had begun to drift to the right of centerline but not enough to warrant a missed approach. At 150 feet, now fully committed to touchdown because of activation of limited-duration photographic and data-collection systems, he attempted to center the flight path with a left aileron input, which resulted in a lateral oscillation.

The Boeing 720B struck the ground 285 feet short of the planned impact point, with the left outboard engine contacting the ground first.

82. Merlin, Boeing 720B Controlled Impact Demonstration flight logs, July 1998, DFRC Historical Reference Collection, NASA Dryden Flight Research Center, Edwards, CA.

83. Edwin L. Fasanella, Emilio Alfaro-Bou, and Robert J. Hayduk, "Impact Data From a Transport Aircraft During a Controlled Impact Demonstration," NASA TP-2589 (Sept. 1986), NASA Headquarters, Washington, DC.

NASA and the FAA conducted a Controlled Impact Demonstration with a remotely piloted Boeing 720 aircraft. NASA.

This caused the airplane to yaw during the slide, bringing the right inboard engine into contact with one of the wing openers and releasing large quantities of degraded (i.e., highly flammable) AMK and exposing them to a high-temperature ignition source. Other obstructions sliced into the fuselage, permitting fuel to enter beneath the passenger cabin. The resulting fireball was spectacular.[84]

To casual observers, this might have made the CID project appear a failure, but such was not the case. The conditions prescribed for the AMK test were very narrow and failed to account for a wide range of variables, some of which were illustrated during the flight test. The results were sufficient to cause FAA officials to abandon the idea of forcing U.S. airlines to use AMK, but the CID provided researchers with a wide range of data for improving transport-aircraft crash survivability.

The experiment also provided significant information for improving RPV technology. The 14 test flights leading up to the final demonstration gave researchers an opportunity to verify analytical models, simulation techniques, RPV control laws, support software, and hardware. The remote pilot assessed the airplane's handling qualities, allowing

84. Ibid.

programmers to update the simulation software and validate the control laws. All onboard systems were thoroughly tested, including AMK degraders, autopilot, brakes, landing gear, nose wheel steering, and instrumentation systems. The CID team also practiced emergency procedures, such as the ability to abort the test and land on a lakebed runway under remote control, and conducted partial testing of an uplinked flight termination system to be used in the event that control of the airplane was lost. Several anomalies—intermittent loss of uplink signal, brief interruption of autopilot command inputs, and failure of the uplink decoder to pass commands—cropped up during these tests. Modifications were implanted, and the anomalies never recurred.[85] Handling qualities were generally good. The ground pilot found landings to be a special challenge as a result of poor depth perception (because of the low-resolution television monitor) and lack of peripheral vision. Through flight tests, the pilot quickly learned that the CID profile was a high-workload task. Part of this was due to the fact that the tracking radar used in the guidance system lacked sufficient accuracy to meet the impact parameters. To compensate, several attempts were made to improve the ground pilot's performance. These included changing the flight path to give the pilot more time to align his final trajectory, improving ground markings at the impact site, turning on the runway lights on the test surface, and providing a frangible 8-foot-high target as a vertical reference on the centerline. All of these attempts were compromised to some degree by the low-resolution video monitor. After the impact flight, members of the control design team agreed that some form of head-up display (HUD) would have been helpful and that more of the piloting tasks should have been automated to alleviate pilot workload.[86] In terms of RPRV research, the project was considered highly successful. The remote pilots accumulated 16 hours and 22 minutes of RPV experience in preparation for the impact mission, and the CID showed the value of comparing predicted results with flight-test data. U.S. Representative William Carney, ranking minority member of the House Transportation, Aviation, and Materials Subcommittee, observed the CID test. "To those who were disappointed with the outcome," he later wrote, "I can only say that the

85. Timothy W. Horton and Robert W. Kempel, "Flight Test Experience and Controlled Impact of a Remotely Piloted Jet Transport Aircraft," NASA TM-4084 (Nov. 1988), NASA Ames Research Center, Dryden Flight Research Facility, Edwards, CA.
86. Ibid.

results dramatically illustrated why the tests were necessary. I hope we never lose sight of the fact that the first objective of a research program is to learn, and failure to predict the outcome of an experiment should be viewed as an opportunity, not a failure."[87]

The British Invasion: CHIRP and HIRM Support the Tornado

In 1981, researchers at NASA Dryden assisted with the first of several series of tests for the British Royal Aircraft Establishment (RAE) under an international agreement to collect data relevant to the Panavia Tornado jet fighter, a large-scale NATO acquisition program. The variable-wing-sweep Tornado eventually became a major deep-strike attack aircraft used by the British, then–West German, and Italian air forces. Britain's Royal Air Force flew an interceptor variant as well. During the 6-week Cooperative High Incidence Research Program (CHIRP), 4 25-percent-scale Tornado models of varying configurations were used to conduct 10 drop tests. Six of these flights were undertaken to gather unaugmented stability and control data to improve RAE engineers' mathematical model of Tornado aerodynamics. The remaining 4 drops (totaling 130 seconds of flight time) were allocated to evaluating a Spin Prevention and Incidence-Limiting System (SPILS) in support of a modification program for the full-scale operational Tornado fleet.

In February and March 1981, NASA and RAE officials met to discuss support requirements for the project. Once details had been decided, Walter B. Olstad of NASA's Office of Aeronautics and Space Technology and R.J.E. Glenny of the British RAE signed a Memorandum of Agreement. The first Tornado model arrived at Dryden in a British Royal Air Force C-130 transport May 11. Edward "Ted" Jeffries and Owen Forder of the RAE arrived a week later to assemble the model and install NASA telemetry equipment. Three more Tornado models arrived at the end of July.[88] The quarter-scale models were constructed of fiberglass, wood, and metal. Each was equipped with a rudder and an all-moving tailplane with differential deflection. Instrumentation included transducers, telemetry, servo systems, and radar transponder equipment. To reduce complexity and cost, the models were not equipped with landing gear. Instead, recovery parachutes

87. Letter from Congressman William Carney to Martin Knutson, Director, Flight Operations, NASA Dryden Flight Research Facility, Dec. 3, 1984, Timothy W. Horton collection via Terri Horton, Lancaster, CA.
88. Merlin, Tornado/HIRM Flight Log, Feb. 2000, DFRC Historical Reference Collection, NASA Dryden Flight Research Center, Edwards, CA.

were provided to allow for a soft landing in the desert. Each model weighed approximately 661 pounds and was towed aloft beneath a helicopter, using a 98-foot cable with an electromechanical release system. A small drogue chute stabilized the model prior to drop in order to maintain proper heading, and it separated at launch. An onboard, preprogrammed controller actuated the model's control surfaces. From a launch altitude of 11,900 feet, each model had a maximum gliding range of about 4.7 miles.[89] The British team, consisting of Jeffries, Forder, Charles O'Leary, Geraldine F. Edwards, and Jim Taylor, had the first model ready for flight by August 25. Dubbed ADV-B—reflecting its shape, which was that of the so-called long-nose Air Defense Variant (ADV) of the Tornado design—the model was carried aloft August 31 beneath a UH-1H on loan from NASA Ames Research Center. The helicopter was piloted by Army Maj. Ron Carpenter and NASA research pilot Donald L. Mallick, with O'Leary as observer. Following release from its tow cable, the Tornado model glided to a landing on the Precision Impact Range Area, east of Rogers Dry Lake.

Tornado model ADV-C was dropped the next day, and ADV-D followed with a test on September 3. Five days later, the fourth model—called IDS-I for Interdiction Strike configuration (the snub-nose surface attack variant of the Tornado)—was successfully dropped over the PIRA. By September 22, the ADV-B and ADV-D models had each flown three more times.[90] Although three of the models were unserviceable at the completion of the tests because of damage sustained during recovery, CHIRP constituted an outstanding success. Previous flights had been made at test ranges near Larkhill, U.K., and Woomera, Australia, but with less impressive results, so much less so that the data acquired during testing at Dryden was equivalent to that collected during 5 years of earlier tests at other locations.

A second test series involving the three Tornado variants previously flown, along with two High-Incidence Research Model (HIRM) vehicles, took place in 1983. The HIRM shape included a boxy fuselage, conventional tail configuration, and close-coupled canards in front of the wings. On July 6, the first of two HIRM models flew once at Larkhill to test all systems and basic aerodynamics.

Following arrival of the test team at Dryden, the first model was ready for flight by September 23, but the mission was canceled because

89. Notes on Safety Aspects of Testing Free-Flight Models of Tornado, n.d., Roy Bryant files, DFRC Historical Reference Collection, NASA Dryden Flight Research Center, Edwards, CA.
90. Merlin, Tornado/HIRM Flight Log.

of adverse weather. ADV-D was successfully dropped 4 days later. The following day, the IDS-I model was flown but was damaged during landing and did not fly again. Two more flights each were made with the ADV-D and ADV-B models in October.[91] The remaining sorties were flown using the two HIRM models, dubbed "Hirmon" and "Hermes." Unlike the Tornado models, these did not resemble an operational aircraft type. Rather, they represented an entirely new research aircraft configuration. The HIRM models were equipped with an active control system capable of maintaining bank angles below 30 degrees.

The first drop of Hirmon at Dryden was terminated after just 22 seconds of flight, when an overspeed sensor triggered the vehicle's parachute recovery system. Hermes flew several days later, but the mission was terminated immediately after launch because of failure of a barometric switch in the recovery system. Successful flights of both HIRM vehicles commenced October 14 and continued through the end of the month, when the test models were packed for shipping back to the United Kingdom.

Of the 20 flights scheduled at Dryden during a 6-week period, 5 were eventually canceled. Fifteen flights were completed successfully. The British team worked punishing 12-hour days and 6-day weeks to sustain the flight rate. Three models remained flyable at the conclusion of the project. One Tornado sustained repairable fuselage damage requiring an alignment fixture not available at Dryden, and a second Tornado sustained minor but extensive damage as the result of being dragged through a small tree after a successful parachute landing. The HIRM models were used in 10 of the flights in this series.[92] A third test series was conducted in 1986 under a joint agreement among NASA, the U.S. Department of Defense, and the British Ministry of Defence. A four-person test team traveled from the U.K. and was joined by five Ames-Dryden project team members who provided management and support-services coordination. The Air Force Flight Test Center and U.S. Army Aviation Engineering Flight Activity group at Edwards provided additional support. Typically, an Army UH-1H helicopter carried the test model to an altitude of between 10,000 to 11,500 feet and released it over the PIRA at 72 to 78 knots indicated airspeed.

Three Tornado and the two HIRM models arrived at Dryden in October. Hirmon and Hermes were flown 12 times, logging a total of 24.48 minutes of flight time. The Tornado models were not used, and

91. Ibid.
92. Ibid.

Hermes flew only once. Two flights resulted in no useful data. Five were canceled because of adverse weather, four because of helicopter unavailability, and five more because of range unavailability. Manual recovery had to be initiated during the third drop test. Both models survived the test series with minimal damage.[93]

Working with Sandia—Avocet and SHIRP

Low-cost RPRVs have contributed to the development of hypersonic vehicle concepts and advanced cruise-missile technology. The first such project undertaken at Dryden originated with the Sandia Winged Energetic Reentry Vehicle (SWERVE).

Sandia National Laboratories developed the SWERVE under an exploratory tactical nuclear weapon program. With a slender cone-shaped body and small triangular fins that provided steering, the SWERVE was capable of maneuvering in the range from Mach 2 to Mach 14. Several flight tests in the late 1970s and early 1980s demonstrated maneuverability at high speeds and high angles of attack. Three SWERVE vehicles of two sizes were lofted to altitudes of 400,00 to 600,000 feet on a Strypi rocket and reentered over the Pacific Ocean. The SWERVE 3 test in 1985 included a level flight-profile segment to extend the vehicle's range. Because technologies demonstrated on SWERVE were applicable to development of such hypersonic vehicles as the proposed X-30 National Aero-Space Plane (NASP), Sandia offered to make a SWERVE-derived vehicle available to defense contractors and Government agencies for use as a hypersonic testbed.[94] During the early 1980s, NASA's Office of Aeronautics and Space Technology (OAST) began studying technologies that would enable development of efficient hypersonic aircraft and aerospace vehicles. As part of the program, OAST officials explored the possibility of a joint NASA–Sandia flight program using a SWERVE-derived vehicle to provide hypersonic entry and flight data. Planners wanted to use the capabilities of both NASA and Sandia to refine the existing SWERVE configuration to enable data measurement in specific flight regimes of interest to NASA engineers.[95] The SWERVE

93. Ibid.

94. William B. Scott, "Vehicle Used in Nuclear Weapon Program Offered as Advanced Hypersonic Testbed," *Aviation Week & Space Technology* (Aug. 6, 1996).

95. Preliminary Draft—Engineering Study for a Joint NASA/Sandia Hypersonic Flight Test Program, Dec. 1985, DFRC Historical Reference Collection, NASA Dryden Flight Research Center, Edwards, CA.

shape was optimized for hypersonic performance, but for a transatmospheric vehicle to be practical, it had to be capable of subsonic operation during the approach and landing phases of flight. In 1986, Sandia and NASA officials agreed to participate in a joint project involving an unpowered, radio-controlled model called Avocet. Based on the SWERVE shape, the model retained the slender conical fuselage but featured the addition of narrow-span delta wings. It was approximately 9 feet long and weighed about 85 pounds, including instrumentation. For flight tests, the Avocet vehicle was dropped from a Piper PA-18-150 Super Cub owned by Larry G. Barrett of Tehachapi, CA. The test plan called for 30 to 40 flights to collect data on low-speed performance, handling qualities, and stability and control characteristics.[96] Dryden engineers Henry Arnaiz and Robert Baron managed the Avocet project. R. Dale Reed worked with Dan Garrabrant and Ralph Sawyer to design and build the model. Principal investigators included Ken Iliff, Alex Sim, and Al Bowers. Larry Schilling developed a simulation for pilot training. James B. Craft, Jr., and William Albrecht served as systems and operations engineers, respectively. Robert Kempel and Bruce Powers developed the flight control system. Eloy Fuentes provided safety and quality assurance. Ed Schneider served as primary project pilot, with Einar Enevoldson as backup.[97] All tests were conducted at the China Lake Naval Weapons Center, about 40 miles northeast of Edwards. The model was carried to an altitude of about 8,000 feet beneath the wing of the Super Cub and released above a small dry lakebed. Schneider piloted the vehicle from a ground station, using visual information from an onboard television camera. After accomplishing all test points on the flight plan, Schneider deployed a parachute to bring the vehicle gently to Earth. Testing began in spring 1986 and concluded November 2. Results indicated the configuration had an extremely low lift-to-drag ratio, probably unacceptable for the planned National Aero-Space Plane then being considered in beginning development studies.[98] In 1988, Sandia officials proposed a follow-on project to study the Avocet configuration's cruise and landing characteristics. Primary objectives included demonstration of powered flight and landing characteristics,

96. Personal diary of Timothy W. Horton, 1980–1986, DFRC Historical Reference Collection, NASA Dryden Flight Research Center, Edwards, CA.
97. Avocet Offsite Operations Plan, (Draft) April 1986, DFRC Historical Reference Collection, NASA Dryden Flight Research Center, Edwards, CA.
98. Ibid.

determination of the long-range cruise capabilities of a SWERVE-type vehicle, and the use of Avocet flight data to determine the feasibility of maneuvering and landing such a vehicle following a hypersonic research flight. The new vehicle, called Avocet II, was a lightweight, radio-controlled model weighing just 20 pounds. Significant weight reduction was made possible, in part, through the use of an advanced miniature instrumentation system weighing 3 pounds—one-tenth the weight of the instrumentation used in Avocet I. Powered by two ducted-fan engines, the Avocet II was capable of taking off and landing under its own power.

NASA Dryden officials saw several potential benefits to the projects. First was the opportunity to flight-test an advanced hypersonic configuration that had potential research and military applications. Second, continued work with Sandia offered access to a wealth of hypersonic experience and quality information. Third, Avocet II expanded the NASA–Sandia SWERVE program that had become the heart of NASA's Generic Hypersonic Program, a research project initiated at Dryden and managed by Dr. Isaiah Blankson at NASA Headquarters. Finally, the small-scale R/C model effort served as an excellent training project for young Dryden engineers and technicians. Moreover, total costs for vehicle, instrumentation, flight-test operations, miscellaneous equipment, data analysis, and travel were estimated to be $237,000, truly a bargain by aeronautical research standards.[99] In 1989, a team of researchers at Dryden began work on Avocet II under the direction of Robert Baron. Many of the original team members were back, including William Albrecht, Henry Arnaiz, R. Dale Reed, Alex Sim, Eloy Fuentes, and Al Bowers. They were joined by engineers Gerald Budd, Mark Collard, James Murray, Greg Noffz, and James Yamanaka. Charles Baker provided additional project management oversight. Others included ground pilot Ronald Gilman, crew chief David Neufeld, model builder Robert Violett, and instrumentation engineer Phil Hamory. James Akkerman built and supplied twin ducted-fan engines for the model.[100] For flight operations, the team traveled to the remote test site in a travel trailer equipped with all tools and supplies necessary for onsite maintenance and repair of the model. After setting up camp on the edge of a dry lakebed, technicians unloaded, preflighted,

99. NASA/Sandia Powered SWERVE Landing Configuration Project briefing, 1988, DFRC Historical Reference Collection, NASA Dryden Flight Research Center, Edwards, CA.
100. AVOCET Program Organization chart, 1988, DFRC Historical Reference Collection, NASA Dryden Flight Research Center, Edwards, CA.

and fueled the model. If the configuration had been changed since the previous flight, an engineer performed a weight-and-balance survey prior to takeoff. When the crew chief was satisfied that the vehicle was ready, the flight-test engineer reviewed all pertinent test cards to ensure that each crewmember was aware of his responsibilities during each phase of flight. The ground pilot followed a structured sequence of events outlined in the test cards in order to optimize the time available for research maneuvers.

Typically, the pilot flew a figure-eight ground track that produced the longest-possible steady, straight-line flight segment between turns at each end of the test range. The ground pilot controlled the Avocet II using a commercially available nine-channel, digital pulse-code modulation radio-control system. Since loss of the vehicle was considered an acceptable risk, there was no redundant control system. Software permitted preprogrammed mixing of several different control functions, greatly simplifying vehicle operation. After landing, recorded test data were downloaded to a personal computer for later analysis.[101] Initial taxi tests revealed that the model lacked sufficient thrust to achieve takeoff. Modifications to the inlet solved the problem, but the model had a very low lift-to-drag ratio, which made it difficult to maneuver. The turning radius was so large that it was nearly impossible to keep the model within visual range of the ground pilot, so the flight-test engineer provided verbal cues regarding heading and attitude while observing the model through binoculars. The pilot executed each research maneuver several times to ensure data quality.[102] The first flight took place November 18, 1989, and lasted just 2 minutes. Ron Gilman lost sight of the model in the final moments of its steep descent, resulting in a hard landing. Over the course of 10 additional flights through February 1991, Gilman determined the vehicle's handling qualities and longitudinal stability, while engineers attempted to define local flow-interference areas using tufts and ground-based high-speed film.[103] The instrumentation system in the Avocet II vehicle, consisting of a Tattletale Model 4 data

101. Gerald D. Budd, Ronald L. Gilman, and David Eichstedt, "Operational and Research Aspects of a Radio-Controlled Model Flight Test Program," presented at the *AIAA 31st Aerospace Sciences Meeting and Exhibit, Reno, NV, Jan. 11–14, 1993.* Also published in *Journal of Aircraft*, vol. 32, no. 3 (May–June 1995). Also published as NASA TM-104266 (1993).

102. Ibid.

103. Avocet II Flight Log, compiled by Merlin, Oct. 2008, DFRC Historical Reference Collection, NASA Dryden Flight Research Center, Edwards, CA.

logger with 32 kilobytes of onboard memory, provided research-quality quantitative analysis data on such performance parameters as lift-curve slope, lift-to-drag ratio, and trim curve. An 11-channel, 10-bit analog-to-digital converter capable of operating at up to 600 samples per second measured analog signals. The 2.2-ounce device, measuring just 3.73 by 2.25 by 0.8 inches, also featured a 128-kilobyte memory expansion board to increase data-storage capability.

The pilot quantified aircraft performance by executing a quasistatic pushover/pull-up (POPU) maneuver. Properly executed, a single POPU maneuver could simultaneously characterize all three of the desired flight-test parameters over a wide angle-of-attack range. Structural vibration at high-power settings—such as those necessary to execute a POPU maneuver—caused interference with onboard instrumentation. Attempts to use different mounting techniques and locations for both engines and accelerometers failed to alleviate the problem. Eventually, engineers developed a POPU maneuver that could be flown in a steep dive with the engines at an idle setting. In this condition, the accelerometers provided usable data.[104] Researchers at Dryden teamed up with Sandia again for the Royal Amber Model (RAM) project, later renamed the Sandia Hybrid Inlet Research Program (SHIRP). This project included tests of subscale and full-scale radio-controlled models of an advanced cruise missile shape designed by Sandia under the Standoff Bomb Program. The goal of the SHIRP experiments was to provide flight-test data on an experimental inlet configuration for use in future weapons, such as the Joint Air-to-Surface Standoff Missile, then under development. Sandia engineers designed an engine inlet to be "stealthy"—not detectable by radar—yet still capable of providing good performance characteristics such as a uniform airflow with no separation. Airflow exiting the inlet and entering the turbine had to be uniform as well. The design of the new inlet was complex. Instead of a standard rectangular channel, the cross-sectional area of the inlet varied from a high aspect ratio V-shape at the front to an almost circular outlet at the back end.[105] Sandia

104. Philip J. Hamory and James E. Murray, "Flight Experience With Lightweight, Low-Power Miniaturized Instrumentation Systems," NASA TM-4463, Mar. 1993, NASA Dryden Flight Research Center, Edwards, CA.

105. Jim Nelsen, "Sandia Uses CFD Software with Adaptive Meshing Capability to Optimize Inlet Design," Journal Articles No. JA060, Fluent, Inc., *www.fluent.com/solutions/articles/ja060.pdf*, 1999, accessed July 1, 2009.

funded Phase I flight tests of a 40-percent-scale RAM from August 1990 through August 1991. Because the project was classified at the time, flight operations could not take place at Dryden. Instead, the test team used secure range areas at Edwards Air Force Base North Base and China Lake Naval Weapons Center.[106] The first flight took place in August 1990 at China Lake. Typically, the model was released from the R/C mother ship at an altitude of about 600 feet. The ground pilot performed a series of gliding and turning maneuvers, followed by a controlled pullup prior to impact. Results from the first four flights indicated good longitudinal and directional stability and neutral lateral stability.

The next three flights took place in February 1991 at North Base, just a few miles northeast of Dryden. During the first of these, a recovery parachute deployed at 150 feet but came loose from the vehicle. The ground pilot made a horizontal landing on the runway centerline. On the next flight, the vehicle exhibited good controllability and stability in both pitch and yaw axes at airspeeds between 35 and 80 miles per hour (mph). The pilot elected to land on the runway rather than use the recovery parachute. The final 10 flights took place at China Lake, ending July 13, 1991.[107] During fall 1991 and early 1992, researchers proposed tasks and milestones for the second phase of testing, and in February 1992, RAM Phase II was reorganized as the unclassified SHIRP project. During spring 1992, however, conditions arose at both Sandia and Dryden that required modification of the proposed schedule.

In support of a Sandia initiative to conduct a prototype flight demonstration program, the stabilizing and lifting surfaces for the baseline Standoff Bomb were reevaluated based on the most recent wind tunnel data and taking into account the current mass properties and flight profiles. This revised geometry was used for the definition of wind tunnel models to collect data on static aerodynamics, diffuser distortion, and total pressure loss. In order to use the revised definition for the SHIRP flight-test models, the schedule had to be compromised.[108] An initial flight-test series in December 1992 involved launching a subscale model called Mini-SHIRP from the R/C Mothership. The team also constructed

106. Merlin, Sandia Hybrid Inlet Research Program Flight Log, Oct. 2008, DFRC Historical Reference Collection, NASA Dryden Flight Research Center, Edwards, CA.
107. Merlin, Sandia Hybrid Inlet Research Program Flight Log.
108. SHIRP project files, 1991–1993, DFRC Historical Reference Collection, NASA Dryden Flight Research Center, Edwards, CA.

two full-scale vehicles, each 14 feet long and weighing about 52 pounds. SHIRP-1 was uninstrumented, unpowered, and lacked inlets. SHIRP-2 featured the experimental inlet configuration and was powered by two electric ducted-fan engines to extend the glide range and provide short periods of level flight (10–15 seconds). The ground pilot controlled the vehicle through a fail-safe pulse-code modulation radio-uplink system. The test vehicles were equipped with deployable wings and pneumatically deployable recovery parachutes. The two full-scale vehicles, tested in 1993, were launched from the modified Rans S-12 (also known as "Ye Better Duck") remotely piloted ultralight aircraft.

Flight operations began with takeoff of the mother ship from North Base followed by launch and landing of the test article in the vicinity of Runway 23 on the northern part of Rogers Dry Lake. The SHIRP flights demonstrated satisfactory lateral, longitudinal, and directional static and dynamic stability. The vehicle had reasonable control authority, required only minimal rudder deflection, and had encouraging wing-stall characteristics.[109] NASA project personnel included Don Bacon, Jerry Budd, Bob Curry, Alex Sim, and Tony Whitmore. Contractors from PRC, Inc., included Dave Eichstedt, Ronald Gilman, R. Dale Reed, B. McCain, and Dave Richwine. Todd M. Sterk, Walt Rutledge, Walter Gutierrez, and Hank Fell of Sandia worked with NASA and PRC personnel to analyze and document the various test data. In a September 1992 memorandum, Gutierrez noted that Sandia personnel recognized the SHIRP effort as "an opportunity to learn from the vast flight-test experience available at Dryden in the areas of experimental testing and data analysis."

In acknowledging the excellent teaming opportunity for both Sandia and NASA, he added that, "Dryden has an outstanding reputation for parameter estimation of aerodynamic characteristics of flight-test vehicles."[110]

Toward Precision Autonomous Spacecraft Recovery

From October 1991 to December 1996, a research program known as the Spacecraft Autoland Project was conducted at Dryden to determine the feasibility of autonomous spacecraft recovery using a ram-air parafoil

109. Ibid.

110. Memorandum dated Sept. 14, 1992, regarding "Status of SHIRP Radio Control Flight Tests at Dryden Flight Research Facility," SHIRP project files, 1991–1993, DFRC Historical Reference Collection, NASA Dryden Flight Research Center, Edwards, CA.

system for the final stages of flight, including a precision landing. The latter characteristic was the focus of a portion of the project that called for development of a system for precision cargo delivery. NASA Johnson Space Center and the U.S. Army also participated in various phases of the program, with the Charles Stark Draper Laboratory of Cambridge, MA, developing Precision Guided Airdrop Software (PGAS) under contract to the Army.[111] Four generic spacecraft models (each called a Spacewedge, or simply Wedge) were built to test the concept's feasibility. The project demonstrated precision flare and landing into the wind at a predetermined location, proving that a flexible, deployable system that entailed autonomous navigation and landing was a viable and practical way to recover spacecraft.

Key personnel included R. Dale Reed, who participated in flight-test operations. Alexander Sim managed the project and documented the results. James Murray served as the principal Dryden investigator and as lead for all systems integration for Phases I and II. He designed and fabricated much of the instrumentation for Phase II and was the lead for flight data retrieval and analysis in Phases II and III. David Neufeld performed mechanical integration for the Wedge vehicles' systems during all three phases and served as parachute rigger, among other duties. Philip Hattis of the Charles Stark Draper Laboratory served as the project technical director for Phase III. For the Army, Richard Benney was the technical point of contact, while Rob Meyerson served as the technical point of contact for NASA Johnson and provided the specifications for the Spacewedges.[112] The Spacewedge configuration consisted of a flattened biconic airframe joined to a ram-air parafoil with a custom harness. In the manual control mode, the vehicle was flown using radio uplink. In the autonomous mode, it was controlled using a small computer that received inputs from onboard sensors. Selected sensor data were recorded onto several onboard data loggers.

Two Spacewedge shapes, resembling half cones with a flattened bottom, were used for four airframes that represented generic hypersonic vehicle configurations. Wedge 1 and Wedge 2 had sloping sides, and the underside of the nose sloped up slightly. Wedge 3 had flattened sides, to

111. Merlin, Spacewedge fact sheet (draft), FS-045, Apr. 1998, NASA Dryden Flight Research Center, Edwards, CA.

112. Merlin, Spacewedge fact sheet, FS-2002-09-045, Sept. 2002, NASA Dryden Flight Research Center, Edwards, CA.

9

create a larger internal volume for instrumentation. The Spacewedge vehicles were 48 inches long, 30 inches wide, and 21 inches in height. The basic weight was 120 pounds, although various configurations ranged from 127 to 184 pounds during the course of the test program. Wedge 1 had a tubular steel structure, covered with plywood on the rear and underside that could withstand hard landings. It had a fiberglass-covered wooden nose and removable aluminum upper and side skins. Wedge 2, originally uninstrumented, was later configured with instrumentation. It had a fiberglass outer shell, with plywood internal bulkheads and bottom structure. Wedge 3 was constructed as a two-piece fiberglass shell, with a plywood and aluminum shelf for instrumentation.[113] A commercially available 288-square-foot ram-air parafoil of a type commonly used by sport parachutists was selected for Phase I tests. The docile flight characteristics, low wing loading, and proven design allowed the project team to concentrate on developing the vehicle rather than the parachute. With the exception of lengthened control lines, the parachute was not modified. Its large size allowed the vehicle to land without flaring and without sustaining damage. For Phase II and III, a smaller (88 square feet) parafoil was used to allow for a wing loading more representative of space vehicle or cargo applications.

Spacewedge Phase I and II instrumentation system architecture was driven by cost, hardware availability, and program evolution. Essential items consisted of the uplink receiver, Global Positioning System (GPS) receiver and antenna, barometric altimeter, flight control computer, servo-actuators, electronic compass, and ultrasonic altimeter. NASA technicians integrated additional such off-the-shelf components as a camcorder, control position transducers, a data logger, and a pocket personal computer. Wedge 3 instrumentation was considerably more complex in order to accommodate the PGAS system.[114] Spacewedge control systems had programming, manual, and autonomous flight modes. The programming mode was used to initialize and configure the flight control computer. The

113. Philip D. Hattis, Robert J. Polutchko, Brent D. Appleby, Timothy M. Barrows, Thomas J. Fill, Peter M. Kachmar, and Terrence D. McAteer, "Final Report: Development and Demonstration Test of a Ram-Air Parafoil Precision Guided Airdrop System," vols. 1 to 4 (Report CSDL-R-2752, Oct. 1996) and Addendum (Report CSDL-R-2771, Dec. 1996).

114. James E. Murray, Alex G. Sim, David C. Neufeld, Patrick K. Rennich, Stephen R. Norris, and Wesley S. Hughes, "Further Development and Flight Test of an Autonomous Precision Landing System Using a Parafoil," NASA TM-4599, July 1994.

manual mode incorporated a radio-control model receiver and uplink transmitter, configured to allow the ground pilot to enter either brake (pitch) or turn (yaw) commands. The vehicle reverted to manual mode whenever the transmitter controls were moved, even when the autonomous mode was selected. Flight in the autonomous mode included four primary elements and three decision altitudes. This mode allowed the vehicle to navigate to the landing point, maintain the holding pattern while descending, enter the landing pattern, and initiate the flare maneuver. The three decision altitudes were at the start of the landing pattern, the turn to final approach, and the flare initiation.

NASA researchers initially launched Wedge 1 from a hillside near the town of Tehachapi, in the mountains northwest of Edwards, to evaluate general flying qualities, including gentle turns and landing flare. Two of these slope soar flights were made April 23, 1992, with approximately 15-knot winds, achieving altitudes of 10 to 50 feet. The test program was then moved to Rogers Dry Lake at Edwards and to a sport parachute drop zone at California City.[115] A second vehicle (known as Inert Spacewedge, or Wedge 2) was fabricated with the same external geometry and weight as Wedge 1. It was initially used to validate parachute deployment, harness design, and drop separation characteristics. Wedge 2 was inexpensive, lacked internal components, and was considered expendable. It was first dropped from a Cessna U-206 Stationair on June 10, 1992. A second drop of Wedge 2 verified repeatability of the parachute deployment system. The Wedge 2 vehicle was also used for the first drop from a Rans S-12 ultralight modified as a RPV on August 14, 1992. Wedge 2 was later instrumented and used for ground tests while mounted on top of a van, becoming the primary Phase I test vehicle.[116] Thirty-six flight tests were conducted during Phase I, the last taking place February 12, 1993. These flights, 11 of which were remotely controlled, verified the vehicle's manual and autonomous landing systems. Most were launched from the Cessna U-206 Stationair. Only two flights were launched from the Rans S-12 RPV.

Phase II of the program, from March 1993 to March 1995, encompassed 45 flights using a smaller parafoil for higher wing loading

115. Alex G. Sim, James E. Murray, David C. Neufeld, and R. Dale Reed, "The Development and Flight Test of a Deployable Precision Landing System for Spacecraft Recovery," NASA TM-4525, Sept. 1993.

116. Sim, Murray, Neufeld, and Reed, "Development and Flight Test of a Deployable Precision Landing System," *AIAA Journal of Aircraft*, vol. 31, no. 5 (Sept. 1994), pp. 1101–1108.

(2 lb/ft^2) and incorporating a new guidance, control, and instrumentation system developed at Dryden. The remaining 34 Phase III flights evaluated the PGAS system using Wedge 3 from June 1995 to December 1996. The software was developed by the Charles Stark Draper Laboratory under contract to the U.S. Army to develop a guidance system to be used for precision offset cargo delivery. The Wedge 3 vehicle was 4 feet long and was dropped at weights varying from 127 to 184 pounds.[117] Technology developed in the Spacewedge program has numerous civil and military applications. Potential NASA users for a deployable, precision, autonomous landing system include proposed piloted vehicles as well as planetary probes and booster-recovery systems. Military applications of autonomous gliding-parachute systems include recovery of aircraft ejection seats and high-altitude, offset delivery of cargo to minimize danger to aircraft and crews. Such a cargo delivery system could also be used for providing humanitarian aid.[118] In August 1995, R. Dale Reed incorporated a 75-square-foot Spacewedge-type parafoil on a 48-inch-long, 150-pound lifting body model called ACRV-X. During a series of 13 flights at the California City drop zone, he assessed the landing characteristics of Johnson Space Center's proposed Assured Crew Return Vehicle design (essentially a lifeboat for the International Space Station). The instrumented R/C model exhibited good flight control and stable ground slide-out characteristics, paving the way for a larger, heavyweight test vehicle known as the X-38.[119]

Models and Mother Ships—Utility RPRV and Ultralight RPRV

By the mid-1990s, it was clear to NASA researchers that use of unpiloted vehicles for research and operational purposes was expanding dramatically. R. Dale Reed and others at Dryden proposed development of in-house, hands-on expertise in flight-testing experimental UAVs to guide and support anticipated research projects. They suggested that lower risks and higher mission-success rates could be achieved by applying lessons learned from flight-test experience and crew training. Additionally, they recommended that special attention be paid to human factors by standardizing ground control consoles and UAV operational procedures.

117. Merlin, Spacewedge fact sheet, (draft).

118. Ibid.

119. R. Dale Reed, "The Flight Test of a 1/6 Scale Model of the ACRV-X Space Craft Using a Parafoil Recovery Parachute System," NASA Dryden Flight Research Center, Nov. 1995.

To meet these goals, Reed recommended using two types of low-cost expendable UAVs. The first was a radio-controlled model airplane weighing less than 50 pounds but capable of carrying miniature downlink television cameras, autopilot, and GPS guidance systems. Requirements for flight termination systems and control redundancy for such an aircraft would be much less stringent than those for larger UAVs, and the model would require much less airspace for flight operations. Reed felt the R/C model could serve as a basic trainer for UAV pilots because the same skills and knowledge are required regardless of vehicle size. Additionally, the R/C model could provide flight research results at very low cost.[120] Second, Reed felt the modified Rans S-12 ("Ye Better Duck") should be returned to flight status since an ultra-light-type vehicle could duplicate the size and flying characteristics of planned high-altitude RPRVs then being developed. He saw the S-12 as an advanced trainer for NASA UAV crews. The S-12 had not been flown since January 1994 and required a thorough inspection of airframe and engine, as well as replacement of batteries in several of its systems. Reed recommended that Tony Frackowiak of the Dryden Physics Lab be given the task of preparing the "Ye Better Duck" for flight status and then serving as primary checkout pilot.[121] Reed submitted his proposals to Dryden director Ken Szalai with a recommendation to develop a Utility UAV as a mother ship for small experimental models. Jenny Baer-Riedhart and John Del Frate, Project Manager and Assistant Project Manager, respectively, for the Environmental Research Aircraft and Sensor Technology (ERAST) program, were willing to support the project plan if the Dryden Operations Division provided a requirement and also pledged strong support for the plan. Research pilots Dana Purifoy, Tom McMurtry, and Steve Ishmael were enthusiastic about the project. Ishmael immediately saw a potential application for the Utility UAV to drop a subscale aerodynamic model of the planned X-33 spacecraft. Project personnel included Reed as Utility UAV project engineer, research pilot Purifoy, crew chief/project pilot Tony Frackowiak, UAV systems technician Howard Trent, and UAV backup pilot Jerry Budd.[122]

120. Dryden Utility UAV Development outline, Reed files, n.d. (circa 1996), NASA DFRC Historical Reference Collection, NASA Dryden Flight Research Center, Edwards, CA.

121. Ibid.

122. Letter to Ken Szalai from Reed and Utility UAV Project organizational chart, Reed files, Mar. 10, 1997, NASA DFRC Historical Reference Collection, NASA Dryden Flight Research Center, Edwards, CA.

During this time, Reed reactivated the old R/C Mothership that had been used to launch lifting body models in the 1960s. Frackowiak removed and overhauled its engines, cleaned the exhaust system, replaced throttle servos, and made other repairs. During six checkout flights November 25, 1996, the Mothership underwent checkout and demonstrated a 20-pound payload capability. It was subsequently used as a launch aircraft for a model of a hypersonic wave rider and a 5-percent-scale model of the Pegasus satellite booster.[123] Meanwhile, Reed had pressed on with plans for the larger Utility UAV. For systems development, Frackowiak acquired a Tower Hobbies Trainer-60 R/C model and modified it to accept several different gyro and autopilot configurations. The Trainer 60 was 57 inches long, had a 69-inch wingspan, and weighed just 8 pounds. Frackowiak conducted more than a dozen test flights with the model in March 1997.[124] In April 1997, the Mothership was equipped with a video camera and telemetry system that would also be used on the Utility UAV. The first three test flights took place at Rosamond Dry Lake on the morning of April 10, with one pilot inside a control van watching a video monitor and another outside directly observing the aircraft. For the first flight, Frackowiak served as outside pilot—controlling takeoff and landing—while Reed familiarized himself with pitch and roll angles in climb, cruise, and descent. On the third flight, they switched positions so Reed could make a low approach to familiarize Frackowiak with the view from the camera. They found that it helped to have a ground marking (such as a runway edge stripe) on the lakebed as a visual reference during touchdown. Other areas for improvement included the reduction of glare on the video monitor, better uplink antenna orientation, and stabilization of pitch and roll rate gyros to help less-experienced pilots more easily gain proficiency.[125] In May 1997, Dana Purifoy began familiarization and training with the Mothership. In August, the aircraft was again used to launch the Pegasus model (for deep-stall tests) as well as a Boeing–UCLA Solar-Powered Formation Flight (SPFF) vehicle.

123. Mothership Notebook, Reed files, 1996–1997, DFRC Historical Reference Collection, NASA Dryden Flight Research Center, Edwards, CA.

124. Tony Frackowiak, Tower Trainer-60 Gyro and Autopilot Test Notes, Mar. 1997, Reed files, 1996–1997, NASA DFRC Historical Reference Collection, NASA Dryden Flight Research Center, Edwards, CA.

125. Utility UAV Report, Apr. 1997, Reed files, 1996–1997, NASA DFRC Historical Reference Collection, NASA Dryden Flight Research Center, Edwards, CA.

A radio controlled model aircraft, acting as a miniature mother ship, carries aloft a radiocontrolled model of the X-33. NASA.

On August 5, Reed piloted the Mothership, while Frackowiak flew the SPFF model.

In September 1997, Frackowiak modified the Mothership's launch hook to accept a scale model of the Lockheed Martin X-33 lifting body vehicle. The X-33 Mini-RPRV was, like the SPFF model, equipped with its own set of radio controls. Initial drop flights took place September 30 at a sod farm near Palmdale, with John Howell piloting the X-33 model.

Following a series of SPFF flights in October, the Mothership was taken to Air Force Plant 42 in Palmdale for more X-33 Mini-RPRV drops. On February 12, 1998, interference led to loss of control. The Mothership crashed, sustaining severe but repairable damage to wing and nose.[126]

While the Mothership was undergoing repairs, Frackowiak completed construction of the 30-pound Utility UAV in April 1998. On April 24, he took the airplane to Tailwinds Field, a popular R/C model airstrip in Lancaster, for its first flight. Takeoff at partial power was uneventful. After gaining 300 feet altitude, Frackowiak applied full power to check the trim then checked controllability in slow flight before bringing the Utility UAV in for a smooth landing.

By the end of June, the aircraft had been cleared to carry payloads weighing up to 20 pounds. Three months later, the Utility UAV was

126. Mothership Notebook.

modified to carry the X-33 Mini-RPRV. On September 10, Reed and John Redman began a series of captive flights at Rosamond Dry Lake. Drop testing at Rosamond began 4 days later, with 4 successful free flights made over a 2-day span to evaluate higher X-33 model weights and a dummy nose boom.[127] On October 1, 1998, the Utility UAV made its 20th flight, and the X-33 model was released for the 5th time at Rosamond. Piloted by Frackowiak, the lifting body's steep descent ended with a flawless landing, but disaster lurked in wait for the drop plane. As Redman maneuvered the Utility UAV toward final approach, he watched it suddenly roll to the left and plunge into the clay surface of the lakebed, sustaining major damage.[128] Further testing of the X-33 Mini-RPRV was undertaken using the repaired Mothership. Several successful drops were made in early October, as well as a familiarization flight for research pilot Mark Stucky. Reed noted in his log: "The Mothership has again proven the practicality of its design, as it has been flawless during these launches. And it is very good to see it flying and performing useful missions again."[129]

Riding the Wave with LoFLYTE

The Low-Observable Flight Test Experiment (LoFLYTE) program was a joint effort among researchers at NASA Langley and the Air Force Research Laboratory with support from NASA Dryden and the 445th Flight Test Squadron at Edwards Air Force Base. Accurate Automation, Corp., of Chattanooga, TN, received a contract under NASA's Small Business Innovation Research program to explore concepts for a stealthy hypersonic wave rider aircraft. The Navy and the National Science Foundation provided additional funding. A wave rider derives lift and experiences reduced drag because of the effects of riding its bow shock wave. Applications for wave rider technology include transatmospheric vehicles, high-speed passenger transports, missiles, and military aircraft.

The LoFLYTE vehicle was designed to serve as a testbed for a variety of emerging aerospace technologies. These included rapid prototyping, instrumentation, fault diagnosis and isolation techniques, real-time

127. Utility UAV Notebook, Reed files, 1997–1998, NASA DFRC Historical Reference Collection, NASA Dryden Flight Research Center, Edwards, CA.

128. Merlin and Tony Moore, *X-Plane Crashes—Exploring Secret, Experimental, and Rocket Plane Crash Sites* (Specialty Press, 2008) and personal log of Merlin, vol. 1, June 1997–Sept. 2006, entry for Oct. 1, 1998.

129. Mothership Notebook.

data acquisition and control, miniature telemetry systems, optimum antenna placement, electromagnetic interference minimization, advanced exhaust nozzle concepts, trajectory control techniques, advanced landing concepts, free-floating wingtip ailerons (called tiperons), and adaptive compensation for pilot-induced oscillations.[130] Most important of all, LoFLYTE was eventually to be equipped with neural network flight controls. Such a system employs a network of control nodes that interact in a similar fashion to neurons in the human brain. The network "learns," altering the aircraft's flight controls to optimize performance and take pilot responses into consideration. This would be particularly useful in situations in which a pilot needed to make decisions quickly and land a damaged aircraft safely, even if its controls are partially destroyed. Researchers also expected that neural network controls would be useful for flying unstable configurations, such as those necessary for efficient hypersonic-flight vehicles. The computing power of Accurate Automation's neural network was provided by 16,000 parallel neurons making 1 billion decisions per second, giving it the capability to adjust to changing flight conditions faster than could a human pilot.[131] The LoFLYTE model was just 100 inches long, with a span of 62 inches and a height of 24 inches. It weighed 80 pounds and was configured as a narrow delta planform with two vertical stabilizer fins. The shell of the model, made from fiberglass, foam, and balsa wood, was constructed at Mississippi State University's Raspet Flight Research Laboratory and then shipped to SWB Turbines in Appleton, WI, for installation of radio control equipment and a 42-pound-thrust microturbine engine.[132] The first flight took place at Mojave Airport, CA, on December 16, 1996. The vehicle was not yet equipped with a neural network and relied instead on conventional computerized stabilization and control systems. All went well as the LoFLYTE climbed to an altitude of about 150 feet and the pilot began a 180-degree turn. At that point—about 34 seconds into the flight—the ground pilot was forced to land the craft wheels-up in the

130. NASA's LoFLYTE Program Flown, NASA FS-1997-07-29-LaRC, NASA Langley Research Center, Hampton, VA, July 1997.

131. Jim Skeen, "Edwards to perform tests on aircraft with control system which 'learns as it flies,'" *Los Angeles Daily News*, Antelope Valley edition, Los Angeles, CA (Aug. 3, 1996). Additional information from Ian Sheppard, "Towards hypersonic flight," Flight International (Nov. 1997).

132. NASA News Release No. 96-126—LoFLYTE, NASA Headquarters, Washington, DC, Aug. 2, 1996.

sand beside the runway because of control difficulties. The model suffered only minor damage, and researchers generally considered the flight a success because it was the first time a wave-rider–concept vehicle had taken off under its own power.[133] Testing resumed in June 1997 with several flights from the Edwards North Base runway. This gave researchers the opportunity to verify the subsonic airworthiness of the wave rider shape and analyze basic handling characteristics. The results showed that a full-scale vehicle would be capable of taking off and landing at normal speeds (i.e., those comparable to such high-speed aircraft as the SR-71). Flight tests of the neural network control system began in December 1997 and continued into 1998. These included experiments to verify the system's ability to handle changes in airframe configuration (such as removal of vertical stabilizers) and simulated damage to control surfaces.[134]

X-36 Tailless Fighter Agility Demonstration

In 1989, engineers from NASA Ames Research Center and the Phantom Works, a division of McDonnell-Douglas—and later Boeing, following a merger of the two companies—began development of an agile, tailless aircraft configuration. Based on results of extensive wind tunnel testing and computational fluid dynamics (CFD) analysis, designers proposed building the X-36—a subscale, remotely piloted demonstrator—to validate a variety of advanced technologies. The X-36 project team consisted of personnel from the Phantom Works, Ames, and Dryden. NASA and Boeing were full partners in the project, which was jointly funded under a roughly fifty-fifty cost-sharing arrangement. Combined program cost for development, fabrication, and flight-testing of two aircraft was approximately $21 million. The program was managed at Ames, while Dryden provided flight-test experience, facilities, infrastructure, and range support during flight-testing.

The X-36 was a 28-percent-scale representation of a generic advanced tailless, agile, stealthy fighter aircraft configuration. It was about 18 feet long and 3 feet high, with a wingspan of just over 10 feet. A single

133. "LoFLYTE makes its maiden flight," *Antelope Valley Press*, Lancaster, CA (Jan. 2, 1997). Additional information from Andreas Parsch, "Directory of U.S. Military Rockets and missiles, Appendix 4: Undesignated Vehicles," 2004, *http://www.designation-systems.net/dusrm/app4/loflyte.html*, accessed June 9, 2009; source material includes Kenneth Munson, ed., *Jane's Unmanned Aerial Vehicles and Targets*, Issue 15, Jane's Information Group, Alexandria, VA, 2000.
134. Parsch, "Directory of U.S. Military Rockets and missiles, Appendix 4: Undesignated Vehicles."

Technicians push the X-36 into a hangar at NASA Dryden Flight Research Center. NASA.

Williams International F112 turbofan engine provided about 700 pounds of thrust. Fully fueled, the X-36 weighed about 1,250 pounds.

The vehicle's small size helped reduce program costs but increased risk because designers sacrificed aircraft system redundancy for lower weight and complexity. The subscale vehicle was equipped with only a single-string flight control system rather than a multiply redundant system more typical in larger piloted aircraft. Canards on the forward fuselage, split ailerons on the trailing edges of the wings, and an advanced thrust-vectoring nozzle provided directional control as well as speed brake and aerobraking functions. Because the X-36 was aerodynamically unstable in both pitch and yaw, an advanced single-channel digital fly-by-wire control system was required to stabilize the aircraft in flight.[135] Risks were mitigated by using a pilot-in-the-loop approach, to eliminate the need for expensive and complex autonomous flight control systems and the risks associated with such systems' inability to correct for unknown or unforeseen phenomena once in flight. Situational-awareness data were provided to the pilot's ground station through a video camera mounted in the vehicle's nose, a standard fighter-type head-up display, and a moving-map representation of the vehicle's position.

135. "X-36 Tailless Fighter Agility Research Aircraft," NASA Fact Sheet, NASA Dryden Flight Research Center, Edwards, CA, 1999.

Boeing project pilot Laurence A. Walker was a strong advocate for the advantages of a full-sized ground cockpit. When an engineer designs a control station for a subscale RPRV, the natural tendency might be to reduce the cockpit control and display suite, but Walker demonstrated that the best practice is just the opposite. In any ground-based cockpit, the pilot will have fewer natural sensory cues such as peripheral vision, sound, and motion. Re-creating motion cues was impractical, but audio, visual, and HUD cues were re-created in order to improve situational awareness comparable to that of a full-sized aircraft.[136] The X-36 Ground Control Station included a full-size stick, rudder pedals and their respective feel systems, throttle, and a full complement of modern fighter-style switches. Two 20-inch monitors provided visual displays to the pilot. The forward-looking monitor provided downlinked video from a canopy-mounted camera, as well as HUD overlay with embedded flight-test features. The second monitor displayed a horizontal situation indicator, engine and fuel information, control surface deflection indicators, yaw rate, and a host of warnings, cautions, and advisories. An audio alarm alerted the pilot to any new warnings or cautions. A redundant monitor shared by the test director and GCS engineer served as a backup, should either of the pilot's monitors fail.[137] To improve the pilot's ability to accurately set engine power and to further improve situational awareness, the X-36 was equipped with a microphone in what would have been the cockpit area of a conventional aircraft. Downlinked audio from this microphone proved to be a highly valuable cue and alerted the team, more than once, to problems such as screech at high-power settings and engine stalls before they became serious.

The X-36 had a very high roll rate and a mild spiral divergence. Because of its size, it was also highly susceptible to gusty wind conditions. As a result, the pilot had to spend a great deal of time watching the HUD, the sole source of attitude cues. Without kinesthetic cues to signal a deviation, anything taking the pilot's focus away from the HUD (such as shuffling test cards on a kneeboard) was a dangerous distraction. To resolve the problem, the X-36 team designed a tray to hold test cards at the lower edge of the HUD monitor for easy viewing.[138] Walker

136. Laurence A. Walker, "Flight Testing the X-36—The Test Pilot's Perspective," NASA CR-198058, NASA Dryden Flight Research Center, Edwards, CA, 1997.

137. "X-36 Tailless Fighter Agility Research Aircraft."

138. Walker, "Flight Testing the X-36—The Test Pilot's Perspective."

piloted the maiden flight May 17, 1997. The X-36 flight-test envelope was limited to 160 knots to avoid structural failure in the event of a flight control malfunction. If a mishap occurred, an onboard parachute was provided to allow safe recovery of the X-36 following an emergency flight termination. Fortunately, the initial flight was a great success with no obvious discrepancies.

The second flight, however, presented a significant problem as the video and downlink signals became weak and intermittent while the X-36 was about 10 miles from the GCS at 12,000 feet altitude. As programmed to do, the X-36 went into lost-link autonomous operation, giving the test team time to initiate recovery procedures to regain control. The engineers were concerned, as each intermittent glimpse of the data showed the vehicle in a steeper angle of bank, well beyond what had yet been flown. Eventually, Walker regained control and made an uneventful landing. The problem was later traced to a temperature sensitivity problem in a low-noise amplifier.[139] Phase I of the X-36 program provided a considerable amount of data on real-time stability margin and parameter identification maneuvers. Automated maneuvers, uplinked to the aircraft, greatly facilitated envelope expansion, and handling qualities were found to be remarkably good.

Phase II testing expanded the flight envelope and demonstrated new software. New control laws and better derivatives improved stability margins and resulted in improved flying qualities. The final Phase II flight took place November 12, 1997. During a 25-week period, 31 safe and successful research missions had been made, accumulating a total of 15 hours and 38 minutes of flight time and using 4 versions of flight control software.[140] In a follow-on effort, the Air Force Research Laboratory (ARFL) contracted Boeing to fly AFRL's Reconfigurable Control for Tailless Fighter Aircraft (RESTORE) software as a demonstration of the adaptability of a neural-net algorithm to compensate for in-flight damage or malfunction of aerodynamic control surfaces. Two RESTORE research flights were flown in December 1998, with the adaptive neural-net software running in conjunction with the original proven control laws. Several in-flight simulated failures of control surfaces were introduced as issues for the reconfigurable control algorithm to address. Each time, the software correctly compensated

139. Ibid.

140. Ibid.

for the failure and allowed the aircraft to be safely flown in spite of the degraded condition.[141] The X-36 team found that having a trained test pilot operate the vehicle was essential because the high degree of aircraft agility required familiarity with fighter maneuvers, as well as with the cockpit cues and displays required for such testing. A test pilot in the loop also gave the team a high degree of flexibility to address problems or emergencies in real time that might otherwise be impossible with an entirely autonomous system. Design of the ground cockpit was also critical, because the lack of normal pilot cues necessitated development of innovative methods to help replace the missing inputs. The pilot also felt that it was vital for flight control systems for the subscale vehicle to accurately represent those of a full-scale aircraft.

Some X-36 team members found it aggravating that, in the minds of some upper-level managers, the test vehicle was considered expendable because it was didn't carry a live crewmember. Lack of redundancy in certain systems created some accepted risk, but process and safety awareness were key ingredients to successful execution of the flight-test program. Accepted risk as it extended to the aircraft and onboard systems did not extend to processes that included qualification testing of hardware and software.[142] The X-36 demonstrator program was aimed at validating technologies proposed by McDonnell-Douglas (and later Boeing) for early concepts of a Joint Strike Fighter (JSF) design, as well as unmanned combat air vehicle (UCAV) proposals. Results were immediately applicable to the company's X-45 UCAV demonstrator project.[143]

ERAST: High-Altitude, Long-Endurance Science Platforms

In the early 1990s, NASA's Earth Science Directorate received a solicitation for research to support the Atmospheric Effects of Aviation project. Because the project entailed assessment of the potential environmental impact of a commercial supersonic transport aircraft, measurements were needed at altitudes around 85,000 feet. Initially, Aurora Flight Sciences of Manassas, VA, proposed developing the Perseus A and Perseus B remotely piloted research aircraft as part of NASA's Small High-Altitude Science Aircraft (SHASA) program.

141. "X-36 Tailless Fighter Agility Research Aircraft."

142. Walker, "Flight Testing the X-36—The Test Pilot's Perspective."

143. Merlin, "Testing Tailless Technology Demonstrators," draft copy of proposed AIAA paper, 2009.

The SHASA effort expanded in 1993 as NASA teamed with industry partners for what became known as the Environmental Research Aircraft and Sensor Technology project. Goals for the ERAST project included development and demonstration of unpiloted aircraft to perform long-duration airborne science missions. Transfer of ERAST technology to an emerging UAV industry validated the capability of unpiloted aircraft to carry out operational science missions.

The ERAST project was managed at Dryden, with significant contributions from Ames, Langley, and Glenn Research Centers. Industry partners included such aircraft manufacturers as AeroVironment, Aurora Flight Sciences, General Atomics Aeronautical Systems, Inc., and Scaled Composites. Thermo-Mechanical Systems, Hyperspectral Sciences, and Longitude 122 West developed sensors to be carried by the research aircraft.[144] The ERAST effort resulted in a diverse fleet of unpiloted vehicles. Perseus A, built in 1993, was designed to stay aloft for 5 hours and reach altitudes around 82,000 feet. An experimental, closed-system, four-cylinder piston engine recycled exhaust gases and relied on stored liquid oxygen to generate combustion at high altitudes. Aurora built two Perseus A vehicles, one of which crashed because of an autopilot malfunction. By that time, the airplane had only reached an altitude of 50,000 feet.

Aurora engineers designed the Perseus B to remain aloft for 24 hours. The vehicle was equipped with a triple-turbocharged engine to provide sea-level air pressure up to 60,000 feet. In the 2 years following its maiden flight in 1994, Perseus B experienced some technical difficulties and a few hard landings that resulted in significant damage. As a result, Aurora technicians made numerous improvements, including extending the wingspan from 58.5 feet to 71.5 feet. When flight operations resumed in 1998, the Perseus B attained an unofficial altitude record of 60,280 feet before being damaged in a crash in October 1999. Despite such difficulties, experience with the Perseus vehicles provided designers with useful data regarding selection of instrumentation for RPRVs and identifying potential failures resulting from feedback deficiencies in a ground cockpit.[145] Aurora Flight Sciences also built a larger UAV named Theseus that was funded by NASA through the Mission To Planet Earth environmental observation program. Aurora and its

144. "ERAST: Environmental Research and Sensor Technology Fact Sheet," NASA Dryden Flight Research Center, Edwards, CA, 2002.
145. Hallion and Gorn, *On the Frontier*, pp. 310–311.

partners, West Virginia University and Fairmont State College, built the Theseus for NASA under an innovative, $4.9-million fixed-price contract. Dryden hosted the Theseus program, providing hangar space and range safety. Aurora personnel were responsible for flight-testing, vehicle flight safety, and operation of the aircraft.

With the potential to carry 700 pounds of science instruments to altitudes above 60,000 feet for durations of greater than 24 hours, the Theseus was intended to support research in areas such as stratospheric ozone depletion and the atmospheric effects of future high-speed civil transport aircraft engines. The twin-engine, unpiloted vehicle had a 140-foot wingspan and was constructed primarily from composite materials. Powered by two 80-horsepower, turbocharged piston engines that drove twin 9-foot-diameter propellers, it was designed to fly autonomously at high altitudes, with takeoff and landing under the active control of a ground-based pilot.

Operators from Aurora Fight Sciences piloted the maiden flight of the Theseus at Dryden on May 24, 1996. The test team conducted four additional checkout flights over the next 6 months. During the sixth flight, the vehicle broke apart and crashed while beginning a descent from 20,000 feet.[146] Innovative designers at AeroVironment in Monrovia, CA, took a markedly different approach to the ERAST challenge. In 1983, the company had built and tested the High-Altitude Solar (HALSOL) UAV using battery power only. Now, NASA scientists were anxious to see how it would perform with solar panels powering its six electrically driven propellers. The aircraft was a flying wing configuration with a rectangular planform and two ventral pods containing landing gear. Its structure consisted of a composite framework encased in plastic skin. In 1993 and 1994, researchers at Dryden flew it using a combination of battery and solar power, in a program sponsored by the Ballistic Missile Defense Organization that sought to develop a long-endurance surveillance platform. By now renamed Pathfinder, the unusual craft joined the ERAST fleet in 1995, where it soon attained an altitude of 50,500 feet, a record for solar-powered aircraft.[147] After additional upgrades and checkout flight at Dryden, ERAST team members transported the Pathfinder to the U.S. Navy's Pacific Missile Range Facility (PMRF) at Barking Sands,

146. Aurora's Theseus remotely piloted aircraft crashes, Release 96-63, NASA Dryden Flight Research Center, Edwards, CA, Nov. 12, 1996.
147. "Pathfinder Solar-Powered Aircraft," FS-034, NASA Dryden Flight Research Center, Edwards, CA, 2001.

Kauai, HI, in April 1997. Predictable weather patterns, abundant sunlight, available airspace and radio frequencies, and the diversity of terrestrial and coastal ecosystems for validating scientific imaging applications made Kauai an optimum location for testing. During one of seven high-altitude flights from the PMRF, the Pathfinder reached a world altitude record for propeller-driven as well as solar-powered aircraft at 71,530 feet.[148] In 1998, technicians at AeroVironment modified the vehicle to include two additional engines and extended the wingspan from 98 feet to 121 feet. Renamed Pathfinder Plus, the craft had more efficient silicon solar cells developed by SunPower, Corp., of Sunnyvale, CA, that were capable of converting almost 19 percent of the solar energy they received to useful electrical energy to power the motors, avionics, and communication systems. Maximum potential power was boosted from about 7,500 watts on the original configuration to about 12,500 watts. This allowed the Pathfinder Plus to reach a record altitude of 80,201 feet during another series of developmental test flights at the PMRF.[149] NASA research teams, coordinated by the Ames Research Center and including researchers from the University of Hawaii and the University of California, used the Pathfinder/Pathfinder Plus vehicle to carry a variety of scientific sensors. Experiments included detection of forest nutrient status, observation of forest regrowth following hurricane damage, measurement of sediment and algae concentrations in coastal waters, and assessment of coral reef health. Several flights demonstrated the practical utility of using high-flying, remotely piloted, environmentally friendly solar aircraft for commercial purposes. Two flights, funded by a Japanese communications consortium and AeroVironment, emphasized the vehicle's potential as a platform for telecommunications relay services. A NASA-sponsored demonstration employed remote-imaging techniques for use in optimizing coffee harvests.[150] AeroVironment engineers ultimately hoped to produce an autonomous aircraft capable of flying at altitudes around 100,000 feet for weeks—or even months—at a time through use of rechargeable solar power cells. Building on their experience with the Pathfinder/ Pathfinder Plus, they subsequently developed the 206-foot-span Centurion. Test flights at Dryden in 1998, using only battery power to drive 14 propellers, demonstrated the aircraft's

148. Ibid.
149. Ibid.
150. Ibid.

The solar-electric Helios Prototype was flown from the U.S. Navy's Pacific Missile Range Facility. NASA.

capability for carrying a 605-pound payload. The vehicle was then modified to feature a 247-foot-span and renamed the Helios Prototype, with a performance goal of 100,000 feet altitude and 96 hours mission duration.

As with its predecessors, a ground pilot remotely controlled the Helios Prototype, either from a mobile control van or a fixed ground station. The aircraft was equipped with a flight-termination system—required on remotely piloted aircraft flown in military restricted airspace—that included a parachute system plus a homing beacon to aid in determining the aircraft's location.

Flights of the Helios Prototype at Dryden included low-altitude evaluation of handling qualities, stability and control, response to turbulence, and use of differential motor thrust to control pitch. Following installation of more than 62,000 solar cells, the aircraft was transported to the PMRF for high-altitude flights. On August 13, 2001, the Helios Prototype reached an altitude of 96,863 feet, a world record for sustained horizontal flight by a winged aircraft.[151]

During a shakedown mission June 26, 2003, in preparation for a 48-hour long-endurance flight, the Helios Prototype aircraft encountered atmospheric turbulence, typical of conditions expected by the test crew, causing abnormally high wing dihedral (upward bowing of both wingtips). Unobserved mild pitch oscillations began but quickly

151. "Helios Prototype: The forerunner of 21st century solar-powered atmospheric satellites," FS-068, NASA Dryden Flight Research Center, Edwards, CA, 2002.

diminished. Minutes later, the aircraft again experienced normal turbulence and transitioned into an unexpected, persistent high wing-dihedral configuration. As a result, the aircraft became unstable, exhibiting growing pitch oscillations and airspeed deviations exceeding the design speed. Resulting high dynamic pressures ripped the solar cells and skin off the upper surface of the outer wing panels, and the Helios Prototype fell into the Pacific Ocean. Investigators determined that the mishap resulted from the inability to predict, using available analysis methods, the aircraft's increased sensitivity to atmospheric disturbances, such as turbulence, following vehicle configuration changes required for the long-duration flight demonstration.[152] Scaled Composites of Mojave, CA, built the remotely piloted RAPTOR Demonstrator-2 to test remote flight control capabilities and technologies for long-duration (12 to 72 hours), high-altitude vehicles capable of carrying science payloads. Key technology development areas included lightweight structures, science payload integration, engine development, and flight control systems. As a result, it had only limited provisions for a scientific payload. The D-2 was unusual in that it was optionally piloted. It could be flown either by a pilot in an open cockpit or by remote control. This capability had been demonstrated in earlier flights of the RAPTOR D-1, developed for the Ballistic Missile Defense Organization in the early 1990s.

D-2 flight tests began August 23, 1994. In late 1996, technicians linked the D-2 to NASA's Tracking and Data Relay Satellite system in order to demonstrate over-the-horizon communications capabilities between the aircraft and ground stations at ranges of up to 2,000 miles. The D-2 resumed flights in August 1998 to test a triple-redundant flight control system that would allow remotely piloted high-altitude missions.[153] General Atomics of San Diego, CA, produced several vehicles for the ERAST program based on the company's Gnat and Predator UAVs. The first two, called Altus (Latin for "high") and Altus 2, looked similar to the company's Gnat 750. Altus was 23.6 feet long and featured long, narrow, high aspect ratio wings spanning 55.3 feet. Powered by a rear-mounted, turbocharged, four-cylinder piston engine rated at 100 horsepower, the vehicle was capable of cruising at 80 to 115 mph and attaining altitudes

152. "Helios mishap report released," NASA Dryden Flight Research Center, Edwards, CA, Sept. 3, 2004.

153. RAPTOR Demonstrator project files, 1993–1999, DFRC Historical Reference Collection, NASA Dryden Flight Research Center, Edwards, CA.

of up to 53,000 feet. Altus could accommodate up to 330 pounds of sensors and scientific instruments.

NASA Dryden personnel initially operated the Altus vehicles as part of the ERAST program. The Altus 2, the first of the two aircraft to be completed, made its first flight May 1, 1996. During subsequent developmental tests, it reached an altitude of 37,000 feet. In late 1996, researchers flew the Altus 2 in an atmospheric-radiation-measurement study sponsored by the Department of Energy's Sandia National Laboratory for the purpose of collecting data on radiation/cloud interactions in Earth's atmosphere to better predict temperature rise resulting from increased carbon dioxide levels. During the course of the project, Altus 2 set a single-flight endurance record for remotely operated aircraft, remaining aloft for 26.18 hours through a complete day-to-night-to-day cycle.[154] The multiagency program brought together capabilities available among Government agencies, universities, and private industry. Sandia provided technical direction, logistical planning and support, data analysis, and a multispectral imaging instrument. NASA's Goddard Space Flight Center and Ames Research Center, Lawrence Livermore National Laboratory, Brookhaven National Laboratory, Colorado State University, and the University of California Scripps Institute provided additional instrumentation. Scientists from the University of Maryland, the University of California at Santa Barbara, Pennsylvania State University, the State University of New York, and others also participated.[155] In September 2001, the Altus 2 carried a thermal imaging system for the First Response Experiment (FiRE) during a demonstration at the General Atomics flight operations facility at El Mirage, CA. A sensor developed for the ERAST program and previously used to collect images of coffee plantations in Hawaii was modified to provide real-time, calibrated, geolocated, multispectral thermal imagery of fire events. This scientific demonstration showcased the capability of an unmanned aerial system (UAS) to collect remote sensing data over fires and relay the information to fire management personnel on the ground.[156] A larger vehicle called Altair, based on the Predator B (Reaper) UAV, was designed to perform

154. "ALTUS II—How High is High?" NASA FS-1998-12-058 DFRC, NASA Dryden Flight Research Center, Edwards, CA, 1998.

155. W.R. Bolton, "Measurements of Radiation in the Atmosphere," *NASA Tech Briefs*, DRC-98-32.

156. Vincent Ambrosia, "Remotely Piloted Vehicles as Fire Imaging Platforms: The Future is Here!" NASA Ames Research Center, Moffett Field, CA, 2002.

a variety of ERAST science missions specified by NASA's Earth Science enterprise. In the initial planning phase of the project, NASA scientists established a stringent set of requirements for the Altair that included mission endurance of 24 to 48 hours at an altitude range of 40,000 to 65,000 feet with a payload of at least 660 pounds. The project team also sought to develop procedures to allow operations from conventional airports without conflict with piloted aircraft. Additionally, the Altair had to be capable of demonstrating command and control beyond-line-of-sight communications via satellite link, undertake see-and-avoid operations relative to other air traffic, and demonstrate the ability to communicate with FAA air traffic controllers. To accomplish this, the Altair was equipped with an automated collision-avoidance system and a voice relay to allow air traffic controllers to talk to ground-based pilots. As the first UAV to meet FAA requirements for operating from conventional airports, with piloted aircraft in the national airspace, the aircraft also had to meet all FAA airworthiness and maintenance standards. The final Altair configuration was designed to fly continuously for up to 32 hours and was capable of reaching an altitude of approximately 52,000 feet with a maximum range of about 4,200 miles. It was designed to carry up to 750 pounds of sensors, radar, communications, and imaging equipment in its forward fuselage.[157] Although the ERAST program was formally terminated in 2003, research continued with the Altair. In May 2005, the National Oceanic and Atmospheric Administration (NOAA) funded the UAV Flight Demonstration Project in cooperation with NASA and General Atomics. The experiment included a series of atmospheric and oceanic research flights off the California coastline to collect data on weather and ocean conditions, as well as climate and ecosystem monitoring and management. The Altair was the first UAV to feature triple-redundant controls and avionics for increased reliability, as well as a fault-tolerant, dual-architecture flight control system.

Science flights began May 7 with a 6.5-hour flight to the Channel Islands Marine Sanctuary west of Los Angeles, a site thought ideal for exploring NOAA's operational objectives with a digital camera system and electro-optical/infrared sensors. The Altair carried a payload of instruments for measuring ocean color, atmospheric composition and temperature, and surface imaging during flights at altitudes of up to 45,000

157. "Altair/Predator B—An Earth Science Aircraft for the 21st Century," NASA FS-073, NASA Dryden Flight Research Center, Edwards, CA, 2001.

feet. Objectives of the experiment included evaluation of an unmanned aircraft system for future scientific and operational requirements related to NOAA's oceanic and atmospheric research, climate research, marine sanctuary mapping and enforcement, nautical charting, and fisheries assessment and enforcement.[158] In 2006, personnel from NASA, NOAA, General Atomics, and the U.S. Forest Service teamed for the Altair Western States Fire Mission (WSFM). This experiment demonstrated the combined use of an Ames-designed thermal multispectral scanner integrated on a large-payload capacity UAV, a data link telemetry system, near-real-time image geo-rectification, and rapid Internet data dissemination to fire center and disaster managers. The sensor system was capable of automatically identifying burned areas as well as active fires, eliminating the need to train sensor operators to analyze imagery. The success of this project set the stage for NASA's acquisition of another General Atomics UAV called the Ikhana and for future operational UAS missions in the national airspace.[159]

Ikhana: Awareness in the National Airspace

Military UAVs are easily adapted for civilian research missions. In November 2006, NASA Dryden obtained a civilian version of the General Atomics MQ-9 Reaper that was subsequently modified and instrumented for research. Proposed missions included supporting Earth science research, fabricating advanced aeronautical technology, and developing capabilities for improving the utility of unmanned aerial systems.

The project team named the aircraft Ikhana, a Native American Choctaw word meaning intelligent, conscious, or aware. The choice was considered descriptive of research goals NASA had established for the aircraft and its related systems, including collecting data to better understand and model environmental conditions and climate and increasing the ability of unpiloted aircraft to perform advanced missions.

The Ikhana was 36 feet long with a 66-foot wingspan and capable of carrying more than 400 pounds of sensors internally and over 2,000 pounds in external pods. Driven by a 950-horsepower turboprop engine, the aircraft has a maximum speed of 220 knots and is capable of reaching

158. Beth Hagenauer, "NOAA and NASA Begin California UAV Flight Experiment," Press Release 05-20, NASA Dryden Flight Research Center, Edwards, CA, 2005.

159. "Altair Western States Fire Mission," *http://ntrs.nasa.gov/archive/nasa/casi.ntrs.nasa.gov/200.700.31044_200.703.2019.pdf*, accessed June 10, 2009.

Research pilot Mark Pestana flies the Ikhana from a Ground Control Station at NASA Dryden Flight Research Center. NASA.

altitudes above 40,000 feet with limited endurance.[160] Initial experiments included the use of fiber optics for wing shape and temperature sensing, as well as control and structural loads measurements. Six hairlike fibers on the upper surfaces of the Ikhana's wings provided 2,000 strain measurements in real time, allowing researchers to study changes in the shape of the wings during flight. Such sensors have numerous applications for future generations of aircraft and spacecraft. They could be used, for example, to enable adaptive wing-shape control to make an aircraft more aerodynamically efficient for specific flight regimes.[161] To fly the Ikhana, NASA purchased a Ground Control Station and satellite communication system for uplinking flight commands and downlinking aircraft and mission data. The GCS was installed in a mobile trailer and, in addition to the pilot's remote cockpit, included computer workstations for scientists and engineers. The ground pilot was linked to the aircraft through a C-band line-of-sight (LOS) data link at ranges up to 150 nautical miles. A Ku-band satellite link allowed for over-the-horizon control. A remote video terminal provided real-time imagery from

160. "Ikhana Unmanned Science and Research Aircraft System," NASA FS-097, NASA Dryden Flight Research Center, Edwards, CA, 2007.
161. Jay Levine, "Measuring up to the Gold Standard," *X-tra,* NASA Dryden Flight Research Center, Edwards, CA, 2008.

the aircraft, giving the pilot limited visual input.[162] Two NASA pilots, Hernan Posada and Mark Pestana, were initially trained to fly the Ikhana. Posada had 10 years of experience flying Predator vehicles for General Atomics before joining NASA as an Ikhana pilot. Pestana, with over 4,000 flight hours in numerous aircraft types, had never flown a UAS prior to his assignment to the Ikhana project. He found the experience an exciting challenge to his abilities because the lack of vestibular cues and peripheral vision hinders situational awareness and eliminates the pilot's ability to experience such sensations as motion and sink rate.[163]

Building on experience with the Altair unpiloted aircraft, NASA developed plans to use the Ikhana for a series of Western States Fire Mission flights. The Autonomous Modular Sensor (AMS), developed by Ames, was key to their success. The AMS is a line scanner with a 12-band spectrometer covering the spectral range from visible to the near infrared for fire detection and mapping. Digitized data are combined with navigational and inertial sensor data to determine the location and orientation of the sensor. In addition, the data are autonomously processed with geo-rectified topographical information to create a fire intensity map.

Data collected with AMS are processed onboard the aircraft to provide a finished product formatted according to a geographical information systems standard, which makes it accessible with commonly available programs, such as Google Earth. Data telemetry is downlinked via a Ku-band satellite communications system. After quality-control assessment by scientific personnel in the GCS, the information is transferred to NASA Ames and then made available to remote users via the Internet.

After the Ikhana was modified to carry the AMS sensor pod on a wing pylon, technicians integrated and tested all associated hardware and systems. Management personnel at Dryden performed a flight readiness review to ensure that all necessary operational and safety concerns had been addressed. Finally, planners had to obtain permission from the FAA to allow the Ikhana to operate in the national airspace.[164]

162. "Ground Control Stations Fact Sheet," General Atomics Aeronautical Systems Company, San Diego, CA, 2007.
163. Author's interview with Mark Pestana, NASA Dryden Flight Research Center, Aug. 13, 2008.
164. Philip Hall, Brent Cobleigh, Greg Buoni, and Kathleen Howell, "Operational Experience with Long Duration Wildfire Mapping UAS Missions over the Western United States," presented at the *Association for Unmanned Vehicle Systems International Unmanned Systems North America Conference, San Diego, CA, June 2008.*

The first four Ikhana flights set a benchmark for establishing criteria for future science operations. During these missions, the aircraft traversed eight western U.S. States, collecting critical fire information and relaying data in near real time to fire incident command teams on the ground as well as to the National Interagency Fire Center in Boise, ID. Sensor data were downlinked to the GCS, transferred to a server at Ames, and autonomously redistributed to a Google Earth data visualization capability—Common Desktop Environment (CDE)—that served as a Decision Support System (DSS) for fire-data integration and information sharing. This system allowed users to see and use data in as little as 10 minutes after it was collected.

The Google Earth DSS CDE also supplied other real-time fire-related information, including satellite weather data, satellite-based fire data, Remote Automated Weather Station readings, lightning-strike detection data, and other critical fire-database source information. Google Earth imagery layers allowed users to see the locations of man-made structures and population centers in the same display as the fire information. Shareable data and information layers, combined into the CDE, allowed incident commanders and others to make real-time strategy decisions on fire management. Personnel throughout the U.S. who were involved in the mission and imaging efforts also accessed the CDE data. Fire incident commanders used the thermal imagery to develop management strategies, redeploy resources, and direct operations to critical areas such as neighborhoods.[165] The Western States UAS Fire Missions, carried out by team members from NASA, the U.S. Department of Agriculture Forest Service, the National Interagency Fire Center, the NOAA, the FAA, and General Atomics Aeronautical Systems, Inc., were a resounding success and a historic achievement in the field of unpiloted aircraft technology.

In the first milestone of the project, NASA scientists developed improved imaging and communications processes for delivering near-real-time information to firefighters. NASA's Applied Sciences and Airborne Science programs and the Earth Science Technology Office developed an Airborne Modular Sensor with the intent of demonstrating its capabilities during the WSFM and later transitioning

165. "Completed Missions," Wildfire Research and Applications Partnership (WRAP), *http://geo.arc.nasa.gov/sge/WRAP/current/com_missions.html*, 2008, accessed Aug. 27, 2009.

those capabilities to operational agencies.[166] The WSFM project team repeatedly demonstrated the utility and flexibility of using a UAS as a tool to aid disaster response personnel through the employment of various platform, sensor, and data-dissemination technologies related to improving near-real-time wildfire observations and intelligence-gathering techniques. Each successive flight expanded capabilities of the previous missions for platform endurance and range, number of observations made, and flexibility in mission and sensing reconfiguration.

Team members worked with the FAA to safely and efficiently integrate the unmanned aircraft system into the national airspace. NASA pilots flew the Ikhana in close coordination with FAA air traffic controllers, allowing it to maintain safe separation from other aircraft.

WSFM project personnel developed extensive contingency management plans to minimize the risk to the aircraft and the public, including the negotiation of emergency landing rights agreements at three Government airfields and the identification and documentation of over 300 potential emergency landing sites.

The missions included coverage of more than 60 wildfires throughout 8 western States. All missions originated and terminated at Edwards Air Force Base and were operated by NASA crews with support from General Atomics. During the mission series, near-real-time data were provided to Incident Command Teams and the National Interagency Fire Center.[167] Many fires were revisited during some missions to provide data on time-induced fire progression. Whenever possible, long-duration fire events were imaged on multiple missions to provide long-term fire-monitoring capabilities. Postfire burn-assessment imagery was also collected over various fires to aid teams in fire ecosystem rehabilitation. The project Flight Operations team built relationships with other agencies, which enabled real-time flight plan changes necessary to avoid hazardous weather, to adapt to fire priorities, and to avoid conflicts with multiple planned military GPS testing/jamming activities.

166. "Western States Fire Mission Team Award for Group Achievement," NASA Ames Research Center Honor Awards ceremony, NASA Ames Research Center, Mountain View, CA, Sept. 20, 2007, and Status Report, "NASA's Ikhana UAS Resumes Western States Fire Mission Flights," NASA Dryden Flight Research Center, Edwards, CA, Sept. 19, 2008.
167. Ibid.

Critical, near-real-time fire information allowed Incident Command Teams to redeploy fire-fighting resources, assess effectiveness of containment operations, and move critical resources, personnel, and equipment from hazardous fire conditions. During instances in which blinding smoke obscured normal observations, geo-rectified thermal-infrared data enabled the use of Geographic Information Systems or data visualization packages such as Google Earth. The images were collected and fully processed onboard the Ikhana and transmitted via a communications satellite to NASA Ames, where the imagery was served on a NASA Web site and provided in the Google Earth–based CDE for quick and easy access by incident commanders.

The Western States UAS Fire Mission series also gathered critical, coincident data with satellite sensor systems orbiting overhead, allowing for comparison and calibration of those resources with the more sensitive instruments on the Ikhana. The Ikhana UAS proved a versatile platform for carrying research payloads. Since the sensor pod could be reconfigured, the Ikhana was adaptable for a variety of research projects.[168]

Lessons Learned—Realities and Recommendations

Unmanned research vehicles have proven useful for evaluating new aeronautical concepts and providing precision test capability, repeatable test maneuver capability, and flexibility to alter test plans as necessary. They allow testing of aircraft performance in situations that might be too hazardous to risk a pilot on board yet allow for a pilot in the loop through remote control. In some instances, it is more cost-effective to build a subscale RPRV than a full-scale aircraft.[169] Experience with RPRVs at NASA Dryden has provided valuable lessons. First and foremost, good program planning is critical to any successful RPRV project. Research engineers need to spell out data objectives in as much detail as possible as early as possible. Vehicle design and test planning should be tailored to achieve these objectives in the most effective way. Definition of operational techniques—air launch versus ground launch, parachute recovery versus horizontal landing, etc.—are highly dependent on research objectives.

168. Ibid.

169. Terrence W. Rezek, "Unmanned Vehicle Systems Experience at the Dryden Flight Research Facility," NASA TM-84913, NASA Ames Research Center, Dryden Flight Research Facility, Edwards, CA, June 1983.

One advantage of RPRV programs is flexibility in regard to matching available personnel, facilities, and funds. Almost every RPRV project at Dryden was an experiment in matching personnel and equipment to operational requirements. As in any flight-test project, staffing is very important. Assigning an operations engineer and crew chief early in the design phase will prevent delays resulting from operational and maintainability issues.[170] Some RPRV projects have required only a few people and simple model-type radio-control equipment. Others involved extremely elaborate vehicles and sophisticated control systems. In either case, simulation is vital for RPRV systems development, as well as pilot training. Experience in the simulator helps mitigate some of the difficulties of RPRV operation, such as lack of sensory cues in the cockpit. Flight planners and engineers can also use simulation to identify significant design issues and to develop the best sequence of maneuvers for maximizing data collection.[171] Even when built from R/C model stock or using model equipment (control systems, engines, etc.), an RPRV should be treated the same as any full-scale research airplane. Challenges inherent with RPRV operations make such vehicles more susceptible to mishaps than piloted aircraft, but this doesn't make an RPRV expendable. Use of flight-test personnel and procedures helps ensure safe operation of any unmanned research vehicle, whatever its level of complexity.

Configuration control is extremely important. Installation of new software is essentially the same as creating a new airplane. Sound engineering judgments and a consistent inspection process can eliminate potential problems.

Knowledge and experience promote safety. To as large a degree as possible, actual mission hardware should be used for simulation and training. People with experience in manned flight-testing and development should be involved from the beginning of the project.[172] The critical role of an experienced test pilot in RPRV operations has been repeatedly demonstrated. A remote pilot with flight-test experience can adapt to changing situations and discover system anomalies with greater flexibility and accuracy than an operator without such experience.

170. Reed, "Flight Research Techniques Utilizing Remotely Piloted Research Vehicles."

171. Ibid.

172. Scaled Composites presentation, *UAV Flight Test Lessons Learned Workshop, NASA Dryden Flight Research Center, Dec. 18, 1996*, DFRC Historical Reference Collection, NASA Dryden Flight Research Center, Edwards, CA.

The need to consider human factors in vehicle and ground cockpit design is also important. RPRV cockpit workload is comparable to that for a manned aircraft, but remote control systems fail to provide many significant physical cues for the pilot. A properly designed Ground Control Station will compensate for as many of these shortfalls as possible.[173] The advantages and disadvantages of using RPRVs for flight research sometimes seem to conflict. On one hand, the RPRV approach can result in lower program costs because of reduced vehicle size and complexity, elimination of man-rating tests, and elimination of the need for life-support systems. However, higher program costs may result from a number of factors. Some RPRVs are at least as complex as manned vehicles and thus costly to build and operate. Limited space in small airframes requires development of miniaturized instrumentation and can make maintenance more difficult. Operating restrictions may be imposed to ensure the safety of people on the ground. Uplink/downlink communications are vulnerable to outside interference, potentially jeopardizing mission success, and line-of-sight limitations restrict some RPRV operations.[174] The cost of designing and building new aircraft is constantly rising, as the need for speed, agility, stores/cargo capacity, range, and survivability increases. Thus, the cost of testing new aircraft also increases. If flight-testing is curtailed, however, a new aircraft may reach production with undiscovered design flaws or idiosyncrasies. If an aircraft must operate in an environment or flight profile that cannot be adequately tested through wind tunnel or computer simulation, then it must be tested in flight. This is why high-risk, high-payoff research projects are best suited to use of RPRVs. High data-output per flight—through judicious flight planning—and elimination of physical risk to the research pilot can make RPRV operations cost-effective and worthwhile.[175] Since the 1960s, remotely piloted research vehicles have evolved continuously. Improved avionics, software, control, and telemetry systems have led to development of aircraft capable of operating within a broad range of flight regimes. With these powerful research tools, scientists and engineers at NASA Dryden continue to explore the aeronautical frontier.

173. Rezek, "Unmanned Vehicle Systems Experience at the Dryden Flight Research Facility."
174. Reed, "Flight Research Techniques Utilizing Remotely Piloted Research Vehicles."
175. Rezek, "Unmanned Vehicle Systems Experience at the Dryden Flight Research Facility."

Recommended Additional Reading

Reports, Papers, Articles, and Presentations:

James W. Akkerman, "Hydrazine Monopropellant Reciprocating Engine Development," *Journal of Engineering for Industry*, vol. 101, no. 4 (Nov. 1979).

Vincent Ambrosia, "Remotely Piloted Vehicles as Fire Imaging Platforms: The Future is Here!" NASA Ames Research Center, Moffett Field, CA (2002).

Henry H. Arnaiz and Paul C. Loschke, "Current Overview of the Joint NASA/USAF HiMAT Program," NASA CP-2162 (1980).

Jennifer L. Baer-Riedhart, "The Development and Flight Test Evaluation of an Integrated Propulsion Control System for the HiMAT Research Airplane," AIAA-1981-2467 (1981).

Daniel D. Baumann and Benjamin Gal-Or, "Thrust Vectoring Fighter Aircraft Agility Research Using Remotely Piloted Vehicles," SAE Paper 921015 (Apr. 1992).

Donald R. Bellman and David A. Kier, "HiMAT—A New Approach to the Design of Highly Maneuverable Aircraft," SAE Paper 740859 (1974).

Al Bowers, "The Dirty, the Boring, and the Dangerous: Uninhabited Aerial Vehicles—An Engineers' Perspective," *ITEA Journal* (2009), vol. 30, International Test and Evaluation Association (Mar. 2009).

L.E. Brown, Jr., M.H. Roe, and R.A. Quam, "HiMAT Systems Development Results and Projections," SAE Paper 801175, *Society of Automotive Engineers Aerospace Congress and Exposition, Los Angeles, CA, Oct. 1980.*

L.E. Brown, M. Roe, and C.D. Wiler, "The HiMAT RPRV System," AIAA Paper 78-1457, *AIAA Aircraft Systems and Technology Conference, Los Angeles, CA, Aug. 1978.*

Gerald D. Budd, Ronald L. Gilman, and David Eichstedt, "Operational and Research Aspects of a Radio-Controlled Model Flight Test Program," *AIAA 31st Aerospace Sciences Meeting and Exhibit, Reno, NV, Jan. 1993*.

Michael V. DeAngelis, "In-Flight Deflection Measurement of the HiMAT Aeroelastically Tailored Wing," AIAA-1981-2450 (1981).

Dwain A. Deets and Lewis E. Brown, "Experience with HiMAT Remotely Piloted Research Vehicle—An Alternate Flight Test Approach," AIAA-1986-2754 (1986).

Dwain A. Deets and Carl A. Crother, "Highly Maneuverable Aircraft Technology," AGARD Paper AG-234 (1978).

Dwain A. Deets and Dana Purifoy, "Operational Concepts for Uninhabited Tactical Aircraft," NASA TM-206549 (1998).

John H. Del Frate and Gary B. Consentino, "Recent Flight Test Exeprience with Uninhabited Aerial Vehicles at the NASA Dryden Flight Research Center," NASA TM-206546 (1998).

Richard DeMeis, "What Really Happened in Safe Fuel Test," *Aerospace America* (July 1985).

E.L. Duke, F.P. Jones, and R.B. Roncoli, "Development of a Flight Test Maneuver Autopilot for a Highly Maneuverable Aircraft," AIAA Paper 83-0061 (Jan. 1983).

H.J. Dunn, "Realizable optimal control for a remotely piloted research vehicle," NASA TP-1654 (May 1980).

John W. Edwards, "Flight Test Results of an Active Flutter Suppression System Installed on a Remotely Piloted Vehicle," NASA TM-83132 (May 1981).

Martha B. Evans and Lawrence J. Schilling, "Simulations Used in the Development and Flight Test of the HiMAT Vehicle," AIAA 1983-2505 (1983).

Edwin L. Fasanella, Emilio Alfaro-Bou, and Robert J. Hayduk, "Impact Data From a Transport Aircraft During a Controlled Impact Demonstration," NASA TP-2589 (Sept. 1986).

Shu W. Gee, Peter C. Carr, William R. Winter, and John A. Manke, "Development of systems and techniques for landing an aircraft using onboard television," NASA TP-1171 (Feb. 1978).

David L. Grose, "The Development of the DAST I Remotely Piloted Research Vehicle for Flight Testing an Active Flutter Suppression Control System," NASA CR-144881, University of Kansas (Feb. 1979).

Philip Hall, Brent Cobleigh, Greg Buoni, and Kathleen Howell, "Operational Experience with Long Duration Wildfire Mapping UAS Missions over the Western United States," presented at the *Association for Unmanned Vehicle Systems International Unmanned Systems North America Conference, San Diego, CA, June 2008.*

Philip J. Hamory and James E. Murray, "Flight Experience With Lightweight, Low-Power Miniaturized Instrumentation Systems," NASA TM-4463, NASA Dryden Flight Research Center, Edwards, CA (Mar. 1993).

Paul F. Harney, "Diversity Techniques for Omnidirectional Telemetry Coverage of the HiMAT Research Vehicle," NASA TP-1830 (1981).

Paul F. Harney, James B. Craft, Jr., and Richard G. Johnson, "Remote Comtrol of an Impact Demonstation Vehicle," NASA TM-85925, NASA Ames Research Center, Dryden Flight Research Facility, Edwards, CA (Apr. 1985).

Philip D. Hattis, Robert J. Polutchko, Brent D. Appleby, Timothy M. Barrows, Thomas J. Fill, Peter M. Kachmar, and Terrence D. McAteer, "Final Report: Development and Demonstration Test of a Ram-Air Parafoil Precision Guided Airdrop System," vols. 1 to 4, Report CSDL-R-2752, and Addendum, Report CSDL-R-2771 (1996).

Michael J. Hirschberg, David M. Hart, and Thomas J. Beutner, "A Summary of a Half-Century of Oblique Wing Research," paper No. AIAA-2007-150, *45th AIAA Aerospace Sciences Meeting, Reno, NV, Jan. 2007.*

Timothy W. Horton and Robert W. Kempel, "Flight Test Experience and Controlled Impact of a Remotely Piloted Jet Transport Aircraft," NASA TM-4084, NASA Ames Research Center, Dryden Flight Research Facility, Edwards, CA (Nov. 1988).

Carl E. Hoyt, Robert W. Kempel, and Richard R. Larson, "Backup Flight Control System for a Highly Maneuverable Remotely Piloted Research Vehicle," AIAA-1980-1761 (1980).

Steve Jacobsen, Tony Frackowiak, and Gordon Fullerton, "Gordon's Unmanned Aerial Vehicle Trainer: A Flight-Training Tool for Remotely Piloted Vehicle Pilots," *2006 Engineering Annual Report*, NASA TM-2007-214622 (Aug. 2007).

Thomas L. Jordan, William M. Langford, and Jeffrey S. Hill, "Airborne Subscale Transport Aircraft Research Testbed: Aircraft Model Development; Final Report," AIAA-2005-6432 (2005).

Michael W. Kehoe, "Highly Maneuverable Aircraft Technology (HiMAT) Flight Flutter Test Program," NASA TM-84907 (1984).

Robert W. Kempel, "Flight Experience with a Backup Flight-Control System for the HiMAT Research Vehicle," AIAA-1982-1541 (1982).

Robert W. Kempel and Michael R. Earls, "Flight Control Systems Development and Flight Test Experience with the HiMAT Research Vehicles," NASA TP-2822 (1988).

A. Kotsabasis, "The DAST-1 remotely piloted research vehicle development and initial flight testing," NASA CR-163105 (Feb. 1981).

Neil W. Matheny and George N. Panageas, "HiMAT Aerodynamic Design and Flight Test Experience," AIAA-1981-2433 (1981).

James E. Murray, Alex G. Sim, David C. Neufeld, Patrick K. Rennich, Stephen R. Norris, and Wesley S. Hughes, "Further Development and Flight Test of an Autonomous Precision Landing System Using a Parafoil," NASA TM-4599 (July 1994).

H.N. Murrow and C.V. Eckstrom, "Drones for Aerodynamic and Structural Testing (DAST)—A Status Report," *AIAA Aircraft Systems and Technology Conference, Los Angeles, CA, Aug. 21–23, 1978.*

Kevin L. Petersen, "Flight Control Systems Development of Highly Maneuverable Aircraft Technology (HiMAT) Vehicle," AIAA-1979-1789 (1979).

Richard F. Porter, David W. Hall, Joe H. Brown, Jr., and Gerald M. Gregorek, "Analytical Study of a Free-Wing/Free-Trimmer Concept," Battelle Columbus Laboratories, Columbus, OH (Dec. 7, 1977).

Terrill W. Putnam and M.R. Robinson, "Closing the Design Loop on HiMAT (Higly Maneuverable Aircraft Technology)," NASA TM-85923 (1984).

R. Dale Reed, "Flight Research Techniques Utilizing Remotely Piloted Research Vehicles," *UAV Flight Test Lessons Learned Workshop, NASA Dryden Flight Research Center, Edwards, CA, Dec. 18, 1996.*

R. Dale Reed, "The Flight Test of a 1/6 Scale Model of the ACRV-X Space Craft Using a Parafoil Recovery Parachute System," NASA Dryden Flight Research Center (Nov. 1995).

Terrence W. Rezek, "Unmanned Vehicle System Experiences at the Dryden Flight Research Facility," NASA TM-84913 (1983).

Ralph B. Roncoli, "A Flight Test Maneuver Autopilot for a Highly Maneuverable Aircraft," NASA TM-81372 (1982).

Shahan K. Sarrafian, "Simulator Evaluation of a Remotely Piloted Vehicle Lateral Landing Task Using a Visual Display," NASA TM-85903 (Aug. 1984).

Alex G. Sim, James E. Murray, David C. Neufeld, and R. Dale Reed, "Development and Flight Test of a Deployable Precision Landing System," *AIAA Journal of Aircraft,* vol. 31, no. 5 (Sept. 1994).

Alex G. Sim, James E. Murray, David C. Neufeld, and R. Dale Reed, "The Development and Flight Test of a Deployable Precision Landing System for Spacecraft Recovery," NASA TM-4525 (Sept. 1993).

Dean S. Smith and Jack L. Bufton, "The Remotely Piloted Vehicle as an Earth Science Research Aircraft," *4th Airborne Science Workshop Proceedings*, NASA (Jan. 1991).

Kenneth J. Szalai, "Role of Research Aircraft in Technology Development," NASA TM-85913 (1984).

Laurence A. Walker, "Flight Testing the X-36—The Test Pilot's Perspective," NASA CR-198058, NASA Dryden Flight Research Center, Edwards, CA (1997).

Books and Monographs:

Richard P. Hallion and Michael H. Gorn, *On the Frontier: Experimental Flight at NASA Dryden* (Washington, DC: Smithsonian Books, 2002).

Peter W. Merlin, *Ikhana Unmanned Aircraft System Western States Fire Missions*, NASA SP-4544 (Washington, DC: NASA, 2009).

R. Dale Reed, with Darlene Lister, *Wingless Flight: The Lifting Body Story*, NASA SP-4220 (Washington, DC: NASA, 1997).

NASA Dryden Flight Research Center's SR-71A, DFRC Aircraft 844, banks away over the Sierra Nevada mountains after air refueling from a USAF tanker during a 1997 flight. NASA.

CASE 10

NASA and Supersonic Cruise

William Flanagan

For an aircraft to attain supersonic cruise, or the capability to fly faster than sound for a significant portion of time, the designer must balance lift, drag, and thrust to achieve the performance requirements, which in turn will affect the weight. Although supersonic flight was achieved over 60 years ago, successful piloted supersonic cruise aircraft have been rare. NASA has been involved in developing the required technology for those rare designs, despite periodic shifting national priorities.

IN THE 1930S AND EARLY 1940S, investigation of flight at speeds faster than sound began to assume increasing importance, thanks initially to the "compressibility" problems encountered by rapidly rotating propeller tips but then to the dangerous trim changes and buffeting encountered by diving aircraft. Researchers at the National Advisory Committee for Aeronautics (NACA) began to focus on this new and troublesome area. The concept of Mach number (ratio of a body's speed to the speed of sound in air at the body's location) swiftly became a familiar term to researchers. At first, the subject seemed heavily theoretical. But then, with the increasing prospect of American involvement in the Second World War, NACA research had to shift to shorter-term objectives of improving American warplane performance, notably by reducing drag and refining the Agency's symmetrical low-drag airfoil sections. But with the development of fighter aircraft with engines exhibiting 1,500 to 2,000 horsepower and capable of diving in excess of Mach 0.75, supersonic flight became an issue of paramount military importance. Fighter aircraft in steep power on-dives from combat altitudes over 25,000 feet could reach 450 mph, corresponding to Mach numbers over 0.7. Unusual flight characteristics could then manifest themselves, such as severe buffeting, uncommanded increasing dive angles, and unusually high stick forces.

The sleek, twin-engine, high-altitude Lockheed P-38 showed these characteristics early in the war, and a crash effort by the manufacturer aided by NACA showed that although the aircraft was not "supersonic," i.e., flying faster than the speed of sound at its altitude, the airflow at the thickest part of the wing was at that speed, producing shock waves that were unaccounted for in the design of the flight control surfaces. The shock waves were a thin area of high pressure, where the supersonic airflow around the body began to slow toward its customary subsonic speed. This shock region increased drag on the vehicle considerably, as well as altered the lift distribution on the wing and control surfaces. An expedient fix, in the form of a dive flap to be activated by the pilot, was installed on the P-38, but the concept of a "critical Mach number" was introduced to the aviation industry: the aircraft flight speed at which supersonic flow could be present on the wing and fuselage. Newer high-speed, propeller-driven fighters, such as the P-51D with its thin laminar flow wing, had critical Mach numbers of 0.75, which allowed an adequate combat envelope, but the looming turbojet revolution removed the self-governing speed limit of reduced thrust because of supersonic propeller tips. Investigation of supersonic aircraft was no longer a theoretical exercise.[1]

Early Transonic and Supersonic Research Approaches

The NACA's applied research was initially restricted to wind tunnel work. The wind tunnels had their own problems with supersonic flow, as shock waves formed and disturbed the flow, thus casting doubt on the model test results. This was especially true in the transonic regime, from Mach 0.8 to 1.2, at which the shock waves were the strongest as the supersonic flow slowed to subsonic in one single step; this was called a "normal" shock, referring to the 90-degree angle of the shock wave to the vehicle motion. Free air experiments were necessary to validate and improve wind tunnel results. John Stack at NACA Langley developed a slotted wind tunnel that promised to reduce some of the flow irregularities. The Collier Trophy was awarded for this accomplishment, but validation of the supersonic tunnel results was still lacking. Pending the development of higher-powered engines for full-scale in-flight experiments, initial experimentation included attaching small wing shapes to NACA P-51 Mustangs, which then performed high-speed dives to and beyond their critical Mach numbers, allowing seconds of transonic

1. Roger E. Bilstein, *Orders of Magnitude: A History of the NACA and NASA, 1915–1990,* NASA SP-4406 (Washington, DC: GPO, 1989), ch. 3.

data collection. Heavy streamlined bomb shapes were released from NACA B-29s, the shapes going supersonic during their 30–45-second trajectories, sending pressure data to the ground via telemetry before impact.[2] Supersonic rocket boosters were fired from the NACA facility at Wallops Island, VA, carrying wind tunnel–sized models of wings and proposed aircraft configurations in order to gain research data, a test method that remained fruitful well into the 1960s. The NACA and the United States Air Force (USAF) formed a joint full-scale flight-test program of a supersonic rocket-powered airplane, the Bell XS-1 (subsequently redesignated the X-1), which was patterned after a supersonic 0.50-caliber machine gun projectile with thin wings and tail surfaces. The program culminated October 14, 1947, with the demonstration of a controllable aircraft that exceeded the speed of sound in level flight. The news media of the day hailed the breaking of the "sound barrier," which would lead to ever-faster airplanes in the future. Speed records popularized in the press since the birth of aviation were "made to be broken"; now, the speed of sound was no longer the limit.

But the XS-1 flight in October was no more a practical solution to supersonic flight than the Wright brothers' flights at Kitty Hawk in December 1903 were a director predecessor to transcontinental passenger flights. Rockets could produce the thrust necessary to overcome the drag of supersonic shock waves, but the thrust was of limited duration. Rocket motors of the era produced the greatest thrust per pound of engine, but they were dangerous and expensive, could not be throttled directly, and consumed a lot of fuel in a short time. Sustained supersonic flight would require a more fuel-efficient motor. The turbojet was an obvious choice, but in 1947, it was in its infancy and was relatively inefficient, being heavy and producing only (at most) several thousand pounds of static thrust. Military-sponsored research continued on improving the efficiency and the thrust levels, leading to the introduction of afterburners, which would increase thrust from 10–30 percent, but at the expense of fuel flows, which doubled to quadrupled that of the more normal subsonic cruise settings. The NACA and manufacturers looked at another form of jet propulsion, the ramjet, which did away with the complex rotating compressors and turbines and relied on forward speed of the vehicle to compress the airflow into an inlet/diffuser, where fuel

2. James Schulz, *Winds of Change* (Washington, DC: NASA, 1992), pp. 56–57.

would then be injected and combusted, with the exhaust nozzle further increasing the thrust.

Gathering the Data for Supersonic Airplane Design

NACA supersonic research after 1947 concentrated on the practical problems of designing supersonic airplanes. Basic transonic and low supersonic test data were collected in a series of experimental aircraft that did not suffer from the necessary compromises of operational military aircraft. The test programs were generally joint efforts with the Air Force and/or Navy, which needed the data in order to make reasonable decisions for future aircraft. The X-1 (USAF) and D-558-1 and D-558-2 (Navy) gathered research data on aerodynamics and stability and control in the transonic regime as well as flight Mach numbers to slightly above 2. The D-558-1 was a turbojet vehicle with a straight wing; as a result, although it had longer mission duration, it could not achieve supersonic flight and instead concentrated on the transonic regime. For supersonic flights, the research vehicles generally used rocket engines, with their corresponding short-duration data test points. Other experimental vehicles used configurations that were thought to be candidates for practical supersonic flight. The D-558-2 used a swept wing and was able to achieve Mach 2 on rocket power. The XF-92A explored the pure delta wing high-speed shape, the X-4 explored a swept wing that dispensed with horizontal tail surfaces, the X-5 configuration had a swept wing that could vary its sweep in flight, and the X-3 explored a futuristic shape with a long fuselage with a high fineness ratio combined with very low aspect ratio wings and a double-diamond cross section that was intended to reduce shock wave drag at supersonic speeds. The Bell X-2 was a NACA–USAF–sponsored rocket research aircraft with a swept wing intended to achieve Mach 3 flight.[3]

Valuable basic data were collected during these test programs applicable to development of practical supersonic aircraft, but sustained supersonic flight was not possible. The limited-thrust turbojets of the era limited the speeds of the aircraft to the transonic regime. The X-3 was intended to explore flight at Mach 2 and above, but its interim engines made that impossible; in a dive with afterburners, it could only reach Mach 1.2. The XF-92A delta wing showed promise for supersonic

3. Richard P. Hallion, *On the Frontier: Flight Research at Dryden, 1946–1981* (Washington, DC: GPO, 1984), pp. 47–85.

NACA stable of experimental aircraft. The X-3 is in the center; around it, clockwise, from lower left: X-1A, D-558-1, XF-92, X-5, D-558-2, and X-4. NASA.

designs but could not go supersonic in level flight.[4] This was unfortunate, as the delta winged F-102—built by Convair, which also manufactured the XF-92—was unable to achieve its supersonic design speeds and required an extensive redesign. This redesign included the "area rule" concept developed by the NACA's Richard Whitcomb.[5] The area rule principle, published in 1952, required a smooth variation in an aircraft's cross-section profile from nose to tail to minimize high drag normal shock wave formation, at which the profile has discontinuities. Avoiding the discontinuities, notably where the wing joined the fuselage, resulted in the characteristic "Coke bottle" or "wasp waist" fuselage adjacent to the wing. This was noticeable in supersonic fighter designs of the late 1950s, which still suffered from engines of limited thrust, afterburner being necessary even for low supersonic flight with the resultant

4. Donald Baals and William Corliss, *Wind Tunnels of NASA*, NASA SP-440 (Washington, DC: GPO, 1981), ch. 5-10: "Area Rule and the F-102."

5. Richard T. Whitcomb, "A Study of the Zero-Lift Drag Rise Characteristics of Wing-Body Combinations Near the Speed of Sound," NACA RM-L52H08 (1952).

short range and limited duration. The rocket-powered swept wing X-2 Mach 3 test program was not productive, with only one flight to Mach 3, ending in loss of the aircraft and its pilot, Capt. Milburn "Mel" Apt.[6]

Feeling the "Need for Speed": Military Requirements in the Atomic Age

In the 1950s and into the 1960s, the USAF and Navy demanded supersonic performance from fighters in level flight. The Second World War experience had shown that higher speed was productive in achieving superiority in fighter-to-fighter combat, as well as allowing a fighter to intercept a bomber from the rear. The first jet age fighter combat over Korea with fighters having swept wings had resulted in American air superiority, but the lighter MiG-15 had a higher ceiling and better climb rate and could avoid combat by diving away. When aircraft designers interviewed American fighter pilots in Korea, they specified, "I want to go faster than the enemy and outclimb him."[7] The advent of nuclear-armed jet bombers meant that destruction of the bomber by an interceptor before weapon release was critical and put a premium on top speed, even if that speed would only be achievable for a short time.

Similarly, bomber experience in World War II had shown that loss rates were significantly lower for very fast bombers, such as the Martin B-26 and the de Havilland Mosquito. The prewar concept of the slow, heavy-gun-studded "flying fortress," fighting its way to a target with no fighter escort, had been proven fallacious in the long run. The use of B-29s in the Korean war in the MiG-15 jet fighter environment had resulted in high B-29 losses, and the team switched to night bombing, where the MiG-15s were less effective. Hence, the ideal jet bomber would be one capable of flying a long distance, carrying a large payload, and capable of increased speed when in a high-threat zone. The length of the high-speed (and probably supersonic) dash might vary on the threat, combat radius, and fuel capacity of the long-range bomber, but it would likely be a longer distance than the short-legged fighter was capable of at supersonic flight. The USAF relied on the long-range bomber as a primary reason for its independent status and existence; hence, it was

6. The first was lost earlier in an explosion during a captive carry flight, resulting in the deaths of two crewmen. However, the EB-50 launch aircraft returned safely to base, thanks to the remarkable airmanship of its two pilots.

7. Interview with Lockheed test pilot Bob Gilliland by author, Western Museum of Flight, May 16, 2009.

interested in using the turbojet to improve bomber performance and survivability. But supersonic speeds seemed out of the question with the early turbojets, and the main effort was on wringing long range from a jet bomber. Swept thin wings promised higher subsonic cruise speed and increased fuel efficiency, and the Boeing Company took advantage of NACA swept wing research initiated by Langley's R.T. Jones in 1945 to produce the B-47 and B-52, which were not supersonic but did have the long range and large payloads.[8]

The development of more fuel-efficient axial-flow turbojets such as the General Electric J47 and Pratt & Whitney J57 (the first mass-produced jet engine to develop over 10,000 pounds static sea level non-afterburning thrust) were another needed element. Aerial refueling had been tried on an experimental basis in the Second World War, but for jet bombers, it became a priority as the USAF sought the goal of a large-payload jet bomber with intercontinental range to fight the projected atomic third World War. The USAF began to look at a supersonic dash jet bomber now that supersonic flight was an established capability being used in the fighters of the day. Just as the medium-range B-47 had served as an interim design for the definitive heavy B-52, the initial result was the delta wing Convair B-58 Hustler. The initial designs had struggled with carrying enough fuel to provide a worthwhile supersonic speed and range; the fuel tanks were so large, especially for low supersonic speeds with their high normal shock drag, that the airplane was huge with limited range and was rejected. Convair adopted a new approach, one that took advantage of its experience with the area rule redesign of the F-102. The airplane carried a majority of its fuel and its atomic payload in a large, jettisonable shape beneath the fuselage, allowing the actual fuselage to be extremely thin. The fuselage and the fuselage/tank combination were designed in accordance with the area rule. The aircraft employed four of the revolutionary J79 engines being developed for Mach 2 fighters, but it was discovered that with the increased fuel capacity, high installed thrust, and reduced drag at low supersonic Mach numbers, the aircraft could sustain Mach 2 for up to 30 minutes, giving it a supersonic range over 1,000 miles, even retaining the centerline store. It could be said that the B-58, although intended to be a

8. James R. Hansen, *Engineer in Charge*, NASA SP-4305 (Washington, DC: GPO, 1987), pp. 276–280; for further discussion on swept wing evolution, see the companion case study by Richard P. Hallion in this volume.

supersonic dash aircraft, became the first practical supersonic cruise aircraft. The B-58 remained in USAF service for less than 10 years for budgetary reasons and its notoriously unreliable avionics. The safety record was not good either, in part because of the difficulty in training pilots to change over from the decidedly subsonic (and huge) B-52 with a crew of six to a "hot ship" delta wing, high-landing-speed aircraft with a crew of three (but only one pilot). Nevertheless, the B-58 fleet amassed thousands of hours of Mach 2 time and set numerous world speed records for transcontinental and intercontinental distances, most averaging 1,000 mph or higher, including the times for slowing for aerial refueling. Examples included 4 hours 45 minutes for Los Angeles to New York and back, averaging 1,045 mph, and Los Angeles to New York 1 way in 2 hours 1 minute, at an average speed of 1,214 mph, with 1 refueling over Kansas.

The later record flight illustrated one of the problems of a supersonic cruise aircraft: heat.[9] The handbook skin temperature flight limit on the B-58 was 240 degrees Fahrenheit (°F). For the speed run, the limit was raised to 260 degrees to allow Mach 2+, but it was a strict limit; there was concern the aluminum honeycomb skin would debond above that temperature. Extended supersonic flight duration meant that the aircraft structure temperature would rise and eventually stabilize as the heat added from the boundary layer balanced with radiated heat from the hot airplane. The stabilization point was typically reached 20–30 minutes after attaining the cruise speed. The B-58's Mach 2 speed at 45,000–50,000 feet had reached a structural limit for its aluminum material; the barrier now was "the thermal thicket"—a heat limit rather the sound barrier.

Airlines and the Jet Age

In the 1930s, the NACA had conducted research on engine cowlings that improved cooling while reducing drag. This led to improvements in airliner speed and economy, which in turn led to increased capacity and more acceptance by the traveling public; airliners were as fast as the fighters of the early Depression era. In World War II, the NACA shifted research focus to military needs, the most challenging being the turbojet,

9. Interviews with B-58 record flight crewmember Capt. Robert Walton on "Operation Heat Rise" by USAF Museum, location and date unknown, *http://www.wvi.com/~sr71webmaster/b58.htm*, accessed June 30, 2009.

and almost doubled potential top speeds. In civil aviation, postwar propeller-driven airliners could span the continent and the oceans, but at 300 mph. Initial attempts to install turbojets in straight winged airliners failed because of the fuel inefficiency of the jets and the increased drag at jet speeds; the loss of life in the mysterious crashes of three British jet-propelled Comets did not instill confidence. Practical airliners had to wait for more efficient engines and a better understanding of high subsonic speeds at high altitudes. NACA aeronautical research of the early 1950s helped provide the latter; the drive toward higher speed in military aircraft provided the impetus for the engine improvements. Boeing's business gamble in funding the 367-80 demonstrator, which first flew in 1954, triggered the avalanche of jet airliner designs. Airlines began to buy the prospective aircraft by the dozens; because the Civil Aeronautics Board (CAB) mandated all ticket prices in the United States, an airline could not afford to be left behind if its competitors offered travel time significantly less than its propeller-driven fleet. Once passengers were exposed to the low vibration and noise levels of the turbine powerplants, compared to the dozens of reciprocating cylinders of the piston engines banging away combined with multiple noisy propellers, the outcome was further cemented. By the mid 1950s, the jet revolution was imminent in the civil aviation world.

In late 1958, commercial transcontinental and transatlantic jet service began out of New York City, but it was not an easy start. Turbojet noise to ground bystanders during takeoff and landings was not a concern to the military; it was to the New York City airport authorities. "Organ pipe" sound suppressors were mandated, which reduced engine performance and cost the airlines money; even with them, special flight procedures were required to minimize residential noise footprints, requiring numerous flight demonstrations and even weight limitations for takeoffs. The 707 was larger than the newly redesigned British Comet and hence noisier; final approval to operate the 707 from Idlewild was given at the last minute, and the delay helped give the British aircraft "bragging rights" on transatlantic jet service.[10]

10. Robert H. Cook, "Pyle Says Jet Noise Still Major Problem" *Aviation Week and Space Technology,* vol. 69, no. 4 (July 28, 1958), p. 30; "Comet Takes Idlewild Noise Test as Step to Transatlantic Service," *Aviation Week and Space Technology,* vol. 69, no. 7 (Aug. 18, 1958), pp. 41–42; Glenn Garrison, "Modified 707 Starts Pan Am Cargo Runs," *Aviation Week and Space Technology,* vol. 69, no. 9 (Sept. 1, 1958), pp. 28–31.

Other jet characteristics were also a concern to operators and air traffic control (ATC) alike. Higher jet speeds would give the pilots less time to avoid potential collisions if they relied on visual detection alone. A high-profile midair collision between a DC-7 and Constellation over the Grand Canyon in 1956 highlighted this problem. Onboard collision warning systems using either radar or infrared had been in development since 1954, but no choice had been made for mandatory use. Long-distance jet operations were fuel critical; early jet transatlantic flights frequently had to make unplanned landings en route to refuel. Jets could not endure lengthy waits in holding patterns; hence, ATC had to plan on integrating increasingly dense traffic around popular destinations, with some of the traffic traveling at significantly higher speeds and potentially requiring priority. A common solution to the traffic problems was to provide ground radar coverage across the country and to better automate the ATC sequencing of flight traffic. This was being introduced as the jet airliner was introduced; a no-survivors midair collision between a United Airlines DC-8 jetliner and a Constellation, this time over Staten Island, NY, was widely televised and emphasized the importance of ATC modernization.[11]

NACA research by Richard Whitcomb that led to the area rule had been used by Convair in reducing drag on the F-102 so it would go supersonic. It was also used to make the B-58 design more efficient so that it had a significant range at Mach 2, propelled by four afterburning General Electric J79 turbojets. Convair had been busy with these military projects and was late in the jet airliner market. It decided that a smaller, medium-range airliner could carve out a niche. An initial design appeared as the Convair 880 but did not attract much interest. The decision was made to develop a larger aircraft, the Convair 990, which employed non-afterburning J79s with an added aft fan to reap the developing turbofan engines' advantages of increased fuel efficiency and decreased sideline noise. Furthermore, the aircraft would employ Whitcomb's area rule concepts (including so-called shock bodies on its

11. Cook, "CAA Studies Terminal Air Control Plan," *Aviation Week and Space Technology*, vol. 69, no. 12 (Sept. 22, 1958), pp. 40–43; Phillip J. Klass, "Anti-Collision Device Tests Promising," *Aviation Week and Space Technology*, vol. 69, no. 9 (Sept. 1, 1958), pp 32–33; Klass, "Collision Avoidance Progress," *Aviation Week and Space Technology*, vol. 73, no. 26 (Dec. 26, 1960), pp. 26–27; Klass, "New Techniques Aimed at Jet Control," *Aviation Week and Space Technology*, vol. 68, no. 9 (Mar. 3, 1958), pp. 219–223.

wings, something it shared with the Soviet Union's Tupolev bombers) to allow it to efficiently cruise some 60–80 mph faster than the 707 and the DC-8, leading to a timesavings on long-haul routes. The aircraft had a higher cruise speed and some limited success in the marketplace, but the military-derived engine had poor fuel economics even with a fan and without an afterburner, was still very noisy, and generated enough black smoke on approach that casual observers often thought the aircraft was on fire (something it shared with its military counterpart, which generated so much smoke that McDonnell F-4 Phantoms often had their position given away by an accusing finger of sooty smoke). The potential trip timesavings was not adequate to compensate for those shortcomings. The lesson the airline industry learned was that, in an age of regulated common airline ticket prices, any speed increase would have to be sufficiently great to produce a significant timesavings and justify a ticket surcharge. The latter was a double-edged sword, because one might lose market share to non–high-speed competitors.[12]

The Quest for Long-Range Supersonic Cruise

Two users were looking to field airplanes in the 1960s with long range at high speeds. One organization's requirement was high profile and the object of much debate: the United States Air Force and its continuing desire to have an intercontinental range supersonic bomber. The other organization was operating in the shadows. It was the Central Intelligence Agency (CIA), and it was aiming to replace its covert subsonic high-altitude reconnaissance plane (the Lockheed U-2). The requirement was simple; the fulfillment would be challenging, to say the least: a mission radius of 2,500 miles, cruising at Mach 3 for the entire time, at altitudes up to 90,000 feet. The payload was to be on the order of 800 pounds, as it was on the U-2.

The evolution of both supersonic cruise aircraft was involved, much more so for the highly visible USAF aircraft that eventually appeared as the XB-70. The B-58 had given the USAF experience with a Mach 2 bomber, but bombing advocates (notably Gen. Curtis LeMay) wanted long range to go with the supersonic performance. As demonstrated in the classic Breguet range equation, range is a direct function of

12. J.S. Butz, Jr., "Industry Studies Transport Area Rule," *Aviation Week and Space Technology*, vol. 69, no. 2 (July 14, 1958), pp. 48–52; Richard Sweeney, "Area Rule Fits Convair 600 to Meet Jet Age Problems," *Aviation Week and Space Technology*, vol. 69, no. 10 (Sept. 8, 1958), pp. 50–57.

lift-to-drag (L/D) ratio. The high drag at supersonic speeds reduced that ratio to the point where large fuel tanks were necessary, increasing the weight of the vehicle, requiring more lift, more drag, and more fuel. Initial designs weighed 750,000 pounds and looked like a "3-ship formation." NACA research on the XF-92 had suggested a delta wing design as an efficient high-speed shape; now, a paper written by Alfred Eggers and Clarence Syvertson of Ames published in 1954 studied simple shapes in the supersonic wind tunnels. They noted that, by mounting a wing atop a half cylindrical shape, they could use the pressure increase behind the shape's shock wave to increase the effective lift of the wing.[13] A lift increase of up to 30 percent could be achieved. This concept was dubbed "compression lift"; more recently, it is referred to as the "wave rider" concept. Using compression lift principles, North American Aviation (NAA) proposed a 6-engined aircraft weighing 500,000 pounds loaded that could cruise at Mach 2.7 to 3 for 5,000 nautical miles. The aircraft would have a delta wing, with a large underslung shape housing the propulsion system, weapons bay, landing gear, and fuel tanks. A canard surface behind the cockpit would provide trim lift at supersonic speeds. To provide additional directional stability at high speeds, the outer wingtips would fold to either 25 or 65 degrees down. Although reducing effective wing lifting surface, it would have an additional benefit of further increasing compression lift caused by wingtip shocks reflecting off the underside of the wing. Because of the 900–1,100-degree sustained skin temperature at such high cruise speeds, the aircraft would be made of titanium and stainless steel, with stainless steel honeycomb being used in the 6,300-square-foot wing to save weight.[14]

Original goals were for the XB-70, as it was designated, to make its first flight in December 1961, after contract award to NAA in January 1958. But the development of the piloted bomber was colliding with the missile and space age. The NACA now became the National Aeronautics and Space Administration (NASA), and the research organization gained the mission of directing the Nation's civilian space program, as well as its traditional aeronautics advancement focus. For military aviation, the development of reliable intercontinental ballistic missiles (ICBM)

13. Edwin P. Hartman, *Adventures in Research: A History of Ames Research Center 1940–1965*, NASA SP-4302 (Washington, DC: NASA, 1970), pp. 249–250.

14. J.W. Ross and D.B. Rogerson, "Technological Advancements of XB-70," AIAA Paper 83-1048 (1983), p. 21.

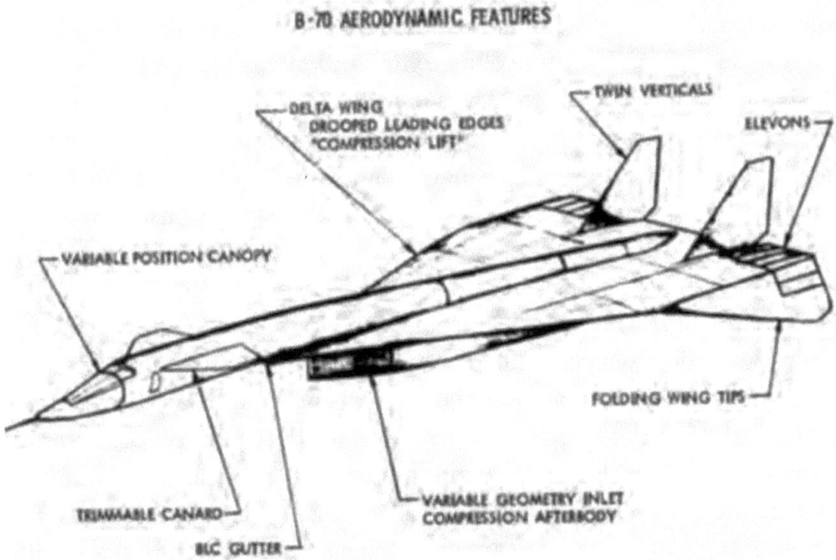

North American Aviation (NAA) XB-70 Valkyrie. NASA.

promised delivery of atomic payloads in 30 minutes from launch. The deployment by the Soviet Union of supersonic interceptors armed with supersonic air-to air missiles and belts of Mach 3 surface-to-air missiles (SAM) increasingly made the survivability of the unescorted bomber once again in doubt. The USAF clung to the concept of the piloted bomber, but in the face of delays in manufacturing the airframe with its new materials, increasing program costs, and the concerns of the new Secretary of Defense Robert S. McNamara, the program was scaled back to an experimental program with only four (later three, then two) aircraft to be built. The Air Force's loss was NASA's gain; a limited test program of 180 hours was to be flown, with the USAF and NASA sharing the cost and the data. At last, a true supersonic cruise aircraft would be available for the NACA's successor to study in the sky. The long-awaited first flight of XB-70 No. 1 occurred before a large crowd at Palmdale, CA, on September 21, 1964. But the other shadow supersonic cruise aircraft had already stolen a march on the star of the show.

In February 1964, President Lyndon Johnson revealed to the world that the United States was operating an aircraft that cruised at Mach 3 at latitudes over 70,000 feet. Describing a plane called the A-11, the initial press release was misleading—deliberately so. The A-11 name was a misnomer; it was a proposed design for the CIA spy plane that was never built, as it had too large a radar cross section. The photograph released was of a slim, long aircraft with two huge wing-mounted engines: the two-seat USAF interceptor version, known as the YF-12. Only three were built, and they were not put into production. The "A-11" that was flying was actually known as the A-12 and was the single-seat low-radar cross-section design plane built in secret by the Lockheed team led by Kelly Johnson, designer of the original U-2. Built almost exclusively of titanium, the aircraft had to be extremely light to achieve its altitude goal; its long range also dictated a high fuel fraction. The twin J58 turbojets had to remain in afterburner for the cruise portion, which dictated even higher-temperature materials than titanium and unique attention to the thermal environment of the vehicle.[15]

The USAF ordered a two-seat reconnaissance version of the A-12, designated the SR-71 and duly announced by the President in summer 1964, before the Presidential election. The single-seat A-12 existence was kept secret for another 20 years at CIA insistence, which had a significant impact on NASA's flight test of the only other Mach 3 piloted aircraft besides the XB-70. Later known collectively known as Blackbirds, a fleet of 50 Mach 3 cruise airplanes were built in the 1960s and operated for over 25 years. But the labyrinth of secrecy surrounding them severely hampered acquisition by NASA of an airplane for research, much less investigating their technical details and publishing reports. This was unfortunate, as now the United States was committed to not only a space race, but also a global race for a new landmark in aviation technology: a practical supersonic jet airliner, more popularly known as the Supersonic Transport (SST). The emerging NASA would be a major participant in this race, and in 1964, the other runners had a headstart.

Civilian Supersonic Cruise: The National SST Effort

The fascination for higher speeds of the 1950s and the new long-range comfortable jet airliners combined to create an interest in a supersonic airliner. The dominance of American aircraft manufacturers designs in the long-range subsonic jet airliner market meant that European manufacturers turned their sights on that goal. As early as 1959, when jet traffic was just commencing, Sir Peter Masefield, an influential aviation figure, said that a supersonic airliner should be a national goal for Britain. Development of such an airplane would contribute to national

15. Peter W. Merlin, *Mach 3+ NASA/USAF YF-12 Flight Research 1969–1979* (Washington, DC: NASA, 2002), pp. 1–6.

prestige, enhance the national technology skill level, and contribute to a favorable trade balance by foreign sales. He recognized that the undertaking would be expensive and that the government would have to support the development of the aircraft. The possibility was also suggested of a cooperative design effort with the United States. Meanwhile, the French aviation industry was pursuing a similar course. Eventually, in 1962, Britain and France merged their efforts to produce a joint European aircraft cruising at Mach 2.2.[16]

A Supersonic Transport had also been envisioned in the United States, and low-level studies had been initiated at NACA Langley in 1956, headed by John Stack. But the European initiatives triggered an intensification of American efforts, for essentially the same reasons listed by Masefield. In 1960, Convair proposed a new 52-seat modified-fuselage version of its Mach 2 B-58, preceded by a testbed B-58 with 5 intrepid volunteers in airline seats in the belly pod (windows and a life-support system were to be installed).[17] The influential magazine *Aviation Week* reflected the tenor of the American feeling by proposing that the United States make SST a national priority, akin to the response to Sputnik.[18] Articles appeared outlining the technology for supersonic cruise speeds up to Mach 4 with existing technology. The USAF's Wright Air Development Division convened a conference in late 1960 to discuss the SST for military as well as civilian use.[19] And in 1961, the newly created Federal Aviation Agency (FAA) began to work with the newly created NASA and the Air Force in Project Horizon to study an American SST program. One of the big questions was whether the design cruise speed should be Mach 2, as the Europeans were striving for, or closer to Mach 3.[20]

16. John Tunstall, "British Weigh Entering Supersonic Race," *Aviation Week and Space Technology*, vol. 70, no. 18 (May 4, 1959), pp. 55–56. For the Anglo-French program, see Kenneth Owen, *Concorde: Story of a Supersonic Pioneer* (London: Science Museum, 2001). The American program is treated in Mel Horwitch, *Clipped Wings: The American SST Conflict* (Cambridge: The MIT Press, 1982), and Erik M. Conway, *High-Speed Dreams: NASA and the Technopolitics of Supersonic Transportation, 1945–1999* (Baltimore: The Johns Hopkins University Press, 2005).

17. "B-58A Proposed for Transport Research," *Aviation Week and Space Technology*, vol. 73, no. 20 (Nov. 14, 1960), pp. 54–61.

18. Robert Hotz, "Supersonic Transport Race," *Aviation Week and Space Technology*, vol. 73, no. 20 (Nov. 14, 1960), p. 21.

19. "WADD Conference on Supersonic Transport," *Aviation Week and Space Technology*, vol. 73, no. 11 (Sept. 12, 1960), p. 53.

20. As detailed in Owen, *Concorde*.

NASA SST baseline configurations in 1963. Clockwise from bottom: SCAT 15, 16, 17, and 4. NASA.

Both Langley and Ames had been engaged in large supersonic aircraft design studies for years and had provided technical support for the Air Force WS-110 program that became the Mach 3 cruise B-70.[21] Langley had also pioneered work on variable-sweep wings, in part drawing upon variable wing sweep technology as explored by the Bell X-5 in NACA testing, to solve the problem of approach speeds for heavy airplanes with highly swept wings for supersonic cruise but also required to operate from existing jet runways. Langley embarked upon developing baseline configurations for a theoretical Supersonic Commercial Air Transport (SCAT), with Ames also participating. Clinton Brown and F. Edward McLean at Langley developed the so-called arrow wing, with highly swept leading and trailing edges, that promised to produce higher L/D at supersonic cruise speeds. In June 1963, the theoretical research

21. For example, D.D. Baals, O.G. Morris, and T.A. Toll, "Airplane Configurations for Cruise at a Mach Number of 3," NASA LRC, *Conference on High Speed Aerodynamics*, NTIS 71N5324 (1958); and M.M. Carmel, A.B. Carraway, and D.T. Gregory, "An Exploratory Investigation of a Transport Configuration Designed for Supersonic Cruise Flight Near a Mach Number of 3," NASA TM-X-216 (1960).

became more developmental, as President John F. Kennedy announced that the United States would build an SST with Government funding of up to $1 billion provided to industry to aid in the development.

In September 1963, NASA Langley hosted a conference for the aircraft industry presenting independent detailed analyses by Boeing and Lockheed of four NASA-developed configurations known as SCAT 4 (arrow wing), 15 (arrow wing with variable sweep), 16 (variable sweep), and 17 (delta with canard). Langley research had produced the first three, and Ames had produced SCAT 17.[22] Additionally, papers on NASA research on SST technology were presented. The detailed analyses by both contractors of the baselines concluded that a supersonic transport was technologically feasible, and that the specified maximum range of 3,200 nautical miles would be possible at Mach 3 but not at Mach 2.2. The economic feasibility of an SST was not evaluated directly, although each contractor commented on operating cost comparisons with the Boeing 707. Although the initial FAA SST specification called for Mach 2.2 cruise, the conference baseline was Mach 3, with one of the configurations also being evaluated at Mach 2.2. The results and the need to make the American SST more attractive to airlines than the European Concorde shifted the SST baseline to a Mach 2.7 to Mach 3 cruise speed. This speed was similar to that of the XB-70, so the results of its test program could be directly applicable to development of an SST. As the 1963 conference report stated, "Significant research will be required in the areas of aerodynamic performance, handling qualities, sonic boom, propulsion, and structural fabrication before the supersonic transport will be a success."[23]

NASA's Valkyrie Supersonic Cruise Flight-Test Program

Although the XB-70 test program was only budgeted for 180 hours, Air Force Category 1 testing with the contractor took first priority. That testing included verification of basic airworthiness and the achievement of the contractually required speed of Mach 3 for an extended cruise period. This proved to be harder than was thought, as the first XB-70 turned out to be almost a jinxed aircraft, as prototypes often are.

22. D.D. Baals, "Summary of Initial NASA SCAT Airframe and Propulsion Concepts," in NASA LRC, *Proceedings of NASA Conference on Supersonic-Transport Feasibility Studies and Supporting Research*, NTIS N67-31606 (1963), pp. 2–21.

23. John G. Lowry, "Summary and Assessment of Feasibility Studies," in NASA, *Proceedings of NASA Conference on Supersonic-Transport Feasibility Studies and Supporting Research* (1963), p. 52.

It was not until the 17th flight, 13 months after 1st flight, that Mach 3 was attained. Earlier flights had been plagued by landing gear problems, in-flight shutdowns of the new GE J93 engines (the most powerful in the world, at 30,000 pounds of thrust each in afterburner), and, most seriously, in-flight shedding of pieces of the stainless steel skin. The stainless steel honeycomb covering much of the wing had proven to be difficult to fabricate, requiring a brazing technique in an inert atmosphere to attach the skins. This process unfortunately resulted in numerous pinholes in the skin welds, which would allow the nitrogen inerting atmosphere required for fuel tanks with fuel heated to over 300 °F to leak away. Correcting this problem delayed the first aircraft by almost a year. The No. 5 fuel tank could never be sealed and was flown empty, further shortening the duration of test sorties on the two prototype aircraft, which had no aerial refueling capability.[24]

Aside from the mechanical difficulties that often shortened test sorties, the design features providing supersonic cruise worked well. The two-pilot XB-70 was initially the heaviest airplane in the world, at 500,000-pound takeoff weight, as well as designed to be the fastest. It was stable, maneuverable, and, aside from the unusually high attitude of the cockpit on takeoff and landing, easy to fly. The folding wingtips (each the size of a B-58 wing) worked flawlessly. The propulsion system of inlets and turbojets, when properly functioning, provided the thrust to reach Mach 3, and handling qualities at that speed were generally satisfactory, although the high speed meant that small pitch changes produced large changes in vertical velocity; it was difficult to maintain level flight manually. Mach 3 cruise in a large SST-size airplane seemed to be technologically achievable.[25]

The inlets for the six engines were another story for complexity, criticality, and pilot workload. An air inlet control system used moving ramps and doors to control the geometry of the inlet to position shock waves in the inlet above flight speed of Mach 1.6.[26] The final shock wave in the inlet was a strong normal shock in the narrow "throat," where the airflow became subsonic downstream of the shock. Proper positioning of the normal shock was vital; if downstream pressure was too high,

24. North American Rockwell, Space Division, "B-70 Aircraft Study Final Report," SD 72-SH-0003, vol. 2 (1972), pp. II-237–238.
25. Ibid., vol. 2, pp. II-278–284.
26. Ibid., vol. 4, pp. IV-16–64.

the normal shock might "pop out" of the inlet, losing the inlet pressure buildup, which actually provided net thrust to the airplane, and causing compressor stalls in the turbojet, as it now received air that was still supersonic. This was known as an inlet " unstart" and usually was corrected by opening bypass doors in the inlet to relieve the pressure and resetting the inlet geometry to allow the normal shock to resume its correct position. Unstarts usually were announced by a loud bang, a rapid yaw in the direction of the inlet that had unstarted because of the lack of thrust, and often by an unstart of the other inlet because of airflow disturbance caused by the yaw. Pilots considered unstarts to be exciting (" breathtaking," as NAA test pilot Al White described it), with motion varying from mild to severe, depending on flight conditions, but not particularly dangerous and usually easily corrected.[27] Although the inlet control system was designed to be automatic, for the first XB-70 (also known as "Ship 1"), the copilot became the flight engineer and manually manipulated the ramps and doors as a function of Mach number and normal shock position indicator. There were two inlets on the aircraft, with each feeding three engines. There had been some concern that problems with one engine might spread to the other two fed by the same inlet, but this did not seem to usually be the case. One exception was on the 12th flight, on May 7, 1965, when a piece of stainless steel skin went down the right inlet at Mach 2.6, damaging all 3 engines, one seriously. The mismanagement of the right inlet doors, because of time pressure and lack of knowledge of the nature of the emergency, led to inlet "duct buzz" pressure fluctuations caused by shock oscillation. This vibration at 2½ cycles per second was near the duct's resonant frequency, which could cause destruction of the duct. The vibration also fed into the highly flexible vehicle fuselage. This in turn led to the pilot reverting to turning the yaw dampers off, with subsequent development of a divergent Dutch roll oscillation. All three engines on the right side were eventually shut down. Fortunately, the flight control anomalies were cleared up, and the pilot performed a successful "3 and ½ engine" landing on the Rogers dry lakebed, touching down at 215 knots. This 5-minute inlet emergency generated a 33-page analytical report and presented some cautionary notes. The author commented in his closing that: "The seriousness of the interaction of the inlet conditions with

27. Ibid., vol. 2, p. II-280.

vehicle performance and handling characteristics tends to be accentuated for high-supersonic aircraft. Bypass-door settings are critical on mixed-compression inlets to maintain efficient inlet conditions."[28]

This observation would prove even more relevant for the Mach 3 Blackbird aircraft that followed the XB-70 in NASA supersonic cruise research. Test crews soon discovered that, as Blackbird researchers ruefully noted, "Around Mach 3, when things go wrong, they also get worse at a rate of Mach 3."[29] Crews who flew the secret twin-engine Blackbird often experienced this fact of life, sometimes with a less happy ending.

XB-70 Early Flight-Testing Experience

A byproduct of this and other incidents was that Ship 1 was eventually limited to Mach 2.5 because of flight safety concerns of the skin shedding. But Ship 2 made its first flight July 17, 1965, and it had numerous improvements. Skin bonding had been improved, an automated air inlet control system had been installed, wing dihedral had been increased to 5 degrees to improve lateral directional stability, and fuel tank No. 5 could now be filled. NASA planned to use Ship 2 for its research program; an extensive instrumentation package recording over 1,000 parameters such as temperature, pressure, and accelerations was installed in the weapons bay for use when NASA took over the direction of the flight-test program. Ship 2 still had some of the gremlins that seemed to haunt the XB-70, mainly connected to the complex landing gear. Flight 37 on AV-2 resulted in the pilots having to do some in-flight maintenance when the nose gear door position prevented proper retraction or extension of the nose gear. The activity was widely advertised as the pilot using "a paperclip" to short an electrical circuit to allow extension (actually, there were no paperclips on board; USAF pilot Joseph Cotton fashioned the device from a wire on his oxygen mask). But AV-2 showed that the high-speed skin-shedding problem had indeed been solved. Beginning in March 1966, AV-2 routinely spent 50 minutes to 1 hour at speeds from Mach 2.5 to Mach 2.9. And on May 19, AV-2 reached the (contractual) holy grail of 32 minutes at Mach 3 (actually up to 3.06). Skin stagnation temperature was over 600 °F. With accomplishment of that goal, NASA moved to put a new pilot in the program.

28. Chester H. Wolowicz, "Analysis of an Emergency Deceleration and Descent of the XB-70-1 Airplane Due to Engine Damage Resulting from Structural Failure," NASA TM-X-1195 (1966).
29. Author's recollection.

NASA X-15 veteran test pilot Joe Walker had been undergoing delta wing training and preparation to fly the B-70 as the program moved to the second stage of flight test. National Sonic Boom Program (NSBP) tests were flown June 6, 1966, to prepare for the official change over to NASA on June 15, but on June 8, disaster struck, dramatically changing the program.

That day, AV-2 took off on a planned flight-test mission that would include a photo session at the end of the sortie with a number of other aircraft powered by engines made by General Electric.[30] One of the aircraft was a Lockheed F-104N Starfighter flown by Joe Walker, who was observing the mission as he prepared to fly the B-70 on the next sortie. During the photo shoot, which required close formation flight, his F-104 was seen to fly within 30–50 feet of the Valkyrie's right wingtip, which had been lowered to the 20-degree intermediate droop position. As the photo session ended, the F-104 tail struck the XB-70 wingtip, causing the F-104 to roll violently to the left and pass inverted over the top of the bomber, shearing off most of the twin vertical tails and causing the Starfighter to erupt in flames, killing Walker. The XB-70 subsequently entered an inverted spin, from which recovery was impossible. Company test pilot Joe Cotton ejected using the complex encapsulated ejection seat and survived; USAF copilot Carl Cross did not eject and died in the ensuing crash. The accident was not related to the Valkyrie design itself; nevertheless, the loss of the improved Ship 2 and its comprehensive instrumentation package meant that AV-1 would now have to become the NASA research aircraft. A new instrumentation package was installed in AV-1, but the Mach 2.5 speed limit imposed on AV-1 for the skin shedding problem and the workload-intensive manual inlets meant the program orientation could be less of an analog for the national SST program, which was now approaching the awarding of contracts for an SST with speeds of Mach 2.7 to 3.

XB-70 Supersonic Cruise Program Takes to the Air

Despite the AV-1 aircraft limitations, the XB-70 test program proceeded, now with NASA directing the effort with USAF support. Eleven flights were flown under NASA direction as Phase II of the original XB-70 planned flight-test program, ending January 31, 1967. Nine of the

30. Dennis R. Jenkins and Tony R. Landis, *Valkyrie: North American's Mach 3 Superbomber* (North Branch, MN: Specialty Press, 2004), pp. 153–164.

flights were primarily dedicated to the NSBP. As the XB-70 was the only aircraft in the world with the speed, altitude capability, and weight of the U.S. SST, priority was given to aspects that supported that program. The sonic boom promised to be a factor that was drastically different from current jet airliner operations and one whose initial impact was underrated. It was thought that a rapid climb to high altitude before going supersonic would muffle the initial strong normal shock; once at high altitude, even at higher Mach numbers, the boom would be sufficiently attenuated by distance from the ground and the shock wave inclination "lay back" as Mach number increased, to not be a disturbance to ground observers. This proved not to be the case, as overflights by B-58s and the XB-70 proved. Another case study in this volume provides details on sonic boom research by NASA. Overpressure measurements on the ground during XB-70 overflights as well as the observer questionnaires and measurements in instrumented homes constructed at Edwards AFB indicated that overland supersonic cruise would produce unacceptable annoyance to the public on the ground. Overpressure beneath the flight path reached values of 1.5 to 2 pounds per square foot. A lower limit goal of not more than 0.5 pounds per foot to preclude ground disturbance seemed unachievable with current designs and technology.[31]

Supersonic cruise test missions proved challenging for pilots and flight-test engineers alike. Ideally, the test conductor on the ground would be in constant contact with the test pilots to assist in most efficient use of test time. But with an aircraft traveling 25–30 miles per minute, the aircraft rapidly disappeared over the horizon from test mission control. Fortunately, NASA had installed a 450-mile "high range" extending to Utah, with additional tracking radars, telemetry receivers, and radio relays for the hypersonic X-15 research rocket plane. The X-15 was typically released from the B-52 at the north end of the range and was back on the ground within 15 minutes. The high range provided extended mission command and control and data collection but was not optimized for the missions flown by the XB-70 and YF-12.

31. Donald S. Findley, Vera Huckel, and Herbert R. Henderson, "Vibration Responses of Test Structure no. 1 During the Edwards AFB Phase of the National Sonic Boom Program," NASA TM-72706 (1975); Domenic J. Maglieri, David A. Hilton, and Norman J. McLeod, "Summary of Variations of Sonic Boom Signatures Resulting from Atmospheric Effects," NASA TM-X-59633 (1967).

The XB-70 ground track presented a different problem for mission planners. The author flew the SR-71 Blackbird from Southern California for 5 years and faced the same problems in establishing a test ground track. The test aircraft would take over 200 miles to get to the test cruise speed and altitude. Then it would remain at test conditions, collecting data for 30–40 minutes. It then required an additional 200–250 miles to slow to "normal" subsonic flight. Ground tracks had to be established that would provide data collection legs while flying straight or performing planned turning maneuvers, and avoiding areas that would be sensitive to the increasingly contentious sonic booms. Examples of the areas included built-up cities and towns; the "avoidance radius" was generally 30 nautical miles. Less obvious areas included mink farms and large poultry ranches, as unexplained sudden loud noises could apparently interfere with breeding habits and egg-laying practices. The Western United States fortunately had a considerably lower population density than the area east of the Mississippi River, and test tracks could be established on a generally north-south orientation.

The presence of Canada to the north and Mexico to the south, not to mention the densely populated Los Angeles/San Diego corridor and the "island" of Las Vegas, set further bounding limits. Planning a test profile that accounted for the limits/avoidance areas could be a challenge, as the turn radius of a Mach 3 aircraft at 30 degrees of bank was over 65 nautical miles. Experience and the sonic boom research showed that a sonic boom laid down by a turning or descending supersonic aircraft would "focus" the boom on the ground, decreasing the area affected but increasing the overpressure on the ground within a smaller region. Because planning ground tracks was so complicated and arduous, once a track was established, it tended to be used numerous times. This in turn increased the frequency of residents being subjected to sudden loud noises, and complaints often appeared only after a track had been used several times. The USAF 9th Reconnaissance Wing operating the Mach 3+ SR-71 at Beale Air Force Base near Sacramento, CA, had the same problem as NASA flight-testing for developing training routes (but without the constraints of maintaining telemetry contact with a test control), and it soon discovered another category for avoidance areas: congressional complaints relayed from the Office of the Secretary of the Air Force.

For the limited XB-70 test program, a ground track was established that remained within radio and telemetry range of Edwards. As a result, the aircraft at high Mach would only fly straight and level for 20 minutes at best, requiring careful sequencing of the test points. The profile included California, Nevada, and Utah.[32]

This planning experience was a forerunner of what problems a fleet of Supersonic Transports would face on overland long-distance flights if they used their design speed. A factor to be overcome in supersonic cruise flight test, it would be critical to a supersonic airliner. Pending development of sonic boom reduction for an aircraft, the impact of off-design-speed operation over land would have to be factored into SST designs. This would affect both range performance and economics.

The flight tests conducted on the XB-70 missions collected data on many areas besides sonic boom impact. The research data were generally focused on areas that were a byproduct of the aeronautical technology inherent in a large airplane designed to go very fast for a long distance with a large payload. An instrumentation package was developed to record research data.[33] Later, boundary layer rakes were installed to measure boundary layer growth on the long fuselage at high Mach at 70,000 feet altitude; this would influence the drag and hence the range performance of a design. The long flexible fuselage of the XB-70 produced some interesting aeroelastic effects when in turbulence, not to mention taxing over a rough taxiway, similar to the pilot being on a diving board. Two 8-inch exciter vane "miniature canards" were mounted near the cockpit as part of the Identically Located Acceleration and Force (ILAF) experiment for the final XB-70 flight-test sorties. These vanes could be programmed to oscillate to induce frequencies in the fuselage to explore its response. Additionally, frequencies could be produced to cancel accelerations induced by turbulence or gusts, leading to a smoother ride for pilots and ultimately SST passengers. This system was demonstrated to be effective.[34] A similar system was employed in the Rockwell B-1 Lancer bomber, the Air Force bomber eventually built instead of the B-70.

Inlet performance would have a critical effect on the specific fuel consumption performance, which had a direct effect on range achieved. In addition to collecting inlet data on all supersonic cruise sorties,

32. "B-70 Aircraft Study Final Report," vol. 2, p. II-292.

33. Ibid., vol. 2, p. II-25.

34. John H. Wykes, et al., "XB-70 Structural Mode Control System Design and Performance Analyses," NASA CR-1557 (1970).

numerous test sorties involved investigating inlet unstarts deliberately induced by pilot action, as well as the "unplanned" events. This was important for future aircraft, as the Valkyrie used a two-dimensional (rectangular) inlet with mixed external (to the inlet)/internal compression, with one inlet feeding multiple engines. As a comparison, the A-12/SR-71 used an axisymmetric (round) inlet, also with external/internal compression feeding a single engine. There was a considerable debate in the propulsion community in general and the Boeing and Lockheed competitive SST designers in particular as to which configuration was better. Theoretical values of pressure recovery had been tested in propulsion installations in wind tunnels, but the XB-70 presented an opportunity to collect data and verify wind tunnel results in extended supersonic free-flight operations, including "off-design" conditions during unstart operations. These data were also important as an operational SST factor, as inlet unstarts were disconcerting to pilots, not to mention prospective passengers.

Traditional aircraft flight-test data on performance, stability, control, and handling qualities were collected, although AV-1 was limited to Mach 2.5 and eventually Mach 2.6. Data to Mach 3 were sometimes also available from AV-2 flights. As USAF–NASA test pilot Fitzhugh Fulton reported in a paper presented to the Society of Automotive Engineers (SAE) in 1968 in Anaheim, CA, on test results as applied to SST operations, the XB-70 flew well, although there were numerous deficiencies that would have to be corrected.[35] The airplane's large size and delta wing high-incidence landing attitude required pilot adjustments in takeoff, approach, and landing techniques but nothing extraordinary. High Mach cruise was controllable, but the lack of an autopilot in the XB-70 and the need of the pilot to "hand-fly" the airplane brought out another pilot interface problem; at a speed of nearly 3,000 feet per second, a change in pitch attitude of only 1 degree would produce a healthy climb or descent rate of 3,000 feet per minute (50 feet per second). Maintaining a precise altitude was difficult. Various expanded instrument displays were used to assist the task, but the inherent lag in Pitot-static instruments relying on measuring tiny pressure differentials (outside static pressure approximately 0.5 pounds per square inch [psi]) to indicate altitude change meant the pilot was often playing catchup.

35. Fitzhugh L. Fulton, Jr., "Lessons from the XB-70 as Applied to the Supersonic Transport," NASA TM-X-56014 (1968).

10

High Mach cruise at 70,000 feet may have become routine, but it required much more careful flight planning than do contemporary subsonic jet operations. The high fuel flows at high Mach numbers meant that fuel reserves were critical in the event of unplanned excursions in flight. Weather forecasts at the extreme altitudes were important, as temperature differences at cruise had a disproportionate influence on fuel flows at a given Mach and altitude; 10 °F hotter than a standard day at altitude could reduce range, requiring an additional fuel stop, unless it was factored into the flight plan. (Early jet operations over the North Atlantic had similar problems; better weather forecasts and larger aircraft with larger fuel reserves rectified this within several years.) Supersonic cruise platforms traveling at 25–30 miles per minute had an additional problem. Although the atmosphere is generally portrayed as a "layer cake," pilots in the XB-70 and Mach 3 Blackbird discovered it was more like a "carrot cake," as there were localized regions of hot and cold air that were quickly traversed by high Mach aircraft This could lead to range performance concerns and autopilot instabilities in Mach hold because of the temperature changes encountered. The increase in stagnation temperatures on a hot day could require the aircraft to slow because of engine compressor inlet temperature (CIT) limitations, further degrading range performance.

Fuel criticality and the over 200 miles required to achieve and descend from the optimum cruise conditions meant that the SST could brook no air traffic control delays, so merging SST operations with subsonic traffic would stress traffic flow into SST airports. Similar concerns about subsonic jet airliner traffic in the mid-1950s resulted in revamping the ATC system to provide nationwide radar coverage and better automate traffic handoffs. To gather contemporary data on this problem for SST concerns, NASA test pilots flew a Mach 2 North American A-5A (former A3J-1) Vigilante on supersonic entry profiles into Los Angeles International Airport. The limited test program flying into Los Angeles showed that the piloting task was easy and that the ATC system was capable of integrating the supersonic aircraft into the subsonic flow.[36]

One result mentioned in test pilot Fulton's paper had serious implications not only for the SST but also supersonic research. The XB-70

36. Donald L. Hughes, Bruce G. Powers, and William H. Dana, "Flight Evaluation of Some Effects of the Present Air Traffic Control System on Operation of a Simulated Supersonic Transport," NASA TN-D-2219 (1964).

had been designed using the latest NASA theories (compression lift) and NASA wind tunnels. Nevertheless, the XB-70 as flown was deficient in achieving its design range by approximately 25 percent. What was the cause of the deficiency? Some theorized the thermal expansion in such a large aircraft at cruise Mach, unaccounted for in the wind tunnels, increased the size of the aircraft to the point where the reference areas for the theoretical calculations were incorrect. Others thought the flexibility of the large aircraft was unaccounted for in the wind tunnel model configuration. Another possibility was that the skin friction drag on the large surface area at high Mach was higher than estimated. Yet another was that the compression lift assumption of up to 30-percent enhancement of lift at cruise speed was incorrect.

The limited duration of the XB-70 test program meant that further flight tests could not be flown to investigate the discrepancy. Flight-test engineer William Schweikhard proposed a reverse investigation. He structured a program that would use specific flight-test conditions from the program and duplicate them in wind tunnels using high-fidelity models of the XB-70 built to represent the configuration of the aircraft as it was estimated to exist at Mach 2.5. The flight-test data would thus serve as a truth source for the tunnel results.[37] This comparison showed good correlation between the flight-test data and the wind tunnel, with the exception of a 20-percent-too-low transonic drag estimate, mainly caused by an incorrect estimate of the control surface deflection necessary to trim the aircraft at transonic speeds. It was doubtful that that would account for the range discrepancy, because the aircraft spent little time at that speed.

The NASA test program with the XB-70 extended from June 16, 1966, to January 22, 1969, with the final flight being a subsonic flight to the Air Force Museum at Wright-Patterson Air Force Base in Dayton, OH. Thirty-four sorties were flown during the program. The original funding agreement with the USAF to provide B-58 chase support and maintenance was due to expire at the end of 1968, and the XB-70 would require extensive depot level maintenance as envisioned at the end of the 180-hour test program. NASA research program goals had essentially been

37. H.H. Arnaiz, J.B. Peterson, Jr., and J.C. Daugherty, "Wind-tunnel/flight correlation study of aerodynamic characteristics of a large flexible supersonic cruise airplane (XB-70-1)," pt. 3: "A comparison between characteristics predicted from wind-tunnel measurements and those measured in flight," NASA TP-1516 (1980).

reached, and because of the high costs of operating a one-aircraft fleet, the program was not extended. The X-15 program was also terminated at this time.

The legacy of the XB-70 program was in the archived mountains of data and the almost 100 technical reports written using that data. As late as 1992, the sonic boom test data generated in the NSBP flights were transferred to modern digital data files for use by researchers of high-speed transports.[38] But it was fitting that the XB-70's final supersonic test sortie included collecting ozone data at high altitudes. The United States SST program that would use supersonic cruise research data was about to encounter something that the engineers had not considered: the increasing interest of both decision makers and the public in the social consequences of high technology, exemplified by the rise of the modern environmental movement. This would have an impact on the direction of NASA supersonic cruise research. Never again in the 20th century would such a large aircraft fly as fast as the Valkyrie.

The American SST Program: Competition, Selection, and Demise

NASA participated extensively in plans to develop an American SST. President Kennedy had committed the U.S. Government to contribute funding for 75 percent of the aircraft's development cost, with a $1-billion upper limit. Industry would contribute the rest of the cost, with the Government money to be repaid via royalty payments as aircraft were sold. This Government backing was a response to the 1962 announcement of a joint government-backed program between France (Sud Aviation) and England (British Aircraft Corporation) companies to develop a Mach 2.2, 100-passenger transport, which emerged as the graceful Concorde. The FAA, NASA, and the Department of Defense would manage the American program and select a final contractor to make the SST a reality.[39] The competition aspect of the program gained even more of a Cold War aspect when the Soviet Union announced in June 1965 that it also was developing a Mach 2.2 SST, which would fly in 1968. The United States was still deciding on a contractor and design to be given the go-ahead.

38. Maglieri, et al., "A Summary of XB-70 Sonic Boom Signature Data," NASA CR-189630 (1992).
39. Glenn Garrison, "Supersonic Transport May Aim at Mach 3," *Aviation Week and Space Technology*, vol. 70, no. 5 (Feb. 2, 1959), pp. 38–40; J.S. Butz, Jr., "FAA, NASA Study Supersonic Transport," *Aviation Week and Space Technology*, vol. 72, no. 18 (May 2, 1960).

Desktop models of American Boeing and Lockheed SST finalist designs. William Flanagan.

The finalist contractors selected in May 1964 were Lockheed and Boeing, after rival Douglas and NAA designs (the latter based on B-70 technology) were eliminated. Although the initial submissions had a speed requirement of Mach 2.2+ with 160 passengers, the selected initial designs were a double delta Lockheed Model 2000 Mach 3 aircraft and a Boeing Model 733 Mach 2.7 variable sweep aircraft reminiscent of the NASA SCAT 16 design. Both finalist contractors had done analyses of the NASA SCAT designs in 1963. They had reached the conclusion that at Mach 2.2, the range specification could not be achieved, so they opted for the higher Mach cruises. FAA Administrator Najeeb Halaby had favored the higher cruise speed with larger capacity to preempt the Concorde in the international airliner marketplace. General Electric and Pratt & Whitney were the engine contractors chosen to develop engines for the SST. Both had developed 30,000-pound thrust engines for supersonic cruise airplanes (GE J93 for the XB-70 and Pratt & Whitney J58 for the A-12/SR-71), but the SST would require four 60,000-pound thrust engines.

The selection was announced on the last day of 1966. The Lockheed configuration had remained relatively unchanged, while the Boeing fuselage had been made longer and the engine position had shifted from under the wing to under the tail. Even the name had been changed, to

Boeing 2707. Both contractors built impressive full-scale mockups that were as much publicity props as engineering tools. (Unfortunately, the impressive mockups would prove to be the only airplanes built.) With fuselage lengths around 300 feet to accommodate up to 300 passengers and the fuel for ranges of over 3,000 miles, the mockups represented a new dawn in civil aviation. But the Boeing design and the Pratt & Whitney engine were chosen as the United States' entry in the supersonic airliner derby. (Details of the Boeing design showed that the variable sweep wing was unachievable because of weight and complexity; the Boeing design had 59 control surfaces, versus the Lockheed design's 16). Eventually, the Boeing design evolved to a fixed double delta with a small horizontal tail and four underwing engine nacelles with axisymmetric inlets. American flag carriers placed $100,000 deposits to reserve delivery positions on the production line with an order book of 120 aircraft by 1969, and work began on the first prototype.[40]

Controversy, Confrontation, and Cancellation

The American involvement in combat operations in Vietnam escalated by the late 1960s to something that was not called a war by the Government but actually was. The public turned against the war as casualties and costs escalated; by 1968, a sense of distrust of the Government and all its programs also affected a significant portion of the populace. The Apollo program was about to achieve President Kennedy's goal of landing on the Moon, but people were beginning to question its value. A youth-oriented cultural shift had not only a pro-peace stance but also an anti-technology bent, and environmental movements such as the Sierra Club were becoming increasingly influential. Nuclear powerplants and nuclear weapons were increasingly cited as being harmful to the environment, and many people wanted them limited or banned. The United States SST program was a high-visibility target and opportunity for environmental movements. Initially, the arguments focused on the sonic booms to be produced by an SST fleet. Dr. William Shurcliff, a Harvard University physicist, formed the Citizens League against the Sonic Boom to argue against the SST. As the *SST and Sonic Boom Handbook* (published in 1970 by an environmental activist organization) stated on its jacket, "This book demonstrates that the SST is an incredible, unnecessary insult to

40. Bill Yenne, "America's Supersonic Transports," *Flightpath*, vol. 3 (2004), pp 146–157.

the living environment, and an albatross around the neck of whatever nations seek to promote it."[41]

American legislators were not deaf to this increasing clamor. The sonic boom problem remained real and apparently technically unsolvable. An operational solution was to ban overland supersonic flights by the SST. This had a serious impact on the economic case for an airplane that would have to fly at off-design cruise speeds for significant amounts of time if flying on anything other than transatlantic or transpacific flights. Transcontinental flights in the United States had always been a prime revenue generator for airlines. A further noise problem was represented by New York City airports surrounded by densely populated communities. Subsonic jets with 10,000-pound thrust engines had difficulty meeting local noise standards; engines with 30,000 and even 60,000 pounds of thrust promised to be even more intractable. But even solving these problems would not satisfy some later environmental concerns. Water vapor from SST exhaust at high altitude having the potential to damage the protective ozone layer was a major concern, which later proved to be unwarranted, even if 500 SSTs had been built (although this was not so for fluorocarbons in spray cans at sea level). At a hearing before the United States Senate in May 1970, a member of the President's Environmental Quality Council referred to the SST as "the most significant unresolved environmental problem."[42] By late 1970, polls showed that American voters were 85 percent in favor of ending Federal funding for the SST program. In May 1971, the Senate voted to withhold further Government funding. Boeing, already developing the 737 and 747 at its own expense, said it could not proceed without $500 million in Federal funding. The American SST program of the 1960s was over.

The international SST race ended with only one horse crossing the finish line, albeit at a walk rather than a gallop. Although the Soviet SST flew first at the end of 1968, it required redesign and never was commercially successful. Two were lost in crashes during trials, and it only flew limited cargo flights from Moscow to Siberia over the sparsely populated Russian landmass. The SST field was left to the Anglo-French Concorde, which finally entered service on the transatlantic run in 1976 and flew for over 25 years. Only 13 aircraft entered service with the British and French national airlines, and most of their traffic was on the transatlantic run

41. William A. Shurcliff, *SST and Sonic Boom Handbook* (New York: Ballantine Books, 1970).
42. Yenne, "America's SST," p. 155.

for which the aircraft had initially been sized. The fuel price increase that started in 1974 and the de facto ban on overland supersonic flight meant that there was no move to improve or expand the Concorde fleet. It essentially remained a limited capacity first-class-only means of quickly getting from the United States to Paris or London while experiencing supersonic flight. Boeing's privately financed gamble on the 747 jumbo jet turned out to be the winning hand, as it revolutionized air traveling habits, especially after airline deregulation began in the United States in 1978.[43]

Shaping NASA Supersonic Cruise Research for the Post-SST Era

With the demise of the national SST program and the popular shift away from a supersonic airliner, a principal reason for funding supersonic research disappeared. Nevertheless, NASA's mission to advance aeronautics research dictated that a program should continue, although not necessarily at the previous urgency. It was obvious from the XB-70 flight test and the SST debate that the integration of the elements of a supersonic cruiser was more critical than for a subsonic aircraft. The shape of the aircraft dictated external shock wave formation, which not only changed drag but also had a major effect on the sonic boom footprint on the ground. The shock waves within the air-breathing engine inlet had a major impact on propulsive efficiency, which affected range and had operational impact if the inlet was not operating at optimal efficiency. To integrate these elements in the design process required better knowledge of how accurate the engineers design tools were. Wind tunnel fidelity in predicting results when the models did not necessarily have the temperature or aeroelastic characteristics of a full-sized aircraft at the high-temperature cruise state required investigation. The same could be said for propulsion wind tunnel models.

NASA developed a research program known initially as Advanced Supersonic Technology (AST), which lasted from 1972 to 1981. (Because of political sensitivities, the name was changed to Supersonic Cruise Aircraft Research [SCAR] in 1974; the "Aircraft" word was deleted in 1979, to avoid connection with the contentious SST label, so it became SCR.)[44]

43. Joe Sutter, with Jay Spenser, *747: Creating the World's First Jumbo Jet and Other Adventures from a Life in Aviation* (Washington, DC: Smithsonian Books, 2006), pp. 202–224.
44. Joseph R. Chambers, *Innovations in Flight: Research of the NASA Langley Flight Research Center on Revolutionary Concepts for Advanced Aeronautics* (Washington DC: NASA, 2005), pp 39–48.

The idea was to research the technical problems that had appeared during the SST development program and the XB-70 flight test. NASA Langley concentrated on refining a configuration for an advanced supersonic cruise aircraft, and it was postulated to have a cruise speed of Mach 2.2, matching the Concorde that was entering service. Its size, payload, and range performance were also reduced in comparison with the 1963 configurations. The configuration often shown for the SCR research resembled the Boeing 2707-300, but Langley continued to favor a refinement of the SCAT 15F, with the arrow wing as the optimum high-speed shape. Unfortunately, without variable sweep, the arrow wing was initially one of the worst low-speed shapes. Lewis Research Center studies focused on a variable cycle engine (VCE) to study optimizing engine performance, including internal aerodynamics for various phases of flight, engine noise, and exhaust emission problems. Dryden and Ames did simulator studies mainly, with some uses of the XB-70 test data followed by the YF-12 test program data.

Funding for AST–SCAR–SCR was limited, mainly because of the SST fallout; nevertheless, SCAR conferences in 1976 and 1979 were well-attended and produced almost 1,000 papers.[45] Nor was the only target application a transport. The USAF was exploring the possibility of a fighter aircraft using supersonic cruise for "global persistence" to operate deep behind the battlefront over the Central European battlefield; thus, the Air Force and its contractors were interested in optimum supersonic performance in an aircraft with limited fuel. A conference at the Air Force Academy hosted by the Air Force Flight Dynamics Laboratory in February 1976 included papers on NASA research results and contractor studies that used NASA's arrow wing to satisfy the supersonic mission requirements. The arrow wing was shown to have superior maximum L/D over the delta wing, to the point that Lockheed studies switched to it from their SST double delta configuration.[46]

45. Sherwood Hoffman, "Bibliography of Supersonic Cruise Aircraft Research (SCAR) Program, 1972-mid 1977," NASA RP-1003 (1977); Hoffman, "Supersonic Cruise Research (SCR) Program Publications FY 1977–1979: Preliminary Bibliography," NASA TM-80184 (1979).

46. Barrett L. Schout, et al., "Review of NASA Supercruise Configuration Studies," presented at *Supercruise Military Aircraft Design Conference, U.S. Air Force Academy, Colorado Springs, CO, Feb. 17–20, 1976;* B.R. Wright, F. Bruckman, et al., "Arrow Wings for Supersonic Cruise Aircraft," *Journal of Aircraft*, vol. 15, no. 12 (Dec. 1978), pp. 829–836.

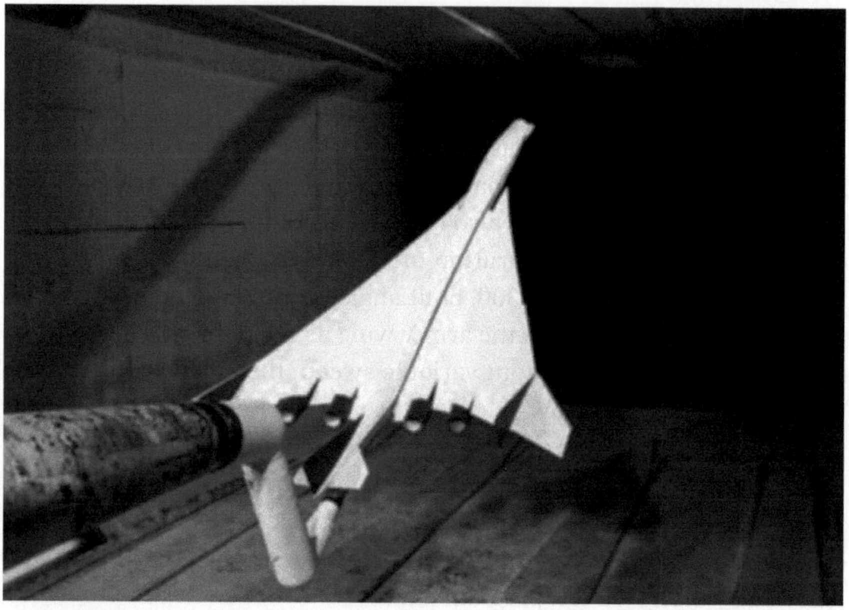

The SCAT 15F sharply swept fixed wing of 1964 promised a high supersonic lift-to-drag ratio of 9.6 at Mach 2.6. NASA.

General Dynamics—later Lockheed Martin Tactical Aircraft Systems (LMTAS)—worked with NASA Langley in the early 1970s in the development of its highly maneuverable F-16 Fighting Falcon fighter.[47] As interest developed in a supersonic cruise fighter in 1977, the company teamed with Langley researchers again to design an arrow wing for its F-16. Known as the Supersonic Cruise and Maneuverability Program (SCAMP), it resulted in the construction of two company-funded aircraft designated F-16XL, which first flew in 1982. Development of an arrow wing aircraft provided an opportunity to develop the features necessary to make the design practical, especially with regard to its low-speed and high-angle-of-attack characteristics. Although USAF interest shifted to the air-to-ground mission, resulting in purchase of the larger McDonnell-Douglas (later Boeing) F-15E Strike Eagle, the two shapely F-16XLs were the first flying testbeds for the arrow wing. (NASA shared in the data from the test program and wisely put the two aircraft in storage for possible future use, for they were later used to accomplish

47. Chambers, *Partners in Freedom: Contributions of the Langley Research Center to U.S. Military Aircraft of the 1990s*, No. 19 in the *Monographs in Aerospace History* series (Washington, DC: NASA, 2000), pp. 156–158.

F-16 with arrow wing SCAMP configuration. NASA.

notable work in refined aerodynamic studies, including supersonic laminar flow control.) The AST–SCAR–SCR program had essentially ended in 1981, as funding for NASA aeronautical research was cut because of the needs of the approaching Space Transportation System (STS, the Space Shuttle). Flight test for supersonic aircraft was too expensive. But one supersonic NASA flight-test program of the 1970s proved to be a spectacular success, one that contributed across a number of technical disciplines: the Blackbirds.

YF-12 Flight Test: NASA's Major Supersonic Cruise Study Effort

The XB-70 test program had focused on SST research, as it was the only large aircraft capable of high Mach cruise. In the 1970s, flight data collection could focus on a smaller aircraft but one that had already demonstrated routine flight at Mach 3. Lockheed's Mach 3 Blackbird was no longer as secret as it had been in its CIA A-12 initial stages, and the USAF was operating a fleet of acknowledged Mach 3 aircraft, although details of its missions, top speeds, and altitudes remained military secrets. NASA had requested Blackbirds as early as 1968 for flight research, but the Agency was rejected as being too open for the CIA's liking. Later, as the NASA flight-test engineers Bill Schweikhard and Gene Matranga, both XB-70 test program veterans, assisted the USAF in SR-71 flight-test data analysis, the atmosphere changed, and the USAF was more willing to provide the aircraft but without compromising the secrecy of the details of the SR-71. The YF-12s were in storage, as the USAF had decided not to buy any further aircraft. Because the YF-12 had a different fuselage and earlier model J58s than the SR-71, a joint test program was proposed, with the USAF providing aircraft and crew support and NASA paying operational costs. Phase I of the program would concentrate on USAF desires to evaluate operational tactics against a high Mach target (such as the new Soviet MiG-25 Foxbat). NASA would instrument the aircraft and collect basic research data, as well as conduct Phase II of the test program with applied research that would benefit from a Mach 3, 80,000-foot altitude supersonic cruise platform.[48]

Between 1969 and 1979, flight-test crews flew 298 flights with 2 YF-12 Blackbird aircraft. The first of these was a modified YF-12A interceptor. The second, which replaced another YF-12A lost from a fire during an Air Force test mission, was a nonstandard SR-71A test aircraft given a fictitious "YF-12C" designation and serial number to mask its spy plane origins. YF-12 supersonic cruise–related test results included isolation of thermal effects on aircraft loads from aerodynamic effects. The instrumented aircraft collected loads and temperature data in flight. It was then heated in purpose-built form-fit ovens on the ground—the High Temperature Loads Laboratory (HTTL)—so the thermal strains and loads could be differentiated from the aerodynamic.[49] For a high-temperature aircraft of the future, separation of these stresses could be crucial in the event that underestimation could lead to skin failures, as experienced by XB-70 AV-1. One byproduct of this research was to correct the factoid still quoted into the 21st century that the airplane expanded in length by 30 inches because of heat. The actual figure was closer to 12 inches, with the difference appearing in structural stresses.

48. Merlin, *Mach 3+*, pp. 7–11.

49. Jerald M. Jenkins and Robert D. Quinn, "A Historical Perspective of the YF-12A Thermal Loads and Structures Program," NASA TM-104317 (1996).

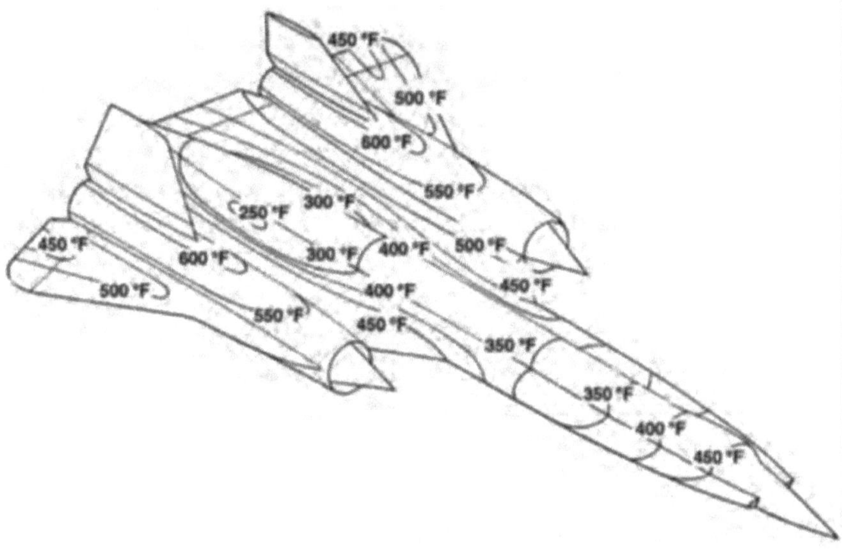

YF-12 representative cruise temperature measurements (upper surfaces). NASA.

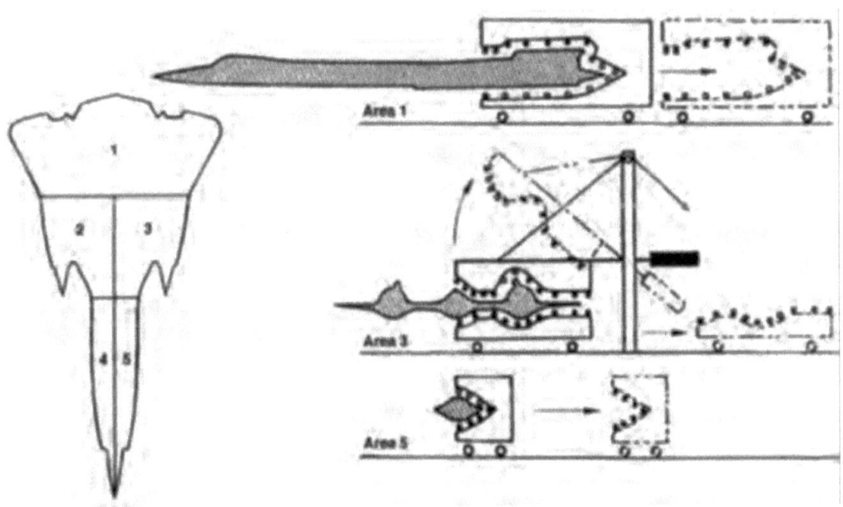

YF-12 thermal zones and High Temperature Loads Laboratory setup. NASA.

Lockheed's masterful Blackbird relied not only on lightweight titanium and a high fuel fraction for its long range but also a finely tuned propulsion system. At Mach 3, over 50 percent of the nacelle net thrust came from the inlet pressure rise, with the engine thrust being only on the order of 20 percent; the remainder came from the accelerated flow exiting the nozzle (given the small percentage of thrust from the engine, Lockheed designer Kelly Johnson used to good-naturedly joke that Pratt & Whitney's superb J58 functioned merely as an air pump for the nacelle at Mach 3 and above). It is not necessarily self-evident why the nozzle should produce such a large percentage of thrust, while the engine's contribution seems so little. The nozzle produces so much thrust because it accelerates the combined flow from the engine and inlet bypass air as it passes through the constricted nozzle throat at the rear of the nacelle. Engine designers concentrate only on the engine, regarding the nacelle inlet and exhaust as details that the airframe manufacturer provides. The low percentage numbers for the jet engine are because it produces less absolute net thrust the faster and higher it goes. Therefore, at the same time the engine thrust goes down (as the plane climbs to high altitude and accelerates to high Mach numbers), the percentage of net thrust from nonengine sources increases drastically (mainly because of inlet pressure buildup). Thus, static sea level thrust is the highest thrust an engine can produce. Integration of the propulsion system (i.e., matching the nacelle with the engine for optimum net thrust) is critical for efficient and economical supersonic cruise, as opposed to accelerating briefly through the speed of sound, which can be achieved by using (as the early Century series did) a "brute force" afterburner to boost engine power over airframe drag.

Air had to be bypassed around the engine to position shock waves properly for the pressure recovery. This bypass led to the added benefit of cooling the engine and to the system being referred to as a turbo ramjet. The NASA YF-12 inlets were instrumented, and much testing was devoted to investigation of the inlet/shock wave/engine interaction. Inlet unstarts in the YF-12 were even more noticeable and critical than they were in the B-70, as the nacelles were close to mid-span on the wing and the instantaneous loss of the inlet thrust led to violent yaws induced in the less massive aircraft. The Blackbirds had automatic inlet controls, unlike XB-70 AV-1, but they were analog control devices and were often not up to the task; operational crewmembers spent much time in the simulator practicing emergency manual control of the inlets. The NASA test sorties revealed that the inlet affected the flight performance of the aircraft during restart recovery. The excess spillage drag airflows from the unstarted inlets induced uncommanded rolling moments. This could result in a "falling leaf" effect at extreme altitudes, as the inlet control systems attempted to reposition the shock

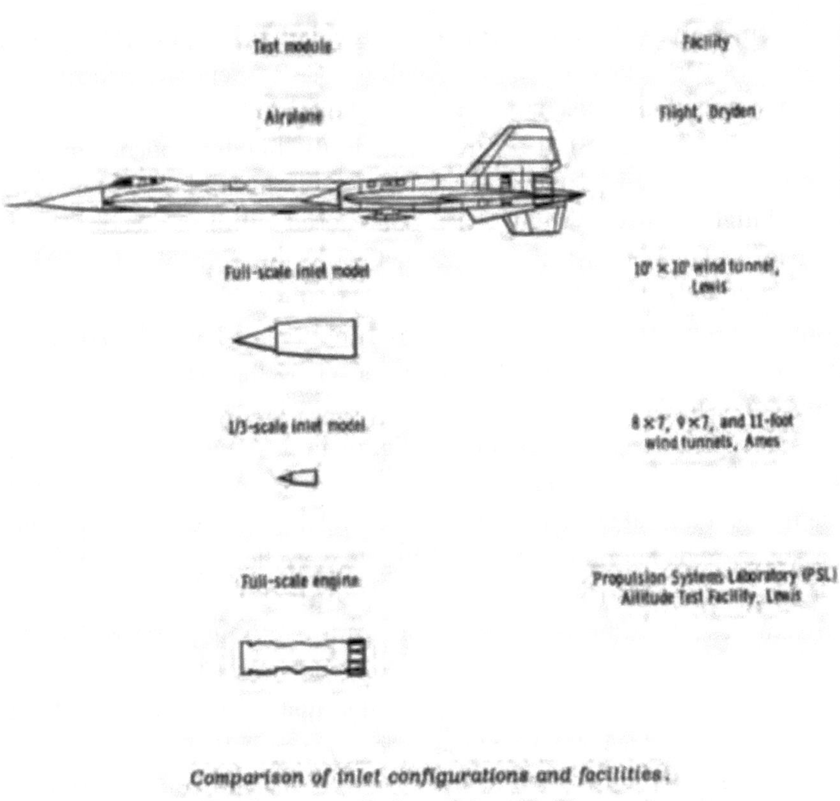

YF-12 NASA propulsion research assets. NASA.

wave properly by spike positioning and door movements.[50] This was an illustration of the strong interaction for a supersonic cruiser between aerodynamics and propulsion.

To further investigate this interaction, much research was dedicated to the inlet system of the YF-12.[51] A salvaged full-scale inlet was tested in a supersonic wind tunnel and a one-third-scale inlet. Using this approach, flight-test data could be compared with wind tunnel data to validate the tunnel data and adjust them as required.

The digital computer era was appearing, and NASA led the way in applying it to aeronautics. In addition to the digital fly-by-wire F-8 flight

50. Donald T. Berry and Glenn B. Gilyard, "A Review of Supersonic Cruise Flight Path Control," in *Proceedings of NASA Aircraft Safety and Operating Problems, Oct. 18–20, 1976,* NASA SP-416 (1977), pp. 147–164.

51. James A. Albers, "Status of YF-12 Propulsion Research," NASA TM-X-56039 (1976).

research, the YF-12 was also employing the digital computer. Originally, the Central Airborne Performance Analyzer (CAPA) general performance digital computer was used to monitor the behavior of the YF-12 Air Inlet Control System (AICS). It behaved well in the harsh airborne environment and provided excellent data. Based upon this and the progress in digital flight control systems, NASA partnered with Lockheed in 1975 to incorporate a Cooperative Airframe/Propulsion Control System computer on the YF-12C (the modified SR-71) that would perform the flight control system and propulsion control functions.[52] This was delivered in 1978, only shortly before the end of the YF-12 flight-test era. The system requirements dictated that pilot interface be transparent between the standard analog aircraft and the digital aircraft and that the aircraft system performance be duplicated digitally. The development included the use of a digital model of the flight controls and propulsion system for software development. Only 13 flights were flown in the 4 months remaining before program shutdown, but the flights were spectacularly successful. After initial developmental "teething problems" early in the program, the aircraft autopilot behavior was 10 times more precise than it was in the analog system, inlet unstarts were rare, and the aircraft exhibited a 5–7-percent increase in range.[53] Air Force test pilots flew the YF-12C on three occasions and were instrumental in persuading the USAF Logistics Command to install a similar digital system on the entire SR-71 fleet. The triple-redundant operational system—called Digital Automatic Flight and Inlet Control System (DAFICS)—was tested and deployed between 1980 and 1985 and exhibited similar benefits.

For the record, the author himself was a USAF SR-71 flight-test engineer and navigator/back-seater for the developmental test flights of DAFICs on the SR-71 and during the approximately 1-year test program experienced some 85 inlet unstarts! Several NASA research papers speculated on the effect of inlet unstarts on passenger, using anecdotal flight-test data from XB-70 and YF-12 flights. The author agreed with the comments in 1968 of XB-70 test pilot Fulton (who also flew the YF-12 for NASA) that he thought paying passengers in an SST would put up with an unstart exactly once. The author

52. Donald T. Berry and William G. Schweikhard, "Potential Benefits of Propulsion and Flight Control Integration for Supersonic Cruise Vehicles," in NASA TM-X-3409 (1974), pp. 433–452.
53. D.L. Anderson, G.F. Connolly, F.M. Mauro, and P.J. Reukauf, "YF-12 Cooperative Airframe/Propulsion Control System Program," vol. 1, NASA CR-163099 (1980), pp. 4-39–4-49.

also has several minutes of supersonic glider time because of dual inlet unstarts followed by dual engine flameouts, accompanied by an unrelated engine mechanical problem inhibiting engine restart. During a dual inlet unstart at 85,000 feet and subsequent emergency single engine descent to 30,000 feet, he experienced the "falling leaf" mode of flight, as the inlets cycled trying to recapture the shock waves within the inlet while the flight Mach number also oscillated. The problems during the test program indicated the sensitivity of the integration of propulsion with the airframe for a supersonic cruise aircraft. Once the in-flight "debugging" of the digital system had been accomplished, however, operational crews never experienced unstarts, except in the event of mechanical malfunction. One byproduct of the digital system was that the inlet setting software could be varied to account for differences in individual airframe inlets because of manufacturing tolerances. This even allowed inlet optimization by tail number.

Other YF-12 research projects were more connected with taking advantage of its high speed and high altitude as platforms for basic research experiments. One measured the increase in drag caused by an aft-facing "step" placed within the Mach 3 boundary layer. As well, researchers measured the thickness and flow characteristics of this turbulent region. The coldwall experiment was the most famous (or infamous).[54] It was a thermodynamics heat transfer experiment that took an externally mounted, insulated, cryogenically cooled cylinder to Mach 3 cruise and then exposed it to the high-temperature boundary layer by explosively stripping the insulation. Basic handling qualities investigations with the cylinder resulted in loss of the carrier YF-12A folding ventral fin. It was replaced with a newer material, producing a bonus materials experiment. When the experiment was finally cleared for deployment, it resulted in sending debris into the left inlet of the YF-12 carrier and unstarts in both inlets, not to mention multiple unstarts of the YF-12C chase aircraft. Both aircraft were grounded for over 6 weeks for inspections and repairs. Fortunately, the next 2 deployments with fewer explosives were more routine. An implicit lesson learned was that at Mach 3, the seemingly routine flight-test techniques may require careful review to ensure that they really are routine.

54. Merlin, *Mach 3+*, pp. 34–36.

Supersonic Cruise in the 1990s: SCR, Tu-144LL, F-16XL, and SR-71

NASA essentially resumed in 1990 what had ended in 1981 with the termination of the SCR program. Enough time had elapsed since the U.S. SST political firestorm to suggest the possibility of developing a practical aircraft.[55] Ironically, one of the justifications was concern that not only the Europeans but also the Japanese were studying a second-generation SST, one that could exploit reduced travel times to the Pacific rim countries, where U.S. overland sonic boom restrictions would not be such an economic handicap. A Presidential finding in 1986 during the Reagan Administration stated that research toward a supersonic commercial aircraft should be conducted. A consortium of NASA Research Centers continued research in conjunction with airframe manufacturers to work toward development of a High-Speed Civil Transport (HSCT), which would essentially become the 21st century SST. The development would incorporate lessons learned from previous SSTs and research conducted since 1981 and would be environmentally friendly. A test concept aircraft (TCA) configuration was established as a baseline for technology development studies. Cruise Mach number was to be Mach 2 to 2.5, and design range was to be 5,000 nautical miles, in deference to the Pacific Ocean traffic. Phase I of the SCR was to last 6 years, while concentrating on such environmental issues as studies on ozone layer impact of an SST fleet and sideline community noise levels. Both areas required extensive propulsion system studies and probable advances in engine technology. Studies of the economics of an HSCT showed that the concept would be more practical if there were a reduction of a sonic boom footprint to the point where overland flight was permissible in some corridors. The Concorde boom average was 2 pounds per square foot, which was deemed unacceptable; the questions were what would be acceptable and how to achieve that level. Phase II was to be focused on development of specific technologies leading to HSCT as a practical commercial aircraft. The initial goal was for a 2006 development decision target date.

The digital revolution has had a major impact on supersonic technology. The nonlinear physics of supersonic flow shock waves made control of a system difficult. But the advent of high-speed computer technology changed that. The improvement in the SR-71 fleet performance shown by the DAFICS, pioneered by NASA in the YF-12 program, showed the

55. Chambers, *Innovation in Flight,* pp. 40–61.

Baseline High-Speed Civil Transport (HSCT) for NASA SCR. NASA.

operational benefits of digital controls. But in SCR, much effort centered on using the computational fluid dynamics (CFD) codes being developed to perform design tasks that traditionally required massive wind tunnel testing.[56] CFD could also be used to predict sonic boom propagation for configurations, once the basic physics of that propagation was better

56. T. Edwards, et al., "Sonic Boom Prediction and Minimization Using Computational Fluid Dynamics," in NASA, *First Annual High Speed Research Workshop, May 14–16, 1991, Williamsburg, VA* (1991).

Tupolev Tu-144LL flying laboratory. NASA.

understood. Another case study in this book addresses the details of the research that was conducted to provide that data. Flight tests included flights by an SR-71 over an instrumented ground array of microphones as in the 1960s that were also accompanied by instrumented chase aircraft that recorded the shock wave characteristics in free space at various distances from the supersonic aircraft. These data were to be used to develop and validate the CFD predictions, just as supersonic flight-test data has traditionally been used to validate supersonic wind tunnel predictions.

Another flight research program devoted to SCR included a post–Cold War cooperative venture with Russia's Central Institute of Aerohydromechanics (TsAGI) to resurrect and fly the Tu-144 SST of the 1970s.[57] Equipped with new engines with more powerful turbofans, the Tu-144LL (the modified designation reflecting the Cyrillic abbreviation for flying laboratory) flew a 2-phase, 26-flight-test program in 1998 and 1999 at cruise Mach numbers to 2.15. All the flights were flown from Zhukovsky Flight Research Center outside Moscow, and NASA pilots flew on 3 of the sorties.[58] Experiments investigated handling qualities, boundary layer characteristics, ground cushion effects of the large delta wing, cabin aerodynamic noise, and sideline engine noise.

57. Tu-144LL testing is the subject of a subsequent case study in this volume.

58. Timothy H. Cox and Alisa Marshall, "Longitudinal Handling Qualities of the Tu-144LL Airplane and Comparisons With Other Large, Supersonic Aircraft," NASA TM-2000-209020 (2000).

NASA F-16XL modified for Supersonic Laminar Flow Control program. The right wing is the normal arrow wing configuration, while the left wing has the LFC glove extending from the fuselage to the mid-span sweep "kink." NASA.

Another flight research program of the 1990s was the NASA use of the arrow wing F-16XL. Flown over 13 months in 1995–1996, the 90-hour, 45-flight-test program was known as the Supersonic Laminar Flow Control program.[59] A glove was fitted over the left wing of the aircraft, which had millions of microscopic laser-drilled holes. A suction system drew the turbulent supersonic boundary layer through the holes to attempt to create a laminar boundary layer with less friction drag. Flight Mach numbers up to Mach 2 showed that the concept was indeed effective at creating laminar flow. This was a significant finding for an HSCT, for which drag reduction at cruise conditions is so critical.[60]

The USAF had taken the SR-71 fleet out of service in 1990 because of cost concerns and opinions that its reconnaissance mission could be

59. Bianca T. Anderson and Marta Bohn-Meyer, "Overview of Supersonic Laminar Flow Control Research on F-16XL Ships 1 and 2," NASA TM-104257 (1992).

60. A.G. Powell "Supersonic LFC: Challenges and Opportunities," in NASA, *First Annual High Speed Research Workshop, May 14–16, 1991, Williamsburg, VA* (1991), p. 1824.

better accomplished by other platforms, including satellites. This freed a number of Mach 3 cruise platforms equipped with advanced digital control systems for possible use by NASA in the SCR effort. Dryden Flight Research Center was allocated two SR-71As for research use and the sole SR-71B airframe for pilot checkout training. The crew-training simulator was also installed at Dryden. It was being updated to new computer technology when the financial ax fell yet again. Some research relevant to supersonic cruise was performed on the SR-71s. Handling qualities and cruise performance using the updated configuration were evaluated. Despite the digital system, the use of an inertial vertical velocity indicator at Mach 3 was still found to be superior to the air-data-driven vertical velocity for precise altitude control.[61] An experimental air-data system using lasers to sense angle of attack and sideslip rather than differential air pressure was also tested to confirm that it would function at the 80,000-foot cruise altitude. Several Sonic Boom Research Program flights were flown, as mentioned earlier, for in-flight sonic boom shock wave measurements. The SR-71 had to slow and descend from its normal cruise levels to accommodate the instrumented chase aircraft. Like the YF-12, the SR-71 was again used as a platform for experiments. Several devices planned for satellite Earth observations were carried in the sensor bays of the SR-71 for observations from above 95 percent of the Earth's atmosphere. An ultraviolet camera funded by the Jet Propulsion Laboratory conducted celestial observations from the same vantage point.

The program that mainly funded retention of the airplanes was actually in support of a proposed (later canceled) reuseable space launch vehicle, the Lockheed-Martin X-33. It would employ a revolutionary rocket engine, the Linear Aerospike Rocket Engine (LASRE). The engine used shock waves to contain the exhaust and increase thrust at a comparatively light structural weight. For risk reduction, the SR-71 would have a fixture mounted atop the fuselage, on which would be installed a 12-percent model of the X-33 with engine for aerial tests. The fixture was installed, but the increased drag of the fixture plus the LASRE limited the maximum Mach number attainable to around Mach 2. The installation was carried on several flights, but insuperable flight safety issues

61. Timothy H. Cox, and Dante Jackson, "Supersonic Flying Qualities Experience Using the SR-71," NASA TM-4800 (1997), pp. 4–5.

meant that the engine never was fired on the aircraft.[62] Funding ended with the demise of the SCR program. The final flight of the world's only Mach 3 supersonic cruise fleet occurred as an overflight of the Edwards Air Force Base Open House on October 9, 1999. The staff of the Russian Test Pilot School furnished an indication of the unique cachet of the aircraft when they visited the USAF Test Pilot School at Edwards as part of a reciprocal exchange in the mid-1990s. They had earlier hosted the Americans in Moscow and allowed them to fly current Russian fighters. When asked what they would like to fly at Edwards, the response was the SR-71. They were told that was unfortunately impossible because of cost and because the SR was a NASA asset, but that a simulator flight might be arranged. Even so, these experienced test pilots welcomed the opportunity to sample the SR-71 simulator.

By 1999, much research work had been performed in support of the HSCT.[63] Nevertheless, no breakthrough seemed to have been made that answered all the issues raised on a practical HSCT development decision. One of the major contractor contributors had been McDonnell-Douglas, which became Boeing in the defense industry implosion of the 1990s.[64] In 1999, Boeing withdrew further major financial support, as it saw no possibility of an HSCT before 2020. Also in 1999, NASA Administrator Daniel S. Goldin cut $600 million from the aeronautics budget to provide support for the International Space Station. These two actions essentially ended the SCR for the time being.

62. Stephen Corda, Bradford A. Neal, Timothy R. Moes, Timothy H. Cox, Richard C. Monaghan, Leonard S. Voelker, Griffin P. Corpening, and Richard R. Larson "Flight Testing the Linear Aerospike SR-71 Experiment (LASRE)," NASA TM-1998-206567 (1998).
63. M. Leroy Spearman, "The Evolution of the High-Speed Civil Transport," NASA TM-109089 (1994) is an excellent configuration survey. See also A. Warner Robins, et al., "Concept Development of a Mach 3.0 High-Speed Civil Transport," NASA TM-4058 (1988); P.G. Parikh and A.L. Nagel, "Application of Laminar Flow Control to Supersonic Transport Configurations," NASA CR-181917 (1990); Christopher D. Domack, et al., "Concept Development of a Mach 4 High-Speed Civil Transport," NASA TM-4223 (1990); T. Edwards, et al., "Sonic Boom Prediction and Minimization Using Computational Fluid Dynamics," in NASA, *First Annual High Speed Research Workshop, May 14–16, 1991, Williamsburg, VA* (1991); A.G. Powell, "Supersonic LFC: Challenges and Opportunities," in NASA, *First Annual High Speed Research Workshop, May 14–16, 1991, Williamsburg, VA* (1991); Bianca T. Anderson and Marta Bohn-Meyer, "Overview of Supersonic Laminar Flow Control Research on F-16XL Ships 1 and 2," NASA TM-104257 (1992).
64. National Science and Technology Council, Executive Office of the President, "Report to the Congress: Impact of the Termination of NASA's High Speed Research Program and The Redirection of NASA's Advanced Subsonic Technology Program," p. 6.

Into the 21st Century

In 2004, NASA Headquarters Aeronautics Research Mission Directorate (ARMD) formed the Vehicle Systems Program (VSP) to preserve core supersonic research capabilities within the Agency.[65] As the program had limited funding, much of the effort concentrated on cooperation with other organizations, notably the Defense Advanced Research Projects Agency (DARPA) and the military. Likely configuration studies pointed toward business jets as being a more likely candidate for supersonic travelers than full-size airliners. More effort was devoted to cooperation with DARPA on the sonic boom problem. An earlier joint program resulted in the shaped sonic boom demonstration of 2003, when a Northrop F-5 fighter with a forward fuselage modified to reduce the type's characteristic sonic boom signature demonstrated that the modification worked.[66]

Military aircraft have traversed the sonic regime so frequently that one can hardly dignify it with the name "frontier" that it once had.

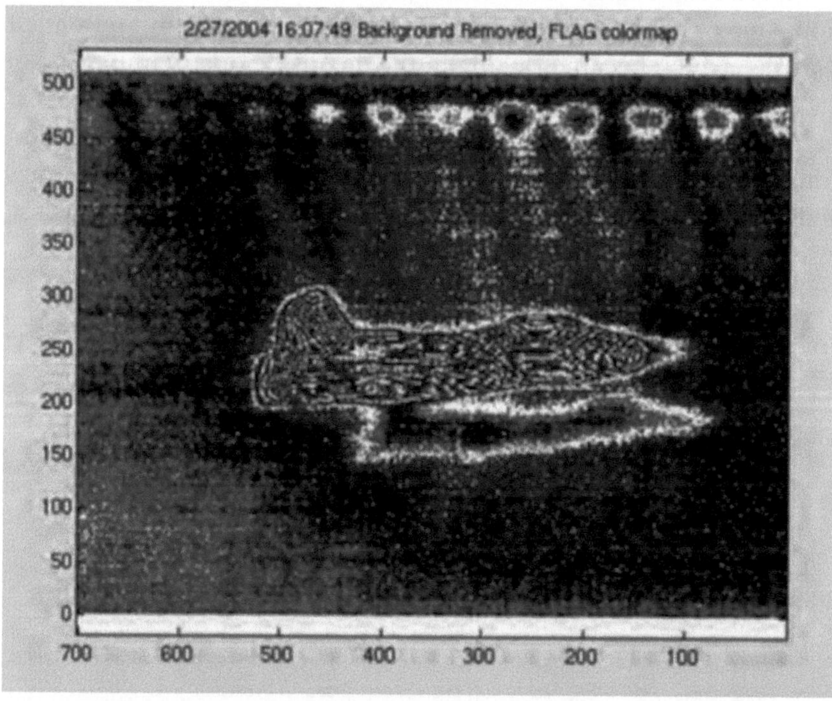

Among the supersonic cruise flight-test research tools, circa 2007, was thermal imagery. NASA.

65. Chambers, *Innovation in Flight,* pp. 62–68.

66. The experiment is detailed in another case study within this volume.

In-flight Schlieren imagery. NASA.

10

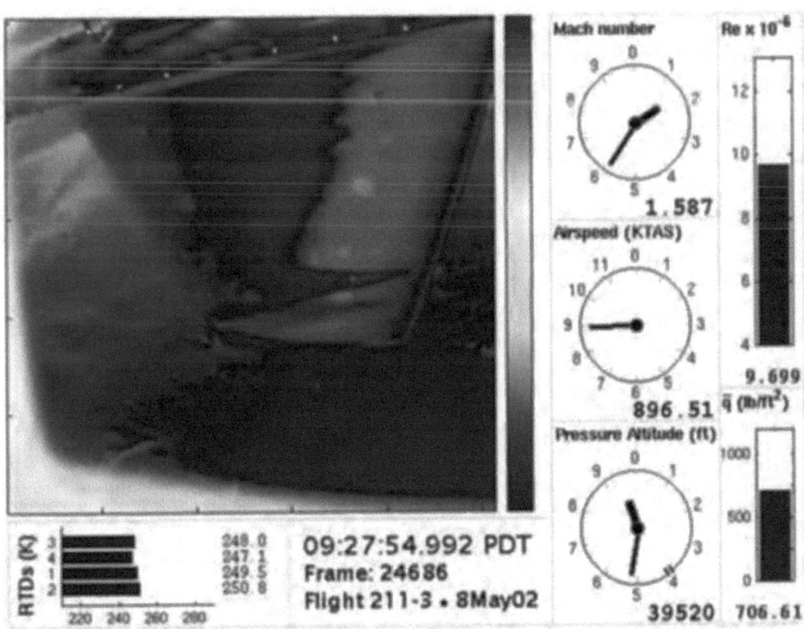

In-flight thermography output. NASA.

Nevertheless, there have been few supercruising aircraft: the SR-71, the Concorde, the Tu-144, and the F-22A constituting notable exceptions. The operational experience gained with the SR-71 fleet with its DAFICS in the 1980s, and the more recent Air Force experience with the low-observable supercruising Lockheed-Martin F-22A Raptor, indicate that a properly designed aircraft with modern digital systems makes high Mach supersonic cruise now within reach technologically. Indeed, at a November 2007 Langley Research Center presentation at the annual meeting of the Aeronautics Research Mission Directorate reflected that although no supersonic cruise aircraft is lying, digital simulation capabilities, advanced test instrumentation, and research tools developed in support of previous programs are nontrivial legacies of the supersonic cruise study programs, positioning NASA well for any nationally identified supersonic cruise aircraft requirement. Whether that will occur in the near future remains to be seen, just as it has since the creation of NASA a half century ago, but one thing is clear: the more than three decades of imaginative NASA supersonic cruise research after cancellation of the SST have produced a technical competency permitting, if needed, design for routine operation of a high Mach supersonic cruiser.[67]

67. Dan Banks, "Overview of Experimental Capabilities," in *NASA Fundamental Aeronautics 2007 Annual Meeting, Oct. 30–Nov. 1, 2007, New Orleans, LA.*

Recommended Additional Reading

Reports, Papers, Articles, and Presentations:

James A. Albers, "Status of YF-12 Propulsion Research," NASA TM-X-56039 (1976).

R.D. Allan and W. Joy, "Advanced Supersonic Propulsion Study, Phases 3 and 4," NASA CR-13526 (1978).

Bianca T. Anderson and Marta Bohn-Meyer, "Overview of Supersonic Laminar Flow Control Research on F-16XL Ships 1 and 2," NASA TM-104257 (1992).

D.L. Anderson, G.F. Connolly, F.M. Mauro, and P.J. Reukauf, "YF-12 Cooperative Airframe/Propulsion Control System Program," vol. 1, NASA CR-163099 (1980).

H.H. Arnaiz, J.B. Peterson, Jr., and J.C. Daugherty, "Wind-tunnel/flight correlation study of aerodynamic characteristics of a large flexible supersonic cruise airplane (XB-70-1). Part 3: A comparison between characteristics predicted from wind-tunnel measurements and those measured in flight," NASA TP-1516 (1980).

D.D. Baals, "Summary of Initial NASA SCAT Airframe and Propulsion Concepts," in NASA LRC, *Proceedings of NASA Conference on Supersonic-Transport Feasibility Studies and Supporting Research*, NTIS N67-31606 (1963).

D.D. Baals, O.G. Morris, and T.A. Toll, "Airplane Configurations for Cruise at a Mach Number of 3," NASA LRC, *Conference on High Speed Aerodynamics*, NTIS 71N5324 (1958).

Donald T. Berry and Glenn B. Gilyard, "A Review of Supersonic Cruise Flight Path Control," *Proceedings of NASA Aircraft Safety and Operating Problems, Oct. 18–20, 1976*, NASA SP-416 (1977).

Donald T. Berry and William G. Schweikhard, "Potential Benefits of Propulsion and Flight Control Integration for Supersonic Cruise Vehicles," NASA TM-X-3409 (1974).

10

E. Bonner, et al., "Influence of Propulsion System Size, Shape, and Location on Supersonic Aircraft Design," NASA CR-132544 (1974).

S.C. Brown, "Computer Simulation of Aircraft Motions and Propulsion System Dynamics for the YF-12 Aircraft at Supersonic Cruise Conditions," NASA TM-X-62245 (1973).

J.S. Butz, Jr., "FAA, NASA Study Supersonic Transport," *Aviation Week and Space Technology* (May 2, 1960).

J.S. Butz, Jr., "Industry Studies Transport Area Rule," *Aviation Week and Space Technology* (July 14, 1958), pp. 48–52.

M.M. Carmel, A.B. Carraway, and D.T. Gregory, "An Exploratory Investigation of a Transport Configuration Designed for Supersonic Cruise Flight Near a Mach Number of 3," NASA TM-X-216 (1960).

P.A. Cooper and R.R. Heldenfels, "NASA Research on Structures and Materials for Supersonic Cruise Aircraft," NASA TM-X-72790 (1976).

Stephen Corda, et al. "Flight Testing the Linear Aerospike SR-71 Experiment (LASRE)," NASA TM-1998-206567 (1998).

Timothy H. Cox and Dante Jackson, "Supersonic Flying Qualities Experience Using the SR-71," NASA TM-4800 (1997).

Timothy H. Cox and Alisa Marshall, "Longitudinal Handling Qualities of the Tu-144LL Airplane and Comparisons With Other Large, Supersonic Aircraft," NASA TM-2000-209020 (2000).

Christopher D. Domack, et al., "Concept Development of a Mach 4 High-Speed Civil Transport," NASA TM-4223 (1990).

T. Edwards, et al., "Sonic Boom Prediction and Minimization Using Computational Fluid Dynamics," in NASA, *First Annual High Speed Research Workshop, May 14–16, 1991, Williamsburg, VA* (1991).

L.J. Ehernberger, "High Altitude Turbulence for Supersonic Cruise Vehicles," NASA TM-88285 (1987).

D.E. Fetterman, Jr., "Preliminary Sizing and Performance Evaluation of Supersonic Cruise Aircraft," NASA TM-X-73936 (1976).

Donald S. Findley, Vera Huckel, and Herbert R. Henderson, "Vibration Responses of Test Structure No. 1 During the Edwards AFB Phase of the National Sonic Boom Program," NASA TM-72706 (1975).

J.E. Fischler, "Structural Design of Supersonic Cruise Aircraft," NTRS 77N18041 (1976).

Fitzhugh L. Fulton, Jr., "Lessons from the XB-70 as Applied to the Supersonic Transport," NASA TM-X-56014 (1968).

Glenn Garrison, "Supersonic Transport May Aim at Mach 3," *Aviation Week and Space Technology* (Feb. 2, 1959), pp. 38–40.

G.B. Gilyard and J.J. Burken, "Development and Flight Test Results of an Autothrottle Control System at Mach 3 Cruise," NASA TP-1621 (1980).

Sherwood Hoffman, "Bibliography of Supersonic Cruise Aircraft Research (SCAR) Program, 1972-mid 1977," NASA RP-1003 (1977).

Sherwood Hoffman, "Bibliography of Supersonic Cruise Research (SCR) Program from 1980 to 1983," NASA RP-1117 (1984).

Sherwood Hoffman, "Supersonic Cruise Research (SCR) Program Publications FY 1977–1979: Preliminary Bibliography," NASA TM-80184 (1979).

Robert Hotz, "Supersonic Transport Race," *Aviation Week and Space Technology* (Nov. 14, 1960), p. 21.

Donald L. Hughes, Bruce G. Powers, and William H. Dana, "Flight Evaluation of Some Effects of the Present Air Traffic Control System on Operation of a Simulated Supersonic Transport," NASA TN-D-2219 (1964).

Jerald M. Jenkins and Robert D. Quinn, "A Historical Perspective of the YF-12A Thermal Loads and Structures Program," NASA TM-104317 (1996).

C.M. Lee and R.W. Niedzwiecki, "High-Speed Commercial Transport Fuels Considerations and Research Needs," NASA TM-102535 (1989).

John G. Lowry, "Summary and Assessment of Feasibility Studies," in NASA LRC, *Proceedings of NASA Conference on Supersonic-Transport Feasibility Studies and Supporting Research,* NTIS N67-31606 (1963).

Domenic J. Maglieri. et al., "A Summary of XB-70 Sonic Boom Signature Data," NASA CR-189630 (1992).

Domenic J. Maglieri, David A. Hilton, and Norman J. McLeod, "Summary of Variations of Sonic Boom Signatures Resulting from Atmospheric Effects," NASA TM-X-59633 (1967).

V.R. Mascitti, "Aerodynamic Performance Studies for Supersonic Cruise Aircraft," NASA TM-X-73915 (1976).

V.R. Mascitti, "Systems Integration Studies for Supersonic Cruise Aircraft," NASA TM-X-72781 (1975).

Eugene A. Morelli, "Low Order Equivalent System Identification for the Tu-144LL Supersonic Transport Aircraft," AIAA Paper 2000-3902 (2000).

NASA, *Proceedings of the SCAR [Supersonic Cruise Aircraft Research] Conference*, pt. 1, NASA LRC, NASA CP-001-PT-1 (1976).

NASA, *Proceedings of the SCAR [Supersonic Cruise Aircraft Research] Conference*, pt. 2, NASA LRC, NASA CP-001-PT-2 (1976).

NASA, *Supersonic Cruise Research 1979*, pt. 1, NASA CP-2108-PT-1 (1980).

NASA, *Supersonic Cruise Research 1979*, pt. 2, NASA CP-2108-PT-2 (1980).

NASA, "High-Speed Civil Transport Study," NASA CR-4233 (1989).

NASA, "High-Speed Civil Transport Study: Summary," NASA CR-4234 (1989)

NASA, "Study of High-Speed Civil Transports," NASA CR-4235 (1989).

North American Rockwell, Space Division, "B-70 Aircraft Study Final Report," SD 72-SH-0003, vols. 2 and 4 (1972).

P.G. Parikh and A.L. Nagel, "Application of Laminar Flow Control to Supersonic Transport Configurations," NASA CR-181917 (1990).

A.G. Powell "Supersonic LFC: Challenges and Opportunities," in NASA, *First Annual High Speed Research Workshop, May 14–16, 1991, Williamsburg, VA* (1991), p. 1824.

S.M.B. Rivers, R.A. Wahls, and L.R. Owens, "Reynolds Number Effects on Leading Edge Radius Variations of a Supersonic Transport at Transonic Conditions," AIAA Paper 2001-2462 (2001).

A. Warner Robins, et al., "Concept Development of a Mach 3.0 High-Speed Civil Transport," NASA TM-4058 (1988).

J.W. Ross and D.B. Rogerson, "Technological Advancements of XB-70," AIAA Paper 83-1048 (1983).

I.F. Sakata and G.W. Davis, "Evaluation of Structural Design Concepts for an Arrow-Wing Supersonic Cruise Aircraft," NASA CR-2667 (1977).

Barrett L. Schout, et al., "Review of NASA Supercruise Configuration Studies," paper presented at *Supercruise Military Aircraft Design Conference, U.S. Air Force Academy, Colorado Springs, CO, Feb. 17–20, 1976.*

R.J. Shaw, "Overview of Supersonic Cruise Propulsion Research," NASA Glenn [Lewis] Research Center, NTRS 91N20088 (1991).

C.L. Smith and L.J. Williams, "An Economic Study of an Advanced Technology Supersonic Cruise Vehicle," NASA TM-X-62499 (1975).

M. Leroy Spearman, "The Evolution of the High-Speed Civil Transport," NASA TM-109089 (1994).

William C. Strack and Shelby J. Morris, Jr., "The Challenges and Opportunities of Supersonic Transport Propulsion Technology," NASA TM-100921 (1988).

R.W. Sudderth, et al., "Development of Longitudinal Handling Qualities Criteria for Large Advanced Supersonic Aircraft," NASA CR-137635 (1975).

John Tunstall, "British Weigh Entering Supersonic Race," *Aviation Week and Space Technology* (May 4, 1959).

M.J. Turner and D.L. Grande, "Study of Advanced Composite Structural Design Concepts for an Arrow Wing Supersonic Cruise Configuration," NASA CR-2825 (1978).

Richard T. Whitcomb, "A Study of the Zero-Lift Drag Rise Characteristics of Wing-Body Combinations Near the Speed of Sound," NACA RM-L52H08 (1952).

Chester H. Wolowicz, "Analysis of an Emergency Deceleration and Descent of the XB-70-1 Airplane Due to Engine Damage Resulting from Structural Failure," NASA TM-X-1195 (1966).

B.R. Wright, F. Bruckman, et. al., "Arrow Wings for Supersonic Cruise Aircraft," *Journal of Aircraft*, vol. 15, no. 12 (Dec. 1978), pp. 829–836.

John H. Wykes, et al., "XB-70 Structural Mode Control System Design and Performance Analyses," NASA CR-1557 (1970).

Bill Yenne, "America's Supersonic Transports," *Flightpath*, vol. 3 (2004), pp. 146–157.

Books and Monographs:

Donald Baals and William Corliss, *Wind Tunnels of NASA*, NASA SP-440 (Washington, DC: GPO, 1981).

Roger E. Bilstein, *Orders of Magnitude: A History of the NACA and NASA, 1915–1990*, NASA SP-4406 (Washington, DC: GPO, 1989).

Joseph R. Chambers, *Innovations in Flight: Research of the NASA Langley Flight Research Center on Revolutionary Concepts for Advanced Aeronautics*, NASA SP-2005-4539 (Washington, DC: GPO 2005), pp. 39–48.

Joseph R. Chambers, *Partners in Freedom: Contributions of the Langley Research Center to U.S. Military Aircraft of the 1990s*, NASA SP-2000-4519 (Washington, DC: GPO, 2000).

Erik M. Conway, *High-Speed Dreams: NASA and the Technopolitics of Supersonic Transportation, 1945–1999* (Baltimore: The Johns Hopkins University Press, 2005).

Richard P. Hallion, *On the Frontier: Flight Research at Dryden, 1946–1981*, NASA SP-4303 (Washington, DC: GPO, 1984).

Mel Horwitch, *Clipped Wings: The American SST Conflict* (Cambridge: The MIT Press, 1982).

James R. Hansen, *Engineer in Charge*, NASA SP-4305 (Washington, DC: GPO, 1987).

Edwin P. Hartman, *Adventures in Research: A History of Ames Research Center 1940–1965*, NASA SP-4302 (Washington, DC: GPO, 1970).

Dennis R. Jenkins and Tony R. Landis, *Valkyrie: North American's Mach 3 Superbomber* (North Branch, MN: Specialty Press, 2004).

Laurence K. Loftin, Jr., *Quest for Performance: The Evolution of Modern Aircraft*, NASA SP-468 (Washington, DC: GPO, 1965).

F. Edward McLean, *Supersonic Cruise Technology*, NASA SP-472 (Washington, DC: GPO, 1985).

Peter W. Merlin, *Mach 3+ NASA/USAF YF-12 Flight Research 1969–1979*, NASA SP-2001-4525 (Washington, DC: GPO, 2002).

Kenneth Owen, *Concorde: Story of a Supersonic Pioneer* (London: Science Museum, 2001).

William A. Shurcliff, *SST and Sonic Boom Handbook* (New York: Ballantine Books, 1970).

Joe Sutter, with Jay Spenser, *747: Creating the World's First Jumbo Jet and Other Adventures from a Life in Aviation* (Washington, DC: Smithsonian Books, 2006).

THIS PAGE INTENTIONALLY BLANK

NASA synthetic vision research promises to increase flight safety by giving pilots perfect positional and situation awareness, regardless of weather or visibility conditions. Richard P. Hallion.

CASE 11

Introducing Synthetic Vision to the Cockpit

Robert A. Rivers

The evolution of flight has witnessed the steady advancement of instrumentation to furnish safety and efficiency. Providing revolutionary enhancements to aircraft instrument panels for improved situational awareness, efficiency of operation, and mitigation of hazards has been a NASA priority for over 30 years. NASA's heritage of research in synthetic vision has generated useful concepts, demonstrations of key technological breakthroughs, and prototype systems and architectures.

THE CONNECTION OF THE NATIONAL AERONAUTICS AND SPACE ADMINISTRATION (NASA) to improving instrument displays dates to the advent of instrument flying, when James H. Doolittle conducted his "blind flying" experiments with the Guggenheim Flight Laboratory in 1929, in the era of the Ford Tri-Motor transport.[1] Doolittle became the first pilot to take off, fly, and land entirely by instruments, his visibility being totally obscured by a canvas hood. At the time of this flight, Doolittle was already a world-famous airman, who had earned a doctorate in aeronautical engineering from the Massachusetts Institute of Technology and whose research on accelerations in flight constituted one of the most important contributions to interwar aeronautics. His formal association with the National Advisory Committee for Aeronautics (NACA) Langley Aeronautical Laboratory began in 1928. In the late 1950s, Doolittle became the last Chairman of the NACA and helped guide its transition into NASA.

The capabilities of air transport aircraft increased dramatically between the era of the Ford Tri-Motor of the late 1920s and the jetliners

1. The author gratefully acknowledges the assistance provided by Louis J. Glaab of Langley and Jeffrey L. Fox of Johnson in providing notes, documents, and interviews. Other valuable assistance was provided by Langley's Lynda J. Kramer, Jarvis J. Arthur, III, and Monica F. Hughes, and by Michael F. Abernathy of Rapid Imaging Software, Inc. This chapter honors the numerous dedicated NASA researchers and technicians whose commitment to the ideals of NASA aeronautics has resulted in profound advancements in the aviation industry.

of the late 1960s. Passenger capacity increased thirtyfold, range by a factor of ten, and speed by a factor of five.[2] But little changed in one basic area: cockpit presentations and the pilot-aircraft interface. As NASA Ames Research Center test pilot George E. Cooper noted at a seminal November 1971 conference held at Langley Research Center (LaRC) on technologies for future civil aircraft:

> Controls, selectors, and dial and needle instruments which were in use over thirty years ago are still common in the majority of civil aircraft presently in use. By comparing the cockpit of a 30-year-old three-engine transport with that of a current four-engine jet transport, this similarity can be seen. However, the cockpit of the jet transport has become much more complicated than that of the older transport because of the evolutionary process of adding information by more instruments, controls, and selectors to provide increased capability or to overcome deficiencies. This trend toward complexity in the cockpit can be attributed to the use of more complex aircraft systems and the desire to extend the aircraft operating conditions to overcome limitations due to environmental constraints of weather (e.g., poor visibility, low ceiling, etc.) and of congested air traffic. System complexity arises from adding more propulsion units, stability and control augmentation, control automation, sophisticated guidance and navigation systems, and a means for monitoring the status of various aircraft systems.[3]

Assessing the state of available technology, human factors, and potential improvement, Cooper issued a bold challenge to NASA and the larger aeronautical community, noting: "A major advance during the 1970s must be the development of more effective means for systematically evaluating the available technology for improving the pilot-aircraft interface if major innovations in the cockpit are to be obtained during

2. See statistics for the Ford Tri-Motor in Kenneth Munson, *Airliners Between the Wars, 1919–1939* (New York: The Macmillan Co., 1972), pp. 54 and 140; and statistics for the Boeing 747 in Kenneth Munson, *Airliners Since 1946* (New York: The Macmillan Co., 1972), pp. 95 and 167.
3. George E. Cooper, "The Pilot-Aircraft Interface," in NASA LRC, *Vehicle Technology for Civil Aviation: The Seventies and Beyond,* NASA SP-292 (1971), pp. 271–272.

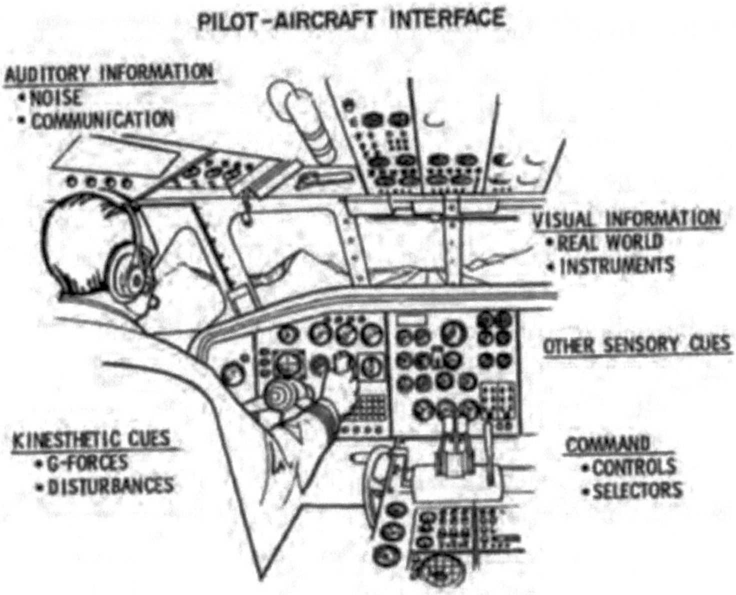

The pilot-aircraft interface, as seen by NASA pilot George E. Cooper, circa 1971. Note the predominance of gauges and dials. NASA.

Cooper's concept of an advanced multifunction electronic cockpit. Note the flightpath "highway in the sky" presentation. NASA.

the 1980s."[4] To illustrate his point, Cooper included two drawings, one representative of the dial-intensive contemporary jetliner cockpit presentation and the other of what might be achieved with advanced multifunction display approaches over the next decade.

11

At the same conference, Langley Director Edgar M. Cortright noted that, in the 6 years from 1964 through 1969, airline costs incurred by congestion-caused terminal area traffic delays had risen from less than $40 million to $160 million per year. He said that it was "symptomatic of the inability of many terminals to handle more traffic," but that "improved ground and airborne electronic systems, coupled with acceptable aircraft characteristics, would improve all-weather operations, permit a wider variety of approach paths and closer spacing, and thereby increase airport capacity by about 100 percent if dual runways were provided."[5] Langley avionics researcher G. Barry Graves noted the potentiality of revolutionary breakthroughs in cockpit avionics to improve the pilot-aircraft interface and take aviation operations and safety to a new level, particularly the use of "computer-centered digital systems for both flight management and advanced control applications, automated communications, [and] systems for wide-area navigation and surveillance."[6]

But this early work generated little immediate response from the aviation community, as requisite supporting technologies were not sufficiently mature to permit their practical exploitation. It was not until the 1980s, when the pace of computer graphics and simulation development accelerated, that a heightened interest developed in improving pilot performance in poor visibility conditions. Accordingly, researchers increasingly studied the application of artificial intelligence (AI)

4. Ibid., p. 277.

5. E.M. Cortright, "Vehicle Technology for Civil Aviation: The Seventies and Beyond—Keynote Address," in NASA LRC, *Vehicle Technology for Civil Aviation*, p. 1, and Figure 4, p. 8. The value of $40 million in 1964 is approximately $279 million in 2009, and $160 million in 1969 is approximately $942 million in 2009.

6. G. Barry Graves, Jr., "Advanced Avionic Systems," in NASA LRC, *Vehicle Technology for Civil Aviation*, p. 287; see also J.P. Reeder, "The Airport-Airplane Interface: The Seventies and Beyond," in the same work, pp. 259–269. The idea of dynamic and intelligent flight guidance displays began with early highway-in-the-sky research by the United States Navy's George W. Hoover in the 1950s; see Joseph R. Chambers, *Innovation in Flight: Research of the NASA Langley Research Center on Revolutionary Advanced Concepts for Aeronautics*, NASA SP-2005-4539 (2005), p. 99. Chambers presents a thorough summary of the history of SVS research at NASA Langley through the end of 2005.

to flight deck functions, working closely with professional pilots from the airlines, military, and flight-test community. While many exaggerated claims were made—given the relative immaturity of the computer and AI field at that time—researchers nevertheless recognized, as Sheldon Baron and Carl Feehrer wrote, "one can conceive of a wide range of possible applications in the area of intelligent aids for flight crew."[7] Interviews with pilots revealed that "descent and approach phases accounted for the greatest amounts of workload when averaged across all system management categories," stimulating efforts to develop what was then popularly termed a "pilot's associate" AI system.[8]

In this growing climate of interest, John D. Shaughnessy and Hugh P. Bergeron's Single Pilot Instrument Flight Rules (SPIFR) project constituted a notable first step, inasmuch as SPIFR's novel "follow-me box" showed promise as an intuitive aid for inexperienced pilots flying in instrument conditions. Subsequently, Langley's James J. Adams conducted simulator evaluations of the display, confirming its potential.[9] Building on these "follow-me box" developments, Langley's Eric C. Stewart developed a concept for portraying an aircraft's current and future desired positions. He created a synthetic display similar, to the scene a driver experiences while driving a car, combining it with

7. Sheldon Baron and Carl Feehrer, "An Analysis of the Application of AI to the Development of Intelligent Aids for Flight Crew Tasks," NASA CR-3944 (1985). See also Richard M. Hueschen and John W. McManus, "Application of AI Methods to Aircraft Guidance and Control," in *Proceedings of the 7th American Control Conference, June 15–17, 1988*, vol. 1 (New York: IEEE, 1988), pp. 195–201.
8. Ibid. For early research, see J.J. Adams, et al., "Description and Preliminary Studies of a Computer Drawn Instrument Landing Approach Display," NASA TM-78771 (1978); D. Warner, "Flight Path Displays," USAF Flight Dynamics Laboratory, Report AFFDL-TR-79-3075 (1979); A.J. Grunwald, et al., "Evaluation of a Computer-Generated Perspective Tunnel Display for Flight Path Following," NASA TP-1736 (1980); Richard M. Hueschen, et al., "Guidance and Control System Research for Improved Terminal Area Operations," in Joseph W. Stickle, ed., *1980 Aircraft Safety and Operating Problems*, NASA CP-2170, pt. 1 (1981), pp. 51–61; Adams, "Simulator Study of a Pictorial Display for General Aviation Instrument Flight," NASA TP-1963 (1982); Adams, "Flight-Test Verification of a Pictorial Display for General Aviation Instrument Approach," NASA TM-83305 (1982); Adams, "Simulator Study of Pilot-Aircraft-Display System Response Obtained with a Three-Dimensional-Box Pictorial Display," NASA TP-2122 (1983); J. Atkins, "Prototypical Knowledge for Expert Systems," *Artificial Intelligence*, vol. 20, no. 2 (Feb. 1983), pp. 163–210; F. Hayes-Roth, et al., *Building Expert Systems* (Reading, MA: Addison-Wesley, 1983); and P. Winston and K. Prendergast, eds., *The AI Business: Commercial Uses of Artificial Intelligence* (Cambridge: The MIT Press, 1984).
9. Chambers, *Innovation in Flight*, p. 99.

highway-in-the-sky (HITS) displays.[10] This so-called "E-Z Fly" project was incorporated into Langley's General-Aviation Stall/Spin Program, a major contemporary study to improve the safety of general-aviation (GA) pilots and passengers. Numerous test subjects, from nonpilots to highly experienced test pilots, evaluated Stewart's concept of HITS implementation. NASA flight-test reports illustrated both the challenges and the opportunities that the HITS/E-Z Fly combination offered.[11]

E-Z Fly decoupled the flight controls of a Cessna 402 twin-engine, general-aviation aircraft simulated in Langley's GA Simulator, and HITS offered a system of guidance to the pilot. This decoupling, while making the simulated airplane "easy to fly," also reduced its responsiveness. Providing this level of HITS technology in a low-end GA aircraft posed a range of technical, economic, implementation, and operational challenges. As stated in a flight-test report, "The concept of placing inexperienced pilots in the National Airspace System has many disadvantages. Certainly, system failures could have disastrous consequences."[12] Nevertheless, the basic technology was sound and helped set the stage for future projects. NASA Langley was developing the infrastructure in the early 1990s to support wide-ranging research into synthetically driven flight deck displays for GA, commercial and business aircraft (CBA), and NASA's High-Speed Civil Transport (HSCT).[13] The initial limited idea of easing the workload for low-time pilots would lead to sophisticated display systems that would revolutionize the flight deck. Ultimately, in 1999, a dedicated, well-funded Synthetic Vision Systems Project was created, headed by

10. Eric C. Stewart, "A Simulation Study of Control and Display Requirements for Zero-Experience General Aviation Pilots," NASA LRC, *Workshop on Guidance, Navigation, Controls, and Dynamics for Atmospheric Flight*, Report N94-25102 (1993); Eric C. Stewart, "A Piloted Simulation Study of Advanced Controls and Displays for General Aviation Airplanes," NASA TM-111545 (1994).
11. Robert A. Rivers, "GA E-Z Fly," NASA Langley Flight Test Report (July 10, Oct. 20, and Dec. 1, 1992). NASA Langley flight-test reports were informal documents written by Langley research pilots describing their work on flight, simulation, or ground tests. Though these reports were written immediately after the flight test and were not peer-reviewed or corrected, they were for the most part extremely detailed and followed a rigorous format. The reports were provided researchers valuable input for final comprehensive reports, such as those referenced in this document. Unfortunately, these documents were not archived by NASA Langley, and most have been lost over the years. The author has relied extensively in this chapter on his reports retained in his personal files. Only one other report from another pilot was located.
12. Rivers, "GA E-Z Fly," NASA Langley Flight Test Report (Oct. 20, 1992), p. 3.
13. M.K. Kaiser, "The Surface Operations Research and Evaluation Vehicle (SOREV)—A testbed for HSCT taxi issues," AIAA Paper 1998-5558, 1998.

Daniel G. Baize under NASA's Aviation Safety Program (AvSP). Inspired by Langley researcher Russell V. Parrish, researchers accomplished a number of comprehensive and successful GA and CBA flight and simulation experiments before the project ended in 2005. These complex, highly organized, and efficiently interrelated experiments pushed the state of the art in aircraft guidance, display, and navigation systems.

Significant work on synthetic vision systems and sensor fusion issues was also undertaken at the NASA Johnson Space Center (JSC) in the late 1990s, as researchers grappled with the challenge of developing displays for ground-based pilots to control the proposed X-38 reentry test vehicle. As subsequently discussed, through a NASA-contractor partnership, they developed a highly efficient sensor fusion technique whereby real-time video signals could be blended with synthetically derived scenes using a laptop computer. After cancellation of the X-38 program, JSC engineer Jeffrey L. Fox and Michael Abernathy of Rapid Imaging Software, Inc., (RIS, which developed the sensor fusion technology for the X-38 program, supported by a small business contract) continued to expand these initial successes, together with Michael L. Coffman of the Federal Aviation Administration (FAA). Later joined by astronaut Eric C. Boe and the author (formerly a project pilot on a number of LaRC Synthetic Vision Systems (SVS) programs), this partnership accomplished four significant flight-test experiments using JSC and FAA aircraft, motivated by a unifying belief in the value of Synthetic Vision Systems technology for increasing flight safety and efficiency.

Synthetic Vision Systems research at NASA continues today at various levels. After the SVS project ended in 2005, almost all team members continued building upon its accomplishments, transitioning to the new Integrated Intelligent Flight Deck Technologies (IIFDT) project, "a multi-disciplinary research effort to develop flight deck technologies that mitigate operator-, automation-, and environment-induced hazards."[14] IIFDT constituted both a major element of NASA's Aviation Safety Program and a crucial underpinning of the Next Generation Air Transportation System (NGATS), and it was itself dependent upon the maturation of SVS begun within the project that concluded in 2005. While

14. For example, J.J. "Trey" Arthur, III, Lawrence Prinzel, III, Kevin Shelton, Lynda J. Kramer, Steven P. Williams, Randall E. Bailey, and Robert M. Norman, "Design and Testing of an Unlimited Field of Regard Synthetic Vision Head-Worn Display for Commercial Aircraft Surface Operations," NASA LRC and Boeing Phantom Works, NTRS Rept. LAR-17290-1 (2007).

much work remains to be done to fulfill the vision, expectations, and promise of NGATS, the principles and practicality of SVS and its application to the cockpit have been clearly demonstrated.[15] The following account traces SVS research, as seen from the perspective of a NASA research pilot who participated in key efforts that demonstrated its potential and value for professional civil, military, and general-aviation pilots alike.

Synthetic Vision: An Overview

NASA's early research in SVS concepts almost immediately influenced broader perceptions of the field. Working with NASA researchers who reviewed and helped write the text, the Federal Aviation Administration crafted a definition of SVS published in Advisory Circular 120-29A, describing it as "a system used to create a synthetic image (e.g., typically a computer generated picture) representing the environment external to the airplane." In 2000, NASA Langley researchers Russell V. Parrish, Daniel G. Baize, and Michael S. Lewis gave a more detailed definition as "a display system in which the view of the external environment is provided by melding computer-generated external topography scenes from on-board databases with flight display symbologies and other information from on-board sensors, data links, and navigation systems. These systems are characterized by their ability to represent, in an intuitive manner, the visual information and cues that a flight crew would have in daylight Visual Meteorological Conditions (VMC)."[16] This definition can

15. For a sampling of this research, see Randall E. Bailey, et al., "Crew and Display Concepts Evaluation for Synthetic/Enhanced Vision Systems," *SPIE Defense and Security Symposium 2006, Apr. 2006*; Z. Rahman, et al., "Automated, On-Board Terrain Analysis for Precision Landings," *Proceedings of SPIE Visual Information Processing XIV*, vol. 6246 (Apr. 2006); J.J. Arthur, III, et al., "Design and Testing of an Unlimited-Field-of-Regard Synthetic Vision Head-worn Display for Commercial Aircraft Surface Operations," *Proceedings of SPIE Enhanced and Synthetic Vision*, vol. 6559 (2007); Bailey, et al., "Fusion of Synthetic and Enhanced Vision for All-Weather Commercial Aviation Operations," (NATO RTO-HFM-141), in *NATO Human Factors and Medicine Symposium on Human Factors and Medical Aspects of Day/Night All Weather Operations: Current Issues and Future Challenges*, pp. 11-1–11-18 (2007); P.V. Hyer, et al., "Cockpit Displays for Enhancing Terminal-Area Situational Awareness and Runway Safety," NASA CR-2007-214545 (2007).

16. Russell V. Parrish, et al., "Aspects of Synthetic Vision Display Systems and the Best Practices of the NASA's SVS Project," NASA TP-2008-215130 (May 2008), p. 2. The referenced definition of SVS can be found in the following source: Parrish, et al., "Synthetic Vision," *The Avionics Handbook* (Boca Raton: CRC Press, 2000), pp. 16-1–16-8. Parrish, et al.'s TP-2008-215130 is a definitive source of the work of the NASA Langley SVS project from 1999 to 2005. Additionally, "Best Practices" contains a bibliography of over 230 articles, technical reports, journal articles, and books on NASA's SVS work.

be expanded further to include sensor fusion. This provides the capability to blend in real time in varying percentages all of the synthetically derived information with video or infrared signals. The key requirements of SVS as stated above are to provide the pilot with an intuitive, equivalent-to-daylight VMC capability in all-weather conditions at any time on a tactical level (with present and near future time and position portrayed on a head-up display [HUD] or primary flight display [PFD]) and far improved situation awareness on a strategic level (with future time and position portrayed on a navigation display [a NAV display, or ND]).

In the earliest days of proto-SVS development during the 1980s and early 1990s, the state of the art of graphics generators limited the terrain portrayal to stroke-generated line segments forming polygons to represent terrain features. Superimposing HITS symbology on these displays was not difficult, but the level of situational awareness (SA) improvement was somewhat limited by the low-fidelity terrain rendering. In fact, the superposition of HITS projected flight paths to include a rectilinear runway presentation at the end of the approach segment on basic PFD displays inspired the development of improved terrain portrayal by suggesting the simple polygon presentation of terrain. The development of raster graphics generators and texturing capabilities allowed these simple polygons to be filled, producing more realistic scenes. Aerial and satellite photography providing "photo-realistic" quality images emerged in the mid-1990s, along with improved synthetic displays enhanced by constantly improving databases. With vastly improved graphics generators (reflecting increasing computational power), the early concept of co-displaying the desired vertical and lateral pathway guidance ahead of the airplane in a three-dimensional perspective has evolved from the crude representations of just two decades ago to the present examples of high-resolution, photo-realistic, and elevation-based three-dimensional displays, replete with overlaid pathway guidance, providing the pilot with an unobstructed view of the world. Effectively, then, the goal of synthetically providing the pilot with an effective daylight, VMC view in all-weather has been achieved.[17]

Though the expressions Synthetic Vision Systems, External Vision Systems (XVS), and Enhanced Vision Systems (EVS) have often been used interchangeably, each is distinct. Strictly speaking, SVS has come

17. Parrish, et al., "Aspects of Synthetic Vision," NASA TP-2008-215130, p. 9.

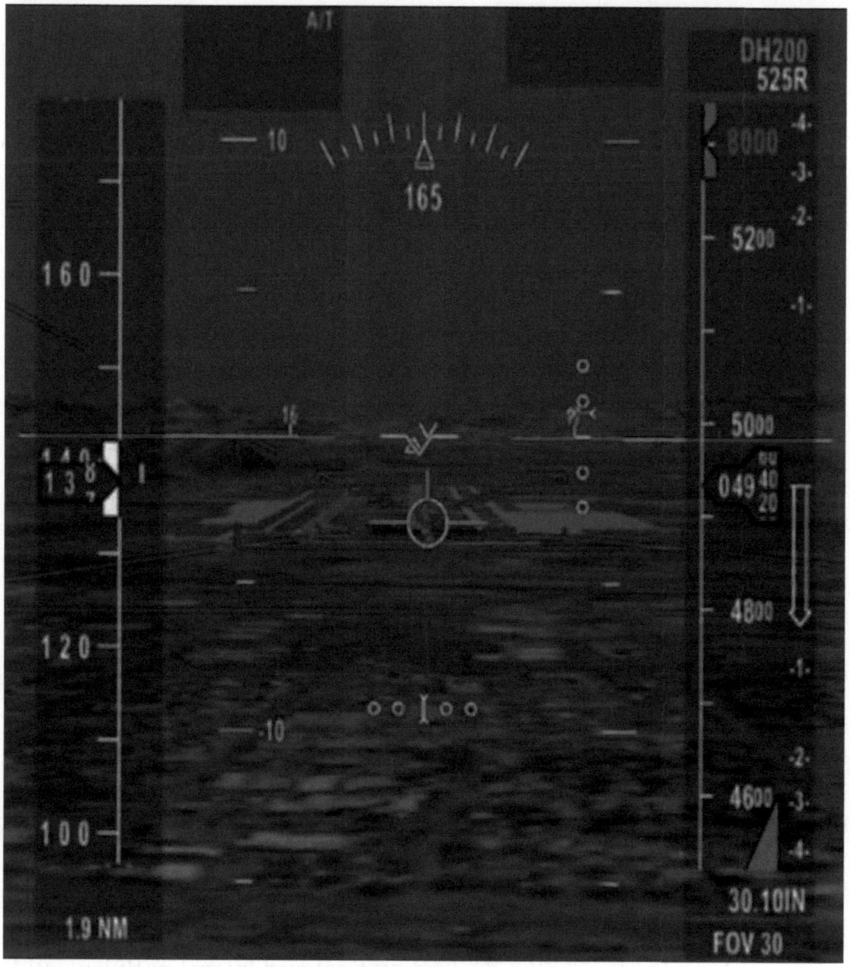

Elevation-based generic primary flight display used on a NASA SVS test in 2005. NASA.

to mean computer-generated imagery from onboard databases combined with precise Global Positioning System (GPS) navigation. SVS joins terrain, obstacle, and airport images with spatial and navigational inputs from a variety of sensor and reference systems to produce a realistic depiction of the external world. EVS and XVS employ imaging sensor systems such as television, millimeter wave, and infrared, integrated with display symbologies (altitude/airspeed tapes on a PFD or HUD, for example) to permit all-weather, day-night operations.[18]

18. Parrish, et al., "Description of 'Crow's Foot,'" p. 6.

Confusion in terminology, particularly in the early years, has characterized the field, including use of multiple terms. For example, in 1992, the FAA completed a flight test investigating millimeter wave and infrared sensors for all-weather operations under the name "Synthetic Vision Technology Demonstration."[19] SVS and EVS are often combined as one expression, SVS/EVS, and the FAA has coined another term as well, EFVS, for Enhanced Flight Vision System. Industry has undertaken its own developments, with its own corporate names and nuances. A number of avionics companies have implemented various forms of SVS technologies in their newer flight deck systems, and various airframe manufacturers have obtained certification of both an Enhanced Vision System and a Synthetic Vision System for their business and regional aircraft. But much still remains to be done, with NASA, the FAA, and industry having yet to fully integrate SVS and EVS/EFVS technology into a comprehensive architecture furnishing Equivalent Visual Operations (EVO), blending infrared-based EFVS with SVS and millimeter-wave sensors, thereby creating an enabling technology for the FAA's planned Next Generation Air Transportation System.[20]

The underlying foundation of SVS is a complete navigation and situational awareness system. This Synthetic Vision System consists mainly of integration of worldwide terrain, obstacle, and airport databases; real-time presentation of immediate tactical hazards (such as weather); an Inertial Navigation System (INS) and GPS navigation capability; advanced sensors for monitoring the integrity of the data base and for object detection; presentation of traffic information; and a real-time synthetic vision display, with advanced pathway or equivalent guidance, effectively affording the aircrew a projected highway-in-the-sky ahead of them.[21] Two enabling technologies were necessary for SVS to be developed: increased computer storage capacity and a global, real-

19. Parrish, et al., "Aspects of Synthetic Vision," NASA TP-2008-215130, p. 9.
20. For example, avionics concerns such as Thales, which certified an EFVS system with EASA, the FAA, Transport Canada, and Rockwell Collins, with its EFVS-4860 system; airframe manufacturers include Gulfstream for its G-IV, G-V, G-300, G-400/450, and G-500/550; Bombardier for its Global Express XRS and Global 5000; Dassault for its Falcon 900EX/DX and 2000EX/DX; and Embraer for its ECJ-190; regarding future EVO, see Capt. Bob Moreau, "Fed Ex HUD/EFVS Overview and LED Replacement for Airport and Approach Lighting Structures," paper presented at the *Illuminating Engineering Society of America Aviation Lighting Conference, 2009*, at *http://www.iesalc.org/docs/FedEx_HUD-EFVS_Overview.pdf*, accessed Dec. 7, 2009.
21. Parrish, et al., "Aspects of Synthetic Vision," NASA TP-2008-215130, p. 2.

time, highly accurate navigation system. The former has been steadily developing over the past four decades, and the latter became available with the advent of GPS in the 1980s. These enabling technologies utilized or improved upon the Electronic Flight Information System (EFIS), or glass cockpit, architecture pioneered by NASA Langley in the 1970s and first flown on Langley's Boeing 737 Advanced Transport Operating System (ATOPS) research airplane. It should be noted that the research accomplishments of this airplane—Boeing's first production 737—in its two decades of NASA service are legendary. These included demonstration of the first glass cockpit in a transport aircraft, evaluation of transport aircraft fly-by-wire technology, the first GPS-guided blind landing, the development of wind shear detection systems, and the first SVS-guided landings in a transport aircraft.[22]

The development of GPS satellite navigation signals technology enabled the evolution of SVS. GPS began as an Air Force–Navy effort to build a satellite-based navigation system that could meet the needs of fast-moving aircraft and missile systems, something the older TRANSIT system, developed in the late 1950s, could not. After early studies by a variety of organizations—foremost of which was the Aerospace Corporation—the Air Force formally launched the GPS research and development program in October 1963, issuing hardware design contracts 3 years later. Known initially as the Navstar GPS system, the concept involved a constellation of 24 satellites orbiting 12,000 nautical miles above Earth, each transmitting a continual radio signal containing a

22. See, for example, W.F. White and L.V. Clark, "Flight Performance of the TCV B-737 Airplane at Kenney Airport Using TRSB/MLS guidance," NASA TM-80148 (1979); White and Clark, "Flight Performance of the TCV B-737 Airplane at Jorge Newbery Airport, Buenos Aires, Argentina, Using TRSB/MLS Guidance," NASA TM-80233 (1980); White and Clark, "Flight Performance of the TCV B-737 Airplane at Montreal Dorval International Airport, Montreal, Canada, Using TRSB/MLS Guidance," NASA TM-81885 (1980); Richard M. Hueschen, J.F. Creedon, W.T. Bundick, and J.C. Young, "Guidance and Control System Research for Improved Terminal Area Operations," in Joseph W. Stickle, ed., *1980 Aircraft Safety and Operating Problems*, NASA CP-2170, pt. 1 (1981), pp. 51–61; J.A. Houck, "A Simulation Study of Crew performance in Operating an Advanced Transport Aircraft in an Automated Terminal Area Environment," NASA TM-84610 (1983); and John J. White, "Advanced Transport Operating Systems Program," SAE Paper 90-1969, presented at the *Society of Automotive Engineers Aerospace Technology Conference and Exposition, Long Beach, CA, Oct. 1–4, 1990*. See also Lane E. Wallace, *Airborne Trailblazer: Two Decades with NASA Langley's 737 Flying Laboratory* (NASA SP 4216), pp. 27–33, for further details on the ATOPS research aircraft.

precise time stamp from an onboard atomic clock. By recording the time of each received satellite signal of a required 4 satellites and comparing the associated time stamp, a ground receiver could determine position and altitude to high accuracies. The first satellite was launched in 1978, and the constellation of 24 satellites was complete in 1995. Originally intended only for use by the Department of Defense, GPS was opened for civilian use (though to a lesser degree of precision) by President Ronald Reagan after a Korean Air Lines Boeing 747 commercial airliner was shot down by Soviet interceptors in 1983 after it strayed miles into Soviet territory. The utility of the GPS satellite network expanded dramatically in 2000, when the United States cleared civilian GPS users to receive the same level of precision as military forces, thus increasing civilian GPS accuracy tenfold.[23]

Database quality was essential for the development of SVS. The 1990s saw giant strides taken when dedicated NASA Space Shuttle missions digitally mapped 80 percent of Earth's land surface and almost 100 percent of the land between 60 degrees north and south latitude. At the same time, radar mapping from airplanes contributed to the digital terrain database, providing sufficient resolution for SVS in route and specific terminal area requirements. The Shuttle Endeavour Radar Topography Mission in 2000 produced topographical maps far more precise than previously available. Digital terrain databases are being produced by commercial and government organizations worldwide.[24]

With the maturation of the enabling technologies in the 1990s and its prior experience in developing glass cockpit systems, NASA Langley was poised to develop the concept of SVS as a highly effective tool for pilots to operate aircraft more effectively and safely. This did not happen directly but was the culmination of experience gained by Langley research engineers and pilots on NASA's Terminal Area Productivity (TAP)

23. George W. Bradley, III, "Origins of the Global Positioning System," in Jacob Neufeld, George M. Watson, Jr., and David Chenoweth, *Technology and the Air Force: A Retrospective Assessment* (Washington, DC: Air Force History and Museums Program, 1997), pp. 245–253; Ivan A. Getting, *All in a Lifetime: Science in the Defense of Democracy* (New York: Vantage Press, 1989), pp. 574–597; John L. McLucas, with Kenneth J. Alnwick and Lawrence R. Benson, *Reflections of a Technocrat: Managing Defense, Air, and Space Programs During the Cold War* (Maxwell AFB, AL: Air University Press, 2006), pp. 295–297; and Randy James, "A Brief History of GPS," *Time.com*, May 26, 2009, *http://www.time.com/time/nation/article/0,8599,1900862,00.html*, accessed Dec. 7, 2009.

24. Chambers, *Innovation in Flight*, p. 95.

and High-Speed Research (HSR) programs in the mid- to late 1990s. By 1999, when the SVS project of the AvSP was initiated and funded, Langley had an experienced core of engineers and research pilots eager to push the state of the art.

TAP, HSR, and the Early Development of SVS

In 1993, responding to anticipated increases in air travel demand, NASA established a Terminal Area Productivity program to increase airliner throughput at the Nation's airports by at least 12 percent over existing levels of service. TAP consisted of four interrelated subelements: air traffic management, reduced separation operations, integration between aircraft and air traffic control (ATC), and Low Visibility Landing and Surface Operations (LVLASO).[25]

Of the four Agency subelements, the Low Visibility Landing and Surface Operations project assigned to Langley held greatest significance for SVS research. A joint research effort of Langley and Ames Research Centers, LVLASO was intended to explore technologies that could improve the safety and efficiency of surface operations, including landing rollout, turnoff, and inbound and outbound taxi; making better use of existing runways; and thus making obvious the need for expensive new facilities and the rebuilding and modification of older ones.[26] Steadily increasing numbers of surface accidents at major airports imparted particular urgency to the LVLASO effort; in 1996, there had been 287 incidents, and the early years of the 1990s had witnessed 5 fatal accidents.[27]

LVLASO researchers developed a system concept including two technologies: Taxiway Navigation and Situational Awareness (T-NASA) and Rollout Turnoff (ROTO). T-NASA used the HUD and NAV display moving map functions to provide the pilot with taxi guidance and data link air traffic control instructions, and ROTO used the HUD to guide the pilot in braking levels and situation awareness for the selected run-

25. Steven D. Young and Denise R. Jones, "Flight Demonstration of Integrated Airport Surface Movement Technologies," NASA TM-1998-206283 (1998).
26. Denise R. Jones and Steven D. Young, "Airport Surface Movement Technologies—Atlanta Demonstration Overview," NASA LRC (1997), NTRS Document ID 200.401.10268, p. 1.
27. Steven D. Young and Denise R. Jones, "Flight Testing of an Airport Surface Guidance, Navigation, and Control System," paper presented at the *Institute of Navigation National Technical Meeting, Jan. 21–23, 1998*, p. 1.

way turnoff. LVLASO also incorporated surface surveillance concepts to provide taxi traffic alerting with cooperative, transponder-equipped vehicles. LVLASO connected with potential SVS because of its airport database and GPS requirements.

In July and August 1997, NASA Langley flight researchers undertook two sequential series of air and ground tests at Atlanta International Airport, using a NASA Boeing 757-200 series twin-jet narrow-body transport equipped with Langley-developed experimental cockpit displays. This permitted surface operations in visibility conditions down to a runway visual range (RVR) of 300 feet. Test crews included NASA pilots for the first series of tests and experienced airline captains for the second. All together, it was the first time that SVS had been demonstrated at a major airport using a large commercial jetliner.[28]

LVLASO results encouraged Langley to continue its research on integrating surface operation concepts into its SVS flight environment studies. Langley's Wayne H. Bryant led the LVLASO effort, assisted by a number of key researchers, including Steven D. Young, Denise R. Jones, Richard Hueschen, and David Eckhardt.[29] When SVS became a focused project under AvSP in 1999, these talented researchers joined their colleagues from the HSR External Vision Systems project.[30] While LVLASO technologies were being developed, NASA was in the midst of one of the largest aeronautics programs in its history, the High-Speed Research Program. SVS research was a key part of this program as well.

After sporadic research at advancing the state of the art in high-speed aerodynamics in the 1970s, the United States began to look at both supersonic and hypersonic cruise technologies more seriously in the mid-1980s. Responding to a White House Office of Science and Technology Policy call for research into promoting long-range, high-speed aircraft, NASA awarded contracts to Boeing Commercial Airplanes and Douglas Aircraft Company in 1986 for market and technology feasibility studies

28. Young and Jones, "Flight Demonstration of Integrated Airport Surface Movement Technologies," and Jones and Young, "Airport Surface Movement Technologies—Atlanta Demonstration Overview."
29. Chambers, *Innovation in Flight*, pp. 103–104.
30. XVS is used as an abbreviation for External Vision Systems. However, under NASA's High-Speed Research (HSR) program, XVS was also shorthand for "eXternal [sic] Visibility System," yet another example of how acronyms and designations evolved over the length of SVS–XVS–EVS studies. See NASA LRC, "NASA's High-Speed Research Program: The eXternal Visibility System Concept," NASA Facts on Line, FS-1998-09-34-LaRC (Sept. 1998), *http://oea.larc.nasa.gov/PAIS/HSR-Cockpit.html*, accessed Dec. 7, 2009.

of a potential High-Speed Civil Transport. The speed spectrum for these studies spanned the supersonic to hypersonic regions, and the areas of study included economic, environmental, and technical considerations. At the same time, LaRC conducted its own feasibility studies led by Charles M. Jackson, Chief of the High-Speed Research Division; his deputy, Wallace C. Sawyer; Samuel M. Dollyhigh; and A. Warner Robbins. These and follow-on studies by 1988 concluded that the most favorable candidate considering all factors investigated was a Mach 2 to Mach 3.2 HSCT with transpacific range.[31]

NASA created the High-Speed Research program in 1990 to investigate technical challenges involved with developing a Mach 2+ HSCT. Phase I of the HSR program was to determine if major environmental obstacles could be overcome, including ozone depletion, community noise, and sonic boom generation. NASA and its industry partners determined that the state of the art in high-speed design would allow mitigation of the ozone and noise issues, but sonic boom mitigation remained elusive.[32]

Buoyed by these assessments, NASA commenced Phase II of the HSR program in 1995, in partnership with Boeing Commercial Airplane Group, McDonnell-Douglas Aerospace, Rockwell North American Aircraft Division, General Electric Aircraft Engines, and Pratt & Whitney as major industry participants. A comprehensive list of technical issues was slated for investigation, including sonic boom effects, ozone depletion, aero acoustics and community noise, airframe/propulsion integration, high lift, and flight deck design. One of the earliest identified issues was forward visibility. Unlike the Concorde and the Tupolev Tu-144 Supersonic Transports, the drooping of the nose to provide forward visibility for takeoff and landing was not a given. By leaving the nose undrooped, engineers could make the final design thousands of pounds lighter. Unfortunately, to satisfy supersonic fineness ratio requirements, the postulated undrooped nose would completely obstruct the pilots' forward vision. A solution had to be found, and the new disciplines of advanced cockpit electronic displays and high-fidelity sensors, in

31. See the Flanagan and Benson cases on supersonic cruise and sonic boom research in these volumes; as well, Chambers provides an informative historical treatment of the HSR program in *Innovations in Flight,* covering the broad areas of research into acoustics, environmental impacts, flight controls, and aerodynamics (to name a few) that are beyond the scope of this chapter.
32. Erik Conway, *High-Speed Dreams: NASA and the Technopolitics of Supersonic Transportation* (Baltimore: The Johns Hopkins University Press, 2005), pp. 213–299.

combination with Langley's HITS development, suggested an answer. A concept known as the External Vision System was developed, which was built around providing high-quality video signals to the flight deck to be combined with guidance and navigation symbology, creating a virtual out-the-window scene.[33]

With the extensive general-aviation highway-in-the-sky experience at Langley, researchers began to expand their focus in the early 1990s to include more sophisticated applications to commercial and business aircraft. This included investigating the no-droop nose requirements of the conceptual High-Speed Civil Transport, which lacked side windows and had such a forward-placed cockpit in relation to the nose wheel of the vehicle—over 50 feet separated the two—as to pose serious challenges for precise ground maneuvering. As the High-Speed Research program became more organized, disciplines became grouped into Integrated Technology Development (ITD) Teams.[34] An XVS element was established in the Flight Deck ITD Team, led by Langley's Daniel G. Baize. Because the HSR program contained so many member organizations, each with its own prior conceptions, it was thought that the ITD concept would be effective in bringing the disparate organizations together. This did not always lead to an efficient program or rapid progress. Partly, this was due to the requirement that consensus must be reached on all ITD Team decisions, a Skunk Works process in reverse. In the case of the XVS element, researchers from NASA Langley and NASA Ames Research Centers joined industry colleagues from Boeing, Douglas, Calspan, and others in designing a system from the bottom up.[35]

Different backgrounds led to different choices for system design from the group. For example, at Langley, the HITS concept was favored with a traditional flight director, while at Ames, much work had been

33. NASA, "NASA's High Speed Research Program: The eXternal Visibility System," in NASA Facts FS-1998-09-34, LaRC, Hampton VA (1998).
34. Parrish, et al., "Aspects of Synthetic Vision Display Systems," p. 17; see also M. Yang, T. Gandhi, R. Kasturi, L. Coraor, O. Camps, and J. McCandless, "Real-time Obstacle Detection System for High Speed Civil Transport Supersonic Aircraft," paper presented at the *IEEE National Aerospace and Electronics Conference, Oct. 2000*; and Mary K. Kaiser, "The Surface Operations Research and Evaluation Vehicle (SOREV): A testbed for HSCT taxi issues," AIAA Paper 1998-5558 (1998).
35. Calspan has for decades been an exemplar of excellence in flight simulation. Originally known as the Cornell Aeronautical Laboratory, in the 1990s, the company was variously known as Arvin Calspan and Veridian. In this chapter, Calspan is used throughout, as that name has the largest recognition among the aerospace professional research community.

devoted to developing a "follow me" aircraft concept developed by Ames researcher Richard Bray, in which an iconic aircraft symbol portrayed the desired position of the aircraft 5–30 seconds in the future. The pilot would then attempt to use the velocity vector to "follow" the leader aircraft. Subsequent research would show that choices of display symbology types profoundly coupled with the type of control law selected. Certain good display concepts performed poorly with certain good control law implementations. As the technology in both flight displays and digital fly-by-wire control laws advanced, one could not arbitrarily select one without considering the other. Flight tests in the United States Air Force (USAF)/Calspan Total In-Flight Simulator (TIFS) aircraft had shown that flightpath guidance cues could lead to pilot-induced oscillations (PIOs) in the flare when control was dependent upon a flight control system employing rate command control laws. For this reason, the Flight Deck and Guidance and Flight Controls (GFC) ITD Teams worked closely together, at times sharing flight tests to ensure that good concert existed between display and flight control architecture. To further help the situation, several individuals served on both teams simultaneously.

From 1994 to 1996, Langley hosted a series of workshops concerning concepts for commercial transports, including tunnel-, pathway-, and highway-in-the-sky concepts.[36] The first two workshops examined potential display concepts and the maturity of underlying technologies, with attendees debating the merits of approaches and their potential utility. The final workshop, the Third XVS Symbology Workshop (September 4–5, 1996), focused on XVS applications for the HSCT. Led by the Flight Deck Integrated Display Symbology Team of Dr. Terrence Abbott and Russell Parrish, from Langley, and Andrew Durbin, Gordon Hardy, and Mary Kaiser, from Ames, the workshop provided an opportunity for participants from related ITD Teams to exchange ideas. Because the sensor image would be the primary means of traffic separation in VMC, display clutter was a major concern. The participants developed the minimal symbology set for the XVS displays to include the virtual out-the-window display and the head-down PFD. The theme of the workshop became, "Less is best, lest we obscure the rest."[37]

36. Russell V. Parrish, ed., "Avionic Pictorial Tunnel-/Pathway-/Highway-In-The-Sky Workshops," NASA CP-2003-212164, presents the proceedings of these interactive workshops.
37. "Minutes of the Third XVS Symbology Workshop," *Third XVS Symbology Workshop, NASA Langley Research Center, Hampton, VA, Sept. 5, 1996.*

As flight tests would troublingly demonstrate, display clutter (excess symbology) would be one of several significant problems revealed while evaluating the utility of displays for object (traffic) detection.

First Steps in Proving XVS: A View from the Cockpit

From 1995 through 1999, the XVS element conducted a number of simulator and flight tests of novel concepts using NASA Langley's and NASA Ames' flight simulators as well as the Calspan–Air Force Research Laboratory's NC-131H Total In-Flight Simulator (an extensively modified Convair 580 twin-turboprop transport, with side force controllers, lift flaps, computerized flight controls, and an experimental cockpit) and Langley's ATOPS Boeing 737.[38]

In 1995, the first formal test, TSRV.1, was conducted in Langley's fixed-base Transport Systems Research Vehicle (TSRV) simulator, which replicated the Research Flight Deck (RFD) in Langley's ATOPS B-737. Under the direction of Principal Investigator Randall Harris, the test was a parametric evaluation of different sensor and HUD presentations of a proposed XVS. A monitor was installed over the copilot's glare shield to provide simulated video, forward-looking infrared (FLIR), and computer-generated imagery (CGI) for the evaluation. The author had the privilege of undertaking this test, and the following is from the report he submitted after its conclusion:

> Approach, flare, and touchdown using 1 of 4 available sensors (2 were FLIR sensors with a simulated selection of the "best" for the ambient conditions) and 1 of 3 HUD presentations making a 3 X 3 test matrix for each scheduled hour long session. Varying the runways and direction of base to final turns resulted in a total matrix of 81 runs. Each of the 3 pilots completed 63 of the 81 possible runs in the allotted time.[39]

38. Like NASA's pioneer 737, the TIFS airplane had its own remarkable history, flying extensively in support of numerous aircraft development programs until its retirement in 2008. For its background and early capabilities, see David A. Brown, "In-Flight Simulator Capabilities Tested," *Aviation Week and Space Technology*, Aug. 9, 1971.

39. Robert A. Rivers, "HSR XVS TSRV Sim 1 FTR," NASA Langley Flight Test Report (Aug. 31, 1995), p. 1.

Commenting on the differences between the leader aircraft flight director and the more traditional HUD/Velocity Vector centered flight director, the author continued:

> Some experimentation was performed to best adapt the amount of lead of the leader aircraft. It was initially agreed that a 25 to 15 sec lead worked best for the TSRV simulator. The 5 sec lead led to a too high gain task for the lateral axis control system and resulted in chasing the leader continuously in a roll PIO state. Adjusting the amount of lead for the leader may need to be revisited in the airplane. A purely personal opinion is that the leader aircraft concept is a higher workload arrangement than a HUD mounted velocity vector centered flight director properly tuned.[40]

At the same time, a team led by Russ Parrish was developing its own fixed-based simulator intended to support HSR XVS research and development. Known as Virtual Imaging Simulator for Transport Aircraft Systems (VISTAS), this simulator allowed rapid plug-and-play evaluation of various XVS concepts and became a valuable tool for XVS researchers and pilots. Over the next 5 years, this simulator evolved through a series of improvements, leading to the definitive VISTAS III configuration. Driven by personal computers rather than the Langley simulation facility mainframe computers, and not subject to as stringent review processes (because of its fixed-base, low-cost concept), this facility became extremely useful and highly productive for rapid prototyping.[41]

From the ground-based TSRV.1 test, XVS took to the air with the next experiment, HSR XVS FL.2, in 1996. Using Langley's venerable ATOPS B-737, FL.2 built upon lessons learned from TSRV.1. FL.2 demonstrated for the first time that a pilot could land a transport aircraft using only XVS imagery, with the Langley research pilots flying the aircraft with no external windows in the Research Flight Deck. As well, they landed using only synthetically generated displays, foreshadowing future SVS work. Two candidate SVS display concepts were evaluated for the first time: elevation-based generic (EBG) and photorealistic. EBG relied on a detailed database to construct a synthetic image of the

40. Ibid., p. 3.

41. Interview of Louis J. Glaab by Rivers, NASA Langley Research Center, June 1, 2009.

NASA Langley's Advanced Transport Operating Systems B-737 conducting XVS guided landings. NASA.

terrain and obstacles. Photorealistic, on the other hand, relied on high-resolution aerial photographs and a detailed database to fuse an image with near-high-resolution photographic quality. These test points were in anticipation of achieving sensor fusion for the HSCT flight deck XVS displays, in which external sensor signals (television, FLIR, etc.) would be seamlessly blended in real time with synthetically derived displays to accommodate surmised varying lighting and visibility conditions. This sensor fusion technology was not achieved during the HSR program, but it would emerge from an unlikely source by the end of the decade.

The second flight test of XVS concepts, known as HSR XVS FL.3, was flown in Langley's ATOPS B-737 in April 1997 and is illustrative of the challenges in perfecting a usable XVS. Several experiments were accomplished during this flight test, including investigating the effects of nonconformality of the artificial horizon portrayed on the XVS forward display and the real-world, out-the-side-window horizon as well as any effects of parallax when viewing the XVS display with a close design eye point rather than viewing the real-world forward scene focused at infinity. Both the Research and Forward Flight Decks (FFD) of the B-737 were highly modified for this test, which was conducted at NASA Wallops Flight Facility (WFF) on Virginia's Eastern Shore, just south of the Maryland border. Located on the Atlantic coast of the Delmarva

Peninsula, Wallops was situated within restricted airspace and immediately adjoining thousands of square miles of Eastern Seaboard warning areas. The airport was entirely a NASA test and rocket launch facility, complete with sophisticated radar- and laser-tracking capability, control rooms, and high-bandwidth telemetry receivers. Langley flight operations conducted the majority of their test work at Wallops. Every XVS flight test would use WFF.

The modifications to the FFD were summarized in the author's research notes as follows:

> The aircraft was configured in one of the standard HSR XVS FL.3 configurations including a 2X2 tiled Elmo lipstick camera array, a Kodak (Megaplus) high resolution monochrome video camera (1028 x 1028 pixels) mounted below the nose with the tiled camera array, an ASK high resolution (1280 x 1024 pixels) color video projector mounted obliquely behind the co-pilot seat, a Silicon Graphics 4D-440VGXT Skywriter Graphics Workstation, and a custom Honeywell video mixer. The projector image was focused on a 24 inch by 12 inch white screen mounted 17.5 inches forward of the right cockpit seat Design Eye Position (DEP). Ashtech Differential GPS receivers were mounted on both the 737 and a Beechcraft B-200 target aircraft producing real time differential GPS positioning information for precise inter-aircraft navigation.[42]

An interesting digression here involves the use of Differential GPS (DGPS) for this experiment. NASA Langley had been a leader in developing Differential GPS technologies in the early 1990s, and the ATOPS B-737 had accomplished the first landing by a transport aircraft using Differential GPS guidance. Plane-plane Differential GPS had been perfected by Langley researchers in prior years and was instrumental in this and subsequent XVS experiments involving traffic detection using video displays in the flight deck. DGPS could provide real-time relative positions of participating aircraft to centimeter accuracy.

With the conformality and parallax investigations as a background, Langley's Beechcraft B-200 King Air research support aircraft was

42. Rivers, "HSR XVS FL.3 Flight Test," NASA Langley Flight Test Report (Apr. 21, 1997), p. 1.

employed for image object detection as a leader aircraft on multiple instrument approaches and as a random traffic target aircraft. FL.3 identified the issue about which a number of XVS researchers and pilots had been concerned about at the XVS Workshop the previous fall: the challenges of seeing a target aircraft in a display. Issues such as pixel per degree resolution, clutter, brightness, sunlight readability, and contrast were revealed in FL.3. From the flight-test report:

> Unfortunately, the resolution and clarity of the video presentation did not allow the evaluation pilot to be able to see the leader aircraft for most of the time. Only if the 737 was flown above the B-200, and it was flying with a dark background behind it, was it readily visible in the display. We closed to 0.6 miles in trail and still had limited success. On final, for example, the B-200 was only rarely discernible against the runway environment background. The several times that I was able to acquire the target aircraft, the transition from forward display to the side window as I tracked the target was seamless. Most of the time the target was lost behind the horizon line or velocity vector of the display symbology or was not visible due to poor contrast against the horizon. Indeed, even with a bright background with sharp cloud boundaries, the video presentation did not readily distinguish between cloud and sky. . . . Interestingly, the landings are easier this time due, in my opinion, to a perceived wider field of view due to the geometry of the arrangement in the Forward Flight Deck (FFD) and to the peripheral benefits of the side window. Also, center of percussion effects may have caused false motion cues in the RFD to the extent that it may have affected the landings. The fact that the pilot is quite comfortable in being very confident of his position due to the presence of the side window may have had an effect in reducing the overall mental workload. The conformality differences were not noticeable at 4 degrees, and at 8 degrees, though noticeable and somewhat limiting, successful landings were possible. By adjusting eye height position the pilot could effectively null the 0 and 4 degree differences.[43]

11

43. Ibid., pp. 4–6.

Convair NC-131H Total In-Flight Simulator used for SVS testing. USAF.

Another test pilot on this experiment, Dr. R. Michael Norman, discussed the effects of rain and insects on the XVS sensors and displays. His words also illustrate the great risks taken by the modern test pilot in the pursuit of knowledge:

> Aerodynamics of the flat, forward facing surface of the camera mount enclosure resulted in static positioning of water droplets which became deposited on the aperture face. The relative size of the individual droplets was large and obtrusive, and once they were visible, they generally stayed in place. Just prior to touchdown, a large droplet became visually superimposed with the velocity vector and runway position, which made lineup corrections and positional situational awareness extremely difficult. Discussions of schemes to prevent aperture environmental contamination should continue, and consideration of incorporation in future flight tests should be made. During one of the runs, a small flying insect appeared in the cockpit. The shadow of this insect amplified its apparent size on the screen, and was somewhat distracting. Shortly thereafter, it landed on the screen, and continued to be distracting. The presence of flying insects in the cockpit is an issue with front projected displays.[44]

44. Dr. R. Michael Norman, "Flight Test Notes, R-806 and R-9807," NASA Langley Flight Test Report (Apr. 21, 1997), p. 3.

Clearly, important strides toward a windowless flight deck had been achieved by FL.3, but new challenges had arisen as well. Recognizing the coupling between flight control law development and advanced flight displays, the GFC and Flight Deck ITD Teams planned a joint test in 1998 on a different platform, the Air Force Research Laboratory-Calspan NC-131H Total In-Flight Simulator aircraft.

TIFS, which was retired to the Air Force Museum a decade afterward, was an exotic-looking, extensively modified Convair 580 twin-engine turboprop transport that Calspan had converted into an in-flight simulator, which it operated for the Air Force. Unique among such simulators, the TIFS aircraft had a simulation flight deck extending in front of and below the normal flight deck. Additionally, it incorporated two large side force controllers on each wing for simulation fidelity, modified flaps to permit direct lift control, and a main cabin with computers and consoles to allow operators and researchers to program models of different existing or proposed aircraft for simulation. TIFS operated on the model following concept, in which the state vector of TIFS was sampled at a high rate and compared with a model of a simulated aircraft. If TIFS was at a different state than the model, the flight control computers on TIFS corrected the TIFS state vector through thrust, flight controls, direct lift control, and side force control to null all the six degree-of-freedom errors. The Simulation Flight Deck (SFD) design was robust and allowed rapid modification to proposed design specifications.

Undertaken from November 1998 through February 1999, the FL.4 HSR experiment combined XVS and GFC experimental objectives. The SFD was configured with a large cathode ray tube mounted on top of the research pilot's glare shield, simulating a notional HSR virtual forward window. Head-down PFD and NAV display completed the simulated HSR flight deck. XVS tests for FL.4 included image object detection and display symbology evaluation. The generic HSR control law was used for the XVS evaluation. A generic XVS symbology suite was used for the GFC experiments flown out of Langley and Wallops. Langley researchers Lou Glaab and Lynda Kramer led the effort, with assistance from the author (who served as Langley HSR project pilot), Calspan test pilot Paul Deppe (among others), and Boeing test pilot Dr. Michael Norman (who was assigned to NASA Langley as a Boeing interface for HSR).

The success of FL.4, combined with some important lessons learned, prepared the way for the final and most sophisticated of the HSR flight tests: FL.5, flown at Langley, Wallops, and Asheville, NC,

USAF/Calspan NC-131H The Total In-Flight Simulator on the ramp at Asheville, NC, with the FL.5 crew. Note the simulation flight deck in the extended nose. NASA.

from September through November 1999. Reprising their FL.4 efforts, Langley's Lou Glaab and Lynda Kramer led FL.5, with valuable assistance from Calspan's Randall E. Bailey, who would soon join the Langley SVS team as a NASA researcher. Russell Parrish also was an indispensable presence in this and subsequent SVS tests. His imprint was felt throughout the period of focused SVS research at NASA.

With the winding down of the HSR program in 1999, the phase of SVS research tied directly to the needs of a future High-Speed Civil Transport came to an end. But before the lights were turned out on HSR, FL.5 provided an apt denouement and fitting climax to a major program that had achieved much. FL.4 had again demonstrated the difficulty in image object detection using monitors or projected displays. Engineers surmised that a resolution of 60 pixels per degree would be necessary for acceptable performance. The requirement for XVS to be capable of providing traffic separation in VMC was proving onerous. For FL.5, a new screen was used in TIFS. This was another rear projection device, providing a 50-degree vertical by 40-degree horizontal field of view (FOV). Adequate FOV parameters had been and would continue to be a topic of study. A narrow FOV (30 degrees or less), while providing good resolution, lacked accommodation for acceptable situation awareness. As FOVs became wider, however, distortion was inevitable, and resolution became an issue. The FL.5 XVS display, in addition to its impressive FOV, incorporated a unique feature: a high-resolution (60 pixels per degree) inset in the center of the display, calibrated appropriately along an axis to provide the necessary resolution for the flare task and traffic detection. The XVS team pressed on with various preparatory checkouts and tests before finally moving on to a terrain avoidance and traffic detection test with TIFS at Asheville, NC.

Asheville was selected because of the terrain challenges it offered and the high-fidelity digital terrain database of the terminal area provided by the United States Geological Survey. These high-resolution terminal area databases are more common now, but in 1999, they were just becoming available. This database allowed the TIFS XVS to provide high-quality head-down PFD SVS information. This foreshadowed the direction Langley would take with FL.5, when XVS gave way to SVS displays incorporating the newer databases. In his FL.5 research notes of the time, the author reflected on the XVS installation, which was by then quite sophisticated:

> The Primary XVS Display (PXD) consisted of three tiled projections, an upper, a lower, and a high resolution inset display. The seams between each projection were noticeable, but were not objectionable. The high resolution inset was designed to approach a resolution of about 60 pixels per degree in the vertical axis and somewhat less than that in the horizontal axis. It is my understanding that this degree of resolution was not actually achieved. The difference in resolution between the high resolution inset and the surround views was not objectionable and did not detract from the utility of any of the displays. Symbology was overwritten on all the PXD displays, but at times there was not a perfect match between the surrounds and the high resolution inset resulting in some duplicated symbology or some occulted symbology. An inboard field-of-view display (IFOV) was also available to the pilot with about the same resolution of the surround views. This also had symbology available.
>
> The symbology consisted of the down selected HSR minimal symbology set and target symbology for the PXD and a

> horizon line, heading marks, and target symbology for the IFOV display. The target symbology consisted of a blue diamond with accompanying digital information (distance, altitude and altitude trend) placed in the relative position on the PXD or IFOV display that the target would actually be located. Unfortunately, due to several unavoidable transport delays, the target symbology lagged the actual target, especially in high track crossing angle situations. For steady relative bearing situations, the symbology worked well in tracking the target accurately. Occasionally, the target symbology would obscure the target, but a well conceived PXD declutter feature corrected this.
>
> The head down displays available to the pilot included a fairly standard electronic Primary Flight Display (PFD) and Navigation Display (ND). The ND was very useful to the pilot from a strategic perspective in providing situation awareness (SA) for target planning. The PXD provided more of a tactical SA for traffic avoidance. TCAS, Radar, Image Object Detection (IOD), and simulated ADSB targets were displayed and could be brought up to the PXD or IFOV display through a touch screen feature. This implementation was good, but at times was just a little difficult to use. Variable ranges from 4 to 80 miles were pilot selectable through the touch screen. In the past sunlight intrusion in the cockpit had adversely affected both the head up and head down displays. The addition of shaded window liners helped to correct this problem, and sun shafting occurrences washing out the displays were not frequent.[45]

The accompanying figure shows the arrangement of XVS displays in the SFD of the TIFS aircraft for the FL.5 experiment.

The author's flight-test report concluded:

> Based on XVS experience to date, it is my opinion that the current state of the art for PXD technologies is insufficient to justify a "windowless" cockpit. Improvements in contrast, resolution, and fields of view for the PXD are required before this concept can be implemented. . . . A visual means of verifying

45. Rivers, "HSR FL.5 Flight Test Report," NASA Langley Flight Test Report (Oct. 2, 1999), pp. 1–2.

the accuracy of the navigation and guidance information presented to the pilot in an XVS configured cockpit seems mandatory. That being said, the use of symbology on the PXD and Nav Display for target acquisition provides the pilot with a significant increase in both tactical and strategic situation awareness. These technologies show huge potentials for use both in the current subsonic fleet as well as for a future HSCT.[46]

The XVS head-up and head-down displays used in the FL.5 flight test. NASA.

Though falling short of fully achieving the "windowless cockpit" goal by program's end, the progress made over the previous 4 years on HSR XVS anticipated the future direction of NASA's SVS research. Much had been accomplished, and NASA had an experienced, motivated team of researchers ready to advance the state of the art as the 20th century closed, stimulated by visions of fleetwide application of Synthetic and Enhanced Vision Systems to subsonic commercial and general-aviation

46. Ibid., p. 4.

aircraft and the need for database integrity monitoring. Meanwhile, a continent away, other NASA researchers, unaware of the achievements of HSR XVS, struggled to develop their own XVS and solved the challenge of sensor fusion along the way.

Sensor Fusion Arrives

Integrating an External Vision System was an overarching goal of the HSR program. The XVS would include advanced television and infrared cameras, passive millimeter microwave radar, and other cutting-edge sensors, fused with an onboard database of navigation information, obstacles, and topography. It would thus furnish a complete, synthetically derived view for the aircrew and associated display symbologies in real time. The pilot would be presented with a visual meteorological conditions view of the world on a large display screen in the flight, deck simulating a front window. Regardless of actual ambient meteorological conditions, the pilot would thus "see" a clear daylight scene, made possible by combining appropriate sensor signals; synthetic scenes derived from the high-resolution terrain, navigation, and obstacle databases; and head-up symbology (airspeed, altitude, velocity vector, etc.) provided by symbol generators. Precise GPS navigation input would complete the system. All of these inputs would be processed and displayed in real time (on the order of 20–30 milliseconds) on the large "virtual window" displays. During the HSR program, Langley did not develop the sensor fusion technology before program termination and, as a result, moved in the direction of integration of the synthetic database derived view with sophisticated display symbologies, redefining the implementation of the primary flight display and navigation display. Part of the problem with developing the sensor fusion algorithms was the perceived need for large, expensive computers. Langley continued on this path when the Synthetic Vision Systems project was initiated under NASA's Aviation Safety Program in 1999 and achieved remarkable results in SVS architecture, display development, human factors engineering, and flight deck integration in both GA and CBA domains.[47]

Simultaneously with these efforts, JSC was developing the X-38 unpiloted lifting body/parafoil recovery reentry vehicle. The X-38 was a technology demonstrator for a proposed orbital crew rescue vehicle

47. This information comes from the author's notes and recollection from that time period.

that could, in an emergency, return up to seven astronauts to Earth, a veritable space-based lifeboat. NASA planners had forecasted a need for such a rescue craft in the early days of planning for Space Station Freedom (subsequently the International Space Station). Under a Langley study program for the Space Station Freedom Crew Emergency Rescue Vehicle (CERV, later shortened to CRV), Agency engineers and research pilots had undertaken extensive simulation studies of one candidate shape, the HL-20 lifting body, whose design was based on the general aerodynamic shape of the Soviet Union's BOR-4 subscale spaceplane.[48] The HL-20 did not proceed beyond these tests and a full-scale mockup. Instead, Agency attention turned to another escape vehicle concept, one essentially identical in shape to the nearly four-decade-old body shape of the Martin SV-5D hypersonic lifting reentry test vehicle, sponsored by NASA's Johnson Space Center. The Johnson configuration spawned its own two-phase demonstrator research effort: the X-38 program, for a series of subsonic drop-shapes air-launched from NASA's NB-52B Stratofortress, and the second, for an orbital reentry shape to be test-launched from the Space Shuttle from a high-inclination orbit. But while tests of the former did occur at the NASA Dryden Flight Research Center (DFRC) in the late 1990s, the fully developed orbital craft did not proceed to development and orbital test.[49]

To remotely pilot this vehicle during its flight-testing at Dryden, project engineers were developing a system displaying the required navigation and control data. Television cameras in the nose of the X-38 provided a data link video signal to a control flight deck on the ground. Video signals alone, however, were insufficient for the remote pilot to perform all the test and control maneuvers, including "flap turns" and "heading hold" commands during the parafoil phase of flight. More information on the display monitor would be needed. Further complications arose because of the design of the X-38: the crew would be lying on its

48. The author was a participant in a series of HL-20 simulations; for details of these, see Robert A. Rivers, E. Bruce Jackson, and W.A. Ragsdale, "Piloted Simulator Studies of the HL-20 Lifting Body," paper presented at the *35th Symposium of the Society of Experimental Test Pilots, Beverly Hills, CA, Sept. 1991.*

49. Scott A. Berry, Thomas J. Horvath, K. James Weilmuenster, Stephan J. Alter, and N. Ronald Merski, "X-38 Experimental Aeroheating at Mach 10," AIAA Paper 2001-2828 (2000), p. 1; see also Jay Miller, *The X-Planes: X-1 to X-45* (Hinckley, England: Midland Publishing, 2001), pp. 378–383.

backs, looking at displays on the "ceiling" of the vehicle. Accordingly, a team led by JSC X-38 Deputy Avionics Lead Frank J. Delgado was tasked with developing a display system allowing the pilot to control the X-38 from a perspective 90 degrees to the vehicle direction of travel. On the cockpit design team were NASA astronauts Rick Husband (subsequently lost in the Columbia reentry disaster), Scott Altman, and Ken Ham, and JSC engineer Jeffrey Fox.

Delgado solicited industry assistance with the project. Rapid Imaging Software, Inc., a firm already working with imaginative synthetic vision concepts, received a Phase II Small Business Innovation Research (SBIR) contract to develop the display architecture. RIS subsequently developed LandForm VisualFlight, which blended "the power of a geographic information system with the speed of a flight simulator to transform a user's desktop computer into a 'virtual cockpit.'"[50] It consisted of "symbology fusion" software and 3-D "out-the-window" and NAV display presentations operating using a standard Microsoft Windows–based central processing unit (CPU). JSC and RIS were on the path to developing true sensor fusion in the near future, blending a full SVS database with live video signals. The system required a remote, ground-based control cockpit, so Jeff Fox procured an extended van from the JSC motor pool. This vehicle, officially known as the X-38 Remote Cockpit Van, was nicknamed the "Vomit Van" by those poor souls driving around lying on their backs practicing flying a simulated X-38. By spring 2002, JSC was flying the X-38 from the Remote Cockpit Van using an SVS NAV Display, an SVS out-the-window display, and a video display developed by RIS. NASA astronaut Ken Ham judged it as furnishing the "best seat in the house" during X-38 glide flights.[51]

Indeed, during the X-38 testing, a serendipitous event demonstrated the value of sensor fusion. After release from the NASA NB-52B Stratofortress, the lens of the onboard X-38 television camera became partially covered in frost, occluding over 50 percent of the FOV. This would have proved problematic for the pilot had orienting symbology

50. Quoted in NASA JSC, "A New Definition of Ground Control," *Spinoff 2002* (Houston: NASA JSC, 2002), pp. 132–133; see also Frank J. Delgado, "Simulation for the X-38/CRV Parafoil and Re-Entry Phases," AIAA Paper 2000-4085 (2000); Frank Delgado and Mike Abernathy, "A Hybrid Synthetic Vision System for the Tele-operation of Unmanned Vehicles," NASA JSC (2004), NTIS Document 200.502.17300; and interview of Michael Abernathy by Robert A. Rivers, Albuquerque, NM, June 11, 2009.
51. NASA JSC, "A New Definition of Ground Control," p. 132.

not been available in the displays. Synthetic symbology, including spatial entities identifying keep-out zones and runway outlines, provided the pilot with a synthetic scene replacing the occluded camera image. This foreshadowed the concept of sensor fusion, in which, for example, blossoming as the camera traversed the Sun could be "blended" out, and haze obscuration could be minimized by adjusting the degree of synthetic blend from 0 to 100 percent.[52]

But then, on April 29, 2002, faced with rising costs for the International Space Station, NASA canceled the X-38 program.[53] Surprisingly, the cancellation did not have the deleterious impact upon sensor fusion development that might have been anticipated. Instead, program members Jeff Fox and Eric Boe secured temporary support via the Johnson Center Director's discretionary fund to keep the X-38 Remote Cockpit Van operating. Mike Abernathy, president of RIS, was eager to continue his company's sensor fusion work. He supported their efforts, as did Patrick Laport of Aerospace Applications North America (AANA). For the next 2 years, Fox and electronics technician James B. Secor continued to improve the van, working on a not-to-interfere basis with their other duties. In July 2004, Fox secured further Agency funding to convert the remote cockpit, now renamed, at Boe's suggestion, the Advanced Cockpit Evaluation System (ACES). It was rebuilt with a single, upright seat affording a 180-degree FOV visual system with five large surplus monitors. An array of five cameras was mounted on the roof of the van, and its input could be blended in real time with new RIS software to form a complete sensor fusion package for the wraparound monitors or a helmet-mounted display.[54] Subsequently, tests with this van demonstrated true sensor fusion. Now, the team looked for another flight project it could use to demonstrate the value of SVS.

Its first opportunity came in November 2004, at Creech Air Force Base in Nevada. Formerly known as Indian Springs Auxiliary Air Field, a backwater corner of the Nellis Air Force Base range, Creech had risen

52. Interview of Jeffrey L. Fox by Robert A. Rivers, NASA Johnson Space Center, Aug. 15, 2008, and June 2, 2009. The author has also relied upon notes, e-mails, memos, and recollections of Jeffrey Fox.
53. Etienne Prandini, "ISS Partnership in Crisis," *Interavia Business & Technology* (May 2002); Mark Carreau, "Project's Cancellation Irks NASA," *Houston Chronicle*, June 9 2002.
54. Eric C. Boe, Jeffrey L. Fox, Francisco J. Delgado, Michael F. Abernathy, Michael Clark, and Kevin Ehlinger, "Advanced Cockpit Evaluation System Van," *2005 Biennial Research and Technology Report*, University Research and Affairs Office, NASA Johnson Space Center (2005).

to prominence after the attacks of 9/11, as it was the Air Force's center of excellence for unmanned aerial vehicle (UAV) operations. It used, as its showcase, the General Atomics Predator UAV. The Predator, modified as a Hellfire-armed attack system, had proven a vital component of the global war on terrorism. With UAVs increasing dramatically in their capabilities, it was natural that the UAV community at Nellis would be interested in the work of the ACES team. Traveling to Nevada to demonstrate its technology to the Air Force, the JSC team used the ACES van in a flight-following mode, receiving downlink video from a Predator UAV. That video was then blended with synthetic terrain database inputs to provide a 180-degree FOV scene for the pilot. The Air Force's Predator pilots found the ACES system far superior to the narrow-view perspective they then had available for landing the UAV.

In 2005, astronaut Eric Boe began training for a Shuttle flight and left the group, replaced by the author, who had spent years over 10 years at Langley as a project or research pilot on all of that Center's SVS and XVS projects. The author transferred to JSC from Langley in 2004 as a research pilot and learned of the Center's SVS work from Boe. The author's involvement with the JSC group linked Langley and JSC's SVS efforts, for he provided the JSC group with his experience with Langley's SVS research.

That spring, a former X-38 cooperative student—Michael Coffman, now an engineer at the FAA's Mike Monroney Aeronautical Center in Oklahoma City—serendipitously visited Fox at JSC. They discussed using the sensor fusion technology for the FAA's flight-check mission. Coffman, Fox, and Boe briefed Thomas C. Accardi, Director of Aviation Systems Standards at FAA Oklahoma City, on the sensor fusion work at JSC, and he was interested in its possibilities. Fox seized this opportunity to establish a memorandum of understanding (MOU) among the Johnson Space Center, the Mike Monroney Aeronautical Center, RIS, and AANA. All parties would work on a quid pro quo basis, sharing intellectual and physical resources where appropriate, without funding necessarily changing hands. Signed in July 2005, this arrangement was unique in its scope and, as will be seen, its ability to allow contractors and Government agencies to work together without cost. JSC and FAA Oklahoma City management had complete trust in their employees, and both RIS and AANA were willing to work without compensation, predicated on their faith in their product and the likely potential return on their investment, effectively a Skunk Works approach taken to the extreme. The stage was set for major SVS accomplishments, for during this same period, huge strides in SVS development had been made at Langley, which is where this narrative now returns.[55]

Langley Transitions SVS into a New Century of Flight

In 1997, in response to a White House Commission on Aviation Safety and Security, NASA created the Aviation Safety Program. SVS fit perfectly within the goals of this program, and the NASA established a SVS project under AvSp, commencing on October 1, 1999. Daniel G. Baize, who had led the XVS element of the Flight Deck ITD Team during the HSR program, continued in this capacity as Project Manager for SVS under AvSP. He wasn't the only holdover from HSR: most of the talented researchers from HSR XVS moved directly to similar roles under AvSP and were joined by their Langley LVLASO colleagues. Funding for FL.5 transitioned from HSR to AvSP, effectively making FL.5 the first of many successful AvSP SVS flight tests.

Langley's SVS research project consisted of eight key technical areas: database rendering, led by Jarvis "Trey" Arthur, III, and Steve Williams; pathway concepts, led by Russell Parrish, Lawrence "Lance" Prinzel, III, Lynda Kramer, and Trey Arthur; runway incursion prevention systems, led by Denise R. Jones and Steven D. Young; controlled flight into terrain (CFIT) avoidance using SVS, led by Trey Arthur; loss of control avoidance using SVS, led by Douglas T. Wong and Mohammad A. Takallu; database integrity, led by Steven D. Young; SVS sensors development, led by Steven Harrah; and SVS database development, led by Robert A. Kudlinski and Delwin R. Croom, Jr. These individuals were supported by numerous NASA and contractor researchers and technicians, and by a number of dedicated industry and academia partners.[56] By any measure, SVS development was moving forward along a broad front at the turn of the 21st century.

The first flight test undertaken under the SVS project occurred at the Dallas-Fort Worth International Airport (DFW) in September and October 2000. It constituted the culmination of Langley's LVLASO project,

55. Interview of Fox by Rivers, NASA Johnson Space Center, Aug. 15, 2008, and June 2, 2009.
56. Chambers, *Innovation in Flight*, pp. 98, 104. The SVS project is well documented, with a number of excellent technical reports. Chambers provides a good overview of the flight tests and personnel, and Parrish, et al., *Aspects of Synthetic Vision* provides detail of the technologies being evaluated. Both sources should be consulted for additional details. For GA SVS research, see Louis J. Glaab, Russell V. Parrish, Monica F. Hughes, and Mohammad A. Takallu, "Effectively Transforming IMC Flight into VMC Flight: An SVS Case Study," *Proceedings of the 25th Digital Avionics System Conference* (2006).

demonstrating the results of 7 years of research into surface display concepts for reduced-visibility ground operations. Because funding for the LVLASO experiment had transitioned to the AvSP, SVS Project Manager Dan Baize decided to combine the LVLASO elements of the test with continued SVS development. SVS was by now bridging the ground operation/flight operation regimes into one integrated system, although at DFW, each was tested separately.

Reduced ground visibility has always constituted a risk in aircraft operations. On March 27, 1977, an experienced KLM 747 flight crew holding for takeoff clearance at Los Rodeos Airport, Tenerife, fell victim to a fatal combination of misunderstood communications and reduced ground visibility. Misunderstanding tower communications, the crewmembers began their takeoff roll and collided with a Pan American 747 still taxiing on landing rollout on the active runway. This accident claimed 578 lives, including all aboard the KLM aircraft and still constitutes the costliest accident in aviation history.[57] Despite the Tenerife disaster, runway incursions continued to rise, and the potential for further tragedies large and small was great. Incursions rose from 186 in 1993 to 431 in 2000, a 132-percent increase. In the first 5 months of 2000, the FAA and National Transportation Safety Board (NTSB) logged 158 incursions, an average of more than 1 runway incursion incident each day.[58]

Recognizing the emphasis on runway incursion accident prevention, researchers evaluated a Runway Incursion Prevention System (RIPS), the key element in the DFW test. RIPS brought together advanced technologies, including surface communications, navigation, and surveillance systems for both air traffic controllers and pilots. RIPS utilized both head-down moving map displays for pilot SA and data link communication and an advanced HUD for real-time guidance. While

57. Spain, Secretary of Civil Aviation, *Report on Tenerife Crash, KLM B-747 Ph-BUF and Pan Am B-747 N736, Collision at Tenerife Airport, Spain, on 27 March 1977* (Oct. 1978), prep. by Harro Ranter, Aircraft Accident Digest, *ICAO Circular* 153-AN/56), pp. 22–68, *http://www.panamair.org/accidents/victor.htm*, accessed Oct. 24, 2009. Complicating the Tenerife accident were unusual traffic pressures on the respective crews after diversion of their flights from Las Palmas to Los Rodeos Airport, Tenerife, caused by a terrorist bombing of the terminal building at Las Palmas. Tenerife was heavily crowded as a result, and the stress was extreme upon both air and ground crews, including ATC personnel.

58. Denise R. Jones, Cuong C. Quach, and Steven D. Young, "Runway Incursion Prevention System—Demonstration and Testing at the Dallas/Fort Worth International Airport," paper presented at the *20th Digital Avionics Systems Conference, Daytona Beach, FL, Oct. 14–18, 2001*, p. 1.

RIPS research was occurring on the ground, SVS concepts were being evaluated in flight for the first time in a busy terminal environment. This evaluation included a Langley-developed opaque HUD concept. Due to the high capacity of flight operations during normal hours at DFW, all research flights occurred at night. HSR veterans Lou Glaab, Lynda Kramer, Jarvis "Trey" Arthur, Steve Harrah, and Russ Parrish managed the SVS experiments, while LVLASO researchers Denise Jones and Richard Hueschen led the RIPS effort.[59]

The successor to Langley's remarkable ATOPS B-737 was a modified Boeing 757, the Aries research airplane. Aries—a name suggested by Langley operations engineer Lucille Crittenden in an employee suggestion campaign—stood for Airborne Research Integrated Experiments System. For all its capabilities, Aries had a somewhat checkered history. Like many new research programs, it provided systemic challenges to researchers that they had not encountered with the B-737. Indeed, Langley's research pilot staff had favored a smaller aircraft than the 757, one that would be less costly and demanding to support. Subsequently, the 757 did prove complex and expensive to maintain, impacting the range of modifications NASA could make to it. For example, Aries lacked the separate mid-fuselage Research Flight Deck that had proven so adaptable and useful in the ATOPS 737. Instead, its left seat of the cockpit (traditionally the "captain's seat" in a multipilot aircraft) of was modified to become a Forward Flight Deck research station. This meant that, unlike the 737, which had two safety pilots in the front cockpit while a test crew was using the Research Flight Deck, the 757 was essentially a "one safety pilot at the controls" aircraft, with the right-seat pilot performing the safety role and another NASA pilot riding in the center jump seat aft and between both the research and safety pilot. This increased the workload of both the research and safety pilots.[60]

As configured for the DFW tests, Aries had an evaluation pilot in the left seat, a NASA safety pilot/pilot-in-command in the right seat, a secondary NASA safety pilot in the center jump seat, and the principal investigator in the second jump seat. The safety pilot monitored two communication frequencies and an intercom channel connected to the numerous engineers and technicians in the cabin. Because the standard B-757 flight deck instrumentation did not support the SVS displays, the SVS

59. Ibid; see also Chambers, *Innovation in Flight*, pp. 108–110.
60. Author's recollections from his experiences flying the Aries aircraft.

researchers developed a portable SVS primary flight display that would be temporarily mounted over the pilot's instrument panel. An advanced HUD was installed in the left-seat position as well, for use during final approach, rollout, turnoff, and taxi. The HUD displayed symbology relating runway and taxiway edge and centerline detail, deceleration guidance, and guidance to gates and hold-short points on the active runway. As well, the Aries aircraft had multifunction display capability, including an electronic moving map (EMM) that could be "zoomed" to various scales and that could display the DFW layout, locations of other traffic, and ATC instructions (the latter displayed both in text and visual formats). Additionally, a test van outfitted with an Automatic Dependent Surveillance-Broadcast (ADS-B) Mode S radar transponder, an air traffic control Radio Beacon System (ATCRBS) transponder, a Universal Access Transceiver (UAT) data link, and a differential GPS was deployed to test sites and used to simulate an aircraft on the ground that could interact during various scenarios with the Aries test aircraft.[61]

The DFW tests occurred in October 2000, with the Aries 757 interacting with the surrogate "airliner" van, and with the airport equipped on its east side with a prototype FAA ground surveillance system developed under the Agency's runway incursion reduction program. Researchers were encouraged by the test results, and industry and Government evaluation pilots agreed that SVS technologies showed remarkable potential, reflecting the thorough planning of the test team and the skill of the flight crew. The results were summarized by Denise R. Jones, Cuong C. Quach, and Steven D. Young as follows:

> The measured performance of the traffic reporting technologies tested at DFW do meet many of the current requirements for surveillance on the airport surface. However, this is apparently not sufficient for a robust runway incursion alerting function with RIPS. This assessment is based on the observed rats of false alerts and missed detections. All false alerts and missed detections at DFW were traced to traffic data that was inaccurate, inconsistent, and/or not received in a timely manner.... All of the subject pilots were complimentary of the RIPS tested at DFW. The pilots stated that the system has the potential

61. Jones, Quach, and Young, "Runway Incursion Prevention System," p. 2, et. seq., details the system.

> to reduce or eliminate runway incursions, although human factors issues must still be resolved. Several suggestions were made regarding the alerting symbology which will be incorporated into future simulation studies. The audible alert was the first display to bring the pilots' attention to the incursion. The EMM would generally be viewed by the non-flying pilot at the time of an incursion since the flying pilot would remain heads up. The pilots stated that two-stage alerting was not necessary and they would take action on the first alert regardless. This may be related to the fact that this was a single pilot operation and the subject pilot did not have the benefit of co-pilot support. In general, after an incursion alert was received, the subject pilots stated they would not want maneuver guidance during final approach or takeoff roll but would like guidance on whether to stop or continue when taxiing across a runway. All of the pilots stated that, in general, the onboard alerts were generated in a timely manner, allowing sufficient time to react to the potential conflict. They all felt safer with RIPS onboard.[62]

Almost exactly a year later, the SVS project deployed to a remote location for a major integrated flight test and demonstration of the Aries B-757, the third year in a row that the team had deployed for an offsite test. This time, the location was the terrain-challenged Eagle County Regional Airport near Vail, CO. Eagle-Vail is situated in a valley with mountains on three sides of the runway. It is also at an elevation of 6,540 feet, giving it a high-density altitude on hot summer days, which is not conducive to airplane performance. Langley's Aries B-757 was configured with two HUDs and four head-down concepts developed by NASA and its industry partner, Rockwell Collins. Enhanced Vision Systems were evaluated as well for database integrity monitoring and imaging the runway environment. Three differently sized head-down PFDs were examined: a "Size A" system, measuring 5.25 inches wide by 5 inches tall, such as flown on a conventional B-757-200 series aircraft; a "Size D" 6.4-inch-wide by 6.4-inch-tall display, such as employed on the B-777 family; and an experimental "Size X," measuring 9 inches wide by 8 inches tall, such as might be flown on a future advanced aircraft.

62. Ibid., p. 9.

Additionally, multiple radar altimeters and differential GPS receivers gathered absolute altitude data to be used in developing database integrity monitoring algorithms.[63]

Randy Bailey was NASA's Principal Investigator, joined by Russ Parrish, Dan Williams, Lynda Kramer, Trey Arthur, Steve Harrah, Steve Young, Rob Kudlinski, Del Croom, and others. Seven pilots from NASA, the FAA, the airline community, and Boeing evaluated the SVS concepts, with particular attention to the terrain-challenged approaches. While fixed-base simulation had indicated that SVS could markedly increase flight safety in terrain-challenged environments, flight-test data had not yet been acquired under such conditions, aside from the limited experience of the Air Force–Calspan TIFS NC-131H trials at Asheville, NC, in September 1999. Of note was the ability of the B-757 to fly circling approaches under simulated instrument meteorological conditions (IMC) using the highly developed SVS displays. Until this test, commercial jet airplanes had not made circling approaches to Eagle-Vail under IMC.[64] SVS were proving their merit in the most challenging of arenas, something evident in the comments of one evaluation pilot, who noted afterward:

> I often commented to people over the years that I never ever flew a circling approach in the -141 [Lockheed C-141 Starlifter] that I was ever comfortable with, particularly at night. It always demanded a lot of attention. This was the first time I ever had an occasion of circling an approach with the kind of information I would love to have in a circling approach. Keeping me safe, I could see the terrain, taking me where I want to go, getting me all types of information in terms to where I am relative to the end of the runway. I mean it's the best of all possible worlds in terms of safety.[65]

Unfortunately, this proved to be the last major flight-test program flown on NASA's B-757 aircraft. An incident during the Eagle-Vail testing had profound effects on its future, illustrating the weakness of not

63. Lynda J. Kramer, Lawrence J. Prinzell, III, Randall E. Bailey, Jarvis J. Arthur, III, and Russell V. Parrish, "Flight Test Evaluation of Synthetic Vision Concepts at a Terrain Challenged Airport," NASA TP-2004-212997 (2004), p. 1.

64. Chambers, *Innovation in Flight*, pp. 110–112.

65. Kramer, et al., "Flight Test Evaluation . . . at a Terrain Challenged Airport," p. 57.

having an independent Research Flight Deck separated from the Forward Flight Deck, which could be occupied by a team of "full-time" safety pilots. After the B-757 missed its approach at Eagle-Vail following a test run, its auto throttles disconnected, without being noticed by the busy flight crew. The aircraft became dangerously slow in the worst possible circumstances: low to the ground and at a high-density altitude. In the subsequent confusion during recovery, the evaluation pilot, unaware that Aries lacked the kind of Full Authority Digital Electronic Control (FADEC) for its turbofan engines on newer B-757s, inadvertently over-boosted both powerplants, resulting in an in-flight abort. The incident reflected as well the decision to procure the B-757 without FADEC engine controls and insufficient training of evaluation pilots before their sorties into the nuances of the non-FADEC airplane. The busy flight deck caused by the FFD design likely also played a role in this incident, as it likely did in previous, less serious events. Safety concerns raised by pilots over this and other issues resulted in the grounding of the B-757 in June 2003. Subsequent examination revealed that it had overloaded floor beams, necessitating costly repairs. Though these repairs were completed during a 12-month period in 2004–2005, NASA retired it from service in 2005, bringing its far-too-brief operational career to an end.[66]

In 2001, NASA Langley's SVS project was organized into two areas: commercial and business aircraft and general-aviation. Randy Bailey had come to NASA from Calspan and became a Principal Investigator for CBA tests, and Lou Glaab assumed the same role for GA. Monica Hughes, Doug Wong, Mohammad Takallu, Anthony P. Bartolome, Francis G. McGee, Michael Uenking, and others joined Glaab in the GA program, while most of the other aforementioned researchers continued with CBA. Glaab and Hughes led an effort to convert Langley's Cessna 206-H Stationaire into a GA SVS research platform. A PFD and NAV display were installed on the right side of the instrument panel, and an instrumentation pallet in the cabin contained processors to drive the displays and a sophisticated data acquisition system.[67]

SVS was particularly important for general aviation, in which two kinds of accidents predominated: controlled flight into terrain and loss

66. Much controversy accompanied the grounding and the pilots' concerns. See David Schleck, "Fear of Reprisals: NASA Langley Pilots Struggle with Safety Versus Silence," *Daily Press* (Apr. 17, 2005).
67. Louis J. Glaab and Monica F. Hughes, "Terrain Portrayal for Head-Down Displays Flight Test," *Proceedings of the 22nd Digital Avionics Systems Conference* (2003).

The modified cockpit of Langley's C-206 research aircraft, showing SVS PFD and ND on the right side of instrument panel. NASA.

of horizon reference (followed by loss of aircraft control and ground impact).[68] To develop a candidate set of GA display concepts, Glaab conceived a General-Aviation Work Station (GAWS) fixed-base simulator, similar to the successful Virtual Imaging Simulator for Transport Aircraft Systems simulator. Doug Wong and other team members helped bring the idea to reality, and GAWS allowed the GA researchers and evaluation pilots to design and validate several promising GA SVS display sets. The GA implementation differed from the previous and ongoing CBA work, in that SVS for the GA community would have to be far lower in cost, computational capability, and weight. A HUD was deemed too expensive, so the PFD would assume added importance. An integrated simulation and flight-test experiment using GAWS and the Cessna 206 known as Terrain Portrayal for Head-Down Displays (TP-HDD) was commenced in summer 2002. The flight test spanned August through October at Newport News and Roanoke, two of Virginia's regional airports.[69]

Both EBG and photorealistic displays were evaluated, and results indicated that equivalent performance across the pilot spectrum could

68. Parrish, et al., "Aspects of Synthetic Vision Display," p. 3.

69. Glaab and Hughes, "Terrain Portrayal for Head-Down Displays Flight Test."

be produced with the less computationally demanding EBG concepts. This was a significant finding, especially for the computationally and economically challenged low-end GA fleet.[70]

The SVS CBA team had planned a comprehensive flight test using the Aries B-757 for summer 2003 at the terrain-challenged Reno-Tahoe International Airport. This flight test was to have included flight and surface runway incursion scenarios and operations using integrated SVS displays, including an SVS HUD and PFD, RIPS symbology, hazard sensors, and database integrity monitoring in a comparative test with conventional instruments. The grounding of Aries ended any hope of completing the Reno-Tahoe test in 2003. Set back yet undeterred, the SVS CBA researchers looked for alternate solutions. Steve Young and his Database Integrity Monitoring Experiment (DIME) team quickly found room on NASA Ames's DC-8 Airborne Science Platform in July and August for database integrity monitoring and Light Detection and Ranging (LIDAR) elevation data collection.[71] At the same time, managers looked for alternate airframes and negotiated an agreement with Gulfstream Aerospace to use a G-V business jet with Gulfstream's Enhanced Vision System. From July to September 2004 at Wallops and Reno-Tahoe International Airport, the G-V with SVS CBA researchers and partners from Rockwell Collins, Gulfstream, Northrop Grumman, Rannoch Corporation, Jeppesen, and Ohio University evaluated advanced runway incursion technologies from NASA–Lockheed Martin and Rannoch Corporation and SVS display concepts from Langley and Rockwell Collins. Randy Bailey again was project lead. Lynda Kramer and Trey Arthur were Principal Investigators for the SVS display development, and Denise Jones led the runway incursion effort. Steve Young and Del Croom managed the DIME investigations, and Steve Harrah continued to lead sensor development.[72]

11

70. Rivers, Glaab interview.

71. Parrish, et al., "Aspects of Synthetic Vision," NASA TP-2008-215130, p. 7; M. Uijt de Haag, Steven D. Young, and J. Campbell, "An X-Band Radar Terrain Feature Detection Method for Low-Altitude SVS Operations and Calibration Using LiDAR," *SPIE Defense and Security Symposium, Apr. 12–16, 2004.*

72. Jarvis J. Arthur, III, L. Kramer, and Randall E. Bailey, "Flight Test Comparison between Enhanced Vision (FLIR) and Synthetic Vision Systems," in Jacques G. Verly, ed., Proceedings of SPIE, Enhanced and Synthetic Vision 2005, vol. 5802-03 (2005); E. Cooper and Steven D. Young, "Database Integrity Monitoring for Synthetic Vision Systems Using Machine Vision and SHADE," paper presented at the *SPIE Defense and Security Symposium, Enhanced and Synthetic Vision Conference, Mar. 28–Apr. 1, 2005*; and Chambers, *Innovation in Flight*, pp. 112–117.

The Reno flight test was a success. SVS technologies had been shown to provide a significant improvement to safe operations in reduced visibility for both flight and ground operations. SVS CBA researchers and managers, moreover, had shown a tenacity of purpose in completing project objectives despite daunting challenges. The last SVS flight test was approaching, and significant results awaited.

In August and September 2005, Lou Glaab and Monica Hughes led their team of SVS GA researchers on a successful campaign to argue for the concept of equivalent safety for VMC operations and SVS in IMC. Russ Parrish, at the time retired from NASA, returned to lend his considerable talents to this final SVS experiment. Using the Langley Cessna 206 from the TP-HDD experiment of 2002, Glaab and Hughes employed 19 evaluation pilots from across the flight-experience spectrum to evaluate three advanced SVS PFD and NAV display concepts and a baseline standard GA concept to determine if measured flight technical error (FTE) from the low-experience pilots could match that of the highly experienced pilots. Additionally, the question of whether SVS displays could provide VMC-like performance in IMC was explored. With pathway-based guidance on SVS terrain displays, it was found that the FTE of low-time pilots could match that of highly experienced pilots. Furthermore, for the more experienced pilots, it was observed that with advanced SVS displays, difficult IMC tasks could be done to VMC performance and workload standards. The experiment was carefully designed to allow the multivariate discriminant analysis method to precisely quantify the results. Truly, SVS potential for providing equivalent safety for IMC flight to that of VMC flight had been established. The lofty goals of the SVS project established 6 years previously had been achieved.[73]

After the Reno SVS CBA flight test and spanning the termination of the SVS project in 2005, Randy Bailey, Lynda Kramer, Lance Prinzel and others investigated the integration of SVS and EVS capabilities in a comprehensive simulation test using Langley's fixed-base Integrated Intelligent Flight Deck Technologies simulator, a modified Boeing 757 flight deck. Twenty-four airline pilots evaluated a HUD and auxiliary head-down display with integrated SVS and EVS presentations, where forward-looking infrared video was used as the enhanced vision signal.

73. Louis J. Glaab, Russell V. Parrish, Monica F. Hughes, and Mohammad A. Takallu, "Effectively Transforming IMC Flight into VMC Flight: An SVS Case Study," *Proceedings of the 25th Digital Avionics System Conference* (2006), pp. 3–14.

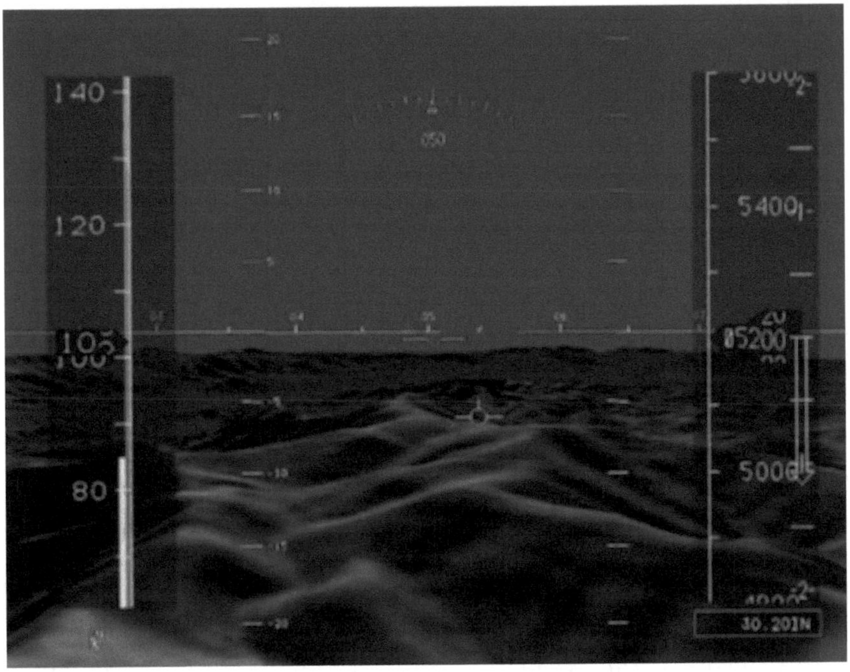

Elevation-based generic primary flight display for SVS GA experiments. NASA.

The fusion here involved blending a synthetic database with the FLIR signal at eight discrete steps selectable by the pilot. Both FLIR and SV signals were imagery generated by the simulation computers. The results showed an increase in SA for all of the subject pilots. Surprisingly, obstacle runway incursion detection did not show significant improvement in either the SV, EV, or fused displays.[74]

The SVS project formally came to an end September 30, 2005. Despite many challenges, the dedicated researchers, research pilots, and technicians had produced an enviable body of work. Numerous technical papers would soon document the results, techniques employed, and lessons learned. From SPIFR's humble beginnings, NASA Langley had designed an SVS display and sensor system that could reliably transform night, instrument conditions to essentially day VMC for commer-

74. Randall E. Bailey, Lynda J. Kramer, and Lawrence J. Prinzel, III, "Fusion of Synthetic and Enhanced Vision for All-Weather Commercial Aviation Operations," *NATO HFM-141 Symposium on Human Factors of Day/Night All-Weather Operations, Heraklion, Greece, Apr. 23–25, 2007.* The paper from this conference contains an excellent bibliography of many of the papers produced after the end of the SVS project.

cial airliners to single-engine, piston-powered GA aircraft. Truly, this was what NASA aeronautics was all about. And now, as the former SVS team transitioned to IIFDT, the researchers at JSC were once again about to take flight.

JSC, the FAA, and Targets of Opportunity

As 2005 drew to a close, Michael Coffman at FAA Oklahoma City had convinced his line management that a flight demonstration of the sensor fusion technology would be a fine precursor to further FAA interest. FAA Oklahoma City had a problem: how best to protect its approaches of flight-check aircraft certifying instruments for the Department of Defense in combat zones. Coffman and Fox had suggested sensor fusion. If onboard video sensors in a flight-check aircraft could image a terminal approach corridor with a partially blended synthetic approach corridor, any obstacle penetrating the synthetic corridor could be quickly identified. Coffman, using the MOU with NASA JSC signed just that July, suggested that an FAA Challenger 604 flight-check aircraft based at FAA Oklahoma City could be configured with SVS equipment to demonstrate the technology to NASA and FAA managers. Immediately, Fox, Coffman, Mike Abernathy of RIS, Patrick Laport and Tim Verborgh of AANA, and JSC electronics technician James Secor began discussing how to configure the Challenger 604. Fox tested his ability to scrounge excess material from JSC by acquiring an additional obsolete but serviceable Embedded GPS Inertial Navigation System (EGI) navigation processor (identical to the one used in the ACES van) and several processors to drive three video displays. Coffman found some FAA funds to buy three monitors, and Abernathy and RIS wrote the software necessary to drive three monitors with SVS displays with full-sensor fusion capability, while Laport and Verborgh developed the symbology set for the displays. The FAA bought three lipstick cameras, JSC's Jay Estes designed a pallet to contain the EGI and processors, and a rudimentary portable system began to take shape.[75]

The author, now a research pilot at the Johnson Space Center, became involved assisting AANA with the design of a notional instrument procedure corridor at JSC's Ellington Field flight operations base. He also obtained permission from his management chain to use JSC's Aircraft

75. The material in the beginning of this section comes from the personal notes, e-mails, and recollection of Jeffrey L. Fox and the author.

Operations Division to host the FAA's Challenger and provide the jet fuel it required. Verborgh, meanwhile, surveyed a number of locations on Ellington Field with the author's help, using a borrowed portable DGPS system to create by hand a synthetic database of Ellington Field, the group not having access to expensive commercial databases. The author and JSC's Donald Reed coordinated the flight operations and air traffic control approvals, Fox and Coffman handled the interagency approvals, and by March 2006, the FAA Challenger 604 was at Ellington Field with the required instrumentation installed and ready for the first sensor fusion–guided instrument approach demonstration. Fox had borrowed helmet-mounted display hardware and a kneeboard computer to display selected sensor fusion scenes in the cabin, and five demonstration flights were completed for over a dozen JSC Shuttle, Flight Crew Operations, and Constellation managers. In May, the flights were completed to the FAA's satisfaction. The sensor fusion software and hardware performed flawlessly, and both JSC and FAA Oklahoma City management gained confidence in the team's capabilities, a confidence that would continue to pay dividends. For its part, JSC could not afford a more extensive, focused program, nor were Center managers uniformly convinced of the applicability of this technology to their missions. The team, however, had greatly bolstered confidence in its ability to accomplish critically significant flight tests, demonstrating that it could do so with "shoestring" resources and support. It did so by using a small-team approach, building strong interagency partnerships, creating relationships with other research organizations and small businesses, relying on trust in one another's professional abilities, and following rigorous adherence to appropriate multiagency safety reviews.

The success of the approach demonstrations allowed the team members to continue with the SVS work on a not-to-interfere basis with their regularly assigned duties. Fox persuaded his management to allow the ACES van to remotely control the JSC Scout simulated lunar rover on three trips to Meteor Crater, AZ, in 2005–2006 using the same sensor fusion software implementation as that on the Challenger flight test. Throughout the remainder of 2006, the team discussed other possibilities to demonstrate its system. Abernathy provided the author with a kneeboard computer, a GPS receiver, and the RIS's LandForm software (for which JSC had rights) with a compressed, high-resolution database of the Houston area. On NASA T-38 training flights, the author evaluated the performance of the all-aspect software in anticipation of an official

WB-57F High-Altitude Research Aircraft being prepared for a WAVE/sensor fusion test. NASA.

evaluation as part of a potential T-38 fleet upgrade. The author had conversations with Coffman, Fox, and Abernathy regarding the FAA's idea of using a turret on flight-check aircraft to measure in real time the height of approach corridor obstacles. The conservations and the portability of the software and hardware inspired the author to suggest a flight test using one of JSC's WB-57F High-Altitude Research Airplanes with the WB-57F Acquisition Validation Experiment (WAVE) sensor as a proof of concept. The WB-57F was a JSC high-altitude research airplane capable of extended flight above 60,000 feet with sensor payloads of thousands of pounds and dozens of simultaneous experiments. The WAVE was a sophisticated, 360-degree slewable camera tracking system developed after the Columbia accident to track Space Shuttle launches and reentries.[76]

The author flew the WB-57F at JSC, including WAVE Shuttle tracking missions. Though hardly the optimal airframe (the sensor fusion proof of concept would be flown at only 2,000 feet altitude), the combination of a JSC airplane with a slewable, INS/GPS–supported camera system was hard to beat. The challenges were many. The two WB-57F airframes at JSC were scheduled years in advance, they were expensive for a single

76. Rivers and Fox, "Synthetic Vision System Flight Testing for NASA's Exploration Space Vehicles," paper presented at the *52nd Symposium of the Society of Experimental Test Pilots, Anaheim, CA, Sept. 2008.*

experiment when designed to share costs among up to 40 simultaneous experiments, and the WAVE sensor was maintained by Southern Research Institute (SRI) in Birmingham, AL. Fortunately, Mike Abernathy of RIS spoke directly to John Wiseman of SRI, and an agreement was reached in which SRI would integrate RIS's LandForm software into WAVE for no cost if it were allowed to use it for other potential WAVE projects.

The team sought FAA funding on the order of $30,000–$40,000 to pay for the WB-57F operation and integration costs and transport the WAVE sensor from Birmingham to Houston. In January 2007, the team invited Frederic Anderson—Manager of Aero-Nav Services at FAA Oklahoma City—to visit JSC to examine the ACES van, meet with the WB-57F Program Office and NASA Exploration Program officials, and receive a demonstration of the sensor fusion capabilities. Anderson was convinced of the potential of using the WB-57/WAVE to prove that an object on the ground could be passively, remotely measured in real time to high accuracy. He was willing to commit $40,000 of FAA money to this idea. With one challenge met, the next challenge was to find a hole in the WB-57F's schedule.

In mid-March 2007, the author was notified that a WB-57F would be available the first week in April. In 3 weeks Fox, pushed through a Space Act Agreement to get FAA Oklahoma City funds transferred to JSC, with pivotal help from the JSC Legal Office. RIS and AANA, working nonstop with SRI, integrated the sensor fusion software into the WAVE computers. Due to a schedule slip with the WB-57, the team only had a day and a half to integrate the RIS hardware into the WB-57, with the invaluable help of WB-57 engineers. Finally, on April 6, on a 45-minute flight from Ellington Field, the author and WAVE operator Dominic Del Rosso of JSC for the first time measured an object on the ground (the JSC water tower) in flight in real time using SVS technology. The video signal from the WAVE acquisition camera was blended with synthetic imagery to provide precise scaling.

The in-flight measurement was within 0.5 percent of the surveyed data. The ramifications of this accomplishment were immediate and profound: the FAA was convinced of the power of the SVS sensor fusion technology and began incorporating the capability into its planned flight-check fleet upgrade.[77]

77. Ibid., pp. 44–51.

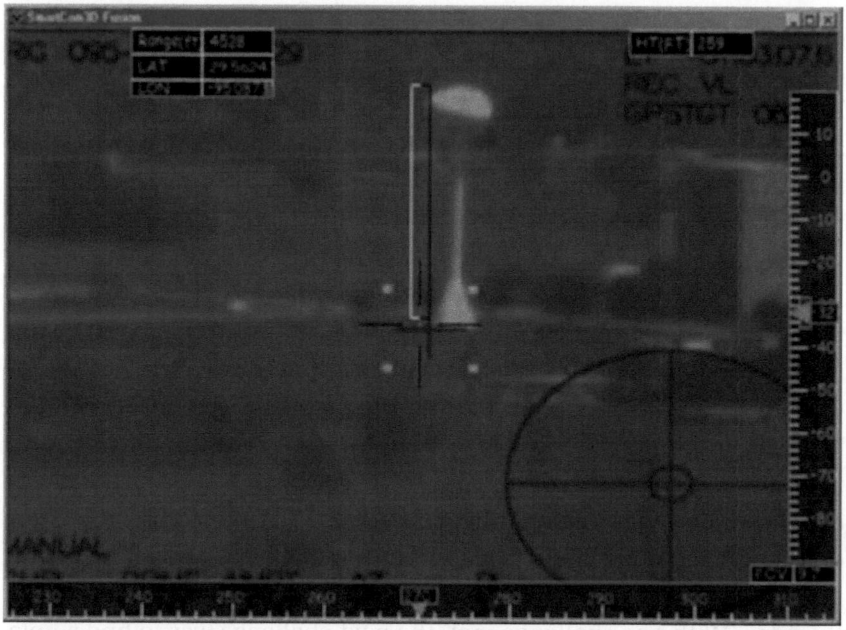

Historic first in-flight measurement of an object on the ground using SVS technology. NASA.

Building on this success, Fox, Coffman, Abernathy, and the author looked at new ways to showcase sensor fusion. In the back of their minds had been the concept of simulating a lunar approach into a virtual lunar base anchored over Ellington Field. The thought was to use the FAA Challenger 604 with the SVS portable pallet installed as before. The problem was money. A solution came from a collaboration between the author and his partner in a NASA JSC aircraft fleet upgrade study, astronaut Joseph Tanner. They had extra money from their fleet study budget, and Tanner was intrigued by the proposed lunar approach simulation because it related to a possible future lunar approach training aircraft. The two approached Brent Jett, who was Director of Flight Crew Operations and the sponsor of their study, in addition to overseeing the astronauts and flight operations at JSC. Jett was impressed with the idea and approved the necessary funds to pay the operational cost of the Challenger 604. FAA Oklahoma City would provide the airplane and crew at its expense.

Once again, RIS and AANA on their own modified the software to simulate a notional lunar approach designed by the author and Fox, derived from the performance of the Challenger 604 aircraft. Coffman was able to retrieve the monitors and cameras from the approach flight tests of 2006. Jim Secor spent a day at FAA Oklahoma City reinstalling the SVS pallet and performing the necessary integration with Michael Coffman. The author worked with Houston Approach Control to gain approval for this simulated lunar approach with a relatively steep flightpath into Ellington Field and within the Houston Class B (Terminal Control Area) airspace. The trajectory commenced at 20,000 feet, with a steep power-off dive to 10,000 feet, at which point a 45-degree course correction maneuver was executed. The approach terminated at 2,500 feet at a simulated 150-feet altitude over a virtual lunar base anchored overhead Ellington Field. Because there was no digital database available for any of the actual proposed lunar landing sites, the team used a modified database for Meteor Crater as a simulated lunar site. The team switched the coordinates to Ellington Field so that the EGI could still provide precise GPS navigation to the virtual landing site anchored overhead the airport.

In early February 2008, all was ready. The ACES van was used to validate the model as there was no time (or money) to do it on the airplane. One instrumentation checkout flight was flown, and several anomalies were corrected. That afternoon, for the first time, an aircraft was used to simulate a lunar approach to a notional lunar base. Sensor fusion was demonstrated on one of the monitors using the actual ambient conditions to provide Sun glare and haze challenges. These were not

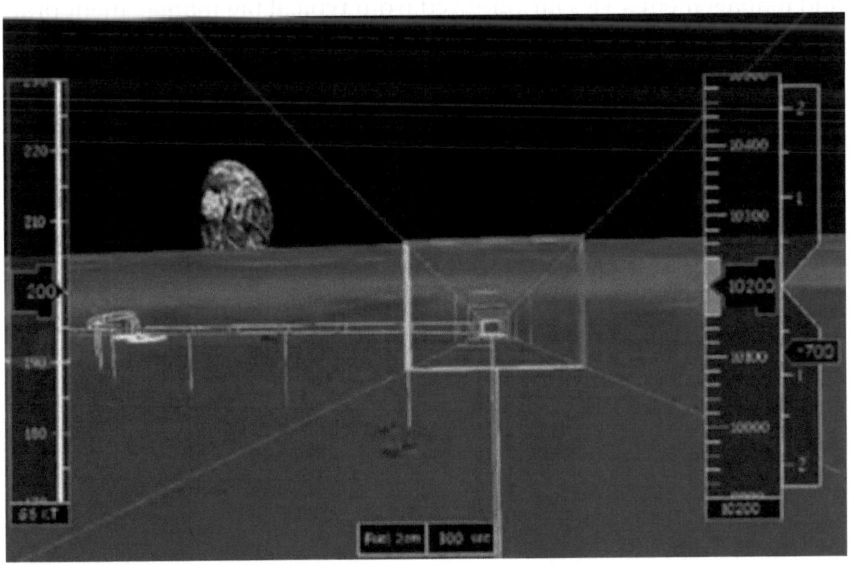

Screen shot of the SVS simulated lunar approach PFD. NASA.

representative of actual lunar issues but were indicative of the benefit of sensor fusion to mitigate the postulated 1-degree Sun angles of the south lunar pole. The second monitor showed the SVS Meteor Crater digital terrain database simulating the lunar surface and perfectly matched to the Houston landscape over which the Challenger 604 was flying.[78] This and two more flights demonstrated the technology to a dozen astronauts and to Constellation and Orion program managers.

Four flight test experiments and three trips to Meteor Crater were completed in a 3-year period to demonstrate the SVS sensor fusion technology. The United States military is using evolved versions of the original X-38 SVS and follow-on sensor fusion software with surveillance sensors on various platforms, and the FAA has contracted with RIS to develop an SVS for its flight-check fleet. Constellation managers have shown much interest in the technology, but by 2009, no decision has been reached regarding its incorporation in NASA's space exploration plans.[79]

The Way Ahead for SVS

NASA's long heritage of research in synthetic vision has generated useful concepts, demonstrations of key technological breakthroughs, and prototype systems and architectures that have influenced both the private and public sectors. Much of this work has been accomplished by small teams of dedicated researchers, often using creative approaches and management styles far removed from typical big management practices. As this book goes to press, synthetic vision and advanced flight path guidance constitutes a critical piece of the Agency's future work on Integrated Intelligent Flight Deck Technologies and related activities aimed at fulfilling the promise of better air transportation and military airpower. While long-range institutional and national budgetary circumstances add greater uncertainties to the challenge of forecasting the future, it is clear that as the advent of blind-flying instrumentation transformed aviation safety and utility in the interwar years, the advent of synthetic vision will accomplish the same in the first years of the 21st century, furnishing yet another example of the enormous and continuing contributions of NASA and its people to the advancement of aeronautics.

78. Ibid., pp. 51–57.

79. As this work goes to press, Agency space exploration plans, of course, are very much a "work in progress" with debates over destinations, types of missions, and mission protocols.

Recommended Additional Readings

Reports, Papers, Articles, and Presentations:

T.S. Abbott and G.G. Steinmetz, "Integration of Altitude and Airspeed Information into a Primary Flight Displays," NASA TM-89064 (1987).

J.J. Adams, "Flight-Test Verification of a Pictorial Display for General Aviation Instrument Approach," NASA TM-83305 (1982).

J.J. Adams, "Simulator Study of a Pictorial Display for General Aviation Instrument Flight," NASA TP-1963 (1982).

J.J. Adams, "Simulator Study of Pilot-Aircraft-Display System Response Obtained with a Three-Dimensional-Box Pictorial Display," NASA TP-2122 (1983).

J.J. Adams and F.J. Lallman, "Description and Preliminary Studies of a Computer Drawn Instrument Landing Approach Display," NASA-TM-78771 (1978).

A.L. Alexander and C.D. Wickens, "Synthetic Vision Systems: Flightpath Tracking, Situation Awareness, and Visual Scanning in an Integrated Hazard Display," paper presented at the *13th International Symposium on Aviation Psychology, 2005.*

Jarvis J. Arthur, III, L. Kramer, and Randall E. Bailey, "Flight Test Comparison between Enhanced Vision (FLIR) and Synthetic Vision Systems," in Jacques G. Verly, ed., *Proceedings of SPIE, Enhanced and Synthetic Vision 2005*, vol. 5802-03 (2005).

Jarvis J. Arthur, III, Lawrence J. Prinzel, III, Lynda J. Kramer, Randall E. Bailey, and Russell V. Parrish, "CFIT [Controlled Flight into Terrain-ed.] Prevention using Synthetic Vision," in Jacques G. Verly, ed., *Proceedings of SPIE, Enhanced and Synthetic Vision 2003*, vol. 5081 (2003), pp. 146–157.

11

Jarvis J. Arthur, III, Lawrence J. Prinzel, III, Lynda J. Kramer, Russell V. Parrish, and Randall E. Bailey, "Flight Simulator Evaluation of Synthetic Vision Display Concepts to Prevent Controlled Flight into Terrain (CFIT)," NASA TP-2004-213008 (2004).

Jarvis J. Arthur, III, Lawrence J. Prinzel, III, K.J. Shelton, Lynda J. Kramer, Steven P. Williams, Randall E. Bailey, and R.M. Norman, "Design and Testing of an Unlimited-Field-of-Regard Synthetic Vision Head-Worn Display for Commercial Aircraft Surface Operations," NASA LRC and Boeing Phantom Works, NTRS Report LAR-17290-1 (2007).

Jarvis J. Arthur, III, Lawrence J. Prinzel, III, Steven P. Williams, and Lynda J. Kramer, "Synthetic Vision Enhanced Surface Operations and Flight Procedures Rehearsal Tool," *SPIE Defense and Security Symposium, Apr. 2006.*

Jarvis J. Arthur, III, Steven P. Williams, Lawrence J. Prinzel, III, Lynda J. Kramer, and Randall E. Bailey, "Flight Simulator Evaluation of Display Media Devices for Synthetic Vision Concepts," in Jacques G. Verly, ed., *Proceedings of SPIE, Enhanced and Synthetic Vision 2004*, vol. 5424 (2004).

J. Atkins, "Prototypical Knowledge for Expert Systems," *Artificial Intelligence*, vol. 20, no. 2 (Feb. 1983), pp. 163–210.

Randall E. Bailey, J.J. Arthur, III, Lawrence J. Prinzel, III, and Lynda J. Kramer, "Evaluation of Head-Worn Display Concepts for Commercial Aircraft Taxi Operations," *Proceedings of SPIE Enhanced and Synthetic Vision*, vol. 6559 (Apr. 2007).

Randall E. Bailey, Jarvis J. Arthur, III, and Steven P. Williams, "Latency Requirements for Head-Worn E/SVS Applications," in Jacques G. Verly, ed., *Proceedings of SPIE, Enhanced and Synthetic Vision 2004*, vol. 5424 (2004).

Randall E. Bailey, Lynda J. Kramer, and Lawrence J. Prinzel, III, "Crew and Display Concepts Evaluation for Synthetic/Enhanced Vision Systems," *Proceedings of SPIE, The International Society For Optical Engineering*, vol. 6226: *Enhanced and Synthetic Vision* (2006).

Randall E. Bailey, Lynda J. Kramer, and Lawrence J. Prinzel, III, "Fusion of Synthetic and Enhanced Vision for All-Weather Commercial Aviation Operations," NATO RTO/HFM-141, paper presented at the *NATO Symposium on Human Factors and Medicine Symposium on Human Factors of Day/Night All Weather Operations, Heraklion, Greece, Apr. 23–25, 2007.*

Randall E. Bailey, Russell V. Parrish, Jarvis J. Arthur, III, Louis J. Glaab, and Lynda J. Kramer, "Flight Test Evaluations (DFW and EGE) of Tactical Synthetic Vision Display Concepts," *SAE Guidance and Control Meeting, Mar. 2002.*

Randall E. Bailey, Russell V. Parrish, Jarvis J. Arthur, III, and R.M. Norman, "Flight Test Evaluation of Tactical Synthetic Vision Display Concepts in a Terrain-Challenged Operating Environment," *SPIE 16th Annual International Symposium on Aerospace/Defense Sensing, Simulation, and Controls, Apr. 2002.*

Randall E. Bailey, Russell V. Parrish, Lynda J. Kramer, S.D. Harrah, and Jarvis J. Arthur, III, "Technical Challenges in the Development of a NASA Synthetic Vision System Concept," *Proceedings of the North Atlantic Treaty Organization (NATO) Symposium on Enhanced and Synthetic Vision Systems* (Sept. 2002).

D. Baize and C. Allen, "Synthetic Vision Systems Project Plan, Version IIId," NASA Langley Research Center (Nov. 2001).

Sheldon Baron and Carl Feehrer, "An Analysis of the Application of AI to the Development of Intelligent Aids for Flight Crew Tasks," NASA CR-3944 (1985).

Anthony P. Bartolone, Louis J. Glaab, Monica F. Hughes, and Russell V. Parrish, "Initial Development of a Metric to Describe the Level of Safety Associated with Piloting an Aircraft with Synthetic Vision Systems (SVS) Displays," in Jacques G. Verly, *Proceedings of SPIE, The International Society For Optical Engineering, vol. 5802: Enhanced and Synthetic Vision 2005* (2005).

A.P. Bartolone, Monica F. Hughes, D.T. Wong, and Mohammad A. Takallu, "Symbology Development for General Aviation Synthetic Vision Primary Flight Displays for the Approach and Missed-Approach Modes of Flight" paper presented at the *Human Factors and Ergonomics Society 48th Annual Meeting, Sept. 2004.*

Scott A. Berry, Thomas J. Horvath, K. James Weilmuenster, Stephan J. Alter, and N. Ronald Merski, "X-38 Experimental Aeroheating at Mach 10," AIAA Paper 2001-2828 (2000).

11

S.O. Beskenis, D.F. Green, P.V. Hyer, and E.J. Johnson, Jr., "Integrated Display System for Low Visibility Landing and Surface Operations," NASA CR-1998-208446 (1998).

Boeing, *Boeing Statistical Summary of Commercial Jet Aircraft Accidents, Worldwide Operations, 1959–1997* (Seattle, WA: Airplane Safety Engineering, Boeing Commercial Airplane Group, 1998).

A. Both, J. Klein, S. Koczo, and T. Lamb, "Preliminary System Requirements for Synthetic Vision," Rockwell Collins Report NCA1-125.11.10.4 (Dec. 1998).

G. Boucek, Jr., "Candidate Concept Descriptions for SVS/EVS Retrofit in Airplanes with CRT Type Primary Flight Instrumentation," Research Triangle Institute Report No. RTI/7473/034-01S (Sept. 2001).

M.A. Burgess, "Synthetic Vision for Low-Visibility Aircraft Operations: What We Know and What We Do Not Know," in J.G. Verly and S.S. Welch, eds., *Proceedings of the SPIE Conference on Sensing, Imaging, and Vision for Control and Guidance of Aerospace Vehicles*, vol. 2220 (July 1994) pp. 206–217.

J.L. Campbell, and M. Uijt de Haag, "Assessment of Radar Altimeter Performance When Used for Integrity Monitoring in a Synthetic Vision System," *Proceedings of the 20th Digital Avionics Systems Conference* (June 2001).

J.L. Campbell, A. Vadlamani, M. Uijt de Haag, and Steven D. Young, "The Application of LiDAR to Synthetic Vision System Integrity," paper presented at the *22nd AIAA/IEEE Digital Avionics Systems Conference, Oct. 2003.*

R. Cassell, C. Evers, and J. Esche, "Safety Benefits of PathProx™—A Runway Incursion Alerting System," paper presented at the *Proceedings of the AIAA/IEEE 22nd Digital Avionics Systems Conference* (Oct. 2003).

R. Cassell, C. Evers, J. Esche, and B. Sleep, "NASA Runway Incursion Prevention System (RIPS) Dallas-Fort Worth Demonstration Performance Analysis," NASA CR-2002-211677 (2002).

R. Cassell, C. Evers, B. Sleep, and J. Esche, "Initial Test Results of PathProx™—A Runway Incursion Alerting System," in *Proceedings of the 20th Digital Avionics Systems Conference* (Oct. 2001).

J.R. Comstock, Jr., Louis J. Glaab, Lawrence J. Prinzel, III, and D.M. Elliott, "Can Effective Synthetic Vision System Displays Be Implemented on Limited Size Display Spaces?" *Proceedings of the 11th International Symposium on Aviation Psychology* (Mar. 2001).

J.R. Comstock, Jr., L.C. Jones, and A.T. Pope, "The Effectiveness of Various Attitude Indicator Display Sizes and Extended Horizon Lines on Attitude Maintenance in a Part-Task Simulation," paper presented at the *Human Factors and Ergonomics Society 2003 Meeting, Oct. 2003.*

E. Cooper and Steven D. Young, "Database Integrity Monitoring for Synthetic Vision Systems Using Machine Vision and SHADE," paper presented at the *SPIE Defense and Security Symposium, Enhanced and Synthetic Vision Conference, Mar. 28–Apr. 1, 2005.*

George E. Cooper, "The Pilot-Aircraft Interface," in NASA LRC, *Vehicle Technology for Civil Aviation: The Seventies and Beyond,* NASA SP-292 (1971).

G. Craig, S. Jennings, N. Link, and R. Kruk, "Flight Test of a Helmet-Mounted, Enhanced and Synthetic Vision System for Rotorcraft Operations," paper presented at the *American Helicopter Society 58th Annual Forum, June 2002.*

O. Dieffenbach, "Autonomous Precision Approach and Landing System," in *Proceedings of SPIE Air Traffic Control Technologies*, vol. 2464 (Apr. 1995), pp. 158–164.

H.J. Dudfield, "Pictorial Displays in the Cockpit: A Literature Review," Royal Aircraft Establishment (U.K.) Technical Memo FS (F) 690 (1988).

Stephen R. Ellis, Mary K. Kaiser, and Arthur Grunwald, eds., *Spatial Displays and Spatial Instruments*, NASA CP-10032 (1989).

Tim Etherington, T. Vogl, M. Lapis, and J. Razo, "Synthetic Vision Information System," in *Proceedings of the 19th Digital Avionics Systems Conference* (Oct. 2000).

Delmar M. Fadden, Rolf Braune, and John Wiedemann, "Spatial Displays as a Means to Increase Pilot Situational Awareness," in Stephen R. Ellis, Mary K. Kaiser, and Arthur Grunwald, eds., *Spatial Displays and Spatial Instruments*, NASA CP-10032 (1989).

D.C. Foyle, A. Ahumada, J. Larimer, and B. Townsend-Sweet, "Enhanced/Synthetic Vision Systems: Human Factors Research and Implications for Future Systems," paper presented at the *Society of Automotive Engineers SAE Aerotech 1992, Anaheim, CA, Oct. 1992.*

D.C. Foyle, A. Andre, B. Hooey, and R. McCann, "T-NASA Taxi Test at Atlanta Airport," NASA TM-112240 (1998).

G. French and T. Schnell, "Terrain Awareness and Pathway Guidance for Head-Up Displays (TAPGUIDE); A Simulator Study of Pilot Performance," in *22nd Digital Avionics Systems Conference, Oct. 2003.*

G. Ganoe and Steven D. Young, "Utilization of GPS Surface Reflected Signals to Provide Aircraft Altitude Verification for SVS," in *SPIE Defense and Security Symposium, Enhanced and Synthetic Vision Conference, Mar. 28–Apr. 1, 2005.*

Louis J. Glaab and Monica F. Hughes, "Terrain Portrayal for Head-Down Displays Flight Test," *Proceedings of the 22nd Digital Avionics Systems Conference* (2003).

Louis J. Glaab, Lynda J. Kramer, Jarvis J. Arthur, III, and Russell V. Parrish, *Flight Test Evaluation of Synthetic Vision Display Concepts at Dallas/Fort-Worth Airport*, NASA 2003-TP212177 (2003).

Louis J. Glaab, Russell V. Parrish, Monica F. Hughes, and Mohammad A. Takallu, "Effectively Transforming IMC Flight into VMC Flight: An SVS Case Study," *Proceedings of the 25th Digital Avionics System Conference* (2006).

Louis J. Glaab, Russell V. Parrish, Monica F. Hughes, and Mohammad A. Takallu, "Transforming IMC Flight into VMC Flight: An SVS Case Study," *25th Digital Avionics Systems Conference, Oct. 2006.*

Louis J. Glaab and Mohammad A. Takallu, "Preliminary Effect of Synthetic Vision Systems Displays to Reduce Low-Visibility Loss of Control and Controlled Flight Into Terrain Accidents," in *2002 SAE General Aviation Technology Conference and Exhibition, Apr. 2002.*

G. Barry Graves, Jr., "Advanced Avionic Systems," in NASA LRC, *Vehicle Technology for Civil Aviation, NASA SP-292 (1971).*

D.F. Green, Jr., "Runway Safety Monitor Algorithm for Runway Incursion Detection and Alerting," NASA CR-2002-211416 (2002).

D.F. Green, Jr., "Runway Safety Monitor Algorithm for Single and Crossing Runway Incursion Detection and Alerting," NASA CR-2006-214275 (2006).

11

A.J. Grunwald, "Improved Tunnel Display for Curved Trajectory Following: Control Considerations and Experimental Evaluation," *Journal of Guidance, Control, and Dynamics*, vol. 19, no. 2 (Mar.–Apr. 1996), pp. 370–384.

A.J. Grunwald, J.B. Robertson, and J.J. Hatfield, "Evaluation of a Computer-Generated Perspective Tunnel Display for Flight Path Following," NASA TP-1736 (1980).

S. Harrah, W. Jones, C. Erickson, and J. White, "The NASA Approach to Realize a Sensor Enhanced-Synthetic Vision System (SE-SVS)," in *Proceeding of the 21st Digital Avionics Systems Conference* (Oct. 2002).

R.L. Harris, Sr., and Russell V. Parrish, "Piloted Studies of Enhanced or Synthetic Vision Display Parameters," paper presented at the *Society of Automotive Engineers SAE Aerotech 1992, Anaheim, CA, Oct. 1992.*

Stella V. Harrison, Lynda J. Kramer, Randall E. Bailey, Denise R. Jones, Steven D. Young, Steven D. Harrah, Jarvis J. Arthur, III, and Russell V. Parrish, "Initial SVS Integrated Technology Evaluation Flight Test Requirements and Hardware Architecture," NASA TM-2003-212644 (2003).

S. Hasan, R.B. Hemm, Jr., and S. Houser, "Analysis of Safety Benefits of NASA Aviation Safety Program Technologies," LMI NS112S2 (Dec. 2002).

S. Hasan, R.B. Hemm, Jr., and S. Houser, "Preliminary Results of an Integrated Safety Analysis of NASA Aviation Safety Program Technologies: Synthetic Vision and Weather Accident Prevention," in *Proceedings of the 20th Digital Avionics Systems Conference* (Mar. 2002).

J.J. Hatfield and Russell V. Parrish, "Advanced Cockpit Technology for Future Civil Transport Aircraft," paper presented at the *11th IEEE/AESS Dayton Chapter Symposium, Nov. 1990.*

C.C. Hawes and M.F. DiBenedetto, "The Local Area Augmentation System: An Airport Surveillance Application Supporting the FAA Runway Incursion Reduction Program Demonstration at the Dallas-Fort Worth International Airport," in *Proceedings of the 20th Digital Avionics Systems Conference* (Oct. 2001).

G. He, T. Feyereisen, K. Conner, S. Wyatt, J. Engels, A. Gannon, and B. Wilson, "EGPWS on Synthetic Vision Primary Flight Display," in *Proceedings of SPIE Enhanced and Synthetic Vision*, vol. 6559 (Apr. 2007).

R.B. Hemm, Jr., "Benefit Estimates of Synthetic Vision Technology," Logistics Management Institute, Report NS002S1 (June 2000).

R.B. Hemm, Jr., and S. Houser, "A Synthetic Vision Preliminary Integrated Safety Analysis," NASA CR-2001-21128 (2000).

R.B. Hemm, Jr., D. Lee, V. Stouffer, and A. Gardner, "Additional Benefits of Synthetic Vision Technology," Logistics Management Institute Report NS014S1 (June 2001).

Glenn D. Hines, Zia-ur Rahman, Daniel J. Jobson, and Glenn A. Woodell, "A Real-Time Enhancement, Registration, and Fusion for a Multi-Sensor Enhanced Vision System," in *SPIE Defense and Security Symposium, Apr. 2006.*

Glenn D. Hines, Zia-ur Rahman, Daniel J. Jobson, and Glenn A. Woodell, "DSP Implementation of the Multiscale Retinex Image Enhancement Algorithm," in *Proceedings of SPIE Visual Information Processing XIII* (Apr. 2004).

Glenn D. Hines, Zia-ur Rahman, Daniel J. Jobson, Glenn A. Woodell, and S.D. Harrah, "Real-time Enhanced Vision System," in *SPIE Defense and Security Symposium 2005, Mar. 2005.*

J.A. Houck, "A Simulation Study of Crew performance in Operating an Advanced Transport Aircraft in an Automated Terminal Area Environment," NASA TM-84610 (1983).

Richard M. Hueschen, J.F. Creedon, W.T. Bundick, and J.C. Young, "Guidance and Control System Research for Improved Terminal Area Operations," in Joseph W. Stickle, ed., *1980 Aircraft Safety and Operating Problems*, NASA CP-2170, pt. 1 (1981), pp. 51–61.

Richard M. Hueschen, W. Hankins, and L.K. Barker, "Description and Flight Test of a Rollout and Turnoff Head-Up Display Guidance System," in *Proceedings of the AIAA/IEEE/SAE 17th Digital Avionics System Conference*, Seattle, WA (Oct. 1998).

Richard M. Hueschen and John W. McManus, "Application of AI Methods to Aircraft Guidance and Control," in *Proceedings of the 7th American Control Conference, June 15–17, 1988*, vol. 1 (New York: Institute of Electrical and Electronics Engineers, 1988), pp. 195–201.

Monica F. Hughes and Louis J. Glaab, "Terrain Portrayal for Head-Down Displays Simulation Experiment Results," *Proceedings of the 22nd Digital Avionics Systems Conference* (2003).

Monica F. Hughes, and Louis J. Glaab, "Terrain Portrayal for Synthetic Vision Systems Head-Down Displays Evaluation Results," NASA TP-2007-214864 (2007).

P.V. Hyer, "Demonstration of Land and Hold Short Technology at the Dallas-Fort Worth International Airport," NASA CR-2002-211642 (2002).

P.V. Hyer and S. Otero, "Cockpit Displays for Enhancing Terminal-Area Situational Awareness and Runway Safety," NASA CR-2007-214545 (2007).

Daniel J. Jobson, Zia-ur Rahman, and Glenn A. Woodell, "The Spatial Aspect of Color and Scientific Implications of Retinex Image Processing," in *Proceedings of the SPIE International Symposium on AeroSense, Conference on Visual Information Processing X* (Apr. 2001).

Daniel J. Jobson, Zia-ur Rahman, and Glenn A. Woodell, "The Statistics of Visual Representation," in *Proceedings of SPIE Visual Information Processing XI*, vol. 4736 (Apr. 2002).

Daniel J. Jobson, Zia-ur Rahman, Glenn A. Woodell, and Glenn D. Hines, "A Comparison of Visual Statistics for the Image Enhancement of FORESITE Aerial Images with Those of Major Image Classes," in *SPIE Defense and Security Symposium* (Apr. 2006).

E.J. Johnson and P.V. Hyer, "Roll-Out and Turn-Off Display Software for Integrated Display System," NASA CR-1999-209731 (1999).

Denise R. Jones, "Runway Incursion Prevention System Testing at the Wallops Flight Facility," in *SPIE Defense and Security Symposium, Mar. 28–Apr. 1, 2005.*

Denise R. Jones and Lawrence J. Prinzel, III, "Runway Incursion Prevention for General Aviation Operations," in *Proceeding of the 25th Digital Avionics Systems Conference* (Oct. 2006).

Denise R. Jones, Cuong C. Quach, and Steven D. Young, "Runway Incursion Prevention System—Demonstration and Testing at the Dallas/Fort Worth International Airport," paper presented at the *20th Digital Avionics Systems Conference, Daytona Beach, FL, Oct. 14–18, 2001.*

Denise R. Jones and J.M. Rankin, "A System for Preventing Runway Incursions," *Journal of Air Traffic Control*, July 2002.

M. Junered, S. Esterhuizen, D. Akos, and P.A. Axelrad, "Modular GPS Remote Sensing Software Receiver for Small Platforms," in *Proceedings of the Institute of Navigation GPS/GNSS Conference* (2006).

Mary K. Kaiser, "The Surface Operations Research and Evaluation Vehicle (SOREV): A Testbed for HSCT taxi issues," AIAA Paper 1998-5558 (1998).

Mary K. Kaiser, "Surface Operations Research and Evaluation Vehicle," NASA ARC, *Research and Technology 1997* (Sept. 1998), pp. 81–82.

Rangachar Kasturi, Octavia Camps, and Lee Coraor, "Performance Characterization of Obstacle Detection Algorithms for Aircraft Navigation," NASA ARC, TR-CSE-00-002, Jan. 2000.

Rangachar Kasturi, Octavia Camps, Y. Huang, A. Narasimhamurthy, and N. Pande, "Wire Detection Algorithms for Navigation," NAG2-1487, Jan. 2002.

Rangachar Kasturi, Sadashiva Devadiga, and Yuan-Liang Tang, "A Model-Based Approach for Detection of Runways and Other Objects in Image Sequences Acquired Using an On-Board Camera," NASA CR-196424 (1994).

M. Keller, T. Schnell, Louis J. Glaab, and Russell V. Parrish, "Pilot Performance as a Function of Display Resolution and Field of View in a Simulated Terrain Following Flight Task Using a Synthetic Vision System," in *Proceeding of the 22nd Digital Avionics Systems Conference* (Oct. 2003).

Lynda J. Kramer, ed., *Synthetic Vision Workshop 2*, NASA CP-1999-209112 (Mar. 1999).

Lynda J. Kramer, Jarvis J. Arthur, III, Randall E. Bailey, and Lawrence J. Prinzel, III, "Flight Testing an Integrated Synthetic Vision System," in Jacques G. Verly, ed., *Proceedings of SPIE Defense and Security Symposium, Enhanced and Synthetic Vision*, vol. 5802-01 (2005).

Lynda J. Kramer and R. Michael Norman, "High-Speed Research Surveillance Symbology Assessment Experiment," ASA TM-2000-210107 (Apr. 2000).

Lynda J. Kramer, Lawrence J. Prinzel, III, Jarvis J. Arthur, III, and Randall E. Bailey, "Advanced Pathway Guidance Evaluations on a Synthetic Vision Head-Up Display," NASA TP-2005-213782 (2005).

Lynda J. Kramer, Lawrence J. Prinzel, III, Jarvis J. Arthur, III, and Randall E. Bailey, "Pathway Design Effects on Synthetic Vision Head-Up Displays," in Jacques G. Verly, ed., *Proceedings of SPIE, Enhanced and Synthetic Vision 2004*, vol. 5424 (Apr. 12, 2004).

Lynda J. Kramer, Lawrence Prinzel, Randall E. Bailey, and Jarvis Arthur, "Synthetic Vision Enhances Situation Awareness and RNP Capabilities for Terrain-Challenged Approaches," *Proceedings of the AIAA Third Aviation Technology, Integration, and Operations Technical Forum*, AIAA 2003-6814 (2003).

Lynda J. Kramer, Lawrence J. Prinzel, III, Randall E. Bailey, Jarvis J. Arthur, III, and Russell V. Parrish, "Flight Test Evaluation of Synthetic Vision Concepts at a Terrain Challenged Airport," NASA TP-2004-212997 (2004).

Lynda J. Kramer, Steven P. Williams, Randall E. Bailey, and Louis J. Glaab, "Synthetic Vision Systems—Operational Considerations Simulation Experiment," *Proceedings of SPIE Enhanced and Synthetic Vision*, vol. 6559 (Apr. 2007).

J. Larimer, M. Pavel, A.J. Ahumada, and B.T. Sweet, "Engineering a Visual System for Seeing Through Fog," paper presented at the *22nd International Conference on Environmental Systems, Seattle, WA, SAE SP-1130, July 1992.*

A. Lechner, P. Mattson, and K. Ecker, "Voice Recognition—Software Solutions in Real Time ATC Workstations," paper presented at the *Proceedings of the 20th Digital Avionics Systems Conference* (Oct. 2001).

K. Lemos and T. Schnell, "Synthetic Vision Systems: Human Performance Assessment of the Influence of Terrain Density and Texture," in *Proceedings of 22nd Digital Avionics Systems Conference* (Oct. 2003).

K. Lemos, T. Schnell, Tim Etherington, and D. Gordon, "'Bye-Bye Steam Gauges, Welcome Glass:' A Review of New Display Technology for General Aviation Aircraft," in *Proceedings of 21st Digital Avionics Systems Conference* (Oct. 2002).

D. Masters, D. Akos, S. Esterhuizen, and E. Vinande, "Integration of GNSS Bistatic Radar Ranging into an Aircraft Terrain Awareness and Warning System," in *Proceedings of the Institute of Navigation GPS/GNSS Conference* (2005).

D. Masters, P. Axelrad, V. Zavorotny, S.J. Katzberg, and F. Lalezari, "A Passive GPS Bistatic Radar Altimeter for Aircraft Navigation," in *Proceedings of the Institute of Navigation GPS/GNSS Conference* (2001).

R.S. McCann, "Building the Traffic, Navigation, and Situation Awareness System (T-NASA) for Surface Operations," NASA CR-203032 (1996).

R.S. McCann, B.L. Hooey, B. Parke, D.C. Foyle, A.D. Andre, and B. Kanki, "An Evaluation of the Taxiway Navigation and Situation Awareness (T-NASA) System in High-Fidelity Simulation," *SAE Transactions: Journal of Aerospace*, No. 107 (1998), pp. 1612–1625.

D. McKay, M. Guirguis, R. Zhang, and R. Newman, "Evaluation of EVS for Approach Hazard Detection," paper presented at the *NATO RTO SET Workshop on Enhanced and Synthetic Vision Systems, NATO RTO-MP-107, Ottawa, Canada, Sept. 2002.*

S. Merchant, Y.T. Kwon, T. Schnell, Tim Etherington, and T. Vogl, "Evaluation of Synthetic Vision Information System (SVIS) Displays Based on Pilot Performance," *Proceedings of the 20th Digital Avionics Systems Conference* (Oct. 2001).

S. Merchant, T. Schnell, and Y. Kwon, "Assessing Pilot Performance in Flight Decks Equipped with Synthetic Vision Information System," in *Proceedings of the 11th International Symposium on Aviation Psychology* (Mar. 2001).

V. Merrick and J. Jeske, "Flightpath Synthesis and HUD Scaling for V/STOL Terminal Area Operations," NASA TM-110348 (1995).

M. Morici, "Aircraft Position Validation using Radar and Digital Terrain Elevation Database," U.S. Patent No. 6,233,522, U.S. Patent Office, May 15, 2001.

R. Mueller, K. Belamqaddam, S. Pendergast, and K. Krauss, "Runway Incursion Prevention System Concept Verification: Ground Systems and STIS-B Link Analysis," in *Proceedings of the 20th Digital Avionics Systems Conference* (Oct. 2001).

NASA, Langley Research Center, *Vehicle Technology for Civil Aviation: The Seventies and Beyond,* NASA SP-292 (Washington, DC: GPO, 1971).

NASA, "NASA's High Speed Research Program: The eXternal Visibility System," in *NASA Facts FS-1998-09-34*, LaRC, Hampton, VA (1998).

Mark Nataupsky, Timothy L. Turner, Harold Land, and Lucille Crittenden, "Development of a Stereo 3-D Pictorial Primary Flight Display," in Stephen R. Ellis, Mary K. Kaiser, and Arthur Grunwald, eds., *Spatial Displays and Spatial Instruments*, NASA CP-10032 (1989).

R.L. Newman, "HUDs, and SDO: A Problem or a Bad Reputation?" paper presented at the *Recent Trends in Spatial Disorientation Conference, Nov. 2000.*

D. Nguyen, P. Chi, S. Harrah, and W. Jones, "Flight Test of IR Sensors on NASA 757 at Newport News/Williamsburg International Airport (PHF)," in *Proceeding of the 21st Digital Avionics Systems Conference* (Oct. 2002).

R.M. Norman, "Synthetic Vision Systems (SVS) Description of Candidate Concepts Document," NASA LRC, Report NAS1-20342 (2001).

Russell V. Parrish, "Avionic Pictorial Tunnel-/Pathway-/Highway-In-The Sky Workshops," NASA CP-2003-212164 (2003).

Russell V. Parrish, Randall E. Bailey, Lynda J. Kramer, Denise R. Jones, Steven D. Young, Jarvis J. Arthur, III, Lawrence J. Prinzel, III, Louis J. Glaab, and Steven D. Harrah, "Aspects of Synthetic Vision Display Systems and the Best Practices of the NASA's SVS Project," NASA TP-2008-215130 (May 2008).

Russell V. Parrish, D.G. Baize, and M.S. Lewis, "Synthetic Vision," in Cary R. Spitzer, *The Digital Avionics Handbook* (Boca Raton, FL: CRC Press, 2000).

Russell V. Parrish, Anthony M. Busquets, and Steven P. Williams, "Recent Research Results in Stereo 3-D Pictorial Displays at Langley Research Center," in *Proceedings of the 9th IEEE/AIAA/NASA Digital Avionics Systems Conference, Virginia, VA, Oct. 15–18, 1990* (New York: IEEE, 1990), pp. 529–539.

Russell V. Parrish, Anthony M. Busquets, Steven P. Williams, and D. Nold, "Evaluation of Alternate Concepts for Synthetic Vision Flight Displays with Weather-Penetrating Sensor Image Inserts During Simulated Landing Approaches," NASA TP-2003-212643 (2003).

Russell V. Parrish, Anthony M. Busquets, Steven P. Williams, and D. Nold, "Spatial Awareness Comparisons between Large-Screen, Integrated Pictorial Displays and Conventional EFIS Displays during Simulated Landing Approaches," NASA TP-3467 (1994).

Russell V. Parrish, Steven P. Williams, Jarvis J. Arthur, III, Lynda J. Kramer, Randall E. Bailey, Lawrence J. Prinzel, III, and R. Michael Norman, "A Description of the 'Crow's Foot' Tunnel Concept," NASA TM-214311 (2006).

Lawrence J. Prinzel, III, Jarvis J. Arthur, III, Lynda J. Kramer, and Randall E. Bailey, "Pathway Concepts Experiment for Head-Down Synthetic Vision Displays," in Jacques G. Verly, ed., *Proceedings of SPIE, Enhanced and Synthetic Vision 2004*, vol. 5424 (Apr. 12, 2004).

Lawrence J. Prinzel, III, Randall E. Bailey, J.R. Comstock, Jr., Lynda J. Kramer, Monica F. Hughes, and Russell V. Parrish, "NASA Synthetic Vision EGE Flight Test," paper presented at the *Human Factors and Ergonomics Society 2002 Meeting, Jan. 2002*.

Lawrence J. Prinzel, III, J.R. Comstock, Jr., Louis J. Glaab, Lynda J. Kramer, Jarvis J. Arthur, III, and J.S. Barry, "Comparison of Head-Up and Head-Down 'Highway-In-The-Sky' Tunnel and Guidance Concepts for Synthetic Vision Displays," in the *Proceedings of the Human Factors and Ergonomics Society* (2004).

Lawrence J. Prinzel, III, J.R. Comstock, Jr., Louis J. Glaab, Lynda Kramer, Jarvis J. Arthur, III, and J.S. Barry, "The Efficacy of Head-Down and Head-Up Synthetic Vision Display Concepts for Retro- and Forward-Fit of Commercial Aircraft," in the *International Journal of Aviation Psychology*, Feb. 2004.

Lawrence J. Prinzel, III, J. Comstock, C. Wickens, M. Endsley, G. French, Tim Etherington, M. Snow, and K. Corker, "Human Factors Issues in Synthetic Vision Displays: Government, Academic, Military, and Industry Perspectives," in *Proceedings of the Human Factors and Ergonomics Society* (2004).

Lawrence J. Prinzel, III, M.F. Hughes, Jarvis J. Arthur, III, Lynda J. Kramer, Louis J. Glaab, Randall E. Bailey, Russell V. Parrish, and M.D. Uenking, "Synthetic Vision CFIT Experiments for GA and Commercial Aircraft: A Picture Is Worth A Thousand Lives," paper presented at the *Human Factors and Ergonomics Society 2003 Meeting, Oct. 2003.*

Lawrence J. Prinzel, III, M. Hughes, Lynda J. Kramer, and Jarvis J. Arthur, III, "Aviation Safety Benefits of NASA Synthetic Vision: Low Visibility Loss-of-Control Incursion Detection, and CFIT Experiments," in *Proceedings of the Human Performance, Situation Awareness, and Automation Technology Conference, Embry-Riddle Aeronautical University, Daytona Beach, FL, Mar. 2004.*

Lawrence J. Prinzel, III, and Denise R. Jones, "Cockpit Technology for Prevention of General Aviation Runway Incursions," in *Proceedings of the 14th International Symposium on Aviation Psychology, Dayton, OH, Apr. 2007.*

Lawrence J. Prinzel, III, and Lynda J. Kramer, "Synthetic Vision Systems" in Waldemar Karwowski, ed., *International Encyclopedia of Ergonomics and Human Factors* (Boca Raton, FL: CRC Press—Taylor and Francis Publishers, 2006 ed.).

Lawrence J. Prinzel, III, Lynda J. Kramer, Jarvis J. Arthur, III, and Randall E. Bailey, "Evaluation of Tunnel Concepts for Advanced Aviation Displays," *Human Performance, Situation Awareness, and Automation Technology Conference, Mar. 2004.*

Lawrence J. Prinzel, III, Lynda J. Kramer, Jarvis J. Arthur, III, Randall E. Bailey, and J.R. Comstock, Jr., "Flight Test Evaluation of Situation Awareness Benefits of Integrated Synthetic Vision System Technology for Commercial Aircraft," *International Symposium on Aviation Psychology, Apr. 18–21, 2005.*

Lawrence J. Prinzel, III, Lynda J. Kramer, Jarvis J. Arthur, III, Randall E. Bailey, and J.L. Sweeters, "Development and Evaluation of 2-D and 3-D Exocentric Synthetic Vision Navigation Display Concepts for Commercial Aircraft," in Jacques G. Verly, ed., *Proceedings of SPIE Defense and Security Symposium, Enhanced and Synthetic Vision*, vol. 5802 (2005).

Lawrence J. Prinzel, III, Lynda J. Kramer, and Randall E. Bailey, "Going Below Minimums The Efficacy of Display Enhanced Synthetic Vision Fusion for Go-Around Decisions during Non-Normal Operations," in *Proceedings of the 14th International Symposium on Aviation Psychology, Dayton, OH* (Apr. 2007).

Lawrence J. Prinzel, III, Lynda J. Kramer, Randall E. Bailey, Jarvis J. Arthur, III, Steven P. Williams, and J. McNabb, "Augmentation of Cognition and Perception through Advanced Synthetic Vision Technology," paper presented at the *1st International Conference on Augmented Cognition, July 22–27, 2005.*

Lawrence J. Prinzel, III, and M. Risser, "Head-Up Displays and Attention Capture," NASA TM-2004-213000 (2004).

Zia-ur Rahman, Daniel J. Jobson, and Glenn A. Woodell, "Retinex Processing for Automatic Image Enhancement," in *Proceedings of SPIE Human Vision and Electronic Imaging VII, Symposium on Electronic Imaging* (Apr. 2002).

Zia-ur Rahman, Daniel J. Jobson, and Glenn A. Woodell, "Retinex Processing for Automatic Image Enhancement," *Journal of Electronic Imaging*, Jan. 2004.

Zia-ur Rahman, Daniel J. Jobson, Glenn A. Woodell, and Glenn D. Hines, "Automated, On-Board Terrain Analysis for Precision Landings," in *Proceedings of SPIE Visual Information Processing XIV* (Apr. 2006).

Zia-ur Rahman, Daniel J. Jobson, Glenn A. Woodell, and Glenn D. Hines, "Image Enhancement, Image Quality, and Noise," in *Proceedings of SPIE Photonic Devices and Algorithms for Computing VII* (Apr. 2005).

Zia-ur Rahman, Daniel J. Jobson, Glenn A. Woodell, and Glenn D. Hines, "Impact of Multi-scale Retinex Computation on Performance of Segmentation Algorithms," in *SPIE Defense and Security Symposium, Apr. 2004*.

Zia-ur Rahman, Daniel J. Jobson, Glenn A. Woodell, and Glenn D. Hines, "Multi-Sensor Fusion and Enhancement Using the Retinex Image Enhancement Algorithm," in *Proceedings of SPIE Visual Information Processing XI* (Apr. 2002).

C.R. Rate, A. Probert, D. Wright, W.H. Corwin, and R. Royer, "Subjective Results of a Simulator Evaluation Using Synthetic Terrain Imagery Presented on a Helmet-Mounted Display," in R. Lewandowski, W. Stephens, and L. Haworth, eds., *SPIE Proceedings Helmet and Head-Mounted Display and Symbology Design Requirements*, vol. 2218 (Apr. 1994), pp. 306–315.

J.P. Reeder, "The Airport-Airplane Interface: The Seventies and Beyond," in NASA LRC, *Vehicle Technology for Civil Aviation*, NASA SP-292 (1971).

Robert A. Rivers and Jeffrey L. Fox, "Synthetic Vision System Flight Testing for NASA's Exploration Space Vehicles," paper presented at the *52nd Symposium Proceedings, Society of Experimental Test Pilots, Anaheim, CA, Sept. 2008*.

Robert A. Rivers, E. Bruce Jackson, and W.A. Ragsdale, "Piloted Simulator Studies of the HL-20 Lifting Body," paper presented at the *35th Symposium of the Society of Experimental Test Pilots, Beverly Hills, CA, Sept. 1991.*

Society of Automotive Engineers (SAE), "Aerospace Recommended Practice (ARP) for Transport Category Airplane Head-Up Display (HUD) Systems," ARP 5288, May 2001.

Society for Automotive Engineers (SAE), "Aerospace Recommended Practice (ARP) of Human Engineering Considerations for Design and Implementation of Perspective Flight Guidance Displays," ARP 5589, Jan. 2005.

J. Schiefele, D. Howland, J. Maris, and P. Wipplinger, "Human Factors Flight Trial Analysis for 2D Situation Awareness and 3D Synthetic Vision Displays," paper presented at the *Human Factors and Ergonomics Society Meeting, Oct. 2003.*

T. Schnell and Tim Etherington, "Evaluation of Synthetic Vision Information Systems in the Simulator and in a Flight Test," *AIAA Journal of Guidance, Control, and Dynamics*, 2002.

T. Schnell and Tim Etherington, "Simulation and Field Testing of a Synthetic Vision Information System for Commercial Flight Decks,: *Proceedings of the 16th Biennial Symposium on Visibility and Simulation* (June 2002).

T. Schnell, Tim Etherington, T. Vogl, and A. Postnikov, "Field Evaluation of a Synthetic Vision Information System onboard the NASA Aries 757 at Eagle County Regional Airport," in *Proceedings of 21st Digital Avionics Systems Conference* (Oct. 2002).

T. Schnell, K. Lemos, and Tim Etherington, "Terrain Sampling Density, Texture, and Shading Requirements for SVIS," in *Final Report to the Iowa Space Grant Consortium*, Dec. 2002.

T. Schnell, Y. Kown, S. Merchant, and Tim Etherington, "Improved Flight Technical Performance in Flight Decks Equipped with Synthetic Vision Information System Displays," *International Journal of Aviation Psychology*, vol. 14 (2004), pp. 79–102.

T. Schnell, Y. Kwon, S. Merchant, Tim Etherington, and T. Vogl, "Reduced Workload and Improved Situation Awareness in Flight Decks Equipped with Synthetic Vision Information System Displays," in *International Journal of Aviation Psychology* (2002).

M.P. Snow and G.A. French, "Human Factors In Head-Up Synthetic Vision Displays," in *Proceedings of the SAE 2001 World Aviation Safety Conference* (2001), pp. 2641–2652.

R.W. Sommer and R.E. Dunhum, Jr., "Evaluation of a Contact-Analog Display in Landing Approaches with a Helicopter," NASA TN-D-5241 (1969).

J.M. Stark, "Investigating Display Integration in Candidate Synthetic Vision System Displays: Creating Endless VMC Conditions to Reduce Limited Visibility Incidents in Commercial Aviation," paper presented at the *Human Factors and Ergonomics Society 2003 Meeting, Oct. 2003*.

J.M. Stark, J.R. Comstock, Lawrence J. Prinzel, III, D.W. Burdette, and M.W. Scerbo, "A Preliminary Examination of Pilot Performance and Situation Awareness within a Synthetic Vision," paper presented at the *Human Factors and Ergonomics Society 2001 Meeting, Oct. 2001*.

Joseph W. Stickle, ed., *1980 Aircraft Safety and Operating Problems*, NASA CP-2170, pts. 1 and 2 (1981).

Mohammad A. Takallu, Louis J. Glaab, Monica F. Hughes, Douglas T. Wong, and Anthony P. Bartolome, *Piloted Simulation of Various Synthetic Vision Systems Terrain Portrayal and Guidance Symbology Concepts for Low Altitude En-Route Scenario*, NASA TP-2008-215127 (2008).

Mohammad A. Takallu, Douglas Wong, Anthony Bartolome, Monica Hughes, and Louis Glaab, "Interaction Between Various Terrain Portrayals and Guidance/Tunnel Symbology Concepts for General Aviation Synthetic Vision Displays During a Low En-Route Scenario," *Proceedings of the 23rd Digital Avionics Systems Conference* (2004).

Mohammad A. Takallu, D.T. Wong, and M.D. Uenking, "Synthetic Vision Systems in GA Cockpit-Evaluation of Basic Maneuvers Performed by Low Time GA Pilots During Transition from VMC to IMC," paper presented at the *International Advanced Aviation Conference, Aug. 2002.*

E. Theunissen, F.D. Roefs, R.M. Rademaker, and Tim Etherington, "Integration of Information in Synthetic Vision Displays: Why, To What Extent and How?" paper presented at *Human Factors and Ergonomics Society 2003 Meeting, Oct. 2003.*

R.J. Thomas and M.F. DiBenedetto, "The Local Area Augmentation System: An Airport Surface Guidance Application Supporting the NASA Runway Incursion Prevention System Demonstration at the Dallas/Fort Worth International Airport" in *Proceedings of the 20th Digital Avionics Systems Conference* (Oct. 2001).

C.L.M. Tiana, J.R. Kerr, and S.D. Harrah, "Multispectral Uncooled Infrared Enhanced-Vision for Flight Test," in Jacques G. Verly, ed., *Proceedings of SPIE Enhanced and Synthetic Vision 2001*, vol. 4363 (Aug. 2001), pp. 231–236.

J. Timmerman, "Runway Incursion Prevention System—ADS-B and DGPS Data Link Analysis, Dallas-Fort Worth International Airport," NASA CR-2001-211242 (2001).

M.D. Uenking and M.F. Hughes, "The Efficacy of Using Synthetic Vision Terrain-Textured Images to Improve Pilot Situation Awareness," in *SAE World Aviation Congress and Display* (Oct. 2002).

M. Uijt de Haag, J. Campbell, and R. Gray, "A Terrain Database Integrity Monitor for Synthetic Vision Systems," *Proceedings of the 19th Digital Avionics Systems Conference* (Oct. 2000).

M. Uijt de Haag, J. Sayre, J. Campbell, Steven D. Young, and R. Gray, "Flight Test Results of a Synthetic Vision Elevation Database Integrity Monitor," *SPIE 15th Annual International Symposium on Aerospace/ Defense Sensing, Simulation, and Controls—AeroSense, Apr. 2001.*

M. Uijt de Haag, J. Sayre, J. Campbell, Steven D. Young, and R. Gray, "Terrain Database Integrity Monitoring for Synthetic Vision Systems," *IEEE Transactions on Aerospace and Electronic Systems, Dec. 2001.*

M. Uijt de Haag, J. Sayre, J. Campbell, Steven D. Young, and R. Gray, "Terrain Database Integrity Monitoring for Synthetic Vision Systems," paper presented at the *IEEE Transactions on Aerospace and Electronic Systems symposium, Mar. 2005.*

M. Uijt de Haag, R. Thomas, and J. Rankin, "Initial Flight Test Results of Ohio University's 3-Dimensional Cockpit Display of Traffic Information," *2002 IEEE Position Location and Navigation Symposium, Apr. 2002.*

M. Uijt de Haag, Steven D. Young, and J. Campbell, "An X-Band Radar Terrain Feature Detection Method for Low-Altitude SVS Operations and Calibration Using LiDAR," *SPIE Defense and Security Symposium, Apr. 12–16, 2004.*

M. Uijt de Haag, Steven D. Young, and R. Gray, "A Performance Evaluation of Elevation Database Integrity Monitors for Synthetic Vision Systems," *8th International Conference of Integrated Navigation Systems, May 2001.*

U.S. Aircraft Owners and Pilots Association, "Accident Trends and Factors for 1996" (1997).

U.S. Aircraft Owners and Pilots Association, "General Aviation Accident Trends and Factors for 1999" (2000).

U.S. Aircraft Owners and Pilots Association, "General Aviation Accident Trends and Factors for 2001" (2002).

11

U.S. Department of Transportation, Federal Aviation Administration, *Advisory Circular 120-29A: Criteria for Approval of Category I and Category II Weather Minima for Approach*, Aug. 12, 2002.

U.S. Department of Transportation, Federal Aviation Administration, *Enhanced Flight Vision Systems; Final Rule. 14 CFR Parts 1, 91, et al.* Jan. 2004.

U.S. Department of Transportation, Federal Aviation Administration, *FAA NPRM Docket No. NM185, Notice 25-01-02-SC, Special Conditions: Enhanced Vision System (EVS) for Gulfstream Model G-V Airplane*, June 2001.

U.S. Department of Transportation, Federal Aviation Administration, *FAA Runway Safety Report,* July 2003.

U.S. Department of Transportation, Federal Aviation Administration, *FAA Runway Safety Report*, Aug. 2004.

U.S. Department of Transportation, Federal Aviation Administration, *FAA Runway Safety Report*, Aug. 2005.

U.S. Department of Transportation, Federal Aviation Administration, *Synthetic Vision and Pathway Depictions on the Primary Flight Display*, AC 23-26, Dec. 2005.

U.S. Department of Transportation, Federal Aviation Administration, *Technical Standard Order (TSO) TSO-C151b, Terrain Awareness and Warning System*, Dec. 17, 2002.

U.S. National Transportation Safety Board, "Most Wanted Transportation Safety Improvements," Nov. 2005, at *http://www.ntsb.gov/recs/mostwanted/index.htm*, accessed Dec. 7, 2009.

U.S. National Transportation Safety Board, Safety Recommendation A-00-66, July 2000, at *http://www.ntsb.gov/recs/mostwanted/index.htm*, accessed Dec. 7, 2009.

A. Vadlamani and M. Uijt de Haag, "Improving the Detection Capability of Spatial Failure Modes Using Downward-Looking Sensors in Terrain Database Integrity Monitors," in *Proceedings of the 22nd Digital Avionics Systems Conference* (Oct. 2003).

D. Warner, "Flight Path Displays," USAF Flight Dynamics Laboratory, Report AFFDL-TR-79-3075 (1979).

John J. White, "Advanced Transport Operating Systems Program," SAE Paper 90-1969, presented at the *Society of Automotive Engineers Aerospace Technology Conference and Exposition, Long Beach, CA, Oct. 1–4, 1990*.

D. Williams, M. Waller, J. Koelling, D.W. Burdette, T. Doyle, W. Capron, J. Barry, and R. Gifford, "Concept of Operations for Commercial and Business Aircraft Synthetic Vision Systems, Version 1.0," NASA TM-2001-211058 (2001).

Steven P. Williams, Jarvis J. Arthur, III, Kevin J. Shelton, Lawrence J. Prinzel, III, and R. Michael Norman, "Synthetic Vision for Lunar Landing Vehicles," in *SPIE Defense and Security Symposium, Orlando, FL, Mar. 16–20, 2008*.

D.T. Wong, Mohammad A. Takallu, Monica F. Hughes, A.P. Bartolone, and Louis J. Glaab, "Simulation Experiment for Developing the Symbology for the Approach and Missed Approach Phases of Flight of Head-Down Synthetic Vision Systems Displays," paper presented at the *AIAA Modeling and Simulation Technologies Conference, Aug. 2004*.

Glenn A. Woodell, Daniel J. Jobson, Zia-ur Rahman, and Glenn D. Hines, "Advanced Image Processing of Aerial Imagery," in *Proceedings of SPIE Visual Information Processing XIV*, vol. 6246 (Apr. 2006).

Glenn A. Woodell, Daniel J. Jobson, Zia-ur Rahman, and Glenn D. Hines, "Enhancement of Imagery in Poor Visibility Conditions," in *Proceedings of SPIE Sensors, and Command, Control, Communications, and Intelligence (C3I) Technologies for Homeland Security and Homeland Defense IV* (Apr. 2005).

M. Yang, T. Gandhi, R. Kasturi, L. Coraor, O. Camps, and J. McCandless, "Real-time Obstacle Detection System for High Speed Civil Transport Supersonic Aircraft," paper presented at the *IEEE National Aerospace and Electronics Conference, Oct. 2000.*

S. Yang, T. Schnell, and K. Lemos, "Spatial Image Content Bandwidth Requirements for Synthetic Vision Displays," in *Proceedings of the 22th Digital Avionics Systems Conference* (Oct. 2003).

Steven D. Young, "On the Development of In-flight Autonomous Integrity Monitoring of Stored Geo-Spatial Data Using Forward-Looking Remote Sensing Technology," Ph.D. dissertation, Ohio State University, Mar. 2005.

Steven D. Young, S. Harrah, and M. Uijt de Haag, "Real-time Integrity Monitoring of Stored Geo-spatial Data using Forward-looking Remote Sensing Technology," in *Proceeding of the 21st Digital Avionics Systems Conference* (Oct. 2002).

Steven D. Young and Denise R. Jones, "Flight Demonstration of Integrated Airport Surface Movement Technologies," NASA TM-1998-206283 (1998).

Steven D. Young and Denise R. Jones, "Flight Testing of an Airport Surface Guidance, Navigation, and Control System," paper presented at the *Institute of Navigation National Technical Meeting, Jan. 21–23, 1998.*

Steven D. Young and Denise R. Jones, "Runway Incursion Prevention: A Technology Solution," paper presented at the *Flight Safety Foundation 54th Annual International Air Safety Seminar, Athens, Greece, Nov. 5–8, 2001.*

Steven D. Young and Denise R. Jones, "Runway Incursion Prevention Using an Advanced Surface Movement Guidance and Control System (A-SMGCS)," in *Proceedings of the 19th Digital Avionics Systems Conference* (Oct. 2000).

Steven D. Young, S. Kakarlapudi, and M. Uijt de Haag, "A Shadow Detection and Extraction Algorithm Using Digital Elevation Models and X-Band Weather Radar Measurements," in *International Journal of Remote Sensing*, vol. 26, no. 8 (Apr. 20, 2005), pp. 1531–1549.

Steven D. Young, M. Uijt de Haag, and J. Campbell, "An X-band Radar Terrain Feature Detection Method for Low-Altitude SVS Operations and Calibration Using LiDAR," in *SPIE Defense and Security Symposium, Apr. 2004.*

Steven D. Young, M. Uijt de Haag, and J. Sayre, "Using X-band Weather Radar Measurements to Monitor the Integrity of Digital Elevation Models for Synthetic Vision Systems," in Jacques G. Verly, ed., *Proceedings of SPIE* 2003 *AeroSense Conference, Enhanced and Synthetic Vision*, vol. 5081 (2003), pp. 66–76.

Steven D. Young and M. Uijt de Haag, "Detection of Digital Elevation Model Errors Using X-band Weather Radar," *AIAA Journal of Aerospace Computing, Information, and Communication* (July 2005).

Books and Monographs:

Joseph R. Chambers, *Innovation in Flight: Research of the NASA Langley Research Center on Revolutionary Advanced Concepts for Aeronautics*, NASA SP-2005-4539.

W.E.L. Grimson, *Object Recognition by Computer: The Role of Geometric Constraints* (Cambridge: MIT Press, 1990).

F. Hayes-Roth, D.A. Waterman, and D.B. Lenat, eds., *Building Expert Systems* (Reading, MA: Addison-Wesley Publishing Co., Inc., 1983).

Waldemar Karwowski, ed., *International Encyclopedia of Ergonomics and Human Factors* (Boca Raton, FL: CRC Press—Taylor and Francis Publishers, 2006 ed.).

Cary R. Spitzer, *The Digital Avionics Handbook* (Boca Raton, FL: CRC Press, 2000).

Lane E. Wallace, *Airborne Trailblazer: Two Decades with NASA Langley's 737 Flying Laboratory*, NASA SP-4216.

P. Winston and K. Prendergast, eds., *The AI Business: Commercial Uses of Artificial Intelligence* (Cambridge: MIT Press, 1984).

THIS PAGE INTENTIONALLY BLANK

Ice formation on aircraft poses a serious flight safety hazard. Here a NASA technician measures ice deposits on a test wing in NASA's Icing Research Tunnel, Lewis (now Glenn) Research Center, Ohio. NASA.

CASE 12

Aircraft Icing: The Tyranny of Temperature

By James Banke

The aerospace environment is a realm of extremes: low to high pressures, densities, and temperatures. Researchers have had the goal of improving flight efficiency and safety. Aircraft icing has been a problem since the earliest days of flight and, historically, researchers have artfully blended theory, ground-and-flight research, and the use of new tools such as computer simulation and software modeling codes to ensure that travelers fly in aircraft well designed to confront this hazard.

ONE FEBRUARY EVENING in the late 1930s, a young copilot strode across a cold ramp of the Nashville airport under a frigid moonlit sky, climbing into a chilled American Airlines DC-2. The young airman was Ernest Gann, later to gain fame as a popular novelist and aviation commentator, whose best-remembered book, The High and the Mighty, became an iconic aviation film. His captain was Walter Hughen, already recognized by his peers as one of the greats, and the two men worked swiftly to ready the sleek twin-engine transport for flight. Behind them, eight passengers settled in, looked after by a flight attendant. They were bound for New York, along AM-23, an air route running from Nashville to New York City. Preparations complete, they taxied out and took off on what should have been a routine 4-hour flight in favorable weather. Instead, almost from the moment the airliner's wheels tucked into the plane's nacelles, the flight began to deteriorate. By the time they reached Knoxville, they were bucking an unanticipated 50-mile-per-hour headwind, the Moon had vanished, and the plane was swathed in cloud, its crew flying by instruments only. And there was something else: ice. The DC-2 was picking up a heavy load of ice from the moisture-laden air, coating its wings and engine cowlings, even its propellers, with a wetly glistening and potentially deadly sheen.[1]

1. Ernest K. Gann, *Fate is the Hunter* (New York: Simon & Schuster, 1961), pp. 79–87.

Suddenly there was "an erratic banging upon the fuselage," as the propellers began flinging ice "chunks the size of baseballs" against the fuselage. In the cockpit, Hughen and Gann desperately fought to keep their airplane in the air. Its leading edge rubber deicing boots, which shattered ice by expanding and contracting, so that the airflow could sweep it away, were throbbing ineffectively: the ice had built up so thick and fast that it shrouded them despite their pulsations. Carburetor inlet icing was building up on each engine, causing it to falter, and only deliberately induced back-firing kept the inlets clear and the engines running. Deicing fluid spread on the propellers and cockpit glass had little effect, as did a hot air hose rigged to blow on the outside of the windshield. Worst of all, the heavy icing increased the DC-2's weight and drag, slowing it down to near its stall point. At one point, the plane began "a sudden, terrible shudder," perilously on the verge of a fatal stall, before Hughen slammed the throttles full-forward and pushed the nose down, restoring some margin of flying speed.[2]

After a half hour of desperate flying that "had the smell of eternity" about it, the battered DC-2 and its drained crew entered clear skies. The weather around them was still foreboding, and so, after trying to return to Nashville, finding it was closed, and then flying about for hours searching for an acceptable alternate, they turned for Cincinnati, Hughen and Gann anxiously watching their fuel consumption. Ice—some as thick as 4 inches—still swathed the airplane, so much so that Gann thought, "Where are the engineers again? The wings should somehow be heated." The rudder was frozen in place, and the elevators and ailerons (controlling pitch and roll) moveable only because of Hughen and Gann's constant control inputs to ensure they remained free. At dawn they reached Cincinnati, where the plane, burdened by its heavy load of ice, landed heavily. "We hit hard," Gann recalled,"and stayed earth-bound. There is no life left in our wings for bouncing." Mechanics took "two hours of hard labor to knock the ice from our wings, engine cowlings, and empennage." Later that day, Hughen and Gann completed the flight to New York, 5 hours late. In the remarks section of his log, explaining the delayed arrival, Gann simply penned "Ice."[3]

2. Ibid., pp. 88–93.
3. Ibid., pp. 94–107.

Gann, ever after, regarded the flight as marking his seasoning as an airman, "forced to look disaster directly in the face and stare it down."[4]

Many others were less fortunate. In January 1939, *Cavalier,* an Imperial Airways S.23 flying boat, ditched heavily in the North Atlantic, breaking up and killing 3 of its 13 passengers and crew; survivors spent 10 cold hours in heaving rafts before being rescued. Carburetor icing while flying through snow and hail had suffocated two of its four engines, leaving the flying boat's remaining two faltering at low power.[5] In October 1941, a Northwest Airlines DC-3 crashed near Moorhead, MN, after the heavy weight of icing prevented its crew from avoiding terrain; this time 14 of 15 on the plane died.[6]

Even when nothing went wrong, flying in ice was unsettling. Trans World Airlines Captain Robert "Bob" Buck, who became aviation's most experienced, authoritative, and influential airman in bad weather flying, recalled in 2002 that

> A typical experience in ice meant sitting in a cold cockpit, windows covered over in a fan-shaped plume from the lower aft corner toward the middle front, frost or snow covering the inside of the windshield frames, pieces as large as eight inches growing forward from the windshield's edges outside, hunks of ice banging against the fuselage and the airplane shaking as the tail swung left and right, right and left, and the action was transferred to the rudder pedals your feet were on so you felt them saw back and forth beneath you The side winds were frosted, but you could wipe them clear enough for a look out at the engines. The nose cowlings collected ice on their leading edge, and I've seen it so bad that the ice built forward until the back of the propeller was shaving it! But still the airplane flew. The indicated airspeed would slow, and

4. Ibid., p. 79.

5. Harold Penrose, *Wings Across the World* (London: Cassell, 1980), p. 114; R.E.G. Davies, *British Airways: An Airline and its Aircraft*, v. 1: *The Imperial Years* (McLean, VA: Paladwr Press, 2005), pp. 94, 96.

6. U.S. Civil Aeronautics Board, Bureau of Safety Investigation, *Comparative Safety Statistics in United States Airline Operations*, Pt. 1: *Years 1938–1945* (Washington, DC: CAB BSI Analysis Division, 15 August 1953), p. 29.

> you'd push up the throttles for more power to overcome the loss but it didn't always take, and the airspeed sometimes went down to alarming numbers approaching stall.[7]

Icing, as the late aviation historian William M. Leary aptly noted, has been a "perennial challenge to aviation safety."[8] It's a chilling fact that despite a century of flight experience and decades of research on the ground and in the air, today's aircraft still encounter icing conditions that lead to fatal crashes. It isn't that there are no preventative measures in place. Weather forecasting, real-time monitoring of conditions via satellite, and ice prediction software are available in any properly equipped cockpit to warn pilots of icing trouble ahead. Depending on the size and type of aircraft, there are several proven anti-icing and de-icing systems that can help prevent ice from building up to unsafe levels. Perhaps most importantly, pilot training includes information on recognizing icing conditions and what to do if an aircraft starts to ice up in flight. Unfortunately the vast majority of icing-related incidents echo a theme in which the pilot made a mistake while flying in known icing conditions. And that shows that in spite of all the research and technology, it's still up to the pilot to take advantage of the experience base developed by NASA and others over the years.

In the very earliest days of aviation, icing was not an immediate concern. That all changed by the end of the First World War, by which time airplanes were operating at altitudes above 10,000 feet and in a variety of meteorological conditions. Worldwide, the all-weather flying needs of both airlines and military air service, coupled with the introduction of blind-flying instrumentation and radio navigation techniques that enabled flight in obscured weather conditions, stimulated study of icing, which began to take a toll on airmen and aircraft as they increasingly operated in conditions of rain, snow, and freezing clouds and sleet.[9]

---

7. Bob Buck [Robert N. Buck], *North Star Over My Shoulder: A Flying Life* (New York: Simon & Schuster, 2002), p. 113. Once America's youngest licensed pilot, Buck authored two influential books on aviation safety, *Weather Flying* (New York: The Macmillan Co., 1978); and *The Pilot's Burden: Flight Safety and the Roots of Pilot Error* (Ames, IA: Iowa State University Press, 1994).
8. William M. Leary, "A Perennial Challenge to Aviation Safety: Battling the Menace of Ice," in Roger D. Launius and Janet R. Daly Bednarek, eds., *Reconsidering a Century of Flight* (Chapel Hill, NC: University of North Carolina Press, 2003), pp. 132–151.
9. See, for example, Wesley L. Smith, "Weather Problems Peculiar to the New York-Chicago Airway," *Monthly Weather Review* vol. 57, no. 12 (Dec. 1929), pp. 503–506.

The NACA's interest in icing dated to the early 1920s, when America's aviation community first looked to the Agency for help. By the early 1930s, both in America and abroad, researchers were examining the process of ice formation on aircraft and means of furnishing some sort of surface coatings that would prevent its adherence, particularly to wings, acquiring data both in actual flight test and by wind tunnel studies. Ice on wings changed their shape, drastically altering their lift-to-drag ratios and the pressure distribution over the wing. An airplane that was perfectly controllable with a clean wing might prove very different indeed with just a simple change to the profile of its airfoil.[10] Various mechanical and chemical solutions were tried. The most popular mechanical approach involved fitting the leading edges of wings, horizontal tails, and, in some cases, vertical fins with pneumatically operated rubber "de-icing" boots that could flex and crack a thin coating of ice. As Gann and Buck noted, they worked at best sporadically. Other approaches involved squirting de-icing fluid over leading edges, particularly over propeller blades, and using hot-air hoses to de-ice cockpit windshields.

Lewis A. "Lew" Rodert—the best known of ice researchers—was a driven and hard-charging NACA engineer who ardently pursued using heat as a means of preventing icing of wings, propellers, carburetors, and windshields.[11] Under Rodert's direction, researchers extensively instrumented a Lockheed Model 12 light twin-engine transport for icing research and, later, a larger and more capable Curtiss C-46 transport. Rodert and test pilot Larry Clausing, both Minnesotans, moved the NACA's ice research program from Ames Aeronautical Laboratory (today the NASA Ames Research Center) to a test site outside Minneapolis. There, researchers took advantage of the often-formidable weather conditions to assemble a large database on icing and icing conditions, and

10. Thomas Carroll and William H. McAvoy, "The Formation of Ice Upon Airplanes in Flight," NACA TN-313 (1929); Montgomery Knight and William C. Clay, "Refrigerated Wind Tunnel Tests on Surface Coatings for Preventing Ice Formation," NACA TN-339 (1930); W. Bleeker, "The Formation of Ice on Aircraft," NACA TM-1027 (1942) [trans. of "Einige Bermerkungen über Eisansatz an Flugzeugen," *Meteorologische Zeitschrift* (Sep. 1932), pp. 349–354.

11. For samples of Rodert's work, see Lewis A. Rodert, "An Investigation of the Prevention of Ice on the Airplane Windshield," TN-754 (1940); Lewis A. Rodert and Alun R. Jones, "A Flight Investigation of Exhaust-Heat De-Icing," NACA TN-783 (1940); Lewis A. Rodert, "The Effects of Aerodynamic Heating on Ice Formations on Airplane Propellers," TN-799 (1941); Lewis A. Rodert and Richard Jackson, "Preliminary Investigation and Design of an Air-Heated Wing for Lockheed 12A Airplane," NACA ARR A-34 (1942).

on the behavior of various modifications to their test aircraft. These tests complemented more prosaic investigations looking at specific icing problems, particularly that of carburetor icing.[12]

The war's end brought Rodert a richly deserved Collier Trophy, American aviation's most prestigious award, for his thermal de-icing research, particularly the development and validation of the concept of air-heated wings.[13] By 1950, a solid database of NACA research existed on icing and its effects upon propeller-driven airplanes.[14] This led many to conclude that the "heroic era" of icing research was in the past, a judgment that would prove to be wrong. In fact, the problems of icing merely changed focus, and NACA engineers quickly assessed icing implications for the civil and military aircraft of the new gas turbine and transonic era.[15] New high-performance interceptor fighters, expected to accelerate quickly and climb to high altitudes, had icing problems of their own, typified by inlet icing that forced performance limitations and required imaginative solutions.[16] When first introduced into service, Bristol's otherwise-impressive Britannia turboprop long-range transport had persistent problems caused by slush ice forming in the induction system of its Proteus turboprop engines. By the time the NACA evolved into the

12. George W. Gray, *Frontiers of Flight: The Story of NACA Research* (New York: Alfred A. Knopf, 1948), pp. 308–316; and Henry A. Essex, "A Laboratory Investigation of the Icing Characteristics of the Bendix-Stromberg Carburetor Model PD-12F5 with the Pratt and Whitney R-1830-C4 Intermediate Rear Engine Section," NACA-WR-E-18 (1944); William D. Coles, "Laboratory Investigation of Ice Formation and Elimination in the Induction System of a Large Twin-Engine Cargo Aircraft," NACA TN-1427 (1947).

13. Edwin P. Hartman *Adventures in Research: A History of Ames Research Center 1940-1965*, NASA SP-4302 (Washington, DC: GPO, 1970), pp. 69–73; and Glenn E. Bugos, "Lew Rodert, Epistemological Liaison, and Thermal De-Icing at Ames," in Pamela E. Mack, ed., *From Engineering Science to Big Science: The NACA and NASA Collier Trophy Research Project Winners*, NASA SP-4219 (Washington, DC: NASA, 1998), pp. 29–58.

14. For example, G. Merritt Preston and Calvin C. Blackman, "Effects of Ice Formations on Airplane Performance in Level Cruising Flight," NACA TN-1598 (1948); Alun R. Jones and William Lewis, "Recommended Values of Meteorological Factors to be Considered in the Design of Aircraft Ice-Prevention Equipment," NACA TN-1855 (1949); Carr B. Neel, Jr., and Loren G. Bright, "The Effect of Ice Formations on Propeller Performance, NACA TN-2212 (1950).

15. James P. Lewis, Thomas F. Gelder, Stanley L. Koutz, "Icing Protection for a Turbojet Transport Airplane: Heating Requirements, Methods of Protection, and Performance Penalties," NACA TN-2866 (1953).

16. Porter J. Perkins, "Icing Frequencies Experienced During Climb and Descent by Fighter-Interceptor Aircraft," NACA TN-4314 (1958); and James P. Lewis and Robert J. Blade, "Experimental Investigation of Radome Icing and Icing Protection," NACA RM-E52J31 (1953).

National Aeronautics and Space Administration in 1958, the fundamental facts concerning the types of ice an aircraft might encounter and the major anti-icing techniques available were well understood and widely in use. In retrospect, as impressive as the NACA's postwar work in icing was, it is arguable that the most important result of NACA work was the establishment of ice measurement criteria, standards for ice-prevention systems, and probabilistic studies of where icing might be encountered (and how severe it might be) across the United States. NACA Technical Notes 1855 (1949) and 2738 (1952) were the references of record in establishing Federal Aviation Administration (FAA) standards covering aircraft icing certification requirements.[17]

NASA and the Aircraft Icing Gap

At a conference in June 1955, Uwe H. von Glahn, the NASA branch chief in charge of icing research at the then-Lewis Research Center (now Glenn Research Center) in Cleveland boldly told fellow scientific investigators: "Aircraft are now capable of flying in icing clouds without difficulty . . . because research by the NACA and others has provided the engineering basis for ice-protection systems."[18]

That sentiment, in combination with the growing interest and need to support a race to the Moon, effectively shut down icing research by

17. As referenced in U.S. National Transportation Safety Board, "In-Flight Icing Encounter and Loss of Control Simmons Airlines, d.b.a. American Eagle Flight 4184 Avions de Transport Regional (ATR) Model 72-212, N401AM, Roselawn, Indiana October 31, 1994," NTSB/AAR-96/01 (Washington, DC: NTSB, 1996), pp. 97–99.

18. William M. Leary, *"We Freeze to Please": A History of NASA's Icing Research Tunnel and the Quest for Flight Safety*, NASA SP-2002-4226 (Washington, DC: NASA, 2002), p. 60. Glahn did much notable work in icing research; see Uwe H. von Glahn and Vernon H. Gray, "Effect of Ice and Frost Formations on Drag of NACA 65-212 Airfoil for Various Modes of Thermal Ice Protection," NACA TN-2962 (1953); Uwe H. von Glahn and Vernon H. Gray, "Effect of Ice Formations on Section Drag of Swept NACA 63A-009 Airfoil with Partial Span Leading Edge Slat for Various Modes of Thermal Ice Protection," NACA RM-E53J30 (1954); Uwe H. von Glahn, Edmund E. Callaghan, and Vernon H. Gray, "NACA Investigation of Icing-Protection Systems for Turbojet-Engine Installations," NACA RM-E51B12 (1951); Uwe H. von Glahn, Thomas F. Gelder, and William H. Smyers, Jr., "A Dye-Tracer Technique for Experimentally Obtaining Impingement Characteristics of Arbitrary Bodies and Method for Determining Droplet Size Distribution," NACA TN-3338 (1955); Vernon H. Gray and Uwe H. von Glahn, "Aerodynamic Effects Caused by Icing of an Unswept NACA 65A004 Airfoil," NACA TN-4155 (1958); Vernon H. Gray, Dean T. Bowden, and Uwe H. von Glahn, "Preliminary Results of Cyclical De-Icing of a Gas-Heated Airfoil," NACA RM-E51J29 (1952).

the NACA, although private industry continued to use Government facilities for their own cold-weather research and certification activities, most notably the historic Icing Research Tunnel (IRT) that still is in use today at the Glenn Research Center (GRC). The Government's return to icing research began in 1972 at a meeting of the Society of Automotive Engineers in Dallas, during which an aeronautics-related panel was set up to investigate ice accretion prediction methods and define where improvements in related technologies could be made. Six years later the panel concluded that little progress in understanding icing had been accomplished since the NACA days. Yet since the formation of NASA in 1958, 20 years earlier, aircraft technology had fundamentally changed. Commercial aviation was flying larger jet airliners and being asked to develop more fuel-efficient engines, and at the same time the U.S. Army was having icing issues operating helicopters in icy conditions in Europe. The Army's needs led to a meeting with NASA and the FAA, followed by a July 1978 conference with 113 representatives from industry, the military, the U.S. Government, and several nations. From that conference sparked the impetus for NASA restarting its icing research to "update the applied technology to the current state of the art; develop and validate advanced analysis methods, test facilities, and icing protection concepts; develop improved and larger testing facilities; assist in the difficult process of standardization and regulatory functions; provide a focus to the presently disjointed efforts within U.S. organizations and foreign countries; and assist in disseminating the research results through normal NASA distribution channels and conferences."[19]

While icing research programs were considered, proposed, planned, and in some cases started, full support from Congress and other stakeholders for the return of a major, sustained icing research effort by NASA did not come until after an Air Florida Boeing 737 took off from National Airport in Washington, DC, in a snowstorm and within seconds crashed on the 14th Street Bridge. The 1982 incident killed 5 people on the bridge, as well as 70 passengers and 4 crewmembers. Only five people survived the crash, which the National Transportation Safety Board blamed on a number of factors, assigning issues related to icing as a major cause of the preventable accident. Those issues included faulting

19. Leary, "*We Freeze to Please*", p. 72.

the flight crew for not activating the twin engine's anti-ice system while the aircraft was on the ground and during takeoff, for taking off with snow and ice still on the airfoil surfaces of the Boeing aircraft, and for the lengthy delay between the final time the aircraft was de-iced on the tarmac and the time it took the crew to be in position to receive takeoff clearance from the control tower and get airborne. While all this was happening the aircraft was exposed to constant precipitation that at various times could be described as rain or sleet or snow.[20]

The immediate aftermath of the accident—including the dramatic rescue of the five survivors who had to be fished out of the Potomac River—was all played out on live television, freezing the issue of aircraft icing into the national consciousness. Proponents of NASA renewing its icing research efforts suddenly had shocking and vivid proof that additional research for safety purposes was necessary in order to deal with icing issues in the future. Approval for a badly needed major renovation of the IRT at GRC was quickly given, and a new, modern era of NASA aircraft icing investigations began.[21]

Baby, It's Cold Out There

Not surprisingly, ice buildup on aircraft is bad. If it happens on the ground, then pilots and passengers alike must wait for the ice to be removed, often with hazardous chemicals and usually resulting in flight delays that can trigger a chain reaction of schedule problems across the Nation's air system. If an aircraft accumulates ice in the air, depending on the severity of the situation, the results could range from mild annoyance that a de-icing switch has to be thrown to complete aerodynamic failure of the wing, accompanied by total loss of control, a spiraling dive from high altitude, a premature termination of the flight and all lives on board, followed by the reward of becoming the lead item on the evening news.

Icing is a problem for flying aircraft not so much because of the added weight, but because of the way even a tiny amount of ice can begin to disrupt the smooth airflow over the wings, wreaking havoc with the wing's ability to generate lift and increasing the amount of drag, which

20. National Transportation Safety Board, *Collision with 14th Street Bridge Near Washington National Airport, Air Florida Flight 90, Boeing 737-222, N62AF, Washington, D.C., January 13, 1982*, NTSB/AAR-82-8 (1982).

21. Leary, *"We Freeze to Please"*, p. 82.

slows the aircraft and pitches the nose down. This prompts the pilot to pull the nose up to compensate for the lost lift, which allows even more ice to build up on the lower surface of the wing. And the vicious circle continues, potentially leading to disaster. Complicating the matter is that even with options for clearing the wing of ice—discussed shortly—ice buildup can remain and/or continue on other aircraft surfaces such as antennas, windshields, wing struts, fixed landing gear, and other protrusions, all of which can still account for a 50-percent increase in drag even if the wing is clean.[22]

From the earliest experience with icing during the 1920s and on through the present day, researchers have observed and understood there to be three primary categories of aircraft ice: clear, rime, and mixed. Each one forms for slightly different reasons and exhibits certain properties that influence the effectiveness of available de-icing measures.[23]

Clear ice is usually associated with freezing rain or a special category of rain that falls through a region of the atmosphere where temperatures are far below the normal freezing point of water, yet the drops remain in a liquid state. These are called super-cooled drops.

Such drops are very unstable and need very little encouragement to freeze. When they strike a cold airframe they begin to freeze, but it is

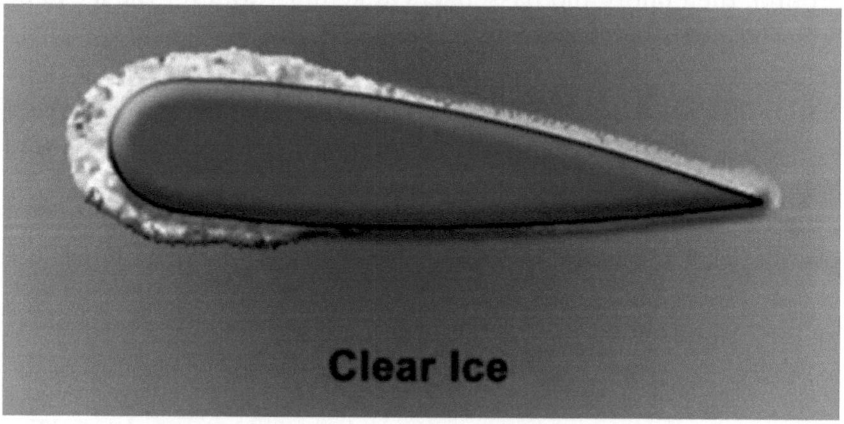

A graphic depicting clear ice buildup on an airfoil.

22. R.J. Ranaudo, K.L. Mikkelsen, R.C. McKnight, P.J. Perkins, Jr., "Performance Degradation of a Typical Twin Engine Commuter Type Aircraft in Measured Natural Icing Conditions," NASA TM-83564 (1984).

23. R. John Hansman, Kenneth S. Breuer, Didier Hazan, Andrew Reehorst, Mario Vargas, "Close-Up Analysis of Aircraft Ice Accretion," NASA TM-105952 (1993).

not an instant process. The raindrop freezes as it spreads out and continues to make contact with an aircraft surface whose skin temperature is at or below 32 degrees Fahrenheit (0 degrees Celsius). The slower the drop freezes, the more time it will have to spread out evenly and create a sheet of solid, clear ice that has very little air enclosed within. This flowback phenomenon is greatest at temperatures right at freezing. Because of its smooth surface, clear ice can quickly disrupt the wing's ability to generate lift by ruining the wing's aerodynamic shape. This type of ice is quite solid in the sense that if any of it does happen to loosen or break off, it tends to come off in large pieces that have the ability to strike another part of the aircraft and damage it.[24]

Rime ice proves size makes a difference. In this case the supercooled liquid water drops are smaller than the type that produces

A graphic depicting rime ice buildup on an airfoil.

clear ice. When these tiny drops of water strike a cold aircraft surface, most of the liquid drops instantly freeze and any water remaining is not enough to create a sheet of ice. Instead, the result is a brittle ice that looks milky white, is opaque, has a rough surface due to its makeup of ice crystals and trapped air, and doesn't accumulate as quickly as clear ice. It does not weigh as much, either, and tends to stick to the leading edge of the wing and the cowl of the engine intakes on a jet, making rime ice just as harmful to the airflow and aerodynamics of the aircraft.[25]

Naturally, when an aircraft encounters water droplets of various sizes, a combination of both clear and rime ice can form, creating the

24. Civil Aviation Authority of New Zealand, "Aircraft Icing Handbook" (2000), p. 2.

25. R.C. McKnight, R.L. Palko, and R.L. Humes, "In-flight Photogrammetric Measurement of Wing Ice Accretions," NASA TM-87191 (1986).

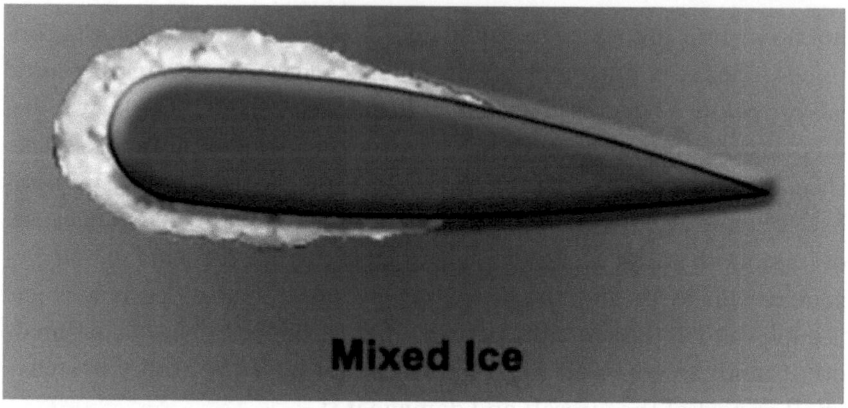

A graphic depicting mixed ice buildup on an airfoil.

third category of icing called mixed ice. The majority of ice encountered in aviation is of this mixed type.[26]

Aircraft must also contend with snow, avoiding the wet, sticky stuff that makes great snowballs on the ground but in the air can quickly accumulate not only on the wings—like ice, a hazard in terms of aerodynamics and weight—but also on the windshield, obscuring the pilot's view despite the best efforts of the windshield wipers, which can be rendered useless in this type of snow. And on the ground, frost can completely cover an aircraft that sits out overnight when there is a combination of humid air and subfreezing temperatures. Frost can also form in certain flying conditions, although it is not as hazardous as any of the ices.[27]

Melting Your Troubles Away

As quickly as the hazards of aircraft icing became known in the early days of aviation, inventive spirits applied themselves to coming up with ways to remove the hazard and allow the airplane to keep flying. These ideas at first took the form of understanding where and when icing occurs and then simply not flying through such conditions, then ways to prevent ice from forming in the first place—proactive anti-icing—were considered, and at the same time options for removing ice once it

26. John R. Hansman, Jr., "The Influence of Ice Accretion Physics on the Forecasting of Aircraft Icing Conditions," NASA Joint University Program for Air Transportation Research, NASA NTRS 90N20928 (1990).

27. M. Dietenberger, P. Kumar, and J. Luers, "Frost Formation on an Airfoil: A Mathematical Model 1," NASA-CR-3129 (1979).

had formed—reactive de-icing—were suggested and tested in the field, in the air, and in the wind tunnel. Of all the options available, the three major ones are the pneumatic boot, spraying chemicals onto the aircraft, and channeling hot bleed air.[28]

A King Air equipped with a de-icing boot on its wing leading edge shows how the boot removes some ice, but not on areas behind the boot.

The oldest of the de-icing methods in use is the pneumatic boot system, invented in 1923 by the B.F. Goodrich Corporation in Akron, OH. The general idea behind the boot has not changed nearly a century later: a thick rubber membrane is attached to the leading edge of a wing airfoil. Small holes in the wing behind the boot allow compressed air to blow through, ever so slightly expanding the boot's volume like a balloon. Any time that ice builds up on the wing, the system is activated, and when the boot expands, it essentially breaks the ice into pieces, which are quickly blown away by the relative wind of the moving aircraft. Again, although the general design of the boot system has not

28. John J. Reinmann, Robert J. Shaw, and W.A. Olsen, Jr., "NASA Lewis Research Center's Program on Icing Research," NASA TM-83031 (1983).

changed, there have been improvements in materials science and sensor technology, as well as changes in the shape of wings used in various sizes and types of aircraft. In this manner, NASA researchers have been very active in coming up with new and inventive ways to enhance the original boot concept and operation.[29]

One way to ensure there is no ice on an aircraft is to remove it before the flight gets off the ground. The most common method for doing this is to spray some type of de-icing fluid onto the aircraft surface as close to takeoff as possible. The idea was first proposed by Joseph Halbert and used by the United Kingdom Royal Air Force in 1937 on the large flying boats then operated by Imperial Airways.[30] Today, the chemicals used in these fluids usually use a propylene glycol or ethylene glycol base and may include other ingredients that might thicken the fluid, help inhibit corrosion on the aircraft, or add a color to the mixture for easier identification. Often water is added to the mixture, which although counterintuitive makes the liquid more effective. Of the two glycols, propylene is more environmentally friendly.[31]

The industry standard for this fluid is set by the aeronautics division of the Society of Automotive Engineers, which has published standards for four types of de-icing fluids, each with slightly different properties and intentions for use. Type I has a low viscosity and is usually heated and sprayed on aircraft at high pressure to remove any snow, ice, or frost. Due to its viscosity, it runs off the aircraft very quickly and provides little to no protection as an anti-icing agent as the aircraft is exposed to snowy or icy conditions before takeoff. Its color is usually orange.[32] Type II fluid has a thickening agent to prevent it from running very quickly off the aircraft, leaving a film behind that acts as an anti-icing agent until the aircraft reaches a speed of 100 knots, when the fluid breaks down from aerodynamic stress. The fluid is usually light yellow. Type III fluid's properties fall in between Type I and II, and it is intended for smaller,

29. A.E. Albright, D.L. Kohlman, W.G. Schweikhard, and P. Evanich, "Evaluation of a Pneumatic Boot Deicing System on a General Aviation Wing," NASA TM-82363 (1981).

30. "The Early Years—1930s," Killfrost, Inc. of Coral Springs, FL (2009).

31. J. Love, T. Elliott, G.C. Das, D.K. Hammond, R.J. Schwarzkopf, L.B. Jones, and T.L. Baker, "Screening and Identification of Cryopreservative Agents for Human Cellular Biotechnology Experiments in Microgravity," 2004 ASGSB Meeting, Brooklyn, NY, Nov. 2004.

32. Society of Automotive Engineers, "Deicing/Anti-icing Fluid, Aircraft, SAE Type 1," AMS 1224 (Rev. J) (2009).

slower aircraft. It is popular in the regional and business aviation markets and is usually dyed light yellow. Type IV fluids are only applied after a Type I fluid is sprayed on to remove all snow, ice, and frost. The Type IV fluid is designed to leave a film on the aircraft that will remain for 30 to 80 minutes, serving as a strong anti-icing agent. It is usually green.[33]

A Type 4 de-icing solution is sprayed on a commercial airliner before takeoff.

NASA researchers have worked with these fluids for many years and found uses in other programs, including the International Space Station. And during the late 1990s, a team of engineers from the Ames Research Center (ARC) at Moffett Field, CA, came up with an anti-icing fluid that was nontoxic—so much so that it was deemed "food grade" because its ingredients were approved by the U.S. Government for use in food—namely ice cream—and promised to last longer as an anti-icing agent for aircraft, as well as work as an effective de-icing agent. Although it

33. Society of Automotive Engineers, "Fluid, Aircraft Deicing/Anti-Icing, Non-Newtonian (Pseudoplastic), SAE Types II, III, and IV," AMS 1228 (Rev. G) (2009).

has not found wide use in the aviation industry, NASA did issue a license to a commercial firm who now sells the product to consumers as "Ice Free," a spray for automobile windshields that can provide protection from snow or ice forming on a windshield in temperatures down to 20 degrees Fahrenheit (-7 degrees Celsius).[34]

The third common technique for dealing with ice accretion is the hot bleed air method. In this scheme, hot air is channeled away from the aircraft engines and fed into tubes that run throughout the aircraft near the areas where ice is most likely to form and do the most damage. The hot air warms the aircraft skin, melting away any ice that is there and discouraging any ice from forming. The hot gas can also be used as the source of pressurized air that inflates a rubber boot, if one is present. While the idea of using hot bleed air became most practical with the introduction of jet engines, the basic concept itself dates back to the 1930s, when NACA engineers proposed the idea and tested it in an open-air-cockpit, bi-wing airplane. The in-flight experiments showed that "a vapor-heating system which extracts heat from the exhaust and distributes it to the wings is an entirely practical and efficient method for preventing ice formation."[35]

As for melting ice that can accrete on or in other parts of an aircraft, such as windshields, protruding Pitot tubes, antennas, and carburetors on piston engines, electrically powered heaters of one kind or another are employed. The problem of carburetor ice is especially important and the one form of icing most prevalent and dangerous for thousands of General Aviation pilots. NASA has studied carburetor ice for engines and aircraft of various configurations through the years[36] and in 1975 surveyed the accident database and found that between 65 and 90 accidents each year involve carburetor icing as the probable cause. And when there are known carburetor icing conditions, between 50 and 70 percent of engine failure accidents are due to carburetor icing. Researchers found the problem to be particularly acute for pilots

34. "Preventing Ice Before it Forms," *Spinoff* 2006 (Washington, DC: NASA, 2006), pp. 46–47.
35. Theodore Theodorsen and William C. Clay, "Ice Prevention on Aircraft by Means of Engine Exhaust Heat and a Technical Study of Heat Transmission from a Clark Y Airfoil," NACA TR-403 (1933).
36. William D. Coles, "Laboratory Investigation of Ice Formation and Elimination in the Induction System of a Large Twin-Engine Cargo Aircraft," NACA TN-1427 (1947). Henry A. Essex, "A Laboratory Investigation of the Icing Characteristics of the Bendix-Stromberg Carburetor Model PD-12F5 with the Pratt and Whitney R-1830-C4 Intermediate Rear Engine Section," NACA WR-E-18 (1944).

with less than 1,000 hours of total flying time and overall exposed about 144 persons to death or injury each year.[37]

Icing's Electromagnetic Personality

Influenced by increasing fuel prices, the search for more profitability in every way, and a growing environmental movement, NASA's aeronautics researchers during the 1980s sought to meet all of those needs in terms of propulsion, airframe design, air traffic control, and more. On the subject of aircraft icing, all three of the traditional de-icing methods provided some drawbacks. The pneumatic boot added weight and disrupted the intended aerodynamics of an otherwise unequipped wing airfoil. Spraying chemicals onto the aircraft, whether on the ground or seeped through the leading edge in flight, contributed toxins to the environment. And bleeding off hot air to warm the interior of the wing and other aircraft cavities reduced the performance of the engines and added to the empty weight of the aircraft. Based on an idea first suggested in 1937 by Rudolf Goldschmidt, a German national living in London, NASA researchers investigated an Electro-Impulse De-Icing (EIDI) system that promised applications both on fixed-wing aircraft and on helicopters.[38]

First tested during the 1970s, the EIDI system researched during the 1980s consisted of flat-wound coils of copper ribbon wire positioned near the skin inside the leading edge of a wing, but leaving a tiny gap between the skin and the coil. The coils were then connected a high-voltage bank of capacitors. When energy was discharged through the wiring, it created a rapidly forming and collapsing electromagnetic field, which in turn set up a sort of a vibration that rippled across the wing, creating a repulsive force of several hundred pounds for just a fraction of a second at a time. The resulting force "shattered, de-bonded and expelled ice instantaneously."[39]

Ground tests in GRC's IRT and flight tests on aircraft such as NASA's Twin Otter and Cessna 206 during 1983 and 1984 conclusively proved the EIDI system would work. The results set up a 1985 symposium with

37. R.W. Obermayer and T.W. Roe, "A Study of Carburetor/Induction System Icing in General Aviation Accidents," NASA CR-143835 (1975).
38. G.W. Zumwalt, R.L. Schrag, W.D. Bernhart, and R.A. Friedberg, "Analyses and Tests for Design of an Electro-Impulse De-Icing System," NASA CR-174919 (1985).
39. G.W. Zumwalt and R.A. Friedberg, "Designing an Electro-Impulse De-Icing System," AIAA Paper 86-0545 (1986).

more than 100 people in attendance representing 10 companies and several Government agencies. As participants observed test runs in the GRC IRT, program engineers stressed that EIDI operated on low energy (in some cases with less power than required to power landing lights), caused no aerodynamic penalties, required minimum maintenance, and compared favorably in terms of weight and cost with existing de-icing systems. Although it was hailed as the de-icing system of the future, the EIDI never found widespread acceptance or lived up to its expectations.[40]

However, in 1988 an ARC engineer by the name of Leonard A. Haslim won NASA's Inventor of the Year Award by coming up with the Electro-Expulsive Separation System (EESS), an apparent combination of the best of the EIDI and traditional rubber boot de-icing systems. In this configuration, the electrically conducting copper ribbons are embedded into the boot with tiny slits in the boot separating each conductor. When a burst of energy is discharged through the system, each conductor pair repels one another in an instant and causes the slits in the boot to expand explosively, instantly breaking free any ice on the wing. In addition to the advantages the EIDI system offers, the EESS can remove ice when it is only as thin as a layer of frost, preventing the possibility of larger chunks of ice breaking free of the leading edge and then causing damage if the ice strikes the tail or tail-mounted engines. With applications for removing ice from large ship superstructures or bridges, the EESS was licensed to Dataproducts New England, Inc. (DNE), to make the product available commercially.[41]

Tail Plane Icing Program

Following the traumatic loss of TWA Flight 800 in 1996, then-President Clinton put together a commission on aviation safety, from which NASA in 1997 began an Aviation Safety Program to address very specific areas of flying in a bid to reduce the accident rate, even as air traffic was anticipated to grow at record rates. The emphasis on safety came at a time when a 4-year program led by NASA with the help of the FAA to understand the phenomenon known as ice-contaminated tail plane stall, or ICTS, was a year away from wrapping up. The successful Tail Plane Icing Program provided immediate benefits to the aviation community and today is considered by veteran NASA

40. G.W. Zumwalt, "Electro-Impulse De-Icing: A Status Report," AIAA Paper 88-0019 (1988).
41. "Breaking the Ice," Spinoff 1989 (Washington, DC: NASA, 1989) pp. 64–65.

researchers as one of the Agency's most important icing-related projects ever conducted.[42]

According to a 1997 fact sheet prepared by GRC, the ICTS phenomenon is "characterized as a sudden, often uncontrollable aircraft nose down pitching moment, which occurs due to increased angle-of-attack of the horizontal tail plane resulting in tail plane stall. Typically, this phenomenon occurs when lowering the flaps during final approach while operating in or recently departing from icing conditions. Ice formation on the tail plane leading edge can reduce tail plane angle-of-attack range and cause flow separation resulting in a significant reduction or complete loss of aircraft pitch control." At the time the program began there had been a series of commuter airline crashes in which icing was suspect or identified as a cause. And while there was a great deal of knowledge about the effects of icing on the primary wing of an aircraft and how to combat it or recover from it, there was little information about the effect of icing on the tail or how pilots could most effectively recover from a tail plane stall induced by icing. As the popularity of the smaller, regional commuter jets grew following airline deregulation in 1978, the incidents of tail plane icing began to grow at a relatively alarming rate. By 1991, when the FAA first had the notion of initiating a review of all aspects of tail plane icing, there had been 16 accidents involving turboprop-powered transport and commuter-class airplanes, resulting in 139 fatalities.[43]

42. Interview of Jaiwon Shin, Associate Administrator for NASA's Aeronautics Research Mission Directorate, by Jim Banke, Orlando, FL, 5 Jan. 2010 Shin's own contributions to the study of aircraft icing have been substantial. For a sampling of his work, see Jaiwon Shin, "Characteristics of Surface Roughness Associated with Leading Edge Ice Accretion," NASA TM-106459 (1994); Jaiwon Shin, "The NASA Aviation Safety Program: Overview," NASA TM-2000-209810 (2000); Jaiwon Shin and Thomas H. Bond, "Results of an Icing Test on a NACA 0012 Airfoil in the NASA Lewis Icing Research Tunnel," NASA TM-105374 (1992); Jaiwon Shin, Hsun H. Chen, and Tuncer Cebeci, "A Turbulence Model for Iced Airfoils and Its Validation," NASA TM-105373 (1992); Jaiwon Shin, Brian Berkowitz, Hsun H. Chen, and Tuncer Cebeci, "Prediction of Ice Shapes and their Effect on Airfoil Performance," NASA TM-103701 (1991); Jaiwon Shin, Peter Wilcox, Vincent Chin, and David Sheldon, "Icing Test Results on an Advanced Two-Dimensional High-Lift Multi-Element Airfoil," NASA TM-106620 (1994); Thomas H. Bond and Jaiwon Shin, "Results of Low Power Deicer Tests on the Swept Inlet Component in the NASA Lewis Icing Research Tunnel," NASA TM-105968 (1993); and Thomas H. Bond, Jaiwon Shin, and Geert A. Mesander, "Advanced Ice Protection Systems Test in the NASA Lewis Icing Research Tunnel," NASA TM-103757 (1991).

43. Dale Hiltner, Michael McKee, Karine La Noé, and Gerald Gregorek, "DHC-6 Twin Otter Tail Plane Airfoil Section Testing in the Ohio State University 7x10 Wind Tunnel," NASA-CR-2000-2099921/VOL1 (2000).

Following a review of all available data on tail plane icing and incidents of the tail stalling on turboprop-powered commuter airplanes as of 1991, the FAA requested assistance from NASA in managing a full-scale research program into the characteristics of ICTS. And so an initial 4-year program began to deal with the problem and propose solutions. More specifically the goals of the program were to collect detailed aerodynamic data on how the tail of a plane contributed to the stability of an aircraft in flight, and then take the same measurements with the tail contaminated with varying severity of ice, and from that information develop methods for predicting the effects of tail plane icing and recovering from them. To accomplish this, a series of wind tunnel tests were performed with a tail section of a De Havilland of Canada DHC-6 Twin Otter aircraft (a design then widely used for regional transport), both in dry air conditions and with icing turned on in the tunnel. Flight tests of a full Twin Otter were made to complement the ground-based studies.[44]

As is typical with many research programs, as new information comes in and questions get answered, the research results often generate additional questions that demand even more study to find solutions. So following the initial tail plane icing research that concluded in 1997, a year later NASA's Ohio-based Field Center initiated a second multiphase program to continue the icing investigations. This time the work was assigned to Wichita State University in Kansas, which would coordinate its activities with support from the Bombardier/Learjet Company. The main goal was of the combined Government/industry/university effort was to expand on the original work with the Twin Otter by coming up with methods and criteria for testing multiple tail plane configurations in a wind tunnel, and then actually conduct the tests to generate a comprehensive database of tail plane aerodynamic performance with and without ice contamination for a range of tail plane/airfoil configurations. The resulting database would then be used to support development and verification of future icing analysis tools.[45]

From this effort pilots were given new tools to recognize the onset of tail plane icing and recover from any disruptions to the aircraft's

44. Gerald Gregorek, John J. Dresse, and Karine La Noé, "Additional Testing of the DHC-6 Twin Otter Tail Plane Airfoil Section Testing in the Ohio State University 7x10 Low Speed Wind Tunnel," NASA-CR-2000-29921/VOL2 (2000).

45. Judith Foss Van Zante, and Thomas P. Ratvasky, "Investigation of Dynamic Flight Maneuvers with an Iced Tail Plane," NASA TM-1999-208849 (1999).

aerodynamics, including a full stall. As part of the education process, a Guest Pilot Workshop was held to give aviators firsthand experience with tail plane icing via an innovative "real world" simulation in which the pilots flew with a model of a typical ice buildup attached to the tail surface of a Twin Otter. The event provided a valuable exchange between real-world pilots and laboratory researchers, which in turn resulted in the collaboration on a 23-minute educational video on tail plane icing that is still used today.[46]

Predicting an Icy Future

With its years of accumulated research about all aspects of icing—i.e., weather conditions that produce it, types of ice that form under various conditions, de-icing and anti-icing measures and when to employ them—NASA's data would be useless unless they were somehow packaged and made available to the aviation community in a convenient manner so that safety could be improved on a daily basis. And so with desktop computers becoming more affordable, available, and increasingly powerful enough to crunch fairly complex datasets, in 1983, NASA researchers at what was still named the Lewis Research Center began developing a computer program that would at first aid NASA's in-house researchers, but would grow to become a tool that would aid pilots, air traffic controllers, and any other interested party in the flight planning process through potential areas of icing. The software was dubbed LEWICE, and version 0.1 originated in 1983 as a research code for in-house use only. As of the beginning of 2010, version 2.0 is the official current version, although a version 3.2.2 is in development, as is the first 0.1 version of GlennICE, which is intended to accurately predict ice growth under any weather conditions for any aircraft surface.[47]

LEWICE, which spelled out is the Lewis Ice Accretion Program, is a freely available desktop software program used by hundreds of people in the aviation community for purposes of predicting the amount, type, and shape of ice an aircraft might experience given a particular weather forecast, as well as what kind of anti-icing heat requirements may be necessary to prevent any buildup of ice from beginning. The software

46. The video is available online via YouTube at *http://www.youtube.com/watch?v=_ifKduc1hE8&feature=PlayList&p=18B9F75B0B7A3DB9&playnext=1&playnext_from=PL&index=3.*

47. William B. Wright, Mark G. Potapczuk, and Laurie H. Levinson, "Comparison of LEWICE and GlennICE in the SLD Regime," NASA TM-2008-215174 (2008).

runs on a desktop PC and provides its analysis of the input data within minutes, fast enough that the user can try out some different numbers to get a range of possible icing experiences in flight. All of the predictions are based on extensive research and real-life observations of icing collected through the years both in flight and in icing wind tunnel tests.[48]

At its heart, LEWICE attempts to predict how ice will grow on an aircraft surface by evaluating the thermodynamics of the freezing process that occurs when supercooled droplets of moisture strike an aircraft in flight. Variables considered include the atmospheric parameters of temperature, pressure, and velocity, while meteorological parameters of liquid water content, droplet diameter, and relative humidity are used to determine the shape of the ice accretion. Meanwhile, the aircraft surface geometry is defined by segments joining a set of discrete body coordinates. All of that data are crunched by the software in four major modules that result in a flow field calculation, a particle trajectory and impingement calculation, a thermodynamic and ice growth calculation, and an allowance for changes in the aircraft geometry because of the ice growth. In processing the data, LEWICE applies a time-stepping procedure that runs through the calculations repeatedly to "grow" the ice. Initially, the flow field and droplet impingement characteristics are determined for the bare aircraft surface. Then the rate of ice growth on each surface segment is determined by applying the thermodynamic model. Depending on the desired time increment, the resulting ice growth is calculated, and the shape of the aircraft surface is adjusted accordingly. Then the process repeats and continues to predict the total ice expected based on the time the aircraft is flying through icing conditions.[49]

The basic functions of LEWICE essentially account for the capabilities of the software up through version 1.6. Version 2.0 was the next release, and although it did not change the fundamental process or models involved in calculating ice accretion, it vastly improved the robustness and accuracy of the software. The current version was extensively tested on different computer platforms to ensure identical results and also incorporated the very latest and complete datasets based on the most

48. Jaiwon Shin, Brian Berkowitz, Hsun Chen, and Tuncer Cebeci, "Prediction of Ice Shapes and their Effect on Airfoil Performance," NASA TM-103701 (1991).
49. William B. Wright and Adam Rutkowski, "Validation Results for LEWICE 2.0," NASA CR-1999-208690 (1999).

recent research available, while also having its prediction results verified in controlled laboratory tests using the Glenn IRT. Version 3.2—not yet released to date—will add the ability to account for the presence and use of anti-icing and de-icing systems in determining the amount, shape, and potential hazard of ice accretion in flight. Previously these variables could be calculated by reading LEWICE output files into other software such as ANTICE 1.0 or LEWICE/Thermal 1.6.[50]

According to Jaiwon Shin, the current NASA Associate Administrator for the Aeronautics Research Mission Directorate, the LEWICE software is the most significant contribution NASA has made and continues to make to the aviation industry in terms of the topic of icing accretion. Shin said LEWICE continues to be used by the aviation community to improve safety, has helped save lives, and is an incredibly useful tool in the classroom to help teach future pilots, aeronautical engineers, traffic controllers, and even meteorologists about the icing phenomenon.[51]

Learning to Fly with SLDs

From the earliest days of aviation, the easiest way for pilots to avoid problems related to weather and icing was to simply not fly through clouds or in conditions that were less than ideal. This made weather forecasting and the ability to quickly and easily communicate observed conditions around the Nation a top priority of aviation researchers. Working with the National Oceanic and Atmospheric Administration (NOAA) during the 1960s, NASA orbited the first weather satellites, which began equipped with black-and-white television cameras and

50. For these programs and their validation, see William B. Wright, "Users Manual for the Improved NASA Lewis Ice Accretion Code LEWICE 1.6," NASA CR-1995-198355 (1995); William B. Wright, "User Manual for the NASA Glenn Ice Accretion Code LEWICE Version 2.0," NASA CR-1999-209409 (1999); William B. Wright, "User Manual for the NASA Glenn Ice Accretion Code LEWICE Version 2.2.2," NASA CR-2002-211793 (2002); William B. Wright, "Further Refinement of the LEWICE SLD Model," NASA CR-2006-214132 (2006); William B. Wright, "User's Manual for LEWICE Version 3.2," NASA CR-2008-214255 (2008); William B. Wright and James Chung, "Correlation Between Geometric Similarity of Ice Shapes and the Resulting Aerodynamic Performance Degradation-A Preliminary Investigation Using WIND," NASA CR-1999-209417 (1999); William B. Wright, R.W. Gent, and Didier Guffond, "DRA/NASA/ONERA Collaboration on Icing Research Pt. II—Prediction of Airfoil Ice Accretion," NASA CR-1997-202349 (1997); William B. Wright, Mark G. Potapczuk, and Laurie H. Levinson, "Comparison of LEWICE and GlennICE in the SLD Regime," NASA TM-2008-215174 (2008).

51. Banke, Shin interview.

have since progressed to include sensors capable of seeing beyond the range of human eyesight, as well as lasers capable of characterizing the contents of the atmosphere in ways never before possible.[52]

Post-flight image shows ice contamination on the NASA Twin Otter airplane as a result of encountering Supercooled Large Droplet (SLD) conditions near Parkersburg, WV.

Our understanding of weather and the icing phenomenon, in combination with the latest navigation capabilities—robust airframe manufacturing, anti- and de-icing systems, along with years of piloting experience—has made it possible to certify airliners to safely fly through almost any type of weather where icing is possible (size of the freezing rain is generally between 100 and 400 microns). The exception is for one category in which the presence of supercooled large drops (SLDs) are detected or suspected of being there. Such rain is made up of water droplets that are greater than 500 microns and remain in a liquid state even though its temperature is below freezing. This makes the drop very unstable, so it will quickly freeze when it comes into contact with a cold object such as the leading edge of an airplane. And while some

52. Andrew Reehorst, David J. Brinker, and Thomas P. Ratvasky, "NASA Icing Remote Sensing System Comparisons from AIRS II," NASA TM-2005-213592 (2005).

of the SLDs do freeze on the wing's leading edge, some remain liquid long enough to run back and freeze on the wing surfaces, making it difficult, if not impossible, for de-icing systems to properly do their job. As a result, the amount of ice on the wing can build up so quickly, and so densely, that a pilot can almost immediately be put into an emergency situation, particularly if the ice so changes the airflow over the wing that the behavior of the aircraft is adversely affected.

This was the case on October 31, 1994 when American Eagle Flight 4184, a French-built ATR 72-212 twin-turboprop regional airliner carrying a crew of 4 and 64 passengers, abruptly rolled out of control and crashed in Roselawn, IN. During the flight, the crew was asked to hold in a circling pattern before approaching to land. Icing conditions existed, with other aircraft reporting rime ice buildup. Suddenly the ATR 72 began an uncommanded roll; its two pilots heroically attempted to recover as the plane repeatedly rolled and pitched, all the while diving at high speed. Finally, as they made every effort to recover, the plane broke up at a very low altitude, the wreckage plunging into the ground and bursting into flame. An exhaustive investigation, including NASA tests and tests of an ATR 72 flown behind a Boeing NKC-135A icing tanker at Edwards Air Force Base, revealed that the accident was all the more tragic for it had been completely preventable. Records indicated that the ATR 42 and 72 had a marked propensity for roll-control incidents, 24 of which had occurred since 1986 and 13 of which had involved icing. The National Transportation Safety Board (NTSB) report concluded:

> The probable cause of this accident were the loss of control, attributed to a sudden and unexpected aileron hinge moment reversal that occurred after a ridge of ice accreted beyond the deice boots because: 1) ATR failed to completely disclose to operators, and incorporate in the ATR 72 airplane flight manual, flightcrew operating manual and flightcrew training programs, adequate information concerning previously known effects of freeing precipitation on the stability and control characteristics, autopilot and related operational procedures when the ATR 72 was operated in such conditions; 2) the French Directorate General for Civil Aviation's (DGAC's) inadequate oversight of the ATR 42 and 72, and its failure to take the necessary corrective action to ensure continued

12

> airworthiness in icing conditions; and 3) the DGAC's failure to provide the FAA with timely airworthiness information developed from previous ATR incidents and accidents in icing conditions, as specified under the Bilateral Airworthiness Agreement and Annex 8 of the International Civil Aviation Organization.
>
> Contributing to the accident were; 1) the Federal Aviation Administration's (FAA's) failure to ensure that aircraft icing certification requirements, operational requirements for flight into icing conditions, and FAA published aircraft icing information adequately accounted for the hazards that can result from light in freezing rain and other icing conditions not specified in 14 Code of Federal Regulations 9CFR) part 25, Appendix C; and 2) the FAA's inadequate oversight of the ATR 42 and 72 to ensure continued airworthiness in icing conditions.[53]

This accident focused attention on the safety hazard associated with SLD and prompted the FAA to seek a better understanding of the atmospheric characteristics of the SLD icing condition in anticipation of a rule change regarding certifying aircraft for flight through SLD conditions, or at least long enough to safely depart the hazardous zone once SLD conditions were encountered. Normally a manufacturer would demonstrate its aircraft's worthiness for certification by flying in actual SLD conditions, backed up by tests involving a wind tunnel and computer simulations. But in this case such flight tests would be expensive to mount, requiring an even greater reliance on ground tests. The trouble in 1994 was lack of detailed understanding of SLD precipitation that could be used to recreate the phenomenon in the wind tunnel or program computer models to run accurate simulations. So a variety of flight tests and ground-based research was planned to support the decision-making process on the new certification standards.[54]

53. National Transportation Safety Board, *In-Flight Icing Encounter and Loss of Control Simmons Airlines, d.b.a. American Eagle Flight 4184 Avions de Transport Regional (ATR) Model 72-212, N401AM Roselawn, Indiana October 31, 1994*, v. 1: *Safety Board Report*, NTSB/AAR-96/01 (Washington, DC: NTSB, 1996), p. 210.
54. Dean R. Miller, Mark G. Potapczuk, and Thomas H. Bond, "Update on SLD Engineering Tools Development," NASA TM-2004-213072 (2004).

One interesting approach NASA took in conducting basic research on the behavior of SLD rain was to employ high-speed, close-up photography. Researchers wanted to learn more about the way an SLD strikes an object: is it more of a direct impact, and/or to what extent does the drop make a splash? Investigators also had similar questions about the way ice particles impacted or bounced when used during research in an icing wind tunnel such as the one at GRC. With water droplets less than 1 millimeter in diameter and the entire impact process taking less than 1 second in time, the close-up, high-speed imaging technique was the only way to capture the sought-after data. Based on the results from these tests, follow-on tests were conducted to investigate what effect ice particle impacts might have on the sensing elements of water content measurement devices.[55]

NASA's Twin Otter ice research aircraft, based at the Glenn Research Center in Cleveland, is shown in flight.

Another program to understand the characteristics of SLDs Supercooled Large Droplets involved a series of flight tests over the Great Lakes during the winter of 1996–1997. GRC's Twin Otter icing research aircraft was flown in a joint effort with the FAA and the National Center for Atmospheric Research (NCAR). Based on weather forecasts

55. Dean R. Miller, Christopher J. Lynch, and Peter A. Tate, "Overview of High Speed Close-Up Imaging in an Icing Environment," NASA TM-2004-212925 (2004).

and real-time pilot reports of in-flight icing coordinated by the NCAR, the Twin Otter was rushed to locations where SLD conditions were likely. Once on station, onboard instrumentation measured the local weather conditions, recorded any ice accretion that took place, and registered the aerodynamic performance of the aircraft in response to the icing. A total of 29 such icing research sorties were conducted, exposing the flight research team to all the sky has to offer—from normal-sized precipitation and icing to SLD conditions, as well as mixed phase conditions. Results of the flight tests added to the database of knowledge about SLDs and accomplished four technical objectives that included characterization of the SLD environment aloft in terms of droplet size distribution, liquid water content, and measuring associated variables within the clouds containing SLDs; development of improved SLD diagnostic and weather forecasting tools; increasing the fidelity of icing simulations using wind tunnels and icing prediction software (LEWICE); and providing new information about SLD to share with pilots and the flying community through educational outreach efforts.[56]

Thanks in large measure to the SLD research done by NASA in partnership with other agencies—an effort NASA Associate Administrator Jaiwon Shin ranks as one of the top three most important contributions to learning about icing—the FAA is developing a proposed rule to address SLD icing, which is outside the safety envelope of current icing certification requirements. According to a February 2009 FAA fact sheet: "The proposed rule would improve safety by taking into account supercooled large-drop icing conditions for transport category airplanes most affected by these icing conditions, mixed-phase and ice-crystal conditions for all transport category airplanes, and supercooled large drop, mixed phase, and ice-crystal icing conditions for all turbine engines."[57]

As of September 2009, SLD certification requirements were still in the regulatory development process, with hope that an initial, draft rule would be released for comment in 2010.[58]

56. Dean R. Miller, Thomas Ratvasky, Ben Bernstein, Frank McDonough, and J. Walter Strapp, "NASA/FAA/NCAR Supercooled Large Droplet Icing Flight Research: Summary of Winter 1996-1997 Flight Operations," NASA TM-1998-206620 (1998).

57. Laura Brown, "FAA Icing Fact Sheet: Flying in Icing Conditions," NTSB Docket No. SA-533, Exhibit No. 2-GGG (2009).

58. "FAA Presentation—Icing Requirements and Guidance," NTSB Docket No. SA-533, Exhibit No. 2-JJJ (2009).

Flaming Out on Ice

And just when the aircraft icing community thought it had seen everything—clear ice, rime ice, glazed ice, SLDs, tail plane icing, and freezing rain encountered within the coldest atmospheric conditions possible—a new icing concern was recently discovered in the least likely of places: the interior of jet engines, where parts are often several hundred degrees above freezing. Almost nothing is known about the mechanism behind engine core ice accretion, except that the problem does cause loss of power, even complete flameouts. According to data compiled by Boeing and cited in a number of news media stories and Government reports, there have been more than 100 dramatic power drops or midair engine stoppages since the mid 1990s, including 14 instances since 2002 of dual-engine flameouts in which engine core ice accretion turned a twin-engine jetliner into a glider. "It's not happening in one particular type of engine and it's not happening on one particular type of airframe," said Tom Ratvasky, an icing flight research engineer at GRC. "The problem can be found on aircraft as big as large commercial airliners, all the way down to business-sized jet aircraft."[59]

The problem came to light in 2004, when the first documented dual-engine flameout occurred with a U.S. business jet due to core ice accretion. The incident was noted by the NTSB, and during the next 2 years Jim Hookey, an NTSB propulsion expert, watched as two more Beechjets lost engine power despite no evidence of mechanical problems or pilot error. One of those incidents took place over Florida in 2005, when both engines failed within 10 seconds of each other at 38,000 feet. Despite three failed attempts to restart the engines the pilots were able to safely glide in to a Jacksonville airport, dodging thunderstorms and threatening clouds all the way down. Hookey took the unusual step of interviewing the pilots and became convinced the cause of the power failures was due to an environmental condition. It was shortly after that realization that both the NTSB and the FAA began pursuing icing as a cause.[60]

Hookey employed some commonsense investigative techniques to find commonality among the incidents he was aware of and others that were suspect. He contacted the engine manufacturers to request they take another look at the detailed technical reports of engines that had failed

59. Phone interview of Tom Ratvasky by Jim Banke, Cape Canaveral, FL, 7 April 2009.
60. Andy Pasztor, "Airline Regulators Grapple with Engine-Shutdown Peril," *The Wall Street Journal*, Page A1, 7 April 2008.

and then also look at the archived weather data to see if any patterns emerged. By May 2006, the FAA began to argue that the engine problems were being caused by ice crystals being ingested into the engine. The NTSB concurred and suggested how ice crystals can build up inside engines even if the interior temperatures are way above freezing. The theory is that ice particles from nearby storms melt in the hot engine air, and as more ice is ingested, some of the crystals stick to the wet surfaces, cooling them down. Eventually enough ice accretes to cause a problem, usually without warning. In August 2006, the NTSB sent a letter to the FAA detailing the problem as it was then understood and advising the FAA to take action.[61]

Part of the action the FAA is taking to continue to learn more about the phenomenon, its cause, and potential mitigation strategies is to partner with NASA and others in conducting an in-flight research program. "If we can find ways of detecting this condition and keeping aircraft out of it, that's something we're interested in doing," said Ratvasky, who will help lead the NASA portion of the research program. Considering the number and type of sensors required, the weight and volume of the associated research equipment, the potentially higher loads that may stress the aircraft as it flies in and around fairly large warm-weather thunderstorms, the required range, and the number of people who would like to be on site for the research, NASA won't be able to use its workhorse Twin Otter icing research aircraft. A twin-turbofan Lockheed S-3B Viking aircraft provided to NASA by the U.S. Navy originally was proposed for this icing research program, but the program requirements outgrew the jet's capabilities. As of early 2010, the Agency still was considering its options for a host aircraft, although it was possible that the NASA DC-8 airborne science laboratory based at the Dryden Flight Research Center (DFRC) might be pressed into service. In any case, it's going to take some time to put together the plan, prepare the aircraft, and test the equipment. It may be 2012 before the flight research begins. "It's a fairly significant process to make sure we are going to be doing this program in a safe way, while at the same time we meet all the research requirements. What we're doing right now is getting the instrumentation integrated onto the aircraft and then doing the appropriate testing to qualify the instrumentation before we go fly all the way across the

61. National Transportation Safety Board, Letter to the Federal Aviation Administration, Safety Recommendation A-06-56 through -59 (Washington, DC: NTSB, 2006).

world and make the measurements we want to make," Ratvasky said. In addition to NASA, organizations providing support for this research include the FAA, NCAR, Boeing, Environment Canada, the Australian Bureau of Meteorology, and the National Research Council of Canada.[62]

In the meantime, ground-based research has been underway and safety advisories involving jet engines built by General Electric and Rolls-Royce has resulted in those companies making changes in their design and operations to prevent the chance of any interior ice buildup that could lead to engine failure. Efforts to unlock the science behind internal engine icing also is taking place at Drexel University in Pennsylvania, where researchers are building computer models for use in better understanding the mechanics of how ice crystals can accrete within turbofan engines at high altitude.[63]

While few technical papers have been published on this subject—none yet appear in NASA's archive of technical reports—expect the topic of engine ingestion of ice crystals and its detrimental effect on safe operations to get a lot of attention during the next decade as more is learned, rules are rewritten, and potential design changes in jet engines are ordered, built, and deployed into the air fleet.

Slip, Sliding Away

Before an aircraft can get into the winter sky and safely avoid the threat of icing, it first must take off from what the pilot hopes is a long, wide, dry runway at the beginning of the flight, as well as at the end of the flight. Likewise, NASA's contributions to air safety in fighting the tyranny of temperature included research into ground operations. While NASA did not invent the plow to push snow off the runway, or flame-throwers to melt off any stubborn runway snow or ice, the Agency has been active in studying the benefits of runway grooves since the first civil runway was introduced in the United States at Washington National Airport in December 1965.[64]

Runway grooves are intended to quickly channel water away from the landing strip without pooling on the surface so as to prevent

62. Banke, Ratvasky interview.

63. Manuel A. Rios, Yung I. Cho, "Analysis of Ice Crystal Ingestion as a Source of Ice Accretion," AIAA-2008-4165 (2208).

64. R.C. McGuire, "Report on Grooved Runway Experience at Washington National Airport," NASA Washington Pavement Grooving and Traction Studies (1969).

NASA's Aircraft Landing Dynamics Facility (ALDF) at Langley Research Center. This facility is used to test landing gear and how it acts when they touch the runway at high speed. ALDF achieved a 200-knot design speed.

hydroplaning. The 3-mile-long runway at the Shuttle Landing Facility is probably the most famous runway in the Nation and known for being grooved. Of course, there is little chance of snow or ice accumulating on the Central Florida runway, so when NASA tests runway surfaces for cold weather conditions it turns to the Langley Aircraft Landing Dynamics Facility at NASA's Langley Research Center (LaRC) in Hampton, VA. The facility uses pressurized water to drive a landing-gear-equipped platform down a simulated runway strip, while cameras and sensors keep an eye on tire pressure, tire temperature, and runway friction. Another runway at NASA's Wallops Flight Facility also has been used to test various surface configurations. During the mid-1980s, tests were performed on 12 different concrete and asphalt runways, grooved and non-grooved, including dry; wet; and snow, slush, and ice-covered surface conditions. More than 200 test runs were made with two transport aircraft, and more than 1,100 runs were made with different ground test vehicles. Ground vehicle and B-737 aircraft friction tests were conducted on grooved and non-grooved surfaces under wet conditions. As expected, grooved runway surfaces had significantly greater friction properties than non-grooved surfaces, particularly at higher speeds.[65]

65. Thomas J. Yager, William A. Vogler, and Paul Baldasare, "Evaluation of Two Transport Aircraft and Several Ground Test Vehicle Friction Measurements Obtained for Various Runway Surface Types and Conditions. A Summary of Test Results From Joint FAA/NASA Runway Friction Program," NASA TP-2917 (1990).

NASA's Cool Research Continues

With additional research required on SLDs and engine core ice accretion, new updates always in demand for the LEWICE software, and the still-unknown always waiting to be discovered, NASA maintains its research capability concentrated within the Icing Branch at GRC. The branch performs research activities related to the development of methods for evaluating and simulating the growth of ice on aircraft surfaces, the effects that ice may have on the behavior of aircraft in flight, and the behavior of ice protection and detection systems. The branch is part of the Research and Technology Directorate and works closely with the staff of the Icing Research Tunnel and the Twin Otter Icing Research Aircraft. Its mission is to develop validated simulation methods—for use in both computer programmed and real-world experiments—suitable for use as both certification and design tools when evaluating aircraft systems for operation in icing conditions. The Icing Branch also fosters the development of ice protection and ice detection systems by actively supporting and maintaining resident technical expertise, experimental facilities, and computational resources. NASA's Aircraft Icing Project at GRC is organized into three sections: Design and Analysis Tools, Aircraft Ice Protection, and Education and Training.[66]

Design and Analysis Tools

The Icing Branch has a continuing, multidisciplinary research effort aimed at the development of design and analysis tools to aid aircraft manufacturers, subsystem manufacturers, certification authorities, the military, and other Government agencies in assessing the behavior of aircraft systems in an icing environment. These tools consist of computational and experimental simulation methods that are validated, robust, and well documented. In addition, these tools are supported through the creation of extensive databases used for validation, correlation, and similitude. Current software offerings include LEWICE, LEWICE 3D, and SmaggIce. LEWICE 3D is computationally fast and can handle large problems on workstations and personal computers. It is a diverse, inexpensive tool for use in determining the icing characteristics of arbitrary aircraft surfaces. The code can interface with most

66. Mario Vargas, "Icing Branch Current Research Activities in Icing Physics," in NASA Glenn Research Center staff, *Proceedings of the Airframe Icing Workshop*, NASA CP-2009-215797 (Cleveland, OH: NASA Glenn Research Center, 2009).

3-D flow solvers and can generate solutions on workstations and personal computers for most cases in less than several hours.[67]

SmaggIce is short for Surface Modeling and Grid Generation for Iced Airfoils. It is a software toolkit used in the process of predicting the aerodynamic performance ice-covered airfoils using grid-based Computational Fluid Dynamics (CFD). It includes tools for data probing, boundary smoothing, domain decomposition, and structured grid generation and refinement. SmaggIce provides the underlying computations to perform these functions, a GUI (Graphical User Interface) to control and interact with those functions, and graphical displays of results. Until 3-D ice geometry acquisition and numerical flow simulation become easier and faster for studying the effects of icing on wing performance, a 2-D CFD analysis will have to play an important role in complementing flight and wind tunnel tests and in providing insights to effects of ice on airfoil aerodynamics. Even 2-D CFD analysis, however, can take a lot of work using the currently available general-purpose grid-generation tools. These existing grid tools require extensive experience and effort on the part of the engineer to generate appropriate grids for moderately complex ice. In addition, these general-purpose tools do not meet unique requirements of icing effects study: ice shape characterization, geometry data evaluation and modification, and grid quality control for various ice shapes. So, SmaggIce is a 2-D software toolkit under development at GRC. It is designed to streamline the entire 2-D icing aerodynamic analysis process from geometry preparation to grid generation to flow simulation, and to provide unique tools that are required for icing effects study.[68]

Aircraft Ice Protection

The Aircraft Ice Protection program focuses on two main areas: development of remote sensing technologies to measure nearby icing conditions, improve current forecast capabilities, and develop systems to transfer and display that information to flight crews, flight controllers, and dispatchers; and development of systems to monitor and assess aircraft performance, notify the cockpit crew about the state of the

67. Colin S. Bidwell and Mark G. Potapczuk, "Users Manual for the NASA Lewis Three-Dimensional Ice Accretion Code: LEWICE 3D," NASA TM-105974 (1993).
68. Marivell Baez, Mary Vickerman, and Yung Choo, "SmaggIce User Guide," NASA TM-2000-209793 (2000).

aircraft, and/or automatically alter the aircraft controlling systems to prevent stall or loss of control in an icing environment. Keeping those two focus areas in mind, the Aircraft Ice Protection program is subdivided to work on these three goals:

- Provide flight crews with real-time icing weather information so they can avoid the hazard in the first place or find the quickest way out.[69]
- Improve the ability of an aircraft to operate safely in icing conditions.[70]
- Improve icing simulation capabilities by developing better instrumentation and measurement techniques to characterize atmospheric icing conditions, which also will provide icing weather validation databases, and increase basic knowledge of icing physics.[71]

In terms of remote sensing, the top level goals of this activity are to develop and field-test two forms of remote sensing system technologies that can reduce the exposure of aircraft to in-flight icing hazards. The first technology would be ground based and provide coverage in

69. Richard H. McFarland and Craig B. Parker, "Weather Data Dissemination to Aircraft," NASA, Langley Research Center, Joint University Program for Air Transportation Research, 1988–1989, pp. 119–127.

70. Sharon Monica Jones, Mary S. Reveley, Joni K. Evans, and Francesca A. Barrientos, "Subsonic Aircraft Safety Icing Study," NASA TM-2008-215107 (2008).

71. For simulation, see Laurie H. Levinson, Mark G. Potapczuk, and Pamela A. Mellor, "Software Development Processes Applied to Computational Icing Simulation," NASA TM-1999-208898 (1999); Thomas B. Irvine, John R. Oldenburg, and David W. Sheldon, "New Icing Cloud Simulation System at the NASA Glenn Research Center Icing Research Tunnel," NASA TM-1999-208891 (1999); Mark G. Potapczuk and John J. Reinmann, "Icing Simulation: A Survey of Computer Models and Experimental Facilities," NASA TM-104366 (1991); Mark G. Potapczuk, M.B. Bragg, O.J. Kwon, and L.N. Sankar, "Simulation of Iced Wing Aerodynamics," NASA TM-104362 (1991); Thomas P. Ratvasky, Billy P. Barnhart, and Sam Lee, "Current Methods for Modeling and Simulating Icing Effects on Aircraft Performance, Stability and Control," NASA TM-2008-215453 (2008); Thomas P. Ratvasky, Kurt Blankenship, William Rieke, and David J. Brinker, "Iced Aircraft Flight Data for Flight Simulation Validation," NASA TM-2003-212114 (2003); Thomas P. Ratvasky, Richard J. Ranaudo, Kurt S. Blankenship, and Sam Lee, "Demonstration of an Ice Contamination Effects Flight Training Device," NASA TM-2006-214233 (2006); Thomas P. Ratvasky, Billy P. Barnhart, Sam Lee, and Jon Cooper, "Flight Testing an Iced Business Jet for Flight Simulation Model Validation," NASA TM-2007-214936 (2007).

12

a limited terminal area to protect all vehicles. The second technology would be airborne and provide unrestricted flightpath coverage for a commuter class aircraft. In most cases the icing hazard to aircraft is minimized with either de-icing or anti-icing procedures, or by avoiding any known icing or possible icing areas altogether. However, being able to avoid the icing hazard depends much on the quality and timing of the latest observed and forecast weather conditions. And once stuck in a severe icing hazard zone, the pilot must have enough information to know how to get out of the area before the aircraft's ice protection systems are overwhelmed. One way to address these problem areas is to remotely detect icing potential and present the information to the pilot in a clear, easily understood manner. Such systems would allow the pilot to avoid icing conditions and also allow rapid escape from icing if severe conditions were encountered.[72]

Education and Training

To support NASA's ongoing goal of improving aviation safety, the Education and Training Element of the Aircraft Icing Project continues to develop education and training aids for pilots and operators on the hazards of atmospheric icing. A complete list of current training aids is maintained on the GRC Web site. Education materials are tailored to several specific audiences, including pilots, operators, and engineers. Due to the popularity of the education products, NASA can no longer afford to print copies and send them out. Instead, interested parties can download material from the Web site[73] or check out the latest catalog from Sporty's Pilot Shop, an internationally known source of professional materials and equipment for aviators.[74]

Icing Branch Facilities

NASA's groundbreaking work to understand the aircraft icing phenomenon would have been impossible if not for a pair of assets available at GRC. The more historic of the two is the Icing Research Tunnel (IRT),

72. Andrew Reehorst, David Brinker, Marcia Politovich, David Serke, Charles Ryerson, Andrew Pazmany, and Frederick Solheim, "Progress Towards the Remote Sensing of Aircraft Icing Hazards," NASA TM-2009-215828 (2009).

73. The GRC Web site can be found at *http://icebox.grc.nasa.gov/education/index.html.*

74. Judith Foss Van Zante, "Aircraft Icing Educational and Training Videos Produced for Pilots," Glenn Research Center Research and Technology Report (1999).

which began service in 1944 and, despite the availability of other wind tunnels with similar capabilities, remains one of a kind. The other asset is the DHC-6 Twin Otter aircraft, which calls the main hangar at GRC its home.

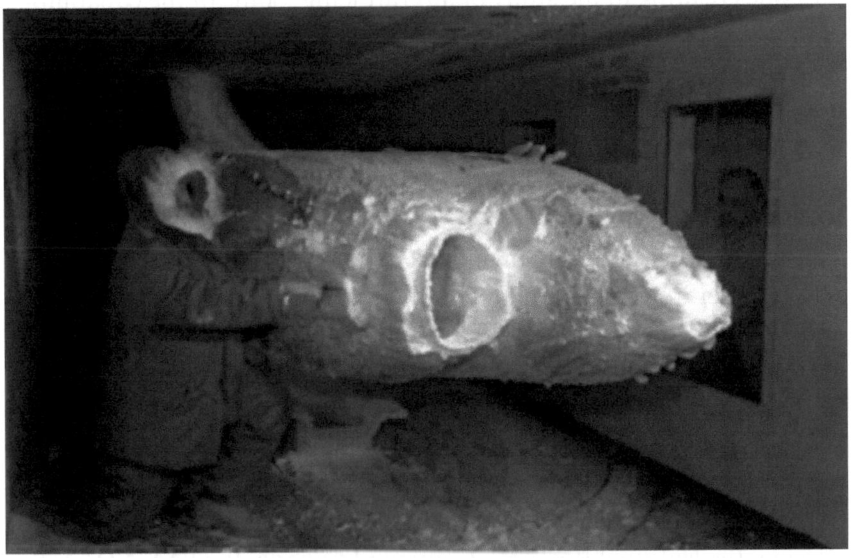

Jack Cotter inspects a Commuter Transport Engine undergoing testing in the Icing Research Tunnel while Ray Soto looks on from the observation window. The Icing Research Tunnel, or IRT, is used to simulate the formation of ice on aircraft surfaces during flight. Cold water is sprayed into the tunnel and freezes on the test model.

For ground-based research it's the IRT, the world's largest refrigerated wind tunnel. It has been used to contribute to flight safety under icing conditions since 1944. The IRT has played a substantial role in developing, testing, and certifying methods to prevent ice buildup on gas-turbine-powered aircraft. Work continues today in the investigation of low-power electromechanical deicing and anti-icing fluids for use on the ground, deicing and anti-icing research on Short Take Off and Vertical Landing (STOVL) rotor systems and certification of ice protection systems for military and commercial aircraft. The IRT is a closed-loop, refrigerated wind tunnel with a 6- by 9-foot test section. It can generate airspeeds from 25 to more than 400 miles per hour. Models placed in the tunnel can be subjected to droplet sprays of varying sizes to produce the natural icing conditions.[75]

75. For a detailed history of the IRT, see the previously cited William Leary, *"We Freeze to Please: A History of NASA's Icing Research Tunnel and the Quest for Flight Safety"*, NASA-SP-2002-4226 (2002).

For its aerial research, the Icing Branch utilizes the capabilities of NASA 607, a DHC-6 Twin Otter aircraft. The aircraft has undergone many modifications to provide both the branch and NASA a "flying laboratory" for issues relating to the study of aircraft icing. Some of the capabilities of this research aircraft have led to development of icing protection systems, full-scale iced aircraft aerodynamic studies, software code validation for ground-based research, development of remote weather sensing technologies, natural icing physics studies, and more.[76]

Partners on Ice

As it is with other areas involving aviation, NASA's role in aircraft icing is as a leader in research and technology, leaving matters of regulations and certifications to the FAA. Often the FAA comes to NASA with an idea or a need, and the Agency then takes hold of it to make it happen. Both the National Center for Atmospheric Research and NOAA have actively partnered with NASA on icing-related projects. NASA also is a major player in the Aircraft Icing Research Alliance (AIRA), an international partnership that includes NASA, Environment Canada, Transport Canada, the National Research Council of Canada, the FAA, NOAA, the National Defense of Canada, and the Defence Science and Technology Laboratory (DSTL)-United Kingdom. AIRA's primary research goals complement NASA's, and they are to

- Develop and maintain an integrated aircraft icing research strategic plan that balances short-term and long-term research needs,
- Implement an integrated aircraft icing research strategic plan through research collaboration among the AIRA members,
- Strengthen and foster long-term aircraft icing research expertise,
- Exchange appropriate technical and scientific information,
- Encourage the development of critical aircraft icing technologies, and
- Provide a framework for collaboration between AIRA members.

76. Thomas P. Ratvasky, Kurt Blankenship, William Rieke, and David J. Brinker, "Iced Aircraft Flight Data for Flight Simulation Validation," NASA TM-2003-212114 (2003).

Finally, among the projects NASA is working with AIRA members includes the topics of ground icing, icing for rotorcraft, characterization of the atmospheric icing environment, high ice water content, icing cloud instrumentation, icing environment remote sensing, propulsion system icing, and ice adhesion/shedding from rotating surfaces—the last two a reference to the internal engine icing problem that is likely to make icing headlines during the next few years.

The NACA-NASA role in the history of icing research, and in searching for means to frustrate this insidious threat to aviation safety, has been one of constant endeavor, constantly matching the growth of scientific understanding and technical capabilities to the threat as it has evolved over time. From crude attempts to apply mechanical fixes, fluids, and heating, NACA and NASA researchers have advanced to sophisticated modeling and techniques matching the advances of aerospace science in the fields of fluid mechanics, atmospheric physics, and computer analysis and simulation. Through all of that, they have demonstrated another constant as well: a persistent dedication to fulfill a mandate of Federal aeronautical research dating to the founding of the NACA itself and well encapsulated in its founding purpose: "to supervise and direct the scientific study of the problems of flight, with a view to their practical solution."

Recommended Additional Readings
Reports, Papers, Articles, and Presentations:

Harold E. Addy, Jr., "Ice Accretions and Icing Effects for Modern Airfoils," NASA TP-2000-210031 (2000).

Harold E. Addy, Jr., "NASA Iced Aerodynamics and Controls Current Research," in NASA Glenn Research Center staff, *Proceedings of the Airframe Icing Workshop,* NASA CP-2009-215797 (Cleveland, OH: NASA Glenn Research Center, 2009).

Harold E. Addy, Jr., and Mark G. Potapczuk, "Full-Scale Iced Airfoil Aerodynamic Performance Evaluated," NASA Glenn Research Center, *2007 Research and Technology Report* (Cleveland, OH: NASA Glenn Research Center, 2008).

Harold E. Addy, Jr., Dean R. Miller, and Robert F. Ide, "A Study of Large Droplet Ice Accretion in the NASA Lewis IRT at Near-Freezing Conditions; Pt. 2," NASA TM 107424 (1996).

Harold E. Addy, Jr., Mark G. Potapczuk, and David W. Sheldon, "Modern Airfoil Ice Accretions," NASA TM-107423 (1997).

A.E. Albright, D.L. Kohlman, W.G. Schweikhard, and P. Evanich, "Evaluation of a Pneumatic Boot Deicing System on a General Aviation Wing," NASA TM-82363 (1981).

David N. Anderson and Jaiwon Shin, "Characterization of Ice Roughness from Simulated Icing Encounters," NASA TM-107400 (1997).

David N. Anderson and Jen-Ching Tsao, "Ice Shape Scaling for Aircraft in SLD Conditions," NASA CR-2008-215302 (2008).

Marivell Baez, Mary Vickerman, and Yung Choo, "SmaggIce User Guide," NASA TM-2000-209793 (2000).

Billy P. Barnhart and Thomas P. Ratvasky, "Effective Training for Flight in Icing Conditions," NASA TM-2007-214693 (2007).

Ben C. Bernstein, Thomas P. Ratvasky, Dean R. Miller, and Frank McDonough, "Freezing Rain as an In-Flight Icing Hazard," NASA TM-2000-210058 (2000).

Colin S. Bidwell and Mark G. Potapczuk, "Users Manual for the NASA Lewis Three-Dimensional Ice Accretion Code: LEWICE 3D," NASA TM-105974 (1993).

Colin S. Bidwell, David Pinella, and Peter Garrison, "Ice Accretion Calculations for a Commercial Transport Using the LEWICE3D, ICEGRID3D and CMARC Programs," NASA TM-1999-208895 (1999).

W. Bleeker, "The Formation of Ice on Aircraft," NACA TM-1027 (1942) [trans. of "Einige Bermerkungen über Eisansatz an Flugzeugen," *Meteorologische Zeitschrift* (1932)], pp. 349–354.

Thomas H. Bond, "FAA Perspective," in NASA Glenn Research Center staff, *Proceedings of the Airframe Icing Workshop,* NASA CP-2009-215797 (Cleveland, OH: NASA Glenn Research Center, 2009).

Thomas H. Bond and Jaiwon Shin, "Results of Low Power Deicer Tests on the Swept Inlet Component in the NASA Lewis Icing Research Tunnel," NASA TM-105968 (1993).

Thomas H. Bond, Jaiwon Shin, and Geert A. Mesander, "Advanced Ice Protection Systems Test in the NASA Lewis Icing Research Tunnel," NASA TM-103757 (1991).

D.T. Bowden, A.E. Gensemer, and C.A. Skeen, "Engineering Summary of Airframe Technical Icing Data," FAA Technical Report ADS-4 (1964).

David J. Brinker, Andrew L. Reehorst, "Web-Based Icing Remote Sensing Product Developed," in NASA Glenn Research Center, *2007 Research and Technology Report* (Cleveland, OH: NASA Glenn Research Center, 2008).

Laura Brown, "FAA Icing Fact Sheet: Flying in Icing Conditions," NTSB Docket No. SA-533, Exhibit No. 2-GGG (2009).

Thomas Carroll and William H. McAvoy, "The Formation of Ice Upon Airplanes in Flight," NACA TN-313 (1929).

Yung K. Choo, John W. Slater, Todd L. Henderson, Colin S. Bidwell, Donald C. Braun, and Joongkee Chung, "User Manual for Beta Version of TURBO-GRD: 2A Software System for Interactive Two-Dimensional Boundary/Field Grid Generation, Modification, and Refinement," NASA TM-1998-206631 (1998).

James J. Chung and Harold E. Addy, Jr., "A Numerical Evaluation of Icing Effects on a Natural Laminar Flow Airfoil," NASA TM-2000-209775 (2000).

Joongkee Chung, Yung K. Choo, Andrew Reehorst, M. Potapczuk, and J. Slater, "Navier-Stokes Analysis of the Flowfield Characteristics of an Ice Contaminated Aircraft Wing," NASA TM-1999-208897 (1999).

William D. Coles, "Laboratory Investigation of Ice Formation and Elimination in the Induction System of a Large Twin-Engine Cargo Aircraft," NACA TN-1427 (1947).

Mark Dietenberger, Prem Kumar, and James Luers, "Frost Formation on an Airfoil: A Mathematical Model 1," NASA CR-3129 (1979).

Edward F. Emery, "Instrument Developed for Indicating the Severity of Aircraft Icing and for Providing Cloud-Physics Measurements for Research," NASA Glenn Research Center, *2007 Research and Technology Report* (Cleveland, OH: NASA Glenn Research Center, 2008).

Henry A. Essex, "A Laboratory Investigation of the Icing Characteristics of the Bendix-Stromberg Carburetor Model PD-12F5 with the Pratt and Whitney R-1830-C4 Intermediate Rear Engine Section," NACA-WR-E-18 (1944).

Thomas F. Gelder, William H. Smyers, Jr., and Uwe H. von Glahn, "Experimental Droplet Impingement on Several Two-Dimensional Airfoils with Thickness Ratios of 6 to 16 Percent," NACA TN-3839 (1956).

Uwe H. von Glahn, Vernon H. Gray, "Effect of Ice and Frost Formations on Drag of NACA 65-212 Airfoil for Various Modes of Thermal Ice Protection," NACA TN-2962 (1953).

Uwe H. von Glahn, Vernon H. Gray, "Effect of Ice Formations on Section Drag of Swept NACA 63A-009 Airfoil with Partial Span Leading Edge Slat for Various Modes of Thermal Ice Protection," NACA RM-E53J30 (1954).

Uwe H. von Glahn, Edmund E. Callaghan, and Vernon H. Gray, "NACA Investigation of Icing-Protection Systems for Turbojet-Engine Installations," NACA RM-E51B12 (1951).

Uwe H. von Glahn, Thomas F. Gelder, and William H. Smyers, Jr., "A Dye-Tracer Technique for Experimentally Obtaining Impingement Characteristics of Arbitrary Bodies and Method for Determining Droplet Size Distribution," NACA TN-3338 (1955).

Vernon H. Gray and Dean T. Bowden, "Comparisons of Several Methods of Cyclic De-Icing of a Gas-Heated Airfoil," NACA RM-E53C27 (1953).

Vernon H. Gray and Uwe H. von Glahn, "Aerodynamic Effects Caused by Icing of an Unswept NACA 65A004 Airfoil," NACA TN-4155 (1958).

Vernon H. Gray, Dean T. Bowden, and Uwe H. von Glahn, "Preliminary Results of Cyclical De-Icing of a Gas-Heated Airfoil," NACA RM-E51J29 (1952).

Gerald Gregorek, John J. Dresse, and Karine La Noé, "Additional Testing of the DHC-6 Twin Otter Tail Plane Airfoil Section Testing in the Ohio State University 7x10 Low Speed Wind Tunnel," NASA CR-2000-29921/VOL2 (2000).

R. John Hansman, Kenneth S. Breuer, Didier Hazan, Andrew Reehorst, and Mario Vargas, "Close-Up Analysis of Aircraft Ice Accretion," NASA TM-105952 (1993).

Julie Haggerty, Frank McDonough, Jennifer Black, Scott Landott, Cory Wolff, Steven Mueller, Patrick Minnis, and William Smith, Jr., "Integration of Satellite-Derived Cloud Phase, Cloud Top Height, and Liquid Water Path into an Operational Aircraft Icing Nowcasting System," *American Meteorological Society, 13th Conference on Aviation, Range, and Aerospace Meteorology, New Orleans, LA, Jan. 2008.*

John R. Hansman, Jr., "The Influence of Ice Accretion Physics on the Forecasting of Aircraft Icing Conditions," NASA Joint University Program for Air Transportation Research (1990).

12

Dale Hiltner, Michael McKee, Karine La Noé, and Gerald Gregorek, "DHC-6 Twin Otter Tail Plane Airfoil Section Testing in the Ohio State University 7x10 Wind Tunnel," NASA-CR-2000-2099921/VOL1 (2000).

Jim Hoppins, "Small Airframe Manufacturer's Icing Perspective," NASA Glenn Research Center staff, *Proceedings of the Airframe Icing Workshop*, NASA CP-2009-215797 (Cleveland, OH: NASA Glenn Research Center, 2009).

Robert F. Ide, David W. Sheldon, "2006 Icing Cloud Calibration of the NASA Glenn Icing Research Tunnel," NASA TM-2008-215177 (2008).

Thomas B. Irvine, John R. Oldenburg, and David W. Sheldon, "New Icing Cloud Simulation System at the NASA Glenn Research Center Icing Research Tunnel," NASA TM-1999-208891 (1999).

Thomas B. Irvine, Susan L. Kevdzija, David W. Sheldon, and David A. Spera, "Overview of the Icing and Flow Quality Improvements Program for the NASA Glenn Icing Research Tunnel," NASA TM-2001-210686 (2001).

Alun R. Jones and William Lewis, "Recommended Values of Meteorological Factors to be Considered in the Design of Aircraft Ice-Prevention Equipment," NACA TN-1855 (1949).

Sharon Monica Jones, Mary S. Reveley, Joni K. Evans, and Francesca A. Barrientos, "Subsonic Aircraft Safety Icing Study," NASA TM-2008-215107 (2008).

Abdollah Khodadoust, Chet Dominik, Jaiwon Shin, and Dean R. Miller, "Effect of In-Flight Ice Accretion on the Performance of a Multi-Element Airfoil," NASA TM-112174 (1995).

Montgomery Knight and William C. Clay, "Refrigerated Wind Tunnel Tests on Surface Coatings for Preventing Ice Formation," NACA TN-339 (1930).

William M. Leary, "A Perennial Challenge to Aviation Safety: Battling the Menace of Ice," in Roger D. Launius and Janet R. Daly Bednarek, eds., *Reconsidering a Century of Flight* (Chapel Hill, NC: University of North Carolina Press, 2003), pp. 132–151.

Laurie H. Levinson, Mark G. Potapczuk, and Pamela A. Mellor, "Software Development Processes Applied to Computational Icing Simulation," NASA TM-1999-208898 (1999).

Laurie H. Levinson and William B. Wright, "IceVal DatAssistant: An Interactive, Automated Icing Data Management System," NASA TM-2008-215158 (2008).

James P. Lewis and Dean T. Bowden, "Preliminary Investigation of Cyclic De-Icing of an Airfoil Using an External Electric Heater," NACA RM-E51J30 (1952).

James P. Lewis and Robert J. Blade, "Experimental Investigation of Radome Icing and Icing Protection," NACA RM-E52J31 (1953). James P. Lewis, Thomas F. Gelder, and Stanley L. Koutz, "Icing Protection for a Turbojet Transport Airplane: Heating Requirements, Methods of Protection, and Performance Penalties," NACA TN-2866 (1953).

William Lewis and Norman R. Bergrun, "A Probability Analysis of the Meteorological Factors Conducive to Aircraft Icing in the United States," NACA TN-2738 (1952).

Richard H. McFarland, and Craig B. Parker, "Weather Data Dissemination to Aircraft," NASA, Langley Research Center, Joint University Program for Air Transportation Research, 1988–1989, pp. 119–127.

R.C. McKnight, R.L. Palko, and R.L. Humes, "In-flight Photogrammetric Measurement of Wing Ice Accretions," NASA TM-87191 (1986).

Dean R. Miller, Harold E. Addy, Jr., and Robert F. Ide, "A Study of Large Droplet Ice Accretion in the NASA Lewis IRT at Near-Freezing Conditions," Pt. 2, NASA TM-107424 (1996).

Dean R. Miller, Thomas P. Ratvasky, Ben C. Bernstein, Frank McDonough, and J. Walter Strapp, "NASA/FAA/NCAR Supercooled Large Droplet Icing Flight Research: Summary of Winter 1996–1997 Flight Operations," NASA TM-1998-206620 (1998).

Dean R. Miller, Christopher J. Lynch, and Peter A. Tate, "Overview of High Speed Close-Up Imaging in an Icing Environment," NASA TM-2004-212925 (2004).

Dean R. Miller, Mark G. Potapczuk, and Thomas H. Bond, "Update on SLD Engineering Tools Development," NASA TM-2004-213072 (2004).

Dean R. Miller, Thomas Ratvasky, Ben Bernstein, Frank McDonough, and J. Walter Strapp, "NASA/FAA/NCAR Supercooled Large Droplet Icing Flight Research: Summary of Winter 1996–1997 Flight Operations," NASA TM-1998-206620 (1998).

Carr B. Neel, Jr., and Loren G. Bright, "The Effect of Ice Formations on Propeller Performance, NACA TN-2212 (1950).

Richard W. Obermayer and William T. Roe, "A Study of Carburetor/Induction System Icing in General Aviation Accidents," NASA CR-143835 (1975).

Michael Papadakis, See-Cheuk Wong, Arief Rachman, Kuohsing E. Hung, Giao T. Vu, and Colin S. Bidwell, "Large and Small Droplet Impingement Data on Airfoils and Two Simulated Ice Shapes," NASA TM-2007-213959 (2007).

Michael Papadakis, Arief Rachman, See-Cheuk Wong, Hsiung-Wei Yeong, Kuohsing E. Hung, Giao T. Vu, and Colin S. Bidwell, "Water Droplet Impingement on Simulated Glaze, Mixed, and Rime Ice Accretions," NASA TM-2007-213961 (2007).

Andy Pasztor, "Airline Regulators Grapple with Engine-Shutdown Peril," *The Wall Street Journal*, 7 April 2008.

Porter J. Perkins, "Icing Frequencies Experienced During Climb and Descent by Fighter-Interceptor Aircraft," NACA TN-4314 (1958).

Porter J. Perkins, Stuart McCullough, and Ralph D. Lewis, "A Simplified Instrument for Recording and Indicating Frequency and Intensity of Icing Conditions Encountered in Flight," NACA RM-E51E16 (1951).

Mark G. Potapczuk, "LEWICE/E: An Euler Based Ice Accretion Code," NASA TM-105389 (1992).

Mark G. Potapczuk, "A Review of NASA Lewis' Development Plans for Computational Simulation of Aircraft Icing," NASA TM-1999-208904 (1999).

Mark G. Potapczuk, "Airframe Icing Research Gaps: NASA Perspective," in NASA Glenn Research Center staff, *Proceedings of the Airframe Icing Workshop*, NASA CP-2009-215797 (Cleveland, OH: NASA Glenn Research Center, 2009).

Mark G. Potapczuk, "NASA Airframe Icing Research Overview Past and Current," in NASA Glenn Research Center staff, *Proceedings of the Airframe Icing Workshop*, NASA CP-2009-215797 (Cleveland, OH: NASA Glenn Research Center, 2009).

Mark G. Potapczuk and Colin S. Bidwell, "Swept Wing Ice Accretion Modeling," NASA TM-103114 (1990).

Mark G. Potapczuk and John J. Reinmann, "Icing Simulation: A Survey of Computer Models and Experimental Facilities," NASA TM-104366 (1991).

Mark G. Potapczuk, M.B. Bragg, O.J. Kwon, and L.N. Sankar, "Simulation of Iced Wing Aerodynamics," NASA TM-104362 (1991).

Mark G. Potapczuk, Kamel M. Al-Khalil, and Matthew T. Velazquez, "Ice Accretion and Performance Degradation Calculations with LEWICE/NS," NASA TM-105972 (1991).

G. Merritt Preston and Calvin C. Blackman, "Effects of Ice Formations on Airplane Performance in Level Cruising Flight," NACA TN-1598 (1948).

Richard J. Ranaudo, Kevin L. Mikkelsen, Robert C. McKnight, Porter J. Perkins, Jr., "Performance Degradation of a Typical Twin Engine Commuter Type Aircraft in Measured Natural Icing Conditions," NASA TM-83564 (1984).

Richard J. Ranaudo, Kevin L. Mikkelsen, Robert C. McKnight, Robert F. Ide, Andrew L. Reehorst, J.L. Jordan, W.C. Schinstock, and S.J. Platz, "The Measurement of Aircraft Performance and Stability and Control After Flight Through Natural Icing Conditions," NASA TM-87265 (1986).

Thomas P. Ratvasky and Richard J. Ranaudo, "Icing Effects on Aircraft Stability and Control Determined from Flight Data, Preliminary Results," NASA TM-105977 (1993).

Thomas P. Ratvasky and Judith Foss Van Zante, "In Flight Aerodynamic Measurements of an Iced Horizontal Tailplane," NASA TM-1999-208901 (1999).

Thomas P. Ratvasky, Judith Foss Van Zante, and James T. Riley, "NASA/FAA Tailplane Icing Program Overview," NASA TM-1999-208901 (1999).

Thomas P. Ratvasky, Billy P. Barnhart, and Sam Lee, "Current Methods for Modeling and Simulating Icing Effects on Aircraft Performance, Stability and Control," NASA TM-2008-215453 (2008).

Thomas P. Ratvasky, Judith Foss Van Zante, and Alex Sim, "NASA/FAA Tailplane Icing Program: Flight Test Report," NASA TP-2000-209908 (2000).

Thomas P. Ratvasky, Kurt Blankenship, William Rieke, and David J. Brinker, "Iced Aircraft Flight Data for Flight Simulation Validation," NASA TM-2003-212114 (2003).

Thomas P. Ratvasky, Richard J. Ranaudo, Kurt S. Blankenship, and Sam Lee, "Demonstration of an Ice Contamination Effects Flight Training Device," NASA TM-2006-214233 (2006).

Thomas P. Ratvasky, Billy P. Barnhart, Sam Lee, and Jon Cooper, "Flight Testing an Iced Business Jet for Flight Simulation Model Validation," NASA TM-2007-214936 (2007).

Andrew L. Reehorst, David J. Brinker, and Thomas P. Ratvasky, "NASA Icing Remote Sensing System Comparisons from AIRS II," NASA TM-2005-213592 (2005).

Andrew L. Reehorst, Mark G. Potapczuk, Thomas P. Ratvasky, and Brenda Gile Laflin, "Wind Tunnel Measured Effects on a Twin-Engine Short-Haul Transport Caused by Simulated Ice Accretions Data Report," NASA TM-107419 (1997).

Andrew L. Reehorst, David J. Brinker, Thomas P. Ratvasky, Charles C. Ryerson, and George G. Koenig, "The NASA Icing Remote Sensing System," NASA TM-2005-213591 (2005).

Andrew L. Reehorst, Marcia K. Politovich, Stephan Zednik, George A. Isaac, and Stewart Cober, "Progress in the Development of Practical Remote Detection of Icing Conditions," NASA TM-2006-214242 (2006).

Andrew L. Reehorst, Joongkee Chung, M. Potapczuk, Yung K. Choo, W. Wright, and T. Langhals, "An Experimental and Numerical Study of Icing Effects on the Performance and Controllability of a Twin Engine Aircraft," NASA TM-1999-208896 (1999).

Andrew L. Reehorst, David Brinker, Marcia Politovich, David Serke, Charles Ryerson, Andrew Pazmany, and Frederick Solheim, "Progress Towards the Remote Sensing of Aircraft Icing Hazards," NASA TM-2009-215828 (2009).

John J. Reinmann, "NASA's Aircraft Icing Technology Program," NASA TM-1991-104518 (1991).

John J. Reinmann, Robert J. Shaw, and W.A. Olsen, Jr., "Aircraft Icing Research at NASA," NASA TM-82919 (1982).

John J. Reinmann, Robert J. Shaw, and W.A. Olsen, Jr., "NASA Lewis Research Center's Program on Icing Research," NASA TM-83031 (1983).

John J. Reinmann, Robert J. Shaw, and Richard J. Ranaudo, "NASA's Program on Icing Research and Technology," NASA TM-101989 (1989).

Manuel A. Rios and Yung I. Cho, "Analysis of Ice Crystal Ingestion as a Source of Ice Accretion," AIAA-2008-4165 (2208).

Lewis A. Rodert, "An Investigation of the Prevention of Ice on the Airplane Windshield," TN-754 (1940).

Lewis A. Rodert, "The Effects of Aerodynamic Heating on Ice Formations on Airplane Propellers," TN-799 (1941).

Lewis A. Rodert and Alun R. Jones, "A Flight Investigation of Exhaust-Heat De-Icing," NACA TN-783 (1940). Lewis A. Rodert and Richard Jackson, "Preliminary Investigation and Design of an Air-Heated Wing for Lockheed 12A Airplane," NACA ARR A-34 (1942).

Charles C. Ryerson, "Remote Sensing of In-Flight Icing Conditions: Operational, Meteorological, and Technological Considerations," NASA CR-2000-209938 (2000).

Charles C. Ryerson, George G. Koenig, Andrew L. Reehorst, and Forrest R. Scott, "Concept, Simulation, and Instrumentation for Radiometric Inflight Icing Detection," NASA TM-2009-215519 (2009).

Robert J. Shaw and John J. Reinmann, "NASA's Rotorcraft Icing Research Program," NASA NTRS N88-16641 (1988).

Robert J. Shaw and John J. Reinmann, "The NASA Aircraft Icing Research Program," NASA NTRS 92N22534 (1990).

Jaiwon Shin, "Characteristics of Surface Roughness Associated with Leading Edge Ice Accretion," NASA TM-106459 (1994).

Jaiwon Shin, "The NASA Aviation Safety Program: Overview," NASA TM-2000-209810 (2000).

Jaiwon Shin and Thomas H. Bond, "Results of an Icing Test on a NACA 0012 Airfoil in the NASA Lewis Icing Research Tunnel," NASA TM-105374 (1992).

Jaiwon Shin, Hsun H. Chen, and Tuncer Cebeci, "A Turbulence Model for Iced Airfoils and Its Validation," NASA TM-105373 (1992).

Jaiwon Shin, Brian Berkowitz, Hsun H. Chen, and Tuncer Cebeci, "Prediction of Ice Shapes and their Effect on Airfoil Performance," NASA TM-103701 (1991).

Jaiwon Shin, Peter Wilcox, Vincent Chin, and David Sheldon, "Icing Test Results on an Advanced Two-Dimensional High-Lift Multi-Element Airfoil," NASA TM-106620 (1994).

Howard Slater, Jay Owns, and Jaiwon Shin, "Applied High-Speed Imaging for the Icing Research Program at NASA Lewis Research Center," NASA TM-104415 (1991).

Wesley L. Smith, "Weather Problems Peculiar to the New York-Chicago Airway," *Monthly Weather Review* vol. 57, no. 12 (Dec. 1929).

Dave Sweet, "An Ice Protection and Detection Systems Manufacturer's Perspective," NASA Glenn Research Center staff, *Proceedings of the Airframe Icing Workshop*, NASA CP-2009-215797 (Cleveland, OH: NASA Glenn Research Center, 2009).

Theodore Theodorsen and William C. Clay, "Ice Prevention on Aircraft by Means of Engine Exhaust Heat and a Technical Study of Heat Transmission from a Clark Y Airfoil," NACA TR-403 (1933).

David S. Thompson and Bharat K. Soni, "ICEG2D (v2.0)—An Integrated Software Package for Automated Prediction of Flow Fields for Single-Element Airfoils with Ice Accretion," NASA CR-2000-209914 (2000).

U.S. Civil Aeronautics Board, Bureau of Safety Investigation, *Comparative Safety Statistics in United States Airline Operations*, Pt. 1: *Years 1938–1945* (Washington, DC: CAB BSI Analysis Division, 15 August 1953).

U.S. National Aeronautics and Space Administration, "Preventing Ice Before it Forms," *Spinoff 2006* (Washington, DC: NASA, 2006), pp. 46–47.

U.S. National Aeronautics and Space Administration, "Deicing System Protects General Aviation Aircraft," *Spinoff 2007* (Washington, DC: NASA, 2007).

U.S. National Aeronautics and Space Administration, NASA Glenn Research Center, *2007 Research and Technology Report* (Cleveland, OH: NASA Glenn Research Center, 2008).

U.S. National Aeronautics and Space Administration, NASA Glenn Research Center staff, *Proceedings of the Airframe Icing Workshop*, NASA CP-2009-215797 (Cleveland, OH: NASA Glenn Research Center, 2009).

U.S. National Transportation Safety Board, *Collision with 14th Street Bridge Near Washington National Airport, Air Florida Flight 90, Boeing 737-222, N62AF, Washington, D.C., January 13, 1982*, NTSB/AAR-82-8 (Washington, DC: NTSB, 1982).

National Transportation Safety Board, *In-Flight Icing Encounter and Loss of Control Simmons Airlines, d.b.a. American Eagle Flight 4184 Avions de Transport Regional (ATR) Model 72-212, N401AM Roselawn, Indiana October 31, 1994*, v. 1: *Safety Board Report*, NTSB/AAR-96/01 (Washington, DC: NTSB, 1996).

U.S. National Transportation Safety Board, Letter to the Federal Aviation Administration, Safety Recommendation A-06-56 through -59 (Washington, DC: NTSB, 2006).

U.S. National Transportation Safety Board, "FAA Presentation—Icing Requirements and Guidance," NTSB Docket No. SA-533, Exhibit No. 2-JJJ (Washington, DC: NTSB, 2009).

Mario Vargas, "Icing Branch Current Research Activities in Icing Physics," in NASA Glenn Research Center staff, *Proceedings of the Airframe Icing Workshop*, NASA CP-2009-215797 (Cleveland, OH: NASA Glenn Research Center, 2009).

Mario Vargas and Eli Reshotko, "Physical Mechanisms of Glaze Ice Scallop Formations on Swept Wings," NASA TM-1998-206616 (1998).

Mario Vargas and Eli Reshotko, "LWC and Temperature Effects on Ice Accretion Formation on Swept Wings at Glaze Ice Conditions," NASA TM-2000-209777 (2000).

Mario Vargas, Howard Broughton, James J. Sims, Brian Bleeze, and Vatanna Gaines, "Local and Total Density Measurements in Ice Shapes," NASA TM-2005-213440 (2005).

Leanne West, Gary Gimmestad, William Smith, Stanislav Kireev, Larry B. Cornman, Philip R. Schaffner, and George Tsoucalas, "Applications of a Forward-Looking Interferometer for the On-board Detection of Aviation Weather Hazards," NASA TP-2008-215536 (2008).

William B. Wright, "Users Manual for the Improved NASA Lewis Ice Accretion Code LEWICE 1.6," NASA CR-1995-198355 (1995).

William B. Wright, "User Manual for the NASA Glenn Ice Accretion Code LEWICE Version 2.0," NASA CR-1999-209409 (1999).

William B. Wright, "User Manual for the NASA Glenn Ice Accretion Code LEWICE Version 2.2.2," NASA CR-2002-211793 (2002).

William B. Wright, "Further Refinement of the LEWICE SLD Model," NASA CR-2006-214132 (2006).

William B. Wright, "User's Manual for LEWICE Version 3.2," NASA CR-2008-214255 (2008).

William B. Wright and James Chung, "Correlation Between Geometric Similarity of Ice Shapes and the Resulting Aerodynamic Performance Degradation-A Preliminary Investigation Using WIND," NASA CR-1999-209417 (1999).

William B. Wright, R.W. Gent, and Didier Guffond, "DRA/NASA/ONERA Collaboration on Icing Research Pt. II—Prediction of Airfoil Ice Accretion," NASA CR-1997-202349 (1997).

William B. Wright and Adam Rutkowski, "Validation Results for LEWICE 2.0," NASA CR-1999-208690 (1999).

William B. Wright, Mark G. Potapczuk, and Laurie H. Levinson, "Comparison of LEWICE and GlennICE in the SLD Regime," NASA TM-2008-215174 (2008).

Thomas J. Yager, William A. Vogler, and Paul Baldasare, "Evaluation of Two Transport Aircraft and Several Ground Test Vehicle Friction Measurements Obtained for Various Runway Surface Types and Conditions. A Summary of Test Results From Joint FAA/NASA Runway Friction Program," NASA TP-2917 (1990).

Judith Foss Van Zante, "A Database of Supercooled Large Droplet Ice Accretions," NASA CR-2007-215020 (2007).

Judith Foss Van Zante and Thomas P. Ratvasky, "Investigation of Dynamic Flight Maneuvers with an Iced Tail Plane," NASA TM-1999-208849 (1999).

G.W. Zumwalt, "Electro-Impulse De-Icing: A Status Report," AIAA Paper 88-0019 (1988).

G.W. Zumwalt, R.A. Friedberg, "Designing an Electro-Impulse De-Icing System," AIAA Paper 86-0545 (1986).

G.W. Zumwalt, R.L. Schrag, W.D. Bernhart, and R.A. Friedberg, "Analyses and Tests for Design of an Electro-Impulse De-Icing System," NASA CR-174919 (1985).

Books and Monographs

David A. Anderton, *NASA Aeronautics*, NASA EP-85 (Washington, DC: NASA, 1982).

Paul F. Borchers, James A. Franklin, and Jay W. Fletcher, *Flight Research at Ames: Fifty-Seven Years of Development and Validation of Aeronautical Technology*, NASA SP-1998-3300 (Moffett Field, CA: Ames Research Center, 1998).

Bob Buck [Robert N. Buck], *North Star Over My Shoulder: A Flying Life* (New York: Simon & Schuster, 2002).

Robert N. Buck, *The Pilot's Burden: Flight Safety and the Roots of Pilot Error* (Ames, IA: Iowa State University Press, 1994).

Robert N. Buck, *Weather Flying* (New York: The Macmillan Company, 1978).

Joseph R. Chambers, *Partners in Freedom: Contributions of the Langley Research Center to U.S. Military Aircraft of the 1990's*, NASA SP-2000-4519 (Washington, DC: GPO, 2000).

Joseph R. Chambers, *Concept to Reality: Contributions of the NASA Langley Research Center to U.S. Civil Aircraft of the 1990s,* NASA SP-2003-4529 (Washington, DC: GPO, 2003).

Joseph R. Chambers, *Innovation in Flight: Research of the NASA Langley Research Center on Revolutionary Advanced Concepts for Aeronautics,* NASA SP-2005-4539 (Washington, DC: GPO, 2005).

Virginia P. Dawson, *Engines and Innovation: Lewis Laboratory and American Propulsion Technology,* NASA SP-4306 (Washington, DC: NASA, 1991).

Ernest K. Gann, *Fate Is the Hunter* (New York: Simon & Schuster, 1961).

Michael H. Gorn, *Expanding the Envelope: Flight Research at NACA and NASA* (Lexington, KY: The University Press of Kentucky, 2001).

George W. Gray, *Frontiers of Flight: The Story of NACA Research* (New York: Alfred A. Knopf, 1948).

James R. Hansen, *Engineer in Charge: A History of the Langley Aeronautical Laboratory, 1917–1958,* NASA SP-4305 (Washington, DC: GPO, 1987).

Edwin P. Hartman, *Adventures in Research: A History of Ames Research Center 1940–1965,* NASA SP-4302 (Washington, DC: GPO, 1970).

Terry T. Lankford, *Aircraft Icing: A Pilot's Guide* (McGraw-Hill Professional, 1999).

Roger D. Launius and Janet R. Daly Bednarek, eds., *Reconsidering a Century of Flight* (Chapel Hill, NC: University of North Carolina Press, 2003).

William M. Leary, *"We Freeze to Please": A History of NASA's Icing Research Tunnel and the Quest for Flight Safety,* NASA SP-2002-4226 (Washington, DC: NASA, 2002).

Pamela E. Mack, ed., *From Engineering Science to Big Science: The NACA and NASA Collier Trophy Research Project Winners,* NASA SP-4219 (Washington, DC: NASA, 1998).

Donald R. Whitnah, *Safer Skyways: Federal Control of Aviation, 1926–1966* (Ames, IA: The Iowa State University Press, 1966).

A drop model of the F/A-18E is released for a poststall study high above the NASA Wallops Flight Center. NASA.

CASE 13

Care-Free Maneuverability At High Angle of Attack

Joseph R. Chambers

Since the airplane's earliest days, maintaining safe flight at low speeds and high angles of attack has been a stimulus for research. As well, ensuring that a military fighter aircraft has good high-angle-of-attack qualities can benefit its combat capabilities. NASA research has provided critical guidance on configuration effects and helped usher in the advent of powerful flight control concepts.

AT THE TIME THAT the National Aeronautics and Space Administration (NASA) absorbed the National Advisory Committee for Aeronautics (NACA), it also inherited one of the more challenging technical issues of the NACA mission: to "supervise and direct the scientific study of the problems of flight with a view to their practical solution." Since the earliest days of heavier-than-air flight, intentional or inadvertent flight at high angles of attack (high alpha) results in the onset of flow separation on lifting surfaces, stabilizing fins, and aerodynamic controls. In such conditions, a poorly designed aircraft will exhibit a marked deterioration in stability, control, and flying qualities, which may abruptly cause loss of control, spin entry, and catastrophic impact with the ground.[1] Stalling and spinning have been—and will continue to be—major areas of research and development for civil and military aircraft. In the case of highly maneuverable military aircraft, high-angle-of-attack characteristics exert a tremendous influence on tactical effectiveness, maneuver options, and safety.

Some of the more notable contributions of NASA to the Nation's military aircraft community have been directed at high-angle-of-attack technology, including the conception, development, and validation of advanced ground- and flight-test facilities; advances in related disciplinary fields, such as aerodynamics and flight dynamics; generation

1. Joseph R. Chambers and Sue B. Grafton, "Aerodynamic Characteristics of Airplanes at High Angles of Attack," NASA TM-74097 (1977).

of high-alpha design criteria and methods; and active participation in aircraft development programs.[2] Applications of these NASA contributions by the industry and the Department of Defense (DOD) have led to a dramatic improvement in high-angle-of-attack behavior and associated maneuverability for the current U.S. military fleet. The scope of NASA activities in this area includes ground-based and flight research at all of its aeronautical field centers. The close association of NASA, industry, and DOD, and the significant advances in the state of the art that have resulted from common objectives, are notable achievements of the Agency's value to the Nation's aeronautical achievements.

The Early Days

Early NACA research on stalling and spinning in the 1920s quickly concluded that the primary factors that governed the physics of stall behavior, spin entry, and recovery from spins were very complicated and would require extensive commitments to new experimental facilities for studies of aerodynamics and flight motions. Over the following 85 years, efforts by the NACA and NASA introduced a broad spectrum of specialized tools and analysis techniques for high-angle-of-attack conditions, including vertical spin tunnels, pressurized wind tunnels to define the impact of Reynolds number on separated flow phenomena, special free-flight model test techniques, full-scale aircraft flight experiments, theoretical studies of aircraft motions, piloted simulator studies, and unique static and dynamic wind tunnel aerodynamic testing capability.[3]

By the 1930s, considerable progress had been made at the NACA Langley Memorial Aeronautical Laboratory on obtaining wind tunnel aerodynamic data on the effectiveness of lateral control concepts at the stall and understanding control effects on motions.[4] A basic understanding began to emerge on the effects of design variables for biplanes of the era, such as horizontal and vertical tail configurations, wing stagger,

2. For detailed discussions of contributions of NASA Langley and its partners to high-angle-of-attack technology for current military aircraft, see Chambers, *Partners in Freedom: Contributions of the NASA Langley Research Center to Military Aircraft of the 1990s*, NASA SP-2000-4519 (2000).

3. D. Bruce Owens, Jay M. Brandon, Mark A. Croom, Charles M. Fremaux, Eugene H. Heim, and Dan D. Vicroy, "Overview of Dynamic Test Techniques for Flight Dynamics Research at NASA LaRC," AIAA Paper 2006-3146 (2006).

4. Fred E. Weick and Robert T. Jones, "The Effect of Lateral Controls in Producing Motion of an Airplane as Computed From Wind-Tunnel Data," NACA TR-570 (1937).

and center-of-gravity location on spinning. Flight-testing of stall characteristics became a routine element of handling quality studies. In the race to conquer stall/spin problems, however, simplistic and regrettable conclusions were frequently drawn.[5]

The sudden onset of World War II and its urgency for aeronautical research and development overwhelmed the laboratory's plodding research environment and culture with high-priority requests from the military services for immediate wind tunnel and flight assessments, as well as problem-solving activities for emerging military aircraft. At that time, the military perspective was that operational usage of high-angle-of-attack capability was necessary in air combat, particularly in classic "dogfight" engagements wherein tighter turns and strenuous maneuvers meant the difference between victory and defeat. Tactical effectiveness and safety, however, demanded acceptable stalling and spinning behavior, and early NACA assessments for new designs prior to industry and military flight-testing and production were required for every new maneuverable aircraft.[6] Spin demonstrations of prototype aircraft by the manufacturer were mandatory, and satisfactory stall characteristics and recoveries from developed spins required extensive testing by the NACA in its conventional wind tunnels and vertical spin tunnel.

The exhausting demands of round-the-clock, 7-day workweeks left very little time for fundamental research, but researchers at Langley's Spin Tunnel, Free-Flight Tunnel, Stability Tunnel, and 7- by 10-Foot Tunnels initiated a series of studies that resulted in advancements in high-angle-of-attack design procedures and analysis techniques.[7]

New Challenges

Arguably, no other technical discipline is as sensitive to configuration features as high-angle-of-attack technology. Throughout World War II, the effects of configuration details such as wing airfoil, wing twist,

5. See, for example, Montgomery Knight, "Wind Tunnel Tests on Autorotation and the Flat Spin," NACA TR-273 (1928), which states, "The results of the investigation indicate that in free flight a monoplane is incapable of flat spinning, whereas an unstaggered biplane has inherent flat-spinning tendencies."
6. Over 100 designs were tested in the Langley Spin Tunnel during World War II.
7. Chambers, *Radical Wings and Wind Tunnels* (Specialty Press, 2008); Anshal I. Neihouse, Walter J. Klinar, and Stanley H. Scher, "Status of Spin Research for Recent Airplane Designs," NASA TR-R-57 (1960).

engine torque, propeller slipstream, and wing placement were critical and, if not properly designed, often resulted in deficient handing qualities accentuated by poor or even vicious stalling behavior. The NACA research staffs at Langley and Ames played key roles in advancing design methodology based on years of accumulated knowledge and lessons learned for straight winged, propeller-driven aircraft. Aberrations of design practice, such as flying wings, had posed new problems such as tumbling, which had also been addressed.[8] However, just as it appeared that the art and science of designing for high-alpha conditions was under control, a wave of unconventional configuration features emerged in the jet aircraft of the 1950s to challenge designers with new problems. Foremost among these radical features was the use of swept-back and delta wings, long pointed fuselages, and the distribution of mass primarily along the fuselage.

Suddenly, topics such as pitch-up, inertial coupling, and directional divergence became the focus of high-angle-of-attack technology. Responding to an almost complete lack of design experience in these areas, the NACA initiated numerous experimental and theoretical studies. One of the more significant contributions to design methods was the development of a predictive criterion that used readily obtained aerodynamic wind tunnel parameters to predict whether a configuration would exhibit a directional divergence (departure) at high angles of attack.[9] Typical of many NACA and NASA contributions, the criterion is still used today by designers of high-performance military aircraft.

As the 1950s progressed, it was becoming obvious that high-alpha maneuverability was becoming a serious challenge. Lateral-directional stability and control were difficult to achieve, and the spin and recovery characteristics of the new breed of fighter aircraft were proving to be extremely marginal. In addition to frequent encounters with unsatisfactory spin recovery, dangerous new posstall motions such as disorienting oscillatory spins and fast flat spins were encountered, which challenged the ability of human pilots to effect recovery.[10]

8. Charles Donlan, "An Interim Report on the Stability and Control of Tailless Airplanes," NACA TR-796 (1944).

9. Martin T. Moul and John W. Paulson, "Dynamic Lateral Behavior of High-Performance Aircraft," NACA RM-L58E16 (1958).

10. Neihouse, Klinar, and Scher, "Status of Spin Research."

Group photo of X-planes at Dryden in 1953 exhibit configuration features that had changed dramatically from the straight winged X-1A and D-558-1, at left, to the delta wing XF-92A, top left, the variable-sweep X-5, the swept wing D-558-2, the tailless X-4, and the slender X-3. The changes had significant effects on high-alpha and spin characteristics. NASA.

Automatic flight control systems were designed to limit the maximum obtainable angle of attack to avoid these high-angle-of-attack deficiencies, but severe degradations in maneuver capability were imposed by this approach for some designs. Researchers considered automatic spin recovery concepts, but such systems required special sensors and control components not used in day-to-day operations at that time. Concerns over the cost, maintenance, and the impact of inadvertent actuation of such systems on safety discouraged interest in the development of automatic spin prevention systems.

As the 1950s came to a close, the difficulty of designing for high-angle-of-attack conditions, coupled with the anticipated dominance of emerging air-to-air missile concepts, resulted in a new military perspective on the need for maneuverability. Under this doctrine, maneuverability required for air-to-air engagements would be built into the missile system, and fighter or interceptor aircraft would be designed as stand-off missile launchers with no need for maneuverability or high-alpha capability. Not only did this scenario result in a minimal analysis of high-angle-of-attack behavior for emerging designs, it resulted in a significant decrease in the advocacy and support for NASA research on stall/spin problems. In the late 1950s, Langley was even threatened with a closure of its spin tunnel.[11]

Revelation and Call to Action

During the Vietnam conflict, U.S. pilots flying F-4 and F-105 aircraft faced highly maneuverable MiG-17 and MiG-19 aircraft, and the unanticipated return of the close-in dogfight demanded maneuverability that had not been required during design and initial entry of the U.S. aircraft into operational service. Unfortunately, aircraft such as the F-4 exhibited a marked deterioration in lateral-directional stability and control characteristics at high angles of attack. Inadvertent loss of control became a major issue, with an alarming number of losses in training accidents. A request for support to the NASA Langley Research Center by representatives of the Air Force Aeronautical Systems Division in 1967 resulted in an extensive analysis of the high-angle-of-attack deficiencies of the aircraft and wind tunnel, free-flight model, and piloted simulator studies.[12]

The F-4 experience is especially noteworthy in NASA's contributions to high-angle-of-attack technology. Based on the successful demonstrations of analysis and design tools by NASA, management within the Air Force, Navy, and NASA strongly supported an active participation by the Agency in high-angle-of-attack technology, resulting in requests for similar NASA involvement in virtually all subsequent DOD high-performance aircraft development programs, which continue to the current day. After the F-4 program, NASA activities at Langley were no longer limited to spin tunnel tests but included conventional and special dynamic wind tunnel tests, analytical studies, and piloted simulator studies.

11. Interview of James S. Bowman, Jr., head of the Langley Spin Tunnel, by author, NASA Langley Research Center, June 5, 1963.

12. For a detailed discussion of NASA contributions to the F-4 high-angle-of-attack issues, see Chambers, *Partners in Freedom*; Chambers and Ernie L. Anglin, "Analysis of Lateral Directional Stability Characteristics of a Twin Jet Fighter Airplane at High Angles of Attack," NASA TN-D-5361 (1969); Chambers, Anglin, and Bowman, "Analysis of the Flat-Spin Characteristics of a Twin-Jet Swept-Wing Fighter Airplane," NASA TN-D-5409 (1969); William A. Newsom, Jr., and Grafton, "Free-Flight Investigation of Effects of Slats on Lateral-Directional Stability of a 0.13-Scale Model of the F-4E Airplane," NASA TM-SX-2337 (1971); and Edward J. Ray and Eddie G. Hollingsworth, "Subsonic Characteristics of a Twin-Jet Swept-Wing Fighter Model with Maneuvering Devices," NASA TN-D-6921 (1973).

The shocking number of losses of F-4 aircraft and aircrews did not, however, escape the attention of senior Air Force leadership. As F-4 stall/spin/out-of-control accidents began to escalate, other aircraft types were also experiencing losses, including the A-7, F-100, and F-111. The situation reached a new level of concern when, on April 26, 1971, Air Force Assistant Secretary for Research and Development (R&D) Grant L. Hansen sent a memorandum to R&D planners within the Air Force noting that during a 5-year period from 1966 through 1970, the service had lost over $200 million in assets in stall/spin/out-of-control accidents while it had spent only $200,000 in R&D.[13] Hansen's memo called for a broad integrated research program to advance the state of the art with an emphasis on "preventing the loss of, rather than recovering, aircraft control." The response of Air Force planners was swift, and in December 1971, a major symposium on stall/poststall/spin technology was held at Wright-Patterson Air Force Base.[14] Presentations at the symposium by Air Force, Navy, and Army participants disclosed that the number of aircraft lost by the combined services to stall/spin/out-of-control accidents during the subject 5-year period was sobering: over 225 aircraft valued at more than $367 million. Some of the aircraft types stood out as especially susceptible to this type of accident—for example, the Air Force, Navy, and Marines had lost over 100 F-4 aircraft in that period.

An additional concern was that valuable test and evaluation (T&E) aircraft and aircrews were being lost in flight accidents during high-angle-of-attack and spin assessments. At the time of the symposium, the Navy had lost two F-4 spin-test aircraft and an EA-6B spin-test vehicle, and the Air Force had lost an F-4 and F-111 during spin-test programs because of unrecoverable spins, malfunctions of emergency spin parachute systems, pilot disorientation, and other spin-related causes. The T&E losses were especially distressing because they were experienced under controlled conditions with a briefed pilot entering carefully planned maneuvers with active emergency recovery systems.

The 1971 symposium marked a new waypoint for national R&D efforts in high-angle-of-attack technology. Spin prevention became a major focus of research, the military services acknowledged the need for controlled flight at high-angle-of-attack conditions, and DOD formally

13. Funds cited in then-year dollars.

14. *Aeronautical System Division/Air Force Flight Dynamics Laboratory Symposium on Stall/Post-Stall/Spin, Wright-Patterson Air Force Base, OH, Dec. 15–17, 1971.*

stated high-angle-of-attack and maneuverability requirements for new high-performance aircraft programs. Collaborative planning between industry, DOD, and NASA intensified for research efforts, including ground-based and flight activities.[15] The joint programs clearly acknowledged the NASA role as a source of corporate knowledge and provider of national facilities for the tasks. With NASA having such responsibilities in a national program, its research efforts received significantly increased funding and advocacy from NASA Headquarters and DOD, thereby reversing the relative disinterest and fiscal doldrums of the late 1950s and 1960s.

One of the key factors in the resurgence of NASA–DOD coupling for high-angle-of-attack research from the late 1960s to the early 1990s was the close working relationships that existed between senior leaders in DOD (especially the Navy) and at NASA Headquarters. With these individuals working on a first-name basis, their mutual interests and priorities assured that NASA could respond in a timely manner with high-priority research for critical military programs.[16]

From a technology perspective, new concepts and challenges were ready for NASA's research and development efforts. For example, at the symposium, Langley presented a paper summarizing recent experimental free-flight model studies of automatic spin prevention concepts along with a perspective that unprecedented opportunities for implementation of such concepts had arrived.[17] Although the paper was highly controversial at the time, within a few months, virtually all high-performance aircraft design teams were assessing candidate systems.

Accelerated Progress

NASA's role in high-angle-of-attack technology rapidly accelerated beginning in 1971. Extensive research was conducted with generic models, simulator techniques for assessing high-alpha behavior were developed, and test techniques were upgraded. Active participation in the F-14, F-15, and B-1 development programs was quickly followed by similar research for the YF-16 and YF-17 Lightweight Fighter prototypes, as

15. The significance of the Dayton symposium cannot be overstated. It was one of the most critical high-angle-of-attack meetings ever held, in view of the national R&D mobilization that occurred in its wake.
16. Key individuals at NASA Headquarters included William S. Aiken, Jr., Gerald G. Kayten, A.J. Evans, and Jack Levine.
17. William P. Gilbert and Charles E. Libby, "Investigation of an Automatic Spin Prevention System for Fighter Airplanes," NASA TN-D-6670 (1972).

well as later efforts for the F-16, F-16XL, F/A-18, X-29, EA-6B, and X-31 programs. Summaries of Langley's contributions in those programs have been documented, and equally valuable contributions from Dryden and Ames will be described herein.[18] Brief highlights of a few NASA contributions and their technical impacts follow.

Spin Prevention: The F-14 Program

Early spin tunnel tests of the F-14 at Langley during the airplane's early development program indicated the configuration would exhibit a potentially dangerous fast, flat spin and that conventional spin recovery techniques would not be effective for recovery from that spin mode—even with the additional deployment of the maximum-size emergency spin recovery parachute considered feasible by Grumman and the Navy. Outdoor radio-controlled models were quickly readied by NASA for drop-testing from a helicopter at a test site near Langley to evaluate the susceptibility of the F-14 to enter the dangerous spin, and when the drop model results indicated marginal spin resistance, Langley researchers conceived an automatic aileron-to-rudder interconnect (ARI) control system that greatly enhanced the spin resistance of the design.[19] The value of NASA participation in the early high-angle-of-attack assessments of the F-14 benefited from the fact that the same Langley personnel had participated earlier in the development of the flight control system for the F-15, which used a similar approach for enhanced spin resistance. Extensive evaluations of the effectiveness of the ARI concept by NASA and Grumman pilots in the Langley Differential Maneuvering Simulator (DMS) air combat simulator reported a dramatic improvement in high-alpha characteristics.

13

However, after the ARI system was conceived by Langley and approved for implementation to the F-14 fleet, a new wing leading-edge maneuver flap concept designed by Grumman was also adopted for retrofit production. Initial flight-testing showed that, when combined, the ARI and maneuver-flap concepts resulted in unsatisfactory

18. For individual details and references for Langley's activities, see Chambers, *Partners in Freedom*.

19. Gilbert, Luat T. Nguyen, and Roger W. Van Gunst, "Simulator Study of Automatic Departure- and Spin-Prevention Concepts to a Variable-Sweep Fighter Airplane," NASA TM-X-2928 (1973). Although the initial concept had been identified in 1972, over 20 years would pass before the concept was implemented in the F-14 fleet. During that time, over 35 F-14s were lost because of departure from controlled flight and the flat spin.

An F-14 used in Dryden's high-alpha flight program extends its foldout nose canards. Spin tunnel tests predicted that the airplane's flat spin would require this modification. NASA.

pilot-induced oscillations and lateral-directional deficiencies in handling qualities at high angles of attack. Meanwhile, NASA had withdrawn from the program, and Grumman's modifications to the ARI to fix the deficiencies actually made the F-14 more susceptible to spins. Made aware of the problem, Langley then revisited the ARI concept and, together with Grumman and Navy participation, corrected the problems. Development and refinement of the ARI system for the F-14 continued for several years.

In the mid-1970s, senior Navy leaders were invited to NASA Headquarters for briefings on the latest NASA technologies that might be of benefit to the F-14. When briefed on the effectiveness of the ARI system, a decision was made to conduct flight evaluations of a new updated NASA version of the system. Joint NASA–Grumman–Navy flight-test assessments of the refined concept took place with a modified F-14 at the NASA Dryden Flight Research Center[20] in 1980. Flight tests of the ARI-equipped aircraft included over 100 flights by 9 pilots over a 2-year period during severe high-angle-of-attack maneuvers at speeds up to low supersonic Mach numbers. Results of the activity were very impressive; however, funding constraints and priorities within

20. Nguyen, Gilbert, Joseph Gera, Kenneth W. Iliff, and Einar K. Enevoldson, "Application of High-Alpha Control System Concepts to a Variable-Sweep Fighter Airplane," AIAA Paper 80-1582 (1980).

the Navy delayed the implementation of the system until an advanced digital flight control system (DFCS) was finally incorporated into fleet airplanes in 1999. The system, designed by a joint GEC-Marconi–Northrop Grumman–Navy team, was essentially a refined version of the concept advanced by Langley over 25 years earlier.[21]

In retrospect, the F-14 experience is a classic example of inadequate followthrough on the technology maturation process for new research concepts. No doubt, if NASA had continued its involvement in the development of the ARI and been tasked to resolve the ARI/maneuver flap issues, the fleet would have benefitted from the concept much earlier.[22]

New Levels of Departure Resistance: The F-15 Program

After its traumatic experiences with the F-4 stability and control deficiencies at high angles of attack, the Air Force encouraged competitors in the F-15 selection process to stress good high-angle-of-attack characteristics for the candidate configurations of their proposed aircraft. As part of the source selection process, an analysis of departure resistance was required based on high Reynolds number aerodynamic data obtained for each design in the NASA Ames 12-Foot Pressure Tunnel. In addition, spin and recovery characteristics were determined during the competitive phase using models in the Langley Spin Tunnel. The source selection team evaluated data from these and other high-angle-of-attack tests and analysis.

In its role as an air superiority fighter, the winning McDonnell-Douglas F-15 design was carefully crafted to exhibit superior stability and departure resistance at high angles of attack. In addition to providing a high level of inherent aerodynamic stability, the McDonnell-Douglas design team devised an automatic control concept to avoid control-induced departures at high angles of attack because of adverse yaw from lateral control (ailerons and differential horizontal tail deflections). By using an automatic aileron washout scheme that reduced the amount of aileron/tail deflections obtainable at high angles of attack and an interconnect system that deflected the rudder for roll control as a

21. For an account of the background and development of the F-14 DFCS, see "F-14 Tomcat Upgrades" GlobalSecurity.Org, *http://www.globalsecurity.org/military/systems/aircraft/f-14-upgrades.htm*, accessed June, 5, 2009.

22. The F-14 ARI scenario is in direct contrast to the beneficial "cradle-to-grave" technology participation that the NACA and NASA enjoyed during the development of the Century series fighters.

13

Radio-controlled drop model of the F-15 undergoing checkout prior to a flight to assess spin susceptibility at a test site near Langley Research Center. The F-15's reluctance to spin was accurately predicted in model tests. NASA.

function of angle of attack within its Command Augmentation System (CAS), the F-15 was expected to exhibit exceptional stability and departure resistance at high angles of attack.

NASA's free-flight model tests of the F-15 in the Langley Full-Scale Tunnel during 1971 verified that the F-15 would be very stable at high-angle-of-attack conditions, in dramatic contrast to its immediate predecessors.[23] During the F-15 development process, spin tunnel testing at Langley provided predictions for spin modes for the basic airplane as well as an extensive number of external stores, and an emergency spin recovery parachute size was determined.

Langley was also requested to evaluate the spin resistance of the F-15 with the outdoor helicopter drop-model technique used at Langley for many previous assessments of spin resistance. During spin entry attempts of the drop model with the CAS operative, it was once again obvious that the configuration was very spin resistant. In fact, an exceptional effort was required by the Langley team to develop a longitudinal

23. Gilbert, "Free-Flight Investigation of Lateral-Directional Characteristics of a 0.10-Scale Model of the F-15 Airplane at High Angles of Attack," NASA TM-SX-2807 (1973).

and lateral-directional control input technique to spin the model. Ultimately, such a technique was identified and demonstrated, although it was successful for a very constrained range of flight variables. This spin entry technique was later used in the full-scale aircraft flight program to promote spins. In 1972, Dryden constructed a larger drop model with a more complete representation of the aircraft flight control system and a larger-scale prediction of the airplanes spin recovery characteristics. Launched from a B-52 and known as the F-15 spin research vehicle (SRV), the remotely piloted vehicle verified the predictions of the smaller model and added confidence to the subsequent flight tests.[24]

Meanwhile, testing in the Spin Tunnel concentrated on one of the more critical spin conditions for the F-15 aircraft—unsymmetrical mass loadings. Model tests showed that the configuration's spin and recovery characteristics deteriorated when lateral unbalance was simulated, as would be the situation for asymmetric weapon store loadings on the right and left wing panels or fuel imbalance between wing tanks. Fuel imbalance can occur during banked turns in strenuous air combat maneuvers when tanks feed at different rates. The results of the spin tunnel tests showed that the spins would be faster and flatter in one direction, and that recovery would not be possible when the mass imbalance exceeded a certain critical value. As frequently happens in the field of spinning and spin recovery, a configuration that was extremely spin resistant in the "clean" configuration suddenly became an unmanageable tiger with mass imbalance.

During its operational service, the F-15 has experienced several accidents caused by unrecoverable spins with asymmetric loadings. At one time, this type of accident was the second greatest cause of F-15 losses, after midair collisions.[25]

Comparison of theoretical predictions, spin tunnel results, drop-model results, and flight results indicated that correlation of a model and airplane results were very good and that risk in the full-scale program had been reduced considerably by the NASA model tests.

24. Euclid C. Holleman, "Summary of Flight Tests to Determine the Spin and Controllability Characteristics of a Remotely Piloted, Large-Scale (3/8) Fighter Airplane Model," NASA TN-D-8052 (1976).
25. Steve Davies and Doug Dildy, *F-15 Eagle Engaged: The World's Most Successful Jet Fighter* (Oxford: Osprey Publishing, 2007), pp. 82–83.

Relaxed Stability Meets High Alpha: The F-16 Program

Initially envisioned as a nimble lightweight fighter with "carefree" maneuverability, the F-16 was designed from the onset with reliance on the flight control system to ensure satisfactory behavior at high-angle-of-attack conditions.[26] By using the concept of relaxed longitudinal stability, the configuration places stringent demands on the flight control system. In addition to extensive static and dynamic wind tunnel testing in Langley's tunnels from subsonic to supersonic speeds and free-flight model studies for high-angle-of-attack conditions and spinning, Langley and its partners from General Dynamics and the Air Force conducted in-depth piloted studies in a Langley simulator. The primary objective of the studies was to assess the ability of the F-16 control system to prevent loss of control and departures for critical dynamic maneuvers involving rapid roll rates at high angles of attack and low airspeeds.[27] General Dynamics used the results of the study to modify gains in the F-16 flight control system and introduce new elements for enhanced departure prevention in production aircraft.

One of the more significant events in NASA's support of the F-16 was the timely identification and solution to a potentially unrecoverable "deep-stall" condition. Analysis of Langley wind tunnel data at extreme angles of attack (approaching 90 degrees) and simulated maneuvers by pilots in the DMS during the earlier YF-16 program indicated that rapid roll maneuvers at high angles of attack could saturate the nose-down aerodynamic control capability of the flight control system, resulting in the inherently unstable airplane pitching up to an extreme angle of attack with insufficient nose-down aerodynamic control to recover to normal flight.[28] The ability of the YF-16 to enter this dangerous condition was demonstrated to General Dynamics and the Air Force, but aerodynamic data obtained in other NASA and industry wind tunnel tests of different YF-16 models did not indicate the existence of such a problem.

26. NASA's participation had begun with the YF-16 program, during which testing in the Full-Scale Tunnel, Spin Tunnel, and 7- by 10-Foot High-Speed Tunnel contributed to the airframe shaping and control system design. For an example, see Newsom, Anglin, and Grafton, "Free-Flight Investigation of a 0.15-Scale Model of the YF-16 Airplane at High Angle of Attack," NASA TM-SX-3279 (1975).
27. Gilbert, Nguyen, and Van Gunst, "Simulator Study of the Effectiveness of an Automatic Control System Designed to Improve the High-Angle-of-Attack Characteristics of a Fighter Airplane."
28. Nguyen, Marilyn E. Ogburn, Gilbert, Kemper S. Kibler, Philip W. Brown, and Perry L. Deal, "Simulator Study of Stall/Post Stall Characteristics of a Fighter Airplane With Relaxed Longitudinal Stability," NASA TP-1538 (1979).

The scope of the ensuing YF-16 flight program was limited and did not allow for exploration of a potential deep-stall problem.

The early production F-16 configuration also indicated a deep-stall issue during Langley tests in the Full-Scale Tunnel, and once again, the data contradicted results from other wind tunnels. As a result, the Langley data were dismissed as contaminated with "scale effects," and concerns over the potential existence of a deep stall were minimal as the aircraft entered flight-testing at Edwards Air Force Base. However, during zoom climbs with combined rolling motions, the specially equipped F-16 high-angle-of-attack test airplane entered a stabilized deep-stall condition, and after finding no effective control for recovery, the pilot was forced to use the emergency spin recovery parachute to recover the aircraft to normal flight. The motions and flight variables were virtually identical to the Langley predictions.

Because Langley's aerodynamic model of the F-16 provided the most realistic inputs for the incident, a joint NASA, General Dynamics, and Air Force team aggressively used the DMS simulator at Langley to develop a piloting strategy for recovery from the deep stall. Under Langley's leadership, the team conceived a "pitch rocker" technique, in which the pilot pumped the control stick fore and aft to set up oscillatory pitching motions that broke the stabilized deep-stall condition and allowed the aircraft to return to normal flight. The concept was demonstrated during F-16 flight evaluations and was incorporated in the early flight control systems as a pilot-selectable emergency mode. Ultimately, the deep stall was eliminated by an increase in size of the horizontal tail (which was done for other reasons) on later production models of the F-16.

The value of Langley's support in the area of high-angle-of-attack behavior for the F-16 represented the first step for advancing methodology for fly-by-wire control systems with special capabilities for severe maneuvers at high angles of attack. The experience demonstrated the advantages of NASA's involvement as a Government partner in development programs and the value of having NASA facilities, technical expertise, and experience available to design teams in a timely manner. The initial objective of carefree maneuverability for the F-16 was provided in a very effective manner by the NASA–industry–DOD team.

Precision Controllability Flight Studies

During the 1970s, NASA Dryden conducted a series of flight assessments of emerging fighter aircraft to determine factors affecting the precision tracking capability of modern fighters at transonic conditions.[29] Although the flight evaluations did not explore the flight envelope beyond stall and departure, they included strenuous maneuvers at high angles of attack and explored typical such handling quality deficiencies as wing rock (undesirable large-amplitude rolling motions), wing drop, and pitch-up encountered during high-angle-of-attack tracking. Techniques were developed for the assessment process and were applied to seven different aircraft during the study. Aircraft flown included a preproduction version of the F-15, the YF-16 and YF-17 Lightweight Fighter prototypes, the F-111A and the F-111 supercritical wing research aircraft, the F-104, and the F-8.

Extensive data were acquired in the flight-test program regarding the characteristics of the specific aircraft at transonic speeds and the impact of configuration features such as wing maneuver flaps and automatic flap deflection schedules with angle of attack and Mach number. However, some of the more valuable observations relative to undesirable and uncommanded aircraft motions provided insight and guidance to the high-angle-of-attack research community regarding aerodynamic and control system deficiencies and the need for research efforts to mitigate such issues. In addition, researchers at Dryden significantly expanded their experience and expertise in conducting high-angle-of-attack flight evaluations and developing methodology to expose inherent handling-quality deficiencies during tactical maneuvers.

Challenging Technology: The X-29 Program

Meetings between Defense Advanced Research Projects Agency (DARPA) and NASA Langley personnel in early 1980 initiated planning for support of an advanced forward-swept wing (FSW) research aircraft project with numerous objectives, including assessments and demonstration of superior high-angle-of-attack maneuverability and departure resistance resulting from the aerodynamic behavior of the FSW at high angles of attack. Langley was a major participant in the subsequent program and conducted high-angle-of-attack wind tunnel tests of models of the competing designs by General Dynamics, Rockwell, and Grumman during 1980 and 1981. When Grumman was selected to develop the X-29

29. For examples, see reports by Thomas R. Sisk and Neil W. Matheny, "Precision Controllability of the F-15 Airplane," NASA TM-72861 (1979), and "Precision Controllability of the YF-17 Airplane," NASA TP-1677 (1980).

The X-29 flies at high angle of attack during studies of the flow-field shed by the fuselage forebody. Note the smoke injected into the flow for visualization and the emergency spin parachute structure on the rear fuselage. NASA.

research aircraft in December 1981, NASA was a major partner with DARPA and initiated several high-angle-of-attack/stall/spin/departure studies of the X-29, including dynamic force-testing and free-flight model tests in the Full-Scale Tunnel, spinning tests in the Spin Tunnel, initial high-angle-of-attack control system concept development and assessment in the DMS, and assessments of spin entry and poststall motions using a radio-controlled drop model.[30]

Early in the test program, Langley researchers encountered an unanticipated aerodynamic phenomenon for the X-29 at high angles of attack. It had been expected that the FSW configuration would maintain satisfactory airflow on the outer wing panels at high angle of attack; however, dynamic wind tunnel testing to measure the aerodynamic roll damping

30. Daniel G. Murri, Nguyen, and Grafton, "Wind-Tunnel Free-Flight Investigation of a Model of a Forward-Swept Wing Fighter Configuration," NASA TP-2230 (1984); David J. Fratello, Croom, Nguyen, and Christopher S. Domack, "Use of the Updated NASA Langley Radio-Controlled Drop-Model Technique for High-Alpha Studies of the X-29A Configuration," AIAA Paper 1987-2559 (1987).

of an X-29 model in the Full-Scale Tunnel indicated that the configuration would exhibit unstable roll damping and a tendency for oscillatory large-amplitude wing-rocking motions for angles of attack above about 25 degrees. After additional testing and analysis, it was determined that the FSW of the aircraft worked as well as expected, but aerodynamic interactions between the vortical flow shed by the fuselage forebody with the wing were the cause of the undesirable wing rock. When the free-flight model was subsequently flown, the wing rock was encountered as predicted by the earlier force test, resulting in large roll fluctuations at high angles of attack. However, the control effectiveness of the wing trailing-edge flapperon used for artificial damping on the full-scale X-29 was extremely powerful, and the model motions quickly damped out when the system was replicated and engaged for the model.

Obtaining reliable aerodynamic data for high-angle-of-attack tests of subscale models at Langley included high Reynolds number tests in the NASA Ames 12-Foot Pressure Tunnel, where it was found that significant aerodynamic differences could exist for certain configurations between model and full-scale airplane test conditions. Wherever possible, artificial devices such as nose strakes were used on the models to more accurately replicate full-scale aerodynamic phenomena. In lieu of approaches to correct Reynolds number effects for all test models, a conservative approach was used in the design of the flight control system to accommodate variability in system gains and logic to mitigate problems demonstrated by the subscale testing.[31]

In the area of spin and recovery, the Langley spin tunnel staff members conducted tests to identify the spin modes that might be exhibited by the X-29 and the size of emergency spin recovery parachute recommended for the flight-test vehicles. They also investigated a growing concern within the airplane development program that the inherently unstable configuration might exhibit longitudinal tumbling during maneuvers involving low speeds and extreme angles of attack (such as during recovery from a "zoom climb" to zero airspeed). This concern was of the general category of ensuring that aircraft motions might overpower the relative ineffectiveness of aerodynamic controls for configurations with relaxed stability at low-speed conditions.

31. Neihouse, et al., "Status of Spin Research" NASA TR-R-57; Fremaux, "Wind-Tunnel Parametric Investigation of Forebody Devices for Correcting Low Reynolds Number Aerodynamic Characteristics at Spinning Attitudes," NASA CR-198321 (1996).

Using a unique, single-degree-of-freedom test apparatus, the research team demonstrated that tumbling might be encountered but that the aft-fuselage strake flaps—intended to be only trimming devices—could be used to prevent uncontrollable tumbling.[32] As a result of these tests, the airplane's control system was modified to use the flaps as active control devices, and with this modification, subsequent flight tests of the X-29 demonstrated a high degree of resistance to tumbling.

In 1987, Langley conducted high-angle-of-attack and poststall assessments of the X-29 using the Langley helicopter drop-model technique that had been applied to numerous configurations since the early 1960s. However, the inherent aerodynamic longitudinal instability and sophisticated flight control architecture of the X-29 required an extensive upgrade to Langley's test technique. The test program was considered the most challenging drop-model project ever conducted by Langley to that time. Among several highlights of the study was a demonstration that the large-amplitude wing rock exhibited earlier by the unaugmented wind tunnel free-flight model also existed for the drop model. In fact, when the angle of attack was increased beyond 30 degrees, the roll oscillations became divergent, and the model exhibited uncontrollable 360 degrees rolls that resulted in severe poststall gyrations. When the active wing-rock roll control system of the airplane was simulated, the roll motions were damped and controllable to extreme angles of attack.[33]

Two X-29 research aircraft conducted joint DARPA–NASA–Grumman flight tests at NASA Dryden from 1984 to 1992.[34] The first aircraft was used to verify the benefits of advanced technologies and expand the envelope to an angle of attack of about 23 degrees and to a Mach number of about 1.5. The second X-29 was equipped with hardware and software modifications for low-speed flight conditions for angles of attack up to about 70 degrees. The test program for X-29 No. 2 was planned and accomplished using collated results from wind tunnel tests, drop-model tests, simulator results, and results obtained from X-29 No. 1 for lower

32. Raymond D. Whipple, Croom, and Scott P. Fears, "Preliminary Results of Experimental and Analytical Investigations of the Tumbling Phenomenon for an Advanced Configuration," AIAA Paper 84-2108 (1984).

33. Fratello, et al., "Updated Radio-Controlled Drop-Model Technique for the X-29A," AIAA Paper 1987-2559.

34. Iliff and Kon-Sheng Charles Wang, "X-29A Lateral-Directional Stability and Control Derivatives Extracted from High-Angle-Of-Attack Flight Data," NASA TP-3664 (1996).

angles of attack. Dryden and the Air Force Flight Test Center designed flight control system modifications, and Grumman made modifications. The high-angle-of-attack flight program included 120 flights between 1989 and 1991. Dryden researchers conducted a series of aerodynamic investigations in mid-1991 to assess the symmetry of flow from the fuselage forebody, the flow separation patterns on the wing as angle of attack was increased, and the flow quality at the vertical tail location.[35] In 1992, the Air Force conducted an additional 60 flights to evaluate the effectiveness of forebody vortex flow control using blowing.

The results of the high-angle-of-attack X-29 program were extremely impressive. Using only aerodynamic controls and no thrust vectoring, X-29 No. 2 demonstrated positive and precise pitch-pointing capability to angles of attack as high as 70 degrees, and all-axis maneuverability for 1 g flight up to an angle of attack of 45 degrees with lateral-directional control maintained. The wing-rock characteristic predicted by the Langley model tests was observed for angles of attack greater than about 35 degrees, but the motions were much milder than those exhibited by the models. It was concluded that the Reynolds number effects observed between model testing and full-scale flight tests were responsible for the discrepancy, as flight-test values were an order of magnitude greater than those of subscale tests.

Cutting Edge: The NASA High-Alpha Program

As the 1970s came to an end, the U.S. military fleet of high-performance fighter aircraft had been transformed from departure-prone designs to new configurations with outstanding stability and departure resistance at high angles of attack. Thanks to the national research and development efforts of industry and Government following the Dayton symposium in 1971, the F-14, F-15, F-16, and F/A-18 demonstrated that the peril of high-angle-of-attack departure exhibited by the previous generation of fighters was no longer a critical concern. Rather, the pilot could exploit high angles of attack under certain tactical conditions without fear of nose slice or pitch-up. At air shows and public demonstrations, the new "supermaneuverable" fighters wowed the crowds with high-angle-of-attack flybys, and more importantly, the high-alpha capabilities provided pilots with new options for air combat. High-angle-of-attack

35. John H. Del Frate and John Saltzman, "In-Flight Flow Visualization Results From the X-29A Aircraft at High Angles of Attack," NASA TM-4430 (1992).

technology had progressed from concerns over stall characteristics to demonstrated spin resistance and was moving into a focus on poststall agility and precision maneuverability.

Reflecting on the advances in high-angle-of-attack technology of the 1970s and concepts yet to be developed, technical managers at Langley, Dryden, and Ames began to advocate for a cohesive, integrated research program focused on technologies and innovative ideas. The Agency was in an excellent position to initiate such a program thanks to the unique ground- and flight-testing capabilities that had been developed and the expertise that had been gathered by interactions of the NASA researchers with the real-world challenges of specific aircraft programs. At Langley, for example, researchers had been intimately involved in high-angle-of-attack/departure/spin activities in the development of all the new fighters and had accumulated in-depth knowledge of the characteristics of the configurations, including aerodynamics, flight control architecture, and handling characteristics at high angles of attack. Technical expertise and facilities at Langley included subscale static and dynamic free-flight model wind tunnel testing, advanced control-law synthesis, and computational aerodynamics. In addition, extensive peer contacts had been made within industry teams and DOD aircraft development offices.

At Dryden, the world-class flight-test facilities and technical expertise for high-performance fighter aircraft had been continually demonstrated in highly successful flight-test programs in which potentially hazardous testing had been handled in a professional manner. The Dryden staff was famous for its can-do attitude and accomplishments, including the conception, development, and routine operation of experimental aircraft; advanced flight instrumentation; and data extraction techniques.

Meanwhile, at Ames, the aeronautical research staff had aggressively led developments in high-performance computing facilities and computational aerodynamics. Computational fluid dynamics (CFD) codes developed at Ames and Langley had shown powerful analysis capability during applications to traditional aerodynamic predictions such as cruise performance and the analysis of flow-field phenomena. In addition to computational expertise, Ames had extensive wind tunnel facilities, including the huge 80-by 120-Foot Tunnel, which had the capability of testing a full-scale fighter aircraft as large as the F/A-18.

From the perspective of the three technical managers, the time was right to bring together the NASA capabilities into a focused program directed toward some of the more critical challenges in high-angle-of-attack technology.[36] The research program that evolved from the planning meetings grew into one of the more remarkable efforts ever undertaken by NASA. The planning, advocacy, and conduct of the program was initiated at the grassroots level and was managed in a most remarkable manner for the duration of the program. Within NASA's aeronautics activities, the program brought an enthusiastic environment of cooperation—not competition—that fostered a deep commitment to team spirit and accomplishments so badly needed in research endeavors. The personal satisfaction of the participants was widely known, and the program has become a model for NASA intercenter relationships and joint programs.[37]

The first task in planning the program was to identify major technical issues facing the high-angle-of-attack community. Foremost among these was the understanding, prediction, and control of aerodynamic phenomena at high-angle-of-attack conditions, especially for aircraft configurations with strong vortical flows. Achieving this goal involved detailed studies of separated flow characteristics; measurement of static and dynamic phenomena in ground-test facilities as well as flight; calibration of flow predictions from CFD methodology, wind tunnels, and flight; and the development of CFD codes for high-angle-of-attack conditions. In addition, the analysis and prediction of aerodynamic phenomena associated with structural fatigue issues for vertical tails immersed in violently fluctuating separated flows at high-angle-of-attack conditions became a major element in the program.

The second research thrust in the proposed program was directed toward an exciting new technology that offered unprecedented levels of controllability at high angles of attack—thrust vectoring. The thrust-vectoring concept had been developed in early rocket control applications by placing vanes in the exhaust of the rocket vehicle, and extensive NASA–industry–DOD studies had been conducted to develop movable nozzle vectoring concepts for aircraft applications. The introduction of the superb fighters of the 1970s had demonstrated a new level of design achievement in stability at high angles of attack, but another nemesis remained—inadequate control at high-angle-of-attack conditions at

36. The advocates and planners for the NASA program were Chambers (Langley), Kenneth J. Szalai (Dryden), and Leroy L. Pressley (Ames).

37. Szalai, "Cooperation, Not Competition," NASA Dryden *X-Press* newspaper, Issue 96-06, June 1996, p. 4.

which conventional aerodynamic control surfaces lose effectiveness because of separated flow. The problem was particularly critical in the lack of ability to create crisp, precise roll control for "nose pointing" at high angles of attack. For such conditions, the ability to roll is dependent on providing high levels of yaw control, which creates sideslip and rolling motion because of dihedral effect. Unfortunately, conventional rudders mounted on vertical tails become ineffective at high angles of attack.

During the early 1980s, researchers at the Navy David Taylor Research Center pursued the application of simple jet-exit vanes to the F-14 for improved yaw control.[38] Teaming with Langley in a joint study in the Langley Differential Maneuvering Simulator, the researchers found that the increased yaw control provided by the vanes resulted in a dramatic improvement in high-angle-of-attack maneuverability and dominance in simulated close-in air combat. Inspired by these results, Langley researchers evaluated the effectiveness of similar vanes on a variety of configurations during free-flight model testing in the Langley Full Scale Tunnel. Following investigations of modified models of the F-16, F/A-18, X-29, and X-31, the researchers concluded that thrust vectoring in yaw provided unprecedented levels of maneuverability and control at high angles of attack. In addition, providing feedback from flight sensors to the vane control system enhanced dynamic stability for the test conditions.

Another technology that had matured to the point of research applications was the control of strong vortical flow shed from the long pointed forebodies of contemporary fighters at high angles of attack. As previously mentioned, Dryden and the Air Force Flight Dynamics Laboratory had conducted a joint program to evaluate the effects of blowing on the nose of the X-29A for enhanced control. Competing concepts for vortical flow control had also received attention during NASA and industry research programs, including investigations at Langley of deflectable forebody strakes that could be used to control flow separation on the forebody for enhanced yaw control.

Perhaps the most contentious issue in planning the integrated NASA high-angle-of-attack program was whether a research aircraft was required and, if so, which aircraft would make the best testbed for research studies. Following prolonged discussions (the Ames

38. Chambers, *Partners in Freedom*, p. 108.

representative did not initially endorse the concept of flight-testing), the planning team agreed that flight-testing was mandatory for the program to be relevant, coordinated, and focused. Consideration was given to the F-15, F-16, X-29, and F/A-18 as potential testbeds, and after discussions, the team unanimously chose the F/A-18, for several reasons. The earlier Navy F/A-18 development program had included extensive support from Langley; therefore, its characteristics were well known to NASA (especially aerodynamic and aeroelastic phenomena, such as vortical flow and vertical tail buffet). During spin-testing for the development program, the aircraft had displayed reliable, stall-free engine operations at high angles of attack and excellent spin recovery characteristics. The F/A-18 was equipped with an advanced digital flight control system that offered the potential for modifications for research flight tests. Finally, the aircraft exhibited a remarkably high-angle-of-attack capability (up to 60 degrees in trimmed low-speed flight)—ideal for aerodynamic tests at extreme angles of attack.

The intercenter planning team presented its integrated research program plan to NASA Headquarters, seeking approval to pursue the acquisition of an F/A-18 from the Navy and for program go-ahead. After Agency approval, the Navy transferred the preproduction F/A-18A Ship 6, which had been used for spin testing at Patuxent River, MD, to NASA Dryden, where it arrived in October 1984. This particular F/A-18A had been stripped of several major airframe and instrument components following the completion of its spin program at Patuxent River, but it was still equipped with a multi-million-dollar emergency spin recovery parachute system and a programmable digital flight control computer ideally suited to NASA's research interests. The derelict aircraft was shipped overland to Dryden and reassembled by a team of NASA and Navy technicians into a unique high-angle-of-attack research airplane known as the F/A-18A High-Alpha Research Vehicle (HARV).[39] The HARV was equipped with several unique research systems, including flow visualization equipment, a thrust-vectoring system using external postexit vanes around axisymmetric nozzles, and deployable nose strakes on a modified fuselage of forebody. Additional aircraft systems included extensive instrumentation, integrated flight research controls with special flight control hardware and software for the thrust-

39. Albion H. Bowers, et al., "An Overview of the NASA F-18 High Alpha Research Vehicle," NASA CP-1998-207676 (1998).

vectoring system, interface controls for the forebody strakes, and safety backup systems including a spin recovery parachute.

The High-Angle-of-Attack Technology program (HATP) was funded and managed under an arrangement that was different from other NASA programs but was extremely efficient and productive. Headquarters provided program management oversight, but recommendations for day-to-day technical planning, distribution of funds, and technical thrusts were provided by an intercenter steering committee consisting of members from each of the participating Centers. In recognition of its technical expertise and accomplishments in high-angle-of-attack technology, Langley was designated the technology lead Center. Dryden was designated the lead Center for flight research and operations of the HARV, and Ames and Langley shared the technical leadership for CFD and experimental aerodynamics. In subsequent years, the NASA Lewis Research Center (now the NASA Glenn Research Center) joined the HATP for experiments on engine inlet aerodynamics for high-angle-of-attack conditions. The HATP included aerodynamics, flight controls, handling qualities, stability and control, propulsion, structures, and thrust vectoring.[40]

The HATP program was conducted in three sequential phases, centering on high-angle-of-attack aerodynamic studies (1987–1989), evaluation of thrust vectoring effects on maneuverability (1990–1994), and forebody flow control (1995–1996), with 383 research flights. In the first activities, aerodynamic characteristics obtained from flight-test results for the baseline HARV (no vectoring) were correlated with wind tunnel and CFD predictions, with emphasis on flow separation predictions and vortical flow behavior on the fuselage forebody and wing-body leading-edge extension (LEX).[41]

The first HARV research flight was April 17, 1987. Flown in its baseline configuration, the HARV provided maximum angles of attack on the order of 55 degrees, limited by aerodynamic control. At the time the flight studies were conducted, CFD had not yet been applied to real

40. Norman Lynn, "High Alpha: Key to Combat Survival?" *Flight International*, Nov. 7, 1987; William B. Scott, "NASA Adds to Understanding of High Angle of Attack Regime," *Aviation Week & Space Technology*, May 22, 1989, pp. 36–42; Gilbert, Nguyen, and Gera, "Control Research in the NASA High-Alpha Technology Program," in NATO, AGARD (Aerodynamics of Combat Aircraft Controls and of Ground Effects) (1990).
41. Farhad Ghaffari, et al., "Navier-Stokes Solutions About the F/A-18 Forebody-LEX Configuration," *NASA Computational Fluid Dynamics Conference*, vol. 1, sessions 1–6, pp. 361–383 (1989).

aircraft shapes at high angles of attack. Rather, researchers had used computational methods to predict flow over simple shapes such as prolate spheroids, and the computation of flow fields, streamlines, and separation phenomena for a modern fighter was extremely challenging. Many leaders in the NASA and industry CFD communities were pessimistic regarding the success of such a venture at the time.

The experimental wind tunnel community was also facing issues on how (or whether) to modify models to better simulate high-angle-of-attack aerodynamics at flight values of Reynolds numbers and to understand the basic characteristics of vortical flows and techniques for the prediction of flow interactions with aircraft structures. Using a highly innovative, Dryden-developed propylene glycol monomethyl ether (PGME) dye flow-visualization technique that emitted colored dye tracers from ports for visualization of surface flows over the HARV forebody and LEX, the team was able to directly compare results, analyze separation phenomena, and modify CFD codes for a valid prediction of the observed on-surface flow characteristics. The ports for the PGME were later modified for pressure instrumentation to provide even more detailed information on flow fields. Additional instrumentation for aerodynamic measurements was initially provided by a nose boom, but evidence of aerodynamic interference from the nose boom caused the team to remove it and replace the boom with wingtip air-data probes. A rotating, foldout flow rake was also used to measure vortical flows shed by the LEX surfaces.[42]

The results of the HATP flight- and ground-based aerodynamic studies provided a detailed perspective of the relative accuracy of computational flow dynamics and wind tunnel testing techniques to predict critical flow phenomena such as surface pressures, separation contours, vortex interaction patterns, and laminar separation bubbles.[43] The scope of correlation included assessments of the impact of Mach and Reynolds numbers on forebody and LEX vortexes as observed in flight with the HARV, the Langley 7- by 10-Foot High-Speed Tunnel, the Langley

42. Bowers, et al., "Overview of the NASA F-18 High Alpha Research Vehicle," NASA CP-1998-207676 (1998).

43. Robert M. Hall, et al., "Overview of HATP Experimental Aerodynamics Data for the Baseline F/A-18 Configuration," NASA TM-112360 (1996); David F. Fisher, et al., "In-Flight Flow Visualization Characteristics of the NASA F-18 High Alpha Research Vehicle at High Angles of Attack," NASA TM-4193 (1991).

An ex–Blue Angel F/A-18 aircraft was tested in the Ames 80- by 120-Foot Tunnel during the NASA HATP program. NASA.

30- by 60-Foot Tunnel, the Navy David Taylor Research Center 7- by 10-Foot Transonic Tunnel, and the Ames 80- by 120-Foot Wind Tunnel.

A wide variety of subscale models of the HARV configuration was tested in the various wind tunnels, and a full-scale F/A-18 aircraft was used for testing in the Ames 80- by 120-Foot Tunnel. The test article was an ex–Blue Angel flight demonstrator, whose life had been exceeded,

13

Shown during a flight to determine on- and off-surface flows at high angles of attack, the F/A-18 HARV provided unprecedented data on phenomena such as aerostructural interactions and vortex physics. NASA.

that had been bailed to NASA for the tests. When the tunnel tests were conducted in 1991 and 1993, the aircraft had both engines, flowthrough inlets, and the wingtip missile launchers removed.

Using extensive instrumentation that had been carefully coordinated between ground and flight researchers gathered an unprecedented wealth of detail on aerodynamic characteristics of a modern fighter at high angles of attack. The effort was successful particularly because it had been planned with common instrumentation locations for pressure ports and flow visualization stations between wind tunnel tests and the flight article. More importantly, the high value of the data obtained was the result of one of the most successful aspects of the program—close communications and working relationships between the flight, wind tunnel, and CFD technical communities.

As NASA neared the end of the aerodynamic phase of testing for the HARV, growing concerns over buffeting of the vertical tail surfaces for military fleet F/A-18 aircraft led the Navy and McDonnell-Douglas to develop vertical longitudinal fences on the upper surfaces of the LEX to extend the service life of the tails of fleet F/A-18s. Although the fences were not installed on HARV during the early aerodynamic studies, they were added during the second and third phases of the program, when extensive wind tunnel and HARV flight studies of the tail buffet phenomenon were conducted. Resulting data were transmitted to the appropriate industry and service organizations for analysis of the F/A-18 specific phenomena as well as for other twin-tail fighter aircraft.

As the second phase of the HATP began, Dryden accepted major program responsibilities for the implementation of a relatively simple and cheap thrust-vectoring system for the HARV aircraft. The objective of NASA's research was not to develop a production-type thrust-vectoring engine/nozzle system, but rather to evaluate the impact of vectoring for high-angle-of-attack maneuvers, assess control-law requirements for high-angle-of-attack applications, and use the control augmentation provided by vectoring to stabilize the aircraft at extreme angles of attack for additional aerodynamic studies. With this philosophy in mind, the program contracted with McDonnell-Douglas to modify the HARV with deflectable external vanes mounted behind the aircraft's two F-404 engines, similar in many respects to the installations used by the Navy F-14 mentioned earlier and the Rockwell X-31 research aircraft.

For the installation, the exhaust nozzle divergent flaps were removed from the engines and replaced with a set of three vanes for each engine, thereby providing both pitch and yaw vectoring capability. The research teams at Dryden and Langley thoroughly studied the specific vane configuration, structural design, and control system modifications required for the project. The scope of activities included measurements of thrust-vane effectiveness for many powered model configurations at Langley, simulator studies of the effectiveness of vectoring on maneuverability and controllability at Langley, and hardware and software development—as well as the integration, checkout, and operations of the system—at Dryden. The implementation of the HARV thrust-vectoring hardware and software modifications proved to be relatively difficult, requiring the NASA research team to participate in the final design of the thrust-vectoring system. The HARV vectoring system followed the HATP objective of providing thrust-vectoring research capability at minimal cost through external airframe modifications rather than a new production-type vectoring engine. With the massive external thrust-vectoring vane actuation system and the emergency spin recovery parachute system both mounted on the rear of the aircraft and necessary ballast added

Ground test of the HARV thrust vectoring system illustrates the deflection of the engine thrust by three vanes in the engine exhaust. NASA.

13

to the nose of the aircraft to maintain balance, the weight of the HARV was increased by about 4,000 pounds.

In the thrust-vectoring phase of the HATP project, the conventional flight control system of the HARV was modified to include a research flight control system (RFCS) to influence control laws. The conventional F/A-18 control laws were used for takeoff, for landing, and as a backup in case of failure of the RFCS, whereas the second set of control laws were for high-angle-of-attack research flights. The design and implementation of the RFCS system was one of the more complex changes to the F/A-18 digital flight control system undertaken at that time.

First flight of the HARV with vectoring engaged occurred in July 1991, a few weeks after the X-31 research aircraft demonstrated pitch-vectoring capability at Edwards, but the HARV conducted the first multiaxis vectoring flights shortly thereafter. Research flight-testing of the HARV equipped with thrust vectoring vividly demonstrated the anticipated benefits at high angles of attack that had been predicted by earlier free-flight model tests and piloted simulator studies. The precision and angular rates available to the pilot were remarkable, and the enhanced stability and control at extreme angles of attack permitted precision aerodynamic studies that had previously been impossible. Angles of attack as high as 70 degrees were flown with complete control in aerodynamic experiments.

During the late 1980s, three NASA–industry–DOD programs had been initiated to explore thrust-vectoring systems for high-angle-of-attack conditions. Each program had different objectives and focused on separate technologies. NASA's HARV aircraft was designed to evaluate fundamental thrust-vectoring system control-law synthesis and use vectoring to stabilize the aircraft at high angles of attack for aerodynamic experiments. The DARPA X-31A aircraft was conceived to demonstrate enhanced fighter maneuverability at poststall angles of attack under simulated tactical conditions. In addition, the Air Force F-16 Variable-Stability In-Flight Simulator Test Aircraft (VISTA) was modified into the F-16 Multi-Axis Thrust Vectoring (MATV) project with an objective of demonstrating the effectiveness of a production-type thrust-vectoring system. All three programs had different goals, and the three research aircraft underwent flight-testing at Edwards in the same time period.

The HATP participants conceived, developed, and assessed several control-law schemes, which included special configurations for longitudinal control at high angles of attack, lateral and directional control mixing strategies, automatic spin prevention, and spin recovery modes. Seventy-five spin attempts (at low power conditions) resulted in 70 fully developed spins with satisfactory recoveries, and the emergency spin recovery parachute was never fired in flight.

As the HARV conducted its thrust-vectoring research program, a critical issue emerged within the advanced fighter design community. With new configurations under consideration having extreme angle-of-attack capability and reduced longitudinal stability for performance and maneuverability enhancements, the issue of providing sufficient nose-down control effectiveness for recovery from high-angle-of-attack excursions became significant. NASA–DOD technical meetings had been held to discuss studies to assess the adequacy of theoretical and wind tunnel predictions, and it appeared that using the HARV flight capability with thrust vectoring would provide highly desirable data for design criteria for future fighters. In view of the urgency of the situation, Langley led a HATP element known as High-Alpha Nosedown Guidelines (HANG), which included extensive simulator studies and flights with the HARV.[44]

44. Marilyn E. Ogburn, et al., "Status of the Validation of High-Angle-of-Attack Nose-Down Pitch Control Margin Design Guidelines," AIAA Paper No. 93-3623 (1993).

Although the main objective of the HARV thrust-vectoring experiments was not air-to-air combat maneuvering, Dryden conducted flight tests to provide validation data for a proposed high-angle-of-attack flying qualities requirement MIL-STD-1797A by using basic fighter maneuvers and limited air combat maneuvering. Six NASA research test pilots from Dryden and Langley provided the major expertise and guidance for the HATP simulator and HARV flight-testing. Other guest pilots from NASA, the Navy, the Canadian Air Force, the United Kingdom, McDonnell-Douglas, and Calspan also participated in flight-test evaluations of the HARV vectoring capabilities.

The third and final phase of the HATP was directed to in-flight assessments of the effectiveness of controlling the powerful vortex flows shed by the fuselage forebody for augmentation of yaw control at high angles of attack. Ground-based research in NASA wind tunnels and simulators had indicated that the most effective method for rolling an aircraft about its flight path for nose pointing at high angles of attack was through the use of yaw control. Unfortunately, conventional rudders suffer a severe degradation and control effectiveness at high angles of attack because of the impingement of low-energy stalled flows only vertical tail surfaces. Years of NASA research had demonstrated that the use of deployable fuselage forebody strakes was a potentially viable concept for yaw control augmentation. With a vast amount of wind tunnel data and pilot opinions derived from air combat simulation, the strake concept was ready for realistic evaluations in flight. Once again, the cohesive nature of the HATP was demonstrated when the strake hardware was designed and fabricated on a special F/A-18 forebody radome in machine shops at Langley and the control laws were developed at Langley and delivered to Dryden, where the flight computer interface and instrumentation were accomplished by the Dryden staff. The project, known as actuated nose strakes for enhanced rolling (ANSER), was evaluated independently and in combination with thrust vectoring.[45]

Implementation of the ANSER concept on the thrust-vectoring-equipped HARV provided three control combinations. The aircraft could be flown with thrust vectoring only, thrust vectoring in longitudinal control with a thrust-vectored and strake-blended mode for lateral control,

45. Murri, Gautam H. Shah, Daniel J. DiCarlo, and Todd W. Trilling, "Actuated Forebody Strake Controls for the F-18 High-Alpha Research Vehicle," *Journal of Aircraft*, vol. 32, no. 3 (1995), pp. 555–562.

and a strake mode with thrust-vectoring control longitudinally and strakes controlling the lateral mode. As was the case for thrust vectoring, the forebody strake flight results demonstrated that a significant enhancement of high-angle-of-attack rolling capability was obtained, particularly at higher subsonic speeds. In fact, at those speeds, the effectiveness of the strakes was comparable to that of thrust vectoring.

Several other subsystems were implemented on the HARV, including an instrumented inlet rake, extensive pressure instrumentation, aeroservoelastic accelerometers, thrust-vectoring vane loads and temperatures, and an emergency power backup system. Notably, although the power backup system was implemented to continue aircraft systems operation in the event of a dual-engine flameout or unrecoverable dual-engine stalls, it was removed later in the program when testing showed excellent high-angle-of-attack engine operations. In fact, 383 high-angle-of-attack flights were made without experiencing an engine stall.

Throughout the HATP program, NASA ensured that results were widely disseminated within industry and DOD. Major HATP technical conferences were held, with at least 200 attendees at Langley in 1990, at Dryden in 1992 and 1994, and a wrap-up conference at Langley in 1996.[46] Hundreds of reports and presentations resulted from the program, and the $74 million (1995 dollars) activity produced cutting-edge technical results that were absorbed into the Nation's latest aircraft, including the F-22, F-35 and F/A-18E.

Supermaneuverability: The X-31 Program

NASA Langley became involved in the X-31 Enhanced Fighter Maneuverability (EFM) program in 1984, when mutual discussions with Rockwell International occurred regarding a fighter configuration capable of highly agile flight at extreme angles of attack. Known as the Super Normal Attitude Kinetic Enhancement (SNAKE) configuration, the design underwent exploratory testing in the Full-Scale Tunnel.[47] The

46. The HATP program produced hundreds of publications covering aerodynamics, control concepts, handling qualities, aerostructural interactions, flight instrumentation, thrust vectoring, wind tunnel test techniques, inlet operation, and summaries of flight investigations. Conference proceedings of the High Alpha Conferences were published as NASA CP-3149 (1990), NASA CP-10143 (1994), and NASA CP-1998-207676 (1998). The volumes are unclassified but limited in distribution by International Traffic in Arms Regulations (ITAR) restrictions.
47. Chambers, *Partners in Freedom*. pp. 216–218.

early cooperative research study later led to a cooperative project using the Langley Full Scale Tunnel, the Langley Spin Tunnel, and the Langley Jet-Exit Test Facility. After DARPA and the West Germany government formally initiated the X-31 program, Langley and Dryden actively participated in the development of the configuration and flight tests of two X-31 demonstrators at Dryden from 1992 to 1995.

In the early SNAKE Langley-Rockwell study, Langley researchers assessed the high-angle-of-attack capabilities of the Rockwell-designed configuration that had been designed using computational methods with minimal use of wind tunnel tests. Preliminary evaluations in the full-scale tunnel disclosed that the configuration was unacceptable, being unstable in pitch, roll, and yaw. Langley's expertise in high-angle-of-attack stability and control contributed to modifications and revisions of the original configuration, eliminating the deficiencies of the SNAKE design.

Simultaneous with the SNAKE activities, several other events contributed to shaping what would become the X-31 program. First, the emerging recognition that thrust vectoring would provide unprecedented levels of control for precision maneuvering at extreme angles of attack had led to joint Langley-Rockwell studies of jet-exit vanes similar to those previously discussed for the Navy F-14 experiments and the NASA F/A-18 HARV vehicle. The tests, which were conducted in the Langley Jet-Exit Test Facility, inspired Rockwell to include multi-axis thrust-vectoring paddles in the SNAKE configuration. Free-flight testing of the revised SNAKE configuration provided impressive proof that the vectoring paddles were extremely effective.

The second major activity was the strong advocacy of the West German Messerschmitt-Bölkow-Blohm (MBB) Company that asserted that high levels of agility for poststall flight conditions provided dominant capabilities for close-in air combat. With the support of DARPA, the X-31 EFM program was initiated in 1986 with a request that Langley be a major participant in the development program. Using the NASA Langley test facility assets for free-flight model testing, spin testing, and drop-model testing uncovered several critical issues for the configuration.

One issue was the general character of inherent poststall motions that might be encountered in the aircraft flight program. Results indicated that the X-31 might have marginal nose-down control for recovery from high-angle-of-attack maneuvers, and that severe unstable wing-rock motions would be exhibited by the configuration, resulting in a violent, disorienting roll departure and an unrecoverable inverted stall

The X-31 demonstrated the tactical effectiveness of extreme maneuvers at high angles of attack during flights at Dryden. NASA.

condition. With these inputs, the X-31 design team worked to configure the flight control system for maximum effectiveness and to prevent the foregoing problems, even without thrust vectoring. The value of these contributions from Langley cannot be understated, but equally important contributions were to come as the drop-model technique maintained operations during the full-scale aircraft flight-test program.

Flight-testing of the two X-31 aircraft began at Dryden in February 1992 under the direction of an International Test Organization (ITO) that included NASA, the U.S. Navy, the U.S. Air Force, Rockwell, the Federal Republic of Germany, and Deutsche Aerospace (formerly MBB). Two issues were encountered in the flight-test program, resulting in additional test requirements from the supporting team of Langley researchers. Early in the flight tests, pilots reported marginal nose-down pitch control and said that significant improvements would be necessary if the aircraft were to be considered an efficient weapon system for close-in combat. In a quick-response mode, Langley conducted evaluations of 16 configuration modifications to improve nose-down control in the Full-Scale Tunnel. From these tests, a decision was made to add strakes to the lower aft fuselage, and pilots of subsequent flight tests with the modified airplane reported that the problem was eliminated.

Another problem encountered in the X-31 flights at extreme angles of attack was the presence of large out-of-trim yawing moments with the potential to overpower corrective inputs from the pilot. After a departure was unexpectedly experienced during a maneuvering flight near an angle of attack of 60 degrees, analysis of the flight records indicated that the departure had been caused by a large asymmetric yawing moment that was much larger than any predicted in subscale wind tunnel testing. The presence of asymmetric moments of this type had been well-known to the aeronautics community, including the fact that the phenomenon might be sensitive to the specific Reynolds number under consideration. Experience had shown that, for some configurations, the out-of-trim moments exhibited during subscale model tests might be larger than those exhibited at the full-scale conditions, and for other configurations, opposite results might occur. In the case of the X-31, the full-scale aircraft exhibited significantly higher values.[48]

The flight-test team sent an urgent request to Langley for solutions to the problem. Once again, tests in the full-scale tunnel were conducted of a matrix of possible airframe modifications, a candidate solution was identified, and real-time recommendations were made to the ITO. In these tunnel tests, a single nose strake was used to predict the maximum level of asymmetry for the airplane, and the solutions worked for that configuration. A pair of nose strakes designed in the tunnel tests was

48. Fisher, et al., "Reynolds Number of Effects at High Angles of Attack," NASA TP-1998-206553 (1998).

implemented and, together with other modifications (grit on the nose boom and slight blunting of the fuselage nose tip), permitted the aircraft flight program to continue. This X-31 experience was noteworthy, in that it demonstrated the need for testing seemingly unimportant details at Reynolds numbers equivalent to flight.

The X-31 EFM program completed an X-plane record of 524 flights with 14 evaluation pilots from the sponsoring organizations.

The New Breed

The intense U.S. research and development programs on high-angle-of-attack technology of the 1970s and 1980s ushered in a new era of carefree maneuvering for tactical aircraft. New options for close-in combat were now available to military pilots, and more importantly, departure/spin accidents were dramatically reduced. Design tools had been sharpened, and the widespread introduction of sophisticated digital flight control systems finally permitted the implementation of automatic departure and spin prevention systems. These advances did not go unnoticed by foreign designers, and emerging threat aircraft were rapidly developed and exhibited with comparable high-angle-of-attack capabilities.[49] As the Air Force and Navy prepared for the next generation of fighters to replace the F-15 and F-14, the integration of superior maneuverability at high angles of attack and other performance- and signature-related capabilities became the new challenge.

Fifth Generation: The F-22 Program

The Air Force initiated its Advanced Tactical Fighter (ATF) program in 1985 as an effort to augment and ultimately replace the F-15. During the competitive phase of the program between the Northrop-led YF-23 and the Lockheed-led YF-22 designs, the Air Force established that each team could draw on the facilities and expertise of NASA for establishing credibility and risk reduction before a competitive fly-off. Lockheed subsequently requested free-flight and spin tests of the YF-22 in the Langley Full-Scale Tunnel and the Langley Spin Tunnel. The relatively

49. The most notable foreign advances in high-alpha technology have come from Russia, where close working relationships between the military and the TsAGI Central Aerodynamic Institute have focused on providing exceptional high-alpha maneuverability for the latest MiG and Sukhoi aircraft. Current products such as the Su-35 employ multiaxis thrust vectoring and carefully tuned high-alpha aerodynamics for outstanding capability.

compressed timeframe of the ATF competition would not permit a feasible schedule for the fabrication and testing of a helicopter drop model of the YF-22.

A joint NASA–Lockheed team conducted conventional tunnel tests in the Full-Scale Tunnel in 1989 to measure YF-22 aerodynamic data for high-angle-of-attack conditions, followed by free-flight model studies to determine the low-speed departure resistance of the configuration. Meanwhile, spin tunnel tests obtained information on spin and recovery characteristics as well as the size and location of an emergency spin recovery parachute for the high-angle-of-attack test airplane. In addition, specialized "rotary-balance" tests were conducted in the spin tunnel to obtain aerodynamic data during simulated spin motions. Lockheed incorporated all of the foregoing results in the design process, leading to an impressive display of capabilities by the YF-22 during the competitive flight demonstrations in 1990.

Lockheed formally acknowledged its appreciation of NASA's participation in the YF-22 program in a letter to NASA, which stated:

> On behalf of the Lockheed YF-22 Team, I would like to express our appreciation of the contribution that the people of NASA Langley made to our successful YF-22 flight test program, and provide some feedback on how well the flight test measurements agreed with the predictions from your wind-tunnel measurements. . . . The highlight of the flight test program was the high-angle-of-attack flying qualities. We relied on aerodynamic data obtained in the full-scale wind tunnel to define the low-speed, high-angle-of-attack static and dynamic aerodynamic derivatives; rotary derivatives from your spin tunnel; and free-flight demonstrations in the full-scale tunnel. We expanded the flight envelope from 20° to 60° angle of attack, demonstrating pitch attitude changes and full-stick rolls about the velocity vector in seven calendar days. The reason for this rapid envelope expansion was the quality of the aerodynamic data used in the control law design and pre-flight simulations.[50]

50. Letter from James A. Blackwell, Lockheed vice president and general manager of the Lockheed ATF Office, to Richard H. Petersen, Director of the NASA Langley Research Center, Mar. 12, 1991.

Free-flight model tests of the YF-22 in the Full-Scale Tunnel accurately predicted the high-alpha maneuverability of the full-scale airplane and provided risk reduction for the F-22 program. NASA.

After the team of Lockheed, Boeing, and General Dynamics was announced as the winner of the ATF competition in April 1991, high-angle-of-attack testing of the final F-22 configuration was conducted in the Full-Scale Tunnel and the Spin Tunnel. Aerodynamic force testing was completed in the Full-Scale Tunnel in 1992, with spin- and rotary-balance tests conducted in 1993. A wind tunnel free-flight model was not fabricated for the F-22 program, but a typical full-scale tunnel model was constructed and used for the aerodynamic tests. A notable contribution from the spin tunnel tests was a relocation of the attachment point for the F-22 emergency spin recovery parachute to clear the exhaust plume of the vectoring engine in 1994. Langley's contributions to the high-angle-of-attack technologies embodied in the F-22 fighter had been completed well in advance of the aircraft's first flight in September 1997.[51]

51. The F-22 has outstanding capabilities at high angles of attack. However, for a number of reasons, the aircraft was designed with thrust vectoring only in pitch. Based on NASA fundamental high-alpha research on many configurations, yaw vectoring would significantly increase high-alpha maneuverability.

New Issues: The F/A-18E/F Program

The U.S. Navy funded the F/A-18E/F Super Hornet program in 1992 to design its next-generation fighter as a replacement for the canceled A-12 aircraft and the earlier legacy F/A-18 versions. Although somewhat similar in configuration to existing F/A-18C aircraft, the new design was a larger aircraft with critical differences in wing design and other features that impact high-angle-of-attack behavior. Two of the first configuration design issues centered on the shape of the wing leading-edge extension and the ability to obtain crisp nose-down control for recovery at extreme angles of attack. Representatives of Langley's high-angle-of-attack specialty areas were participants in a 15-member NASA–industry–DOD team who conducted wind tunnel studies and analyses that provided the basis for the final design of the F/A-18E/F LEX.[52]

Aerodynamic stability and control characteristics for the Super Hornet for high-angle-of-attack conditions were conducted in the Full-Scale Tunnel to develop a database for piloted simulator evaluations using the Langley and Boeing simulators. Once again, the Spin Tunnel was used for identifying spin modes, spin recovery characteristics, an acceptable emergency spin recovery parachute, and measurement of rotational aerodynamic characteristics using the rotary-balance technique. Langley used an extremely large (over 1,000 pounds) drop model for departure susceptibility and poststall testing at the NASA Wallops Flight Facility to provide risk reduction for the subsequent full-scale flight-test program.[53]

One of NASA's more critical contributions to the Super Hornet program began in March 1996, when a preproduction F/A-18E experienced an unacceptable uncommanded abrupt roll-off that randomly occurred at high angles of attack (below maximum lift) at transonic speeds and involved rapid bank angle changes of up to 60 degrees in the heart of the maneuvering envelope. Engineering analyses indicated that the wing drop was caused by a sudden asymmetric loss of lift on the wing, but the fundamental cause of the problem was not well understood. Following the formation of a DOD Blue Ribbon Panel, a research program was recommended to be undertaken to develop design methods to avoid such problems on future fighter aircraft. This recommenda-

52. Chambers, *Partners in Freedom*. pp. 45–46.

53. Croom, Holly M. Kenney, and Murri, "Research on the F/A-18E/F Using a 22%-Dynamically-Scaled Drop Model," AIAA Paper 2000-3913 (2000).

tion was accepted, and a joint NASA and Navy Abrupt Wing Stall (AWS) program was initiated to conduct the research.[54]

Meanwhile, extensive efforts by industry and the Navy were underway to resolve the wing-drop problem through wind tunnel tests and "cut and try" airframe modifications during flight tests. Over 25 potential wing modifications were assessed, and computational fluid dynamics studies were undertaken without a feasible fix identified. Subsequently, the automatically programmed wing leading-edge flaps were examined as a solution. Typical of current advanced fighters, the F/A-18E/F uses flaps with deflection programs scheduled as functions of angle of attack and Mach number. A revised deflection schedule was adopted in 1997 as a major improvement, but the aircraft still exhibited less serious wing drops at many test conditions. As the Navy test and evaluation staff continued to explore further solutions to wing drop, exploratory flight tests with the outer-wing fold fairing removed indicated that the wing drop had been eliminated. However, unacceptable performance and buffet characteristics resulted from removing the fairing.

Langley personnel suggested that passive porosity be examined as a more acceptable treatment of the wing fold area based on NASA's extensive fundamental research. Subsequently evaluated by the Navy flight-test team, the porous fold doors became a feature of the production F/A-18E/F and permitted continued production of the aircraft.

With the F/A-18E/F wing-drop problem resolved, NASA and the Naval Air Systems Command began their efforts in the AWS research program that used a coordinated approach involving static and dynamic tests at Langley in several wind tunnels, piloted simulator studies, and computational fluid dynamics studies conducted by the Navy and NASA. The scope of research focused on the causes and resolution of the unexpected wing drop that had been experienced for the preproduction F/A-18E/F and the wealth of aerodynamic wind tunnel and flight data that had been collected, but the program was intentionally designed to include assessments of other aircraft for validation of conclusions. The studies included the F/A-18C and the F-16 (both of which do not exhibit wing drop) and the AV-8B and the preproduction version of the F/A-18E (which do exhibit wing drop at the extremes of the flight envelope).

54. Robert M. Hall, Shawn H. Woodson, and Chambers, "Accomplishments of the Abrupt Wing Stall Program and Future Research Requirements," AIAA Paper 2003-0927 (2003).

After 3 years of intense research on the complex topic of transonic shock-induced asymmetric stall at high angles of attack, the AWS program produced an unprecedented amount of design information, engineering tools, and recommendations regarding developmental approaches to avoid wing drop for future fighters. Particularly significant output from the program included the development and validation of a single-degree-of-freedom free-to-roll wind tunnel testing technique for detection of wing-drop tendencies, an assessment of advanced CFD codes for prediction of steady and unsteady shock-induced separation at high angles of attack for transonic flight, and a definition of simulator model requirements for assessment and prediction of wing drop. NASA and Lockheed Martin have already applied the free-to-roll concept in the development of the wing geometry for the F-35 fighter.[55]

Opportunities

After the results of the NASA HATP project in 1996 and the F/A-18E/F wing-drop and AWS programs were disseminated, it was widely recognized that computational fluid dynamics had tremendous potential as an additional tool in the designer's toolkit for high-angle-of-attack flight conditions. However, it was also appreciated that the complexity of the physics of flow separation, the enormous computational resources required for accurate predictions, and the fundamental issues regarding representation of key characteristics such as turbulence would be formidable barriers to progress. Even more important, the lack of communication between the experimental test and evaluation community and the CFD community was apparent. More specifically, the T&E community placed its trust in design methods it routinely used for high-angle-of-attack analysis—namely, the wind tunnel and experimental methods. Furthermore, a majority of the T&E engineers were not willing to accept what they regarded as an aggressive "oversell" of CFD capabilities without many examples that the computer could reliably predict aircraft stability and control parameters at high angles of attack. Meanwhile, the CFD community had continued its focus on applications related to aircraft performance, with little or no awareness of the aerodynamic

55. Ibid.; Francis J. Capone, D. Bruce Owens, and Hall, "Development of a Free-to-Roll Transonic Test Capability," AIAA Paper 2003-0749 (2003); Owens, Jeffrey K. McConnell, Jay M. Brandon, and Hall, "Transonic Free-To-Roll Analysis of the F/A-18E and F-35 Configurations," AIAA Paper 2004-5053 (2003).

problems faced by the T&E community for high-angle-of-attack predictions. One example of the different cultures of the communities was that a typical CFD expert was used to striving for accuracies within a few percent for performance-related estimates, whereas the T&E analyst was, in many cases, elated to know simply whether parameters at high angles of attack were positive or negative.

Stimulated to bring these two groups together for discussions, Langley conceived a plan for a project known as Computational Methods for Stability and Control (COMSAC), which could potentially spin off focused joint programs to assess, modify, and calibrate computational codes for the prediction of critical aircraft stability and control parameters for high-angle-of-attack conditions.[56] Many envisioned the start of another HATP-like effort, with similar outlooks for success. In 2004, Langley hosted a COMSAC Workshop, which was well-attended by representatives of the military and civil aviation industries, DOD, and academia. As expected, controversy was widespread regarding the probability of success in applying CFD to high-angle-of-attack stability and control predictions. Stability and control attendees expressed their "show me that it works" philosophy regarding CFD, while the CFD experts were alarmed by the complexity of typical experimental aerodynamic data for high-angle-of-attack flight conditions. Nonetheless, the main objective of establishing communications between the two scientific communities was accomplished, and NASA's follow-on plans for establishing research efforts in this area were eagerly awaited.

Unfortunately, changes in NASA priorities and funding distributions terminated the COMSAC planning activity after the workshop. However, several attendees returned to their organizations to initiate CFD studies to evaluate the ability of existing computer codes to predict stability and control at high angles of attack. Experts at the Naval Air Systems Command have had notable success using the F/A-18E as a test configuration.[57]

---

56. Hall, Fremaux, and Chambers, "Introduction to Computational Methods for Stability and Control (COMSAC)," *http://ntrs.nasa.gov/archive/nasa/casi.ntrs.nasa.gov/200.400.84128_200.408.6282.pdf*, accessed Sept. 10, 2009.

57. Bradford E. Green and James J. Chung, "Transonic Computational Fluid Dynamics Calculations on Preproduction F/A-18E for Stability and Control," *AIAA Journal of Aircraft*, vol. 44, no. 2, pp. 420–426 (2007); Green, "Computational Prediction of Roll Damping for the F/A-18E at Transonic Speeds," AIAA Paper 2008-6379 (2008).

Despite the inability to generate a sustainable NASA research effort to advance the powerful CFD methods for stability and control, the COMSAC experience did inspire other organizations to venture into the area. It appears that such an effort is urgently needed, especially in view of the shortcomings in the design process.

Challenges

As clearly evidenced by U.S. military experiences, the technical area of high-angle-of-attack/departure/spin behavior will continue to challenge design teams of highly maneuverable aircraft. The Nation has been fortunate in assembling and maintaining unique expertise and facilities for the timely identification and resolution of problems that might have had a profound impact on operational capability or program viability. In the author's opinion, several situations are emerging that threaten the traditional partnerships and mutual resources required for advancing the state of the art in high-angle-of-attack technology for military aircraft.

The end of the Cold War has naturally resulted in a significant decrease in new military aircraft programs and the need for continued research in a number of traditional research areas. As technical personnel exit from specialty areas such as high-angle-of-attack and spin behavior, the corporate knowledge and experience base that was the jewel in NASA's crown rapidly erodes, and lessons learned become forgotten.

Of even more concern is the change in traditional working-level relationships between the NASA and DOD communities. During the term of NASA Administrator Daniel S. Goldin in the 1990s, NASA turned its priorities away from its traditional links with military aircraft R&D to the extent that long-time working-level relationships between NASA, industry, and DOD peers were ended. At the same time, aeronautics funding within the Agency was significantly reduced, and remaining aeronautics activities were redirected to civil goals. As a result of those programmatic decisions and commitments, NASA does not even highlight military-related research as part of its current mission. It has become virtually impossible for researchers and their peers in the military, industry, or DOD research laboratories to consider the startup of highly productive, unclassified military-related programs such as the NASA F/A-18 High-Angle-of-Attack Technology program.

Meanwhile, leaders in military services and research organizations have now been replaced with many who are unfamiliar with the traditional NASA–military ties and accomplishments. Without those relationships, the military R&D organizations have turned to hiring their own aeronautical talent and conducting major research undertakings in areas that were previously exclusive to NASA Centers.

Finally, one of the more alarming trends underway has been the massive closures of NASA wind tunnels, which have been the backbone of NASA's ability to explore concepts and ideas and to respond to high-priority military requests and problem-solving exercises in specialty areas such as high-angle-of-attack technology.

In summary, this essay has discussed some of the advances made in high-angle-of-attack technology by NASA, which have contributed to a dramatic improvement in the capabilities of the Nation's first-line military aircraft. Without these contributions, many of the aircraft would have been subject to severe operational restrictions, excessive development costs, significantly increased risk, and unacceptable accidents and safety-of-flight issues. In the current era of relative inactivity for development of new aircraft, it is critical that the resources required to provide such technology be protected and nurtured for future applications.

Recommended Additional Reading

Reports, Papers, Articles, and Presentations:

Seth B. Anderson, "Handling Qualities Related to Stall/Spin Accidents of Supersonic Fighter Aircraft," AIAA Paper 84-2093 (1984).

Seth B. Anderson, Einar K. Enevoldson, and Luat T. Nguyen, "Pilot Human Factors in Stall/Spin Accidents of Supersonic Fighter Aircraft," NASA TM-84348 (1983).

Jeffrey E. Bauer, Robert Clarke, and John J. Burken, "Flight Test of the X-29A at High Angle of Attack: Flight Dynamics and Controls," NASA TP-3537 (1995).

Lisa J. Bjarke, John H. Del Frate, and David F. Fisher, "A Summary of the Forebody High-Angle-of-Attack Aerodynamics Research on the F-18 and the X-29A Aircraft," NASA TM-104261 (1992).

Albion H. Bowers, et al., "An Overview of the NASA F-18 High Alpha Research Vehicle," NASA CP-1998-207676 (1998).

Jay M. Brandon, "Low-Speed Wind-Tunnel Investigation of the Effect of Strakes and Nose Chines on Lateral-Directional Stability of a Fighter Configuration," NASA TM-87641 (1986).

Jay M. Brandon, Daniel G. Murri, and Luat T. Nguyen, "Experimental Study of Effects of Forebody Geometry on High Angle of Attack Static and Dynamic Stability and Control," *ICAS 15th Congress, London, England, Sept. 7–12, 1986, Proceedings*, vol. 1 (New York: American Institute of Aeronautics and Astronautics, Inc., 1986), pp. 560–572.

Jay M. Brandon, James M. Simon, D. Bruce Owens, and Jason S. Kiddy, "Free-Flight Investigation of Forebody Blowing for Stability and Control," AIAA Paper 96-3444 (1996).

Francis J. Capone, D. Bruce Owens, and Robert M. Hall, "Development of a Free-to-Roll Transonic Test Capability," AIAA Paper 2003-0749 (2003).

Peter C. Carr and William P. Gilbert, "Effects of Fuselage Forebody Geometry on Low-Speed Lateral-Directional Characteristics of Twin-Tail Fighter Model at High Angles Of Attack," NASA TP-1592 (1979).

Joseph R. Chambers and Ernie L. Anglin, "Analysis of Lateral Directional Stability Characteristics of a Twin Jet Fighter Airplane at High Angles of Attack," NASA TN-D-5361 (1969).

Joseph R. Chambers, Ernie L. Anglin, and James S. Bowman, Jr., "Analysis of the Flat-Spin Characteristics of a Twin-Jet Swept-Wing Fighter Airplane," NASA TN-D-5409 (1969).

Joseph R. Chambers and Sue B. Grafton, "Aerodynamic Characteristics of Airplanes at High Angles of Attack," NASA TM-74097 (1977).

Brent R. Cobleigh and Mark A. Croom, "Comparison of X-31 Flight and Ground-Based Yawing Moment Asymmetries at High Angles of Attack," NASA TM-210393 (2001).

Mark A. Croom, Holly M. Kenney, and Daniel G. Murri, "Research on the F/A-18E/F Using a 22%-Dynamically-Scaled Drop Model," AIAA Paper 2000-3913 (2000).

John H. Del Frate and John A. Saltzman, "In-Flight Flow Visualization Results From the X-29A Aircraft at High Angles of Attack," NASA TM-4430 (1992).

Charles Donlan, "An Interim Report on the Stability and Control of Tailless Airplanes," NACA TR-796 (1944).

David F. Fisher, et al., "In-Flight Flow Visualization Characteristics of the NASA F-18 High Alpha Research Vehicle at High Angles of Attack," NASA TM-4193 (1991).

David F. Fisher, et al., "Reynolds Number of Effects at High Angles of Attack," NASA TP-1998-206553 (1998).

David F. Fisher, David M. Richwine, and Stephan Landers, "Correlation of Forebody Pressures and Aircraft Yawing Moments on the X-29A Aircraft at High Angles of Attack," AIAA Paper 92-4105 (1992).

John V. Foster, Holly M. Ross, and Patrick A. Ashley, "Investigation of High-Alpha Lateral-Directional Control Power requirements for High-Performance Aircraft," AIAA Paper 93-3647 (1993).

David J. Fratello, Mark A. Croom, Luat T. Nguyen, and Christopher S. Domack, "Use of the Updated NASA Langley Radio-Controlled Drop-Model Technique for High-Alpha Studies of the X-29A Configuration," AIAA Paper 87-2559 (1987).

13

C. Michael Fremaux, "Wind-Tunnel Parametric Investigation of Forebody Devices for Correcting Low Reynolds Number Aerodynamic Characteristics at Spinning Attitudes," NASA CR-198321 (1996).

Joseph Gera, R. Joseph Wilson, Einar K. Enevoldson, and Luat T. Nguyen, "Flight Test Experience with High-Alpha Control System Techniques on the F-14 Airplane," AIAA Paper 81-2505 (1981).

William P. Gilbert, "Free-Flight Investigation of Lateral-Directional Characteristics of a 0.10-Scale Model of the F-15 Airplane at High Angles of Attack," NASA TM-SX-2807 (1973).

William P. Gilbert and Charles E. Libby, "Investigation of an Automatic Spin Prevention System for Fighter Airplanes," NASA TN-D-6670 (1972).

William P. Gilbert, Luat T. Nguyen, and Joseph Gera, "Control Research in the NASA High-Alpha Technology Program," in NATO, AGARD (Aerodynamics of Combat Aircraft Controls and of Ground Effects) (1990).

William P. Gilbert, Luat T. Nguyen, and Roger W. Van Gunst, "Simulator Study of Automatic Departure- and Spin-Prevention Concepts to a Variable-Sweep Fighter Airplane," NASA TM-X-2928 (1973).

William P. Gilbert, Luat T. Nguyen, and Roger W. Van Gunst, "Simulator Study of the Effectiveness of an Automatic Control System Designed to Improve the High-Angle-of-Attack Characteristics of a Fighter Airplane," NASA TN-D-8176 (1976).

Sue B. Grafton, Mark A. Croom, and Luat T. Nguyen, "High-Angle-of-Attack Stability Characteristics of a 3-Surface Fighter Configuration," NASA TM-84584 (1983).

Sue B. Grafton, "Low-Speed Wind-Tunnel Study of the High-Angle-of-Attack Stability and Control Characteristics of a Cranked-Arrow-Wing Fighter Configuration," NASA TM-85776 (1984).

Bradford E. Green, "Computational Prediction of Roll Damping for the F/A-18E at Transonic Speeds," AIAA Paper 2008-6379 (2008).

Bradford E. Green and James J. Chung, "Transonic Computational Fluid Dynamics Calculations on Preproduction F/A-18E for Stability and Control," *Journal of Aircraft*, vol. 44, no. 2 (2007), pp. 420–426.

H. Douglas Greer, "Summary of Directional Divergence Characteristics of Several High-Performance Aircraft Configurations," NASA TN-D-6993 (1972).

Robert M. Hall, et al., "Overview of HATP Experimental Aerodynamics Data for the Baseline F/A-18 Configuration," NASA TM-112360 (1996).

Robert M. Hall, C. Michael Fremaux, and Joseph R. Chambers, "Introduction to Computational Methods for Stability and Control (COMSAC)," at *http://ntrs.nasa.gov/archive/nasa/casi.ntrs.nasa.gov/200.400.84128_200.408.6282.pdf*, accessed Sept. 16, 2009.

Robert M. Hall, Shawn H. Woodson, and Joseph R. Chambers, "Accomplishments of the Abrupt Wing Stall (AWS) Program and Future Research Requirements," AIAA Paper 2003-0927 (2003).

Euclid C. Holleman, "Summary of Flight Tests to Determine the Spin and Controllability Characteristics of a Remotely Piloted, Large-Scale (3/8) Fighter Airplane Model," NASA TN-D-8052 (1976).

Kenneth W. Iliff and Kon-Sheng Charles Wang, "X-29A Lateral-Directional Stability and Control Derivatives Extracted from High-Angle-Of-Attack Flight Data," NASA TP-3664 (1996).

Joseph L. Johnson, Jr., Sue B. Grafton, and Long P. Yip, "Exploratory Investigation of the Effects of Vortex Bursting on the High Angle-of-Attack Lateral-Directional Stability Characteristics of Highly-Swept Wings," AIAA Paper 80-0463 (1980).

Frank L. Jordan, Jr., and David E. Hahne, "Wind-Tunnel Static and Free-Flight Investigation of High-Angle-of-Attack Stability and Control Characteristics of a Model of the EA-6B Airplane," NASA TP-3194 (1992).

Montgomery Knight, "Wind Tunnel Tests on Autorotation and the Flat Spin," NACA TR-273 (1928).

Norman Lynn, "High Alpha: Key to Combat Survival?" *Flight International*, Nov. 7, 1987.

Martin T. Moul and John W. Paulson, "Dynamic Lateral Behavior of High-Performance Aircraft," NACA RM-L58E16 (1958).

Scott M. Murman, Yehia M. Rizk, and Lewis B. Schiff, "Coupled Numerical Simulation of the External and Engine Inlet Flows for the F-18 at Large Incidence," AIAA Paper 92-2621 (1992).

Daniel G. Murri, Luat T. Nguyen, and Sue B. Grafton, "Wind-Tunnel Free-Flight Investigation of a Model of a Forward-Swept Wing Fighter Configuration," NASA TP-2230 (1984).

Daniel G. Murri, Gautam H. Shah, Daniel J. DiCarlo, and Todd W. Trilling, "Actuated Forebody Strake Controls for the F-18 High-Alpha Research Vehicle," *Journal of Aircraft*, vol. 32, no. 3 (1995), pp. 555–562.

NASA, DFRC, *Proceedings of the Fourth High Alpha Conference July 12–14, 1994,* NASA CP-10143 (1994).

NASA, *HATP [High Angle of Attack Technology Program] 1990 Program Conference Proceedings*, NASA CP-3149 (1990).

NASA, *HATP [High Angle of Attack Technology Program] 1994 Program Conference Proceedings*, NASA CP-10143 (1994).

NASA, *HATP [High Angle of Attack Technology Program] 1998 Program Conference Proceedings*, NASA CP-1998-207676 (1998).

Anshal I. Neihouse, Walter J. Klinar, and Stanley H. Scher, "Status of Spin Research for Recent Airplane Designs," NASA TR-R-57 (1960).

William A. Newsom, Jr., and Sue B. Grafton, "Free-Flight Investigation of Effects of Slats on Lateral-Directional Stability of a 0.13-Scale Model of the F-4E Airplane," NASA TM-SX-2337 (1971).

W.A. Newsom, Jr., Ernie L. Anglin, and S.B. Grafton, "Free-Flight Investigation of a 0.15-Scale Model of the YF-16 Airplane at High Angle of Attack," NASA TM-SX-3279 (1975).

Luat T. Nguyen, Ernie L. Anglin, and William P. Gilbert, "Recent Research Related to Prediction of Stall/Spin Characteristics of Fighter Aircraft," Proceedings of *AIAA Atmospheric Flight Mechanics Conference, June 7–9*, pp. 79–91 (1976).

Luat T. Nguyen and William P. Gilbert, "Impact of Emerging Technologies on Future Combat Aircraft Agility," AIAA Paper 90-1304 (1990).

Luat T. Nguyen, William P. Gilbert, Joseph Gera, Kenneth W. Iliff, and Einar K. Enevoldson, "Application of High-Alpha Control System Concepts to a Variable-Sweep Fighter Airplane," AIAA Paper 80-1582 (1980).

Luat T. Nguyen, William P. Gilbert, and Marilyn E. Ogburn, "Control-System Techniques for Improved Departure/Spin Resistance for Fighter Aircraft," NASA TP-1689 (1980).

Luat T. Nguyen, Marilyn E. Ogburn, William P. Gilbert, Kemper S. Kibler, Philip W. Brown, and Perry L. Deal, "Simulator Study of Stall/Post Stall Characteristics of a Fighter Airplane With Relaxed Longitudinal Stability," NASA TP-1538 (1979).

Marilyn E. Ogburn, et al., "Status of the Validation of High-Angle-of-Attack Nose-Down Pitch Control Margin Design Guidelines," AIAA Paper No. 93-3623 (1993).

D. Bruce Owens, Jay M. Brandon, Mark A. Croom, Charles M. Fremaux, Eugene H. Heim, and Dan D. Vicroy, "Overview of Dynamic Test Techniques for Flight Dynamics Research at NASA LaRC," AIAA Paper 2006-3146 (2006).

D. Bruce Owens, Francis Capone, Robert M. Hall, Jay Brandon, and Joseph R. Chambers, "Transonic Free-To-Roll Analysis of Abrupt Wing Stall on Military Aircraft," *Journal of Aircraft*, vol. 41, no. 3 (2004), pp. 474–484.

D. Bruce Owens, Jeffrey K. McConnell, Jay M. Brandon, and Robert M. Hall, "Transonic Free-To-Roll Analysis of the F/A-18E and F-35 Configurations," AIAA Paper 2004-5053 (2003).

Edward J. Ray and Eddie G. Hollingsworth, "Subsonic Characteristics of a Twin-Jet Swept-Wing Fighter Model with Maneuvering Devices," NASA TN-D-6921 (1973).

Victoria Regenie, Donald Gatlin, Robert Kempel, and Neil Matheny, "The F-18 High Alpha Research Vehicle—A High-Angle-of-Attack Testbed Aircraft," AIAA Paper 92-4121 (1992).

David M. Richwine, Robert E. Curry, and Gene V. Tracy, "A Smoke Generator System for Aerodynamic Flight Research," NASA TM-4137 (1989).

William B. Scott, "NASA Adds to Understanding of High Angle of Attack Regime," *Aviation Week & Space Technology*, May 22, 1989.

Thomas R. Sisk and Neil W. Matheny, "Precision Controllability of the F-15 Airplane," NASA TM-72861 (1979).

Thomas R. Sisk and Neil W. Matheny, "Precision Controllability of the YF-17 Airplane," NASA TP-1677 (1980).

Patrick C. Stoliker and John T. Bosworth, "Evaluation of High-Angle-of-Attack Handling Qualities for the X-31A Using Standard Evaluation Maneuvers," NASA TM-104322 (1996).

Kenneth J. Szalai, "Cooperation, Not Competition," NASA Dryden *X-Press* newspaper, issue 96-06 (June 1996).

Kevin R. Walsh, Andrew J. Yuhas, John G. Williams, and William G. Steenken, "Inlet Distortion for an F/A-18A Aircraft during Steady Aerodynamic conditions up to 60 Deg Angle of Attack," NASA TM-104329 (1997).

Fred E. Weick and Robert T. Jones, "The Effect of Lateral Controls in Producing Motion of an Airplane as Computed From Wind-Tunnel Data," NACA TR-570 (1937).

Stephan A. Whitmore and Timothy R. Moes, "The Effects of Pressure Sensor Acoustics on Air Data Derived From a High-Angle-of-Attack Flush Air Data Sensing (HI-FADS) System," NASA TM-101736 (1991).

Raymond D. Whipple, "Rockets for Spin Recovery," NASA CR-159240 (1980).

Raymond D. Whipple, Mark A. Croom, and Scott P. Fears, "Preliminary Results of Experimental and Analytical Investigations of the Tumbling Phenomenon for an Advanced Configuration," AIAA Paper 84-2108 (1984).

Books and Monographs:

Joseph R. Chambers, *Partners in Freedom; Contributions of the NASA Langley Research Center to Military Aircraft of the 1990s,* NASA SP-2000-4519 (Washington, DC: GPO, 2000).

Joseph R. Chambers and Mark A. Chambers, *Radical Wings and Wind Tunnels* (North Branch, MN: Specialty Press, 2008).

Steve Davies and Doug Dildy, *F-15 Eagle Engaged: The World's Most Successful Jet Fighter* (Oxford: Osprey Publishing, 2007).

Richard P. Hallion, *On The Frontier: Flight Research at Dryden 1946–1981*, NASA SP-4303 (Washington, DC: NASA, 1984).

THIS PAGE INTENTIONALLY BLANK

Three important NASA research aircraft representing different approaches to V/STOL flight pass in review over NASA's Ames Research Center. Left to right: the deflected lift QSRA, the tilt rotor XV-15, and the vectored-thrust Harrier. NASA.

CASE 14

On the Up and Up: NASA Takes on V/STOL

G. Warren Hall

The advent of vertical flight required mastery of aerodynamics, propulsion, and flight control technology. In the evolution of flight characterized by progressive development of the autogiro, helicopter, and various convertiplanes, the NACA and NASA have played a predominant role. NASA developed the theoretical underpinning for vertical flight, evaluated requisite technologies and research vehicles, and expanded the knowledge base supporting V/STOL flight technology.

ONE OF THE MAJOR ACCOMPLISHMENTS in the history of aviation has been the development of practical Vertical Take-Off and Landing (VTOL) aircraft, exemplified by the emergence of the helicopter in the 1930s and early 1940s, and the vectored-thrust jet airplane of the 1960s. Here indeed was a major challenge that confronted flight researchers, aeronautical engineers, military tacticians, and civilian planners for over 50 years, particularly those of the National Aeronautics and Space Administration (NASA) and its predecessor, the National Advisory Committee for Aeronautics (NACA). While perhaps not regarded by aviation aficionados as being as glamorous as the experimental craft that streaked to new speeds and altitudes, early vertical flight testbeds were likewise revolutionary at the other end of the performance spectrum, in vertical ascents and descents, low-speed controllability, and hover, areas challenging accepted knowledge and practice in aerodynamics, propulsion, and flight controls and controllability.[1]

The accomplishment of vertical flight was as challenging as inventing the airplane itself. Only four decades after Kitty Hawk were vertical takeoff, hovering, and landing aircraft beginning to enter service. These were, of course, the first helicopters: successors to the interim rotary wing autogiro that relied on a single or multiple rotors to give them Vertical/Short

1. John J. Schneider, "The History of V/STOL Aircraft," (two parts) *Vertiflite*, vol. 29, nos. 3–4 (Mar.–Apr. and May–June, 1983), pp. 22–29, 36–43.

Take-Off and Landing (V/STOL) performance. Before the end of the Second World War, the helicopter had flown in combat, proved its value as a life-saving craft, and shown its adaptability for both land- and sea-based operation.[2] The faded promises of many machines litter the path to the modern V/STOL vehicle. The dedicated research accompanying this work nevertheless led to a class of flight craft that have expanded the use of civil and military aeronautics, saving the lives of nearly a half million people over the last seven decades. The oil rigger in the Gulf going on leave, the yachtsman waiting for rescue, and the infantryman calling in gunships to fend off attack can all thank the flight researchers, particularly those of the NACA and NASA, who made the VTOL aircraft possible.[3]

Helicopters matured significantly during the Korean war, setting the stage for their pervasive employment in the war in Southeast Asia a decade later.[4] Helicopters revolutionized warfare and became the iconic image of the Vietnam war. On the domestic front, outstanding helicopter research was being carried on at NASA Langley. Of particular note were the contributions of researchers and test pilots such as Jack Reeder, John P. Campbell, Richard E. Kuhn, Marion O. McKinney, and Robert H. Kirby. In the late 1950s, military advisers realized how much of the Nation's defense structure depended on a few large airbases and a few large aircraft carriers. Military interests were driven by the objective of achieving operations into and out of unprepared remotely dispersed sites independent of conventional airfields. Meanwhile, commercial air transportation organizations were pursuing ways to cut the amount of real estate required to accommodate new aircraft and long airstrips.[5]

---

2. For NACA work on autogiros, see J.B. Wheatley, "Lift and Drag Characteristics and Gliding Performance of an Autogiro as Determined In Flight," NACA Report 434 (1932); Wheatley, "An Aerodynamic Analysis of the Autogiro Rotor With Comparison Between Calculated and Experimental Results," NACA Report 487 (1934).

3. Michael J. Hirschberg, *The American Helicopter: An Overview of Helicopter Developments in America, 1908–1999* (Arlington, VA: ANSER, 1999); and Col. H.F. Gregory, *Anything a Horse Can Do* (New York: Reynal & Hitchcock, 1944). NACA–NASA contributions to helicopter development are examined in a separate case study in volume one of this work, by John F. Ward.

4. Edgar C. Wood, "The Army Helicopter, Past, Present and Future," *Journal of the American Helicopter Society*, vol. 1, no. 1 (Jan. 1956), pp 87–92; and Lt. Gen. John J. Tolson, *Airmobility, 1961–1971*, a volume in the *U.S. Army Vietnam Studies* series (Washington, DC: U.S. Army, 1973).

5. F.B. Gustafson, "History of NACA/NASA Rotating-Wing Aircraft Research, 1915–1970," *Vertiflite*, Reprint VF-70 (Apr. 1971), pp. 1–27; John F. Ward, "An Updated History of NACA/NASA Rotary-Wing Aircraft Research 1915–1984," *Vertiflite*, vol. 30, no. 4 (May–June 1984), pp. 108–117.

The Vought-Sikorsky V-173 "Flying Flapjack" was an important step on the path to practical V/STOL aircraft. NASA.

Since NASA's inception in 1958, its researchers at various Centers have advanced the knowledge base of V/STOL technology via many specialized test aircraft and flying techniques. Some key discoveries include the realization that V/STOL aircraft must be designed with good Short Take-Off and Landing (STOL) performance capability to be cost-effective, and that, arguably, the largest single obstacle to the implementation of STOL powered-lift technology for civil aircraft is the increasingly objectionable level of aircraft-generated noise at airports close to populated areas.

But NASA interest in fixed wing STOL and VTOL convertiplanes predates formation of the Agency, going back to the unsuccessful combined rotor and wing design by Emile and Henry Berliner tested at College Park Airport, MD, in the early 1920s. In the late 1930s and early 1940s, NACA researcher Charles Zimmerman undertook pioneering research on such craft, his interest leading to the Vought V-173, popularly known as the "Flying Flapjack," because of its peculiar near-circular wing shape. It led to an abortive Navy fighter concept, the Vought XF5U-1, which was built but never flown. The V-173, however, contributed notably to the emerging understanding of V/STOL aircraft challenges and performance. Aside from this sporadic interest, the Agency's research staff did not place great emphasis upon such studies until the postwar era. Then, beginning in the early 1950s, a veritable explosion of interest followed, with a number of design studies and flight-test

The Convair XFY-1 "Pogo" of 1954 was a daring but impractical attempt at developing an operational VTOL naval fighter. U.S. Navy.

programs undertaken at Langley and Ames laboratories (later the NASA Langley and Ames Research Centers). This interest corresponded to rising interest in the military in the possibility of vertical flight vehicles for a variety of missions.

For example, the U.S. Navy sponsored two unsuccessful experimental "Pogo" tail-sitting turboprop-powered VTOL fighters: the Lockheed XFV-1 and the Convair XFY-1. Only the XFY-1 subsequently operated in true VTOL mode, and flight trials indicated that neither represented a reasonable approach to practical VTOL flight. The Air Force developed a pure-jet equivalent: the VTOL delta-winged Ryan X-13. Though widely demonstrated (even outside the Pentagon), it was equally impracticable.[6] The U.S. Army's Transportation and Research Engineering Command sponsored ducted-fan flying jeep and other saucerlike circular flying platforms by Avro and Hiller, with an equivalent lack of success. Overall, the Army's far-seeing V/STOL testbed program, launched in 1956 and undertaken in cooperation with the U.S. Navy's Office of Naval Research, advanced a number of so-called "VZ"-designated research aircraft exploring a range of technical approaches to V/STOL flight.[7] NATO planners envisioned V/STOL close-air support, interdiction, and nuclear attack aircraft. This interest eventually helped spawn the British Aerospace Harrier strike fighter of the late 1960s and other designs that, though they entered flight-testing, did not prove suitable for operational service.[8]

NACA–NASA and Boundary Layer Control, Externally Blown Flap, and Upper Surface Blowing STOL Research

Short Take-Off and Landing flight research was primarily motivated by the desire of military and civil operators to develop transport aircraft with short-field operational capability typical of low-speed airplanes yet the high cruising speed of jets. For Langley and Ames, it was a natural extension of their earlier boundary layer control (BLC) activity undertaken in the 1950s to improve the safety and operational efficiency of military aircraft, such as naval jet fighters that had to land on aircraft carriers, by improving their low-speed controllability and reducing approach and landing speeds.[9] Indeed, as NACA–NASA engineer-

6. See Stephan Wilkinson, "Going Vertical," *Air & Space*, vol. 11, no. 4 (Oct.–Nov. 1996), pp. 50–61; and Ray Wagner, *American Combat Planes* (Garden City, NY: Doubleday & Co., 1982 ed.), pp. 396–397, 515, and 529–530 for details on these aircraft.

7. John J. Schneider, "The History of V/STOL Aircraft," pt. 2, *Vertiflite*, vol. 29, no. 4 (May–June 1983), p. 36.

8. C.H. Zimmerman, Paul R. Hill, and T.L. Kennedy, "Preliminary Experimental Investigation of the Flight of a Person Supported by a Jet Thrust Device Attached to his Feet," NACA RM-L52D10 (1953); Robert H. Kirby, "Flight Investigation of the Stability and Control Characteristics of a Vertically Rising Airplane Research Model with Swept or Unswept Wings and x or + Tails," NACA TN-3312 (1956).

9. Maurice D. White, Bernard A. Schlaff, and Fred J. Drinkwater, III, "A Comparison of Flight-Measured Carrier-Approach Speeds with Values Predicted by Several Different Criteria for 41 Fighter-Type Airplane Configurations," NACA RM-A57L11 (1958).

14

The Stroukoff YC-134A was the first large STOL research aircraft flown at NASA's Ames Research Center. NASA.

historian Edwin Hartman wrote in 1970, "BLC was the first practical step toward achieving a V/STOL airplane."[10] This research had demonstrated the benefits of boundary layer flap-blowing, which eventually was applied to operational high-performance aircraft.[11]

NASA's first large-aircraft STOL flight research projects involved two Air Force–sponsored experimental transports: a Stroukoff Aircraft Corporation YC-134A and a Lockheed NC-130B Hercules. Both aircraft used boundary layer control over their flaps to augment wing lift.

10. Edwin P. Hartman, *Adventures in Research, A History of Ames Research Center, 1940–1965*, NASA SP-4302 (Washington, DC: NASA, 1970), p. 352.

11. L. Stewart Rolls and Robert C. Innis, "A Flight Evaluation of a Wing-Shroud-Blowing Boundary-Layer Control System Applied to the Flaps of an F9F-4 Airplane," NACA RM-A55K01 (1956); Seth B. Anderson, Hervey C. Quigley, and Robert C. Innis, "Flight Measurements of the Low-Speed Characteristics of a 35° Swept-Wing Airplane with Blowing-Type Boundary-Layer Control on the Trailing-Edge Flaps," NACA RM-A56G30 (1956); George E. Cooper and Robert C. Innis, "Effect of Area-Suction-Type Boundary-Layer Control on the Landing-Approach Characteristics of a 35° Swept-Wing Fighter," NACA RM-A55K14 (1957); Hervey C. Quigley, Seth B. Anderson, and Robert C. Innis, "Flight Investigation of the Low-Speed Characteristics of a 45° Swept-Wing Fighter-Type Airplane with Blowing Boundary-Layer Control Applied to the Trailing-Edge Flaps," NACA RM-A58E05 (1958).

The NC-130B boundary layer control STOL testbed just before touchdown at Ames Research Center; note the wing-pod BLC air compressor, drooped aileron, and flap deflected 90 degrees. NASA.

The YC-134A was a twin-propeller radial-engine transport derived on the earlier Fairchild C-123 Provider tactical transport and designed in 1956. It had drooped ailerons and trailing-edge flaps that deflected 60 degrees, together with a strengthened landing gear. A J30 turbojet compressor provided suction for the BLC system. Tested between 1959 and mid-1961, the YC-134A confirmed expectations that deflected propeller thrust used to augment a wing's aerodynamic lift could reduce stall speed. However, in other respects, its desired STOL performance was still limited, indicative of the further study needed at this time.[12]

More promising was the later NC-130B, first evaluated in 1961 and then periodically afterward. Under an Air Force contract, the Georgia

12. Seth B. Anderson, *Memoirs of an Aeronautical Engineer: Flight Testing at Ames Research Center, 1940–1970*, NASA SP-2002-4526, No. 26 in the *Monographs in Aerospace History* series (Washington, DC: GPO, 2002), p. 29; Robert C. Innis and Hervey C. Quigley, "A Flight Examination of Operating Problems of V/STOL Aircraft in STOL-Type Landing and Approach," NASA TN-D-862 (1961); Hartman, *Adventures in Research*, p. 354; Paul F. Borchers, James A. Franklin, and Jay W. Fletcher, *Flight Research at Ames: Fifty-Seven Years of Development and Validation of Aeronautical Technology*, NASA SP-1998-3300 (Moffett Field, CA: Ames Research Center, 1998), Table 8, p. 49.

Division of Lockheed Aircraft Corporation modified a C-130B Hercules tactical transport to a STOL testbed. Redesignated as the NC-130B, it featured boundary layer blowing over its trailing-edge flaps (which could deflect a full 90 degrees down), ailerons (which were also drooped to enhance lift-generation), elevators, and rudder (which was enlarged to improve low-speed controllability). The NC-130 was powered by four Allison T-56-A-7 turbine engines, each producing 3,750 shaft horsepower and driving four-bladed 13.5-foot-diameter Hamilton Standard propellers. Two YT-56-A-6 engines driving compressors mounted in outboard wing-pods furnished the BLC air, at approximately 30 pounds of air per second at a maximum pressure ratio varying from 3 to 5. Roughly 75 percent of the air blew over the flaps and ailerons and 25 percent over the tail surfaces.[13] Thanks to valves and crossover ducting, the BLC air could be supplied by either or both of the BLC engines. Extensive tests in Ames's 40- by 80-foot wind tunnel validated the ability of the NC-130B's BLC flaps to enhance lift at low airspeeds, but uncertainties remained regarding low-speed controllability. Subsequent flight-testing indicated that such concern was well founded. The NC-130B, like the YC-134A before it, had markedly poor lateral-directional control characteristics during low-speed approach and landing. Ames researchers used a ground simulator to devise control augmentation systems for the NC-130B. Flight test validated improved low-speed lateral-directional control.

For a corresponding margin above the stall, the handling qualities of the NC-130B in the STOL configuration were changed quite markedly from those of the standard C-130 airplane. Evaluation pilots found the stability and control characteristics to be unsatisfactory. At 100,000 pounds gross weight, a conventional C-130B stalled at 80 knots; the BLC NB-130B stalled at 56 knots. Approach speed reduced from 106 knots for the unmodified aircraft to between 67 and 75 knots, though, as one NASA report noted, "At these speeds, the maneuvering capability of the aircraft was severely limited."[14] The most seriously affected character-

13. Hervey C. Quigley and Robert C. Innis, "Handling Qualities and Operational Problems of a Large Four-Propeller STOL Transport Airplane," NASA TN-D-1647 (1963), p. 4.

14. Fred J. Drinkwater, III, L. Stewart Rolls, Edward L. Turner, and Hervey C. Quigley, "V/STOL Handling Qualities as Determined by Flight Test and Simulation Techniques," paper presented at the *3rd International Congress of the Aeronautical Sciences, Stockholm, Sweden, Aug. 27–Sept. 1, 1962,* NTIS Report 62N12456 (1962), pp. 12–13.

istics were about the lateral and directional axes, exemplified by problems maneuvering onto and during the final approach, where the pilots found their greatest problem was controlling sideslip angle.[15]

Landing evaluations revealed that the NC-130B did not conform well to conventional traffic patterns, an indication of what could be expected from other large STOL designs. Pilots were surprised at the length of time required to conduct the approach, especially when the final landing configuration was established before turning onto the base leg. Ames researchers Hervey Quigley and Robert Innis noted:

> The time required to complete an instrument approach was even longer, since with this particular ILS system the glide slope was intercepted about 8 miles from touchdown. The requirement to maintain tight control in an instrument landing system (ILS) approach in combination with the aircraft's undesirable lateral-directional characteristics resulted in noticeable pilot fatigue. Two methods were tried to reduce the time spent in the STOL (final landing) configuration. The first and more obvious was suitable for VFR patterns and consisted of merely reducing the size of the pattern, flying the downwind leg at about 900 feet and close abeam, then transitioning to the STOL configuration and reducing speed before turning onto the base leg. Ample time and space were available for maneuvering, even for a vehicle of this size. The other procedure consisted of flying a conventional pattern at high speed (120 knots) with 40° of flap to an altitude of about 500 feet, and then performing a maximum deceleration to the approach angle-of-attack using 70° flap and 30° of aileron droop with flight idle power. Power was then added to maintain the approach angle-of-attack while continuing to decelerate to the approach speed. This procedure reduced the time spent in the approach and generally expedited the operation. The most noticeable adverse effect of this technique was the departure from the original approach path in order to slow down. This effect would compromise its use on a conventional ILS glide path.[16]

14

15. Quigley and Innis, "Handling Qualities and Operational Problems of a Large Four-Propeller STOL Transport Airplane," p. 7.

16. Ibid.

Flight evaluation of the NC-130B offered important experience and lessons for subsequent STOL development. Again, as Quigley and Innis summarized, it clearly indicated that

> The flight control system of an airplane in STOL operation must have good mechanical characteristics (such as low friction, low break-out force, low force gradients) with positive centering and no large non-linearities.
>
> In order to aid in establishing general handling qualities criteria for STOL aircraft, more operational experience was required to help define such items as:
>
> (1) Minimum airport pattern geometry,
> (2) Minimum and maximum approach and climb-out angles,
> (3) Maximum cross wind during landings and take-offs, and
> (4) All-weather operational limits.[17]

Overall, Quigley and Innis found that STOL tests of the NC-130B BLC testbed revealed

> (1) With the landing configuration of 70° of flap deflection, 30° of aileron droop, and boundary-layer control, the test airplane was capable of landing over a 50-foot obstacle in 1,430 feet at a 100,000 pounds gross weight. The approach speed was 72 knots and the flight-path angle 5° for minimum total distance. The minimum approach speed in flat approaches was 63 knots.
>
> (2) Take-off speed was 65 knots with 40° of flap deflection, 30° of aileron droop, and boundary-layer control at a gross weight of 106,000 pounds. Only small gains in take-off distance over a standard C-130B airplane were possible because of the reduced ground roll acceleration associated with the higher flap deflections.
>
> (3) The airplane had unsatisfactory lateral-directional handling qualities resulting from low directional stability and

17. Ibid., p. 15.

> damping, low side-force variation with sideslip, and low aileron control power. The poor lateral-directional characteristics increased the pilots' workload in both visual and instrument approaches and made touchdowns a very difficult task especially when a critical engine was inoperative.
>
> (4) Neither the airplane nor helicopter military handling quality specifications adequately defined stability and control characteristics for satisfactory handling qualities in STOL operation.
>
> (5) Several special operating techniques were found to be required in STOL operations:
>
> > (a) Special procedures are necessary to reduce the time in the STOL configuration in both take-offs and landings.
> > (b) Since stall speed varies with engine power, BLC effectiveness, and flap deflection, angle of attack must be used to determine the margin from the stall.
>
> (6) The minimum control speed with the critical engine inoperative (either of the outboard engines) in both STOL landing and take-off configurations was about 65 knots and was the speed at which almost maximum lateral control was required for trim. Neither landing approach nor take-off speed was below the minimum control speed for minimum landing or take-off distance.[18]

During tests with the YC-134B and the NC-130B, NASA researchers had followed related foreign development efforts, focusing upon two: the French Breguet 941, a four-engine prototype assault transport, and the Japanese Shin-Meiwa UF-XS four-engine seaplane, both of which used deflected propeller slipstream to give them STOL performance. The Shin-Meiwa UF-XS, which a NASA test team evaluated at Omura Naval Air Base in 1964, was built using the basic airframe of a Grumman UF-1 (Air Force SA-16) Albatross seaplane. It was a piloted scale model of a much larger turboprop successor that went on to a

18. Ibid., pp. 15–16.

distinguished career as a maritime patrol and rescue aircraft.[19] However, the Breguet 941 did not, even though both America's McDonnell company and Britain's Short firm advanced it for a range of civil and military applications. A NASA test team was allowed to fly and assess the 941 at the French Centre d'Essais en Vol (the French flight-test center) at Istres in 1963 and undertook further studies at Toulouse and when it came to America at the behest of McDonnell. In conjunction with the Federal Aviation Administration, the team undertook another evaluation in 1972 to collect data for a study on developing civil airworthiness criteria for powered-lift aircraft.[20] The team members found that it had "acceptable performance," thanks largely to its cross-shafted and opposite rotation propellers. The propellers minimized trim changes and asymmetric trim problems in the event of engine failure and ensured no lateral or directional moment changes with variations in airspeed and engine power. But they also found that its longitudinal and lateral-directional stability was "too low for a completely satisfactory rating" and concluded, "More research is required to determine ways to cope with the problem and to adequately define stability and control requirements of STOL airplanes."[21] Their judgment likely matched that of the French, for only four production Breguet 941S aircraft were built; the last of which was retired in 1974. Undoubtedly, however, it was for its time a remarkable and influential aircraft.[22]

Another intriguing approach to STOL design was use of lift-enhancing rotating cylinder flaps. Since the early 1920s, researchers in

19. Curt A. Holzhauser, Robert C. Innis, and Richard F. Vomaske, "A Flight and Simulator Study of the Handling Qualities of a Deflected Slipstream STOL Seaplane Having Four Propellers and Boundary-Layer Control," NASA TN-D-2966 (1965).

20. Barry C. Scott, Charles S. Hynes, Paul W. Martin, and Ralph B. Bryder, "Progress Toward Development of Civil Airworthiness Criteria for Powered-Lift Aircraft," FAA-RD-76-100 and NASA TM-X-73124 (1976), pp. 2–3.

21. Hervey C. Quigley, Robert C. Innis, and Curt A. Holzhauser, "A Flight Investigation of the Performance, Handling Qualities, and Operational Characteristics of a Deflected Slipstream STOL Transport Airplane Having Four Interconnected Propellers," NASA TN-D-2231 (1964), p. 19; Seth B. Anderson, *Memoirs of an Aeronautical Engineer: Flight Testing at Ames Research Center, 1940–1970* (Washington, DC: GPO, 2002), pp. 41–42.

22. NASA test pilots from Langley and Ames participated in an 11-hour flight-test program in the German Dornier DO-31, a 10–jet-engine, 50,000-pound VTOL transport. The tests concentrated on transition, approaches, and vertical landing. Though a vectored and direct-lift thrust system, it is included here to show the international sweep of NASA research in the V/STOL field.

Europe and America had recognized that the Magnus effect produced by a rotating cylinder in an airstream could be put to use in ships and airplanes.[23] Germany's Ludwig Prandtl, Anton Flettner, and Kurt Frey; the Netherland's E.B. Wolff; and NACA Langley's Elliott Reid all examined airflow around rotating cylinders and around wings with spanwise cylinders built into their leading, mid, and trailing sections.[24] All were impressed, for, as Wolff noted succinctly, "The rotation of the cylinder had a remarkable effect on the aerodynamic properties of the wing."[25] Flettner even demonstrated a "Rotorschiff" (rotor-ship) making use of two vertical cylinders functioning essentially as rotating sails.[26] However, because of mechanical complexity, the need for an independent propulsion source to rotate the cylinder at high speed, and the lack of advantage in applying these to aircraft of the interwar era because of their modest performance, none of these systems resulted in more than laboratory experiments. However, that changed in the jet era, particularly as aircraft landing and takeoff speeds rose appreciably. In 1963, Alberto Alvarez-Calderon advocated using a rotating cylinder in conjunction with a flap to increase a wing's lift and reduce its drag. The combination would serve to reenergize the wing's boundary layer without use of the traditional methods of boundary-layer suction or blowing. Advances in propulsion and high-speed rotating shaft systems, he concluded, "indicated to this investigator the need of examining the rotating cylinder as a high lift device for VTOL aircraft."[27]

23. Theodore von Kármán, *Aerodynamics* (New York: McGraw-Hill Book Publishing Co., 1963 ed.), pp. 32–34; Michael Eckert, *The Dawn of Fluid Dynamics* (Weinheim, Germany: Wiley-VCH Verlag, 2006), p. 175; and Jan A. van der Bliek, *75 Years of Aerospace Research in the Netherlands, 1919–1994* (Amsterdam: NLR, 1994), p. 20.

24. Elliott G. Reid, "Tests of Rotating Cylinders," NACA TN-209 (1924); E.B. Wolff "Tests for Determining the Effect of a Rotating Cylinder Fitted into the Leading Edge of an Airplane Wing," NACA TM-354 (1926); Kurt Frey, "Experiments with Rotating Cylinders in Combination with Airfoils," NACA TM-382 (1926).

25. Wolff, "Tests for Determining the Effect of a Rotating Cylinder," p. 1.

26. Jakob Ackeret, *Das Rotorschiff und seine physikalischen Grundlagen* (Göttingen: Vandenhoeck & Ruprecht, 1925), pp. 34–48.

27. Alberto Alvarez-Calderon, "VTOL and the Rotating Cylinder Flap," *Annals of the New York Academy of Sciences*, vol. 107, no. 1 (Mar. 1963), pp. 249–255; see also his later "Rotating Cylinder Flaps for V/STOL: Some Aspects of an Investigation into the Rotating Cylinder Flap High Lift System for V/STOL Aircraft Conducted Jointly by the Peruvian Air Force and The National University of Engineering of Peru," *Aircraft Engineering and Aerospace Technology*, vol. 36, no. 10 (Oct. 1964), pp. 304–309.

The NASA YOV-10A rotating cylinder flap research aircraft. NASA.

In 1971, NASA Ames Program Manager James Weiberg had North American-Rockwell modify the third prototype, YOV-10A Bronco, a small STOL twin-engine light armed reconnaissance aircraft (LARA), with an Alvarez-Calderon rotating cylinder flap system. As well as installing the cylinder, which was 12 inches in diameter, technicians cross-shafted the plane's two Lycoming T53-L-11 turboshaft engines for increased safety, using the drive train from a Canadair CL-84 Dynavert, a twin-engine tilt rotor testbed. The YOV-10A's standard three-bladed propellers were replaced with the four-bladed propellers used on the CL-84, though reduced in diameter so as to furnish adequate clearance of the propeller disk from the fuselage and cockpit. The rotating cylinder, between the wing and flap, energized the plane's boundary layer by accelerating airflow over the flap. The flaps were modified to entrap the plane's propeller slipstream, and the combination thus enabled steep approaches and short landings.[28]

Before attempting flight trials, Ames researchers tested the modified YOV-10A in the Center's 40- by 80-foot wind tunnel, measuring

28. Leonard Roberts and Wallace R. Deckert, "Recent Progress in VSTOL Technology," NASA TM-84238 (1982), pp. 3–4.

changes in boundary layer flow at various rotation speeds. They found that at 7,500 revolutions per minute (rpm), equivalent to a rotational speed of 267.76 mph, the flow remained attached over the flaps even when they were set vertically at 90 degrees to the wing. But in the course of 34 flight-test sorties by North American-Rockwell test pilot Edward Gillespie and NASA pilot Robert Innis, researchers found significant differences between tunnel predictions and real-world behavior. Flight tests revealed that the YOV-10A had a lift coefficient fully a third greater than the basic YOV-10. It could land with approach speeds of 55 to 65 knots, at descent angles up to 8 degrees, and at flap angles up to 75 degrees. Researchers found that

> Rotation angles to flare were quite large and the results were inconsistent. Sometimes most of the sink rate was arrested and sometimes little or none of it was. There never was any tendency to float. The pilot had the impression that flare capability might be quite sensitive to airspeed (C_L)[29] at flare initiation. None of the landings were uncomfortable.[30]

The modified YOV-10A had higher than predicted lift and downwash values, likely because of wind tunnel wall interference effects. It also had poor lateral-directional dynamic stability, with occasional longitudinal coupling during rolling maneuvers, though this was a characteristic of the basic aircraft before installation of the rotating cylinder flap and had, in fact, forced addition of vertical fin root extensions on production OV-10A aircraft. Most significantly, at increasing flap angles, "deterioration of stability and control characteristics precluded attempts at landing,"[31] manifested by an unstable pitch-up, "which required full nose-down control at low speeds" and was "a strong function of flap deflection, cylinder operation, engine power and airspeed."[32]

29. Engineering abbreviation for lift coefficient.
30. James A. Weiberg, Demo Giulianetti, Bruno Gambucci, and Robert C. Innis, "Takeoff and Landing Performance and Noise Measurements of a Deflected Slipstream STOL Airplane with Interconnected Propellers and Rotating Cylinder Flaps," NASA TM-X-62320 (1973), p. 9.
31. Ibid., p. 1.
32. D.R. Cichy, J.W. Harris, and J.K. MacKay, "Flight Tests of A Rotating Cylinder Flap on a North American Rockwell YOV-10 Aircraft," NASA CR-2135 (1972), pp. ii and 17–18; re: addition of vertical fin root extensions, see David Willis, "North American Bronco," *Aeroplane*, vol. 38, no. 1 (Jan. 2010), p. 63.

As David Few subsequently noted, the YOV-10A's rotating cylinder flap-test program constituted the first time that: "a flow-entrainment and boundary-layer-energizing device was used for turning the flow downward and increasing the wing lift. Unlike all or most pneumatic boundary layer control, jet flap, and similar concepts, the mechanically driven rotating cylinder required very low amounts of power; thus there was little degradation to the available takeoff horsepower."[33]

Unfortunately, the YOV-10A did not prove to be a suitable research aircraft. As modified, it could not carry a test observer, had too low a wing loading—just 45 pounds per square foot—and so was "easily disturbed in turbulence." Its marginal stability characteristics further hindered its research utility, so after this program, it was retired.[34]

NASA's next foray in BLC research was a cooperative program between the United States and Canada that began in 1970 and resulted in NASA's Augmentor Wing Jet STOL Research Aircraft (AWJSRA) program. The augmentor wing concept was international in origin, with significant predecessor work in Germany, France, Britain, Canada, and the United States.[35] The augmentor wing included a blown flap on the trailing edge of a wing, fed by bleed air taken from the aircraft's engines, accelerating ambient air drawn over the flap and directing it downward to produce lift, using the well-known Coanda effect. Ames researchers conducted early tunnel tests of the concept using a testbed that used a J85 engine powering two compressors that furnished air to the wind tunnel model.[36] Encouraged, Ames Research Center and Canada's Department of Industry, Trade, and Commerce (DTIC) moved to collaborate in flying

33. D.D. Few, "A Perspective on 15 Years of Proof-of-Concept Aircraft Development and Flight Research at Ames–Moffett by the Rotorcraft and Powered-Lift Flight Projects Division, 1970–1985," NASA RP-1187 (1987), p. 7.

34. Weiberg, Giulianetti, Gambucci, and Innis, "Takeoff and Landing Performance and Noise Measurements of a Deflected Slipstream STOL Airplane," p. 9; Roberts and Deckert, "Recent Progress in VSTOL Technology," p. 4.

35. M.H. Schmitt, "Remarks on the Paper of Mr. D. C. Whittley (Paper no. 13): Some Aspects of Propulsion for the Augmentor Wing Concept," NASA TT-F-14005 (1971), pp. 1–9; W.S. Hindson and G. Hardy, "A Summary of Joint US-Canadian Augmentor Wing Powered-Lift STOL Research programs at the Ames Research Center, NASA, 1975–1980," NASA TM-81215 (1980).

36. W.L. Cook and D.C. Whittley, "Comparison of Model and Flight Test Data for an Augmented Jet Flap STOL Research Aircraft," NASA TM-X-62491 (1975); Few, "A Perspective on 15 Years of Proof-of-Concept Aircraft," pp. 7–8; Roberts and Deckert, "Recent Progress in VSTOL Technology," p. 5.

The C-8A augmentor wing testbed on takeoff. NASA.

a testbed system. Initially, researchers examined putting an augmentor wing on a modified U.S. Army de Havilland CV-7A Caribou twin-piston-engine light STOL transport. But after studying it, they chose instead its bigger turboprop successor, the de Havilland C-8A Buffalo.[37] Boeing, de Havilland, and Rolls-Royce replaced its turboprop engines with Rolls-Royce Spey Mk 801-SF turbofan engines modified to have the rotating lift nozzle exhausts of the Pegasus engine used in the vectored-thrust P.1127 and Harrier aircraft. They also replaced its high aspect ratio wing with a lower aspect ratio wing with spoilers, blown ailerons, augmentor flaps, and a fixed leading-edge slat. Because it was intended strictly as a low-speed testbed, the C-8A was fitted with a fixed landing gear. As well, it had a long proboscis-like noseboom, which, given the fixed gear and classic T-tail high wing configuration of the basic Buffalo from which it was derived, endowed it with a quirky and somewhat thrown-together appearance. The C-8A project was headed by David Few, with techni-

37. E.H. Kemper and D.J. Renselaer, "CV-7A Transport Aircraft Modification to Provide an Augmentor Wing Jet STOL Research Aircraft, v. 1: Design Study," NASA CR-73321 (1969).

cal direction by Hervey Quigley, who succeeded Few as manager in 1973. The NASA pilots were Robert Innis and Gordon Hardy. The Canadian pilots were Seth Grossmith, from the Canadian Ministry of Transport, and William Hindson, from the National Research Council of Canada.[38]

The C-8A augmentor wing research vehicle first flew on May 1, 1972, and subsequently enjoyed great technical success.[39] It demonstrated thrust augmentation ratios of 1.20, achieved a maximum lift coefficient of 5.5, flew approach speeds as low as 50 knots, and took off and landed over 50-foot obstructions in as little as 1,000 feet, with ground rolls of only 350 feet. It benefitted greatly from the cushioning phenomena of ground effect, making its touchdowns "gentle and accurate."[40] Beyond its basic flying qualities, the aircraft also enabled Ames researchers to continue their studies on STOL approach behavior, flightpath tracking, and the landing flare maneuver. The Ames Avionics Research Branch used it to help define automated landing procedures and evaluated an experimental NASA–Sperry automatic flightpath control system that permitted pilots to execute curved steep approaches and landings, both piloted and automatic. Thus equipped, the C-8A completed its first automatic landing in 1975 at Ames's Crows Landing test facility. Ames operated it for 4 years, after which it returned to Canada, where it continued its own flight-test program.[41]

Upper surface blowing (USB) constituted another closely related concept for using accelerated flows as a means of enhancing lift production. Following on the experience with the augmentor wing C-8A testbed, it became NASA's "next big thing" in transport-related STOL

38. Few, "A Perspective on 15 Years of Proof-of-Concept Aircraft," pp. 7–8.

39. Hervey C. Quigley, Robert C. Innis, and Seth Grossmith, "A Flight Investigation of the STOL Characteristics of an Augmented Jet Flap STOL Research Aircraft," NASA TM-X-62334 (1974); Richard F. Vomaske, Robert C. Innis, Brian E. Swan, and Seth W. Grossmith, "A Flight Investigation of the Stability, Control, and Handling Qualities of an Augmented Jet Flap STOL Airplane," NASA TP-1254 (1978); L.T. Dufraimont, "Evaluation of the Augmentor-Wing Jet STOL Aircraft," NASA CR-169853 (1980).

40. Roberts and Deckert, "Recent Progress in VSTOL Technology," p. 5.

41. DeLamar M. Watson, Gordon H. Hardy, and David N. Warner, Jr., "Flight Test of the Glide-Slope Track and Flare-Control Laws for an Automatic Landing System for a Powered-Lift STOL Airplane," NASA TP-2128 (1983); James A. Franklin and Robert C. Innis, "Longitudinal Handling Qualities During Approach and Landing of a Powered Lift STOL Aircraft," NASA TM-X-62144 (1972); Elizabeth A. Muenger, *Searching the Horizon: A History of Ames Research Center, 1940–1976,* NASA SP-4304 (Washington, DC: GPO, 1985), p. 272.

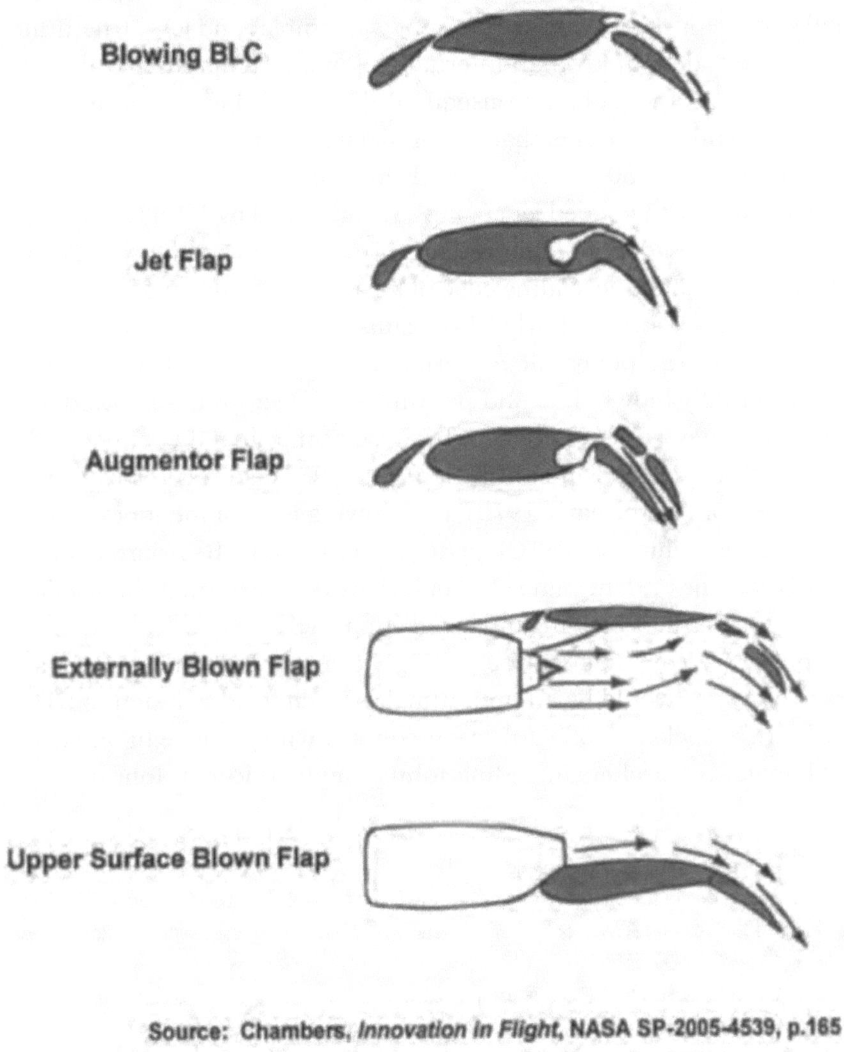

Powered lift concepts. NASA.

aircraft research. Agency interest in USB was an outgrowth of NACA–NASA research at Langley and Ames on BLC and the engine-bleed-air-fed jet flap, exemplified by tests in 1963 at Langley with a Boeing 707 jet airliner modified to have engine compressor air blown over the wing's trailing-edge flaps. An Ames 40-foot by 80-foot tunnel research program in 1969 used a British Hunting H.126, a jet-flap research aircraft flight-tested between 1963 and 1967. It used a complex system of ducts and nozzles to divert over half of its exhaust over its flaps.[42] As a fully external system, the upper surface concept was simpler and less structurally intrusive and complex than internally blown systems such as the augmentor wing and jet flap. Consequently, it enjoyed more success than these and other concepts that NASA had pursued.[43]

In the mid-1950s, Langley's study of externally blown flaps used in conjunction with podded jet engines, spearheaded by John P. Campbell, had led to subsequent Center research on upper surface blowing, using engines built into the leading edge of an airplane's wing and exhausting over the upper surface. Early USB results were promising. As Campbell recalled, "The aerodynamic performance was comparable with that of the externally blown flap, and preliminary noise studies showed it to be a potentially quieter concept because of the shielding effect of the wing."[44] Noise issues meant little in the 1950s, so further work was dropped. But in the early 1970s, the growing environment noise issue and increased interest in STOL performance led to USB's resurrection. In particular, the evident value of Langley's work on externally blown flaps and upper surface blowing intrigued Oran Nicks, appointed as Langley Deputy Director in September 1970. Nicks concluded that upper surface blowing "would be an optimum approach for the design of STOL aircraft."[45] Nicks's strong advocacy, coupled with the insight and drive of Langley researchers including John Campbell, Joseph Johnson, and

42. Albert W. Hall, Kalman J. Grunwald, and Perry L. Deal, "Flight Investigation of Performance Characteristics During Landing Approach of a Large Powered-Lift Jet Transport," NASA TN-D-4261 (1967); T.N. Aiken and A.M. Cook, "Results of Full-Scale Wind tunnel Tests on the H.126 Jet Flap Aircraft," NASA TN-D-7252 (1973); I.M. Davidson, "The Jet Flap," *Journal of the Royal Aeronautical Society*, vol. 60, no. 541 (Jan. 1956), pp. 25–50; and D.A. Spence, "The Flow Past a Thin Wing with an Oscillating Jet Flap," *Philosophical Transactions of the Royal Society of London*, Series A, Mathematical and Physical Sciences, vol. 257, no. 1085 (June 3, 1965), pp. 445–477.
43. John P. Campbell, "Overview of Powered-Lift Technology," in NASA LRC, "Powered-Lift Aerodynamics and Acoustics: A Conference Held at Langley Research Center, Hampton, Virginia, May 24–26, 1976," NASA SP-406 (Washington, DC: NASA, 1976), pp. 2–3.
44. Campbell, "Overview of Powered-Lift Technology," p. 3; for early studies, see Thomas R. Turner, Edwin E. Davenport, and John M. Riebe, "Low-Speed Investigation of Blowing from Nacelles Mounted Inboard and on the Upper Surface of an Aspect-Ratio-7.0 35° Swept Wing With Fuselage and Various Tail Arrangements," NASA Memo 5-1-59L (1959); Domenic J. Maglieri and Harvey H. Hubbard, "Preliminary Measurements of the Noise Characteristics of Some Jet-Augmented-Flap Configurations," NASA Memo 12-4-58L (1959).
45. Joseph R. Chambers, *Innovation in Flight*, NASA SP-2005-4539 (Washington, DC: GPO, 2005), p. 174.

Arthur Phelps, William Letko, and Robert Henderson, swiftly resulted in modification of an existing externally blown flap (EBF) wind tunnel model to a USB one. The resulting tunnel tests, completed in 1971, confirmed that the USB concept could result in a generous augmentation of lift and low noise. Encouraged, Langley researchers expanded their USB studies using the Center's special V/STOL tunnel, conducted tests of a much larger USB model in Langley's Full-Scale Tunnel, and moved on to tests of even larger models derived from modified Cessna 210 and Aero Commander general-aviation aircraft to acquire data more closely matching full-size aircraft. At each stage, wind tunnel testing confirmed that the USB concept offered high lifting properties, warranting further exploration.[46]

Langley's research on EBF and USB technology resulted in application to actual aircraft, beginning with the Air Force's experimental Advanced Medium STOL Transport (AMST) development effort of the 1970s, a rapid prototyping initiative triggered by the Defense Science Board and Deputy Secretary of Defense David Packard. Out of this came the USB Boeing YC-14 and the EBF McDonnell-Douglas YC-15, evaluated in the 1970s in similar fashion to the Air Force's Lightweight Fighter (LWF) competition between the General Dynamics YF-16 and Northrop YF-17. Unlike the other evaluation, the AMST program did not spawn a production model of either the YC-14 or YC-15. NASA research benefited the AMST effort, particularly Boeing's USB YC-14, which first flew in August 1976. It demonstrated extraordinary performance during flight-testing and a 1977 European tour. The merits of YC-14-style USB impressed the engineers of the Soviet Union's Antonov design bureau. They subsequently produced a transport, the An-72/74, which bore a remarkable similarity to the YC-14.[47]

14

46. Chambers details Langley's progression from EBF to USB in his *Innovation in Flight*, pp. 173–181. See also Arthur E. Phelps, III, William Letko, and Robert L. Henderson, "Low Speed Wind-Tunnel Investigation of a Semispan STOL Jet Transport Wing-Body with an Upper-Surface Blown Jet Flap," NASA TN-D-7183 (1973); Phelps and Charles C. Smith, Jr., "Wind-Tunnel Investigation of an Upper Surface Blown Jet-Flap Powered-Lift Configuration," NASA TN-D-7399 (1973); Phelps, "Wind Tunnel Investigation of a Twin-Engine Straight Wing Upper Surface Blown Jet Flap Configuration," NASA TN-D-7778 (1975); William C. Sleeman, Jr., and William C. Hohlweg, "Low-Speed Wind-Tunnel Investigation of a Four-Engine Upper Surface Blown Model Having a Swept Wing and Rectangular and D-Shaped Exhaust Nozzles," NASA TN-D-8061 (1975).

47. E.J. Montoya and A.E. Faye, Jr., "NASA Participation in the AMST Program," in NASA LRC Staff, "Powered-Lift Aerodynamics and Acoustics: A Conference Held at Langley Research Center, Hampton, Virginia, May 24–26, 1976," NASA SP-406 (1976), pp. 465–478.

The Quiet Short-Haul Research Airplane during trials from the USS Kitty Hawk in July 1980. NASA.

In January 1974, NASA launched a study program for a Quiet Short-Haul Research Airplane (QSRA) using USB. The QSRA evolved from earlier proposals by Langley researchers for a quiet STOL transport, the QUESTOL, possibly using a modified Douglas B-66 bomber, an example of which had already served as the basis for an experimental laminar flow testbed, the X-21. However, for the proposed four-engine USB, NASA decided instead to modify another de Havilland C-8, issuing a contract to Boeing as prime contractor for the conversion in 1976.[48] The QSRA thus benefited fortuitously from Boeing's work on the YC-14. Again, as with the earlier C-8 augmentor wing, the QSRA had a fixed landing gear and a long conical proboscis. Four 7,860-pound-thrust Avco Lycoming YF102 turbofans furnished the USB. As the slotted flaps lowered, the exhaust followed their curve via Coanda effect, creating additional propulsive lift. First flown in July 1978, the QSRA could take off and land in less than 500 feet, and its high thrust enabled a rapid climbout while making a steep turn over the point from which it became airborne. On approach, its high drag allowed the QSRA to execute a steep approach, which enhanced both its STOL performance and further reduced its

48. M.D. Shovlin and J.A. Cochrane, "An Overview of the Quiet Short-Haul Research Aircraft Program," NASA TM-78545 (1978); Roberts and Deckert, "Recent Progress in VSTOL Technology," p. 6.

already low noise signature.[49] It demonstrated high lift coefficients, from 5.5 to as much as 11. Despite a moderately high wing-loading of 80 pounds per square foot, it could fly at landing approach speeds as low as 60 knots. Researchers evaluated integrated flightpath and airspeed controls and displays to assess how precisely the QSRA could fly a precision instrument approach, refined QSRA landing performance to the point where it achieved carrier-like precision landing accuracy, and, in conjunction with Air Force researchers, used the QSRA to help support the development of the C-17 transport, with Air Force and McDonnell-Douglas test pilots flying the QSRA in preparation for their flights in the much larger C-17 transport. Lessons from display development for the QSRA were also incorporated in the Air Force's MC-130E Combat Talon I special operations aircraft, and the QSRA influenced Japan's development of its USB testbed, the ASKA, a modified Kawasaki C-1 with four turbofan engines flown between 1985 and 1989.[50]

Not surprisingly, as a result of its remarkable Short Take-Off and Landing capabilities, the QSRA attracted Navy interest in potentially using USB aircraft for carrier missions, such as antisubmarine patrol, airborne early warning, and logistical support. This led to trials of the QSRA aboard the carrier USS Kitty Hawk in 1980. In preparation, Ames researchers undertook a brief QSRA carrier landing flight simulation using the Center's Flight Simulator for Advanced Aircraft (FSAA), and the Navy furnished a research team from the Carrier Suitability Branch at the Naval Air Test Center, Patuxent River, MD. The QSRA did have one potential safety issue: it could slow without any detectable change in control force or position, taking a pilot unawares. Accordingly, before the carrier landing tests, NASA installed a speed indexer light system that the pilot could monitor while tracking the carrier's mirror-landing

49. J.A. Cochrane, D.W. Riddle, and V.C. Stevens, "Quiet Short-Haul Research Aircraft—TheFirst Three Years of Flight Research," AIAA Paper 81-2625 (1981); J.A. Cochrane, D.W. Riddle, V.C. Stevens, and M.D. Shovlin, "Selected Results from the Quite Short-Haul Research Aircraft Flight Research Program," *Journal of Aircraft*, vol. 19, no. 12 (Dec. 1982), pp. 1076–1082; Joseph C. Eppel, Dennis W. Riddle, and Victor C. Stevens, "Flight Measured Downwash of the QSRA," NASA TM-101050 (1988); Jack D. Stephenson and Gordon H. Hardy, "Longitudinal Stability and Control Characteristics of the Quiet Short-Haul Research Aircraft (QSRA)," NASA TP-2965 (1989); Jack D. Stephenson, James A. Jeske, and Gordon H. Hardy, "Lateral-Directional Stability and Control Characteristics of the Quiet Short-Haul Research Aircraft (QSRA)," NASA TM-102250 (1990).
50. Borchers, Franklin, and Fletcher, *Flight Research at Ames*, pp. 187–189; Chambers, *Innovation in Flight*, pp. 188–189.

system Fresnel lens during the final approach to touchdown. The indexer used a standard Navy angle-of-attack indicator modified to show the pilot deviations in airspeed rather than changes in angle of attack. After final reviews, the QSRA team received authorization from both NASA and the Navy to take the plane to sea.

Sea trials began July 10, 1980, with the Kitty Hawk approximately 100 nautical miles southwest of San Diego. Over 4 days, Navy and NASA QSRA test crews completed 25 low approaches, 37 touch-and-go landings, and 16 full-stop landings, all without using an arresting tail hook during landing or a catapult for takeoff assistance. With the carrier steaming into the wind, standard Navy approach patterns were flown, at an altitude of 600 feet above mean sea level (MSL). The initial pattern configuration was USB flaps at 0 degrees and double-slotted wing flaps at 59 degrees. On the downwind leg, abeam of the bow of the ship, the aircraft was configured to set the USB flaps at 30 degrees and turn on the BLC. The 189-degree turn to final approach to the carrier's angled flight deck was initiated abeam the round-down of the flight deck, at the stern of the ship. The most demanding piloting task during the carrier evaluations was alignment with the deck. This difficulty was caused partially by the ship's forward motion and consequent continual lateral displacement of the angle deck to the right with the relatively low QSRA approach speeds. In sum, to pilots used to coming aboard ship at 130 knots in high-performance fighters and attack aircraft, the 60-knot QSRA left them with a disconcerting feeling that the ship was moving, so to speak, out from under them. But this was a minor point compared with the demonstration that advanced aerospace technology had reached the point where a transport-size aircraft could land and takeoff at speeds so remarkably slow that it did not need either a tail hook to land or a catapult for takeoff. Landing distance was 650 feet with zero wind over the carrier deck and approximately 170 feet with a 30-knot wind over the deck. Further, the QSRA demonstrated a highly directional noise signature, in a small 35-degree cone ahead of the airplane, with noise levels of 90 engine-perceived noise decibels at a sideline distance of 500 feet, "the lowest ever obtained for any jet STOL design."[51]

51. Quote from Chambers, *Innovation in Flight*, p. 187; V.C. Stevens, D.W. Riddle, J.L. Martin, and R.C. Innis, "Powered-lift STOL Aircraft Shipboard Operations—a Comparison of Simulation, Land-based and Sea Trial Results for the QSRA," AIAA Paper 81-2480 (1981); Roberts and Deckert, "Recent Progress in VSTOL Technology," p. 7; David D. Few, "A Perspective on 15 Years of Proof-of-Concept Aircraft," p. 11.

The QSRA's performance made it a crowd pleaser at any airshow where it was flown. Most people had never seen an airplane that large fly with such agility, and it was even more impressive from the cockpit. One of the QSRA's noteworthy achievements was appearing at the Paris Air Show in 1983. The flight, from California across Canada and the North Atlantic to Europe, was completed in stages by an airplane having a maximum flying range of just 400 miles. Another was a demonstration landing at Monterey airport, where it landed so quietly that airport monitoring microphones failed to detect it.[52]

By the early 1980s, the QSRA had fulfilled the expectations its creators, having validated the merits of USB as a means of lift augmentation. Simultaneously, another Coanda-rooted concept was under study, the notion of circulation control around a wing (CCW) via blowing sheet of high-velocity air over a rounded trailing edge. First evaluated on a light general-aviation aircraft by researchers at West Virginia University in 1975 and then refined and tested by a David Taylor Naval Ship Research and Development (R&D) Center team under Robert Englar using a modified Grumman A-6A twin-engine attack aircraft in 1979, CCW appeared as a candidate for addition to the QSRA.[53] This resulted in a full-scale static ground-test demonstration of USB and CCW on the QSRA aircraft and a proposal to undertake flight trials of the QSRA using both USB and CCW. This, however, did not occur, so QSRA at last retired in 1994. In its more than 15 years of flight research, it had accrued nearly 700 flight hours and over 4,000 STOL approaches and landings, justifying the expectations of those who had championed the QSRA's development.[54]

52. Author's recollections; Glenn E. Bugos, *Atmosphere of Freedom, Sixty Years at the Ames Research Center*, NASA SP-4314 (Washington, DC: NASA, 2000), p. 139.

53. John L. Loth, "Why Have Only Two Circulation-Controlled STOL Aircraft Been Built and Flown in Years 1974–2004?" in Gregory S. Jones and Ronald D. Joslin, eds., *Proceedings of the 2004 NASA/ONR Circulation Control Workshop*, NASA CP-2005-213509/PT1 (Washington, DC: NASA, 2005), pp. 603–615; Robert J. Englar, "Development of the A-6/Circulation Control Wing Flight Demonstrator Configuration," U.S. Navy DTNSRDC Report ASED-79/01 (1979).

54. Robert J. Englar, "Development of Circulation Control Technology for Powered-Lift STOL Aircraft," and Dennis W. Riddle and Joseph C. Eppel, "A Potential Flight Evaluation of an Upper-Surface-Blowing/Circulation-Control-Wing Concept," both in NASA ARC, "Proceedings of the Circulation Control Workshop, 1986," NASA CP-2432 (1986); R.J. Englar, J.H. Nichols, Jr., M.J. Harris, J.C. Eppel, and M.D. Shovlin, "Circulation Control Technology Applied to Propulsive High Lift Systems," Society of Automotive Engineers, SAE Paper 841497, in Society of Automotive Engineers, *V/STOL: An Update and Overview* (Warrendale, PA: SAE, 1984), pp. 31–43.

NACA–NASA Research on Deflected Slipstream and Tilt Wing V/STOL

In contrast to STOL aircraft systems, which used wing lift generated by forward movement to take off, VTOL aircraft would necessarily have to have some provision for direct vertical propulsive thrust, with the thrust level well in excess of the airplane's operating weight, to lift off the ground. This drove deflected propeller thrust, tilt wing, tilt rotor, and vectored jet thrust technical approaches, all of which NASA researchers intensively studied. In all of this, the researchers' assessment of the system's VTOL control capability was of special interest—for they had to be able to be controlled in pitch, roll, and yaw without any reliance upon the traditional forces imposed upon an airplane by its movement through the air. The first two approaches that NACA–NASA researchers explored were those of deflected propeller flow and pivoted tilt wings.

At the beginning of 1958, the Ryan Company of San Diego unveiled its Model 92, the VZ-3RY Vertiplane. The Vertiplane, a single-seat twin-propeller high wing design with a T-tail, used propeller thrust to attain vertical flight and maintain hover, deflecting the propeller slipstream via a variable-area and variable-camber wing. The wing's trailing edge consisted of large, 40-percent-chord, double-slotted flaps that transformed into a gigantic curved flow channel, with wingtip ventral fins serving to further entrap the air and concentrate its flow vertically below the craft. Roll control in hover came via varying the propeller pitch to achieve changes in slipstream flow. Power to its twin three-bladed propellers was furnished by a single Lycoming T53 turboshaft engine, which also had its exhaust channeled through a tailpipe to a universal-joint nozzle that furnished pitch and yaw control for the airplane when it was in hover mode via deflected jet thrust.[55]

Before the aircraft flew, Ames researchers undertook a series of wind tunnel tests in the 40-foot by 80-foot full-scale wind tunnel to define performance, stability and control, and handling and control characteristics.[56] As a result of these tests, the aircraft's landing gear was changed from a "tail-dragger" to tricycle arrangement, and engineers added a ventral fin to enhance directional stability in conventional flight. Thus modified, the VZ-3RY completed its first flight January 21, 1959, piloted by Ryan test pilot Pete Girard. Less than a month later, it

55. Swanborough, *Vertical Flight Aircraft*, p. 78; Anderson, "Historical Overview of V/STOL," pp. 9-7–9-8.

56. Borchers, Franklin, and Fletcher, *Flight Research at Ames*, p. 56.

The VZ-3RY, in final configuration with fully deflected wing and flaps, and full-span leading-edge slat, at Ames Research Center in California. NASA.

was damaged in a landing accident at the conclusion of its 13th flight, when a propeller pitch control mechanism malfunctioned, leaving the VZ-3RY with insufficient lift to drag (L/D) available to flare for landing. It was late summer before it returned to the air, being delivered to Ames in 1960 for NASA testing. Howard L. Turner oversaw the project, and Glen Stinnett and Fred Drinkwater undertook most of the flying. The aircraft was severely damaged when Stinnett ran out of nose-down control at a low-power setting and the aircraft pitched inverted. Fortunately, Stinnett ejected before it nosed into the salt ponds north of the Moffett Field runway. Despite this seemingly disastrous accident, the aircraft was rebuilt yet again and completed the test program. The addition of full-span wing leading-edge slats to enhance lift production permitted hover out of ground effect (OGE). However, air recirculation effects limited in ground effect (IGE) operation to speeds greater than 10 knots, as marginal turning of the slipstream and random upset disturbances caused by slipstream recirculation prevented true VTOL performance. A static pitch instability was often encountered at high lift coefficients, and large pitch trim changes occurred with flap deflection and power changes. The transition required careful piloting technique to avoid pitch-up. Although adequate, descent performance was limited in the extreme by low roll control power and airflow separation on the wing when power was reduced to descend. Despite these quirks and two accidents, the VZ-3RY demonstrated excellent STOL performance, achieving a maximum lift coefficient of 10, with a moderate to good cruise speed range. Thus, it must be considered a successful research program. Transitions were completed from maximum speed down to 20 knots with "negligible change in longitudinal trim and at rates comparable to those done with a helicopter."[57] Indeed, as Turner and Drinkwater concluded in 1963, "Flight tests with the Ryan VZ-3RY V/STOL deflected-slipstream test vehicle have indicated that the concept has some outstanding advantages as a STOL aircraft where very short take-off and landing characteristics are desired."[58]

As well as pursuing the BLC and deflected slipstream projects, NASA researchers examined tilt wing concepts then being pursued in America and abroad. The tilt wing promised a good blend of moderate low- and high-speed compatibility, with good STOL performance provided by slipstream-induced lift. For takeoff and landing, the wing would pivot so that the engine nacelles and propellers pointed vertically. After takeoff, the wing would be gradually rotated back to the horizontal, enabling conventional flight. Various research aircraft were built to investigate the tilt wing approach to V/STOL flight, notably including the Canadair CL-84, Hiller X-18, the Kaman K-16B, and the joint-service Ling-Temco-Vought XC-142. The first such American aircraft was the Boeing-Vertol VZ-2 (the Vertol Model 76). It was powered by a single Lycoming YT53L1 gas turbine, driving two propellers via extension shafts and small tail fans for low-speed pitch and yaw control. Conceived from a jointly funded U. S. Army–Office of Naval Research study, the VZ-2 first flew in August 1957 and was an important early step in demonstrating the potential of tilt wing V/STOL technology. On July 16, 1958, piloted by Leonard La Vassar, it made the world's first full-conversion of a tilt wing aircraft from vertical to horizontal flight, an important milestone in the history of V/STOL. Vertol completed its testing in September 1959 and then shipped the VZ-2 to Langley Research Center for evaluation by NASA.[59]

57. Drinkwater, Rolls, Turner, and Quigley, "V/STOL Handling Qualities as Determined by Flight Test and Simulation Techniques," p. 15.

58. Turner and Drinkwater, "Some Flight Characteristics of a Deflected Slipstream VSTOL Aircraft," NASA TN-D-1891 (1963), p. 9; Turner and Drinkwater, "Longitudinal-Trim Characteristics of a Deflected Slipstream V/STOL Aircraft During Level Flight at Transition Flight Speeds," NASA TN-D-1430 (1962).

59. Swanborough, *Vertical Flight Aircraft,* p. 19; Schneider, "The History of V/STOL," p. 36; Aircraft Industries Association, *The Aircraft Year Book for 1959* (Washington, DC: American Aviation Publications, Inc., 1959), p. 447.

The ungainly Boeing-Vertol VZ-2, shown here shortly after completion in 1957, made important contributions to early V/STOL tilt wing understanding. NASA.

Subsequent Langley tests confirmed that the tilt wing was undoubtedly promising. However, like many first-generation technological systems, the VZ-2 had a number of limitations. NASA test pilot Don Mallick recalled, "it was extremely difficult to fly," with "lots of cross-coupling between the roll and yaw controls," and that "It took everything I had to keep from 'dinging' or crashing the aircraft."[60] Langley research pilot Jack Reeder found that its VTOL roll control—which, as in a helicopter, was provided by varying the propeller pitch and hence its thrust—was too sensitive. Further, the two ducted fans at the tail responsible for pitch and yaw control furnished only marginal control power. In particular, weak yaw control generated random heading deviations. When slowing into ground effect at a wing tilt angle of 70 degrees, directional instabilities were encountered, though there was no appreciable aerodynamic lift change.[61]

Reflected flows from the ground caused buffeting and unsteady aircraft behavior, resulting in poor hover precision. Because of low

60. Donald L. Mallick, with Peter W. Merlin, *The Smell of Kerosene: A Test Pilot's Odyssey*, NASA SP-4108 (Washington, DC: GPO, 2003), p. 55.

61. John P. Reeder, "Handling Qualities Experience with Several VTOL Research Aircraft," NASA TN-D-735 (1961); Seth B. Anderson, "Historical Overview of V/STOL Aircraft Technology," NASA TM-81280 (1981), pp. 8–9.

pitch control power, lack of a Stability Augmentation System (SAS), and low inherent damping of any pitch oscillations, researchers prudently undertook hover trials only in calm air. Among its positive qualities, good STOL performance was provided by slipstream-induced lift. Transition to wing-supported flight was satisfactory, with little pitch-trim change required. In transitions, as the wing pivoted down to normal flight position, hover controls were phased out. The normal aerodynamic controls were phased in, with the change from propeller to wing-supported flight being judged satisfactory. However, deceleration on descent was severely restricted by wing stall. When power was reduced, lateral-directional damping decreased to unsatisfactory levels. Changes were made to "droop" the leading edge 6 degrees to improve descent performance, and the modification improved behavior and controllability so greatly that Langley test pilot Jack Reeder concluded the "serious stall limitations in descent and level-flight deceleration were essentially eliminated from the range of practical flight operation, at least at incidence angles up to 50°."[62] In spite of this seemingly poor "report card," the awkward-looking VZ-2 contributed greatly to early understanding of the behavior and foibles of V/STOL tilt wing designs. All together, it completed 450 research sorties, including 34 full transitions from vertical to horizontal flight. The VZ-2 flight program proved to be one of the more productive American V/STOL programs, furnishing much information on wing-propeller aerodynamic interactions and basic V/STOL handling qualities.[63]

In addition to the pioneering VZ-2, the Hiller and Kaman companies also pursued the concept, the former for the Air Force and the latter for the Navy, though with significantly less success. Using an off-the-shelf development approach followed by many V/STOL programs, Hiller joined the fuselage and tail section of a Chase YC-122 assault transport to a tilt wing, creating the X-18, the first transport-sized tilt wing testbed. It used two Allison T40 turboprop engines driving three-bladed contra-rotating propellers, plus a Westinghouse J34 to furnish pitch control via a lengthy tailpipe. The sole X-18 made a conventional flight in November 1959 and completed a further 19 test sorties before being

62. John P. Reeder, "Handling Qualities Experience with Several VTOL Research Aircraft," NASA TN-D-735 (1961), p. 8.

63. Robert J. Pegg, "Summary of Flight-Test Results of the VZ-2 Tilt-Wing Aircraft," NASA TN-D-989 (1962); Pegg, "Flight-Test Investigation of Ailerons as a Source of Yaw Control on the VZ-2 Tilt-Wing Aircraft," NASA TN-D-1375 (1962).

grounded. Though it demonstrated wing tilt in flight to an angle of 33 degrees, it never completed a VTOL takeoff and transition. On November 4, 1960, a propeller malfunction led to it entering an inverted spin. Through superb airmanship, test pilots George Bright and Bruce Jones recovered the aircraft and landed safely, but it never flew again.[64] Kaman undertook a similar development program for the Navy, joining a tilt wing with two General Electric T58-GE-2A turboshaft engines to the fuselage and tail section of a Grumman JRF Goose amphibian, creating the K-16B. Tested in Ames's 40-foot by 80-foot tunnel, the K-16B never took to the air.[65]

Despite these failures, confidence in the tilt wing concept had advanced so rapidly that in February 1961, after 2 years of feasibility studies, the Department of Defense issued a joint-service development specification for an experimental VTOL transport that could possibly be developed into an operational military system. After evaluating proposals, the department selected the Vought-Hiller-Ryan Model VHR 447, ordering it into development under the Tri-Service Assault Transport Program as the XC-142A.[66] All three of these companies had previously employed variable position wings, with the F-8 Crusader fighter, the X-18, and the VZ-3RY, though only the last two were V/STOL designs. The XC-142A was powered by four General Electric T64 turboshaft engines, each rated at 3,080 horsepower, driving four-bladed Hamilton Standard propellers, with the propellers cross-linked by drive shafts to prevent a possibly disastrous loss of control during VTOL transitions. The combination of great power and light weight ensured not only that it could take off and land vertically, but also that it would have a high top-end speed of over 400 mph. Piloted by Stuart Madison, the first of five XC-142As completed a conventional takeoff in late September 1964, made its first hover at the end of December 1964, and accomplished its first transition from vertical to horizontal flight January 11, 1965, "with no surprises."[67]

The five XC-142A test aircraft underwent extensive joint-service evaluation, moving a variety of vehicles and troops, undertaking

64. Jay Miller, *The X-Planes: X-1 to X-45* (Hinckley, England: Midland Publishing, 2001 ed.), pp. 219–222.

65. Swanborough, *Vertical Flight Aircraft*, p. 50.

66. More popularly known as the Ling-Temco-Vought (LTV) XC-142A after the merger of Vought with these other partners.

67. Stuart G. Madison, "First Twelve Months of Flight of the XC-142A Program," *Technical Review* (of the Society of Experimental Test Pilots), vol. 7, no. 4 (1965), p. 20.

14

NASA's XC-142A undergoes hover trials at Langley in January 1969. NASA.

simulated recovery of downed aircrew via a recovery sling, landing aboard an aircraft carrier, and even flying a demonstration at the 1967 Paris Air Show. With a payload of 8,000 pounds and a gross weight of 37,500 pounds, the XC-142A had a thrust-to-weight ratio of 1.05 to 1. In STOL mode, with the wing set at 35 degrees and with flaps set at 30 degrees, the XC-142A could almost double this payload yet still clear a 50-foot obstacle after a 200-foot takeoff run.[68] Unfortunately, program costs rose from an estimated $66 million at inception to $115 million (in FY 1963 dollars), resulting in overruns that eventually truncated the aircraft's development.[69] The five aircraft experienced a number of mishaps, most related to shafting and propulsion problems. Sadly, one accident resulted in the death of test pilot Madison and a Ling-Temco-

68. Recollection of USAF XC-142A project test pilot Jesse Jacobs to Richard P. Hallion in an e-mail on Dec. 19, 2009.

69. USAF Scientific Advisory Board, "Report of the USAF Scientific Advisory Board Aerospace Vehicles Panel," Feb. 1968, p. 4, copy in the office files of the SAB, Headquarters USAF, Pentagon, Washington, DC. The amount of $11 million in 1961 is approximately $355 million in 2009; $115 million in 1963 is approximately $813 million in 2009.

Vought (LTV) test crew in May 1967, after a loss of tail rotor pitch control from fatigue failure of a critical part during a hover at low altitude.[70]

NASA Langley took ownership of the fourth XC-142A in October 1968, subsequently flying it until May 1970. The lead pilot was Bob Champine. When these tests concluded, the program came to an end. The Air Force Scientific Advisory Board's Aerospace Vehicle Panel concluded that, "The original premise that the propeller-tilt wing was well within the state of the art and that it was possible to go directly to operational prototypes was essentially a correct one," and that the tilt wing "has remarkable STOL capabilities that should be exploited to the maximum." Indeed, "One of the major advantages of the propeller-tilt wing is the fact that it is a magnificent STOL," but the panel also acknowledged that, on the XC-142A program, "The technical surprises were few, but important."[71]

The results of combined contractor, military, and NASA testing indicated that, as Seth Anderson noted subsequently, despite the XC-142A's clear promise:

Some mechanical control characteristics were unsatisfactory:

> (1) directional friction and breakout forces varied with wing tilt angle,
> (2) non-linear control gearing,
> (3) possibility of control surface hard-over, and
> (4) collective control had to be disengaged manually from the throttles in transition.

Hover handling qualities were good with SAS on, with no adverse flow upsets, resulting in precise spot positioning. Propeller thrust in hover was 12% less than predicted. No adverse lateral-directional characteristics were noted in sideward flight up to 25 knots. In slow forward flight, a long-period (20 sec) oscillation was apparent which could lead to an uncontrollable pitch-up. On one occasion full forward stick did not arrest the pitch-up, whereupon the pilot reduced engine power, the nose fell through, and the aircraft was extensively damaged in a hard landing because the pilot did not add sufficient power to arrest the high sink rate for fear of starting another pitch-up.

70. Ibid., p. 4.

71. USAF SAB, "Report of the USAF Scientific Advisory Board Aerospace Vehicles Panel," pp. 3–4.

STOL performance was not as good as predicted and controllability compromised IGE by several factors:

(1) severe recirculation of the slipstream for wing tilt angles in the range 40° to 80° (speed range 30 to 60 knots) producing large amplitude lateral-directional upsets;
(2) weak positive, neutral, and negative static longitudinal stability; and
(3) low directional control power.

Transition corridor was satisfactory with ample acceleration and deceleration capabilities. Conventional flight performance was less than predicted (11% less) due to large boat-tail drag-cruise.

Stability and control was deficient in several areas:

(1) low to neutral pitch stability,
(2) nonlinear stick force per "g" gradient, and
(3) tendency for a pitch Pilot Induced Oscillation (PIO) during recovery from rolling maneuvers.[72]

A failure of the drive shaft to the tail pitch propeller in low-speed flight caused a fatal crash that essentially curtailed further development of this concept. The experience of Canadair with the CL-84 Dynavert, a twin-engine tilt wing powered by two Lycoming T53 turboshafts, was in many respects similar to that of the XC-142A. In October 1966, NASA Langley pilots Jack Reeder and Bob Champine had evaluated the CL-84 at the manufacturer's plant, finding that, "The flying qualities were considered generally good except for a slow arrest of rate of descent at constant power and airspeed that could be of particular significance during instrument flight."[73] For a while after the conclusion of the XC-142A

72. Quoted from Seth B. Anderson, "Historical Overview of V/STOL Aircraft Technology," NASA TM-81280 (1981), pp. 9-10–9-11; W.P. Nelms and S.B. Anderson, "V/STOL Concepts in the United States—Past, Present, and Future," NASA TM-85938 (1984).

73. Henry L. Kelley, John P. Reeder, and Robert A. Champine, "Summary of a Flight-Test Evaluation of the CL-84 Tilt-Wing V/STOL Aircraft," NASA TM-X-1914 (1970). For the CL-84, see W.S. Longhurst, "Initial Development and Testing of Tilt-Wing V/STOL," *Technical Review* (of the Society of Experimental Test Pilots), vol. 7, no. 4 (1965), pp. 34–41; Frederick C. Phillips, "The Canadair CL-84 Tilt-Wing V/STOL Programme," *The Aeronautical Journal of the Royal Aeronautical Society*, vol. 73, no. 704 (Aug. 1969); Frederick C. Phillips, "Lessons Learned: The Development of the Canadair CL-84 Dynavert Experimental V/STOL Research Aircraft," *Canadian Aviation Historical Society Journal*, vol. 30, no. 3 (fall 1992).

program, the U.S. Navy sponsored further tilt rotor research with the Canadair CL-84, in trials at sea and at the Naval Air Test Center, Patuxent River, MD, looking at combat search and rescue and fleet logistical support missions. Undoubtedly, it was a creative design of great promise and clear potential, marred by a series of mishaps, though fortunately without loss of life. But after 1974, when the CL-84 joined the XC-142 in retirement, whatever merits the tilt wing might have possessed for piloted aircraft were set aside in favor of other technical approaches.

NACA–NASA and Ducted-Fan V/STOL Research Programs

One of the more intriguing forms of aircraft propulsion is the ducted fan: the fan enclosed within a ring and powered by a drive train from an engine typically located elsewhere in the aircraft. Researchers interested in V/STOL flight expended great effort on ducted-fan approaches, and indeed, such an approach is incorporated on the Joint Strike Fighter, the Lockheed Martin F-35 Lightning II. Though ducted-fan propulsion for Conventional Take-Off and Landing aircraft had enjoyed at best a mixed record, those advocating it for V/STOL applications were hopeful it would prove more successful. Ducted-fan options included pivoting fans that could furnish direct lift, like tilt rotors, then pivot to furnish power for wing-borne forward flight, or rely on horizontal fans in a wing or fuselage to generate vertical lift, or combinations of these.

Ducted-fan aircraft intended for conventional flight were tried in many nations. Likewise, ducted-fan V/STOL adherents in various countries had proposed concepts for such craft. But the first two American ducted-fan V/STOL airplanes—that is, ducted-fan aircraft with wings, as opposed to various Hiller and Piasecki flying platforms—were the Doak Model 16, designated the VZ-4, and the Vanguard Omniplane. Each represented a different approach, though, of the two, only the Doak flew.[74]

74. Developed as a private venture by the Vanguard Air and Marine Corporation, the Omniplane had a ducted pusher propeller for forward flight, with two lift fans built into its wings. A Lycoming 0-540 520-horsepower piston engine was geared to run both the fans for vertical lift and the pusher propeller to enable transition and forward flight. The Omniplane was completed in 1959. Tests in the Ames 40-foot by 80-foot tunnel unveiled serious deficiencies, with the Omniplane subsequently receiving a more powerful Lycoming YT53 turboshaft engine and a nose lift fan for pitch control. Ground trials followed in 1961, but a fan driveshaft failure so damaged the vehicle that the program was subsequently abandoned. The technical heir to this approach to ducted lift-fan/pitch-control-fan V/STOL, the Ryan XV-5, is discussed subsequently. See Swanborough, *Vertical Flight Aircraft*, p. 105.

The Doak VZ-4 in flight, showing hover, transition, and cruise. U.S. Army.

The Doak began as an Army research project, first flying in February 1956. A pleasing and imaginative design of conventional straight wing aerodynamic layout, it had a single 860-horsepower Lycoming YT53 turboshaft engine driving two pivoted ducted fans on the wingtips. During hover, variable inlet guide vanes in the ducts furnished roll control, with pitch and yaw control provided by vectored jet exhaust from a variable tail nozzle. The Doak proved that its design approach worked, readily making vertical descents, transitions to conventional flight, and transition back to hover and landing, accelerating in just 17 seconds from 0 to 200 knots. It arrived at the NACA's Langley laboratory in September 1957. Testing revealing a mix of undesirable handling qualities across its flight envelope, with one subsequent NASA study concluding that it

> suffered from low inherent control power about all axes, sensitivity to ground-effect disturbances, large side forces associated with the large ducts, and a large (positive) dihedral effect which restricted operation to calm-air conditions and no crosswinds. No large STOL performance benefit was evident with this design.[75]

During vertical descents, it buffeted with alternate left-and-right wing-dropping. This became so severe as the plane approached stall angle that "roll control was not adequate to keep the aircraft upright," noted pilot Jack Reeder. Large nose-up pitching moments required careful speed and duct-angle management to prevent duct-lip airflow stalling. In hover, the inlet guide vanes were "very inadequate." In ground effect, Reeder noted:

> if lifted clear of the ground by several feet, uncontrollable yawing and persistent lateral upsetting tendencies have been encountered. With the weak yaw control and, particularly the weak roll control, the unindoctrinated pilot may find himself unable to control the aircraft.[76]

The Doak Company closed in 1960, and NASA retired the airplane in 1972. By that time, a ducted-fan successor, the Bell Aerospace Textron X-22A, was already flying. Far more unconventional in appearance than was the Doak, it nevertheless owed a technical debt to the earlier design. As NASA researchers had concluded from the VZ-4's testing, the little Doak had "indicated the feasibility as well as the inherent problems of the tilt-duct concept, which helped the X-22 design which followed."[77]

75. Nelms and Anderson, "V/STOL Concepts in the United States," p. 3.

76. Reeder, "Handling Qualities Experience with Several VTOL Research Aircraft," pp. 2, 4, 6, 8; Anderson, "Historical Overview of V/STOL Aircraft Technology," p. 9-8.

77. Nelms and Anderson, "V/STOL Concepts in the United States," p. 3.

Bell Aerospace Textron's X-22A grew out of company studies in the mid-1950s. Successive examination of Bell-proposed military concepts led to increasing service interest by the Navy, Air Force, Marine Corps, and Army for a range of transport, rescue, and counterinsurgency applications. In late 1962, the U.S. Navy signed a development contract with Bell for a half-scale flying testbed of one of the company's proposed designs. This became the X-22A, two of which were built. Because the loss of any one fan would spell disaster, the X-22A used four General Electric T58 turboshaft engines, interconnected to the ducted fans so that an engine failure would not result in a loss fan power. NASA supported Bell's development with extensive wind tunnel studies at Ames and Langley. Control was exercised by changing the pitch of each of the four propellers and by moving the four elevons. These eight variables were used to control the X-22 in normal flight with ducts horizontal and in hover with ducts vertical. In horizontal flight, pitch and roll were controlled by the elevons and yaw by differential variation of propeller pitch. For hovering flight, propeller pitch adjustments controlled pitch and roll, while elevon movements controlled yaw. During transition, control functions were phased in gradually as a function of duct tilt angle. The pilot was provided with artificial "feel" in yaw during forward flight, but this was removed during transition to hover. Pitch and roll "feel" were provided by a hydroelectric system that applied stick reactions proportional to g-forces. For hover, transition, and low-speed flight, a Stability Augmentation System was used to improve aircraft stability and handling characteristics. The X-22A was equipped with a sophisticated Variable Stability and Control System (VSCS) developed by the Calspan Corp. This allowed it to be programmed to behave like other existing or projected VTOL aircraft for assessment of flight characteristics. The VSCS interacted with the Smiths Industries head-up display (HUD) and the Kaiser Electronics head-down display (HDD). Data inputs to the VSCS included those from a low-speed airspeed sensor, the Linear Omnidirectional Resolving Airspeed System (LORAS), invented by Calspan's Jack Beilman.[78]

The first X-22A flew March 17, 1966, but a hydraulic system failure led to the loss of the aircraft in August of that year during an emergency vertical landing, fortunately without injury to the crew. The second X-22A became one of the more successful research aircraft ever flown,

78. Technical data on the X-22A and its development are well-covered in Miller, *The X-Planes*, pp. 249–255.

The X-22A at altitude. NASA.

completing over 500 flights and over 1,300 transitions, between commencement of its flight-test program at the end of January 1967 through its retirement from flight-testing 17 years later. It hovered at over 8,000 feet altitude and achieved a forward speed of 315 mph, proving conclusively that a tilt duct vehicle could fly faster than could a conventional helicopter. In May 1969, it was turned over to the Navy, which appointed Calspan to continue the flight-test program and fly it as a variable stability research and training aircraft. Eventually, three NASA test pilots flew it, two of whom were formerly at Calspan Corporation—Rogers Smith and G. Warren Hall—and Ron Gerdes from Ames. Assessing the X-22A's place in V/STOL history, NASA researchers concluded:

> Hover operation Out of Ground Effect (OGE) in no wind was rated excellent, with no perceptible hot-gas ingestion. A 12% positive thrust increase was generated In Ground Effect (IGE) by the favorable fountain. Airframe shaking and buffeting occurred at wheel heights up to about 15 ft, and cross-wind effects were quite noticeable because of large side forces generated by the ducts. Vertical cross-wind landings required an excessive bank angle to avoid lateral drift. STOL performance was rated good by virtue of the increased duct-lifting forces. Highspeed performance was limited by inherent high drag associated with the four large ducts. Transition to conventional flight could be made easily because of a wide transition

corridor; however, inherent damping was low. Deceleration and descent at low engine powers caused undesirable duct "buzz" as a result of flow separation on the lower duct lips. Vortex generators appreciably improved this flow-separation problem.[79]

The X-22A proved to be a successful and versatile research tool, flying for at least 17 years and providing much valuable information on ducted VTOL systems and the larger operational issues of VTOL and STOL aircraft.

NASA expended a great deal of study effort examining the benefits of lift-fan technology and various design approaches that might be taken in design of a practical military and civil lift-fan aircraft. In the course of these trials, involving model tests, tests of candidate fan technologies, and examinations of lift-fan aircraft (such as the ill-fated Vanguard Omniplane), researchers studied an experimental Army-Ryan program, the XV-5A Vertifan. The XV-5A was an ill-fated program, like its contemporary, the Army-Lockheed XV-4 Hummingbird, which is discussed subsequently. Between them, the aircraft built of these two types killed three test pilots and nearly a fourth. Powered by two General Electric J85 engines driving in-wing fans and a nose pitch-control fan (like the Vanguard Omniplane) and used for conventional propulsion, the first of two XV-5As flew in 1964 but crashed during a public demonstration at Edwards AFB in August 1965, killing Ryan test pilot Lou Everett. The second fared little better, crashing in October 1966 at Edwards after one lift fan ingested a rescue hoist deployed from the aircraft, causing an asymmetric loss of lift. Air Force test pilot Maj. David Tittle perished while ejecting from the ailing aircraft, which, in sad irony, impacted with surprisingly little damage. Rebuilt as the XV-5B with some changes to its avionics, cockpit layout, ejection seat, and landing gear, it flew again in 1968, flying afterward at Ames Research Center on a variety of NASA investigations led by David Hickey, until its retirement in 1974.[80]

79. Nelms and Anderson, "V/STOL Concepts in the United States," p. 4.

80. Daniel J. March, "VTOL Flat-Risers: Lockheed XV-4 and Ryan/GE XV-5," *International Air Power Review*, vol. 16 (2005), pp. 118–127. See, for example, Adolph Atencio, Jr., Jerry V. Kirk, Paul T. Soderman, and Leo P. Hall, "Comparison of Flight and Wind tunnel Measurements of Jet Noise for the XV-5B Aircraft," NASA TM-X-62182 (1972); D.L. Stimpert, "Effect of Crossflow Velocity on VTOL Lift Fan Blade Passing Frequency Noise Generation," NASA CR-114566 (1973); Ronald M. Gerdes and Charles S. Hynes, "Factors Affecting Handling Qualities of a Lift-Fan Aircraft during Steep Terminal Area Approaches," NASA TM-X-62424 (1975).

The XV-5B in a hover test at Ames Research Center. NASA.

Among these studies were tests in and out of ground effect of various wing and inlet configurations, exit-vane designs, nose fans, and control devices. The research studies focused on problems of transition from vertical to horizontal flight, and on improvements of the lift fans to provide quieter, smaller fans with greater thrust. These studies were funded in part by the U.S. Army Aeronautical Research Laboratory, reflecting the Army's interest in V/STOL aircraft technology and maturation. Ames researchers found that the XV-5B could take off and land vertically from an area the size of a tennis court; hover in midair for several minutes like a helicopter; and fly straight up, down, backward, or to either side at speeds up to 25 mph. As well, it could operate like a conventional jet airplane using a runway, flying up to 525 mph. However, though a NASA summary report on V/STOL concepts concluded, "The lift-fan concept proved to be relatively free of mechanical problems," tests revealed that the XV-4B was still far from a practical vehicle. Hot-gas ingestion degraded engine performance while in ground effect, drag from the fan installations limited STOL performance, combinations of fan overspeed and a nose-up tendency complicated conversions, and the design layout hinted at a potential deep-stall problem characteristic of many T-tail aircraft. NASA concluded: "This configuration has limited high-speed potential because of the relatively thick wing section needed to house the lift fans and vectoring hardware."[81]

As part of an Advanced Short Take-Off and Vertical Landing (ASTOVL) study program that began in 1980, NASA continued detailed studies of ducted lift fans, among other propulsion concepts intended for a supersonic successor to the vectored-thrust AV-8B Harrier II. The outcome of that development effort was validated in the successful flight-testing of a lift fan on the STOVL variant of the Lockheed Martin X-35, the experimental proof-of-concept demonstrator for the F-35 Joint Strike Fighter a quarter century later.[82] In this regard, lessons learned from NASA's various lift-fan programs, particularly the XV-5A and XV-5B, compiled by Ames test pilot and distinguished V/STOL researcher Ronald M. Gerdes, are included as an appendix to this study.

Proving the Tilt Rotor: From XV-3 and X-100 to XV-15 and on to V-22

One V/STOL concept that proved to have enduring appeal was the tilt rotor, which entered production and operational service with the joint-service Bell-Boeing V-22 Osprey. The tilt rotor functioned like a twin-rotor helicopter during lift-off, hover, and landing. But for cruising flight, it tilted forward to operate as high aspect ratio propellers. Such a concept meant that the tilt rotor would necessarily have its rotors pod-mounted on the tips of conventional wings.[83]

Though various designers across the globe envisioned tilt rotor convertiplanes, the first successful one was the Bell Model 200, produced by Bell Helicopter for the Air Force and Army as the XV-3 under a joint Army–USAF "convertiplane" program started in August 1950. Relatively

81. Nelms and Anderson, "V/STOL Concepts in the United States," p. 4.

82. Paul M. Bevilaqua, "Genesis of the F-35 Joint Strike Fighter," *Journal of Aircraft*, vol. 46, no. 6 (Nov.–Dec. 2009), pp. 1826, 1832. Tim Naumowicz, Richard Margason, Doug Wardwell, Craig Hange, and Tom Arledge, "Comparison of Aero/Propulsion Transition Characteristics for a Joint Strike Fighter Configuration," paper presented at the *International Powered Lift Conference, West Palm Beach, FL, Nov. 18–21, 1996*; David W. Lewis, "Lift Fan Nozzle for Joint Strike Fighter Tested in NASA Lewis' Powered Lift Rig," in *NASA Research and Technology 1997* (Apr. 1998).

83. Borchers, Franklin, and Fletcher, *Flight Research at Ames*, p. 58.

The Bell XV-3 tilt rotor shown after transitioning to conventional flight during NASA testing in April 1961. NASA.

streamlined and looking more like an airplane than did many early V/STOL testbeds, the XV-3 had an empty weight of 3,600 pounds and a normal gross weight of just 4,800 pounds, as it was relatively underpowered. A single Pratt & Whitney R985 radial piston engine producing 450 horsepower drove two three-bladed rotors via drive shafts. With this propulsion system, the XV-3 completed its first hover in August 1955, piloted by Floyd Carlson.[84]

Flight-testing over the next year demonstrated flight at progressive levels of rotor tilt, though it had not made a full 90-degree conversion of its rotors to level position before it crashed while landing in October 1956 from a rotor instability. Bell test pilot Dick Stansbury survived but was seriously injured. Afterward, the second XV-3 was equipped with stiffer two-bladed rotors. On December 18, 1958, Bell test pilot Bill Quinlan achieved a full conversion from a helicopter-like ascent to forward flight like an airplane. During the XV-3 flight-test program, the lack of engine power prevented it from hovering out of ground effect. When

84. Swanborough, *Vertical Flight Aircraft*, p. 15; Stu Fitrell and Hank Caruso, eds., *X-traordinary Planes, X-traordinary Pilots: Historic Adventures in Flight Testing* (Lancaster: Society of Experimental Test Pilots Foundation, 2008), p. 48.

it did hover in ground effect, reflected rotor wash caused unpredictable darting, something the tilt rotor V-22 experienced four decades later during its testing. The lack of an SAS further exacerbated pilot hover challenges, and in gusty air, high pilot workload was required to hover. The XV3 transited rapidly from hover to conventional flight, requiring only small pitch changes across the range of speed and angle of attack encompassed by the transition corridor. Pitch and yaw dynamic instability triggered by side forces as blade angle was increased limited maximum cruise speed to 140 knots and pointed to rotor dynamics and flight control challenges that future tilt rotors would have to overcome. In all of this research, Bell blended extensive analytical studies and scale model experiments with tests in the Ames 40-foot by 80-foot wind tunnel.[85]

In May 1959, the surviving XV-3 was delivered to the Air Force Flight Test Center at Edwards Air Force Base (AFB), where it underwent a 3-month Air Force evaluation before being delivered for more extensive testing and research to the Ames Research Center. During Edwards's testing, Maj. Robert Ferry successfully demonstrated a power-off reconversion to a vertical autorotation descent and landing, an important milestone. At Ames, Hervey Quigley carried out the research, and Don Heinle and Fred Drinkwater conducted most of the test flying, in the course of which the XV-3 was modified with a large ventral fin to improve its directional stability. In the Ames tests, flapping of the teetering rotors during maneuvers introduced moments that reduced damping of the longitudinal and lateral-directional oscillations to near zero at speeds approaching 140 knots. Despite these problems and despite being underpowered and limited in payload, the XV-3 proved the capability of the tilt rotor to perform in-flight conversions between the helicopter and the airplane modes, though much work on understanding rotor dynamics and flight control issues needed to be done. The XV-3 flew at Ames until summer 1962, when it began an extensive series of wind tunnel studies in the 40-foot by 80-foot tunnel. In November 1968, during 200-mph tunnel trials, fatigue failure in one wingtip led to separation of the rotors and their pylons from the aircraft, bringing its 13-year test career, at last, to an end. By that time, it had validated the tilt rotor concept, thus influencing—as discussed subsequently—the next step forward in experimental tilt rotor design, the XV-15. That vehicle,

85. Anderson, "Historical Overview of V/STOL Aircraft Technology," p. 9-7; Fitrell and Hank Caruso, eds., *X-traordinary Planes, X-traordinary Pilots*, p. 48.

of course, would exert an even greater influence upon development of its operational successor, the V-22 Osprey.[86]

Before settlement on the tilt rotor as exemplified by Bell's design approach with the XV-3, researchers considered another seemingly closely related concept: the tilt prop. However, the tilt prop idea was different. Researchers had long known that rotating propellers generate a powerful side force, and Curtiss-Wright Corporation engineers envisioned taking advantage of this property by using smaller diameter and lower aspect ratio propellers than tilt rotors that could use this "radial lift force" as a means of lifting a V/STOL airplane vertically. Such a design, they hoped, would have higher top-end speed after conversion than an XV-3-like tilt rotor approach.[87]

The result was the X-100, a small testbed whose twin broad-chord propellers were driven by a single Lycoming YT53-L-1 turboshaft engine. Its jet exhaust vented through an omnidirectional tail nozzle, furnishing low-speed pitch and yaw control. Differential propeller operation furnished roll control during hover. The X-100 underwent testing in Ames 40-foot by 80-foot tunnel and extensive ground trials before making its first flight in March 1960. In August, it underwent a NASA flight evaluation, after which it went to Langley Research Center for further testing, including downwash effects on various kinds of ground surfaces.[88] Langley pilot Jack Reeder found it longitudinally unstable during conversions, something "very undesirable during landing approaches, particularly under instrument conditions." During hover it demonstrated "erratic wing dropping and yawing," necessitating "noticeably large" corrective control inputs to correct, and "weak" yaw control that prevented holding a desired heading. It "settled rapidly toward the ground when upset in bank or pitch attitude" while in ground effect, again, something he found "very undesirable." On the positive side, he found that "The X-100 aircraft suffers no apparent stall problems."[89]

86. Anderson, "Historical Overview of V/STOL Aircraft Technology," p. 9-7.

87. Ibid., p. 9-8; for earlier propeller side force studies, see Herbert S. Ribner, "Formulas for Propellers in Yaw and Charts of the Side-Force Derivative," NACA WR-L-217 (originally ARR-3E19) (1943).

88. Anderson, "Historical Overview of V/STOL Aircraft Technology," p. 9-8; Miller, *The X-Planes*, pp. 225–226; Swanborough, *Vertical Flight Aircraft*, p. 29.

89. Quotes from John P. Reeder, "Handling Qualities Experience with Several VTOL Research Aircraft," NASA TN-D-735 (1961), pp. 4–5, 7, 11.

The Curtiss-Wright X-100 undergoing ground-testing. NASA.

In October 1961, the X-100 was seriously damaged in a hovering accident that, fortunately, did not result in injury to its pilot. Despite its mediocre performance, it had demonstrated the feasibility of the radial-lift propeller concept. Thus, Curtiss-Wright continued pursuing the tilt prop approach but now chose to make a four-propeller craft with equal span fore and aft wings, rather than an X-100-like twin-rotor design. The company subsequently received an Air Force developmental contract for this larger and more powerful design, which became the experimental X-19. Of the two that were built, only the first flew, and it had a brief and troubled flight-test program before crashing in August 1965 at the FAA's National Aviation Facilities Experimental Center (NAFEC) after experiencing a catastrophic gearbox failure. Fortunately, its crew ejected from the now-propless testbed before it plunged to Earth. At the company's request, the X-19 program was terminated the following December. The accident, one NASA authority concluded, "exemplified an inherent deficiency of this VTOL (lift) arrangement: to safely transmit power to the extremities of the planform, very strong (and fatigue-resistant) structures must be incorporated with an obvious weight penalty."[90] The future belonged to the tilt rotor, not tilt prop.

Though tests with the XV-3 had identified numerous challenges in stability and control, handling qualities, and the dynamics of the combined wing-pylon-rotor interactions, the program encouraged tilt rotor proponents to continue their studies. So promising did the tilt rotor appear that the Army and NASA formed a joint project office at Ames to study tilt rotor technology and undertook a number of simulations of such systems to refine project goals and efficiencies.[91] In 1971, Dr. Leonard Roberts of Ames's Aeronautics and Flight Mechanics Directorate established a V/STOL Projects Office under Woody Cook to develop and flight-test new V/STOL aircraft. That same year, in partnership with the Army, NASA launched a competitive development program for the design and fabrication of two tilt rotor research aircraft. Four companies responded, and Boeing and Bell received study contracts in October 1972. After evaluating each proposal, NASA selected Bell's Model D301 for development, issuing Bell a contract at the end of July 1973.[92]

As developed, the XV-15 was an elegant and streamlined technology demonstrator, a two-pilot testbed powered by twin Lycoming T53 turboshafts rated at 1,550 horsepower each, driving 25-foot-diameter three-bladed prop rotors. Bell completed the first XV-15 in October 1976 and, after ground tie-down testing, undertook its first preliminary hover

90. Nelms and Anderson, "V/STOL Concepts in the United States," p. 4; Miller, *The X-Planes*, pp. 227–228.

91. Martin D. Maisel and Lt. Col. Clifford M. McKeithan, "Tilt-Rotor Aircraft," *Army Research, Development & Acquisition Magazine* (May–June 1980), pp. 1–2; Daniel C. Dugan, Ronald G. Erhart, and Laurel G. Schroers, "The XV-15 Tilt Rotor Research Aircraft," NASA TM-81244, AVRADCOM TR-80-A-15 (1980).

92. Few, "A Perspective on 15 Years of Proof-of-Concept Aircraft Development and Flight Research at Ames–Moffett by the Rotorcraft and Powered-Lift Flight Projects Division, 1970–1985," NASA RP-1187 (1987), p. 9. For XV-15 background and development, see Martin D. Maisel, Demo J. Giulianetti, and Daniel C. Dugan, The History of the XV-15 Tilt Rotor Research Aircraft From Concept to Flight, NASA SP-2000-4517, No. 17 in the *Monographs in Aerospace History* series (Washington, DC: GPO, 2000), pp. 26–41. As well, Bell had won a NASA design study contract (NASA V/STOL Tilt-Rotor Aircraft Study, Contract NAS2-6599) for a "representative military and/or commercial tilt-proprotor aircraft." The company designated this the Model D302. It could carry 40 passengers 400 nautical miles at 348 knots and 30,000 feet, or 48 troops or 5 tons of cargo over a 500-nautical-mile radius at up to 370 knots. See Bell Helicopter Co., "V/STOL Tilt-Rotor Study Task I: Conceptual Design (NASA Contract NAS2-6599)," vol. 1, Bell Report 300-099-005, NASA CR-114441 (1973), p. I-1.

14

NASA 703, the second of the elegant XV-15 tilt rotors, in hover during NASA testing at Ames Research Center. NASA.

trials in May 1977, piloted by Ron Erhart and Dorman Cannon. In May 1978, before flight envelope expansion, it went into the Ames 40-foot by 80-foot wind tunnel for extensive stability, performance, and loads tests. Ames's aeronautical facilities greatly influenced the XV-15's development, particularly simulations of anticipated behavior and operational nuances, and tilt rotor performance and dynamic tests in the wind tunnel.[93]

The second XV-15 went to Dryden for contractor flight tests, conducted between April 1979 and July 1980, and was delivered to NASA for research in August.[94] By that time, NASA, Army, and contractor researchers had already concluded:

> The XV-15 tilt rotor has exhibited excellent handling qualities in all modes of flight. In the helicopter mode it is a stable platform that allows precision hover and agility with low pilot workload. Vibration levels are low as are both internal and external noise levels. The conversion procedure is uncomplicated by schedules, and it is easy to perform. During the

93. Borchers, Franklin, and Fletcher, *Flight Research at Ames*, p. 59.

94. Dugan, Erhart, and Schroers, "The XV-15," p. 2; Martin D. Maisel and D.J. Harris, "Hover Tests of the XV-15 Tilt Rotor Research Aircraft," AIAA Paper 81-2501 (1981).

> conversion or reconversion, acceleration or deceleration are impressive and make it difficult for conventional helicopters or airplanes to stay with the XV-15. Handling qualities are excellent within the airplane mode envelope investigated to date; however, gust response is unusual. Although internal noise levels are up somewhat in the airplane mode, external noise levels are very low. Overall the XV-15 is a versatile and unique aircraft which is demonstrating technology that has the potential for widespread civil and military application.[95]

Such belief in the aircraft led to its participation in the 1981 Paris Air Show, the first time NASA had demonstrated one of its research vehicles in an international venue. It was an important vote of confidence in tilt rotor technology, made more evident still by the XV-15's stopover at the Royal Aircraft Establishment at Farnborough, where it demonstrated its capabilities before British aeronautical authorities. In 1995, 14 years after its first Paris appearance, the XV-15 would again fly at Le Bourget, this time in company with its successor, the Bell-Boeing V-22 Osprey.[96]

Over its two-decade test program, the XV-15 was not immune to various mishaps, though fortunately, no one was seriously injured. Both aircraft experienced various emergencies, including forced landings after engine failures, a close call from a bird strike that cracked a wing spar, a tree strike, near-structural failure caused by an unsuitable form of titanium alloy fortuitously discovered before it could do harm, intergranular corrosion that caused potentially dangerous hairline blade cracks, and even one major accident. In August 1991, the first XV-15 crashed while landing after an improperly secured nut separated from a linkage controlling one of the prop rotors. Pilots Ron Erhart and Guy Dabadie were not seriously injured, though the accident destroyed the aircraft.[97]

In retrospect, the XV-15 was the most influential demonstrator aircraft program that Ames ever pursued. For a cost to taxpayers of $50.4 million, NASA and its partners significantly advanced the technology and capability of tilt rotor technology. In over two decades of flight operations, more than 300 guest pilots would fly in the XV-15. As well, it would operate from the New York Port Authority heliport, fly abroad,

95. Dugan, Erhart, and Schroers, "The XV-15," p. 9.
96. Maisel, Giulianetti, and Dugan, *History of the XV-15*, pp. 89–90, 101–102.
97. Ibid., pp. 98–99.

A Bell-Boeing CV-22 during 2009 testing at the Air Force Flight Test Center, Edwards Air Force Base. USAF.

and go to sea, demonstrating its ability to operate from amphibious assault ships. Among the many at Ames who contributed to making the program a success were NASA's Wally Deckert, Mark Kelly, and Demo Giulianetti, and the Army's Paul Yaggy, Dean Borgman, and Kipling Edenborough, who furnished critical guidance and oversight as the project was being established. Dave Few, Army Lt. Col. James Brown, and John Magee served as Program Managers. Principal investigators were Laurel Schroers, Gary Churchill, Marty Maisel, and Jim Weiberg. The project pilots were Daniel Dugan, Ronald Gerdes, George Tucker, Lt. Col. Grady Wilson, and Lt. Col. Rick Simmons. They shepherded the XV-15 through two decades of research on flying qualities and stability and control evaluations, control law development, side stick controller tests, performance evaluations in all flight modes, acoustics tests, flow surveys, and documentation of its loads, structural dynamics, and aeroelastic stability characteristics, generating a useful database that was digitized by Ames and made available to industry and military customers. In sum, the XV-15 did much to advance the V/STOL cause, particularly that of the tilt rotor concept.[98]

98. As is detailed in Maisel, Giulianetti, and Dugan, *History of the XV-15*, and more briefly treated in Borchers, Franklin, and Fletcher, *Flight Research at Ames*, pp. 59–61.

In particular, flight experience with the XV-15 contributed greatly to the development of the joint-service V-22 Osprey tilt rotor.[99] This Bell-Boeing aircraft, now in service with the U.S. Marines, the U.S. Navy, and the U.S. Air Force, fulfills a variety of roles, including combat assault, insertion and support of special operations forces (SOF), combat search and rescue (CSAR), and logistical support. Time will tell whether the V-22 Osprey will come to enjoy the longevity and ubiquity attendant to conventional joint-service fixed and rotary wing transports, such as the legendary Douglas C-47, Lockheed C130, Bell UH-1, and Sikorsky H-53.

NACA–NASA and Thrust Vectored Approaches: from X-14 to YAV-8B

The advent of the gas turbine engine at the end of the 1930s, and its demonstration and incorporation on aircraft in the 1940s, set the stage for a revolution in flight propulsion that affected nearly all powered flying vehicles by the mid-1950s. The pure-jet engine could power aircraft through the speed of sound and transport passengers across global distances. The turbopropeller engine applied to tactical transports, and the turboshaft engine applied to helicopters and V/STOL designs, gave them the power to weight ratios and reliability that earlier piston engines had lacked, enabling generations of far more efficient aircraft typified by the ubiquitous Lockheed C-130 Hercules and the Bell UH-1 "Huey" helicopter.

As well, the jet engine enabled designers to envision STOL and VTOL aircraft taking advantage of its power. Initially, many designers thought that a VTOL aircraft would need to have many small jet engines for vertical lift, coupled with one or more major powerplants for conventional flight. For example, the delta wing Short SC.1, a British low-speed VTOL testbed design that first flew in 1957, had five small jet engines: four to produce a stabilizing "bedpost" of vertical thrust vectors and a fifth to propel it through the air. Other such aircraft, for example, the Dassault Balzac and Dassault Mirage IIIV, followed a generally similar approach (though none, however, entered service).[100]

But other designers wisely rejected the complexity and inherent unreliability of such multiengine conglomerations. Instead, they envisioned a more efficient form of propulsion, vectoring the thrust of a jet engine so that the aircraft could lift vertically and then transition

99. Brenda Forman, "The V-22 Tiltrotor 'Osprey:' The Program That Wouldn't Die," *Vertiflite*, vol. 39, no. 6 (Nov.–Dec. 1993), pp. 20–23.

100. Swanborough, *Vertical Flight Aircraft*, has data on these.

into forward flight. This approach was pursued most successfully with the Hawker P.1127, forerunner of the Harrier fighter family.[101] Within the United States, Lockheed received an Army development contract for two research aircraft, the XV-4A, using a form of vectored thrust, whereby the exhaust of two jet engines buried in the wing roots would be deflected through a central fuselage chamber and mixed with air drawn through a fuselage intake, with this "augmented" exhaust enabling vertical flight. Optimistically named the Hummingbird and first flown in 1962, the XV-4A never enjoyed the success attendant to the P.1127. Most seriously, anticipated augmented flow efficiencies were not achieved, limiting performance. The first aircraft crashed during a VTOL conversion in 1964, killing its pilot. The second was modified with a retrograde propulsion system reminiscent of the SC-1, using four lift jets and two thruster jets, and was redesignated the XV-4B Hummingbird II. It also crashed in 1969, though its pilot ejected safely. In contrast, the P.1127 program went along relatively smoothly both in Britain and the United States. In the U.S., thanks to John Stack of Langley, it received strong technical endorsement, in part because the Agency was already following an important and evolving vectored-thrust study effort: the Bell X-14 program. America's story of vectored-thrust research thus begins not with Langley and Ames's exposure to the streamlined P.1127, but with quite another design: the X-14. Like the XV-15, the X-14 became one of the more successful research aircraft of all time, having flown almost a quarter century and contributing to the success of a variety of other programs.[102]

X-14: A Little Testbed That Could

On May 24, 1958, Bell test pilot David Howe completed a vertical takeoff followed by conventional flight, a transition, and a vertical landing during testing at Niagara Falls Airport, NY. His short foray was a milestone in aviation history, for the flight demonstrated the practicality of using vectored thrust for vertical flight. Howe took off straight up, hovered like a helicopter, flew away at about 160 mph, climbed to 1,000 feet, circled back, approached at about 95 mph, deflected the engine

101. H.C.M. Merewether, *Prelude to the Harrier: P.1127 Prototype Flight Testing and Kestrel Evaluation* (Beirut, Lebanon: HPM Publications, 1998).
102. Schneider, "The History of V/STOL Aircraft," pt. 2, p. 41; March, "VTOL Flat-Risers," pp. 118–127.

thrust (which caused the plane to slow to a hover a mere 10 feet off the ground), made a 180-degree turn, and then settled down, anticipating the behavior and capabilities of future operational aircraft like the British Aerospace Harrier and Soviet Yak-38 Forger.

The plane that he flew into history was the Bell X-14, a firmly subsonic accretion of various aircraft components that proved to have surprising value and utility. Before proceeding with this ungainly creature, company engineers had first built a VTOL testbed: the Bell Model 65 Air Test Vehicle (ATV). The ATV used a mix of components from a glider, a lightplane, and a helicopter, with two Fairchild J44 jet engines attached under its wing. Each engine could be pivoted from horizontal to vertical, and it had a stabilizing tail exhaust furnished by a French Turboméca Palouste compressor as well. Tests with the ATV convinced Bell of the possibility of a jet convertiplane, though not by using that particular approach, and the ATV never attempted a full conversion from VTOL to conventional flight. Accordingly, X-14 differed from all its predecessors because it used a cascade thrust diverter, essentially a venetian-blind–like vane system, to deflect the exhaust from the craft's two small British-built Armstrong-Siddeley Viper ASV 8 engines for vertical lift. Each engine produced 1,900 pounds of thrust. Since the aircraft gross weight was 3,100 pounds, the X-14 had a thrust to weight ratio of 1.226. Compressed-air reaction "controls" kept the craft in balance during takeoff, hovering, and landing, when its conventional aerodynamic control surfaces lacked effectiveness. To simplify construction, the X-14 had an open cockpit, no ejection seat, the wings of a Beech Bonanza, and fuselage and tail of a Beech T-34 Mentor trainer.[103]

14

Early testing revealed that, as completed, the aircraft had numerous deficiencies typical of a first-generation technological system. After Ames acquired the aircraft, its research team operated the X-14A with due caution. Not surprisingly, weight limitations precluded installation of an ejection seat or even a rollover protection bar. The twin engines imparted strong gyroscopic "coupling" forces, these being dramatically illustrated on one flight when the X-14's strong gyroscopic moments generated a severe pitch-up during a yaw, "which resulted in the aircraft

103. Ronald M. Gerdes, "The X-14: 24 Years of V/STOL Flight Testing," *Technical Review* (of the Society of Experimental Test Pilots), vol. 16, no. 2 (1981), p. 3, and p. 11, Table I; W.P. Nelms and S.B. Anderson, "V/STOL Concepts in the United States—Past, Present, and Future," NASA TM-85938 (1984), p. 2; Schneider, "The History of V/STOL Aircraft," pt. 1, p. 29.

performing a loop at zero forward speed."[104] The X-14's hover flight-test philosophy was rooted in an inviolate rule: hover either at 2,500 feet, or at 12–15 feet, but never in between. At the higher altitude, the pilot would have sufficient height to recover from a single engine failure or to bail out. At the lower altitude, he could complete an emergency landing.[105] Close to the ground, the aircraft lost approximately 10 percent of its lift from so-called aerodynamic suck-down while operating in ground effect. During hover operations, the jet engines ingested the hot exhaust gas, degrading their performance. As well, it possessed low control power about all axes, and the lack of an SAS resulted in marginal hover characteristics. Hover flights were often flown over the ramp or at the concrete VTOL area north of the hangar, and typical flights ran from 20 to 40 minutes and within an area close enough to allow for a comfortable glide back to the airfield. Extensive flight-testing investigated a range of flying qualities in hover. Those flights resulted in criteria for longitudinal, lateral, and directional control power, sensitivity, and damping.[106]

By 1960, Ames V/STOL expertise was well-known throughout the global aeronautical community. This led to interaction with aeronautical establishments in many countries pursuing their own V/STOL programs, via the North Atlantic Treaty Organization's (NATO) Advisory Group for Aeronautical Research and Development (AGARD).[107] For example, Dassault test pilot Jacques Pinier flew the X-14 before flying the Balzac. So, too, did Hawker test pilots Bill Bedford and Hugh Merewether before tackling the P.1127. Both arrived at Ames in April 1960 for familiarization sorties in the X-14 to gain experience in a "simple" vectored-thrust airplane before trying the more complex British jet in VTOL, then in final development. Unfortunately, on Merewether's sortie, the X-14 entered an uncontrolled sideslip, touching down hard and breaking up its landing gear, a crash attributed to low roll control power and no SAS. "Though bad for the ego," the British pilot wrote good-naturedly later, "it was probably a blessing in

104. Anderson, *Memoirs of an Aeronautical Engineer: Flight Testing at Ames Research Center, 1940–1970*, No. 26 in the *Monographs in Aerospace History* series (Washington, DC: GPO, 2002), p. 32.
105. Gerdes, "The X-14: 24 Years of V/STOL Flight Testing," p. 5.
106. Anderson, "Historical Overview of V/STOL Aircraft Technology," p. 9-7; Nelms and Anderson, "V/STOL Concepts in the United States—Past, Present, and Future," p. 2.
107. Borchers, Franklin, and Fletcher, *Flight Research at Ames*, p. 11.

The X-14A shown during a hover test flight at Ames Research Center. NASA.

disguise since it brought home to all and sundry the perils of weak reaction controls."[108]

Later that year, Ames technicians refitted the X-14 with more powerful 2,450-pound thrust General Electric J85-5 turbojet engines and modified its flight control system with a response-feedback analog computer controlling servo reaction control nozzles (in addition to its existing manually controlled ones), thus enabling it to undertake variable stability in-flight simulation studies. NASA redesignated the extensively modified craft as the X-14A Variable Stability and Control Research Aircraft (VSCRA). In this form, the little jet contributed greatly to understanding the special roll, pitch, and yaw control power needs of V/STOL vehicles, particularly during hovering in and out of ground effect and at low speeds, where conventional aerodynamic control surfaces lacked effectiveness.[109] It still had modest performance capabilities. Even though its engine power had increased significantly, so had its weight, to 3,970 pounds. Thus, the thrust to weight ratio of the X-14A was only

108. Recollection of Hugh Merewether in Merewether, *Prelude to the Harrier,* p. 13. Sadly, Pinier was later killed in a hovering accident with the Mirage Balzac in 1964.
109. Frank A. Pauli, Daniel M. Hegarty, and Thomas M. Walsh, "A System for Varying the Stability and Control of a Deflected-Jet Fixed-Wing VTOL Aircraft," NASA TN-D-2700 (1965).

marginally better than the X-14.[110] For one handling qualities study, researchers installed a movable exhaust vane to generate a side force so that the X-14A could undertake lateral translations, so they could study how larger V/STOL aircraft, of approximately 100,000 pounds gross weight, could be safely maneuvered at low speeds and altitudes. To this end, NASA established a maneuver course on the Ames ramp. The X-14A, fitted with wire-braced lightweight extension tubes with bright orange Styrofoam balls simulating the wingspan and wingtips of a much larger aircraft, was maneuvered by test pilots along this track in a series of flat turns and course reversals. The results confirmed that, for best low-speed flight control, V/STOL vehicles needed attitude-stabilization, and, as regards wingspan effects, "None of the test pilots could perceive any effect of the increased span, per se, on their tendency to bank during hovering maneuvers around the ramp or in their method of flying the airplane in general."[111]

Attitude control during hover and low-speed flight was normally accomplished in the X-14A through reaction control nozzles in the tail for pitch and yaw and on each wingtip for roll control. Engine compressor bleed air furnished the reaction control moments. For an experimental program in 1969, its wingtip reaction controls were replaced temporarily by two 12.8-inch-diameter lift fans, similar to those on the XV-5B fan-in-wing aircraft, to investigate their feasibility for VTOL roll control. Bleed air, normally supplied to the wingtip reaction control nozzles, drove the tip-turbine-driven fans. Fan thrust was controlled by varying the pressure ratio to the tip turbine and thereby controlling fan speed. Rolling moments were generated by accelerating the rpm of one fan and decelerating the other to maintain a constant net lift.[112]

A number of "lessons learned" were generated as a result of this handling qualities flight-test investigation, as noted by project pilot Ronald M. Gerdes. The fans were so simple, efficient, and reliable that the total bleed air requirement was reduced by about 20 percent from that required

110. Total thrust was 3,800 pounds for the X-14, 4,900 for the X-14A, and 5,500 for the X-14B. Gross weight was 3,100 pounds for the X-14, 3,970 for the X-14A, and 4,270 for the X-14B. Thus, thrust to weight ratio for the X-14 was 1.226, 1.234 for the later X-14A, and 1.288 for the eventual X-14B. Computed from Gerdes, "The X-14: 24 Years of V/STOL Flight Testing," p. 11, Table I.
111. Terrell W. Feistel, Ronald M. Gerdes, and Emmet B. Fry, "An Investigation of a Direct Side-Force Maneuvering System on a Deflected Jet VTOL Aircraft," NASA TN-D-5175 (1969), pp. 13–14.
112. L. Stewart Rolls and Ronald M. Gerdes, "Flight Evaluation of Tip-Turbine-Driven Fans in a Hovering VTOL Vehicle," NASA TN-D-5491 (1969).

using the tip nozzles. As a consequence, the jet engines produced about 4 percent more thrust and could operate at lower temperatures during vertical takeoffs. Despite this, however, during the flight tests, control system lag and increases in the aircraft moment of inertia caused by placement of the fans at the tips negated the increased roll performance that the fans had over the reaction control nozzles and resulted in the pilot having a constant tendency to overcontrol roll-attitude and thus induce oscillations during any maneuver. The wingtip lift-fan control system was thus rated unacceptable, even for emergency conditions, as it scored a Cooper Harper pilot rating of 6½ to 7½ (on a 1–10 scale, where 1 is best and 10 is worst). Finally, Gerdes concluded: "This test also demonstrated a principle that must be kept in mind when considering fans for controls. Even though the time response characteristics of a fan system are capable of improvement by such means as closing the loop with rpm feedback, full authority operation of the control eliminates the fan speed-up capabilities provided by the closed loop, and the fans revert to their open-loop time constants. In the case of the X-14A, its open- and closed-loop first-order time constraints were 0.58 and 0.34 seconds, respectively."[113]

The X-14A flew for two decades for NASA at the Ames Research Center, piloted by Fred Drinkwater and his colleagues on a variety of research investigations. These ranged from evaluating sophisticated electronic control systems to simulating the characteristics of a lunar lander in support of the Apollo effort. In 1965, it was configured to enable simulations of lunar landing approach profiles. The future first man on the Moon, Neil Armstrong, flew the X-14A to evaluate its control characteristics and a visual simulation of the vertical flightpath that the Apollo Lunar Module would fly during its final 1,500-foot descent from the Command Module (CM) to a landing upon the lunar surface.[114]

Another study effort examined soil erosion caused by VTOL operations off unprepared surfaces. In this case, a 5-second hover at 6 feet

113. Gerdes, "Lift-Fan Aircraft—Lessons Learned: the Pilot's Perspective," NASA CR-177620 (1993), p. 9.

114. At NASA's Flight Research Center, researchers built an experimental jet- and rocket-powered lunar landing simulator, the Lunar Landing Research Vehicle (LLRV), which led to a series of Lunar Landing Training Vehicles (LLTV). This constituted yet another example of NASA's creative work in the VTOL field, but, as it was specifically related to the space program, it is not treated further in this essay. See Gene J. Matranga, C. Wayne Ottinger, and Calvin R. Jarvis, with Christian Gelzer, *Unconventional, Contrary, and Ugly: The Lunar Landing Research Vehicle*, NASA SP-2204-4535, No. 35 in the *Monographs in Aerospace History* series (Washington, DC: GPO, 2006).

resulted in chunks of soil and grass being thrown into the air, where they were ingested by the engines, damaging their compressors and forcing subsequent replacement of both engines.[115]

In 1971, under the direction of Richard Greif and Terry Gossett, NASA modified the X-14A a third time, to install a digital variable stability system and up-rated GE J85-19 engines to improve its hover performance. It was redesignated as the X-14B and flown in a program "to establish criteria for pitch and roll attitude command concepts, which had become the control augmentation of choice for precision hover."[116] Unfortunately, in May 1981, a control software design flaw led to saturation of the VSCS autopilot roll control servos, a condition from which the pilot could not recover before it landed heavily. Although NASA contemplated repairing it, the X-14B never flew again.[117]

As a personal aside, having had the opportunity to fly the X-14B near its final flight, I was impressed with its simplicity.[118] For example, one of the more important instruments on the airplane was a 4-inch piece of yarn attached to a small post in the center of the front windshield bow. You never wanted to see the yarn pointed to the front of the airplane. If you did, it meant you were flying backward, and that was a real no-no! The elevator had a nasty tendency to dig in and flip the aircraft over on its back. We aptly named the flip the "Williford maneuver," after J.R. Williford, the first test pilot to inadvertently "accomplish" it. The next most important instrument was the fuel gauge, because the X-14 didn't carry much gas. In retrospect, I consider it a privilege to have flown one of the most successful research aircraft of all time, one that in over 20 years contributed greatly to a variety of other VTOL programs in technical input and piloting training, and to the evolution of V/STOL technology generally.

Vectored V/STOL Comes of Age: The P.1127, Kestrel, and YAV-8B VSRA

In 1957, Britain's Hawker and Bristol firms began development of what would prove to be the most revolutionary V/STOL airplane developed to that point in aviation history, the P.1127. This aircraft program, begun

115. Anderson, *Memoirs of an Aeronautical Engineer*, p. 32.

116. Borchers, Franklin, and Fletcher, *Flight Research at Ames*, p. 57.

117. Gerdes, "The X-14: 24 Years of V/STOL Flight Testing," pp. 8–9.

118. The author had wide-ranging experience in high-performance naval fighters and sophisticated variable stability aircraft before flying the X-14; for his early flying background, see G. Warren Hall, *Demons, Phantoms, and Me: A Love Affair with Flying* (Bloomington, IN: 1st Books Library, 2003)—ed.

The Hawker P.1127 during early hovering trials. NASA.

14

as a private development by two of Britain's more respected companies, was the product of Sir Sidney Camm and Ralph Hooper of Hawker, and Stanley Hooker of Bristol. It eventually spawned a remarkable operational aircraft that fought in multiple wars and served in the air forces and naval air services of many nations. Hawker had an enviable reputation for designing high-performance aircraft, dating to the Sopwith fighters of the First World War, and Bristol had an equally impressive one in the field of aircraft propulsion. NATO's Mutual Weapon Development Project (MWDP) supported the project as it evolved, and it drew heavily upon American support from John Stack of NASA and the Langley Research Center, and from the U.S. Marine Corps. (The P.1127 design was extensively tested in Langley's 30-Foot by 60-Foot Full Scale Tunnel, and the 16-Foot Transonic Tunnel, helping identify and alleviate a potentially serious pitch-up problem exacerbated by power effects during transition upon the original horizontal tail configuration).[119] Powered by a

119. Joseph R. Chambers, *Partners in Freedom: Contributions of the Langley Research Center to U.S. Military Aircraft of the 1990s*, NASA SP-2000-4519 (Washington, DC: GPO, 2000), pp. 13–14. The Air Force had requested NASA tunnel tests of the P.1127; see Charles C. Smith, "Flight Tests of a 1/6-Scale Model of the Hawker P 1127 Jet VTOL Airplane," NASA TM-SX-531 (1961). As a result of these trials, the P.1127 was given pronounced anhedral on its horizontal tail, and ultimately its size was increased as well.

single Bristol Siddeley Pegasus 5 vectored-thrust turbofan of 15,000-pound thrust, the P.1127 completed its first tethered hover in October 1960, an untethered hover the next month, and, after extensive preparation, its first transition from vertical to conventional in September 1961. As with the X-14 and other V/STOL testbeds, bleed air reaction nozzles were used for hover attitude control and, in the P.1127's initial configuration, had no SAS. Low control power, aerodynamic suck-down, and marginal altitude control power made for a high pilot workload for this early Harrier predecessor. Even so, NACA researchers quickly realized that the P.1127 offered remarkable promise. NASA pilots Jack Reeder from Langley and Fred Drinkwater from Ames went to Europe to fly the P.1127 in June 1962, Reeder confiding afterward: "The British are ahead of us again."[120] His flight evaluation report noted:

> The P.1127 is not a testbed aircraft in the usual sense. It is advanced well beyond this stage and is actually an operational prototype, with which it is now possible to study the VSTOL concept in relation to military requirements by actual operation in the field. The aircraft is easily controlled and has safe flight characteristics throughout the range from hover to airplane flight. The performance range is very great; yet, conversions to or from low or vertical flight can be accomplished simply, quickly, and repeatedly.[121]

Camm's P.1127 led to the Hawker Kestrel F.G.A. Mk. 1, an interim "militarized" variant, nine of which undertook operational suitability trials with a NATO tripartite (U.K., U.S., and Federal Republic of Germany) evaluation squadron in 1965. The trials confirmed not only the basic performance of the aircraft, but also its military potential. So the Kestrel, in turn, led directly to a production military derivative, the Hawker Harrier G.R. Mk. 1—or, as known in U.S. Marine Corps service, the AV-8A. Eight of the Kestrel aircraft, designated XV-6A, remained in the United States for follow-on testing. NASA received two Kestrels, flying them in an extensive evaluation program at Langley with pilots

120. Mallick, *The Smell of Kerosene*, p. 55.

121. John P. Reeder, Memorandum for Associate Director (of Langley Research Center), Subject: Flight Evaluation by NASA Pilots of the Hawker P-1127 [*sic*] V/STOL strike-fighter aircraft in England, July 24, 1962, copy in NASA Headquarters Historical Division archives.

Jack Reeder, Lee H. Person, Jr., Robert Champine, and Perry L. Deal, under the supervision of project engineer Richard Culpepper.[122]

Langley tunnel-testing and flight-testing revealed a number of deficiencies, though not of such magnitude as to detract from the impression that the P.1127 was a remarkable accomplishment, and that it had tremendous potential for development. For example, a directional instability was noticed in turning out of the wind, yaw control power was low but not considered unsafe, and pitch-trim changes occurred when leaving ground effect. The usual hot-gas ingestion problem could be circumvented by maintaining a low forward speed in takeoff and landing. A static pitch instability was encountered at alphas greater than approximately 15 degrees, and a large positive dihedral effect limited crosswind operations. Transition characteristics were outstanding, with only small trim changes required. Overall, low- and high-speed performance was excellent. Like any swept wing airplane, the Kestrel's "Dutch roll" lateral-directional damping was low at altitude, requiring provision of a yaw damper. It had good STOL performance when the engine nozzles were deflected between purely vertical and purely horizontal settings. Indeed, this would later become one of the Harrier strike fighter's strongest operational qualities.[123]

Like any operational aircraft, the Harrier went through progressive refinement. Its evolution coincided with the onset of advanced avionics, the emergence of composite structures, and NASA's development of the supercritical wing. All were developments incorporated in the next generation of Harrier, the AV-8B Harrier II, developed at the behest of the U.S

122. Chambers, *Partners in Freedom*, p. 16; details on the Kestrel are from Francis K. Mason, *The Hawker P. 1127 and Kestrel*, No. 198 in the *Profile Publications* series (Leatherhead, Eng.: Profile Publications, Ltd., 1967); Richard P. Hallion, "Hawker XV-6A Kestrel," in Lynne C. Murphy, ed., *Aircraft of the National Air and Space Museum* (Washington, DC: Smithsonian Institution Press, 1976).
123. Smith, "Flight Tests of a 1/6-Scale Model of the Hawker P 1127"; Reeder, "Flight Evaluation by NASA Pilots of the Hawker P-1127"; Samuel A. Morello, Lee H. Person, Jr., Robert E. Shanks, and Richard G. Culpepper, "A Flight Evaluation of A Vectored-Thrust-Jet V/STOL Airplane During Simulated Instrument Approaches Using the Kestrel (XV-6A) Airplane," NASA TN-D-6791 (1972); Richard J. Margason, Raymond D. Vogler, and Matthew M. Winston, "Wind-Tunnel Investigation at Low Speeds of a Model of the Kestrel (XV-6A) Vectored-Thrust V/STOL Airplane," NASA TN-D-6826 (1972); William T. Suit and James L. Williams, "Longitudinal Aerodynamic Parameters of the Kestrel Aircraft (XV-6A) Extracted from Flight Data," NASA TN-D-7296 (1973); Suit and Williams, "Lateral Static and Dynamic Aerodynamic Parameters of the Kestrel Aircraft (XV-6A) Extracted from Flight Data," NASA TN-D-7455 (1974).

Marine Corps and adopted, in slightly different form, as the Harrier Mk. 5 by the Royal Air Force. As well, the AV-8B benefited from Langley research on optimum positioning of engine nozzles, trailing-edge flaps, and the wing, in order to obtain higher propulsive lift. (This jet age work mirrored much earlier work on optimum positioning of propellers, engines, and nacelles undertaken at Langley in the 1920s by the NACA).[124]

Two AV-8A Harriers had been modified to serve as prototypes of the new Harrier II, these being designated YAV-8B. Though deceptively similar to the earlier AV-8A, the YAV-8B relied extensively on graphite epoxy composite structure and had a leading-edge extension at its wing-root and a bigger, supercritical wing. The first made its initial flight in November 1978, joined shortly afterward by the second. A year later, in November 1979, the second YAV-8B crashed after engine failure; its pilot ejected safely. However, flight-testing by contractor and service pilots confirmed that the AV-8B would constitute a significant advance over the earlier AV-8A for, during its evaluation program, "all performance requirements were met or exceeded."[125] Not surprisingly, the AV-8B entered production and squadron service with the U.S. Marine Corps, replacing the older Vietnam-legacy AV-8A.

In 1984, after the AV-8B entered operational service, the U.S. Marine Corps delivered the surviving YAV-8B to Ames so that Ames researchers could investigate advanced controls and flight displays, such as those that might be incorporated on future V/STOL combat systems called upon to conduct vertical envelopment assaults from small assault carriers and other vessels in all-weather conditions. The study effort that followed built upon Ames's legacy of V/STOL simulation studies, using both ground and flight simulators to evaluate a variety of guidance, control, and display concepts, particularly the research of Vernon K. Merrick, Ernesto Moralez, III, Jeffrey A. Schroeder, and their associates.[126] NASA designated the YAV-8B the V/STOL Systems Research Aircraft (VSRA). A team led by Del Watson and John D. Foster modifying it with digital fly-by-wire controls for pitch, roll, yaw, thrust magnitude and thrust deflection, and programmable electronic head-up displays. Researchers subsequently flew the YAV-8B in an extensive evaluation of control

124. Eastman N. Jacobs, "The Drag and Interference of a Nacelle in the Presence of a Wing," NACA TN-320 (1929).

125. K.V. Stenberg, "YAV-8B Flight Demonstration Program," AIAA Paper 83-1055 (1983).

126. Their prolific research is referenced in the recommended readings at the end of this work.

The NASA Ames YAV-8B V/STOL Systems Research Aircraft. NASA.

system concepts and behavior, from decelerations to hover, and then from hover to a vertical landing, assessing flying qualities tradeoffs for each of the various control concepts studied and evaluating advanced guidance and navigation displays as well.[127] In addition to NASA pilots, a range of Marine, Royal Air Force, McDonnell-Douglas, and Rolls-Royce test pilots flew the aircraft. Their inputs, combined with data from Ames's Vertical Motion Simulator, helped researcher Jack Franklin develop flying qualities criteria and control system and display concepts supporting the Joint Strike Fighter program.[128] With the conclusion of the

127. Ernesto Moralez, III, Vernon K. Merrick, and Jeffrey A. Schroeder, "Simulation Evaluation of the Advanced Control Concept for the NASA V/STOL Research Aircraft (VSRA)," AIAA Paper 87-2535 (1987); John D. Foster, Ernesto Moralez, III, James A. Franklin, and Jeffery A. Schroeder, "Integrated Control and Display Research for Transition and Vertical Flight on the NASA V/STOL Research Aircraft (VSRA)," NASA TM-100029 (1987); Paul F. Borchers, Ernesto Moralez, III, Vernon K. Merrick and Michael W. Stortz, "YAV-8B Reaction Control System Bleed and Control Power Usage in Hover and Transition," NASA TM-104021 (1994); D.W. Dorr, Ernesto Moralez, III, and Vernon K. Merrick, "Simulation and Flight Test Evaluation of Head-Up Display Guidance for Harrier Approach Transitions," *AIAA Journal of Aircraft*, vol. 31, no. 5 (Sept.–Oct. 1994), pp. 1089–1094.

128. Borchers, Franklin, and Fletcher, *Flight Research at Ames*, p. 62. The JSF competition pitted two rival designs, the Boeing X-32, and the Lockheed Martin X-35, against each other for a future joint-service USAF–USN–USMC fighter-bomber. The Marine variant would be a V/STOL. Eventually, the X-35 won, and the JSF moved into development as the F-35.

VSRA aircraft program in 1997, NASA Ames's role in V/STOL research came to an end.

In conclusion, in spite of the many challenges revealed in these summaries of V/STOL aircraft, the information accumulated from the design, development, and flight evaluations has provided a useful database for V/STOL designs. It is of interest to note that even though most of the aircraft were deficient, to some degree, in terms of aerodynamics, propulsion systems, or performance, it was always possible to develop special operating techniques to circumvent these problems. For the most part, this review would indicate that performance and handling-qualities limitations severely restricted operational evaluations for all types of V/STOL concepts. It has become quite obvious that V/STOL aircraft must be designed with good STOL performance capability to be cost-effective, a virtue not shared by many of the aircraft researched by NASA. Further, flight experience has shown that good handling qualities are needed, not only in the interest of safety, but also to permit the aircraft to carry out its mission in a cost-effective manner. It was apparent also that SAS was required to some degree for safely carrying out even simple operational tasks. The question of how much control system complexity is needed for various tasks and missions is still unanswered. Another area deserving of increased attention derives from the fact that most of the V/STOL aircraft studied suffered to some degree from adverse ground effects. In this regard, better prediction techniques are needed to avoid costly aircraft modifications or restricted operational use. Finally, there is an important continued need for good testing techniques and facilities to ensure satisfactory performance and control before and during flight-testing.

Today, NASA's investment in V/STOL technology promises to be a key enabling technology in making the airspace system more environmentally friendly and efficient. Cruise Efficient Short Take-Off and Landing Aircraft (CESTOL) and Civil Tilt Rotor (CTR) promise to expand the number of takeoff and landing locations, operating in terminal areas in a simultaneous noninterfering manner (SNI) with conventional traffic, relieving overtaxed hub airports. CESTOL–CTR aircraft avoid the airspace and runways required by commercial aircraft using steeply curved approach and departure paths, thus enabling greater system capacity, reducing delays, and saving fuel. To fulfill this vision, performance penalties associated with STOL capability requires continued NASA research to mitigate.[129] While much still remains to be accomplished, much has already been achieved, and the vision of future V/STOL remains vibrant and exciting. That it is constitutes an accolade to those men and women of NASA, and the NACA before, whose contributions made V/STOL aircraft a practical reality.

129. Craig E. Hange, "Short Field Take-Off and Landing Performance as an Enabling Technology for a Greener, More Efficient Airspace System," Report ARC-E-DAA-TN554, paper presented at the *Green Aviation Workshop, NASA Ames Research Center, Apr. 25–26, 2009.*

Appendix: Lessons from Flight-Testing the XV-5 and X-14 Lift Fans

Note: The following compilation of lessons learned from the XV-5 and X-14 programs is excerpted from a report prepared by Ames research pilot Ronald M. Gerdes based upon his extensive flight research experience with such aircraft and is of interest because of its reference to Supersonic Short Take-Off, Vertical Landing Fighter (SSTOVLF) studies anticipating the advent of the SSTOVLF version of the F-35 Joint Strike Fighter:[130]

> The discussion to follow is an attempt to apply the key issues of "lessons learned" to what might be applicable to the preliminary design of a hypothetical Supersonic Short Take-off and Vertical Landing Fighter/attack (SSTOVLF) aircraft. The objective is to incorporate pertinent sections of the "Design Criteria Summary" into a discussion of six important SSTOVLF preliminary design considerations to form the viewpoint of the writer's lift-fan aircraft flight test experience. These key issues are discussed in the following order: (1) Merits of the Gas-Driven Lift-Fan, (2) Lift-Fan Limitations, (3) Fan-in-Wing Aircraft Handling Qualities, (4) Conversion System Design, (5) Terminal Area Approach Operations, and (6) Human Factors.

14

MERITS OF THE XV-5 GAS-DRIVEN LIFT-FAN

> The XV-5 flight test experience demonstrated that a gas-driven lift-fan aircraft could be robust and easy to maintain and operate. Drive shafts, gear boxes and pressure lubrication systems, which are highly vulnerable to enemy fire, were not required with gas drive. Pilot monitoring of fan machinery health is thus reduced to a minimum which is highly desirable for a single-piloted aircraft such as the SSTOVLF. Lift-fans have proven to be highly resistant to ingestion of foreign objects which is a plus for remote site operations. In one instance an XV-5A wing-fan continued to produce substantial lift despite considerable damage inflicted by the ingestion of a rescue collar weight. All pilots who have flown the XV-5 felt confident in the integrity of the lift-fans, and it was felt that the combat effectiveness of the SSTOVLF would be enhanced by using gas-driven lift-fans.

130. Excerpted from Gerdes, "Lift-Fan Aircraft—Lessons Learned: the Pilot's Perspective," NASA CR-177620 (1993). References to illustrations and figures have been removed.

LIFT-FAN LIMITATIONS

It is recommended that a nose-mounted lift-fan NOT be incorporated into the design of the SSTOVLF for pitch attitude control. XV-5A flight tests demonstrated that although the pitch-fan proved to be effective for pitch attitude control, fan ram drag forces caused adverse handling qualities and reduced the conversion airspeed corridor. It is thus recommended that a reaction control system be incorporated.

The X-14A roll-control lift-fan tests revealed that control of rolling moment by varying fan rpm was unacceptable due to poor fan rpm response characteristics even when closed-loop control techniques were employed. Thus this method should not be considered for the SSTOVLF. However, lift-fan thrust spoiling proved to be successful in the XV-5 and is recommended for the SSTOVLF.

Avoidance of the fan stall boundary placed significant operational limitations on the XV5 and had the potential of doing the same with the SSTOVLF. Fan stall, like wing stall, must be avoided and a well defined safety margin required. Approach to the fan stall boundary proved to be a particular problem in the XV-5B, especially when performing steep terminal area maneuvers during simulated or real instrument landing approaches. The SSTOVLF preliminary designers must account for anticipated fan stall limitations and allow for adequate safety margins when determining SSTOVLF configurations and flight profile specifications.

FAN-IN-WING AIRCRAFT HANDLING QUALITIES

The XV-5 was a proof-of-concept lift-fan aircraft and thus employed a completely "manual" powered-lift flight control system. The lack of an integrated powered-lift system required the pilot to manually control the aircraft flight-path through independent manipulation of stick, engine power, thrust vector angle and collective lift. This lack of an integrated powered-lift management system (and in particular, the conversion controls) was responsible for most of the adverse handling qualities of the aircraft. An advanced digital fly-by-wire control system must provide level one handling qualities, especially for integrated powered-lift management.

CONVERSION SYSTEM DESIGN

The manually operated conversion system was the most exacting, interesting and potentially hazardous flight operation associated with the XV-5. This type of "bang-bang" conversion system should not be considered for the SSTOVLF. Ideally, the conversion should consist of a fully reversible and continuously controllable process. That is, the pilot must be able to continuously control the conversion process. Good examples are the XV-15 Tilt Rotor, the X-22A and the AV-8 Harrier. Furthermore, the conversion of the SSTOVLF with an advanced digital flight control system should be fully decoupled so that the pilot would not have to compensate for lift, attitude or speed changes. The conversion controller should be a single lever or beeper-switch that is safety-interlocked against inadvertent actuation. The conversion airspeed limit corridor must be wide enough to allow for operational flexibility and compensate for single-pilot operation where mission demands can compete for pilot attention.

TERMINAL AREA APPROACH OPERATIONS

The XV-5B demonstrated that lift-fan aircraft are capable of performing steep simulated instrument approaches with up to 20° flight-path angles. Once more, lack of an integrated powered-lift flight control system was the primary cause of adverse handling qualities and operational limitations. The SSTOVLF's integrated powered-lift system must provide decoupled flight path control for glide slope tracking where a single controller, such as a throttle-type lever is used for direct flight-path modulation while airspeed and/or angle-of-attack are held constant. Simulator evaluations of such systems have indicated significant improvements in handling qualities and reductions in pilot workload, an integrated powered-lift system a must in a single-piloted SSTOVF.

Evaluations of the XV-5B's ability to perform simulated instrument landing approaches along a 10° glide slope revealed that pilots preferred to approach with a deck-parallel attitude (near-zero angle-of-attack) instead of using deck-level attitude (near 10° angle-of-attack) instead of 15°. Fan-stall boundary and random aerodynamic lift disturbances were cited as the causes.

SSTOVLF designers should encourage the development of lift-fans with increased angle-of-attack capability which would enhance Instrument Meteorological Conditions (IMC) operational capability and improve safety.

All pilots that flew the XV-5 (the "XV-5 Fan Club") were of the unanimous opinion that the conversion handling qualities of the Vertifan were completely unsatisfactory for IMC operations. Trying to contend with the large power changes, attitude and altitude displacements, and abrupt airspeed changes while trying to fly instruments with the XV-5's "manual" control system was too much to handle. The enhanced operational flexibility requirement laid on the SSTOVLF requires that it have full IMC operational capability.

HUMAN FACTORS

Human factors played a part in some of the key issues that have already been discussed above. Examples are: confidence in lift-fans, concern for approach to the fan-stall boundary, high pilot workload tasks, and conversion controller design.

The human factor issue that concerned the writer the most was that of the cockpit arrangement. An XV-5A and its pilot were probably lost because of the inadvertent actuation of an incorrectly specified and improperly positioned conversion switch. This tragic lesson must not be repeated, and careful human factor studies must be included in the design of modern lift-fan aircraft such as the SSTOVLF. Human factor considerations should be incorporated early in the design and development of the SSTOVLF from the first simulation effort on through the introduction of the production aircraft. It is therefore the writer's hope that SSTOVLF designers will remember the past as they design for the future and take heed of the "Lessons learned."

Fatal Accident #1

One of the two XV-5As being flown at Edwards AFB during an official flight demonstration on the morning of April 27, 1965, crashed onto the lakebed, killing Ryan's Chief Engineering Test Pilot, Lou Everett. The two aircraft were simultaneously demonstrating the high-and low-speed capabilities of the Vertifan.

During a high-speed pass, Everett's aircraft pushed over into a 30° dive and never recovered. The accident board concluded that the uncontrolled dive was the result of an accidental actuation of the conversion switch that took place when the aircraft's speed was far in excess of the safe jet-mode to fan-mode conversion speed limit. The conversion switch (a simple 2-position toggle switch) was, at the time, (improperly) located on the collective for pilot "convenience." It was speculated that the pilot inadvertently hit the conversion switch during the high-speed pass which initiated the conversion sequence: 15° of nose-down stabilizer movement was accompanied by actuation of the diverter valves to the fan-mode. The resulting stabilizer pitching moment created an uncontrollable nose-down flight path. (Note: Mr. Everett initiated a low altitude (rocket) ejection, but tragically, the ejection seat was improperly rigged...another lesson learned!) As a result of this accident, the conversion switch was changed to a lift-lock toggle and relocated on the main instrument panel ahead of the collective lever control.

Fatal Accident #2

The remaining XV-5A was rigged with a pilot-operated rescue hoist, located on the left side of the fuselage just ahead of the wing fan. An evaluation test pilot was fatally injured during the test program while performing a low-speed, steep-descent "pick-up" maneuver at Edwards AFB. The heavily-weighted rescue collar was ingested into the left wing fan as the pilot descended and simultaneously played-out the collar. The damaged fan continued to rotate, but the resultant loss in fan lift caused the aircraft to roll-left and settle toward the ground. The pilot apparently leveled the wings; applied full power and up-collective to correct for the left wing-fan lift loss. The damaged left fan produced enough lift to hold the wings level and somewhat reduce the ensuing descent rate. The pilot elected to eject from the aircraft as it approached the ground in this wings-level attitude. As the pilot released the right-stick displacement and initiated the ejection, the aircraft rolled back to the left which caused the ejected seat trajectory to veer-off to a path parallel to the ground. The seat impacted the ground, and the pilot did not survive the ejection. Post-accident analysis revealed that despite the ingestion of the rescue collar and its weight, the wing-fan continued to operate and produce enough lift force to hold a wings-level roll attitude and reduce descent rate to a value that may have allowed the pilot to survive the ensuing "emergency landing" had he stayed with the aircraft. This was a grim testimony as to the ruggedness of the lift-fan.

Recommended Additional Readings

Reports, Papers, Articles, and Presentations:

Edwin W. Aiken, Robert A. Jacobson, Michelle M. Eshow, William S. Hindson, and Douglas H. Doane, "Preliminary Design Features of the RASCAL—A NASA/Army Rotorcraft In-Flight Simulator," AIAA Paper 92-4175 (1992).

Thomas N. Aiken, "Advanced Augmentor-Wing Research," NASA TM-X-62-250 (1972).

Thomas N. Aiken and A.M. Cook, "Low Speed Aerodynamic Characteristics of a Large Scale STOL Transport Model with an Augmented Jet Flap," NASA TM-X-62017 (1971).

Thomas N. Aiken and A.M. Cook, "Results of Full-Scale Wind tunnel Tests on the H.126 Jet Flap Aircraft," NASA TN-D-7252 (1973).

James A. Albers and John Zuk, "Civil Applications of High-Speed Rotorcraft and Powered-Lift Aircraft Configurations," NASA TM-100035 (1987).

Alberto Alvarez-Calderon, "Rotating Cylinder Flaps for V/STOL: Some Aspects of an Investigation into the Rotating Cylinder Flap High Lift System for V/STOL Aircraft Conducted Jointly by the Peruvian Air Force and The National University of Engineering of Peru," *Aircraft Engineering and Aerospace Technology*, vol. 36, no. 10 (Oct. 1964), pp. 304–309.

Alberto Alvarez-Calderon, "VTOL and the Rotating Cylinder Flap," *Annals of the New York Academy of Sciences*, vol. 107, no. 1 (Mar. 1963), pp. 249–255.

Seth B. Anderson, "An Examination of Handling Qualities Criteria for V/STOL Aircraft," NASA TN-D-331 (1960).

Seth B. Anderson, "Historical Overview of V/STOL Aircraft Technology," NASA TM-81280 (1981).

Seth B. Anderson and G. Warren Hall, "Some Lessons Learned from An Historical Review of Aircraft Operations", in *Aircraft Flight Safety*, the proceedings of an international conference held at Zhukovsky, Russia, in Aug.–Sept. 1993.

Seth B. Anderson, Hervey C. Quigley, and Robert C. Innis, "Flight Measurements of the Low-Speed Characteristics of a 35° Swept-Wing Airplane with Blowing-Type Boundary-Layer Control on the Trailing-Edge Flaps," NACA RM-A56G30 (1956).

Seth B. Anderson, Robert C. Innis, and Hervey C. Quigley, "Stability and Control Characteristics of STOL Aircraft in Low-Speed Landing Approach Portion of Flight Regime," AIAA Paper 65-715 (1965).

Hal Andrews, "Four Ducts and Eleven Gearboxes! The Development of the Bell X-22," *Vertiflite* (summer 2001).

Adolph Atencio, Jr., Jerry V. Kirk, Paul T. Soderman, and Leo P. Hall, "Comparison of Flight and Wind tunnel Measurements of Jet Noise for the XV-5B Aircraft," NASA TM-X-62-182 (1972).

F.J. Bailey, Jr., "A Simplified Theoretical Method of Determining the Characteristics of a Lifting Rotor in Forward Flight," NACA Report 716 (1941).

F.A. Baker, D.N. Jaynes, L.D. Corliss, S. Liden, R.B. Merrick, and D.C. Dugan, "V/STOLAND Avionics System Flight-Test Data on a UH-1H Helicopter," NASA TM-78591 (1980).

Robert C. Ball, "Summary Highlights of the Advanced Rotor Transmission (ART) Program," AIAA Paper 92-3362 (1992).

Paul M. Bevilaqua, "Genesis of the F-35 Joint Strike Fighter," *Journal of Aircraft*, vol. 46, no. 6 (Nov.–Dec. 2009), pp. 1825–1836.

E.A. Bielefeldt, "STOL Aircraft With Mechanical High-Lift Systems Compared with STOL Aircraft with Wings Equipped with Blown Flaps," NASA Translation TT-F-14,895 (of "STOL-Flugzeugen mit mechanischen Hochauftriebsystemen im Vergleich zu STOL-Flugzeugen mit Blasklappenflügeln,") Messerschmitt-Bölkow-Blohm GmbH, Report UH-12-72, presented as Paper 72-057 at the *5th Annual Meeting of the DGLR* (*Deutsche Gesellschaft für Luft-und-Raumfahrt, Oct. 4–6, 1972*) (1973).

Paul F. Borchers, Ernesto Moralez, III, Vernon K. Merrick, and Michael W. Stortz, "YAV-8B Reaction Control System Bleed and Control Power Usage in Hover and Transition," NASA TM-104021 (1994).

G.W. Brooks, "The Application of Models to Helicopter Vibration and Flutter Research," *Proceedings* of the ninth annual forum of the American Helicopter Society (May 1953).

Robert T.N. Chen, William S. Hindson, and Arnold W. Mueller, "Acoustic Flight Tests of Rotorcraft Noise-Abatement Approaches Using Local Differential GPS Guidance," NASA TM-110370 (1995).

J.A. Cochrane, D.W. Riddle, and V.C. Stevens, "Quiet Short-Haul Research Aircraft—The First Three Years of Flight Research," AIAA Paper 81-2625 (1981).

J.A. Cochrane, D.W. Riddle, V.C. Stevens, and M.D. Shovlin, "Selected Results from the Quite Short-Haul Research Aircraft Flight Research Program," *Journal of Aircraft*, vol. 19, no. 12 (Dec. 1982), pp. 1076–1082.

R.P. Coleman, "Theory of Self-Excited Mechanical Oscillations of Hinged Rotor Blades," NACA WR-L-308 (formerly NACA Advanced Restricted Report 3G29) (1943).

Woodrow L. Cook and David H. Hickey, "Aerodynamics of V/STOL Aircraft Powered by Lift Fans," NASA TM-X-60455 (1967).

Woodrow L. Cook and David H. Hickey, "Correlation of Low Speed Wind Tunnel and Flight Test Data for V/STOL Aircraft," NASA TM-X-62-423 (1975).

Woodrow L. Cook and D.C. Whittley, "Comparison of Model and Flight Test Data for an Augmented Jet Flap STOL Research Aircraft," NASA TM-X-62491 (1975).

George E. Cooper, "Understanding and Interpreting Pilot Opinion," *Aeronautical Engineering Review*, vol. 16, no. 3 (Mar. 1957), p. 47–51.

George E. Cooper and Robert C. Innis, "Effect of Area-Suction-Type Boundary-Layer Control on the Landing-Approach Characteristics of a 35° Swept-Wing Fighter," NACA RM-A55K14 (1957).

George E. Cooper and Robert P. Harper, Jr., "The Use of Pilot Rating in the Evaluation of Aircraft Handling Qualities," NASA TN-D-5153 (1969).

I.H. Culver and J.E. Rhodes, "Structural Coupling in the Blades of a Rotating Wing Aircraft," IAS Paper 62-33 (1962).

I.M. Davidson, "The Jet Flap," *Journal of the Royal Aeronautical Society*, vol. 60, no. 541 (Jan. 1956), pp. 25–50.

Mike Debraggio, "The American Helicopter Society—A Leader for 40 Years," *Vertiflite*, vol. 30, no. 4 (May–June 1984).

W.H. Deckert, C.A. Holzhauser, M.W. Kelly, and H.C. Quigley, "Design and Operating Considerations of Commercial STOL Transports," AIAA Paper 64-285 (1965).

D.W. Dorr, Ernesto Moralez, III, and Vernon K. Merrick, "Simulation and Flight Test Evaluation of Head-Up Display Guidance for Harrier Approach Transitions," *AIAA Journal of Aircraft*, vol. 31, no. 5 (Sept.–Oct. 1994), pp. 1089–1094.

Fred J. Drinkwater, III, and L.S. Rolls, "The Application of Flight and Simulation Testing to VTOL Aircraft Handling Qualities Specifications," NASA TM-X-57472 (1963).

Fred J. Drinkwater, III, W.F. Rollins, and L.S. Rolls, "The Use of a Jet-Propelled VTOL Aircraft to Investigate Lunar Landing Trajectories," *Quarterly Review of the Society of Experimental Test Pilots,* vol. 6 (spring 1963), pp. 19–27.

Fred J. Drinkwater, III, L. Stewart Rolls, Edward L. Turner, and Hervey C. Quigley, "V/STOL Handling Qualities as Determined by Flight Test and Simulation Techniques," paper presented at the *3rd International Congress of the Aeronautical Sciences, Stockholm, Sweden, Aug. 27–Sept. 1, 1962,* NTIS Report 62N12456 (1962).

L.T. Dufraimont, "Evaluation of the Augmentor-Wing Jet STOL Aircraft," NASA CR-169853 (1980).

Daniel C. Dugan, Ronald G. Erhart, and Laurel G. Schroers, "The XV-15 Tilt Rotor Research Aircraft," NASA TM-81244, AVRADCOM Technical Report 80-A-15 (1980).

Daniel C. Dugan, "VTOL Aircraft Thrust Control—an Historical Perspective," paper presented at *The Next Generation Vertical Lift Technologies symposium, American Helicopter Society's Southwest Chapter International Specialists Meeting, Dallas, TX, Oct. 15–17, 2008.*

B.J. Dvorscak, "The Hummingbird Experience at Lockheed Georgia," *1991 Report to the Aerospace Profession* (Lancaster, CA: Society of Experimental Test Pilots, 1991), pp. 236–254.

H.K. Edenborough, T.M. Gaffey, and J.A. Weiberg, "Analysis and Tests Confirm Design of Proprotor Aircraft," AIAA Paper 72-803 (1972).

Robert J. Englar, "Circulation Control for High Lift and Drag Generation on a STOL Aircraft," *Journal of Aircraft,* vol. 12, no. 5 (May 1975), pp. 457–463.

Robert J. Englar, "Development of the A-6/Circulation Control Wing Flight Demonstrator Configuration," U.S. Navy DTNSRDC Report ASED-79/01 (1979).

Robert J. Englar and C.A. Applegate, "Circulation Control: A Bibliography of DTNSRDC Research and Selected outside References (Jan. 1969 through Dec. 1983)," U.S. Navy DTNSRDC Report 84/052 (1984).

Robert J. Englar, "Development of Circulation Control Technology for Powered-Lift STOL Aircraft," in NASA ARC, *Proceedings of the Circulation Control Workshop, 1986*, NASA CP-2432 (1986).

Robert J. Englar, J.H. Nichols, Jr., M.J. Harris, J.C. Eppel, and M.D. Shovlin, "Circulation Control Technology Applied to Propulsive High Lift Systems," Society of Automotive Engineers, SAE Paper 841497, in Society of Automotive Engineers, *V/STOL: An Update and Overview* (Warrendale, PA: SAE, 1984), pp. 31–43.

Joseph C. Eppel, Dennis W. Riddle, and Victor C. Stevens, "Flight Measured Downwash of the QSRA," NASA TM-101050 (1988).

J. Eshlemen and R.E. Kuhn, "Ground Effects on V/STOL and STOL Aircraft—A Survey," AIAA Paper 85-4033 (1985).

Terrell W. Feistel, Ronald M. Gerdes, and Emmet B. Fry, "An Investigation of a Direct Side-Force Maneuvering System on a Deflected Jet VTOL Aircraft," NASA TN-D-5175 (1969).

Don Fertman, "The Helicopter History of Sikorsky Aircraft," *Vertiflite*, vol. 30, no. 4 (May–June 1984).

David D. Few, "A Perspective on 15 Years of Proof-of-Concept Aircraft Development and Flight Research at Ames–Moffett by the Rotorcraft and Powered-Lift Flight Projects Division, 1970–1985," NASA RP-1187 (1987).

Brenda Forman, "The V-22 Tiltrotor 'Osprey:' The Program That Wouldn't Die," *Vertiflite*, vol. 39, no. 6 (Nov.–Dec. 1993), pp. 20–23.

John D. Foster, Ernesto Moralez, III, James A. Franklin, and Jeffery A. Schroeder, "Integrated Control and Display Research for Transition and Vertical Flight on the NASA V/STOL Research Aircraft (VSRA)," NASA TM-100029 (1987).

James A. Franklin and Robert C. Innis, "Longitudinal Handling Qualities During Approach and Landing of a Powered Lift STOL Aircraft," NASA TM-X-62144 (1972).

James A. Franklin, "V/STOL Dynamics, Control, and Flying Qualities," NASA TP-2000-209591 (2000).

James A. Franklin, Robert C. Innis, Gordon H. Hardy, and Jack D. Stephenson, "Design Criteria for Flightpath and Airspeed Control for the Approach and Landing of STOL Aircraft," NASA TP-1911 (1982).

James A. Franklin, Michael W. Stortz, Paul F. Borchers, and Ernesto Moralez, III, "Flight Evaluation of Advanced Controls and Displays for Transition and Landing on the NASA V/STOL Systems Research Aircraft," NASA TP-3607 (1996).

Kurt Frey, "Experiments with Rotating Cylinders in Combination with Airfoils," NACA TM-382 (1926).

F. Garren, J.R. Kelly, and R.W. Summer, "VTOL Flight Investigation to Develop a Decelerating Instrument Approach Capability," Society of Automotive Engineers Paper No. 690693 (1969).

Ronald M. Gerdes, "The X-14: 24 Years of V/STOL Flight Testing," *Technical Review* (of the Society of Experimental Test Pilots), vol. 16, no. 2 (1981), pp. 3–21.

Ronald M. Gerdes, "Lift-Fan Aircraft—Lessons Learned: the Pilot's Perspective," NASA CR-177620 (1993).

Ronald M. Gerdes and Charles S. Hynes, "Factors Affecting Handling Qualities of a Lift-Fan Aircraft during Steep Terminal Area Approaches," NASA TM X-62-424 (1975).

B.P. Gupta, A.H. Logan, and E.R. Wood, "Higher Harmonic Control for Rotary Wing Aircraft," AIAA Paper 84-2484 (1984).

F.B. Gustafson, "Effects on Helicopter Performance of Modifications in Profile-Drag Characteristics of Rotor-Blade Airfoil Sections," NACA WR-L-26 (Formerly NACA Advanced Confidential Report ACR L4H05) (1944).

F.B. Gustafson, "A History of NACA Research on Rotating-Wing Aircraft," *Journal of the American Helicopter Society*, vol. 1, no. 1 (Jan. 1956), p. 16.

F.B. Gustafson, "History of NACA/NASA Rotating-Wing Aircraft Research, 1915–1970," *Vertiflite*, Reprint VF-70 (Apr. 1971), pp. 1–27.

Albert W. Hall, Kalman J. Grunwald, and Perry L. Deal, "Flight Investigation of Performance Characteristics During Landing Approach of a Large Powered-Lift Jet Transport," NASA TN-D-4261 (1967).

G. Warren Hall, "Flight Test Research at NASA Ames Research Center: A Test Pilot's Perspective," NASA TM-100025 (1987).

Richard P. Hallion, "Hawker XV-6A Kestrel," in Lynne C. Murphy, ed., *Aircraft of the National Air and Space Museum* (Washington, DC: Smithsonian Institution Press, 1976).

Craig E. Hange, "Short Field Take-Off and Landing Performance as an Enabling Technology for a Greener, More Efficient Airspace System," Report ARC-E-DAA-TN554, paper presented at the *Green Aviation Workshop, NASA Ames Research Center, Apr. 25–26, 2009*.

Robert K. Heffley, Robert L. Stapleford, and Robert C. Rumold, "Airworthiness Criteria Development for Powered-Lift Aircraft," NASA CR-2791, FAA RD-76-195 (1977).

Harry H. Heyson and S. Katzoff, "Induced Velocities Near a Lifting Rotor with Nonuniform Disk Loading," NACA Report 1319 (1957).

David H. Hickey and Jerry V. Kirk, "Survey of Lift-Fan Aerodynamic Technology," NASA CP-177615 (1993).

T.P. Higgins, E.G. Stout, and H.S. Sweet," "Study of Quiet Turbofan STOL Aircraft for Short Haul Transportation," NASA CR-2355 (1973).

Curt A. Holzhauser, Robert C. Innis, and Richard F. Vomaske, "A Flight and Simulator Study of the Handling Qualities of a Deflected Slipstream STOL Seaplane Having Four Propellers and Boundary-Layer Control," NASA TN-D-2966 (1965).

Curt A. Holzhauser, Samuel A. Morello, Robert C. Innis, and James M. Patton, Jr., "A Flight Evaluation of A VTOL Jet Transport Under Visual and Simulated Instrument Conditions," NASA TN-D-6754 (1972) (formerly NASA TM-X-62-083 (1971)).

Robert J. Huston, "An Exploratory Investigation of Factors Affecting the Handling Qualities of a Rudimentary Hingeless Rotor Helicopter," NASA TN-D-3418 (1966).

Robert J. Huston, Robert A. Golub, and James C. Yu, "Noise Considerations for Tilt Rotor," AIAA Paper 89-2359 (1989).

Robert C. Innis and Hervey C. Quigley, "A Flight Examination of Operating Problems of V/STOL Aircraft in STOL-Type Landing and Approach," NASA TN-D-862 (1961).

Robert C. Innis, Curt A. Holzhauser, and Richard P. Gallant, "Flight Tests Under IFR with an STOL Transport Aircraft," NASA TN-D-4939 (1968).

Robert C. Innis, Curt A. Holzhauser, and Hervey C. Quigley, "Airworthiness Considerations for STOL Aircraft," NASA TN-D-5594 (1970).

Karen Jackson, Richard L. Boitnott, Edwin L. Fasanella, Lisa E. Jones, and Karen H. Lyle, "A Summary of DOD-Sponsored Research Performed at NASA Langley's Impact Dynamics Research Facility," *Journal of the American Helicopter Society*, vol. 51, no. 1 (June 2004).

R.A. Jacobsen and F.J. Drinkwater, III, "Exploratory Flight Investigation of Aircraft Response to the Wing Vortex Wake Generated by the Augmentor Wing Jet STOL Research Aircraft," NASA TM-X-62387 (1975).

Wayne Johnson, "Model for Vortex Ring State Influence on Rotorcraft Flight Dynamics," NASA TP-2005-213477 (2005).

Gregory S. Jones and Ronald D. Joslin, eds., *Proceedings of the 2004 NASA/ONR Circulation Control Workshop*, NASA CP-2005-213509/PTI (Washington, DC: NASA, 2005).

Henry L. Kelley, John P. Reeder, and Robert A. Champine, "Summary of a Flight-Test Evaluation of the CL-84 Tilt-Wing V/STOL Aircraft," NASA TM-X-1914 (1970).

J.R. Kelly, F.R. Niessen, J.J. Thibodeaux, K.R. Yenni, and J.F. Garren, Jr., "Flight Investigation of Manual and Automatic VTOL Decelerating Instrument Approaches and Landings," NASA TN-D-7524 (1974).

E.H. Kemper and D.J. Renselaer, "CV-7A Transport Aircraft Modification to Provide an Augmentor Wing Jet STOL Research Aircraft, v. 1: Design Study," NASA CR-73321 (1969).

Robert H. Kirby, "Flight Investigation of the Stability and Control Characteristics of a Vertically Rising Airplane Research Model with Swept or Unswept Wings and x or + Tails," NACA TN-3312 (1956).

Robert M. Kufeld and Paul C. Loschke, "UH-60 Airloads Program—Status and Plans," AIAA Paper 91-3142 (1991).

R.G. Kvaternik and W.G. Walton, Jr., "A Formulation of Rotor-Airframe Coupling for the Design Analysis of Vibrations of Helicopter Airframes," NASA RP-1089 (1982).

R.G. Kvaternik, "The NASA/Industry Design Analysis Methods for Vibration (DAMVIBS) Program—A Government Overview," AIAA Paper 92-2200 (1992).

14

A.W. Linden and M.W. Hellyer, "The Rotor Systems Research Aircraft," AIAA Paper No. 74-1277 (1974).

Mark Liptak, "International Helicopter Study Team (IHST) Overview Briefing," *Helicopter Association International HELI EXPO Meeting, Houston, TX, Feb. 21–23, 2009.*

W.S. Longhurst, "Initial Development and Testing of Tilt-Wing V/STOL," *Technical Review* (of the Society of Experimental Test Pilots), vol. 7, no. 4 (1965), pp. 34–41.

Stuart G. Madison, "First Twelve Months of Flight of the XC-142A Program," *Technical Review* (of the Society of Experimental Test Pilots), vol. 7, no. 4 (1965), pp. 18–25.

Domenic J. Maglieri and Harvey H. Hubbard, "Preliminary Measurements of the Noise Characteristics of Some Jet-Augmented-Flap Configurations," NASA Memo 12-4-58L (1959).

Martin D. Maisel and Lt. Col. Clifford M. McKeithan, "Tilt-Rotor Aircraft," *Army Research, Development & Acquisition Magazine* (May–June 1980), pp. 1–2.

Martin D. Maisel and D.J. Harris, "Hover Tests of the XV-15 Tilt Rotor Research Aircraft," AIAA Paper 81-2501 (1981).

W.R. Mantay, W.T. Yeager, Jr., M.N. Hamouda, R.G. Cramer, Jr., and C.W. Langston, "Aeroelastic model Helicopter Testing in the Langley TDT," NASA TM-86440, USAAVSCOM TM-85-8-5 (1985).

Daniel J. March, "VTOL Flat-Risers: Lockheed XV-4 and Ryan/GE XV-5," *International Air Power Review*, vol. 16 (2005), pp. 118–127.

Richard J. Margason, Raymond D. Vogler, and Matthew M. Winston, "Wind-Tunnel Investigation at Low Speeds of a Model of the Kestrel (XV-6A) Vectored-Thrust V/STOL Airplane," NASA TN-D-6826 (1972).

Ruth M. Martin, "NASA/AHS Rotorcraft Noise Reduction Program: NASA Langley Acoustics Division Contributions," *Vertiflite,* vol. 35, no. 4 (May–June 1989), pp. 48–52.

J.G. McArdle, "Outdoor Test Stand Performance of a Convertible Engine with Variable Inlet Guide Vanes for Advanced Rotorcraft Propulsion," NASA TM-88939 (1986).

M.O. McKinney and W.A. Newsom, "Fan V/STOL Aircraft," NASA TM-X-59739 (1967).

Vernon K. Merrick, "Study of the Application of an Implicit Model-Following Flight Controller to Lift-Fan VTOL Aircraft," NASA TP-1040 (1977).

Vernon K. Merrick, "Simulation Evaluation of Two VTOL Control/Display Systems in IMC Approach and Shipboard Landing," NASA TM-85996 (1984).

Vernon K. Merrick, "Some VTOL Head-Up display Drive-Law Problems and Solutions," NASA TM-104027 (1993).

Vernon K. Merrick and Ronald M. Gerdes, "Design and Evaluation of an Integrated Flight-Control System Concept for Manual IFR VTOL Operations," AIAA Paper 77-601 (1977).

Vernon K. Merrick and Ronald M. Gerdes, "VTOL Controls for Shipboard Operations," Society of Automotive Engineers, SAE Paper 831428 (1983).

Vernon K. Merrick and James A. Jeske, "Flightpath Synthesis and HUD Scaling for V/STOL Terminal Area Operations," NASA TM-110348 (1995).

Vernon K. Merrick, Glenn G. Farris, and Andrejs A. Vanags, "A Head Up Display for Application to V/STOL Aircraft Approach and Landing," NASA TM-102216 (1990).

Vernon K. Merrick, Ernesto Moralez, III, Michel W. Stortz, Gordon H. Hardy, and Ronald M. Gerdes, "Simulation Evaluation of a Speed-Guidance Law for Harrier Approach Transitions," NASA TM-102853 (1991).

Ernesto Moralez, III, Vernon K. Merrick, and Jeffrey A. Schroeder, "Simulation Evaluation of the Advanced Control Concept for the NASA V/STOL Research Aircraft (VSRA)," AIAA Paper 87-2535 (1987).

Samuel A. Morello, Lee H. Person, Jr., Robert E. Shanks, and Richard G. Culpepper, "A Flight Evaluation of A Vectored-Thrust-Jet V/STOL Airplane During Simulated Instrument Approaches Using the Kestrel (XV-6A) Airplane," NASA TN-D-6791 (1972).

M. Mosher and R.L. Peterson, "Acoustic Measurements of a Full-Scale Coaxial Helicopter," AIAA Paper 83-0722 (1983).

Lt. Col. Arthur L. Nalls, Jr., and Michael Stortz, "Assessment of Russian VSTOL Technology: Evaluating the Yak-38 Forger and Yak-141 Freestyle," *1993 Report to the Aerospace Profession* (Lancaster: Society of Experimental Test Pilots, 1993), pp. 40–59.

NASA, ARC staff, "Conference on V/STOL and STOL Aircraft, Ames Research Center, Moffett Field, California April 4–5, 1966," NASA SP-116 (1966).

NASA, ARC staff, "Proceedings of the 1985 NASA Ames Research Center's Ground-Effects Workshop," NASA CP-2462 (1985).

NASA, ARC staff, "Proceedings of the Circulation Control Workshop, 1986," NASA CP-2432 (1986).

NASA, LRC staff, "A Preliminary Study of V/STOL Transport Aircraft and Bibliography of National research in the VTOL-STOL Field," NASA TN-D-624 (1961).

NASA, LRC staff, "Powered-Lift Aerodynamics and Acoustics: A Conference Held at Langley Research Center, Hampton, Virginia, May 24–26, 1976," NASA SP-406 (1976).

Tim Naumowicz, Richard Margason, Doug Wardwell, Craig Hange, and Tom Arledge, "Comparison of Aero/Propulsion Transition Characteristics for a Joint Strike Fighter Configuration," paper presented at the *International Powered Lift Conference, West Palm Beach, FL, Nov. 18–21, 1996.*

W.P. Nelms and S.B. Anderson, "V/STOL Concepts in the United States—Past, Present, and Future," NASA TM-85938 (1984).

Jack N. Nielsen and James C. Biggers, "Recent Progress in Circulation Control Aerodynamics," AIAA Paper 87-0001 (1987).

North Atlantic Treaty Organization, Advisory Group for Aeronautical Research and Development, "Recommendations for V/STOL Handling Qualities," Report 408A (1964), NTIC Report N65-28861.

Frank A. Pauli, Daniel M. Hegarty, and Thomas M. Walsh, "A System for Varying the Stability and Control of a Deflected-Jet Fixed-Wing VTOL Aircraft," NASA TN-D-2700 (1965).

Frank A. Pauli, Tom G. Sharpe, and James R. Rogers, "A Reaction Attitude Control System for Jet VTOL Research Aircraft," NASA TN-D-5715 (1970).

Robert J. Pegg, "Flight-Test Investigation of Ailerons as a Source of Yaw Control on the VZ-2 Tilt-Wing Aircraft," NASA TN-D-1375 (1962).

Robert J. Pegg, "Summary of Flight-Test Results of the VZ-2 Tilt-Wing Aircraft," NASA TN-D-989 (1962).

Arthur E. Phelps, III, "Wind Tunnel Investigation of a Twin-Engine Straight Wing Upper Surface Blown Jet Flap Configuration," NASA TN-D-7778 (1975).

Arthur E. Phelps, III, and Charles C. Smith, Jr., "Wind-Tunnel Investigation of an Upper Surface Blown Jet-Flap Powered-Lift Configuration," NASA TN-D-7399 (1973).

Arthur E. Phelps, III, William Letko, and Robert L. Henderson, "Low Speed Wind-Tunnel Investigation of a Semispan STOL Jet Transport Wing-Body with an Upper-Surface Blown Jet Flap," NASA TN-D-7183 (1973).

Hervey C. Quigley, Seth B. Anderson, and Robert C. Innis, "Flight Investigation of the Low-Speed Characteristics of a 45° Swept-Wing Fighter-Type Airplane with Blowing Boundary-Layer Control Applied to the Trailing-Edge Flaps," NACA RM-A58E05 (1958).

Hervey C. Quigley and David G. Koenig, "A Flight Study of the Dynamic Stability of a Tilting-Rotor Convertiplane," NASA TN-D-778 (1961).

Hervey C. Quigley and Robert C. Innis, "Handling Qualities and Operational Problems of a Large Four-Propeller STOL Transport Airplane," NASA TN-D-1647 (1963).

Hervey C. Quigley, Robert C. Innis, and Curt A. Holzhauser, "A Flight Investigation of the Performance, Handling Qualities, and Operational Characteristics of a Deflected Slipstream STOL Transport Airplane Having Four Interconnected Propellers," NASA TN-D-2231 (1964).

Hervey C. Quigley, S.R. Sinclair, T.C. Nark, Jr., and J.V. O'Keefe, "A Progress Report on the Development of an Augmentor Wing Jet STOL Research Aircraft," Society of Automotive Engineers Paper 710757 (1971).

Hervey C. Quigley, Robert C. Innis, and Seth Grossmith, "A Flight Investigation of the STOL Characteristics of an Augmented Jet Flap STOL Research Aircraft," NASA TM-X-62-334 (1974).

Hervey C. Quigley and J.A. Franklin, "Lift/Cruise Fan VTOL Aircraft," *Astronautics and Aeronautics*, vol. 15, no. 12 (Dec. 1977), pp. 32–37.

Wilmer H. Reed, III, "Review of Propeller-Rotor Whirl Flutter," NASA TR-R-264 (1967).

John P. Reeder, "Handling Qualities Experience with Several VTOL Research Aircraft," NASA TN-D-735 (1961).

Elliott G. Reid, "Tests of Rotating Cylinders," NACA TN-209 (1924).

Herbert S. Ribner, "Formulas for Propellers in Yaw and Charts of the Side-Force Derivative," NACA WR-L-217 (originally ARR-3E19) (1943).

Dennis W. Riddle, R.C. Innis, J.L. Martin, and J.A. Cochrane, "Powered-lift Takeoff Performance Characteristics Determined from Flight Test of the Quiet Short-Haul Research Aircraft (QSRA)," AIAA Paper 81-2409 (1981).

Dennis W. Riddle, "Quiet Short-Haul Research Aircraft—A Summary of Flight Research Since 1981," Society of Automotive Engineers Paper 872315 (1988).

Leonard Roberts and Wallace R. Deckert, "Recent Progress in VSTOL Technology," NASA TM-84238 (1982).

L. Stewart Rolls and Robert C. Innis, "A Flight Evaluation of a Wing-Shroud-Blowing Boundary-Layer Control System Applied to the Flaps of an F9F-4 Airplane," NACA RM-A55K01 (1956).

L. Stewart Rolls and Fred J. Drinkwater, III, "A Flight Determination of the Attitude Control Power and Damping Requirements for a Visual Hovering Task in the Variable Stability and Control X-14A Research Vehicle," NASA TN-D-1328 (1962).

L. Stewart Rolls and Fred J. Drinkwater, III, "A Flight Evaluation of Lunar Landing Trajectories Using a Jet VTOL Test Vehicle," NASA TN-D-1649 (1963).

L. Stewart Rolls and Ronald M. Gerdes, "Flight Evaluation of Tip-Turbine-Driven Fans in a Hovering VTOL Vehicle," NASA TN-D-5491 (1969).

L. Stewart Rolls, Fred J. Drinkwater, III, and Robert C. Innis, "Effects of Lateral Control Characteristics on Hovering a Jet Lift VTOL Aircraft," NASA TN-D-2701 (1965).

L. Stewart Rolls, H.C. Quigley, and R.G. Perkins, Jr., "Review of V/STOL Lift/Cruise Fan Technology," AIAA Paper 76-931 (1976).

Herman R. Salmon, "XFV-1," *Technical Review* (of the Society of Experimental Test Pilots), vol. 13, no. 2 (1976), pp. 238–240.

S. Salmirs and R.J. Tapscott, "The Effects of Various Combinations of Damping and Control Power on Helicopter Handling Qualities During Both Instrument and Visual Flight," NASA TN-D-58 (1959).

John J. Schneider, "The History of V/STOL Aircraft," (two parts) *Vertiflite*, vol. 29, nos. 3–4 (Mar.–Apr. and May–June, 1983), pp. 22–29, 36–43.

Jeffrey A. Schroeder and Vernon K. Merrick, "Flight Evaluations of Several Hover Control and Display Combinations for Precise Blind Vertical landings," AIAA Paper 0-3479 (1990).

Jeffrey A. Schroeder and Vernon K. Merrick, "Control and Display Combinations for Blind Vertical Landings," *Journal of Guidance, Control, and Dynamics*, vol. 15, no. 3 (May–June 1992), pp. 751–760.

Jeffrey A. Schroeder, Ernesto Moralez, III, and Vernon K. Merrick, "Simulation Evaluation of the Control System Command Monitoring Concept for the NASA V/STOL Research Aircraft (VSRA)," AIAA Paper 87-2255 (1987).

Jeffrey A. Schroeder, Ernesto Moralez, III, and Vernon K. Merrick, "Evaluation of a Command Monitoring Concept for a V/STOL Research Aircraft," *Journal of Guidance, Control, and Dynamics*, vol. 12, no. 1 (Jan.–Feb. 1989), pp. 46–53.

J.M. Schuler, R.E. Smith, and J.V. Lebacqz, "An Experimental Investigation of STOL Longitudinal Flying Qualities in the Landing Approach Using the Variable Stability X-22A Aircraft," AHS Preprint No. 642 of a paper presented at the *28th Annual National Forum of the American Helicopter Society, May 17–19, 1972.*

Barry C. Scott, Charles S. Hynes, Paul W. Martin, and Ralph B. Bryder, "Progress Toward Development of Civil Airworthiness Criteria for Powered-Lift Aircraft," FAA-RD-76-100 and NASA TM-X-73124 (1976).

James Sheiman, "A Tabulation of Helicopter Rotor-Blade Differential Pressures, Stresses, and Motions As Measured In Flight," NASA TM-X-952 (1964).

James Sheiman and L.H. Ludi, "Qualitative Evaluation of Effect of Helicopter Rotor Blade Tip Vortex on Blade Airloads," NASA TN-D-1637 (1963).

M.D. Shovlin and J.A. Cochrane, "An Overview of the Quiet Short-Haul Research Aircraft Program," NASA TM-78545 (1978).

William C. Sleeman, Jr., and William C. Hohlweg, "Low-Speed Wind-Tunnel Investigation of a Four-Engine Upper Surface Blown Model Having a Swept Wing and Rectangular and D-Shaped Exhaust Nozzles," NASA TN-D-8061 (1975).

Charles C. Smith, "Flight Tests of a 1/6-Scale Model of the Hawker P 1127 Jet VTOL Airplane," NASA TM-SX-531 (1961).

William J. Snyder, John Zuk, and Hans Mark, "Tilt Rotor Technology Takes Off," AIAA Paper 89-2359 (1989).

D.A. Spence, "The Flow Past a Thin Wing with an Oscillating Jet Flap," *Philosophical Transactions of the Royal Society of London. Series A, Mathematical and Physical Sciences*, vol. 257, no. 1085 (June 3, 1965), pp. 445–477.

R.E. Spitzer, P.C. Rumsey, and H.C. Quigley, "Use of the Flight Simulator in the Design of a STOL Research Aircraft," AIAA Paper 72-762 (1972).

K.V. Stenberg, "YAV-8B Flight Demonstration Program," AIAA Paper 83-1055 (1983).

Jack D. Stephenson and Gordon H. Hardy, "Longitudinal Stability and Control Characteristics of the Quiet Short-Haul Research Aircraft (QSRA)," NASA TP-2965 (1989).

Jack D. Stephenson, James A. Jeske, and Gordon H. Hardy, "Lateral-Directional Stability and Control Characteristics of the Quiet Short-Haul Research Aircraft (QSRA)," NASA TM-102250 (1990).

V.C. Stevens, D.W. Riddle, J.L. Martin, and R.C. Innis, "Powered-lift STOL Aircraft Shipboard Operations—a Comparison of Simulation, Land-based and Sea Trial Results for the QSRA," AIAA Paper 81-2480 (1981).

D.L. Stimpert, "Effect of Crossflow Velocity on VTOL Lift Fan Blade Passing Frequency Noise Generation," NASA CR-114566 (1973).

William T. Suit and James L. Williams, "Longitudinal Aerodynamic Parameters of the Kestrel Aircraft (XV-6A) Extracted from Flight Data," NASA TN-D-7296 (1973).

William T. Suit and James L. Williams, "Lateral Static and Dynamic Aerodynamic Parameters of the Kestrel Aircraft (XV-6A) Extracted from Flight Data," NASA TN-D-7455 (1974).

Howard L. Turner and Fred J. Drinkwater, III, "Longitudinal-Trim Characteristics of a Deflected Slipstream V/STOL Aircraft During Level Flight at Transition Flight Speeds," NASA TN-D-1430 (1962).

Howard L. Turner and Fred J. Drinkwater, III, "Some Flight Characteristics of a Deflected Slipstream VSTOL Aircraft," NASA TN-D-1891 (1963).

Thomas R. Turner, Edwin E. Davenport, and John M. Riebe, "Low-Speed Investigation of Blowing from Nacelles Mounted Inboard and on the Upper Surface of an Aspect-Ratio-7.0, 35° Swept Wing With Fuselage and Various Tail Arrangements," NASA Memo 5-1-59L (1959).

Richard F. Vomaske, Robert C. Innis, Brian E. Swan, and Seth W. Grossmith, "A Flight Investigation of the Stability, Control, and Handling Qualities of an Augmented Jet Flap STOL Airplane," NASA TP-1254 (1978).

John F. Ward, "Helicopter Rotor Periodic Differential Pressures and Structural Response Measured in Transient and Steady-State Maneuvers," *Journal of the American Helicopter Society*, vol. 16, no. 1 (Jan. 1971).

John F. Ward, "An Updated History of NACA/NASA Rotary-Wing Aircraft Research 1915–1984," *Vertiflite*, vol. 30, no. 4 (May–June 1984), pp. 108–117.

William Warmbrodt and J.L. McCloud, II, "A Full-Scale Wind Tunnel Investigation of a Helicopter Bearingless Main Rotor," NASA TM-81321 (1981).

William Warmbrodt, Charles Smith, and Wayne Johnson, "Rotorcraft Research Testing in the National Full-Scale Aerodynamics Complex at NASA Ames Research Center," NASA TM-86687 (1985).

DeLamar M. Watson, Gordon H. Hardy, and David N. Warner, Jr., "Flight Test of the Glide-Slope Track and Flare-Control Laws for an Automatic Landing System for a Powered-Lift STOL Airplane," NASA TP-2128 (1983).

J.B. Wheatley, "An Aerodynamic Analysis of the Autogiro Rotor With Comparison Between Calculated and Experimental Results," NACA Report 487 (1934).

J.B. Wheatley, "Lift and Drag Characteristics and Gliding Performance of an Autogiro as Determined In Flight," NACA Report 434 (1932).

D.E. Wilcox and H.C. Quigley, "V/STOL Aircraft Simulation: Requirements and Capabilities at Ames Research Center," AIAA Paper 8-1515 (1978).

Stephan Wilkinson, "Going Vertical," *Air & Space*, vol. 11, no. 4 (Oct.–Nov. 1996), pp. 50–61.

E.B. Wolff, "Tests for Determining the Effect of a Rotating Cylinder Fitted into the Leading Edge of an Airplane Wing," NACA TM-354 (1926).

Edgar C. Wood, "The Army Helicopter, Past, Present and Future," *Journal of the American Helicopter Society*, vol. 1, no. 1 (Jan. 1956), pp. 87–92.

William T. Yeager, Jr., and Raymond G. Kvaternik, "A Historical Overview of Aeroelasticity Branch and Transonic Dynamics Tunnel Contributions to Rotorcraft Technology and Development," NASA TM-2001-211054, ARL-TR-2564 (2001).

C.H. Zimmerman, "Aerodynamic Characteristics of Several Airfoils of Low Aspect Ratio," NACA TN No. 539 (1935).

C.H. Zimmerman, Paul R. Hill, and T.L. Kennedy, "Preliminary Experimental Investigation of the Flight of a Person Supported by a Jet Thrust Device Attached to his Feet," NACA RM-L52D10 (1953).

Books and Monographs:

Jakob Ackeret, *Das Rotorschiff und seine physikalischen Grundlagen* (Göttingen: Vandenhoeck & Ruprecht, 1925).

Seth B. Anderson, *Memoirs of an Aeronautical Engineer: Flight Testing at Ames Research Center, 1940–1970,* no. 26 in the *Monographs in Aerospace History* series (Washington, DC: GPO, 2002).

Andy Bilimovich, *Jets and Missiles* (New York: Trend Books, Inc., 1957).

Jan A. van der Bliek, *75 Years of Aerospace Research in the Netherlands, 1919–1994* (Amsterdam: NLR, 1994).

Paul F. Borchers, James A. Franklin, and Jay W. Fletcher, *Flight Research at Ames: Fifty-Seven Years of Development and Validation of Aeronautical Technology,* NASA SP-1998-3300 (Moffett Field, CA: Ames Research Center, 1998).

Walter J. Boyne and Donald S. Lopez, eds., *Vertical Flight: The Age of the Helicopter* (Washington, DC: Smithsonian Institution Press, 1984).

Glenn E. Bugos, *Atmosphere of Freedom, Sixty Years at the Ames Research Center,* NASA SP-4314 (Washington, DC: NASA, 2000).

Joseph R. Chambers, *Partners in Freedom: Contributions of the Langley Research Center to U.S. Military Aircraft of the 1990s,* NASA SP-2000-4519 (Washington, DC: GPO, 2000).

Joseph R. Chambers, *Innovation in Flight: Research of the NASA Langley Research Center on Revolutionary Advanced Concepts for Aeronautics,* NASA SP-2005-4539 (Washington, DC: GPO, 2005).

Mark A. Chambers, *Engineering Test Pilot, The Exceptional Career of John P. "Jack" Reeder* (Richmond: Virginia Aeronautical Historical Society, 2007).

Michael Eckert, *The Dawn of Fluid Dynamics: A Discipline Between Science and Technology* (Weinheim, Ger.: Wiley-VCH Verlag GmbH & Co. KGaA, 2006).

Michael H. Gorn, *Expanding the Envelope: Flight Research at NACA and NASA* (Lexington: The University Press of Kentucky, 2001).

Col. H.F. Gregory, *Anything a Horse Can Do* (New York: Reynal & Hitchcock, 1944).

Richard P. Hallion, *Test Pilots: The Frontiersmen of Flight* (Garden City, NY: Doubleday & Co., 1981).

James R. Hansen, *Engineer in Charge: A History of the Langley Aeronautical Laboratory, 1917–1958,* NASA SP-4305 (Washington, DC: GPO, 1987).

James R. Hansen, *Spaceflight Revolution: Langley Research Center From Sputnik to Apollo*, NASA SP-4308 (Washington, DC: GPO, 1995).

Edwin P. Hartman, *Adventures in Research, A History of Ames Research Center, 1940–1965*, NASA SP-4302 (Washington, DC: GPO, 1970).

Michael J. Hirschberg, *The American Helicopter: An Overview of Helicopter Developments in America, 1908–1999* (Arlington, VA: ANSER, 1999).

Theodore von Kármán, *Aerodynamics* (New York: McGraw-Hill Book Publishing Co., 1963 ed.).

Martin D. Maisel, Demo J. Giulianetti, and Daniel C. Dugan, *The History of the XV-15 Tilt Rotor Research Aircraft From Concept to Flight*, NASA SP-2000-4517, No. 17 in the *Monographs in Aerospace History* series (Washington, DC: GPO, 2000).

Donald L. Mallick, with Peter W. Merlin, *The Smell of Kerosene: A Test Pilot's Odyssey,* NASA SP-4108 (Washington, DC: GPO, 2003).

Francis K. Mason, *The Hawker P.1127 and Kestrel*, No. 198 in the *Profile Publications* series (Leatherhead, England: Profile Publications, Ltd., 1967).

Gene J. Matranga, C. Wayne Ottinger, and Calvin R. Jarvis, with Christian Gelzer, *Unconventional, Contrary, and Ugly: The Lunar Landing Research Vehicle,* NASA SP-2204-4535, No. 35 in the *Monographs in Aerospace History* series, (Washington, DC: GPO, 2006).

H.C.M. Merewether, *Prelude to the Harrier: P.1127 Prototype Flight Testing and Kestrel Evaluation* (Beirut, Lebanon: HPM Publications, 1998).

Jay Miller, *The X-Planes: X-1 to X-45* (Hinckley, England: Midland Publishing, 2001 ed.).

Elizabeth A. Muenger, *Searching the Horizon: A History of Ames Research Center, 1940–1976,* NASA SP-4304 (Washington, DC: GPO, 1985).

F.G. Swanborough, *Vertical Flight Aircraft of the World* (Los Angeles, CA: Aero Publishers Inc., 1964).

Lt. Gen. John J. Tolson, *Airmobility, 1961–1971*, a volume in the U.S. Army *Vietnam Studies* series (Washington, DC: U.S. Army, 1973).

Ray Wagner, *American Combat Planes* (Garden City, NY: Doubleday & Co., 1982 ed.).

Tupolev-144 SST on takeoff from Zhukovsky Air Development Center in Russia with a NASA pilot at the controls. NASA.

CASE 15

NASA's Flight Test of the Russian Tu-144 SST

Robert A. Rivers

The aeronautics community has always had a strong international flavor. This case study traces how NASA researches in the late 1990s used a Russian supersonic airliner, the Tupolev Tu-144LL—built as a visible symbol of technological prowess at the height of the Cold War—to derive supersonic cruise and aerodynamic data. Despite numerous technical, organizational, and political challenges, the joint research team obtained valuable information and engendered much goodwill.

ON A COOL, CLEAR, AND GUSTY SEPTEMBER MORNING in 1998, two NASA research pilots flew a one-of-a-kind, highly modified Russian Tupolev Tu-144LL Mach 2 Supersonic Transport (SST) side by side with a Tupolev test pilot, navigator, and flight engineer from a formerly secret Soviet-era test facility, the Zhukovsky Air Development Center 45 miles southeast of Moscow, on the first of 3 flights to be flown by Americans.[1] These flights in Phase II of the joint United States-Russian Tu-144 flight experiments sponsored by NASA's High-Speed Research (HSR) program were the culmination of 5 years of preparation and cooperation by engineers, technicians, and pilots in the largest joint aeronautics program ever accomplished by the two countries. The two American pilots became the first and only non-Russian pilots to fly the former symbol of Soviet aeronautics prowess, the Soviet counterpart of the Anglo-French Concorde SST.

They completed a comprehensive handling qualities evaluation of the Tu-144 while 6 other experiments gathered data from hundreds of onboard sensors that had been painstakingly mounted to the airframe

1. The author gratefully acknowledges the essential and superb support provided by Russ Barber and Glenn Bever, of the Dryden Flight Research Center, and Bruce Jackson, Steve Rizzi, and Donna Amole, of the Langley Research Center. This case study is dedicated to the many diligent professionals in the United States and Russia who made this project a reality and to my wife, Natale, and my sons, Jack and Sam, without whose love and support it could not have been completed.

in the preceding 3 years by NASA, Tupolev, and Boeing engineers and technicians. Only four more flights in the program awaited the Tu-144LL, the last of its kind, before it was retired. With the removal from service of the Concorde several years later, the world lost its only supersonic passenger aircraft and witnessed the end of an amazing era.

This is the story of a remarkable flight experiment involving the United States and Russia, NASA and Tupolev, and the men and women who worked together to accomplish a series of unique flight tests from late 1996 to early 1999 while overcoming numerous technical, programmatic, and political obstacles. What they accomplished in the late 1990s cannot be accomplished today. There are no more Supersonic Transports to be used as test platforms, no more national programs to explore commercial supersonic flight. NASA and Tupolev established a benchmark for international cooperation and trust while producing data of incalculable value with a class of vehicles that no longer exists in a regime that cannot be reached by today's transport airplanes.[2]

HSR and the Genesis of the Tu-144 Flight Experiments

NASA's High-Speed Research program was initiated in 1990 to investigate a number of technical challenges involved with developing a Mach 2+ High-Speed Civil Transport (HSCT). This followed several years of NASA-sponsored studies in response to a White House Office of Science and Technology Policy call for research into promoting long-range, high-speed aircraft. The speed spectrum for these initial studies spanned the supersonic to transatmospheric regions, and the areas of interest included economic, environmental, and technical considerations. The studies suggested a viable speed for a proposed aircraft in the Mach 2 to Mach 3.2 range, and this led to the conceptual model for the HSR program. The initial goal was to determine if major environmental obstacles—including ozone depletion, community noise, and sonic boom generation—could be overcome. NASA selected the Langley Research Center in Hampton, VA, to lead the effort, but all NASA aeronautics Centers became deeply involved in this enormous program. During this

2. Much of the background material for this essay, unless otherwise referenced, comes from the author's extensive notes and papers related to this program. As one of the NASA research pilots participating in this experiment, the author was a firsthand witness to the events described herein. Where published documents do not exist, interviews and the notes from the other participants were used to complete the account.

Phase I period, NASA and its industry partners determined that the state of the art in high-speed design would allow mitigation of the ozone and noise issues, but sonic boom alleviation remained a daunting challenge.[3]

Encouraged by these assessments, NASA began Phase II of the HSR program in 1995 in partnership with Boeing Commercial Airplane Group, McDonnell-Douglas Aerospace, Rockwell North American Aircraft Division, General Electric Aircraft Engines, and Pratt & Whitney. By this time, a baseline concept had emerged for a Mach 2.4 aircraft, known as the Reference H model and capable of carrying 300 passengers non-stop across the Pacific Ocean. A comprehensive list of technical issues was slated for investigation, including sonic boom effects, ozone depletion, aeroacoustics and community noise, airframe/propulsion integration, high lift, and flight deck design. Of high interest to NASA Langley Research Center engineers was the concept of Supersonic Laminar Flow Control (SLFC). Maintaining laminar flow of the supersonic airstream across the wing surface for as long as possible would lead to much higher cruise efficiencies. NASA Langley investigated SLFC using wind tunnel, computational fluid dynamics, and flight-test experiments, including the use of NASA's two F-16XL research aircraft flown at NASA Langley and NASA Dryden Flight Research Centers. Unfortunately, the relatively small size of the unique, swept wing F-16XL led to contamination of the laminar flow by shock waves emanating from the nose and canopy of the aircraft. Clearly, a larger airplane was needed.[4]

That larger airplane seemed more and more likely to be the Tupolev Tu-144 as proposals devolved from a number of disparate sources, and a variety of serendipitous circumstances aligned in the early 1990s to make that a reality. Aware of the HSR program, the Tupolev Aircraft Design Bureau as early as 1990 proposed a Tu-144 as a flying laboratory for supersonic research. In 1992, NASA Langley's Dennis Bushnell discussed with Tupolev this possibility of returning to flight one of the few remaining Tu-144 SSTs as a supersonic research aircraft. Pursuing Bushnell's initial inquiries, Joseph R. Chambers, Chief of Langley's Flight Applications Division, and Kenneth Szalai, NASA's Dryden Flight Research Center

3. Joseph R. Chambers, *Innovation in Flight: Research of the NASA Langley Research Center on Revolutionary Advanced Concepts for Aeronautics*, NASA SP-2005-4539, pp. 49–54. Chambers provides an informative history of supersonic research at NASA from the 1960s through the High-Speed Research program as well as a complete bibliography.

4. Chambers, *Innovation in Flight*, pp. 58–60.

Director, developed a formal proposal for NASA Headquarters suggesting the use of a Tu-144 for SLFC research. Szalai discussed this idea with his friend Lou Williams, of the HSR Program Office at NASA Headquarters, who became very interested in the Tu-144 concept. NASA Headquarters had, in the meantime, already been considering using a Tu-144 for HSR research and had contracted Rockwell North American Aircraft Division to conduct a feasibility study. NASA and Tupolev officials, including Ken Szalai, Lou Williams, and Tupolev chief engineer Alexander Pukhov, first directly discussed the details of a joint program at the Paris Air Show in 1993, after Szalai and Williams had requested to meet with Tupolev officials the previous day.[5] The synergistic force ultimately uniting all of this varied interest was the 1993 U.S.–Russian Joint Commission on Economic and Technological Cooperation. Looking at peaceful means of technological cooperation in the wake of the Cold War, the two former adversaries now pursued programs of mutual interest. Spurred by the Commission, NASA, industry, and Tupolev managers and researchers evaluated the potential benefits of a joint flight experiment with a refurbished Tu-144 and developed a prioritized list of potential experiments. With positive responses from NASA and Tupolev, a cooperative Tu-144 flight research project was initiated and an agreement signed in 1994 in Vancouver, Canada, between Russian Prime Minister Viktor Chernomyrdin and Vice President Al Gore. Ironically, Langley's interest in SLFC was not included in the list of experiments to be addressed in this largest joint aeronautics research project between the two former adversaries.[6] Ultimately, seven flight experiments were funded and accomplished by NASA, Tupolev, and Boeing personnel (Boeing acquired McDonnell-Douglas and Rockwell's aerospace division in December 1996). Overcoming large distances, language and political barriers, cultural differences, and even different approaches to technical and engineering problems, these dedicated researchers, test pilots, and technicians accomplished 27 successful test flights in 2 years.

5. Interview of Marvin R. Barber by author, NASA Dryden Flight Research Center, July 1, 2009. Russ Barber was the NASA Dryden Project Manager for the Tu-144 flight experiment from Sept. 1994 through Dec. 1998. He managed the contractual, budget, and support activities and served as an interface between NASA and the contractors, including Boeing and Tupolev.

6. Barber, "The Tu-144LL Supersonic Flying Laboratory," unpublished, draft NASA study, NASA HD archives, 2000.

The Tu-144 Flight Experiments Project

While negotiations were underway in 1993, leading to the agreement between the United States and Russia to return a Tu-144D to flight status as a supersonic flying laboratory, the HSR Program Office selected NASA Dryden to establish a Project Office for all Tu-144 activities. This initially involved developing a rapport with a British company, IBP, Ltd., which served as the business representative for Tupolev, now known as the Tupolev Aircraft Company (or Tupolev ANTK) after the economic evolution in Russia in the 1990s. Ken Szalai and IBP's Judith DePaul worked to establish an effective business relationship, and this paid dividends in the ensuing complex relationships involving NASA, Rockwell, McDonnell-Douglas, Boeing, Tupolev, and IBP. A degree of cooperation flourished at a level not always observed in NASA–Russian partnerships. Having a business intermediary such as IBP navigate the paths of international business helped ensure the success of the Tu-144 experiment, according to Dryden Tu-144 Project Manager Russ Barber.[7]

Originally, the Tu-144 flight experiment was envisioned as a 6-month, 30-flight program.[8] As events unfolded, the experiment evolved into a two-phase operation. This was due, in part, to the inevitable delays in an enterprise of this magnitude and complexity, to learning from the results of the initial experiments, and to data acquisition issues.[9] By 1995, after two meetings in Russia,the HSR Program Office, Boeing, Rockwell, McDonnell-Douglas, and Tupolev established the requirements for returning a Tu-144D to flight and fabricating an instrumentation system capable of supporting the postulated lineup of experiments.[10] From a list of some 50 proposed experiments, the NASA, industry, and Tupolev officials selected 6 flight experiments for inclusion (a 7th was later added).[11]

---

7. Barber interview.
8. Chambers, *Innovation in Flight*, p. 60.
9. Roy V. Harris, Jr., "Tu-144LL Flight Experiments Review and Critique," NASA HSR Program Office, NASA Langley Research Center (Feb. 9, 1999). This valuable, unpublished document was a report to the HSR Program Office by a former NASA Langley aerodynamicist and Director of Aeronautics. The HSR Program Office tasked him to independently review the Phase I results, data quality, data uniqueness, and program applicability, and to make recommendations regarding use of the Tu-144LL. Harris was an expert on supersonic flight.
10. E-mail interview of Norman H. Princen by author, Apr. 7, 2009.
11. NASA Dryden Flight Research Center, "The Tu-144LL: A Supersonic Flying Laboratory," Internet Fact Sheet.

A somewhat complex international organization developed that, despite the superficial appearance of duplication, ended up working very smoothly. NASA Dryden represented the HSR Program Office as the overseer for all Tu-144 activity. Boeing was contracted to install the instrumentation system, a complex task with over 700 individual pressure transducers, accelerometers, thermocouples, boundary layer rakes, pressure belts, microphones, and other sensors. NASA Dryden installed a complex French-built Damien digital data acquisition system (DAS) for five of the original six experiments.[12] The remaining experiment, a NASA Langley Structure/Cabin Noise experiment, used its own Langley-built DAS.[13] In a sense, traditional roles had to be adjusted, because Boeing, as the contractor, directed NASA, as the Government Agency and supplier, when to provide the necessary sensors and DAS.[14] Boeing and Tupolev would install the sensors, and NASA would then calibrate and test them. The Damien DAS ultimately became problematic and led to some erroneous data recording in Phase I.[15]

Tupolev assumed the role of returning the selected Tu-144D, SSSR-771114, to flight. This was no trivial matter. Even though 771114 had last flown in 1990, the engines were no longer supported and had to be replaced (as discussed in a subsequent section), which necessitated major modifications to the engine nacelles, elevons, and flight deck.[16] As Tupolev was completing this work in 1995 and 1996, IBP acted as its business interface with NASA and Boeing.

In general, the HSR program funded the American effort. The cost to NASA for the Tu-144 flight experiment was $18.3 million for 27 flights. Boeing contributed $3.3 million, and it is estimated that Tupolev spent $25 million.[17] Tupolev gained a fully instrumented and refurbished Tu-144, but unfortunately, after NASA canceled the HSR program in 1999, Tupolev could find no other customers for its airplane.

---

12. Ibid.

13. Stephen A. Rizzi, "Brief Background of Program and Overview of Experiment," unpublished notes, NASA Langley Research Center, June 19, 1998.

14. Barber interview.

15. Harris, "Flight Experiments Review and Critique."

16. Robert A. Rivers, E. Bruce Jackson, C. Gordon Fullerton, Timothy H. Cox, and Norman H. Princen, "A Qualitative Piloted Evaluation of the Tupolev Tu-144 Supersonic Transport," NASA TM-2000-209850 (Feb. 2000), p. 4.

17. Harris, "Flight Experiments Review and Critique."

During the initial program definition and later during the aircraft modification, a number of HSR, Dryden, and Langley personnel made numerous trips to Zhukovsky. HSR managers coordinated program schedules and experiment details, Dryden personnel observed the return to flight efforts as well as the instrumentation modifications and provided flight operations inputs, and Langley instrumentation technicians and researchers assisted with their experiment installation. Among the Dryden visitors to Zhukovsky was NASA research pilot Gordon Fullerton. Fullerton was the NASA pilot interface during these development years and worked with his Tupolev counterparts on flight deck and operational issues. In an interview with the author, he recalled the many contrasts in the program regarding the Russian and American methods of engineering and flight operations. Items worthy of minute detail to the Russians seemed trivial at times to the Americans, while American practices at times resulted in confused looks from the Tupolev personnel. By necessity, because of a lack of computer assets, the Tupolev pilots, engineers, and technicians worked on a "back of the envelope" methodology. Involvement of multiple parties in decisions was thus restricted simply because of a lack of easy means to include them all. Carryovers from the Soviet days were still prevalent in the flightcrew distribution of duties, lack of flight deck instrumentation available to the pilots, and ground procedures that would be viewed as wholly inefficient by Western airlines. Nevertheless, Tupolev produced an elegant airplane that could fly a large payload at Mach 2.[18]

As the American and Russian participants gained familiarity, a spirit of trust and cooperation developed that ultimately contributed to the project's success. The means of achieving this trust were uniquely Russian. As the various American delegations arrived in Moscow or Zhukovsky, they were routinely feted to gala dinners with copious supplies of freely offered vodka. This was in the Russian custom of becoming acquainted over drinks, during which inhibitions that might mask hidden feelings were relaxed. The custom was repeated over and over again throughout the program. Few occasions passed without a celebratory party of some degree: preflight parties, postflight parties, welcoming parties, and farewell parties were all on the agenda. Though at times challenging for some of the American guests who did not drink,

18. Interview of Fullerton by author, Lancaster, CA, July 1, 2009.

these social gatherings were very effective at cementing friendships among two peoples who only a few years before uneasily coexisted, with all of their respective major cities targeted by the other's missiles. To a person, the Americans who participated in this program realized that on a personal level, the Russians were generous hosts, loyal friends, and trusted colleagues. If nothing else, this was a significant accomplishment for this program.

Nineteen flights were completed by early 1998, achieving most of the original program goals. However, some data acquisition problems had rendered questionable some of the data from the six experiments.[19] The HSR Program Office decided that it would be valuable to have United States research pilots evaluate the Tu-144 in order to develop corporate knowledge within NASA regarding SST handling qualities and to ascertain if the adverse handling qualities predicted by the data collected actually existed. Furthermore, there were additional data goals developed since the inception of the program, and a seventh experiment was organized. The resumption of the test flights was scheduled for September 1998. The HSR Program Office and Boeing selected Gordon Fullerton from Dryden and NASA research pilot Robert A. Rivers from Langley as the evaluation pilots. Fullerton had been the Dryden project pilot for the Tu-144 modification and refurbishment, and he was familiar with the Tupolev flightcrews and the airplane. Rivers had been the HSR project pilot for several years, had participated in every HSR flight simulation experiment, served on two HSR integrated test development teams, and had performed an extensive handling qualities evaluation of the Concorde SST the previous year. To accompany them to Zhukovsky were two NASA flight control engineers, Timothy H. Cox from Dryden and E. Bruce Jackson from Langley, and Boeing Tu-144 project handling qualities engineer Norman H. Princen. Jackson had completed extensive work on flight control development for the HSCT Reference H model. During summer 1998, the team members worked together to develop a draft test plan, flew both the Ames and Langley 6-degree-of-freedom motion simulators with the Reference H model, and began studying the Tu-144 systems with the rudimentary information available in the United States at that time. On September 4, they departed for Zhukovsky.

19. Harris, "Flight Experiments Review and Critique."

Members of the United States Pilot Evaluation Team (USPET) and their Russian counterparts in front of the KGB sanitarium in Zhukovsky, Russia. From left to right, Dryden's Tim Cox and Gordon Fullerton, Langley's Rob Rivers, Tupolev's Victor Pedos, Langley's Bruce Jackson, Tupolev's Sergei Borisov, Boeing's Norm Princen, and Russian translator Yuri Tsibulin. NASA.

Onsite in Zhukovsky

The United States Pilot Evaluation Team (USPET)[20] arrived in Moscow on Sunday, September 6, 1998, and was met by Professor Alexander Pukhov and a delegation of Tupolev officials. (Ill fortune had struck the team when NASA Langley research pilot Robert Rivers severely broke his right leg and ankle 2 weeks before departure. Because visas for work in Russia required 60 days' lead time and because no other pilot could be prepared in time, Rivers remained on the team, though it required a great deal of perseverance to obtain NASA approval. Tupolev presented relatively few obstacles, by contrast, to Rivers's participation.) Pukhov was the Tupolev Manager for the Tu-144 experiment and a former engineer on the original design team for the airplane. At Pukhov's insistence, USPET was billeted in Zhukovsky at the former KGB sanitarium. Sanitaria in the Soviet Union were rest and vacation spas for the various professional groups, and the KGB sanitarium was similar to a large hotel. The sanitarium was minutes from the Zhukovsky Air

20. Langley engineer and USPET member Bruce Jackson coined USPET on arrival in Russia as a wordplay between NASA's penchant for acronyms and the Russian language the team was trying to learn. The name stuck, at least informally.

Development Center and saved hours of daily commute time that otherwise might have been wasted had the team been housed in Moscow.

The next day began a very intense training period lasting 2 weeks but was punctuated September 15 by the first flight by American pilots, a subsonic sojourn. The training was complicated by the language differences but was facilitated by highly competent Russian State Department translators. Nevertheless, humorous if not frustrating problems arose when nontechnical translators attempted to translate engineering and piloting jargon with no clear analogs in either language. The training consisted of one-on-one sitdown sessions with various Tu-144 systems experts using manuals and charts written in Russian. There were no English language flight or systems manuals for the Tu-144, and USPET's attempt over the summer to procure a translated Tu-144 flight manual was unsuccessful. Training included aircraft systems, life support, and flight operations. Because flights would achieve altitudes of 60,000 feet and because numerous hull penetrations had occurred to accommodate the instrumentation system, all members of the flightcrew wore partial pressure suits. Because of the experimental nature of the flights, a manual bailout capability had been incorporated in the Tu-144. This involved dropping through a hatch just forward of the mammoth engine inlets. The hope was that the crewmember would pass between the two banks of engines without being drawn into the inboard inlets. Thankfully, this theory was never put to the test.

Much time was spent with the Tupolev flightcrew for the experiment, and great trust and friendship ensued. Tupolev chief test pilot Sergei Borisov was the pilot-in-command for all of the flights. Victor Pedos was the navigator, in actuality a third pilot, and Anatoli Kriulin was the flight engineer. Tupolev's chief flight control engineer, Vladimir Sysoev, spent hours each day with USPET working on the test plan for each proposed flight. Sysoev and Borisov represented Tupolev in the negotiations to perform the maneuvers requested by the various researchers.[21] An effective give-and-take evolved as the mutual trust grew. From Tupolev's perspective, the Tu-144 was a unique asset, into which the fledgling free-market company had invested millions of dollars. It provided badly needed funds at a time when the Russian economy was struggling, and

21. Rivers, et al., "A Qualitative Piloted Evaluation of the Tu-144," NASA TM-2000-209850, p. 18. Complete details of the flight-test planning and preparation for the American-flown flights is provided in that document.

the payments from NASA via Boeing and IBP were released only at the completion of each flight. The Tupolev crewmembers could not afford to risk the airplane. At the same time, they were anxious to be as cooperative as possible. Careful and inventive planning resulted in nearly all of the desired test points being flown.

The Aircraft: Tu-144LL SSSR-77114

The Tu-144 was the world's first Supersonic Transport, when it took off from Zhukovsky Airfield on December 31, 1968. The design of the aircraft had commenced in early 1963, after the Soviet Union selected the Tupolev Design Bureau for the task. The famed Andrei Tupolev named his son Aleksei Tupolev to be chief designer, and over 1,000 staff members from other design bureaus were temporarily assigned to Tupolev for this project of national prestige.[22] For the researchers to evaluate the wing design, a Mig-21 fighter was configured with a scaled model of the wing for in-flight testing. The prototype was completed in the summer of 1968, and in December of that year, Eduard Yelian piloted serial No. SSSR-68001on the Tu-144's first flight. The Tu-144 first exceeded the speed of sound on June 5, 1969 and achieved speeds in excess of Mach 2.0 on May 26, 1970, in every case just beating Concorde.[23]

The prototype was displayed at the Paris Air Show for the first time in June 1971. Tragically, the second production aircraft crashed spectacularly at the 1973 Paris Air Show. This, in combination with range capabilities only about half of what was expected (2,200 miles versus 4,000 miles), led to Aeroflot (the Soviet national airline company) having a diminishing interest in the aircraft. Still, a number of significant modifications to the aircraft occurred in the 1970s. The engine nacelles were move farther outboard, necessitating the relocation of the main landing gear to the center of the nacelles, and the original Kuznetsov NK-144 engines were replaced by Kolesov RD-36-51A variants capable of 44,092 pounds of thrust with afterburner. With these engines, the type was redesignated the Tu-144D, and serial No. SSSR-74105, the fifth

22. Howard Moon, *Soviet SST: The Technopolitics of the Tupolev-144* (New York: Orion Books, 1989), pp. 75–88. Moon provides a detailed history of Tupolev's efforts to build and market the Tu-144 against a background of politics and national pride.
23. Paul Duffy and Andrei Kandalov, *Tupolev: The Man and His Aircraft* (Shrewsbury, England: Airlife Publishing, Ltd., 1996), pp. 153–157. Duffy and Kandalov report on the history of the Tupolev Design Bureau and give an account of the development of the Tu-144.

15

A low pass over Zhukovsky Air Development Center by Tu-144LL SSSR-771114 in September 1998. Note the Russian and American flags on the tail. NASA.

production aircraft, first flew with the new engines in November 1974. Cargo and mail service commenced in December 1975, but Aeroflot crews never commanded a single Tu-144. Only Tupolev test pilots ever flew as pilots-in-command. On November 1, 1977, the Tu-144 received its certificate of airworthiness, and passenger service commenced within the Soviet Union. Ten percent larger than the Concorde, the Tu-144 was configured with 122 economy and 11 first-class passenger seats. Only two production aircraft served on these passenger routes. The service was terminated May 31, 1978, after the first production Tu-144D crashed on a test flight from Zhukovsky while making an emergency landing because of an in-flight fire. After this crash, four more Tu-144s were produced but were used only as research aircraft. Two continued flying until 1990, including SSSR-771114. The fleet of 16 flyable aircraft accumulated 2,556 flights and 4,110 flying hours by 1990.[24]

After the 1994 U.S.–Russian agreement enabling the HSR Tu-144 flight experiments, SSSR-77114 was selected to be refurbished for flight. The final production aircraft, 77114, was built in 1981 and flew only as a research aircraft, before being placed in storage in 1990. Amazingly, it had only accumulated 83 flight hours at that time. Because the RD-36-51A

24. Duffy and Kandalov, *Tupolev: The Man and His Aircraft*, pp. 153–157.

engines were no longer being produced or supported, Tupolev switched to the Kuznetsov NK-321 engines from the Tu-160 Blackjack strategic bomber as powerplants.[25] Redesignated the Tu-144LL, or Flying Laboratory, 77114 first flew under the command of Tupolev test pilot Sergei Borisov on November 29, 1996.[26]

The Tu-144, although it seems outwardly similar to the Concorde, was actually about 10-percent larger, with a different wing and engine configuration, and with low-speed retractable canard control surfaces that the Concorde lacked. It also solved the many challenges to sustained high-altitude, supersonic flight by different means. Where documentation in the West is complete with Concorde systems and operations manuals and descriptions, NASA and Boeing engineers and pilots could find no English counterparts for the Tu-144. This was due in part to the secrecy of the Tu-144 development in the 1960s and 1970s. Therefore, it is worth briefly describing the systems and operation of the Tu-144 in this essay.

This system description will also give insight into the former Soviet design philosophies. It should be noted that many of the systems on the Tu-144LL were designed in the 1960s, and though completely effective, were somewhat dated by the mid to late 1990s.[27]

The Tu-144LL is a delta platform, low wing, four engine Supersonic Transport aircraft. Features of interest included a very high coefficient of lift retractable canard and three position-hinged nose structure. The retractable canard is just aft of the cockpit on top of the fuselage and includes both leading- and trailing-edge flaps that deflect when the canard is deployed in low-speed flight. The only aerodynamic control surfaces are 8 trailing-edge elevons, each powered by two actuators and upper and lower rudder segments. The nominal cockpit crew

25. Rivers, et al., "A Qualitative Piloted Evaluation of the Tu-144," NASA TM-2000-209850, p. 2.

26. Stephen A. Rizzi, Robert G. Rackl, and Eduard V. Andrianov, "Flight Test Measurements from the Tu-144LL Structure/Cabin Noise Experiment," NASA TM-2000-209858 (Jan. 2000), p. 1.

27. A more thorough description can be found in Rivers, et al., "A Qualitative Piloted Evaluation of the Tu-144," NASA TM-2000-209850, pp. 4–15. This technical manuscript contains a detailed systems and operations description of the Tu-144, which, as far as is known, is the only extant English description. The systems information was obtained from the author's extensive notes, taken onsite at the Tupolev test facility in Zhukovsky, Russia, in Sept. 1998. These notes were derived from one-on-one lectures from various Tupolev systems experts, as conveyed through translators. While there may be some minor discrepancies, these systems descriptions should for the most part accurately portray the Tu-144LL. Other than the powerplant and some fuel system modifications, this account describes the generic Tu-144D aircraft as well.

A view of the cockpit of the Tu-144LL from the flight engineer's station looking forward. NASA.

consisted of two pilots, a navigator situated between the two pilots, and a flight engineer seated at a console several feet aft of the navigator on the right side of the aircraft.

The Tu-144LL was 215 feet 6 inches long with a wingspan of 94 feet 6 inches and a maximum height at the vertical stabilizer of 42 feet 2 inches. Maximum takeoff weight was 447,500 pounds, with a maximum fuel capacity of 209,440 pounds.

Quadruple redundant stability augmentation in all axes and an aileron-rudder interconnect characterized a flight control system that provided a conventional aircraft response. Control inceptors included the standard wheel-column and rudder pedals. Pitch and roll rate sensor feedbacks passed through a 2.5-hertz (Hz) structural filter to remove aeroservoelastic inputs from the rate signals. Sideslip angle feedback was used to facilitate directional stability above Mach 1.6 or when the canard or landing gear were extended. Similarly, and aileron-rudder interconnect provided additional coordination in roll maneuvers through first-order lag filters between Mach 0.9 and 1.6 and whenever the canard or landing gear were extended. A yaw rate sensor signal was fed back through a lead-lag filter to oppose random yaw motions and allow steady turn rates.

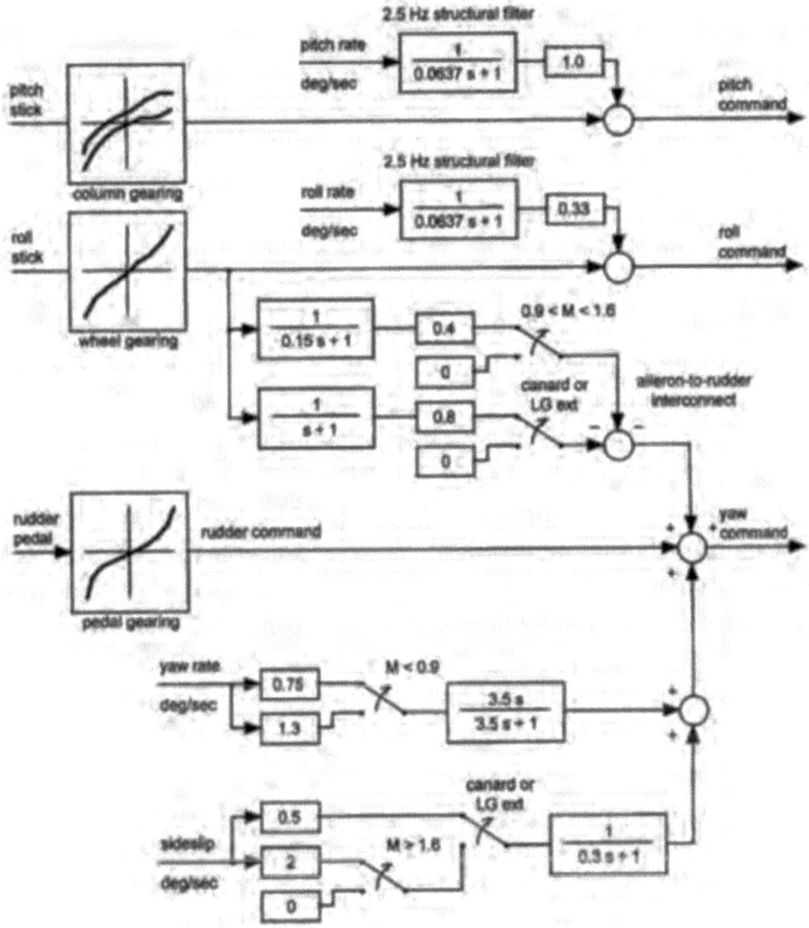

Schematic of the Tu-144LL flight control system as interpreted by NASA flight control engineer Bruce Jackson from conversations with Tupolev engineers in Zhukovsky, Russia. NASA.

Because the elevons provided both pitch and roll control, a mixer logic limited the combined pitch and roll commands to allowable elevon travel while favoring pitch commands in the limit cases. Pitch-roll harmony was moderately objectionable by Western standards because of excessive pitch sensitivity contrasted with very weak roll sensitivity.

The installed Kuznetsov NK-321 engines were rated at 55,000 pounds sea level static thrust in afterburner and 31,000 pounds dry thrust. These engines are 5 feet longer and over 1/3 inch wider than the RD-36-51A engines in the Tu-144D, which necessitated extensive modifications to the engine nacelles and nozzle assemblies. The NK-321 engines were mounted 5 feet farther forward in the nacelles, and to accommodate the larger nozzles, the inboard elevons were modified. The axisymmetric, afterburning, three-stage compressor NK-321 engines were digitally controlled, and this necessitated a redesigned flight engineer's (FE) panel with eight rows of electronic engine parameter displays. The fuel control consisted of a two channel digital electronic control and a backup hydromechanical control. The pilot is only presented with N1 revolutions per minute (rpm) indications and throttle command information, which was used to set the desired thrust through power lever angle in degrees (referred to as throttle alpha by Tupolev). All other engine information, including fuel flows and quantities, oil pressures and temperatures, and exhaust gas temperatures, was displayed on the FE panel, which is not visible to the pilot. The pilot's throttles mounted on the center console had a very high friction level, and in normal situations, the FE set the thrust as commanded by the pilot in degrees throttle alpha. Typical thrust settings in throttle alpha were 72 degrees for maximum dry power, 115 degrees for maximum wet power (afterburner), 98 degrees for Mach 2 cruise, and 59 degrees for supersonic deceleration and initial descent. For takeoff weights less than or equal to 350,000 pounds, 98 degrees throttle alpha was commanded, and for heavier takeoff weights, 115 degrees was used. Operations in the 88- to 95-degree range were avoided for undisclosed reasons.

A fairly unsophisticated, 2-channel autothrottle (A/T) system was available for approach and landing characterized by a 20-second period and an accuracy of plus or minus 4 mph. The A/T control panel was on the center console, with a left/right selector switch, two selectors for channels, and a rocker switch to command the speed bug on the respective pilot's airspeed indicator. A throttle "force" of 45 pounds was needed to override the A/T, or individual A/Ts could be deselected by microswitches in each throttle knob. If two or more were deselected, the system was disconnected. For the system to be engaged, the FE engaged A/T clutches on the FE throttle quadrant. The A/T could be used from 100 mph up to 250 mph indicated airspeed normally or up to 310 mph under test conditions.

The variable geometry inlets were rectangular, with a moderate fore-to-aft rake. An internal horizontal ramp varied from an up position at speeds below Mach 1.25 to full down at Mach 2. Three shocks were contained in the inlet during supersonic flight to slow the inlet flow to subsonic speeds; unlike those of other supersonic aircraft, the Tu-144LL's inlets showed no tendency to experience shock wave–displacing inlet unstart or other undesired responses during supersonic flight. Even when pilots made full rudder deflections while maintaining a steady heading, generating supersonic sideslips at Mach 2, the inlets and engines continued to function normally. Likewise, when they made 30-degree banked turns and moderately aggressive changes in pitch angle, there were no abnormal results from either the engine or the inlet. This contrasted markedly with the Olympus engines in Concorde. While the Olympus engines were more efficient and were designed in conjunction with the inlets and nozzles to provide a complete, interrelated powerplant system, the Tu-144LL's forced use of nonoptimized NK-321 engines required using afterburner to maintain Mach 2 cruise. Of interest was the fact that Concorde's more efficient engines (Mach 2 cruise was sustained without the use of afterburner) were far more susceptible to inlet unstarts and stalls, and as a result, the aggressive engine maneuvers performed in the Tu-144 flight experiment at Mach 2 could not have been accomplished in the Concorde. The RD-36-51A engines did not require afterburner during supersonic cruise, even though the sea level static thrust rating was lower than that of the NK-321 engines. This was due to the optimized engine/inlet/nozzle system, in which 50 percent of the thrust at supersonic cruise was derived from the inlets and nozzles.[28]

The fuel system was comprised of 8 fuel storage areas, including 17 separate tanks. The nomenclature referred to fuel tanks 1 through 8, but only tanks 6, 7, and 8 were single units. Tanks 1, 2, and 8 were balance tanks used to maintain the proper center-of-gravity (CG) location through new, high-capacity fuel transfer jet pumps with peak pressure capacity of 20 atmospheres. These transfer pumps were hydraulically driven and controlled by direct current (DC) power. Fuel boost pumps in each tank were powered by the main alternating current (AC) electrical systems. Tank system No. 4 consisted of 6 tanks, 4 of which provide tank-to-engine fuel. A cross-feed capability was used to control lateral balance. Emergency fuel dumping could be accomplished from all fuel tanks. All fuel system information was displayed on the FE panel, and all fuel system controls were accessible only to the FE. Numerous fuel quantity probes were used to provide individual tank system quantity indications and provide inputs to the CG indicator com-

28. British Aircraft Corporation, Ltd., Commercial Aircraft Division, *An Introduction to the Slender Delta Supersonic Transport* (Bristol, England: Printing and Graphic Services, Ltd., 1975), pp. 15–19.

puter on the FE panel, which continually calculated and displayed the CG location. Proper control of the Tu-144 CG during the transonic through supersonic flight regimes was critical in maintaining aircraft control as the center of lift rapidly changed during sonic transients.

The Tu-144LL incorporated four hydraulic systems, all of which were connected to separate flight control systems. Up to two hydraulic systems could fail without adversely affecting flight control capability. The flight controls consisted of four elevons per wing and an upper and lower rudder. Each control surface had two actuators with two hydraulic channels each so that each hydraulic system partially powered each control surface. The four hydraulic systems were powered by variable displacement engine driven pumps. There were no electrically powered pumps. Engine Nos. 1 and 2 each powered the No. 1 and 2 hydraulic systems, and engine Nos. 3 and 4 each powered the No. 3 and 4 hydraulic systems. Systems No. 1 and 2 and systems No. 3 and 4 shared reservoirs, but dividers in each reservoir precluded a leak in one system from depleting the other. System pressure was nominally between 200 and 220 atmospheres, and a warning was displayed to the pilot if the pressure in a system fell below 100 atmospheres. In the event of the loss of 2 hydraulic systems, an emergency hydraulic system powered by an auxiliary power unit (APU) air-driven pump (or external pneumatic source) was available, but the APU could only be operated below 3-mile altitude (and could not be started above 1.8-mile altitude). For emergency operation of the landing gear (lowering only), a nitrogen system serviced to 150 atmospheres was provided. If one hydraulic system failed, the aircraft was required to decelerate to subsonic speeds. If a second system failed, the aircraft had to be landed as soon as possible.

The landing gear was of the traditional tricycle arrangement, except the Tu-144 had eight wheels on each main truck. Each main landing gear was a single strut with a dual-twin tandem wheel configuration. The landing gear included a ground lock feature that prevented the strut from pivoting about the bogey when on the ground. This resulted in a farther aft ground rotation point, because the aircraft would have to pitch around the aft wheels rather than the strut pivot point, thus preventing the aircraft from tilting back on the tail during loading. The redesign of the Tu-144 in the early 1970s moved the engine nacelles farther out on the wings, placing the main landing gear in the middle of the engine inlet ducting. This issue was solved by having the gear bogey rotate 90 degrees about the strut longitudinal axis before retracting into the tall but narrow wheel well nestled between the adjoining engine inlets.

The wheel brake system was normally powered by the No. 1 hydraulic system, but a capability existed to interconnect to the No. 2 hydraulic system if necessary. An emergency braking capability using nitrogen gas pressurized to 100 atmospheres was provided. Independent braking levers on both the pilot and copilot's forward center console areas allowed differential braking with this system. A locked wheel protection circuit prevented application of the brakes airborne above 110 mph airspeed. On the ground, full brake pressure was available 1.5 seconds after full pedal pressure was applied. Above 110 mph on the ground, the brake pressure was reduced to 70 atmospheres. Below 110 mph, brake pressure was increased to 80 atmospheres. A starting brake was available to hold the aircraft in position during engine runups. This was essential, as the engines had to be run for a minimum of 30 minutes on the ground prior to flight. The brakes had to be "burned in" by holding them while taxiing in order to warm them to a minimum temperature to be effective. Furthermore, the braking capability was augmented by a drag parachute on landing to save wear on the tires and brakes.

The Tu-144 was supplied with main AC power at 115 volts and 400 hertz, secondary AC power at 36 volts and 400 Hz, and DC power at 27 volts. Each engine was connected to its respective Integrated Drive Generator (IDG), rated at 120 kilovolt-amperes (KVA) and providing independent AC power to its respective bus. No parallel generator operation was allowed under normal circumstances. Most systems could be powered from more than one bus, and one generator could provide all of the electrical power requirements, except for the canard and inlet anti-ice. A separate APU generator rated at 60 KVA at 400 Hz and provisions for external AC power were provided. The many fuel tank boost pumps were the main electrical power consumers. Other important AC systems were the canard and the retractable nose. The DC system consisted of 4 transformer/rectifiers (TR) and 4 batteries. The normal DC load was 12 kilowatts, and DC power was used for communication units, relays, and signaling devices.

Fire detection sensors and extinguishing agents were available for all engines, the APU, and the 2 cargo compartments. The extinguishing agent was contained in 6 canisters of 8-liter capacity each. When an overheat condition was detected, an annunciation was displayed on the FE panel showing the affected area. The pilot received only a "fire" light on the forward panel, without seeing which area was affected. In the case of APU fire detection, the extinguishing agent was automatically released into the APU compartment. In the case of an engine fire, the pilot could do nothing, because all engine fire extinguishing and shutdown controls were on the FE panel.

The air-conditioning and pressurization system consisted of identical, independent left and right branches. Any one branch could sustain pressurization during high-altitude operations. Nos. 1 and 2 engines and Nos. 3 and 4 engines shared common ducts for their respective bleed air. The right system provided conditioned air to the cockpit and forward cabin areas, and the left system furnished conditioned air to the mid and aft passenger cabin areas. The pressurization system provided an air exchange rate of 33 pounds per person per hour, and the total air capacity was 9,000 pounds per hour. Air was not recirculated back into the cabin. The pressurization controller maximum change rate was 0.18 millimeters (mm) of mercury (Hg) per second.

Hot engine bleed air was cooled initially to 374 degrees Fahrenheit (°F) by engine inlet bleed air in an air-air heat exchanger, then compressed in an air cycle machine (ACM) to 7.1 atmospheres with an exit temperature of 580 °F, and finally cooled in a secondary heat exchanger to 375 °F or less. If the air temperature were in excess of 200 °F and fuel temperature less than 160 °F, the air would be passed through a fuel-air heat exchanger. Passage through a water separator preceded entry into the expansion turbine of the ACM. Exit temperature from the turbine must be less than or equal to 85 °F, or the turbine would shut down. The FE changed the cockpit and cabin temperature using a hot air mix valve to control the temperature in the supply ducts. An idle descent from high altitude could result in an ACM overheat. In this case, speed must be increased to provide more air for the inlet air heat exchanger. There were four outflow valves on the left side of the fuselage and two on the right. The landing gear and brakes were cooled on the ground with air from the outflow valves. The FE controlled the air-conditioning and pressurization system. Desired cabin pressure was set in mm Hg, with 660 mm nominally being set on the ground. During high-altitude cruise, the ambient cabin altitude was nominally 1.7 to 1.9 miles. Warnings were displayed in the cockpit for cabin altitudes in excess of 2 miles, and 2.5 miles was the maximum.

There was no provision for wing leading-edge anti-icing. Flight-testing of the Tu-144 prototype indicated this was not necessary, because of the high speeds normally flown by the aircraft and the large degree of leading-edge sweep. The canard, however, was electrically heated for anti-ice protection requiring 20 KVA of AC power. No information was available on engine anti-icing, but the inlets were electrically heated for anti-ice protection.

Communication capability consisted of standard frequency band UHF and VHF radios and an Interphone Communication System (ICS). A variety of aural tones and messages were available, including master warning messages, radio altitude calls, and marker beacon tones. The annunciation was in a synthetic female voice format in Russian. Navigation capability consisted of three Inertial Navigation Systems (INS), VOR/DME and ILS receivers, and a Russian version of TACAN. The ILS was not compatible with Western frequency bands. A navigation computer controlled the three INS units. The mutually independent INS units provided attitude and true heading information to the attitude and horizontal situation indicators provided to each pilot. The No. 3 INS provided inputs to the pilot's instruments, No. 2 did the same for the copilot's instruments, and No. 1 could be selected by either pilot if necessary. If the navigation computer failed, the pilot could select raw INS data. Each INS could only accept 20 waypoints. When within 60 miles of the base airport, magnetic heading was used, but outside of that distance, true heading was selected. The crew had the ability to correct the computed position of each INS separately, in 1-mile increments. The Sensitive Pitch Angle Indicator (SPI) mounted above the center glare shield was driven by the No. 3 INS. This provided the pilots with precise pitch angle information necessary for approach and landing. A pilot-designed Vertical Regime Indicator (VRI) was a clever instrument that provided guidance to the pilot for the complex climb and acceleration profiles and descent and deceleration profiles. Concorde, on the other hand, had no such instrument and relied instead charted data.

The autopilot used the same actuators as the manual flight control system and was considered a subsystem of the flight control system. The dampers in all three axes must be operative for the autopilot to be used. The autopilot was a simple two-axis system operated from mode control panels (MCP) on the pilots' control wheels. Autopilot longitudinal and lateral modes included attitude hold, altitude hold, Mach hold, bank-angle hold, heading hold, localizer tracking, and glide-slope tracking. Each mode was selected by pressing a button on the MCP. As an example of the selector logic, for Mach or bank angle hold to be engaged,

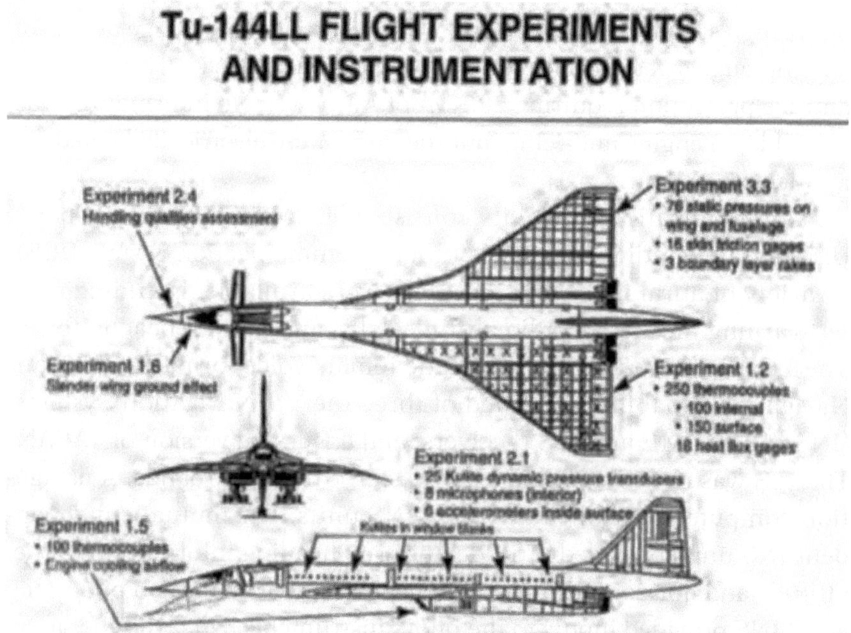

The sensor arrangement for the six Phase I experiments are shown in this three-view drawing of the Tu-144LL. NASA.

attitude hold must first have been selected. Altitude hold could be selected above 1,300-feet altitude but could not be used between 0.85 indicated Mach number (IMN) and 1.2 IMN, because of significant transonic effects. The lateral modes of the autopilot would command roll angles up to 30 degrees, but 25 degrees was the nominal limit. The longitudinal modes operated between 30-degrees nose-up to 11-degrees nose-down and possessed a 10-degree elevon trim range capability. Two autopilot disconnect switches were on each MCP, the left one to disconnect the lateral channel and the right one to disconnect the longitudinal channel. In addition, a red emergency disconnect switch was on each control wheel. The autopilot channels could be manually overridden or disconnected with a 1-inch pitch input or a 15-degree roll input.

The Experiments

The HSR Program Office assigned the six Phase I and one Phase II flight experiments reference numbers.

All six Phase I experiments were continued in Phase II and were identified in their Phase II form by the letter "A" following the number. Only experiment 1.5 changed in nature in Phase II. All of the experiments were assigned Tupolev principal investigator counterparts. The experiments and principal NASA–Boeing investigators are listed below:

- 1.2 Surface/Structure Equilibrium Temperature Verification: Craig Stephens (NASA Dryden).
- 1.5 Propulsion System Thermal Environment: Warren Beaulieu (Boeing).
- 1.5A Fuel System Thermal Database: Warren Beaulieu (Boeing).
- 1.6 Slender Wing Ground Effects: Robert Curry (NASA Dryden).
- 2.1 Structure/Cabin Noise: Stephen Rizzi (NASA Langley) and Robert Rackl (Boeing).
- 2.4 Handling Qualities Assessment: Norman Princen (Boeing).
- 3.3 C_p, C_f, and Boundary Layer Measurement and CFD Comparisons: Paul Vijgen (Boeing).
- 4.1 In-Flight Wing Deflection Measurements: Robert Watzlavick (Boeing).[29]

Because the HSR program was the primary funding source for the Tu-144LL flight experiment, it followed that the relevant HSR Integrated Technology Development (ITD) teams would be the primary customers. Subsequent to Phase I, however, it became apparent that some of the experiments did not have the ITD teams' complete support. The experimenters believed that data analysis would be accomplished by the interested ITD teams, but the ITD teams who had little or no input in the planning and selection of the experiments had no plans to use the data. This was complicated by the cancellation of the HSR program by NASA in April 1999.[30] In retrospect, it appeared that the experiment selection process did not properly consider the ultimate needs of the logical customers in all cases. In deference to the HSR program, however, it should be noted that the joint U.S.–Russian Tu-144 project had political aspects that had to be considered and inputs for data from Tupolev that may not have fit neatly into HSR requirements. Fortunately, the bulk of the raw data from all of

29. Barber, "Tu-144LL Reports, Data, and Documentation Disposition," NASA Dryden Flight Research Center, July 11, 2001.
30. Harris, "Flight Experiments Review and Critique."

the experiments, except Langley's 2.1 and 2.1A, is maintained at NASA Dryden.[31] The data from 2.1 were fully analyzed and reported in several NASA and Boeing reports.[32]

The data from all but experiment 2.1, Structure/Cabin Noise, were collected by the Damien DAS and were for the most part managed in Zhukovsky by Tupolev engineers. Experiment 2.1 had a dedicated DAS and experienced none of the data acquisition problems suffered at times by the other experiments. NASA Dryden's Glenn A. Bever was the NASA onsite engineer and instrumentation engineer for the duration of the program. In this capacity, he supported all of the experiments, except Langley's experiment 2.1, which had its own engineers and technicians. From 1995 to 1999, Bever made 19 trips to Zhukovsky, "a total of 8 months in Russia all told hitting every month of the year at least once."[33] Because Dryden had responsibility for instrumentation, Bever worked with Tupolev instrumentation engineers and technicians directly to ensure that all of the experiments' data other than 2.1 were properly captured. Often, he was the only American in Zhukovsky and found himself the point of contact for all aspects of the project. He "wrote Summaries of Discussion at the end of each trip which tended, we discovered, to act like contracts to direct what work was to happen next and document deliverables and actions."[34] Bever utilized a rather new concept at the time, when he transmitted all of the collected data from the experiments under his purview to Dryden via the Internet. He translated the instrumentation calibration information files into English calibration files, wrote the programs that reduced the data to a manipulative format, applied the calibrations, formatted the data for storage, and archived the data on Dryden's flight data computer and on CDs. One of his final accomplishments was to design the air data sensor system that collected altitude and airspeed information from the Phase II flights flown by the NASA pilots.[35] Langley's instrumentation technician,

31. Barber, "Tu-144LL Reports, Data, and Documentation Disposition."

32. Rizzi, et al., "Structure/Cabin Noise Experiment," NASA TM-2000-209858; Robert G. Rackl and Stephen A. Rizzi, "Structure/Cabin Noise," HSR-AT Contract No. NAS1-20220, TU-144LL Follow On Program, vol. 5 (June 1999). Unfortunately, the data archival effort received a setback when Boeing, for reasons unknown, discarded much of its HSR documentation on pre-2001 efforts. Barber, "Tu-144LL Reports, Data, and Documentation Disposition."

33. E-mail interview of Glenn A. Bever by author, NASA Dryden Flight Research Center, May 28, 2009.

34. Ibid.

35. Ibid.

Donna Amole, and Dryden's Project Manager, Russ Barber, attested to the significant efforts Bever contributed to the project.

Experiment 1.2/1.2A, Surface/Structure Equilibrium Temperature, consisted of 250 thermocouples and 18 heat flux gauges installed on predetermined locations on the left wing, fuselage, and engine nacelles, which measured temperatures from takeoff through landing on Mach 1.6 and 2 test flights.[36] High noise levels and significant zero offsets resulted in poor quality data for the Phase I flights. This was due to problems with the French-built Damien DAS. For Phase II, a Russian-designed Gamma DAS was used, with higher-quality data being recorded. Unfortunately, the HSR program did not analyze the data, because the relevant ITD team did not believe this experiment was justified, based on prior work and preexisting prediction capability at these Mach numbers. The initial poor data quality also did not suggest that further analysis was warranted.[37]

Experiment 1.5, Propulsion System Thermal Environment, sampled temperatures in the engine compartment and inlet and measured accessory section maximum temperatures, engine compartment cooling airflow, and engine temperatures after shutdown. Thirty-two thermocouples on the engine, 35 on the firewall, and 10 on the outboard shield recorded the temperature data.[38] The data provided valuable information on thermal lag during deceleration from Mach 2 flight and on the temperature profiles in the engine compartment after shutdown. Experiment 1.5A in Phase II developed a Thermal Database on the aircraft fuel system using 42 resistance temperature devices and 4 fuel flow meters to collect temperature and fuel flow time histories on engines 1 and 2 and heat rejection data on the engine oil system during deceleration from supersonic speeds. HSR engineers did not fully analyze these data before program cancellation.[39]

Experiments 1.6/1.6A, Slender Wing Ground Effects, demonstrated no evidence of dynamic ground effects on the Tu-144LL. This correlated

36. The Boeing Company, "Volume 2: Experiment 1.2. Surface/Structure Equilibrium Temperature Verification," *Flight Research Using Modified Tu-144 Aircraft, Final Report*, HSR-AT Contract No. NAS1-20220 (May 1998). This and the subsequent volumes of the Boeing contractor report, HSR-AT Contract No. NAS1-20220, provide much detail on the six Phase I experiments, while the Follow-On Program Boeing contractor report of June 1999 reports on the Phase II experiments.
37. Harris, "Flight Experiments Review and Critique."
38. The Boeing Company, "Volume 3: Experiment 1.5, Propulsion System Thermal Environment Database," *Flight Research Using Modified Tu-144 Aircraft, Final Report*, HSR-AT Contract No. NAS1-20220 (May 1998).
39. Harris, "Flight Experiments Review and Critique."

with wind tunnel data and NASA evaluation pilot comments.[40] Effects were determined on lift, drag, and pitching moment with the canard, both retracted and extended. Forty-eight parameters were measured in flight, including inertial parameters, control surface positions, height above the ground, airspeed, and angle of attack. From these, aerodynamic forces and moments were derived, and weight and thrust were computed postflight. A NASA Differential Global Positioning System (DGPS) provided highly precise airspeed and angle-of-attack data and repeatable heights above runway accurate to less than 0.5 feet. Getting this essential DGPS equipment into Russia had been difficult because of Russian import restrictions. In Phase I, 10 good maneuvers from the 19 flights were accomplished, evaluating a range of weights, sink rates, and canard positions. The data quality was excellent, and the results indicated that there is still much to be learned regarding dynamic ground effects for slender, swept wing aircraft.[41]

Langley's Structure/Cabin Noise, experiment 2.1, was unique among the seven flight experiments, in that it used its own Langley-built DAS and had on site its own support personnel for all flights on which data were collected. Another unique feature of this experiment was its direct tie to a specific customer, the HSR structural acoustics ITD team. The two principal investigators, Stephen Rizzi and Robert Rackl, were members of the team, and Rizzi was the team lead. This arrangement allowed the structure of the experiment to be designed directly to meet team requirements.[42] Several datasets, including boundary layer fluctuating pressure measurements, fuselage sidewall vibration and interior noise data, jet noise data, and inlet noise data, were used to update or validate various acoustic models, such as a boundary layer noise source model, a coupled boundary layer/structural interaction model, a near-field jet noise model, and an inlet noise model.[43] The size of the dataset and sampling rates was staggering. The required rate was 40,000 samples per second for each of 32 channels. The Damien DAS was not capable of sampling at these rates, thus necessitating the Langley DAS. Langley, as a result, provided personnel on site to support experiment 2.1. These included

40. The Boeing Company, "Volume 4: Experiment 1.6, Slender Wing Ground Effects," *Flight Research Using Modified Tu-144 Aircraft, Final Report*, HSR-AT Contract No. NAS1-20220 (May 1998).

41. Harris, "Flight Experiments Review and Critique."

42. Rizzi, "Brief Background of Program and Overview of Experiments."

43. Rizzi, et al., "Structure/Cabin Noise Experiment," NASA TM-2000-209858, pp. 1–5.

Rizzi, Rackl, and several instrumentation technicians from Langley's Flight Instrumentation Branch, including Vernie Knight, Keith Harris, and Donna Amole, the only onsite American female on the project. Amole spent about 5 months in Zhukovsky during 8 trips. Her first trip was challenging, to say the least. The Tupolev personnel were not eager to have an American woman working with them. Whether because of superstition (Amole initially was told she could not enter the airplane on flight days), cultural differences, or perhaps a misunderstood fear of potential American sexual harassment issues, Amole for the first 2 weeks was essentially ignored by her Tupolev counterparts. She would not be deterred, however, and won the respect and friendship of her Russian colleagues. Glenn Bever and Stephen Rizzi provided essential support, but many times, she was, like Bever, the only American on site.[44]

Experiment 2.4, Handling Qualities Assessment, suffered in Phase I from poor data quality, which predicted a very poor flying aircraft. The aircraft response to control deflections indicated a 0.25-second delay between control movement and aircraft response. Furthermore, angle-of-attack, angle-of-sideslip, heading, altitude, and airspeed data all were of suspect quality at times.[45] These data issues contributed to the HSR program's desire for U.S. pilots to fly the airplane to evaluate the handling qualities, because access to the Tupolev pilots was limited. Additionally, in Phase II, a new air data sensor from NASA Dryden corrected the nagging air data errors. This experiment will be covered in more detail in the following section on the Tu-144LL Handling Qualities Assessment.

Experiments 3.3/3.3A—C_p, C_f, and Boundary Layer Measurements—collected data on surface pressures, local skin friction coefficients, and boundary layer profiles on the wing and fuselage using 76 static pressure orifices, 16 skin friction gauges consisting of 10 electromechanical balances and 6 hot film sensors, 3 boundary layer rakes, 3 reference probes, 5 full chord external pressure belts consisting of 3 on the wing upper surface and 2 on the lower surface, and angle-of-attack and angle-of-sideslip vanes. Measurements from the 250 thermocouples from experiment 1.2 were used in the aerodynamic data analysis.[46] Data were collected at Mach

44. Interview of Donna Amole by author, Hampton, VA, July 3, 2009.

45. Harris, "Flight Experiments Review and Critique."

46. The Boeing Company, "Volume 7: Experiment 3.3, Cp, Cf, and Boundary Layer Measurements Database," *Flight Research Using Modified Tu-144 Aircraft, Final Report*, HSR-AT Contract No. NAS1-20220 (May 1998).

0.9, 1.6, and 2 and included over 80 minutes of stabilized supersonic flight. Data quality was good, although some calibration problems with the pressure transducers and mechanical skin friction balances arose. On flight 10, the lower wing surface midspan pressure belt detached and was lost, and 4 tubes on the upper midspan belt debonded. Fortunately, the failures occurred after the minimum data requirements had been met. In Phase II, Preston tubes and optical-mechanical sensors developed at Russia's Central Institute of Aerohydromechanics (TsAGI) were implemented for additional skin friction measurements. The HSR program did not fully analyze these data, believing that prior XB-70 data already filled these requirements.[47]

Experiment 4.1A, In-Flight Wing Deflection Measurements, provided a limited verification of the wing geometry under in-flight loads. These data are needed for validating the aeroelastic prediction methodology and providing the in-flight geometry needed in computational fluid dynamics analysis. Boeing's Optitrak active target photogrammetry system was used, and Boeing managed the experiment. The installed system incorporated 24 infrared reflectors mounted on the upper surface of the right wing, each pulsed in sequence. Two cameras captured the reflected signals in order to provide precise x, y, and z coordinates.[48] The system was used on Langley's Boeing 737 in the early 1990s high lift experiment, designed to quantify the precise effect of high-lift devices.

Not listed among the formal experiments was a Phase II independent "piggyback" experiment leveraging off the data collected from experiment 2.4, Handling Qualities Assessment, flown by the NASA research pilots. This involved a new longitudinal, lateral, and directional closed-loop Low-Order Equivalent System (LOES) method of aircraft parameter identification using an equation-error method in the frequency domain. Because the data were accumulated by pilot-in-the-loop frequency sweep and multistep maneuvers, these were added to the test cards for the first four Phase II flights.[49] Langley's Dr. Eugene A. Morrelli requested theses datasets and developed the pilot maneuvers necessary to acquire them. This was a unique example of a researcher taking advantage of his colleagues' work on a once-in-a-lifetime

47. Harris, "Flight Experiments Review and Critique."

48. Ibid.

49. Eugene A. Morrelli, "Low-Order Equivalent System Identification for the Tu-144LL Supersonic Transport Aircraft," *Journal of Guidance, Control, and Dynamics*, vol. 26, no. 2, (Mar.–Apr. 2003), p. 354.

experiment and of the spirit of cooperation among NASA researchers that allowed this opportunity develop.

The Tu-144LL Handling Qualities Assessment

In Phase I, typical flights involved a climb and acceleration to supersonic speeds and cruise altitudes, 15 minutes of stable supersonic cruise, a descent and deceleration to subsonic cruise conditions for subsonic test points, and finally, approach and landing work.[50] All 19 Phase I flights were accomplished by Tupolev crews. Flights 20 through 23 incorporated the NASA pilot evaluations at the beginning of Phase II. The description of the Handling Qualities Assessment Experiment 2.4/2.4A will center on these flights, because they are of more special interest to NASA.[51]

Working with Tupolev chief test pilot Sergei Borisov and project engineer Vladimir Sysoev, USPET developed a set of efficient handling qualities maneuvers to be used on these flights. These maneuver sets were derived from the consensus reached among USPET members regarding the highest-priority tasks from Mach 2 to approach and landing. To assist the pilots, specifically defined maneuvers were repeated for different flight conditions and aircraft configurations. These maneuver sets included

- Integrated test block (ITB): The ITB was a standard block of maneuvers consisting of pitch attitude captures, bank captures, heading captures, steady heading sideslips, and a level acceleration/deceleration.
- Parameter identification (PID) maneuvers: The PID maneuvers generated either a sinusoidal frequency sweep or a timed pulse train in the axis of interest and contributed to the dataset needed for the LOES analysis.
- Simulated engine failure: This consisted of retarding an outboard throttle to minimum setting, stabilizing on a trimmed condition, and performing a heading capture.
- Slow flight: Accomplished in both level and turning flight, this maneuver was flown at minimum airspeed.

50. Harris, "Flight Experiments Review and Critique."
51. Rivers, et al., "A Qualitative Piloted Evaluation of the Tupolev Tu-144 Supersonic Transport," NASA TM-2000-209850, describes flights 20–23 in detail, to include the planning, execution, and results.

NASA and Russian engineers monitoring a U.S. evaluation flight from the Gromov Russian Federation State Scientific Center. NASA.

15

- Structural excitation maneuvers: These maneuvers consisted of sharp raps on each control inceptor to excite and observe any aeroservoelastic response of the aircraft.
- Approaches and landings: Different configurations were specified to include canard retracted, lateral offset, manual throttle, nose retracted (zero forward visibility), simulated engine out, visual, and Instrument Landing System (ILS) approaches.[52]

Flight 20 was flown by an all-Russian crew but was observed from a control room at the Gromov Russian Federation State Scientific Center at the Zhukovsky Air Development Center.

This flight provided USPET with an excellent opportunity to observe Tu-144 planning and operations and prepared the team for the NASA piloted flights. With a better sense of Tupolev operations, USPET was able to develop English checklists and procedures to complement the Russian ones. Fullerton and Rivers learned all of the Russian-labeled

52. Rivers, et al., "A Qualitative Piloted Evaluation of the Tu-144," NASA TM-2000-209850, p. 17.

switches and controls and procedural calls. USPET made bound checklists from the cardboard backs of engineering tablets, because office material was in short supply at that time in Russia. Flight 20 also allowed USPET engineers Jackson, Cox, and Princen and pilots Fullerton and Rivers to develop a working relationship with Tupolev project engineer Sysoev in developing the test cards for the U.S. flights. The stage was set for the first flight of a Tu-144 by a United States pilot.

Flight 21 was scheduled for September 15, 1998. Fullerton and Rivers agreed that Fullerton would pilot this flight and Rivers would observe from the cockpit, taking notes, timing maneuvers, and assisting with the crew coordination. As it turned out, Fullerton's communications failed during the flight, and Rivers had to relay Tupolev pilot Sergei Borisov's comments to Fullerton. Borisov sat in the left seat and Fullerton in the right; Victor Pedos occupied the navigator's seat and Anatoli Kriulin the flight engineer's station. Rivers stood behind Borisov and next to Pedos. Jackson and Cox had seats in the Gromov control room. Flight 21 was to be a subsonic flight with handling qualities maneuvers completed by Fullerton during the climb, Mach 0.9 cruise, descent, low-altitude slow-flight maneuvering, and approach and landing tasks. Because of the shortage of tires, each flight was allowed only one landing. The multiple approaches flown were to low approach (less than 200-feet altitude) only.

The flight is best described by the flight test summary contained in a NASA report titled "A Qualitative Piloted Evaluation of the Tu-144":

> Shortly after take-off a series of ITBs were conducted for the take-off and the clean configurations at 2 km altitude. Acceleration to 700 km/hr was initiated followed by a climb to the subsonic cruise condition of Mach 0.9, altitude 9 km. Another ITB was performed followed by evaluations of a simulated engine failure and slow speed flight. After descent to 2 km, evaluations of slow speed flight in the take-off and landing configurations were conducted as well as an ITB and a simulated engine failure in the landing configuration. Following a descent to pattern altitude three approaches to 60 m altitude were conducted with the following configurations: a canard retracted configuration using the ILS localizer, a nominal configuration with a 100 m offset correction at 140 m altitude, and a nominal configuration using visual cues. The flight ended with a visual approach to touchdown in the nominal configuration.

15

> However, due to unusually high winds the plane landed right at its crosswind limit, necessitating the Russian pilot in command to take control during the landing. Total flight time was approximately 2 hours 40 minutes. The maximum speed and altitude was 0.9 Mach and 9 km.[53]

The flight completed all test objectives. Thorough debriefs ensued, the obligatory postflight party sponsored by Tupolev was held, and USPET began intensive training and planning for the first supersonic flight, to be flown just 3 days later.

September 18 opened cool, clear, and much less gusty than the preceding days. Flight 22 would be a Mach 2 mission to an altitude of 60,000 feet, with at least 20 minutes flight at twice the speed of sound. Rob Rivers was the NASA pilot for this flight. Pukhov's only requirement for Rivers was that he no longer need his crutches by flight day. Two nights before, Bruce Jackson had helped Rivers practice using a cane for over an hour until Rivers was comfortable. At the next day's preflight party, Rivers demonstrated to Pukhov his abilities without crutches, and his approval for the flight was assured. At 11:08 a.m. local time, the Tu-144 became airborne.

The flight is described below in Rivers's original flight test report:

> Flight Profile. The flight profile included takeoff and acceleration to 700 kilometers per hour (km/hr) to intercept the climb schedule to 16.5 kilometers (km) and Mach 2.0. The flight direction was southeast toward the city of Samara on the Volga River at a distance of 700 km from Zhukovsky. Approximately 20 minutes were spent at Mach 2.0 cruise which included an approximately 190 degree course reversal and a cruise climb up to a maximum altitude of 17.3 km. A descent and deceleration to 9 km and Mach 0.9 was followed by a brief cruise period at that altitude and airspeed prior to descent to the traffic pattern at Zhukovsky Airfield for multiple approaches followed by a full stop landing on Runway 30.
>
> Flight Summary. After all preflight checklists had been completed, the evaluation pilot taxied Tu-144LL Serial Number

53. Rivers, et al., "A Qualitative Piloted Evaluation of the Tu-144," p. 20.

77144 onto Runway 12, and the brake burn-in process was accomplished. At 11:08 brakes were released for takeoff, power was set at 98° PLA (partial afterburner), the start brake was released, and after a 30 sec takeoff roll, the aircraft lifted off at approximately 355 km/hr. The landing gear was raised with a positive rate of climb, the canard was retracted out of 120 m altitude, and the nose was raised out of 1000 m altitude. The speed was initially allowed to increase to 600 km/hr and then to 700 km/hr as the Vertical Regime Indicator (VRI) profile was intercepted. Power remained at 72° PLA (maximum dry power) for the climb until Mach 0.95 and CG of 47.5% at which point the throttles were advanced to maximum power, 115° PLA. The climb task was a high workload task due to the sensitivity of the head up pitch reference indicator, the sensitivity of the pitch axis, and the continual change in CG requiring almost continuous longitudinal trim inputs. Also, since the instantaneous center of rotation is located at the pilot station, there are no cockpit motion cues available to the pilot for pitch rate or attitude changes. Significant pitch rates can be observed on the pitch attitude reference indicator that are not sensed by the pilot. During the climb passing 4 km, the first of a repeating series of bank angle captures (±15°) and control raps in all three axes (to excite any aircraft structural modes) was completed. These maneuvers were repeated at 6 km and when accelerating through Mach 0.7, 0.9, 1.1, 1.4, and 1.8. The bank angle captures demonstrated rather high roll forces and relatively large displacements required for small roll angles. A well damped (almost deadbeat) roll mode at all airspeeds up to Mach 2.0 was noted. The control raps showed in general a higher magnitude lower frequency response in all three axes at subsonic speeds and lower magnitude, higher frequency responses at supersonic speeds. The pitch response was in general of lower amplitude and frequency with fewer overshoots (2-3) than the lateral and directional responses (4-5 overshoots) at all speeds. Also of interest was that the axis exhibiting the flexible response was the axis that was perturbed, i.e., pitch raps resulted in essentially only pitch responses. The motions definitely seemed to be aeroservoelastic in nature, and with the strong damping in the lateral and directional axes, normal control inputs resulted in well damped responses. Level off at 16.5 km and Mach 1.95 occurred 19 minutes after takeoff. The aircraft was allowed to accelerate to Mach 2.0 IMN as the throttles were reduced to 98° PLA, and a series of control raps was accomplished. Following this, a portion of the Integrated Test Block set of maneuvers consisting of pitch captures, steady heading sideslips, and a level deceleration was completed. The pitch captures resulted in slight overshoots and indicated a moderate delay between pitch attitude changes and flight path angle changes. The steady heading sideslips showed a slight positive dihedral effect, but no more than approximately 5° angle of bank was required to maintain a constant heading. No unpleasant characteristics were noted. At this point the first set of three longitudinal and lateral/directional parameter identification (PID) maneuvers were completed with no unusual results. By this time a course reversal was necessary, and the bank angle and heading capture portions of the ITB were completed during the over 180° turn which took approximately 7 min to complete at Mach 1.95. During the inbound supersonic leg, two more sets of PID maneuvers with higher amplitude (double the first set) control inputs were completed as were several more sets of control raps. Maximum altitude achieved during the supersonic maneuvering was 17.3 km.

The descent and deceleration from Mach 2.0 and 17 km began with a power reduction from the nominal 98° PLA to 59° and a deceleration to 800 km/hr. During the descent bank angle captures (±30°) and control raps were accomplished at or about Mach 1.8, 1.4, 1.1, and 0.9 with similar results as reported above. The aircraft demonstrated increased pitch sensitivity in the transonic region decelerating through Mach 1.0. The pitch task during descent in following the VRI guidance was fairly high in workload, and the head-up pitch reference indicator was very sensitive and indicated fairly large pitch responses from very small pitch inputs. Since the CG is being transferred aft during supersonic descent, frequent pitch trimming is required. A level off at 9 km at Mach 0.9 was accomplished without difficulty, and an ITB (as described above) was completed. Further descents as directed by air traffic control placed the aircraft in the landing pattern with 32 metric tons of fuel, 6 tons above the planned amount. Five total approaches including the final full stop landing were completed. These included a straight-in localizer only approach with the canard retracted; an offset approach with the nose raised until on final; a manual throttle offset approach; a manual throttle straight-in approach; and a straight-in visual approach to a full stop landing. The first approach with the canard retracted was flown at 360 km/hr due to the loss of about 12 tons of lift from the retracted canards. Pitch control was not as precise in this configuration. There was also a learning curve effect as the evaluation pilot gained experience in making very small, precise pitch inputs which is necessary to properly fly the aircraft on approach and to properly use the pitch reference indicator. After terminating the approach at 60 m, a canard retracted, gear down low pass up the runway at 30-40 m was completed in accordance with a ground effects experiment requirement. The nose-up approach demonstrated the capability to land this aircraft with the nose retracted providing an angling approach with some sideslip is used. The offset approaches were not representative of the normal offset approaches flown in the HSR program since they are to low approach only and do not tax the pilot with the high gain spot landing task out of the corrective turn. No untoward pitch/roll coupling or tendency to overcontrol the pitch or roll axes was noted. The manual approaches were very interesting in that the Tu-144LL, though a back-sided airplane on approach, was not difficult to control even with the high level of throttle friction present. The engine time constant appears reasonable. It was noted that a large pitching moment results from moderate or greater throttle inputs which can lead to overcontrolling the pitch axis if the speed is not tightly controlled and large throttle inputs are required. The full stop landing was not difficult with light braking required due to the decelerating effects of the drag parachutes. The flight terminated with the evaluation pilot taxiing the aircraft clear of the runway to the parking area. 16 tons of fuel remained.

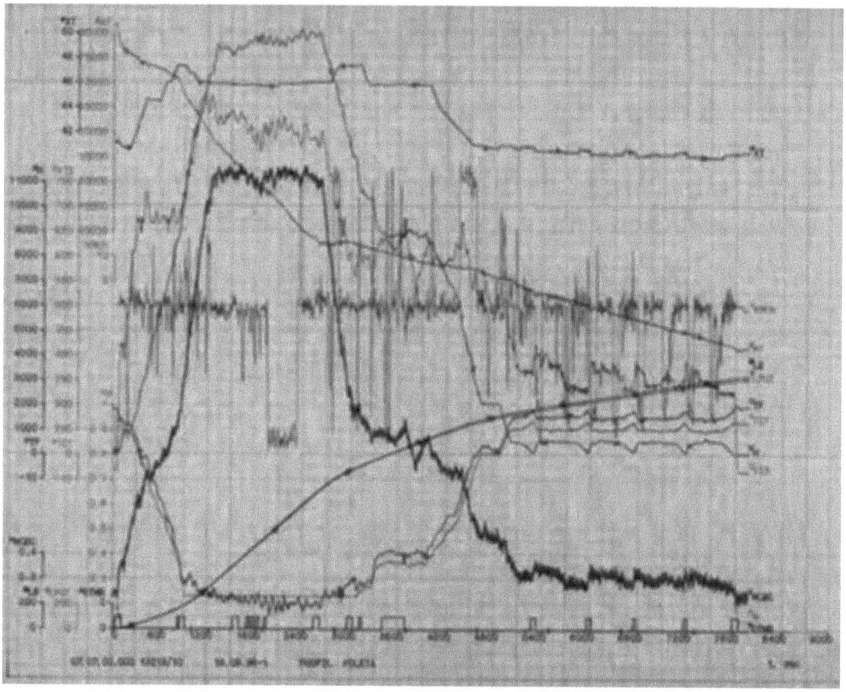

The Tupolev time history plot of flight 22 showing several parameters plotted against time. NASA.

> All test points were accomplished, and several additional optional test points were completed since the flight remained ahead of the planned fuel burn. One additional approach was completed. The planned flight profile was matched very closely, and all flight objectives were achieved.[54]

Onboard recording was used to gather all of the data, because the flight profiles took the Tu-144LL far out of telemetry range. Subsequent to each flight, Tupolev would produce a data time history plot, including over a dozen measured parameters plotted on the vertical axis versus time on the horizontal axis. On one plot, the entire flight could quickly be viewed. From the plotted time histories, much additional data could be ascertained. By comparing fuel quantity expended versus time, for example, fuel flows could be determined. This contrasts with the methods in NASA in which, with paper supplies not of concern, the practice

54. Rivers, et al., "A Qualitative Piloted Evaluation of the Tu-144," pp. 34–36.

The Tupolev and NASA flightcrews after the completion of the last U.S. piloted evaluation flight, with Tu-144LL "Moscow" in the background. NASA.

The Tu-144LL landing, with a NASA pilot at the controls. NASA.

is often to plot individual time histories. USPET members felt that this straightforward Tupolev method showed great merit.

Flight 23 was completed September 24, after several days of weather delays. Gordon Fullerton was the NASA evaluation pilot for this flight, which was very similar to flight 22. The only differences occurred at Mach 2, at which Fullerton simulated an engine failure at the beginning of descent from just over 10-mile altitude and in the landing pattern in which a clean pass was flown for a photographic opportunity, and two simulated engine failure approaches and an additional ILS approach were accomplished. All test objectives were achieved.

The USPET team was feted to a final postflight party and, jokingly, according to Professor Pukhov, was not allowed to leave until a preliminary report was completed. The U.S. team completed the report and departed September 26, with a mutual exchange of best wishes with the Tupolev Tu-144 project staff. Four more Phase II flights were completed with the Tupolev crew to gather more handling qualities data and data for the other six experiments. After Sergei Borisov shut down the engines following the last flight in winter 1999, the Tu-144 never flew again.

NASA TM-2000-209850 thoroughly describes the operational qualities of the Tu-144LL. A brief description will be presented here. The Tu-144 taxied much like a Boeing 747 with mild cockpit accelerations and nominal cockpit overshoots while turning. Throttle friction was extremely high because of the rerouted throttle cables for the retrofitted NK-321 engines. The engines had operational limits and restrictions, some peculiar to a specific engine, but they performed well throughout the flight envelope, were robust and forgiving at Mach 2 cruise, and responded well in the landing pattern. Takeoff acceleration was very rapid, and the takeoff speeds were quite high, as expected with unstick occurring at 220 mph after 30 seconds of ground roll. A very high ambient noise level and moderate buffet were experienced, with the nose drooped to the 11-degree takeoff position and the canard extended. With the nose retracted, the forward view was blocked, and the view through the somewhat distorted and crazed side windows was poor. Because the rate dampers were required to be engaged at all times, the unaugmented characteristics of the aircraft were not investigated. Pitch forces were moderately heavy, and small pitch inputs resulted in significant longitudinal motion, creating a tendency to overcontrol the pitch axis. The lateral forces were high, and large displacements were necessary for small roll rates, resulting in poor pitch-roll harmony. Roll inputs would often couple into undesired pitch inputs. With poor pitch cues because of the visibility issues mentioned earlier, the pilot relied on the Sensitive Pitch Angle Indicator for pitch control. The pitch axis was the high workload axis, and this was exacerbated by the rapid center-of-gravity changes because of fuel transfer balancing in the transonic range. Roll response was very well-damped, with no proverse or adverse yaw, even with large lateral inputs. Precise bank angle captures were easy to accomplish. The aircraft demonstrated positive speed stability. Rudder inputs produced a positive dihedral effect and were well-damped/deadbeat, but rudder pedal forces were very high. Full pedal deflection required 250–300 pounds of force. All of these characteristics were invariant with speed and configuration, except for the slightly degraded handling qualities near Mach 1. With the exception of the heavy control forces (typical of Russian airplanes), the Tu-144 possessed adequate to desirable handling qualities. This result disputed the data taken in Phase I and led engineers to uncover the artificial 0.25-second time delay in the Damien DAS that produced such questionable handling qualities data.

Reflections and Lessons Learned

The HSR, industry, and Tupolev team completed a remarkable project that accentuated the best possibilities of international cooperation. Against a backdrop of extreme challenges in the Russian economy, at a time when the value of the ruble declined over 80 percent from the time USPET arrived in Russia until it departed, all participants worked with a sense of commitment and fraternity. The Tupolev team members were not assured of any pay in those trying days, yet they maintained a cordial and helpful attitude. Typical of their hospitality, Sergei Borisov gave his only video playback receiver and some of his cherished airshow tapes to the Americans for entertainment in the austere KGB sanitarium. Though food shortages existed at that time, one of the translators, Mikhail Melnitchenko, had the USPET team to his apartment for dinner. Borisov, Pukhov, and several other Tupolev officials hosted a visit by the team to the Russian Air Force Museum one weekend. Everywhere the Americans went, they were greeted with hospitality. Scarcity of such basic materials as paper did not affect the Tupolev professionals in the least and left the Americans with a new appreciation for their good fortune. Despite challenges, the Tupolev personnel produced a magnificent airplane.

Many lessons have been presented throughout this essay. Certainly, the political nature of this joint project resulted in some experimental data not being required and never being analyzed. The abrupt cancellation of the HSR program and the climate of NASA in the late 1990s did not allow the proper utilization of this valuable and costly data. The concept of the HSR ITD teams did not contribute to an efficient method of engineering work. Full team consensus was required on decisions, resulting in far too much time being expended to make even minor ones. The size and diversity of the HSR program led to inefficiencies, as each participant had specific interests to consider. It would take a far more in-depth study than presented here to determine if there was a fair return on investment, but there is little doubt that the HSR program did not achieve its primary purpose: to develop technologies leading to a commercially successful HSCT. However, despite this dour assessment, the benefits derived from the HSR program will certainly provide additional payback in the coming years. That payback will likely be seen in technology transfer to the subsonic air transport fleet in the near term and to another HSCT concept or supersonic business jet in the far term. Should the United States ever embark on a national aeronautics program of the scope of HSR, it is hoped that the profound lessons learned from the HSR program will be applied. Regardless of the aerospace community's inability to produce a HSCT design and take it from the drawing board to the flight line, all of those involved in the Tu-144LL flight experiments should take pride in the work they accomplished, when two former adversaries joined to complete their project goals against a background of prodigious challenges.

Recommended Additional Readings

Reports and Papers:

The Boeing Company, "Volume 2: Experiment 1.2. Surface/Structure Equilibrium Temperature Verification," *Flight Research Using Modified Tu-144 Aircraft, Final Report*, HSR-AT Contract No. NAS1-20220 (May 1998).

The Boeing Company, "Volume 3: Experiment 1.5, Propulsion System Thermal Environment Database," *Flight Research Using Modified Tu-144 Aircraft, Final Report*, HSR-AT Contract No. NAS1-20220 (May 1998).

The Boeing Company, "Volume 4: Experiment 1.6, Slender Wing Ground Effects," *Flight Research Using Modified Tu-144 Aircraft, Final Report*, HSR-AT Contract No. NAS1-20220 (May 1998).

The Boeing Company, "Volume 7: Experiment 3.3, C_p, C_f, and Boundary Layer Measurements Database," *Flight Research Using Modified Tu-144 Aircraft, Final Report*, HSR-AT Contract No. NAS1-20220 (May 1998).

Timothy H. Cox and Alisa Marshall, "Longitudinal Handling Qualities of the TU-144LL Airplane and Comparisons with Other Large, Supersonic Aircraft," NASA TM-2000-209020 (May 2000).

Eugene A. Morrelli, "Low-Order Equivalent System Identification for the Tu-144LL Supersonic Transport Aircraft," *Journal of Guidance, Control, and Dynamics*, vol. 26, no. 2, (Mar.–Apr. 2003).

Robert G. Rackl and Stephen A. Rizzi, "Structure/Cabin Noise," HSR-AT Contract No. NAS1-20220, TU-144LL Follow On Program, vol. 5 (June 1999).

Robert A. Rivers, E. Bruce Jackson, C. Gordon Fullerton, Timothy H. Cox, and Norman H. Princen, "A Qualitative Piloted Evaluation of the Tupolev Tu-144 Supersonic Transport," NASA TM-2000-209850 (Feb. 2000).

Stephen A. Rizzi, Robert G. Rackl, and Eduard V. Andrianov, "Flight Test Measurements from the Tu-144LL Structure/Cabin Noise Experiment," NASA TM-2000-209858 (Jan. 2000).

Stephen A. Rizzi, Robert G. Rackl, and Eduard V. Andrianov, "Flight Test Measurements from the Tu-144LL Structure/Cabin Noise Follow-On Experiment," NASA TM-2000-209859 (Feb. 2000).

Robert Watzlavick, "TU-144LL Follow On Program, Volume 8: In Flight Wing Deflection Measurements," HSR-AT Contract No. NAS1-20220 (June 1999).

Books and Monographs:

British Aircraft Corporation, Ltd., Commercial Aircraft Division, *An Introduction to the Slender Delta Supersonic Transport* (Bristol, England: Printing and Graphic Services, Ltd., 1975).

Joseph R. Chambers, *Innovation in Flight: Research of the NASA Langley Research Center on Revolutionary Advanced Concepts for Aeronautics*, NASA SP-2005-4539 (Washington, DC: GPO, 2005).

Paul Duffy and Andrei Kandalov, *Tupolev: The Man and His Aircraft* (Shrewsbury, England: Airlife Publishing, Ltd., 1996).

Vladislav Klein and Eugene A. Morelli, *Aircraft System Identification: Theory and Practice* (Reston, VA: American Institute of Aeronautics and Astronautics, Inc., 2006).

Howard Moon, *Soviet SST: The Technopolitics of the Tupolev-144* (New York: Orion Books, 1989).

THIS PAGE INTENTIONALLY BLANK

THIS PAGE INTENTIONALLY BLANK

Index

D

E

N

T

THIS PAGE INTENTIONALLY BLANK

THIS PAGE INTENTIONALLY BLANK

Printed by Libri Plureos GmbH in Hamburg,
Germany